TWO-FACTOR ANOVA

$$SS_{\text{between treatments}} = \Sigma \frac{T^2}{n} - \frac{G^2}{N}$$

$$df_{\text{between treatments}} = \text{number of cells} - 1$$

$$SS_{\text{within treatments}} = \Sigma SS_{\text{each treatment}}$$

$$df_{\text{within treatments}} = \Sigma df_{\text{each treatment}}$$

$$SS_A = \Sigma \frac{T_{\text{ROW}}^2}{n_{\text{ROW}}} - \frac{G^2}{N}$$

$$df_A = (\text{number of levels of } A) - 1$$

$$SS_B = \Sigma \frac{T_{\text{COL}}^2}{n_{\text{COL}}} - \frac{G^2}{N}$$

$$df_B = (\text{number of levels of } B) - 1$$

$$SS_{A \times B} = SS_{\text{between treatments}} - SS_A - SS_B$$

$$df_{A \times B} = df_{\text{between treatments}} - df_A - df_B$$

$$F_A = \frac{MS_A}{MS_{\text{within}}} \qquad F_B = \frac{MS_B}{MS_{\text{within}}} \qquad F_{A \times B} = \frac{MS_{A \times B}}{MS_{\text{within}}}$$

$$\text{where each } MS = \frac{SS}{df}$$

PEARSON CORRELATION

$$r = \frac{SP}{\sqrt{SS_X SS_Y}}$$

$$\text{where } SP = \Sigma(X - M_X)(Y - M_Y) = \Sigma XY - \frac{(\Sigma X)(\Sigma Y)}{n}$$

SPEARMAN CORRELATION

$$r_s = 1 - \frac{6\Sigma D^2}{n(n^2 - 1)}$$

REGRESSION

$$\hat{Y} = bX + a \qquad \text{where } b = \frac{SP}{SS_X} \quad \text{and} \quad a = M_Y - bM_X$$

$$\text{standard error of estimate} = \sqrt{\frac{SS_{\text{error}}}{df}} = \sqrt{\frac{\Sigma(Y - \hat{Y})^2}{n - 2}}$$

CHI-SQUARE STATISTIC

$$\chi^2 = \Sigma \frac{(f_o - f_e)^2}{f_e}$$

MANN-WHITNEY *U*

$$U_A = n_A n_B + \frac{n_A(n_A + 1)}{2} - \Sigma R_A$$

$$U_B = n_A n_B + \frac{n_B(n_B + 1)}{2} - \Sigma R_B$$

KRUSKAL-WALLIS TEST

$$H = \frac{12}{N(N + 1)}\left(\Sigma \frac{T^2}{n}\right) - 3(N + 1)$$

MEASURES OF EFFECT SIZE

$$\text{Cohen's } d = \frac{\text{mean difference}}{\text{standard deviation}}$$

r^2 and η^2 (Percentage of Variance Accounted For)

$$r^2 = \frac{t^2}{t^2 + df} \qquad \text{(for Independent and Repeated } t\text{)}$$

$$\eta^2 = \frac{SS_{\text{treatment}}}{SS_{\text{treatment}} + SS_{\text{error term}}} \qquad \text{(for Analysis of Variance)}$$

www.wadsworth.com

www.wadsworth.com is the World Wide Web site for Wadsworth and is your direct source to dozens of online resources.

At *www.wadsworth.com* you can find out about supplements, demonstration software, and student resources. You can also send email to many of our authors and preview new publications and exciting new technologies.

www.wadsworth.com
Changing the way the world learns®

About the Authors

Frederick J Gravetter is a Professor of Psychology at the State University of New York College at Brockport. Dr. Gravetter has taught at Brockport since the early 1970s, specializing in statistics, experimental design, and cognitive psychology. He received his bachelor's degree in mathematics from M.I.T. and his Ph.D. in psychology from Duke University. In addition to publishing this textbook and several research articles, Dr. Gravetter co-authored *Research Methods for the Behavioral Science*s and *Essentials of Statistics for the Behavioral Sciences*.

Nicki and Fred

Larry B. Wallnau is Professor of Psychology at the State University of New York College at Brockport. The recipient of grants and awards in both research and teaching, Dr. Wallnau has published numerous research articles, primarily on the effect of psychotropic drugs. With Dr. Gravetter, he co-authored *Essentials of Statistics for the Behavioral Sciences*. He also has provided editorial consulting for publishers such as Allyn and Bacon and Harper & Row and for journals such as *Pharmacology Biochemistry and Behavior, Psychology & Behavior, Journal of Comparative Psychology*, and *Psychobiology*.

Nico and Larry

Statistics for the Behavioral Sciences

Sixth Edition

Frederick J Gravetter
State University of New York College at Brockport

Larry B. Wallnau
State University of New York College at Brockport

THOMSON
WADSWORTH

Australia · Canada · Mexico · Singapore · Spain · United Kingdom · United States

Psychology Editor: *Vicki Knight*
Assistant Editor: *Dan Moneypenny*
Editorial Assistant: *Monica Sarmiento*
Technology Project Manager: *Darin Derstine*
Marketing Manager: *Lori Grebe*
Marketing Assistant: *Laurel Anderson*
Advertising Project Manager: *Brian Chaffee*
Project Manager, Editorial Production: *Kirk Bomont*
Print/Media Buyer: *Jessica Reed*
Permissions Editor: *Sarah Harkrader*

Production Service: *Helen Walden*
Text Designer: *Kim Rokusek/Rokusek Design*
Copy Editor: *Helen Walden*
Illustrator: *Better Graphics, Inc.*
Cover Designer: *Lisa Berman*
Cover Art: *Nel Whatmore/Applejack Licensing*
Cover Printer: *Quebecor World/Taunton*
Compositor: *Better Graphics, Inc.*
Printer: *Quebecor World/Taunton*

For more information about our products, contact us at:
Thomson Learning Academic Resource Center
1-800-423-0563
For permission to use material from this text, contact us by:
Phone: 1-800-730-2214   **Fax:** 1-800-730-2215
Web: http://www.thomsonrights.com

Library of Congress Control Number: 2003103219

Student Edition: ISBN 0-534-60246-0
Instructor's Edition: ISBN 0-534-62203-8
International Student Edition: ISBN 0-534-62204-6

Wadsworth/Thomson Learning
10 Davis Drive
Belmont, CA 94002-3098
USA

Asia
Thomson Learning
5 Shenton Way #01-01
UIC Building
Singapore 068808

Australia/New Zealand
Thomson Learning
102 Dodds Street
Southbank, Victoria 3006
Australia

Canada
Nelson
1120 Birchmount Road
Toronto, Ontario M1K 5G4
Canada

Europe/Middle East/Africa
Thomson Learning
High Holborn House
50/51 Bedford Row
London WC1R 4LR
United Kingdom

Latin America
Thomson Learning
Seneca, 53
Colonia Polanco
11560 Mexico D.F.
Mexico

Spain/Portugal
Paraninfo
Calle/Magallanes, 25
28015 Madrid, Spain

Contents

Preface

Many students in the behavioral sciences view the required statistics course as an intimidating obstacle that has been placed in the middle of an otherwise interesting curriculum. They want to learn about human behavior—not about math and science. However, as long as the behavioral sciences are founded in science, a knowledge of statistics will be necessary. Statistical procedures provide researchers with objective and systematic methods for describing and interpreting their research results. Scientific research is the system that we use to gather information, and statistics are the tools that we use to distill the information into sensible and justified conclusions. The goal of this book is not only to teach the methods of statistics, but also to convey the basic principles of objectivity and logic that are essential for science and valuable in everyday life.

Those of you who are familiar with previous editions of *Statistics for the Behavioral Sciences* will notice that some changes have been made. These changes are summarized in the section entitled "To the Instructor." In revising this text, our students have been foremost in our minds. Over the years, they have provided honest and useful feedback. Their hard work and perseverance has made our writing and teaching most rewarding. We sincerely thank them. Students who are using this edition should please read the section of the preface entitled "To the Student."

ANCILLARIES

Ancillaries for this edition include the following:

- *Study Guide:* Contains chapter overviews, learning objectives, new terms and concepts, new formulas, step-by-step procedures for problem solving, study hints and cautions, self-tests, and review. The Study Guide contains answers to the self-test questions.

- *Instructor's Manual:* Contains test items, as well as solutions to all end-of-chapter problems included in the text.

- *Transparency CD-ROM:* Includes about 90 tables and figures taken directly from the text. Electronic; in Powerpoint.

ACKNOWLEDGMENTS

It takes a lot of good, hard-working people to produce a book. Our friends at Wadsworth have made an enormous contribution to this textbook. We thank our editor, Vicki Knight, who has been most supportive and encouraging; assistant editor, Dan Moneypenny, who also edited the supplements; editorial assistant, Monica Sarmiento; technology manager, Darin Derstine; senior production editor, Kirk Bomont; creative director, Vernon Boes; marketing manager, Lori Grebe; marketing assistant, Laurel Anderson; advertising project manager, Brian Chaffee; print/media buyer, Jessica Reed; and permissions editor, Sarah Harkrader.

Special thanks go to Helen Walden, who shepherded us through production, and to Roxy Peck, who checked our computations and made many helpful suggestions during the preparation of the manuscript.

Reviewers play a very important role in the development of a manuscript. Accordingly, we offer our appreciation to the following colleagues for their thoughtful reviews: Janet Andrews, Vassar College; Garth Bellah, Northwestern State University; Jeffrey Berman, University of Memphis; Mark Calogero, University of Washington; David Feigley, Rutgers—The State University of New Jersey; Kory Floyd, Arizona State University; Michael Gasser, University of Northern Iowa; Bryan Headricks, University of Wisconsin—Madison; Linda Henkel, Fairfield University; Stanley Klein, University of California, Santa Barbara; Meredith Kneavel, Hunter College; Elizabeth Krupinski, University of Arizona; Brian Powell, Indiana University; J. Whittaker, University of Utah; Lynette Zelezny, California State University, Fresno.

TO THE INSTRUCTOR   Those of you familiar with the previous edition of *Statistics for the Behavioral Sciences* will notice a number of changes in the Sixth Edition. A general summary of the revisions is as follows:

- Throughout the book, the topic of measuring effect size has been added as a supplement to hypothesis testing.
- In all chapters for which an SPSS analysis is feasible, there is a brief section at the end of the chapter that describes, step-by-step, how to enter data and run the SPSS analysis.
- The end-of-chapter problem sets have been revised.

The following are examples of the specific and noteworthy revisions.

Chapter 1   New figures illustrate the relationship between samples and populations and the concept of sampling error.

Chapter 2   The material on frequency distribution graphs has been reorganized so that different types of graphs are classified by the scale of measurement that was used. Also, the section on frequency distribution graphs for populations has been expanded and clarified.

Chapter 3   The notation for the sample mean has been changed from $\overline{X}$ (*X*-bar) to *M*. Also, the sections on when to use the median and the mode have been reorganized and clarified.

Chapter 4   The discussion of biased and unbiased statistics has been simplified by focusing on sample variance instead of sample standard deviation. The discussion of degrees of freedom has been expanded, and there is a stronger connection between the concept of degrees of freedom and the concept of an unbiased measure of sample variance.

Chapter 5   A new section demonstrates how *z*-scores can be used as inferential statistics to help evaluate the outcome of a research study. This section looks ahead to hypothesis testing and provides students with another example of the usefulness of *z*-scores.

Chapter 6   The unit normal table has been expanded by adding a column that lists the proportion between the mean and *z*. (Proportions in the body and the tail are still included as in previous editions.) Also, there are new examples demonstrating the use of the unit normal table.

Chapter 7 The section examining the relationship between standard error and sample size has been expanded. Also, a new section illustrates how standard error is used in the context of inferential statistics (looking ahead to hypothesis tests).

Chapter 8 A new section on effect size discusses the limitations of statistical significance and introduces the measurement of effect size as a supplement to the traditional hypothesis test.

Chapter 9 A new section on effect size extends the basic introduction in Chapter 8. The concept of measuring effect size using the percentage of explained variance (r^2) is introduced and the relationship between effect size and sample variance is examined.

Chapter 10 Measures of effect size are introduced for two-sample, independent-measures designs. Hypothesis tests and measures of effect size are examined in terms of how they are influenced by different factors such as sample size, sample variance, and mean difference.

Chapter 11 Measures of effect size are introduced for the repeated-measures t statistic.

Chapter 12 The section on estimation with z-scores has been eliminated so the chapter focuses entirely on estimation using the t statistic.

Chapter 13 A new section demonstrates that the variability between treatments is actually based on the variance (or SS) for the sample means. Also, measuring effect size is introduced for analysis of variance.

Chapter 14 Measuring effect size for the repeated-measures analysis of variance is introduced.

Chapter 15 Notation for the two-factor analysis of variance was changed to emphasize rows and columns in a data matrix rather than factors A and B. Also, measuring effect size is introduced for main effects and for the interaction.

Chapter 16 The z-score formula for a correlation has been clarified to account for situations in which the sample standard deviation is used in place of the population standard deviation. There is an expanded discussion differentiating the strength (or effect size) of a correlation as measured by r^2 and the significance of a correlation as measured by a hypothesis test. A new section also demonstrates the concept of explained variance (r^2) and unexplained variance ($1 - r^2$) in the context of regression.

Chapter 17 A new section discusses measuring effect size for the chi-square test for independence using either a phi-coefficient or Cramér's V, including a discussion on the difference between the strength of a relationship as measured by effect size and the significance of a relationship as measured by a hypothesis test.

Chapter 19 A new section introduces the normal approximation to the Wilcoxon test, paralleling the existing material on the normal approximation to the Mann-Whitney test.

TO THE STUDENT   A primary goal of this book is to make the task of learning statistics as easy and pain-less as possible. Among other things, you will notice that the book provides you with a number of opportunities to practice the techniques you will be learning in the form of learning checks, examples, demonstrations, and end-of-chapter problems. We encourage you to take advantage of these opportunities. Read the text rather than just memorize the formulas. We have taken care to present each statistical procedure in a conceptual context that explains why the procedure was developed and when it should be used. If you read this material and gain an understanding of the basic concepts underlying a statistical formula, you will find that learning the formula and how to use it will be much easier. In the following section, "Study Hints," we provide advice that we give our own students. Ask you instructor for advice as well; we are sure that other instructors will have ideas of their own.

Over the years, the students in our classes and other students using our book have given us valuable feedback. If you have any suggestions or comments about this book, you can write to either Professor Frederick Gravetter or Professor Emeritus Larry Wallnau at the Department of Psychology, SUNY College at Brockport, 350 New Campus Drive, Brockport, New York 14420. You can also contact Professor Gravetter directly at fgravett@brockport.edu.

Study Hints   You may find some of these tips helpful, as our own students have reported.

- The key to success in a statistics course is to keep up with the material. Each new topic builds on previous topics. If you have learned the previous material, then the new topic is just one small step forward. Without the proper background, however, the new topic can be a complete mystery. If you find that you are falling behind, get help immediately.

- You will learn (and remember) much more if you study for short periods several times per week rather than try to condense all of your studying into one long session. For example, it is far more effective to study half an hour every night than to have a single $3\frac{1}{2}$-hour study session once a week. We cannot even work on *writing* this book without frequent rest breaks.

- Do some work before class. Keep a little ahead of the instructor by reading the appropriate sections before they are presented in class. Although you may not fully understand what you read, you will have a general idea of the topic, which will make the lecture easier to follow. Also, you can identify material that is particularly confusing and then be sure the topic is clarified in class.

- Pay attention and think during class. Although this advice seems obvious, often it is not practiced. Many students spend so much time trying to write down every example presented or every word spoken by the instructor that they do not actually understand and process what is being said. Check with your instructor. There may not be a need to copy every example presented in class, especially if there are many examples like it in the text. Sometimes, we tell our students to put their pens and pencils down for a moment and just listen.

- Test yourself regularly. Do not wait until the end of the chapter or the end of the week to check your knowledge. After each lecture, work some of the end-of-chapter problems, and do the Learning Checks. Review the Demonstration Problems, and be sure you can define the Key Terms. If you are having trouble, get your questions answered *immediately* (reread the section, go to your instructor, or ask questions in class). By doing so, you will be able to move ahead to new material.

• Do not kid yourself! Avoid denial. Many students observe their instructor solve problems in class and think to themselves, "This looks easy, I understand it." Do you really understand it? Can you rellly do the problem on your own without having to leaf through the pages of a chapter? Although there is nothing wrong with using examples in the text as models for solving problems, you should try working a problem with your book closed to test your level of mastery.

• We realize that many students are embarassed to ask for help. It is our biggest challenge as instructors. You must find a way to overcome this aversion. Perhaps contacting the instructor directly would be a good starting point, if asking questions in class is too anxiety-provoking. You could be pleasantly surprised to find that your instructor does not yell, scold, or bite! Also, your instructor might know of another student who can offer assistance. Peer tutoring can be very helpful.

Frederick J Gravetter
Larry B. Wallnau

We have divided this book into four parts with the idea that each part covers a general topic area of statistics. The first part, which consists of Chapters 1–4, will provide a broad overview of statistical methods and a more focused presentation of those methods that are classified as *descriptive statistics*. To provide a frame of reference for introducing these topics, we ask you to imagine a psychologist who is studying the relationship between violent/aggressive behavior in children and the violence that they see on TV. Imagine that our psychologist obtains a group of 50 preschool children and divides the children into two groups. One group of 25 children watches a relatively violent one-half hour TV program immediately before playtime each afternoon. The other group of 25 children watches a nonviolent TV program for the same time period. The psychologist then observes the 50 children during their playtime and records the number of aggressive acts committed by each child. At this point, our psychologist has a set of 50 numbers, and the problem is to make some sense of the measurements that have been collected. This is a job for statistics. Specifically, statistical methods will provide our psychologist with a set of mathematical tools that can be used to organize and interpret the results from the research study.

By the time you finish the four chapters in this part you should have a good understanding of the general goals of statistics, and you should be familiar with the basic terminology and notation used in statistics. In addition, you should be familiar with the basic techniques of descriptive statistics. Specifically, you should be able to take a set of scores and organize them into a table or graph that provides an overall picture of the complete set. Also, you should be able to describe a set of scores by reporting the average value (central tendency) and a measure of variability that describes how the individual scores are distributed around the average.

C H A P T E R

1

Introduction to Statistics

Preview

Before we begin our discussion of statistics, we ask you to take a few moments to read the following paragraph, which has been adapted from a psychology experiment reported by Bransford and Johnson (1972).

> The procedure is actually quite simple. First you arrange things into different groups depending on their makeup. Of course, one pile may be sufficient, depending on how much there is to do. If you have to go somewhere else due to lack of facilities, that is the next step; otherwise you are pretty well set. It is important not to overdo any particular endeavor. That is, it is better to do too few things at once than too many. In the short run this may not seem important, but complications from doing too many can easily arise. A mistake can be expensive as well. The manipulation of the appropriate mechanisms should be self-explanatory, and we need not dwell on it here. At first the whole procedure will seem complicated. Soon, however, it will become just another facet of life. It is difficult to foresee any end to the necessity for this task in the immediate future, but then one never can tell.*

You probably find the paragraph a little confusing, and most of you probably think it is describing some obscure statistical procedure. Actually, this paragraph describes the everyday task of doing laundry. Now that you know the topic of the paragraph, try reading it again—it should make sense now.

Why did we begin a statistics textbook with a paragraph about washing clothes? Our goal is to demonstrate the importance of context—when not in the proper context, even the simplest material can appear difficult and confusing. In the Bransford and Johnson experiment, people who knew the topic before reading the paragraph were able to recall 73% more than people who did not know that it was about doing laundry. When you have the appropriate background, it is much easier to fit new material into your memory and to recall it later. In this book, we begin each chapter with a preview. The purpose of the preview is to provide the background or context for the new material in the chapter. As you read each preview section, you should gain a general overview of the chapter content. Remember: All statistical methods were developed to serve a purpose. If you understand why a new procedure is needed, you will find it much easier to learn the procedure.

The objectives for this first chapter are to provide an introduction to the topic of statistics and to give you some background for the rest of the book. We will discuss the role of statistics within the general field of scientific inquiry, and we will introduce some of the vocabulary and notation that are necessary for the statistical methods that follow. In some respects, this chapter serves as a preview section for the rest of the book.

As you read through the following chapters, keep in mind that the general topic of statistics follows a well-organized, logically developed progression that leads from basic concepts and definitions to increasingly sophisticated techniques. Thus, the material presented in the early chapters of this book will serve as a foundation for the material that follows. The content of the first nine chapters, for example, provides an essential background and context for the statistical methods presented in Chapter 10. If you turn directly to Chapter 10 without reading the first nine chapters, you will find the material confusing and incomprehensible. However, if you learn and use the background material, you will have a good frame of reference for understanding and incorporating new concepts as they are presented.

Finally, we cannot promise that learning statistics will be as easy as washing clothes. But if you begin each new topic with the proper context, you should eliminate some unnecessary confusion.

---

*Bransford, J. D., and Johnson, M. K. (1972), Contextual prerequisites for understanding: Some investigations of comprehension and recall. *Journal of Verbal Learning and Verbal Behavior, 11*, 717–726. Copyright by Academic Press. Reprinted by permission of the publisher and M. K. Johnson.

1.1   STATISTICS, SCIENCE, AND OBSERVATIONS

DEFINITIONS OF STATISTICS   By one definition, *statistics* consist of facts and figures such as average income, crime rate, birth rate, average snowfall, and so on. These statistics are usually informative and time-saving because they condense large quantities of information into a few simple figures. Later in this chapter we will return to the notion of calculating statistics (facts and figures) but, for now, we will concentrate on a much broader definition of statistics. Specifically, we will use the term statistics to refer to a set of mathematical procedures. In this case, we are using the term *statistics* as a shortened version of *statistical procedures*. For example, you are probably using this book for a statistics course in which you will learn about the statistical techniques that are used for research in the behavioral sciences.

Research in psychology (and other fields) involves gathering information. To determine, for example, whether violence on television has any effect on children's behavior, you would need to gather information about children's behaviors. When researchers finish the task of gathering information, they typically find themselves with pages and pages of measurements such as IQ scores, personality scores, reaction time scores, and so on. The role of statistics is to help researchers make sense of this information. Specifically, statistics serve two general purposes:

1. Statistics are used to organize and summarize the information so that the researcher can see what happened in the research study and can communicate the results to others.

2. Statistics help the researcher to answer the general questions that initiated the research by determining exactly what conclusions are justified based on the results that were obtained.

DEFINITION   The term **statistics** refers to a set of mathematical procedures for organizing, summarizing, and interpreting information.

Statistical procedures help ensure that the information or observations are presented and interpreted in an accurate and informative way. In somewhat grandiose terms, statistics help researchers bring order out of chaos. In addition, statistics provide researchers with a set of standardized techniques that are recognized and understood throughout the scientific community. Thus, the statistical methods used by one researcher will be familiar to other researchers, who can accurately interpret the statistical analyses with a full understanding of how the analysis was done and what the results signify.

1.2   POPULATIONS AND SAMPLES

WHAT ARE THEY?   Scientific research typically begins with a general question about a specific group (or groups) of individuals. For example, a researcher may be interested in the effect of divorce on the self-esteem of preteen children. Or a researcher may want to examine the attitudes toward abortion for men versus women. In the first example, the researcher is interested in the group of *preteen children*. In the second example, the researcher wants to compare the group of *men* with the group of *women*. In statistical terminology, the entire group that a researcher wishes to study is called a *population*.

DEFINITION

A **population** is the set of all the individuals of interest in a particular study.

As you can well imagine, a population can be quite large—for example, the entire set of women on the planet Earth. A researcher might be more specific, limiting the population for study to women who are registered voters in the United States. Perhaps the investigator would like to study the population consisting of women who are heads of state. Populations can obviously vary in size from extremely large to very small, depending on how the investigator defines the population. The population being studied should always be identified by the researcher. In addition, the population need not consist of people—it could be a population of rats, corporations, parts produced in a factory, or anything else an investigator wants to study. In practice, populations are typically very large, such as the population of fourth-grade children in the United States or the population of small businesses.

Although research questions concern an entire population, it usually is impossible for a researcher to examine every single individual in the population of interest. Therefore, researchers typically select a smaller, more manageable group from the population and limit their studies to the individuals in the selected group. In statistical terms, a set of individuals selected from a population is called a *sample*. A sample is intended to be representative of its population, and a sample should always be identified in terms of the population from which it was selected.

DEFINITION

A **sample** is a set of individuals selected from a population, usually intended to represent the population in a research study.

Just as we saw with populations, samples can vary in size. For example, one study might examine a sample of only 10 children in a preschool program, and another study might use a sample of over 1000 registered voters representing the population of a major city.

So far we have talked about a sample being selected from a population. However, this is actually only half of the full relationship between a sample and its population. Specifically, when a researcher finishes examining the sample the goal is to generalize the results back to the entire population. Remember that the research started with a general question about the population. To answer the question, a researcher studies a sample and then generalizes the results from the sample to the population. The full relationship between a sample and a population is shown in Figure 1.1.

PARAMETERS AND STATISTICS

When describing data it is necessary to distinguish whether the data come from a population or a sample. A characteristic that describes a population—for example, the population average—is called a *parameter*. On the other hand, a characteristic that describes a sample is called a *statistic*. Thus, the average score for a sample is an example of a statistic. Typically, the research process begins with a question about a population parameter. However, the actual data come from a sample and are used to compute sample statistics.

DEFINITION

A **parameter** is a value, usually a numerical value, that describes a population. A parameter may be obtained from a single measurement, or it may be derived from a set of measurements from the population.

FIGURE 1.1

The relationship between a population and a sample.

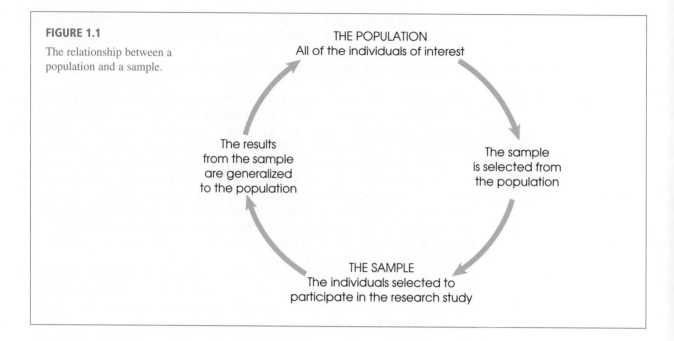

THE POPULATION
All of the individuals of interest

The sample
is selected from
the population

THE SAMPLE
The individuals selected to
participate in the research study

The results
from the sample
are generalized
to the population

DEFINITION

A **statistic** is a value, usually a numerical value, that describes a sample. A statistic may be obtained from a single measurement, or it may be derived from a set of measurements from the sample.

Typically, every population parameter has a corresponding sample statistic, and much of this book is concerned with the relationship between sample statistics and the corresponding population parameters. In Chapter 7, for example, we examine the relationship between the mean obtained for a sample and the mean for the population from which the sample was obtained.

DESCRIPTIVE AND INFERENTIAL STATISTICAL METHODS

The task of answering a research question begins with gathering information. In science, information is gathered by making observations and recording measurements for the individuals to be studied. The measurement or observation obtained for each individual is called a *datum* or, more commonly, a *score* or *raw score*. The complete set of scores or measurements is called the *data set* or simply the *data*. After data are obtained, statistical methods are used to organize and interpret the data.

DEFINITIONS

Data (plural) are measurements or observations. A **data set** is a collection of measurements or observations. A **datum** (singular) is a single measurement or observation and is commonly called a **score** or **raw score**.

Although researchers have developed a variety of different statistical procedures to organize and interpret data, these different procedures can be classified into two general categories. The first category, *descriptive statistics*, consists of statistical procedures that are used to simplify and summarize data.

DEFINITION

Descriptive statistics are statistical procedures used to summarize, organize, and simplify data.

Descriptive statistics are techniques that take raw scores and organize or summarize them in a form that is more manageable. Often the scores are organized in a table or a graph so that it is possible to see the entire set of scores. Another common technique is to summarize a set of scores by computing an average. Note that even if the data set has hundreds of scores, the average provides a single descriptive value for the entire set.

The second general category of statistical techniques is called *inferential statistics*. Inferential statistics are methods that use sample data to make general statements about a population.

DEFINITION

Inferential statistics consist of techniques that allow us to study samples and then make generalizations about the populations from which they were selected.

It usually is not possible to measure everyone in the population. Because populations are typically very large, a sample is selected to represent the population. By analyzing the results from the sample, we hope to make general statements about the population. Typically, researchers use sample statistics as the basis for drawing conclusions about population parameters.

One problem with using samples, however, is that a sample provides only limited information about the population. Although samples are generally *representative* of their populations, a sample is not expected to give a perfectly accurate picture of the whole population. There usually is some discrepancy between a sample statistic and the corresponding population parameter. This discrepancy is called *sampling error*, and it creates another problem to be addressed by inferential statistics (see Box 1.1).

DEFINITION

Sampling error is the discrepancy, or amount of error, that exists between a sample statistic and the corresponding population parameter.

The concept of sampling error is illustrated in Figure 1.2. The figure shows a population of 1000 college students and three samples, each with five students, that have been selected from the population. Notice that each sample contains different individuals who have different characteristics. Because the samples contain different people, the sample statistics vary from one sample to another. In addition, none of the three samples has statistics that are exactly the same as the population parameters. This is the basic concept of sampling error.

BOX

1.1   **THE MARGIN OF ERROR BETWEEN STATISTICS AND PARAMETERS**

One common example of sampling error is the error associated with a sample proportion. For example, in newspaper articles reporting results from political polls, you frequently find statements such as this:

Candidate Brown leads the poll with 51% of the vote. Candidate Jones has 42% approval, and the remaining 7% are undecided. This poll was taken from a sample of registered voters and has a margin of error of plus-or-minus 4 percentage points.

The "margin of error" is the sampling error. In this case, the percentages that are reported were obtained from a sample and are being generalized to the whole population. As always, you do not expect the statistics from a sample to be perfect. There always will be some "margin of error" when sample statistics are used to represent population parameters.

FIGURE 1.2

A demonstration of sampling error. Three samples are selected from the same population. Notice that the sample statistics are different from one sample to another, and all of the sample statistics are different from the corresponding population parameters. The natural differences that exist, by chance, between a sample statistic and a population parameter are called sampling error.

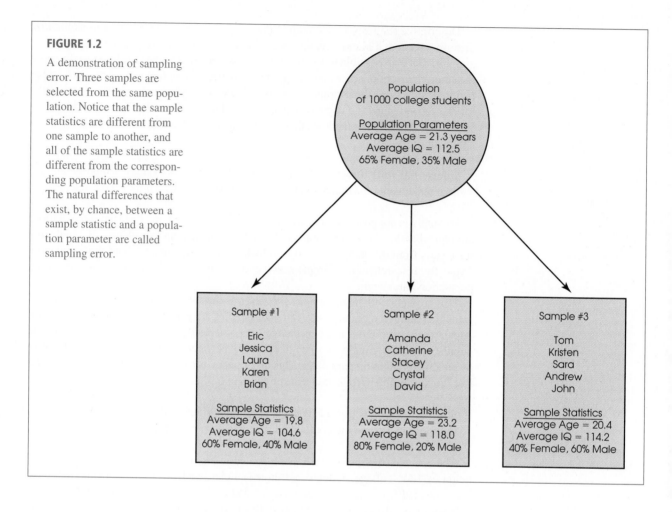

Figure 1.3 shows an overview of a general research situation and demonstrates the roles that descriptive and inferential statistics play. The purpose of the research study is to evaluate the difference between two methods for teaching reading to first-grade children. Two samples are selected from the population of first-grade children. The children in sample A are assigned to teaching method A and the children in sample B are assigned to method B. After 6 months, all of the students are given a standardized reading test. At this point, the researcher has two sets of data: the scores for sample A and the scores for sample B (see Figure 1.3). Now is the time to begin using statistics.

First, descriptive statistics are used to simplify the pages of data. For example, the researcher could draw a graph showing the scores for each sample or compute the average score for each sample. Note that descriptive methods provide a simplified, organized description of the scores. In this example, the students taught by method A averaged 76 on the standardized test, and the students taught by method B averaged only 71.

Once the researcher has described the results, the next step is to interpret the outcome. This is the role of inferential statistics. In this example, the researcher has found a 5-point difference between the two samples. The problem for inferential statistics is to differentiate between the following two interpretations:

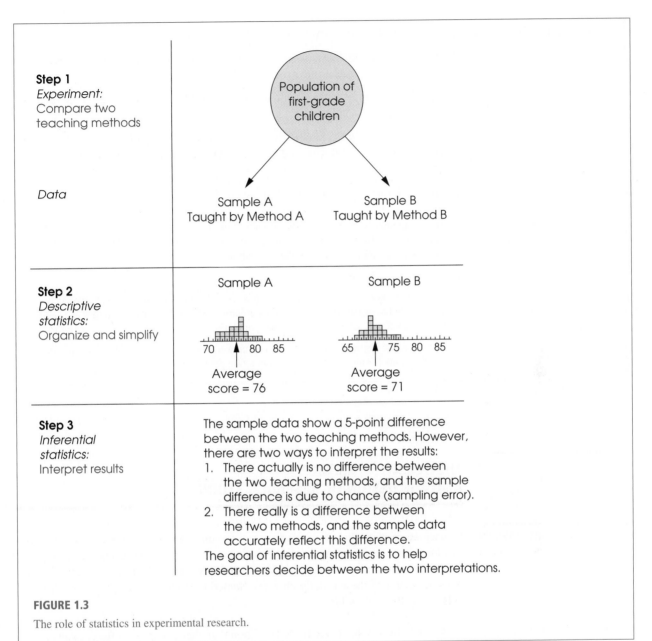

FIGURE 1.3

The role of statistics in experimental research.

1. There is no real difference between the two teaching methods, and the 5-point difference between the samples is just an example of sampling error (like the samples in Figure 1.2).

2. There really is a difference between the two teaching methods, and the 5-point difference between the samples was caused by the different methods of teaching.

In simple English, does the 5-point difference between samples provide convincing evidence of a difference between the two teaching methods or is the 5-point difference just chance? The purpose of inferential statistics is to answer this question.

1. What is a *population* and what is a *sample*?

2. A characteristic that describes a population, such as the population average, is called a _____.

3. The relationship between a population and a parameter is the same as the relationship between a sample and a _____.

4. Statistical techniques are classified into two general categories. What are the two categories called, and what is the general purpose for the techniques in each category?

5. Briefly define the concept of sampling error.

ANSWERS

1. The population is the entire set of individuals of interest for a particular research study. The sample is the specific set of individuals selected to participate in the study. The sample is expected to be representative of the population.

2. parameter   3. statistic

4. The two categories are descriptive statistics and inferential statistics. Descriptive techniques are intended to organize, simplify, and summarize data. Inferential techniques use sample data to reach general conclusions about populations.

5. Sampling error is the error or discrepancy between the value obtained for a sample statistic and the value for the corresponding population parameter.

1.3   THE SCIENTIFIC METHOD AND THE DESIGN OF RESEARCH STUDIES

RELATIONSHIPS BETWEEN VARIABLES

Science attempts to discover orderliness in the universe. Even people of ancient civilizations noted regularity in the world around them—the change of seasons, changes in the moon's phases, changes in the tides—and they were able to make many observations to document these orderly changes. Something that can change or have different values is called a *variable*.

DEFINITION

A **variable** is a characteristic or condition that changes or has different values for different individuals.

Variables may be characteristics that differ from one individual to another, such as height, weight, gender, or personality. Variables also can be environmental conditions that change, such as temperature, time of day, or the size of the room in which the research is being conducted.

When variables are measured, the resulting values are often identified by letters, usually X and Y. For example, the variable height could be identified by the letter X, and shoe size could be identified by Y. It is reasonable to expect a consistent relationship between these two variables. As X changes, Y also changes in a predictable way.

A value that does not change or vary is called a *constant*. For example, an instructor may adjust the exam scores for a class by adding 4 points to each student's score. Because every individual gets the same 4 points, this value is a constant.

DEFINITION

A **constant** is a characteristic or condition that does not vary but is the same for every individual.

In a research study, it is common to control variables by holding them constant. For example, a research study may examine only 10-year-old children who are all tested in the same room at 2 P.M. In this study, the participants' age, the room, and the time of day are all constant.

Researchers also examine relationships between variables. For example, there is a relationship between the amount of rainfall and crop growth. Rainfall is one of the variables. It varies from year to year and season to season. Crop growth is the other variable. Some years, the cornstalks seem short and stunted; other years, they are tall and full. When there is very little rainfall, the crops are short and shriveled. When rain is ample, the crops show vigorous growth. Note that in order to document the relationship, one must make observations—that is, measurements of the amount of rainfall and size of the crops.

There are a variety of research methods for obtaining observations and investigating the relationships between variables. Several of the more commonly used methods are discussed in the following sections.

THE CORRELATIONAL METHOD

The simplest way to look for relationships between variables is to make observations of the two variables as they exist naturally for a set of individuals. This is called the *correlational method*.

DEFINITION

With the **correlational method**, two variables are observed to see whether there is a relationship.

In a correlational study, a researcher measures two different variables for each individual. For example, a researcher might obtain a sample of 10 children and interview each child's parents to obtain a measure of how much television violence the child sees each week. Later, the researcher could observe the children on the school playground to obtain a measure of aggressive behavior for each child. Thus, the researcher has two scores, often called X and Y, for each of the children. Table 1.1 shows a hypothetical set of data for this correlational study. The goal of the study is to determine whether there is any consistent, predictable relationship between TV violence and aggressive behavior. We have constructed the data in Table 1.1 so that there is a relationship and, if you look closely, you should be able to see it. For these data, higher scores for TV violence tend to be associated with higher scores on aggression. Later, in Chapter 16, we will examine the statistical procedures that are used to measure and describe correlational relationships.

When a correlational study shows a relationship between two variables, it is often tempting to conclude that one variable is causing changes in the other variable. For example, it would be tempting to conclude that viewing TV violence causes children to behave violently. However, this conclusion is not justified. A limitation of the correlational method is that it simply describes the relationship—it does not explain the

TABLE 1.1

Hypothetical data showing results from a correlational study evaluating the relationship between exposure to TV violence and aggressive behavior for a sample of 10 children. Note that we have measured two different variables, obtaining two different scores, for each child. The data show a tendency for higher levels of TV violence to be associated with higher levels of aggressive behavior.

CHILD	AMOUNT OF TV VIOLENCE	AGGRESSIVE BEHAVIOR
A	4	11
B	9	18
C	2	7
D	8	15
E	9	17
F	6	14
G	2	4
H	7	14
I	3	8
J	8	17

cause-and-effect mechanism of the relationship. For example, it may be that children who are violent and aggressive simply prefer to watch TV programs that are also violent.

To establish a cause-and-effect relationship, it is necessary to exert a much greater level of control over the variables being studied. This is accomplished by the experimental method. We shall see that the experimental method is a highly structured, systematic approach to the study of relationships between variables.

THE EXPERIMENTAL METHOD

The goal of the experimental method is to establish a cause-and-effect relationship between two variables. That is, the method of observing variables is intended to show that changes in one variable are *caused* by changes in the other variable. To accomplish this goal, the experimental method has two distinguishing characteristics:

1. The researcher *manipulates* one of the variables and observes the second variable to determine whether or not the manipulation causes changes to occur.

2. The researcher must exercise some *control* over the research situation to ensure that other, extraneous variables do not influence the relationship being examined.

To demonstrate these two characteristics, consider an experiment in which a researcher is examining the effects of room temperature on memory performance. The purpose of the experiment is to determine whether changes in room temperature *cause* changes in memory performance.

In more complex experiments, a researcher may systematically manipulate more than one variable and may observe more than one variable. Here we are considering the simplest case, where only one variable is manipulated and only one variable is observed.

The researcher manipulates temperature by creating two or more different treatment conditions. For example, our researcher could set the temperature at 70° for one condition and then change the temperature to 90° for a second condition. The experiment would consist of observing the memory performance for a group of individuals (often called *subjects* or *participants*) in the 70° room and comparing their scores with another group that is tested in the 90° room. The structure of this experiment is shown in Figure 1.4.

To be able to say that differences in memory performance are caused by temperature, the researcher must rule out any other possible explanations for the difference. That is, any other variables that might affect memory performance must be controlled. Researchers typically use two basic techniques to control variables. First, the researcher could use *random assignment* so that each subject has an equal chance of being assigned to each of the treatment conditions. Random assignment helps assure that the subjects in one treatment condition are not substantially different from the subjects in

FIGURE 1.4

The structure of an experiment. Volunteers are randomly assigned to one of two treatment conditions: a 70° room or a 90° room. A list of words is presented and the participants are tested by writing down as many words as they can remember from the list. A difference between groups is attributed to the treatment (the temperature of the room).

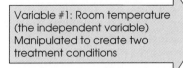

Variable #1: Room temperature (the independent variable) Manipulated to create two treatment conditions

Variable #2: Memory scores (the dependent variable) Measured in each of the treatment conditions

70° Room	90° Room
17	12
19	10
16	14
12	15
17	13
18	12
15	11
16	13

Any difference?

another treatment condition. For example, a researcher should not assign all the young subjects to one condition and all the old subjects to another. If this were done, the researcher could not be sure that the difference in memory performance was caused by temperature; instead, it could be caused by age differences. Second, the experimental method requires that the treatment conditions be identical except for the one variable that is being manipulated. This is accomplished by *controlling* or *holding constant* any other variables that might influence performance. For example, the researcher should test both groups of subjects at the same time of day, in the same room, with the same instructions, and so on. Again, the goal is to eliminate the contribution of all other variables, except for temperature, that might account for the difference in memory performance.

DEFINITION

In the **experimental method**, one variable is manipulated while changes are observed in another variable. To establish a cause-and-effect relationship between the two variables, an experiment attempts to eliminate or minimize the effect of all other variables by using *random assignment* and by *controlling* or *holding constant* other variables that might influence the results.

THE INDEPENDENT AND DEPENDENT VARIABLES

Specific names are used for the two variables that are studied by the experimental method. The variable that is manipulated by the experimenter is called the *independent variable*. It can be identified as the treatment conditions to which subjects are assigned. For the example in Figure 1.4, temperature is the independent variable. The variable that is observed to assess a possible effect of the manipulation is the *dependent variable*.

DEFINITIONS

The **independent variable** is the variable that is manipulated by the researcher. In behavioral research, the independent variable usually consists of the two (or more) treatment conditions to which subjects are exposed. The independent variable consists of the *antecedent* conditions that were manipulated *prior* to observing the dependent variable.

The **dependent variable** is the one that is observed in order to assess the effect of the treatment.

In psychological research, the dependent variable is typically a measurement or score obtained for each subject. For the temperature experiment (Figure 1.4), the dependent variable is the number of words recalled on the memory test. Differences between groups in performance on the dependent variable suggest that the manipulation had an effect. That is, changes in the dependent variable *depend* on the independent variable.

An experimental study evaluates the relationship between two variables by manipulating one variable (the independent variable) and measuring one variable (the dependent variable). Note that in an experiment only one variable is actually measured. You should realize that this is very different from a correlational study, where both variables are measured and the data consist of two separate scores for each individual.

Often an experiment will include a condition where the subjects do not receive any treatment. The scores from these subjects are then compared with scores from subjects who do receive the treatment. The goal of this type of study is to demonstrate that the treatment has an effect by showing that the scores in the treatment condition are substantially different from the scores in the no-treatment condition. In this kind of research, the no-treatment condition is called the *control condition*, and the treatment condition is called the *experimental condition*.

D E F I N I T I O N S

Individuals in a **control condition** do not receive the experimental treatment. Instead, they either receive no treatment or they receive a neutral, placebo treatment. The purpose of a control condition is to provide a baseline for comparison with the experimental condition.

Individuals in the **experimental condition** do receive the experimental treatment.

Note that the independent variable always consists of at least two values. (Something must have at least two different values before you can say that it is "variable.") For the temperature experiment (Figure 1.4), the independent variable is temperature (using values of 90° and 70°). For an experiment with an experimental group and a control group, the independent variable would be treatment versus no treatment.

PROBLEMS THAT CAN ARISE IN EXPERIMENTS

In general, the purpose of an experiment is to show that the manipulated variable, and *only* that variable, is responsible for the changes in the second variable. As noted earlier, this may be accomplished by controlling other variables. However, sometimes an uncontrolled variable can vary along with the independent variable. Any difference that exists between the treatment conditions other than the manipulated variable is called a *confounding variable*. When a researcher allows the treatment conditions to differ with respect to some variable other than the ones being manipulated or controlled, the experiment is flawed, and the researcher cannot interpret the results as demonstrating a cause-and-effect relationship.

D E F I N I T I O N

A **confounding variable** is an uncontrolled variable that is unintentionally allowed to vary systematically with the independent variable.

For example, an instructor would like to assess the effectiveness of computer laboratory exercises in assisting students to grasp the fundamentals of statistics. There are two treatment conditions. One section of statistics is taught three times a week as a lec-

FIGURE 1.5

In this experiment, the effect of instructional method (the independent variable) on test performance (the dependent variable) is examined. However, any difference between groups in performance cannot be attributed to the method of instruction. In this experiment, there is a confounding variable. The instructor teaching the course varies with the independent variable, so that the treatment of the groups differs in more ways than one (instructional method and instructor vary).

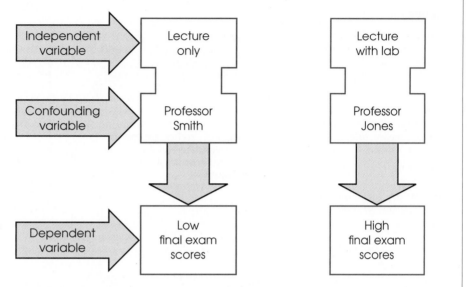

ture course. A second section of students also meets three times a week, but one of these meetings is a laboratory of computer exercises in statistics. Professor Smith teaches the lecture, and Professor Jones teaches the second group of students. At the end of the semester, both groups receive the same final exam, and their performances are compared. The experiment is summarized in Figure 1.5.

Let's assume that a statistical analysis determined that the lecture/lab students performed better on the final exam. Can the instructors conclude that the method of instruction had an effect on learning? The answer is emphatically *no*. The independent variable, method of instruction, is not the only way the groups differ in terms of how they are treated. The instructor was allowed to vary along with the method of instruction. It is possible that the lecture/lab group performed better because Professor Jones is a better instructor, one who motivates students and can explain concepts in understandable terms. That is, the difference between groups in performance might be due to the different instructors rather than the different treatments. One cannot be sure which interpretation is correct. When a study is confounded, it is impossible to make meaningful conclusions about the results. The lesson in this example is that as many variables as possible must be controlled.

OTHER RESEARCH DESIGNS

You should notice that evaluating the results from a correlational study involves looking for patterns *within* a set of data (see Table 1.1). On the other hand, evaluating the results from an experimental study involves looking for differences *between* two or more sets of data (see Figure 1.4). Evaluating the differences between sets of data is a common statistical technique that is used in a wide variety of research situations. In particular, there are a number of different research designs that are not true experiments, but still involve comparing two or more sets of data. Figure 1.6 shows two common examples.

The top part of the figure shows a research study comparing two groups of participants in which the two groups are defined by a participant variable (in this case, gender). A participant variable is a preexisting characteristic such as age or gender that varies from one individual to another. Notice that this study involves comparing two groups of scores (like an experiment). However, you should also notice that the

FIGURE 1.6

Two examples of nonexperimental studies that involve comparing two groups of scores. In (a), a participant variable (gender) is used to create groups, and then the dependent variable (verbal score) is measured in each group. In (b), time is the variable used to define the two groups, and the dependent variable (depression) is measured at each of the two times.

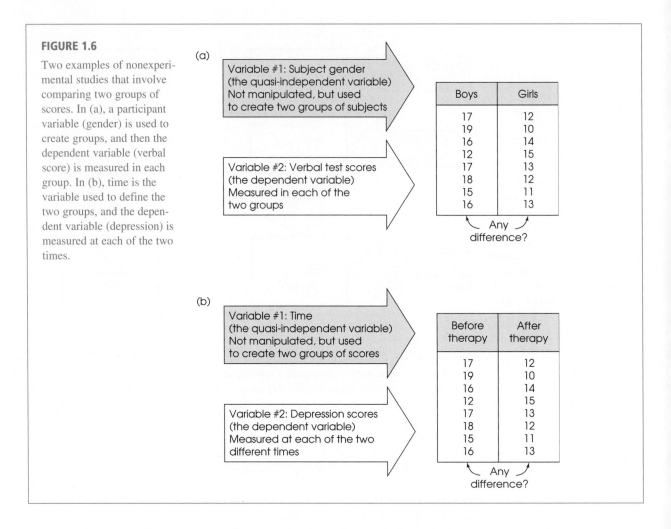

researcher did not manipulate gender to create the two groups. Specifically, the researcher cannot assign individuals to groups and then make one group into females and make the second group into males. Because there is no manipulation and no control over group assignment to create equivalent groups, this study is not a true experiment.

The bottom part of Figure 1.6 also shows a study comparing two groups of scores. The study measures depression scores for a group of participants before therapy and again after therapy. In this case, the two groups of scores are obtained by measuring the same individuals at different points in time. Although the researcher is manipulating a treatment, it is impossible to control or manipulate the passage of time. Also, the researcher has no control over other variables that change with time. For example, the weather could change from dark and gloomy before therapy to bright and sunny after therapy. In this case, the depression scores could improve because of the weather and not because of the therapy. Because the researcher cannot control the passage of time or other variables related to time, this study is not a true experiment.

Although the two research studies shown in Figure 1.6 are not true experiments, you should notice that they produce the same kind of data that are found in an experiment (see Figure 1.4). In each case, one variable is used to create groups, and a second variable is measured to obtain scores within each group. In an experiment, the groups are defined by different values of the independent variable and the scores are the dependent variable. In other types of research, such as the examples shown in Figure 1.6, the variable that defines the groups is called the *quasi-independent variable* and the scores are called the dependent variable.

DEFINITION

Many research studies involve comparing groups that were not created by manipulating an independent variable. Instead, the groups are usually determined by a participant variable (such as male versus female) or a time variable (such as before treatment versus after treatment). In these nonexperimental studies, the variable that determines the groups is called a **quasi-independent variable**.

Most of the statistical procedures presented in this book are designed for research studies that compare sets of scores like the experimental study in Figure 1.4 and the nonexperimental studies in Figure 1.6. Specifically, we will examine descriptive statistics that summarize and describe the scores in each group, and we will examine inferential statistics that will allow us to use the groups, or samples, to generalize to the entire population.

LEARNING CHECK

1. A researcher observes that elderly people who take regular doses of an anti-inflammatory drug tend to have a lower risk of Alzheimer's disease than their peers who do not take the drug. Is this study an example of the correlational method or the experimental method?

2. What two elements are necessary for a research study to be an experiment?

3. The results from an experiment indicate that increasing the amount of indoor lighting during the winter months results in significantly lower levels of depression. Identify the independent variable and the dependent variable for this study.

ANSWERS

1. This is a correlational study. The researcher is simply observing the variables.

2. First, the researcher must manipulate one of the two variables being studied. Second, all other variables that might influence the results must be controlled.

3. The independent variable is the amount of indoor lighting, and the dependent variable is a measure of depression for each individual.

1.4 VARIABLES AND MEASUREMENT

The scores that make up the data from a research study are the result of observing and measuring variables. For example, a researcher may finish a study with a set of IQ scores, or personality scores, or reaction time scores. In this section, we will take a closer look at the variables that are being measured and the process of measurement.

CONSTRUCTS AND OPERATIONAL DEFINITIONS

Many of the variables studied by behavioral scientists are actually hypothetical concepts that are used to help describe and interpret behavior. For example, we say that a child does well in school because he or she is *intelligent*. Or we say that someone is anxious in a social situation because he or she has low *self-esteem*. The concepts of "intelligence" and "self-esteem" are called *constructs*, and because they cannot be directly observed, they are hypothetical.

If constructs are hypothetical and cannot be observed, then how can they possibly be studied? The answer is that we have to *define* the construct so that it can be studied. An *operational definition* defines a construct in terms of an observable and measurable response. This definition should include the process and operations involved in making the observations. For example, an operational definition for emotionality might be stated as the amount of increase in heart rate after a person is insulted. An operational definition of intelligence might be the score on the Wechsler Adult Intelligence Scale.

DEFINITIONS   **Constructs** are hypothetical concepts that are used to help describe and explain behavior.

An **operational definition** defines a construct in terms of specific operations or procedures and the measurements that result from them. Thus, an operational definition consists of two components: First, it describes a set of operations or procedures for measuring a construct. Second, it defines the construct in terms of the resulting measurements.

DISCRETE AND CONTINUOUS VARIABLES

The variables in a study can be characterized by the type of values that can be assigned to them. A *discrete variable* consists of separate, indivisible categories. For this type of variable, there are no intermediate values between two adjacent categories. Consider the values displayed when dice are rolled. Between neighboring values—for example, seven dots and eight dots—no other values can ever be observed.

DEFINITION   A **discrete variable** consists of separate, indivisible categories. No values can exist between two neighboring categories.

Discrete variables are commonly restricted to whole countable numbers—for example, the number of children in a family or the number of students attending class. If you observe class attendance from day to day, you may find 18 students one day and 19 students the next day. However, it is impossible ever to observe a value between 18 and 19. A discrete variable may also consist of observations that differ qualitatively. For example, a psychologist observing patients may classify some as having panic disorders, others as having dissociative disorders, and some as having psychotic disorders. The type of disorder is a discrete variable because there are distinct and finite categories that can be observed.

On the other hand, many variables are not discrete. Variables such as time, height, and weight are not limited to a fixed set of separate, indivisible categories. You can measure time, for example, in hours, minutes, seconds, or fractions of seconds. These variables are called *continuous* because they can be divided into an infinite number of fractional parts.

DEFINITION   For a **continuous variable**, there are an infinite number of possible values that fall between any two observed values. A continuous variable is divisible into an infinite number of fractional parts.

Suppose, for example, that a researcher is measuring weights for a group of individuals participating in a diet study. Because weight is a continuous variable, it can be pictured as a continuous line (see Figure 1.7). Note that there are an infinite number of possible points on the line without any gaps or separations between neighboring points. For any two different points on the line, it is always possible to find a third value that is between the two points.

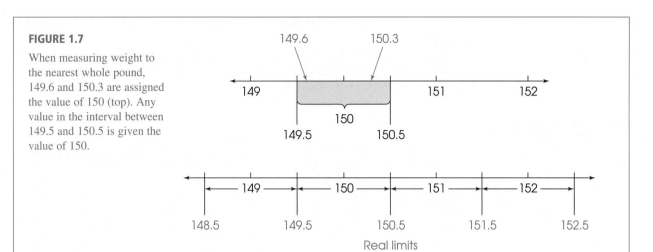

FIGURE 1.7

When measuring weight to the nearest whole pound, 149.6 and 150.3 are assigned the value of 150 (top). Any value in the interval between 149.5 and 150.5 is given the value of 150.

Two other factors apply to continuous variables:

1. When measuring a continuous variable, it should be very rare to obtain identical measurements for two different individuals. Because a continuous variable has an infinite number of possible values, it should be almost impossible for two people to have exactly the same score. If the data show a substantial number of tied scores, then you should suspect that the measurement procedure is very crude or that the variable is not really continuous.

2. When measuring a continuous variable, each measurement category is actually an *interval* that must be defined by boundaries. For example, two people who both claim to weigh 150 pounds are probably not *exactly* the same weight. However, they are both around 150 pounds. One person may actually weigh 149.6 and the other 150.3. Thus, a score of 150 is not a specific point on the scale but instead is an interval (see Figure 1.7). To differentiate a score of 150 from a score of 149 or 151, we must set up boundaries on the scale of measurement. These boundaries are called *real limits* and are positioned exactly halfway between adjacent scores. Thus, a score of $X = 150$ pounds is actually an interval bounded by a lower real limit of 149.5 at the bottom and an upper real limit of 150.5 at the top. Any individual whose weight falls between these real limits will be assigned a score of $X = 150$.

DEFINITION **Real limits** are the boundaries of intervals for scores that are represented on a continuous number line. The real limit separating two adjacent scores is located exactly halfway between the scores. Each score has two real limits. The **upper real limit** is at the top of the interval, and the **lower real limit** is at the bottom.

The concept of real limits applies to any measurement of a continuous variable, even when the score categories are not whole numbers. For example, if you were measuring time to the nearest tenth of a second, the measurement categories would be 31.0, 31.1, 31.2, and so on. Each of these categories represents an interval on the scale that is bounded by real limits. For example, a score of $X = 31.1$ seconds indicates that the actual measurement is in an interval bounded by a lower real limit of 31.05 and an upper real limit of 31.15. Remember that the real limits are always halfway between adjacent categories.*

Later in this book, real limits are used for constructing graphs and for various calculations with continuous scales. For now, however, you should realize that real limits are a necessity whenever you make measurements of a continuous variable.

SCALES OF MEASUREMENT

It should be obvious by now that data collection requires that we make measurements of our observations. Measurement involves either categorizing events (qualitative measurements) or using numbers to characterize the size of the event (quantitative measurement). Several types of scales are associated with measurements. The distinctions among the scales are important because they underscore the limitations of certain types of measurements and because certain statistical procedures are appropriate for data collected on some scales but not on others. If you were interested in people's heights, for example, you could measure a group of individuals by simply classifying them into three categories: tall, medium, and short. However, this simple classification would not tell you much about the actual heights of the individuals, and these measurements would not give you enough information to calculate an average height for the group. Although the simple classification would be adequate for some purposes, you would need more-sophisticated measurements before you could answer more-detailed questions. In this section, we examine four different scales of measurement, beginning with the simplest and moving to the most sophisticated.

THE NOMINAL SCALE

A *nominal scale* of measurement labels observations so that they fall into different categories.

DEFINITION

A **nominal scale** consists of a set of categories that have different names. Measurements on a nominal scale label and categorize observations, but do not make any quantitative distinctions between observations.

The word *nominal* means "having to do with names." Measurements that are made on this scale involve naming things. For example, if we wish to know the sex of a person responding to a questionnaire, it would be measured on a nominal scale consisting of two categories. A product warranty card might have you check the box that best describes your occupation, and it lists "sales," "professional," "skilled trade," and "other (please specify)." Occupations are being measured in terms of category names; therefore, a nominal scale is being used. A researcher observing the behavior of a group of infant monkeys might categorize responses as playing, grooming, feeding, acting aggressive, or showing submissiveness. Again, this instance typifies a nominal scale of measurement.

---

*Technical Note: It is important to distinguish between the *real limits* for an interval and the process of rounding scores. The real limits form boundaries that define an interval on a continuous scale. For example, the real limits 32.5 and 33.5 define an interval that is identified by the score $X = 33$. The real limits, however, are not necessarily a part of the interval. In this example, 32.5 is the lower real limit of the interval. If you are using a rounding process that causes 32.5 to be rounded up, then a measurement of 32.5 would be rounded to 33 and would be included in the 32.5–33.5 interval. On the other hand, if your rounding process causes 32.5 to be rounded down to 32, then this value would not be included in the interval. In general, the question of whether an upper real limit or a lower real limit belongs in an interval is determined by the rounding process that you have adopted.

If two individuals are measured on a nominal scale, it is possible to determine whether the two measurements are the same or different. However, if the measurements are different, you cannot determine how big the difference is, and you cannot say that one measurement is "more" or "less" than the other. A nominal scale consists of *qualitative* differences. It does not provide any information about *quantitative* differences between individuals.

Although the categories on a nominal scale are not quantitative values, they are occasionally represented by numbers. For example, the rooms or offices in a building may be identified by numbers. You should realize that the room numbers are simply names and do not reflect any quantitative information. Room 109 is not necessarily bigger than Room 100 and certainly not 9 points bigger. It also is fairly common to use numerical values as a code for nominal categories when data are entered into computer programs. For example, the data from a survey may code males with a 0 and females with a 1. Again, the numerical values are simply names and do not represent any quantitative difference. The scales that follow do reflect an attempt to make quantitative distinctions.

THE ORDINAL SCALE

With an *ordinal* scale of measurement, the categories that make up the scale not only have separate names (as in a normal scale), but also are ranked in terms of magnitude. Thus, as the word *ordinal* implies, observations measured on an ordinal scale are categorized and arranged in rank order.

DEFINITION

An **ordinal scale** consists of a set of categories that are organized in an ordered sequence. Measurements on an ordinal scale rank observations in terms of size or magnitude.

Often, an ordinal scale consists of a series of ranks (first, second, third, and so on) like the order of finish in a horse race. Occasionally, the categories are identified by verbal labels like small, medium, and large drink sizes at a fast-food restaurant. In either case, the fact that the categories form an ordered sequence means that there is a directional relationship between categories. With measurements from an ordinal scale, you can determine whether two individuals are different and you can determine the direction of difference. However, ordinal measurements do not allow you to determine the magnitude of the difference between two individuals. For example, if Billy is placed in the low-reading group and Tim is placed in the high-reading group, you know that Tim is a better reader, but you do not know how much better.

THE INTERVAL AND RATIO SCALES

An *interval scale* of measurement consists of an ordered set of categories (like an ordinal scale) with the additional requirement that the categories form a series of intervals that are all exactly the same size. The additional feature of equal-sized intervals makes it possible to compute distances between values on an interval scale. On a ruler, for example, a 1-inch interval is the same size at every location on the ruler, and a 4-inch distance is exactly the same size no matter where it is measured on the ruler. Thus, an interval scale allows you to measure differences in the size or amount of events.

Another characteristic of an interval scale is that it has an arbitrary zero point. That is, the value zero is assigned to a particular location on the scale simply as a matter of convenience or reference. Specifically, a value of zero does *not* indicate the total absence of the variable being measured. For example, a temperature of 0 degrees

Fahrenheit does not mean that there is no temperature, and it does not prohibit the temperature from going even lower.

A *ratio scale* of measurement has all the features of an interval scale, but adds an absolute zero point. That is, on a ratio scale, a value of zero indicates none (a complete absence) of the variable being measured. The advantage of an absolute zero is that ratios of numbers on the scale reflect ratios of magnitude for the variable being measured. Thus, a ratio scale not only allows you to measure the difference between two individuals, but also allows you to describe the difference in terms of ratios. For example, you can state that one measurement is three times larger than another, or you can say that one score is only half as large as another. The distinction between an interval scale and a ratio scale is demonstrated in Example 1.1.

DEFINITIONS

An **interval scale** consists of ordered categories that are all intervals of exactly the same size. With an interval scale, equal differences between numbers on the scale reflect equal differences in magnitude. However, ratios of magnitudes are not meaningful.

A **ratio scale** is an interval scale with the additional feature of an absolute zero point. With a ratio scale, ratios of numbers do reflect ratios of magnitude.

EXAMPLE 1.1

A researcher obtains measurements of height for a group of 8-year-old boys. Initially, the researcher simply records each child's height in inches, obtaining values such as 44, 51, 49, and so on. These initial measurements constitute a ratio scale. A value of zero represents no height (absolute zero). Also, it is possible to use these measurements to form ratios. For example, a child who is 80 inches tall is twice as tall as a 40-inch-tall child.

Now suppose that the researcher converts the initial measurement into a new scale by calculating the difference between each child's actual height and the average height for this age group. A child who is 1 inch taller than average now gets a score of +1; a child 4 inches taller than average gets a score of +4. Similarly, a child who is 2 inches shorter than average gets a score of −2. The new scores constitute an interval scale of measurement. A score of zero no longer indicates an absence of height; now it simply means average height.

Notice that both sets of scores involve measurement in inches, and you can compute differences, or intervals, on either scale. For example, there is a 6-inch difference in height between two boys who measure 57 and 51 inches tall on the first scale. Likewise, there is a 6-inch difference between two boys who measure +9 and +3 on the second scale. However, you should also notice that ratio comparisons are not possible on the second scale. For example, a boy who measures +9 is not three times as tall as a boy who measures +3.

For most statistical applications, the distinction between an interval scale and a ratio scale is not particularly important. As we have seen, the process of measuring inches on a ruler can produce either interval scores or ratio scores depending on how the ruler is used. If you begin measuring at ground level (starting at zero) the ruler will produce ratio scores. On the other hand, measuring distances from an arbitrary (nonzero) point produces interval scores. However, the values from either scale allow basic arithmetic operations that permit us to calculate differences between scores, to sum scores, and to calculate average scores. For this reason, most dependent variables we will encounter are measured on either an interval or a ratio scale. Finally, you should realize that the distinction between different scales of measurement is often unclear when considering specific measurements. For example, the scores resulting from an IQ test are usually treated as measurements on an interval scale, but many researchers believe that IQ scores are more accurately described as ordinal data. An IQ score of 105 is clearly greater than a score of 100, but there is some question concerning how much difference in intelligence is reflected in the 5-point difference between these two scores.

LEARNING CHECK

1. The local fast-food restaurant offers small, medium, and large soft drinks. What kind of scale is used to measure the size of the drinks?

2. The Scholastic Achievement Test (SAT) most likely measures aptitude on a(n) _____ scale.

3. In a study on perception of facial expressions, participants must classify the emotions displayed in photographs of people as anger, sadness, joy, disgust, fear, or surprise. Emotional expression is measured on a(n) _____ scale.

4. A researcher studies the factors that determine how many children couples decide to have. The variable, number of children, is a _____ (discrete/continuous) variable.

5. An investigator studies how concept-formation ability changes with age. Age is a _____ (discrete/continuous) variable.

6. **a.** When measuring weight to the nearest pound, what are the real limits for a score of $X = 150$ pounds?
   b. When measuring weight to the nearest $\frac{1}{2}$ pound, what are the real limits for a score of $X = 144.5$ pounds?

ANSWERS   1. ordinal   2. interval   3. nominal   4. discrete   5. continuous

6. **a.** 149.5 and 150.5   **b.** 144.25 and 144.75

1.5 STATISTICAL NOTATION

Measurements of behavior usually will provide data composed of numerical values. These numbers form the basis of the computations that are done for statistical analyses. There is a standardized notation system for statistical procedures, and it is used to identify terms in equations and mathematical operations. Some general mathematical operations, notation, and basic algebra are outlined in the review section of Appendix A. There is also a skills assessment exam (p. A-1) to help you determine whether you need the basic mathematics review. Here we will introduce some statistical notation that is used throughout this book. In subsequent chapters, additional notation will be introduced as it is needed.

SCORES

X		X	Y
37		72	165
35		68	151
35		67	160
30		67	160
25		68	146
17		70	160
16		66	133

Making observations of a dependent variable in a study will typically yield values or scores for each subject. Raw scores are the original, unchanged set of scores obtained in the study. Scores for a particular variable are represented by the letter X. For example, if performance in your statistics course is measured by tests and you obtain a 35 on the first test, then we could state that $X = 35$. A set of scores can be presented in a column that is headed by X. For example, a list of quiz scores from your class might be presented as shown in the margin (the single column on the left).

When observations are made for two variables, there will be two scores for each subject. The data can be presented as two lists labeled X and Y for the two variables. For example, observations for people's height in inches (variable X) and weight in pounds (variable Y) can be presented as shown in the margin (the double column on the right). Each pair X, Y represents the observations made of a single participant.

It is also useful to specify how many scores are in a set. We will use an uppercase letter N to represent the number of scores in a population and a lowercase letter n to represent the number of scores in a sample. Throughout the remainder of the book you will notice that we often will use notational differences to distinguish between samples and populations. For the height and weight data in the preceding table, $n = 7$ for both variables. Note that by using a lowercase letter n, we are implying that these data are a sample.

SUMMATION NOTATION

Many of the computations required in statistics will involve adding a set of scores. Because this procedure is used so frequently, a special notation is used to refer to the sum of a set of scores. The Greek letter *sigma*, or Σ, is used to stand for summation. The expression ΣX means to add all the scores for variable X. The summation sign Σ can be read as "the sum of." Thus, ΣX is read "the sum of the scores." For the following set of quiz scores,

$$10, \quad 6, \quad 7, \quad 4$$

$\Sigma X = 27$ and $N = 4$.

To use summation notation correctly, keep in mind the following two points:

1. The summation sign Σ is always followed by a symbol or mathematical expression. The symbol or expression identifies exactly which values are to be summed. To compute ΣX, for example, the symbol following the summation sign is X, and the task is to find the sum of the X values. On the other hand, to compute $\Sigma(X - 1)^2$, the summation sign is followed by a relatively complex mathematical expression, so your first task is to calculate all of the $(X - 1)^2$ values and then sum the results.

2. The summation process is often included with several other mathematical operations, such as multiplication or squaring. To obtain the correct answer, it is essential that the different operations be done in the correct sequence. Following is a list showing the correct *order of operations* for performing mathematical operations. Most of this list should be familiar, but you should note that we have inserted the summation process as the fourth operation in the list.

More information on the order of operations for mathematics is available in the Math Review appendix, page 665.

Order of Mathematical Operations

1. Any calculation contained within parentheses is done first.

2. Squaring (or raising to other exponents) is done second.

3. Multiplying and/or dividing is done third. A series of multiplication and/or division operations should be done in order from left to right.

4. Summation using the Σ notation is done next.

5. Finally, any other addition and/or subtraction is done.

The following examples demonstrate how summation notation will be used in most of the calculations and formulas we will present in this book.

EXAMPLE 1.2

X	X^2
3	9
1	1
7	49
4	16

A set of four scores consists of values 3, 1, 7, and 4. We will compute ΣX, ΣX^2, and $(\Sigma X)^2$ for these scores. To help demonstrate the calculations, we have constructed a *computational table* showing the original scores (the X values) in the first column. Additional columns can then be used to show additional steps that may be required in the calculation. The table to the left shows the squared scores (the X^2 values) that are needed to compute ΣX^2.

The first calculation, ΣX, does not include any parentheses, squaring, or multiplication, so we go directly to the summation operation. The X values are listed in the first column of the table, and we simply sum the values in this column:

$$\Sigma X = 3 + 1 + 7 + 4 = 15$$

To compute ΣX^2, the correct order of operations is to square each score and then find the sum of the squared values. The computational table shows the original scores and the results obtained from squaring (the first step in the calculation). The second step is to find the sum of the squared values, so we simply add the numbers in the X^2 column.

$$\Sigma X^2 = 9 + 1 + 49 + 16 = 75$$

The final calculation, $(\Sigma X)^2$, includes parentheses, so the first step is to perform the calculation inside the parentheses. Thus, we first find ΣX and then square this sum. Earlier, we computed $\Sigma X = 15$, so

$$(\Sigma X)^2 = (15)^2 = 225$$

EXAMPLE 1.3

We will use the same set of four scores from Example 1.2 and compute $\Sigma(X - 1)$ and $\Sigma(X - 1)^2$. The following computational table will help demonstrate the calculations.

X	$(X - 1)$	$(X - 1)^2$	
3	2	4	The first column lists the
1	0	0	original scores. A second
7	6	36	column lists the $(X - 1)$
4	3	9	values, and a third column
			shows the $(X - 1)^2$ values.

To compute $\Sigma(X - 1)$, the first step is to perform the operation inside the parentheses. Thus, we begin by subtracting one point from each of the X values. The resulting values are listed in the middle column of the table. The next step is to sum the $(X - 1)$ values.

$$\Sigma(X - 1) = 2 + 0 + 6 + 3 = 11$$

The calculation of $\Sigma(X - 1)^2$ requires three steps. The first step (inside parentheses) is to subtract 1 point from each X value. The results from this step are shown in the middle column of the computational table. The second step is to square each of the $(X - 1)$ values. The results from this step are shown in the third column of the table. The final step is to sum the $(X - 1)^2$ values to obtain

$$\Sigma(X - 1)^2 = 4 + 0 + 36 + 9 = 49$$

Notice that this calculation requires squaring before summing. A common mistake is to sum the $(X - 1)$ values and then square the total. Be careful!

EXAMPLE 1.4   In both of the preceding examples, and in many other situations, the summation operation is the last step in the calculation. According to the order of operations, parentheses, exponents, and multiplication all come before summation. However, there are situations in which extra addition and subtraction are completed after the summation. For this example we will use the same scores that appeared in the previous two examples, and we will compute $\Sigma X - 1$.

With no parentheses, exponents, or multiplication, the first step is the summation. Thus, we begin by computing ΣX. Earlier we found $\Sigma X = 15$. The next step is to subtract one point from the total. For these data,

$$\Sigma X - 1 = 15 - 1 = 14$$

EXAMPLE 1.5   For this example, each individual has two scores. The first score is identified as X, and the second score is Y. With the help of the following computational table, we will compute ΣX, ΣY, and ΣXY.

X	Y	XY
3	5	15
1	3	3
7	4	28
4	2	8

To find ΣX, simply sum the values in the X column.

$$\Sigma X = 3 + 1 + 7 + 4 = 15$$

Similarly, ΣY is the sum of the Y values.

$$\Sigma Y = 5 + 3 + 4 + 2 = 14$$

To compute ΣXY, the first step is to multiply X times Y for each individual. The resulting products (XY values) are listed in the third column of the table. Finally, we sum the products to obtain

$$\Sigma XY = 15 + 3 + 28 + 8 = 54$$

SUMMARY

1. The term *statistics* is used to refer to methods for organizing, summarizing, and interpreting data.

2. Scientific questions usually concern a population that is the entire set of individuals that one wishes to study. Usually, populations are so large that it is impossible to examine every individual, so most research is conducted with samples. A sample is a group selected from a population, usually for purposes of a research study.

3. Statistical methods can be classified into two broad categories: descriptive statistics, which organize and summarize data, and inferential statistics, which use sample data to draw inferences about populations.

4. The correlational method looks for interrelationships between variables but cannot determine the cause-and-effect nature of the relationship. The experimental method is able to establish causes and effects in a relationship.

5. In the experimental method, one variable (the independent variable) is manipulated, and another variable (the dependent variable) is observed for changes that may occur as a result of the manipulation. All other variables are controlled or held constant.

6. Constructs are hypothetical concepts used in theories to describe the mechanisms of behavior. Because they are hypothetical, they cannot be observed. Constructs are studied by providing operational definitions for them. An operational definition defines a construct in terms of an observable and measurable response or event.

7. A measurement scale consists of a set of categories that are used to classify individuals. A nominal scale consists of categories that differ only in name and are not differentiated in terms of magnitude or direction. In an ordinal scale, the categories are differentiated in terms of direction, forming an ordered series. An interval scale consists of an ordered series of categories where the categories are all equal-sized intervals. With an interval scale, it is possible to differentiate direction and magnitude (or distance) between categories. Finally, a ratio scale is an interval scale where the zero point indicates none of the variable being measured. With a ratio scale, ratios of measurements reflect ratios of magnitude.

8. A discrete variable is one that can have only a finite number of values between any two values. It typically consists of whole numbers that vary in countable steps. A continuous variable can have an infinite number of values between any two values. For a continuous variable, each score corresponds to an interval. The boundaries that separate intervals are called real limits. The real limits are located exactly halfway between adjacent scores.

9. The letter X is used to represent scores for a variable. If a second variable is used, Y represents its scores. The letter N is used as the symbol for the number of scores in a set.

10. The Greek letter sigma (Σ) is used to stand for summation. Therefore, the expression ΣX is read "the sum of the scores." Summation is a mathematical operation (like addition or multiplication) and must be performed in its proper place in the order of operations; summation occurs after parentheses, exponents, and multiplying/dividing have been completed.

RESOURCES

WebTUTOR

If you are using WebTutor along with this book, you will find a WebTutor section corresponding to this chapter. The WebTutor contains a brief summary of Chapter 1, hints for learning the new material and for avoiding common errors, and sample exam items including solutions.

KEY TERMS

statistics	correlational method	discrete variable
population	experimental method	continuous variable
sample	independent variable	real limits
population parameter	dependent variable	upper real limit
sample statistic	control group	lower real limit
descriptive statistics	experimental group	nominal scale
raw score	confounding variable	ordinal scale
inferential statistics	quasi-independent variable	interval scale
variable	construct	ratio scale
constant	operational definition	

FOCUS ON PROBLEM SOLVING

1. It may help to simplify summation notation if you observe that the summation sign is always followed by a symbol or symbolic expression—for example, ΣX or $\Sigma(X + 3)$. This symbol specifies which values you are to add. If you use the symbol as a column heading and list all the appropriate values in the column, your task is simply to add up the numbers in the column. To find $\Sigma(X + 3)$ for example, start a column headed with $(X + 3)$ next to the column of Xs. List all the $(X + 3)$ values; then find the total for the column.

2. Often, summation notation is part of a relatively complex mathematical expression that requires several steps of calculation. In this case, the best procedure is to use a computational table. Begin with the X values listed in the first column, then create a new column of values for each step in the calculation. For example, computing $\Sigma(X - 1)^2$ requires three steps and will use a computational table with three columns (see Example 1.3). The first column simply lists the original scores, X values.

STEP 1   The first step in the calculation is to subtract 1 from each X. A new column is added to the computational table to list each of the $(X - 1)$ values.

STEP 2   The second step is to square each $(X - 1)$ value. Again, a new column is added to list each of the $(X - 1)^2$ values.

STEP 3   The third step is to sum the $(X - 1)^2$ values, so you simply add the numbers in the $(X - 1)^2$ column.

Note that summation (step 3) produces a single number. All of the other steps before summation produce a set of values corresponding to a new column in the computational table.

DEMONSTRATION 1.1

SUMMATION NOTATION

A set of data consists of the following scores:

> 7   3   9   5   4

For these data, find the following values:
> **a.** ΣX **b.** $(\Sigma X)^2$ **c.** ΣX^2 **d.** $\Sigma X + 5$ **e.** $\Sigma(X - 2)$

Compute ΣX   To compute ΣX, we simply add all of the scores in the group. For these data, we obtain

$$\Sigma X = 7 + 3 + 9 + 5 + 4 = 28$$

Compute $(\Sigma X)^2$ The key to determining the value of $(\Sigma X)^2$ is the presence of parentheses. The rule is to perform the operations that are inside the parentheses first.

STEP 1   Find the sum of the scores, ΣX.

STEP 2   Square the total.
  We have already determined that ΣX is 28. Squaring this total, we obtain

$$(\Sigma X)^2 = (28)^2 = 784$$

Compute ΣX^2   Calculating the sum of the squared scores, ΣX^2, involves two steps.

STEP 1   Square each score.

STEP 2   Sum the squared values.
  These steps are most easily accomplished by constructing a computational table. The first column has the heading X and lists the scores. The second column is labeled X^2 and contains the squared value for each score. For this example, the table is presented in the margin. To find the value for ΣX^2, we sum the X^2 column.

X	X^2
7	49
3	9
9	81
5	25
4	16

$$\Sigma X^2 = 49 + 9 + 81 + 25 + 16 = 180$$

Compute $\Sigma X + 5$ In this expression, there are no parentheses. Thus, the summation sign is applied only to the X values.

STEP 1   Find the sum of X.

STEP 2   Add the constant 5 to the total from step 1.
    Earlier we found that the sum of the scores is 28. For $\Sigma X + 5$, we obtain the following:

$$\Sigma X + 5 = 28 + 5 = 33$$

Compute $\Sigma(X - 2)$ The summation sign is followed by an expression with parentheses. In this case, $X - 2$ is treated as a single expression, and the summation sign applies to the $(X - 2)$ values.

STEP 1   Subtract 2 from every score.

STEP 2   Sum these new values.
    This problem can be done by using a computational table with two columns, headed X and $X - 2$, respectively.

X	$X - 2$
7	5
3	1
9	7
5	3
4	2

To determine the value for $\Sigma(X - 2)$, we sum the $X - 2$ column.

$$\Sigma(X - 2) = 5 + 1 + 7 + 3 + 2 = 18$$

DEMONSTRATION 1.2

SUMMATION NOTATION WITH TWO VARIABLES

The following data consist of pairs of scores (X and Y) for four individuals:

X	Y
5	8
2	10
3	11
7	2

Determine the values for the following expressions:
    a. $\Sigma X \Sigma Y$ **b.** ΣXY

Compute $\Sigma X \Sigma Y$ This expression indicates that we should multiply the sum of X by the sum of Y.

STEP 1   Find the sum of X.

S T E P   2   Find the sum of Y.

S T E P   3   Multiply the results of steps 1 and 2.
First, we find the sum of X.

$$\Sigma X = 5 + 2 + 3 + 7 = 17$$

Next, we compute the sum of Y.

$$\Sigma Y = 8 + 10 + 11 + 2 = 31$$

Finally, we multiply these two totals.

$$\Sigma X \Sigma Y = 17(31) = 527$$

Compute ΣXY   Now we are asked to find the sum of the products of X and Y.

S T E P   1   Find the XY products.

S T E P   2   Sum the products.
The computations are facilitated by using a third column labeled XY.

X	Y	XY
5	8	40
2	10	20
3	11	33
7	2	14

For these data, the sum of the XY products is

$$\Sigma XY = 40 + 20 + 33 + 14 = 107$$

PROBLEMS

*1. Statistical methods are classified into two major categories: descriptive and inferential. Describe the general purpose for the statistical methods in each category.

2. Define the terms population, sample, parameter, and statistic, and explain how these terms are related to each other.

3. A researcher would like to determine whether regular doses of St. John's wort have an effect on memory for college students. A group of 50 students is selected and each student is given a standardized memory test. Then each of the students takes a daily dose of St. John's wort for 6 weeks. At the end of 6 weeks, the students are once again given a memory test. The researcher intends to compare the scores from the two tests to see whether there is any change in memory ability.
   a. For this study, describe the population.
   b. For this study, describe the sample.

4. It is known that the population of laboratory rats reaches an average weight of 510 grams at age 6 months. To test the effectiveness of a new growth hormone, a researcher selects a sample of 10 newborn rats and injects each rat with the hormone. Six months later, the rats in the sample are weighed and the

---

*Solutions for odd-numbered problems are provided in Appendix C.

researcher finds an average weight of 528 grams. Based on these results, can the researcher conclude that the growth hormone had an effect on the rats in the sample? Explain why or why not. (*Hint:* If the hormone had no effect, how can you explain the difference between the sample average and the population average?)

5. Identify the two elements of an experiment that make this type of research study different from other research methods.

6. Define the concept of *sampling error*. Be sure that your definition includes the concepts of *statistic* and *parameter*.

7. In general, research attempts to establish and explain relationships between variables. What terms are used to identify the two variables in an experiment? Define each term.

8. A researcher reports that individuals who survived one heart attack and were given daily doses of aspirin were significantly less likely to suffer a second heart attack than survivors who were given an inactive placebo instead of aspirin. For this study, identify the independent variable and the dependent variable.

9. A developmental psychologist conducts a research study comparing vocabulary skills for 5-year-old boys and 5-year-old girls. Is this study an *experiment*? Explain why or why not.

10. A recent report indicates that elderly drivers (over 65 years old) are more likely to be injured in an automobile accident than are younger drivers. Is this study an *experiment*? Explain why or why not.

11. A researcher is investigating the relationship between sugar consumption and activity level for a group of preschool children. The researcher interviews each child's parents to obtain information about the child's diet (sugar consumption) and then observes each child for 30 minutes during an outdoor play period to obtain a measure of activity level. Is this study an *experiment*? Explain why or why not.

12. A researcher would like to evaluate the claim that large doses of vitamin C can help prevent the common cold. One group of participants is given a large dose of the vitamin (500 mg per day), and a second group is given a placebo (sugar pill). The researcher records the number of colds each individual experiences during the 3-month winter season.
    a. Identify the dependent variable for this study.
    b. Is the dependent variable discrete or continuous?

c. What scale of measurement (nominal, ordinal, interval, or ratio) is used to measure the dependent variable?

13. Four scales of measurement were introduced in this chapter: nominal, ordinal, interval, and ratio.
    a. What additional information is obtained from measurements on an ordinal scale compared to measurements on a nominal scale?
    b. What additional information is obtained from measurements on an interval scale compared to measurements on an ordinal scale?
    c. What additional information is obtained from measurements on a ratio scale compared to measurements on an interval scale?

14. Define and differentiate between a discrete variable and a continuous variable.

15. A researcher studying the effects of environment on mood asks participants to sit alone in a waiting room for 15 minutes at the beginning of an experiment. Half of them are assigned to a room with dark blue walls, and the other half are assigned to a room with bright yellow walls. After 15 minutes in the waiting room, each participant is brought into the lab and given a mood-assessment questionnaire. The scores on the mood questionnaire are then examined.
    a. Identify the independent and dependent variables for this study.
    b. What scale of measurement is used for the independent variable?

16. Three researchers are evaluating taste preferences among three leading brands of cola. After participants taste each brand, the first researcher simply asks each participant to identify his/her favorite. The second researcher asks each participant to identify the most preferred, the second most preferred, and the least preferred. The third researcher asks each participant to rate each of the colas on a 10-point scale, where a rating of 1 indicates "terrible taste" and 10 indicates "excellent taste." Identify the scale of measurement used by each researcher.

17. For the following scores, find the value of each expression:
    a. ΣX
    b. ΣX^2
    c. $(\Sigma X)^2$
    d. $\Sigma(X - 1)$

X
4
2
3
1

18. Two scores, X and Y, are recorded for each of $n = 4$ subjects. For these scores, find the value of each expression.

a. ΣX
b. ΣY
c. ΣXY

SUBJECT	X	Y
A	5	1
B	2	4
C	6	3
D	4	2

19. For the following set of scores, find the value of each expression:

a. ΣX
b. ΣX^2
c. $\Sigma(X + 1)$
d. $\Sigma(X + 1)^2$

X
1
5
0
4

20. For the following set of scores, find the value of each expression:

a. ΣX
b. ΣX^2
c. $\Sigma(X + 3)$

X
5
-2
0
-3

21. For the following set of scores, find the value of each expression:

a. ΣX
b. ΣY
c. ΣXY

X	Y
-3	4
2	5
0	3
4	-2
-3	-3

22. Use summation notation to express each of the following calculations:
   a. Sum the scores, and then add 3 points to the total.
   b. Subtract 2 points from each score, and square the result. Then sum the squared values.
   c. Square each score, and sum the squared values. Then subtract 10 points from the sum.

23. Each of the following summation expressions requires a specific sequence of mathematical operations. For each expression, state in words the sequence of operations necessary to perform the specified calculations. For example, the expression $\Sigma(X + 1)$ instructs you to add 1 point to each score and then sum the resulting values.
   a. $\Sigma X - 4$
   b. ΣX^2
   c. $\Sigma(X + 4)^2$

24. For the following set of scores, find the value of each expression:

a. ΣX^2
b. $(\Sigma X)^2$
c. $\Sigma(X - 3)$
d. $\Sigma(X - 3)^2$

X
9
1
1
1

25. Statistical analyses often use an expression that has the following form: $\Sigma(X - C)^2$.
   a. What is the first step for this computation? The second step? The third?
   b. Find $\Sigma(X - 4)^2$ for the following set of scores:

   8   1   3

   c. Find $\Sigma(X - 4)$ for the same set of scores.

C H A P T E R

2

Frequency Distributions

Tools You Will Need

The following items are considered essential background material for this chapter. If you doubt your knowledge of any of these items, you should review the appropriate chapter or section before proceeding.

- Proportions (math review, Appendix A)
  - Fractions
  - Decimals
  - Percentages

- Scales of measurement (Chapter 1): Nominal, ordinal, interval, and ratio

- Continuous and discrete variables (Chapter 1)

- Real limits (Chapter 1)

Preview

If at first you don't succeed, you are probably not related to the boss.

Did we make you chuckle or, at least, smile a little? The use of humor is a common technique to capture attention and to communicate ideas. Advertisers, for example, will often try to make a commercial funny so that people will notice it and, perhaps, remember the product. After-dinner speakers will always put a few jokes into the speech in an effort to maintain the audience's interest. Although humor seems to capture our attention, does it actually affect our memory?

In an attempt to answer this question, Stephen Schmidt (1994) conducted a series of experiments examining the effects of humor on memory for sentences. Humorous sentences were collected from a variety of sources and then a nonhumorous version was constructed for each sentence. For example, the nonhumorous version of our opening sentence was:

> People who are related to the boss often succeed the very first time.

Participants were then presented with a list containing half humorous and half nonhumorous sentences. Later, each person was asked to recall as many sentences as possible. The researcher measured the number of humorous sentences and the number of nonhumorous sentences recalled by each participant. Hypothetical data similar to the results obtained by Schmidt are shown in Table 2.1. Looking at the numbers in the table, do you see an obvious difference in the scores for the two types of sentence?

TABLE 2.1

Memory scores for a sample of 16 participants. The scores represent the number of sentences recalled from each category.

HUMOROUS SENTENCES				NONHUMOROUS SENTENCES			
4	5	2	4	5	2	4	2
6	7	6	6	2	3	1	6
2	5	4	3	3	2	3	3
1	3	5	5	4	1	5	3

Just by looking at the numbers it is difficult to see any clear pattern. This situation confronts every researcher at the end of a research study. The raw data often do not reveal any obvious patterns. This is why researchers rely on descriptive statistics to summarize and organize their results so that it becomes easier to identify patterns in the data, if any patterns actually exist.

One descriptive technique is to present the data in a table or graph that provides an organized picture of the outcome. For example, the same memory scores that are presented in Table 2.1 have been organized in a graph in Figure 2.1. The graph is an example of a frequency distribution and shows exactly how many people had each score. In Figure 2.1, for example, each individual is represented by a block that is placed directly above that individual's score. The resulting pile of blocks shows a picture of how the individual scores are distributed. For this example, it is now easy to see that the scores for the humorous sentences are generally higher than the scores for the nonhumorous sentences.

In this chapter we present techniques for organizing data into tables and graphs so that an entire set of scores can be presented in a relatively simple display or illustration.

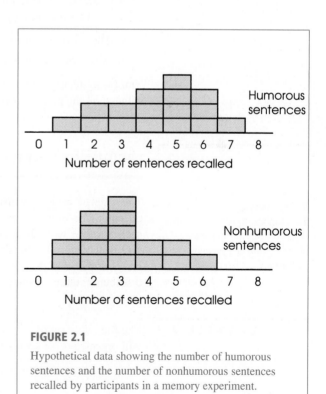

FIGURE 2.1

Hypothetical data showing the number of humorous sentences and the number of nonhumorous sentences recalled by participants in a memory experiment.

2.1   OVERVIEW

When a researcher finishes the data collection phase of an experiment, the results usually consist of pages of numbers. The immediate problem for the researcher is to organize the scores into some comprehensible form so that any trends in the data can be seen easily and communicated to others. This is the job of descriptive statistics: to simplify the organization and presentation of data. One of the most common procedures for organizing a set of data is to place the scores in a frequency distribution.

DEFINITION   A **frequency distribution** is an organized tabulation of the number of individuals located in each category on the scale of measurement.

A frequency distribution takes a disorganized set of scores and places them in order from highest to lowest, grouping together all individuals who have the same score. If the highest score is $X = 10$, for example, the frequency distribution groups together all the 10s, then all the 9s, then the 8s, and so on. Thus, a frequency distribution allows the researcher to see "at a glance" the entire set of scores. It shows whether the scores are generally high or low and whether they are concentrated in one area or spread out across the entire scale and generally provides an organized picture of the data. In addition to providing a picture of the entire set of scores, a frequency distribution allows you to see the location of any individual score relative to all of the other scores in the set.

A frequency distribution can be structured either as a table or as a graph, but in either case the distribution presents the same two elements:

1. The set of categories that make up the original measurement scale.

2. A record of the frequency, or number of individuals in each category.

Thus, a frequency distribution presents a picture of how the individual scores are distributed on the measurement scale—hence the name *frequency distribution*.

2.2   FREQUENCY DISTRIBUTION TABLES

It is customary to list scores from highest to lowest, but this is an arbitrary arrangement. Many computer programs will list scores from lowest to highest.

The simplest frequency distribution table presents the measurement scale by listing the different measurement categories (X values) in a column from highest to lowest. Beside each X value, we indicate the frequency, or the number of times that particular measurement occurred in the data. It is customary to use an X as the column heading for the scores and an f as the column heading for the frequencies. An example of a frequency distribution table follows.

EXAMPLE 2.1   The following set of $N = 20$ scores was obtained from a 10-point statistics quiz. We will organize these scores by constructing a frequency distribution table. Scores:

8, 9, 8, 7, 10, 9, 6, 4, 9, 8,

7, 8, 10, 9, 8, 6, 9, 7, 8, 8

X	f
10	2
9	5
8	7
7	3
6	2
5	0
4	1

1. The highest score is $X = 10$, and the lowest score is $X = 4$. Therefore, the first column of the table will list the categories that make up the scale of measurement (X values) from 10 down to 4. Notice that all of the possible values are listed in the table. For example, no one had a score of $X = 5$, but this value is included. With an ordinal, interval, or ratio scale, the categories are listed in order (usually highest to lowest). For a nominal scale, the categories can be listed in any order.

2. The frequency associated with each score is recorded in the second column. For example, two people had scores of $X = 6$, so there is a 2 in the f column beside $X = 6$.

Because the table organizes the scores, it is possible to see very quickly the general quiz results. For example, there were only two perfect scores, but most of the class had high grades (8s and 9s). With one exception (the score of $X = 4$), it appears that the class has learned the material fairly well.

Notice that the X values in a frequency distribution table represent the scale of measurement, *not* the actual set of scores. For example, the X column lists the value 10 only one time, but the frequency column indicates that there are actually two values of $X = 10$. Also, the X column lists a value of $X = 5$, but the frequency column indicates that no one actually had a score of $X = 5$.

You also should notice that the frequencies can be used to find the total number of scores in the distribution. By adding up the frequencies, you will obtain the total number of individuals:

$$\Sigma f = N$$

OBTAINING ΣX FROM A FREQUENCY DISTRIBUTION TABLE

There may be times when you need to compute the sum of the scores, ΣX, or perform other computations for a set of scores that has been organized into a frequency distribution table. Such calculations can present a problem because many students tend to disregard the frequencies and use only the values listed in the X column of the table. However, this practice is incorrect because it ignores the information provided by the frequency (f) column.

When it is necessary to perform calculations for scores that have been organized into a frequency distribution table, the safest procedure is to take the individual scores out of the table before you begin any computations. Consider the frequency distribution table for Example 2.1. The table shows that the distribution has two 10s, five 9s, seven 8s, and so on. If you simply list all of the individual scores, then you can safely proceed with calculations such as finding ΣX or ΣX^2. Note that the complete set contains $N = \Sigma f = 20$ scores and your list should contain 20 values. For example, to compute ΣX you simply add all 20 of the scores:

$$\Sigma X = 10 + 10 + 9 + 9 + 9 + 9 + 9 + 8 + 8 + 8 + 8 + 8 + 8 + 8 + \cdots$$

For the data in Example 2.1, you should obtain $\Sigma X = 158$. Try it yourself. Similarly, to obtain ΣX^2 you simply square each of the 20 scores and then add the squared values.

$$\Sigma X^2 = 10^2 + 10^2 + 9^2 + 9^2 + 9^2 + 9^2 + 9^2 + 8^2 + 8^2 + 8^2 + \\ 8^2 + 8^2 + 8^2 + 8^2 + \cdots$$

This time you should obtain $\Sigma X^2 = 1288$.

An alternative way to get ΣX from a frequency distribution table is to multiply each X value by its frequency and then add these products. This sum may be expressed in symbols as $\Sigma f X$. The computation is summarized as follows for the data in Example 2.1:

Caution: Doing calculations within the table works well for ΣX but can lead to errors for more complex formulas.

X	f	fX	
10	2	20	(The two 10s total 20)
9	5	45	(The five 9s total 45)
8	7	56	(The seven 8s total 56)
7	3	21	(The three 7s total 21)
6	2	12	(The two 6s total 12)
5	0	0	(There are no 5s)
4	1	4	(The one 4 totals 4)

$$\Sigma f X = 158$$

No matter which method you use to find ΣX, the important point is that you must use the information given in the frequency column in addition to the information in the X column.

PROPORTIONS AND PERCENTAGES

In addition to the two basic columns of a frequency distribution, there are other measures that describe the distribution of scores and can be incorporated into the table. The two most common are proportion and percentage.

Proportion measures the fraction of the total group that is associated with each score. In Example 2.1, there were two individuals with $X = 6$. Thus, 2 out of 20 people had $X = 6$, so the proportion would be $\frac{2}{20} = 0.10$. In general, the proportion associated with each score is

$$\text{proportion} = p = \frac{f}{N}$$

Because proportions describe the frequency (f) in relation to the total number (N), they often are called *relative frequencies*. Although proportions can be expressed as fractions (for example, $\frac{2}{20}$), they more commonly appear as decimals. A column of proportions, headed with a p, can be added to the basic frequency distribution table (see Example 2.2).

In addition to using frequencies (f) and proportions (p), researchers often describe a distribution of scores with percentages. For example, an instructor might describe the results of an exam by saying that 15% of the class earned *A*s, 23% *B*s, and so on. To compute the percentage associated with each score, you first find the proportion (p) and then multiply by 100:

$$\text{percentage} = p(100) = \frac{f}{N}(100)$$

Percentages can be included in a frequency distribution table by adding a column headed with % (see Example 2.2).

EXAMPLE 2.2   The frequency distribution table from Example 2.1 is repeated here. This time we have added columns showing the proportion (p) and the percentage (%) associated with each score.

X	f	$p = f/N$	$\% = p(100)$
10	2	$2/20 = 0.10$	10%
9	5	$5/20 = 0.25$	25%
8	7	$7/20 = 0.35$	35%
7	3	$3/20 = 0.15$	15%
6	2	$2/20 = 0.10$	10%
5	0	$0/20 = 0$	0%
4	1	$1/20 = 0.05$	5%

GROUPED FREQUENCY DISTRIBUTION TABLES

When the scores are whole numbers, the total number of rows for a regular table can be obtained by finding the difference between the highest and the lowest scores and adding 1:

rows = highest − lowest + 1

When a set of data covers a wide range of values, it is unreasonable to list all the individual scores in a frequency distribution table. For example, a set of exam scores ranges from a low of $X = 41$ to a high of $X = 96$. These scores cover a range of over 50 points.

If we were to list all the individual scores from $X = 96$ down to $X = 41$, it would take 56 rows to complete the frequency distribution table. Although this would organize and simplify the data, the table would be long and cumbersome. Remember: The purpose for constructing a table is to obtain a relatively simple, organized picture of the data. This can be accomplished by grouping the scores into intervals and then listing the intervals in the table instead of listing each individual score. For example, we could construct a table showing the number of students who had scores in the 90s, the number with scores in the 80s, and so on. The result is called a *grouped frequency distribution table* because we are presenting groups of scores rather than individual values. The groups, or intervals, are called *class intervals*.

There are several rules that help guide you in the construction of a grouped frequency distribution table. These rules should be considered as guidelines rather than absolute requirements, but they do help produce a simple, well-organized, and easily understood table.

RULE 1   The grouped frequency distribution table should have about 10 class intervals. If a table has many more than 10 intervals, it becomes cumbersome and defeats the purpose of a frequency distribution table. On the other hand, if you have too few intervals, you begin to lose information about the distribution of the scores. At the extreme, with only one interval, the table would not tell you anything about how the scores are distributed. Remember that the purpose of a frequency distribution is to help a researcher see the data. With too few or too many intervals, the table will not provide a clear picture. You should note that 10 intervals is a general guide. If you are constructing a table on a blackboard, for example, you probably will want only 5 or 6 intervals. If the table is to be printed in a scientific report, you may want 12 or 15 intervals. In each case, your goal is to present a table that is relatively easy to see and understand.

RULE 2   The width of each interval should be a relatively simple number. For example, 2, 5, 10, or 20 would be a good choice for the interval width. Notice that it is easy to count by 5s or 10s. These numbers are easy to understand and make it possible for someone to see quickly how you have divided the range.

RULE 3   The bottom score in each class interval should be a multiple of the width. If you are using a width of 10 points, for example, the intervals should start with 10, 20, 30, 40, and so on. Again, this makes it easier for someone to understand how the table has been constructed.

RULE 4   All intervals should be the same width. They should cover the range of scores completely with no gaps and no overlaps, so that any particular score belongs in exactly one interval.

The application of these rules is demonstrated in Example 2.3.

EXAMPLE 2.3   An instructor has obtained the set of $N = 25$ exam scores shown here. To help organize these scores, we will place them in a frequency distribution table. The scores are:

82,   75,   88,   93,   53,   84,   87,   58,   72,   94,   69,   84,   61,

91,   64,   87,   84,   70,   76,   89,   75,   80,   73,   78,   60

Remember, the number of rows is determined by

highest − lowest + 1

The first step is to determine the range of scores. For these data, the smallest score is $X = 53$ and the largest score is $X = 94$, so a total of 42 rows would be needed for a table that lists each individual score. Because 42 rows would not provide a simple table, we will have to group the scores into class intervals.

The best method for finding a good interval width is a systematic trial-and-error approach that uses rules 1 and 2 simultaneously. According to rule 1, we want about 10 intervals; according to rule 2, we want the interval width to be a simple number. For this example, the scores cover a range of 42 points, so we will try several different interval widths to see how many intervals are needed to cover this range. For example, if we try a width of 2, how many intervals would it take to cover the range of scores? With each interval only 2 points wide, we would need 21 intervals to cover the range. This is too many. What about an interval width of 5? What about a width of 10? The following table shows how many intervals would be needed for these possible widths:

WIDTH	NUMBER OF INTERVALS NEEDED TO COVER A RANGE OF 42 POINTS	
2	21	(too many)
5	9	(OK)
10	5	(too few)

Notice that an interval width of 5 will result in about 10 intervals, which is exactly what we want.

The next step is to actually identify the intervals. The lowest score for these data is $X = 53$, so the lowest interval should contain this value. Because the interval should have a multiple of 5 as its bottom score, the interval would be 50 to 54. Notice that the interval contains five values (50, 51, 52, 53, 54) so it does have a width of 5.

The next interval would start at 55 and go to 59. The complete frequency distribution table showing all of the class intervals is presented in Table 2.2.

Once the class intervals are listed, you complete the table by adding a column of frequencies or proportions or percentages. The values in the frequency column indicate the number of individuals whose scores are located in that class interval. For this example, there were three students with scores in the 60–64 interval, so the frequency for this class interval is $f = 3$ (see Table 2.2).

Finally, you should note that after the scores have been placed in a grouped table, you lose information about the specific value for any individual score. For example, Table 2.2 shows that one person had a score between 65 and 69, but the table does not identify the exact value for the score. In general, the wider the class intervals are, the more information is lost. In Table 2.2 the interval width is 5 points, and the table shows that there are three people with scores in the lower 60s and one person with a score in the upper 60s. This information would be lost if the interval width was increased to 10 points. With an interval width of 10, all of the 60s would be grouped together into one interval labeled 60–69. The table would show a frequency of four people in the 60–69 interval, but it would not tell whether the scores were in the upper 60s or the lower 60s.

TABLE 2.2

A grouped frequency distribution table showing the data from Example 2.3. The original scores range from a high of $X = 94$ to a low of $X = 53$. This range has been divided into 9 intervals with each interval exactly 5 points wide. The frequency column (f) lists the number of individuals with scores in each of the class intervals.

X	f
90–94	3
85–89	4
80–84	5
75–79	4
70–74	3
65–69	1
60–64	3
55–59	1
50–54	1

REAL LIMITS AND FREQUENCY DISTRIBUTIONS

Recall from Chapter 1 that a continuous variable has an infinite number of possible values and can be represented by a number line that is continuous and contains an infinite number of points. However, when a continuous variable is measured, the resulting measurements correspond to *intervals* on the number line rather than single points. For example, a score of $X = 8$ for a continuous variable actually represents an interval bounded by the real limits 7.5 and 8.5. Thus, a frequency distribution table showing a frequency of $f = 3$ individuals all assigned a score of $X = 8$ does not mean that all three individuals had exactly the same measurement. Instead, you should realize that the three measurements are simply located in the same interval between 7.5 and 8.5.

The concept of real limits also applies to the class intervals of a grouped frequency distribution table. For example, a class interval of 40–49 contains scores from $X = 40$ to $X = 49$. These values are called the *apparent limits* of the interval because it appears that they form the upper and lower boundaries for the class interval. But $X = 40$ is actually an interval from 39.5 to 40.5. Similarly, $X = 49$ is an interval from 48.5 to 49.5. Therefore, the real limits of the interval are 39.5 (the lower real limit) and 49.5 (the upper real limit). Notice that the next higher class interval is 50–59, which has a lower real limit of 49.5. Thus, the two intervals meet at the real limit 49.5, so there are no gaps in the scale. You also should notice that the width of each class interval becomes easier to understand when you consider the real limits of an interval. For example, the interval 50–54 has real limits of 49.5 and 54.5. The distance between these two real limits (5 points) is the width of the interval.

LEARNING CHECK

1. Place the following scores in a frequency distribution table.

   2, 3, 1, 2, 5, 4, 5, 5, 1, 4, 2, 2

2. A set of scores ranges from a high of $X = 142$ to a low of $X = 65$.

   a. Explain why it would not be reasonable to display these scores in a *regular* frequency distribution table.

   b. Determine what interval width would be most appropriate for a grouped frequency distribution for this set of scores.

   c. What range of values would form the bottom interval for the grouped table?

3. Find the values for N and ΣX for the population of scores in the following frequency distribution table.

X	f
5	1
4	3
3	5
2	2
1	2

4. Using only the frequency distribution table presented in Table 2.2, how many individuals had a score of $X = 73$?

ANSWERS

1.

X	f
5	3
4	2
3	1
2	4
1	2

2. a. It would require a table with 78 rows to list all the individual score values. This is too many rows.

   b. An interval width of 10 points would probably be best. With a width of 10, it would require 8 intervals to cover the range of scores. In some circumstances an interval width of 5 points (requiring 16 intervals) might be appropriate.

   c. With a width of 10, the apparent limits of the bottom interval would be 60–69. The real limits would be 59.5–69.5.

3. There are $N = 13$ scores summarized in the table ($\Sigma f = 13$). The 13 scores sum to $\Sigma X = 38$.

4. After a set of scores has been summarized in a grouped table, you cannot determine the frequency for any specific score. There is no way to determine from the table alone how many individuals had $X = 73$.

2.3 FREQUENCY DISTRIBUTION GRAPHS

A frequency distribution graph is basically a picture of the information available in a frequency distribution table. We will consider several different types of graphs, but all start with two perpendicular lines called *axes*. The horizontal line is called the *X*-axis, or the abscissa. The vertical line is called the *Y*-axis, or the ordinate. The measurement scale (set of *X* values) is listed along the *X*-axis in increasing value from left to right. The frequencies are listed on the *Y*-axis in increasing value from bottom to top. As a general rule, the point where the two axes intersect should have a value of zero for both the scores and the frequencies. A final general rule is that the graph should be constructed so that its height (*Y*-axis) is approximately two-thirds to three-quarters of its length (*X*-axis). Violating these guidelines can result in graphs that give a misleading picture of the data (see Box 2.1).

GRAPHS FOR INTERVAL OR RATIO DATA

The first graphs that we will consider are designed to be used with numerical scores that have been measured on an interval or a ratio scale of measurement. The two types of graph in this category are called histograms and polygons.

Histograms   To construct a histogram, you first list the numerical scores (the categories of measurement) along the *X*-axis. Then you draw a bar above each *X* value so that

a. The height of the bar corresponds to the frequency for that category.

b. The width of the bar extends to the real limits of the category.

Because the bars extend to the real limits for each category, adjacent bars will touch so that there are no spaces or gaps between bars. An example of a histogram is shown in Figure 2.2(a).

When data have been grouped into class intervals, you can construct a frequency distribution histogram by drawing a bar above each interval so that the width of the bar extends to the real limits of the interval (the lower real limit of the lowest score and the upper real limit of the highest score in the interval). This process is demonstrated in Figure 2.2(b).

For the two histograms shown in Figure 2.2, notice that the values on both the horizontal and vertical axes are clearly marked and that both axes are labeled. In these examples, we used the generic term "scores" to label the horizontal axis. In most circumstances, however, a more specific measurement would be identified and often the units of measurement are specified. Figure 2.3, for example, shows a frequency distribution for the heights (measured in inches) of the individuals in a sample. Note also that the horizontal axis in Figure 2.3 does not list all of the possible heights starting from zero and going up to 72 inches. Instead, the graph clearly shows a break between zero and 65, indicating that some scores have been omitted.

A modified histogram   A slight modification to the traditional histogram produces a very easy to draw and simple to understand sketch of a frequency distribution. Instead of drawing a bar above each score, the modification consists of drawing a stack of

FIGURE 2.2

(a) An example of a frequency distribution histogram. The same set of data is presented in a frequency distribution table and in a histogram.

X	f
6	1
5	2
4	2
3	4
2	2
1	1

(b) An example of a frequency distribution histogram for grouped data. The same set of data is presented in a grouped frequency distribution table and in a histogram.

X	f
12–13	4
10–11	5
8–9	3
6–7	3
4–5	2

blocks. Each block represents one individual, so the number of blocks above each score corresponds to the frequency for that score. You may recall that we used this pile-of-blocks display in the Preview section of this chapter. Another example is shown in Figure 2.4.

Note that the number of blocks in each stack makes it very easy to see the absolute frequency for each category. In addition it is easy to see the exact difference in frequency from one category to another. In Figure 2.4, for example, it is easy to see that there are exactly two more people with scores of $X = 2$ than with scores of $X = 1$. Because the frequencies are clearly displayed by the number of blocks, this type of display eliminates the need for a vertical line (the Y-axis) showing frequencies. In general, this kind of display provides a simple and concrete picture of the distribution for a sample of scores. Note that we will be using this kind of display to show sample data throughout the rest of the book. You should also note, however, that this kind of dis-

FIGURE 2.3

A frequency distribution histogram showing the heights for a sample of $n = 20$ adults.

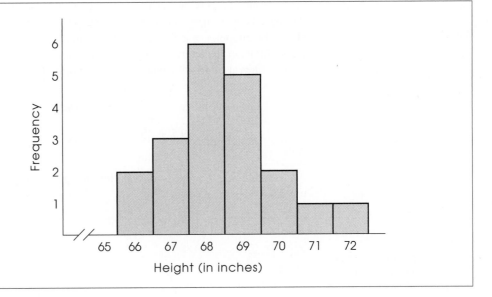

FIGURE 2.4

A frequency distribution in which each individual is represented by a block placed directly above the individual's score. For example, three people had scores of $X = 2$.

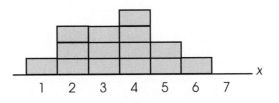

play simply provides a sketch of the distribution and is not a substitute for an accurately drawn histogram with two labeled axes.

Polygons   To construct a polygon, you begin by listing the numerical scores (the categories of measurement) along the X-axis. Then,

    a. A dot is centered above each score so that the vertical position of the dot corresponds to the frequency for the category.

    b. A continuous line is drawn from dot to dot to connect the series of dots.

    c. The graph is completed by drawing a line down to the X-axis (zero frequency) at each end of the range of scores. The final lines are usually drawn so that they reach the X-axis at a point that is one category below the lowest score on the left side and one category above the highest score on the right side.

An example of a polygon is shown in Figure 2.5. A polygon also can be used with data that have been grouped into class intervals. In this case, you position each dot directly above the midpoint of a class interval. The midpoint can be found by averaging the apparent limits of the interval or by averaging the real limits of the interval. For example, a class interval of 40–49 would have a midpoint of 44.5.

$$\text{apparent limits: } \frac{40 + 49}{2} = \frac{89}{2} = 44.5$$

$$\text{real limits: } \frac{39.5 + 49.5}{2} = \frac{89}{2} = 44.5$$

An example of a frequency distribution polygon with grouped data is shown in Figure 2.6.

FIGURE 2.5

An example of a frequency distribution polygon. The same set of data is presented in a frequency distribution table and in a polygon. Note that these data are shown in a histogram in Figure 2.2(a).

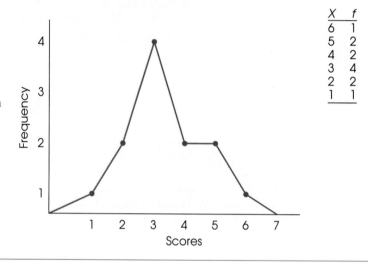

X	f
6	1
5	2
4	2
3	4
2	2
1	1

FIGURE 2.6

An example of a frequency distribution polygon for grouped data. The same set of data is presented in a grouped frequency distribution table and in a polygon. Note that these data are shown in a histogram in Figure 2.2(b).

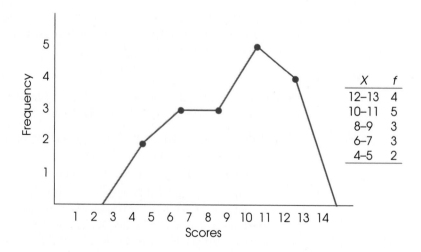

X	f
12–13	4
10–11	5
8–9	3
6–7	3
4–5	2

GRAPHS FOR NOMINAL OR ORDINAL DATA

When the scores are measured on a nominal or ordinal scale (usually nonnumerical values), the frequency distribution can be displayed in a bar graph.

Bar graphs A bar graph is essentially the same as a histogram, except that spaces are left between adjacent bars. For a nominal scale, the space between bars emphasizes that the scale consists of separate, distinct categories. For ordinal scales, separate bars are used because you cannot assume that the categories are all the same size.

To construct a bar graph, you list the categories of measurement along the X-axis and then draw a bar above each category so that the height of the bar correponds to the frequency for the category. An example of a bar graph is shown in Figure 2.7.

FIGURE 2.7

A bar graph showing the distribution of personality types in a sample of college students. Because personality type is a discrete variable measured on a nominal scale, the graph is drawn with space between the bars.

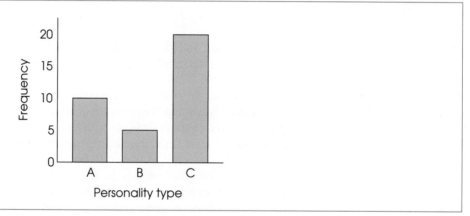

GRAPHS FOR POPULATION DISTRIBUTIONS

Although it is possible to construct graphs showing population frequency distributions, there are two fundamental differences between population graphs and the histograms, polygons, and bar graphs that are used to display sample data. The two differences involve relative frequencies and smooth curves.

Relative frequencies For a sample it is possible to obtain the exact frequency for any score by simply counting the number of individuals who have that score. For a population, however, you typically do not have access to all of the individuals, so it is impossible to count frequencies. For a sample of $n = 50$ people, for example, we could easily determine exactly how many have IQs of $X = 110$. For the entire population (even limited to the population of adults in the United States) it would be impossible to determine exactly how many people have IQ scores of 110.

Although you usually cannot find the absolute frequency for each score in a population, you very often can obtain relative frequencies. For example, you may not know exactly how many fish are in the lake, but after years of fishing you do know that there are twice as many bluegill as there are bass. You can represent these relative frequencies in a bar graph by making the bar above bluegill two times taller than the bar above

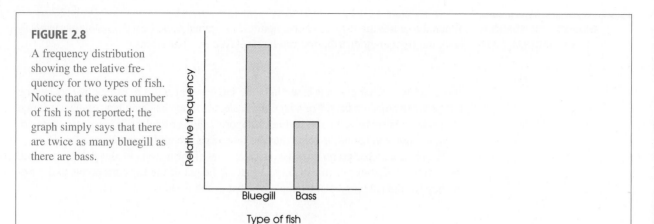

FIGURE 2.8

A frequency distribution showing the relative frequency for two types of fish. Notice that the exact number of fish is not reported; the graph simply says that there are twice as many bluegill as there are bass.

bass (see Figure 2.8). Notice that the graph does not show the absolute number of fish. Instead, it shows the relative number of bluegill and bass.

Smooth curves   When a population consists of numerical scores from an interval or a ratio scale, it is customary to draw the distribution with a smooth curve instead of the jagged, step-wise shapes that occur with histograms and polygons. The smooth curve indicates that you are not connecting a series of dots (real frequencies) but instead are showing the relative changes that occur from one score to the next. One commonly occurring population distribution is the normal curve. The word *normal* refers to a specific shape that can be precisely defined by an equation. Less precisely, we can describe a normal distribution as being symmetrical, with the greatest frequency in the middle and relatively smaller frequencies as you move toward either extreme. A good example of a normal distribution is the population distribution for IQ scores shown in Figure 2.9. Because normal-shaped distributions occur commonly and because this shape is mathematically guaranteed in certain situations, it will receive extensive attention throughout this book.

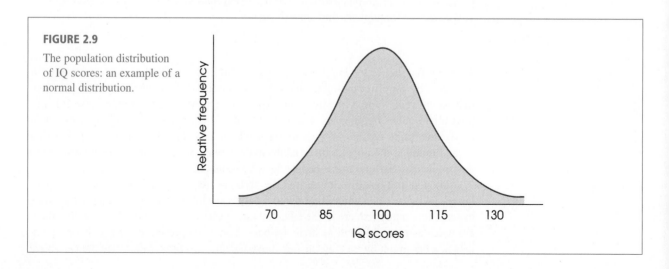

FIGURE 2.9

The population distribution of IQ scores: an example of a normal distribution.

BOX
2.1   THE USE AND MISUSE OF GRAPHS

Although graphs are intended to provide an accurate picture of a set of data, they can be used to exaggerate or misrepresent a set of scores. These misrepresentations generally result from failing to follow the basic rules for graph construction. The following example demonstrates how the same set of data can be presented in two entirely different ways by manipulating the structure of a graph.

For the past several years, the city has kept records of the number of major felonies. The data are summarized as follows:

YEAR	NUMBER OF MAJOR FELONIES
2000	218
2001	225
2002	229

These same data are shown in two different graphs in Figure 2.10. In the first graph, we have exaggerated the height, and we started numbering the Y-axis at 210 rather than at zero. As a result, the graph seems to indicate a rapid rise in the crime rate over the 3-year period. In the second graph, we have stretched out the X-axis and used zero as the starting point for the Y-axis. The result is a graph that shows little change in the crime rate over the 3-year period.

Which graph is correct? The answer is that neither one is very good. Remember that the purpose of a graph is to provide an accurate display of the data. The first graph in Figure 2.10 exaggerates the differences between years, and the second graph conceals the differences. Some compromise is needed. You also should note that in some cases a graph may not be the best way to display information. For these data, for example, showing the numbers in a table would be better than either graph.

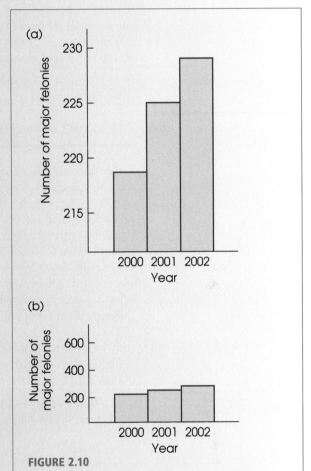

FIGURE 2.10

Two graphs showing the number of major felonies in a city over a 3-year period. Both graphs show exactly the same data. However, the first graph gives the appearance that the crime rate is high and rising rapidly. The second graph gives the impression that the crime rate is low and has not changed over the 3-year period.

In the future, we will be referring to *distributions of scores*. Whenever the term *distribution* appears, you should conjure up an image of a frequency distribution graph. The graph provides a picture showing exactly where the individual scores are located. To make this concept more concrete, you might find it useful to think of the graph as showing a pile of individuals just like we showed a pile of blocks in Figure 2.4. For the population of IQ scores shown in Figure 2.9, the pile is highest at an IQ score around 100 because most people have average IQs. There are only a few individuals piled up at an IQ of 130; it must be lonely at the top.

2.4   THE SHAPE OF A FREQUENCY DISTRIBUTION

Rather than drawing a complete frequency distribution graph, researchers often simply describe a distribution by listing its characteristics. There are three characteristics that completely describe any distribution: shape, central tendency, and variability. In simple terms, central tendency measures where the center of the distribution is located. Variability tells whether the scores are spread over a wide range or are clustered together. Central tendency and variability will be covered in detail in Chapters 3 and 4. Technically, the shape of a distribution is defined by an equation that prescribes the exact relationship between each X and Y value on the graph. However, we will rely on a few less-precise terms that will serve to describe the shape of most distributions.

Nearly all distributions can be classified as being either symmetrical or skewed.

D E F I N I T I O N S

In a **symmetrical distribution**, it is possible to draw a vertical line through the middle so that one side of the distribution is a mirror image of the other (see Figure 2.11).

In a **skewed distribution**, the scores tend to pile up toward one end of the scale and taper off gradually at the other end (see Figure 2.11).

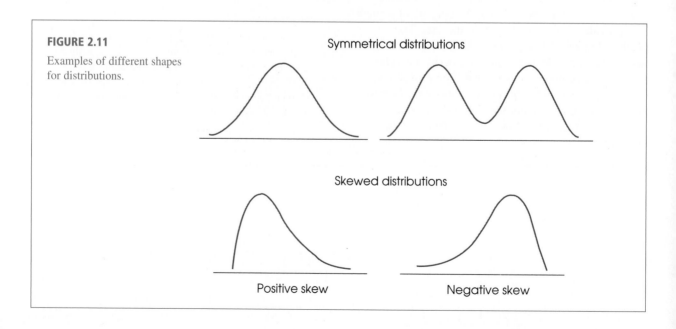

FIGURE 2.11

Examples of different shapes for distributions.

Symmetrical distributions

Skewed distributions

Positive skew

Negative skew

The section where the scores taper off toward one end of a distribution is called the **tail** of the distribution.

A skewed distribution with the tail on the right-hand side is said to be **positively skewed** because the tail points toward the positive (above-zero) end of the *X*-axis. If the tail points to the left, the distribution is said to be **negatively skewed** (see Figure 2.11).

For a very difficult exam, most scores will tend to be low, with only a few individuals earning high scores. This will produce a positively skewed distribution. Similarly, a very easy exam will tend to produce a negatively skewed distribution, with most of the students earning high scores and only a few with low values.

LEARNING CHECK

1. Sketch a frequency distribution histogram and a frequency distribution polygon for the data in the following table:

X	f
5	4
4	6
3	3
2	1
1	1

2. Describe the shape of the distribution in Exercise 1.

3. What type of graph would be appropriate to show the number of gold medals, silver medals, and bronze medals won by the United States during the 2000 Olympics?

4. What shape would you expect for the distribution of salaries for all employees of a major industry?

ANSWERS

1. The graphs are shown in Figure 2.12.

2. The distribution is negatively skewed.

3. A bar graph is appropriate for ordinal data.

4. The distribution probably would be positively skewed, with most employees earning an average salary and a relatively small number of top executives with very large salaries.

FIGURE 2.12

Answers to Learning Check Exercise 1.

2.5   PERCENTILES, PERCENTILE RANKS, AND INTERPOLATION

Although the primary purpose of a frequency distribution is to provide a description of an entire set of scores, it also can be used to describe the position of an individual within the set. Individual scores, or X values, are called raw scores. By themselves, raw scores do not provide much information. For example, if you are told that your score on an exam is $X = 43$, you cannot tell how well you did. To evaluate your score, you need more information such as the average score or the number of people who had scores above and below you. With this additional information, you would be able to determine your relative position in the class. Because raw scores do not provide much information, it is desirable to transform them into a more meaningful form. One transformation that we will consider changes raw scores into percentiles.

DEFINITIONS

The **rank** or **percentile rank** of a particular score is defined as the percentage of individuals in the distribution with scores at or below the particular value.

When a score is identified by its percentile rank, the score is called a **percentile**.

Suppose, for example, that you have a score of $X = 43$ on an exam and that you know that exactly 60% of the class had scores of 43 or lower. Then your score $X = 43$ has a percentile rank of 60%, and your score would be called the 60th percentile. Notice that *percentile rank* refers to a percentage and that *percentile* refers to a score. Also notice that your rank or percentile describes your exact position within the distribution.

CUMULATIVE FREQUENCY
AND CUMULATIVE
PERCENTAGE

To determine percentiles or percentile ranks, the first step is to find the number of individuals who are located at or below each point in the distribution. This can be done most easily with a frequency distribution table by simply counting the number who are in or below each category on the scale. The resulting values are called *cumulative frequencies* because they represent the accumulation of individuals as you move up the scale.

EXAMPLE 2.4

In the following frequency distribution table, we have included a cumulative frequency column headed by *cf*. For each row, the cumulative frequency value is obtained by adding up the frequencies in and below that category. For example, the score $X = 3$ has a cumulative frequency of 14 because exactly 14 individuals had scores in this category or in a lower category.

X	f	cf
5	1	20
4	5	19
3	8	14
2	4	6
1	2	2

The cumulative frequencies show the number of individuals located at or below each score. To find percentiles, we must convert these frequencies into percentages. The resulting values are called *cumulative percentages* because they show the percentage of individuals who are accumulated as you move up the scale.

EXAMPLE 2.5   This time we have added a cumulative percentage column ($c\%$) to the frequency distribution table from Example 2.4. The values in this column represent the percentage of individuals who are located in and below each category. For example, 70% of the individuals (14 out of 20) had scores of $X = 3$ or lower. Cumulative percentages can be computed by

$$c\% = \frac{cf}{N}(100\%)$$

X	f	cf	c%
5	1	20	100%
4	5	19	95%
3	8	14	70%
2	4	6	30%
1	2	2	10%

The cumulative percentages in a frequency distribution table give the percentage of individuals with scores at or below each X value. However, you must remember that the X values in the table are not points on the scale, but rather intervals. A score of $X = 2$, for example, means that the measurement was somewhere between the real limits of 1.5 and 2.5. Thus, when a table shows that a score of $X = 2$ has a cumulative percentage of 30%, you should interpret this as meaning that 30% of the individuals have been accumulated by the time you reach the top of the interval for $X = 2$. Notice that each cumulative percentage value is associated with the upper real limit of its interval. This point is demonstrated in Figure 2.13, which shows the same data that were used in

FIGURE 2.13

The relationship between cumulative frequencies (*cf* values) and upper real limits. Notice that two people have scores of $X = 1$. These two individuals are located between the real limits of 0.5 and 1.5. Although their exact locations are not known, you can be certain that both had scores below the upper real limit of 1.5.

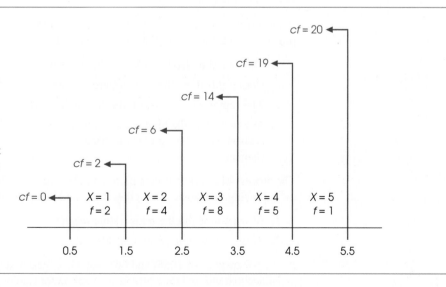

Example 2.5. Figure 2.13 shows that two people, or 10%, had scores of $X = 1$; that is, two people had scores between 0.5 and 1.5. You cannot be sure that both individuals have been accumulated until you reach 1.5, the upper real limit of the interval. Similarly, a cumulative percentage of 30% is reached at 2.5 on the scale, a percentage of 70% is reached at 3.5, and so on.

INTERPOLATION

It is possible to determine some percentiles and percentile ranks directly from a frequency distribution table, provided the percentiles are upper real limits and the ranks are percentages that appear in the table. Using the table in Example 2.5, for example, you should be able to answer the following questions:

1. What is the 95th percentile? (Answer: $X = 4.5$.)
2. What is the percentile rank for $X = 3.5$? (Answer: 70%.)

However, there are many values that do not appear directly in the table, and it is impossible to determine these values precisely. Referring to the table in Example 2.5 again,

1. What is the 50th percentile?
2. What is the percentile rank for $X = 4$?

Because these values are not specifically reported in the table, you cannot answer the questions. However, it is possible to obtain estimates of these intermediate values by using a standard procedure known as interpolation.

Before we apply the process of interpolation to percentiles and percentile ranks, we will use a simple, commonsense example to introduce this method. Suppose that you hear the weather report at 8 A.M. and again at noon. At 8:00, the temperature was 60°, and at noon, it was 68°. What is your estimate of the temperature at 9:00? To make your task a bit easier, we will create a table showing the time and temperature relationships:

TIME	TEMPERATURE
8:00	60
12:00	68

If you estimated the temperature to be 62° at 9:00, you have done interpolation. You probably went through the following logical steps:

1. The total time from 8:00 to 12:00 is 4 hours.
2. During this time, the temperature changed 8°.
3. 9:00 represents 1 hour, or one-fourth of the total time.
4. Assuming that the temperature went up at a constant rate, it should have increased by 2° during the hour because 2° equals one-fourth of the temperature change.

The process of interpolation is pictured in Figure 2.14. Using the figure, try answering the following questions about other times and temperatures:

1. At what time did the temperature reach 64°?
2. What was the temperature at 11:00?

If you got answers of 10:00 and 66°, you have mastered the process of interpolation.

Notice that interpolation provides a method for finding intermediate values—that is, values that are located between two specified numbers. This is exactly the problem we

FIGURE 2.14

The graphic representation of the process of interpolation. The same interval is shown on two separate scales, temperature and time. Only the endpoints of the scales are known—at 8:00, the temperature is 60°, and at 12:00, the temperature is 68°. Interpolation allows you to estimate values within the interval by assuming that fractional portions of one scale correspond to the same fractional portions of the other. For example, it is assumed that halfway through the temperature scale corresponds to halfway through the time scale.

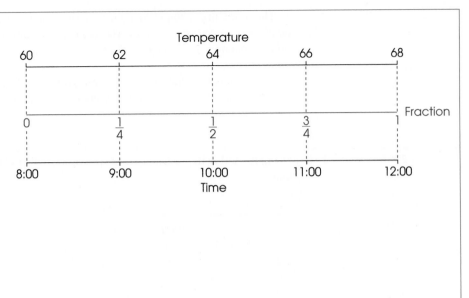

faced with percentiles and percentile ranks. Some values are given in the table, but others are not. Also notice that interpolation only *estimates* the intermediate values. In the time and temperature example, we do not know what the temperature was at 10:00. It may have soared to 80° between 8:00 and noon. The basic assumption underlying interpolation is that the change from one end of the interval to the other is a regular, linear change. We assumed, for example, that the temperature went up consistently at 2° per hour throughout the time period. Because interpolation is based on this assumption, the values we calculate are only estimates. The general process of interpolation can be summarized as follows:

1. A single interval is measured on two separate scales (for example, time and temperature). The endpoints of the interval are known for each scale.

2. You are given an intermediate value on one of the scales. The problem is to find the corresponding intermediate value on the other scale.

3. The interpolation process requires four steps:

   a. Find the width of the interval on both scales.

   b. Locate the position of the intermediate value in the interval. This position corresponds to a fraction of the whole interval:

   $$\text{fraction} = \frac{\text{distance from the top of the interval}}{\text{interval width}}$$

   c. Use this fraction to determine the distance from the top of the interval on the other scale:

   $$\text{distance} = (\text{fraction}) \times (\text{width})$$

   d. Use the distance from the top to determine the position on the other scale.

The following examples demonstrate the process of interpolation as it is applied to percentiles and percentile ranks. The key to successfully working these problems is that each cumulative percentage in the table is associated with the upper real limit of its score interval.

You may notice that in each of these problems we use interpolation working from the *top* of the interval. However, this choice is arbitrary, and you should realize that interpolation can be done just as easily working from the bottom of the interval.

E X A M P L E   2 . 6   Using the following distribution of scores, we will find the percentile rank corresponding to $X = 7.0$:

X	f	cf	c%
10	2	25	100%
9	8	23	92%
8	4	15	60%
7	6	11	44%
6	4	5	20%
5	1	1	4%

Notice that $X = 7.0$ is located in the interval bounded by the real limits of 6.5 and 7.5. The cumulative percentages corresponding to these real limits are 20% and 44%, respectively. These values are shown in the following table:

For interpolation problems, it is always helpful to create a table showing the range on both scales.

SCORES (X)	PERCENTAGES
7.5	44%
7.0 -------------------	?
6.5	20%

S T E P   1   For the scores, the width of the interval is 1 point. For the percentages, the width is 24 points.

S T E P   2   Our particular score is located 0.5 point from the top of the interval. This is exactly halfway down in the interval.

S T E P   3   Halfway down on the percentage scale would be

$$\frac{1}{2}(24 \text{ points}) = 12 \text{ points}$$

S T E P   4   For the percentages, the top of the interval is 44%, so 12 points down would be

$$44\% - 12\% = 32\%$$

This is the answer. A score of $X = 7.0$ corresponds to a percentile rank of 32%.

This same interpolation procedure can be used with data that have been grouped into class intervals. Once again, you must remember that the cumulative percentage values are associated with the upper real limits of each interval. The following example demonstrates the calculation of percentiles and percentile ranks using data in a grouped frequency distribution.

EXAMPLE 2.7   Using the following distribution of scores, we will use interpolation to find the 50th percentile:

X	f	cf	c%
20–24	2	20	100%
15–19	3	18	90%
10–14	3	15	75%
5–9	10	12	60%
0–4	2	2	10%

A percentage value of 50% is not given in the table; however, it is located between 10% and 60%, which are given. These two percentage values are associated with the upper real limits of 4.5 and 9.5, respectively. These values are shown in the following table:

SCORES (X)	PERCENTAGES
9.5	60%
? -----------------	50%
4.5	10%

STEP 1   For the scores, the width of the interval is 5 points. For the percentages, the width is 50 points.

STEP 2   The value of 50% is located 10 points from the top of the percentage interval. As a fraction of the whole interval, this is 10 out of 50, or $\frac{1}{5}$ of the total interval.

STEP 3   Using this same fraction for the scores, we obtain a distance of

$$\frac{1}{5}(5 \text{ points}) = 1 \text{ point}$$

The location we want is 1 point down from the top of the score interval.

STEP 4   Because the top of the interval is 9.5, the position we want is

$$9.5 - 1 = 8.5$$

This is the answer. The 50th percentile is $X = 8.5$.

LEARNING CHECK

1. On a statistics exam, would you rather score at the 80th percentile or at the 40th percentile?

2. For the distribution of scores presented in the following table,

   a. Find the 60th percentile.

   b. Find the percentile rank for $X = 39.5$.

X	f	cf	c%
40–49	4	25	100%
30–39	6	21	84%
20–29	10	15	60%
10–19	3	5	20%
0–9	2	2	8%

3. Using the distribution of scores from Exercise 2 and interpolation,

   a. Find the 40th percentile.

   b. Find the percentile rank for $X = 32$.

ANSWERS

1. The 80th percentile is the higher score.

2. a. $X = 29.5$ is the 60th percentile.   **b.** $X = 39.5$ has a rank of 84%.

3. a. Because 40% is between the values of 20% and 60% in the table, you must use interpolation. The score corresponding to a rank of 40% is $X = 24.5$.

   b. Because $X = 32$ is between the real limits of 29.5 and 39.5, you must use interpolation. The percentile rank for $X = 32$ is 66%.

2.6   STEM AND LEAF DISPLAYS

The general term display *is used because a stem and leaf display combines the elements of a table and a graph.*

In 1977, J.W. Tukey presented a technique for organizing data that provides a simple alternative to a frequency distribution table or graph (Tukey, 1977). This technique, called a *stem and leaf display*, requires that each score be separated into two parts: the first digit (or digits) is called the *stem*, and the last digit (or digits) is called the *leaf*. For example, $X = 85$ would be separated into a stem of 8 and a leaf of 5. Similarly, $X = 42$ would have a stem of 4 and a leaf of 2. To construct a stem and leaf display for a set of data, the first step is to list all the stems in a column. For the data in Table 2.3, for example, the lowest scores are in the 30s and the highest scores are in the 90s, so the list of stems would be

STEMS
3
4
5
6
7
8
9

The next step is to go through the data, one score at a time, and write the leaf for each score beside its stem. For the data in Table 2.3, the first score is $X = 83$, so you would write 3 (the leaf) beside the 8 in the column of stems. This process is continued for the entire set of scores. The complete stem and leaf display is shown with the original data in Table 2.3.

TABLE 2.3

A set of $N = 24$ scores presented as raw data and organized in a stem and leaf display.

DATA			STEM AND LEAF DISPLAY	
83	82	63	3	23
62	93	78	4	26
71	68	33	5	6279
76	52	97	6	283
85	42	46	7	1643846
32	57	59	8	3521
56	73	74	9	37
74	81	76		

COMPARING STEM AND LEAF DISPLAYS WITH FREQUENCY DISTRIBUTIONS

Notice that the stem and leaf display is very similar to a grouped frequency distribution. Each of the stem values corresponds to a class interval. For example, the stem 3 represents all scores in the 30s—that is, all scores in the interval 30–39. The number of leaves in the display shows the frequency associated with each stem. It also should be clear that the stem and leaf display has several advantages over a traditional frequency distribution:

1. The stem and leaf display is very easy to construct. By going through the data only one time, you can construct a complete display.

2. The stem and leaf display allows you to identify every individual score in the data. In the display shown in Table 2.3, for example, you know that there were three scores in the 60s and that the specific values were 62, 68, and 63. A frequency distribution would tell you only the frequency, not the specific values.

3. The stem and leaf display provides both a listing of the scores and a picture of the distribution. If a stem and leaf display is viewed from the side, it is essentially the same as a frequency distribution histogram (see Figure 2.15).

FIGURE 2.15

A grouped frequency distribution histogram and a stem and leaf display showing the distribution of scores from Table 2.3. The stem and leaf display is placed on its side to demonstrate that the display gives the same information provided in the histogram.

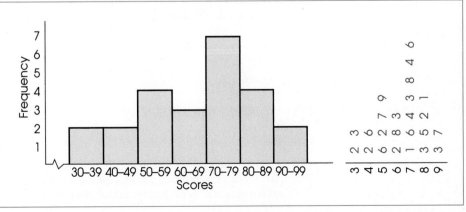

4. Because the stem and leaf display presents the actual value for each score, it is easy to modify a display if you want a more-detailed picture of a distribution. The modification simply requires that each stem be split into two (or more) parts. For example, Table 2.4 shows the same data that were presented in Table 2.3, but now we have split each stem in half. Notice that each stem value is now listed twice in the display. The first half of each stem is associated with the lower leaves (values 0–4), and the second half is associated with the upper leaves (values 5–9). In essence, we have regrouped the distribution using an interval width of 5 points instead of a width of 10 points in the original display.

Although stem and leaf displays are quite useful, they are considered to be only a preliminary means for organizing data. Typically, a researcher would use a stem and leaf display to get a first look at experimental data. The final, published report normally would present the distribution of scores in a traditional frequency distribution table or graph.

TABLE 2.4

A stem and leaf display with each stem split into two parts. Note that each stem value is listed twice: The first occurrence is associated with the lower leaf values (0–4), and the second occurrence is associated with the upper leaf values (5–9). The data shown in this display are taken from Table 2.3.

3	23
3	
4	2
4	6
5	2
5	679
6	23
6	8
7	1434
7	686
8	321
8	5
9	3
9	7

LEARNING CHECK

1. Use a stem and leaf display to organize the following set of scores:

86,   114,   94,   107,   96,   100,   98,   118,   107,

132,   106,   127,   124,   108,   112,   119,   125,   115

2. Explain how a stem and leaf display contains more information than a grouped frequency distribution.

ANSWERS

1. The stem and leaf display for these data is

8	6
9	468
10	70768
11	48295
12	745
13	2

2. A grouped frequency distribution table tells only the number of scores in each interval; it does not identify the exact value for each score. The stem and leaf display gives the individual scores as well as the number in each interval.

SUMMARY

1. The goal of descriptive statistics is to simplify the organization and presentation of data. One descriptive technique is to place the data in a frequency distribution table or graph that shows exactly how many individuals (or scores) are located in each category on the scale of measurement.

2. A frequency distribution table lists the categories that make up the scale of measurement (the X values) in one column. Beside each X value, in a second column, is the frequency or number of individuals in that category. The table may include a proportion column showing the relative frequency for each category:

$$\text{proportion} = p = \frac{f}{n}$$

The table may include a percentage column showing the percentage associated with each X value:

$$\text{percentage} = p(100) = \frac{f}{n}(100)$$

3. It is recommended that a frequency distribution table have a maximum of 10 to 15 rows to keep it simple. If the scores cover a range that is wider than this suggested maximum, it is customary to divide the range into sections called class intervals. These intervals are then listed in the frequency distribution table along with the frequency or number of individuals with scores in each interval. The result is called a grouped frequency distribution. The guidelines for constructing a grouped frequency distribution table are as follows:
   a. There should be about 10 intervals.
   b. The width of each interval should be a simple number (e.g., 2, 5, or 10).
   c. The bottom score in each interval should be a multiple of the width.
   d. All intervals should be the same width, and they should cover the range of scores with no gaps.

4. A frequency distribution graph lists scores on the horizontal axis and frequencies on the vertical axis. The type of graph used to display a distribution depends on the scale of measurement used. For interval or ratio scales, you should use a histogram or a polygon. For a histogram, a bar is drawn above each score so that the height of the bar corresponds to the frequency. Each bar extends to the real limits of the score, so that adjacent bars touch. For a polygon, a dot is placed above the midpoint of each score or class interval so that the height of the dot corresponds to the frequency; then lines are drawn to connect the dots. Bar graphs are used with nominal or ordinal scales. Bar graphs are similar to histograms except that gaps are left between adjacent bars.

5. Shape is one of the basic characteristics used to describe a distribution of scores. Most distributions can be classified as either symmetrical or skewed. A skewed distribution that tails off to the right is said to be positively skewed. If it tails off to the left, it is negatively skewed.

6. The cumulative percentage is the percentage of individuals with scores at or below a particular point in the distribution. The cumulative percentage values are associated with the upper real limits of the corresponding scores or intervals.

7. Percentiles and percentile ranks are used to describe the position of individual scores within a distribution. Percentile rank gives the cumulative percentage associated with a particular score. A score that is identified by its rank is called a percentile.

8. When a desired percentile or percentile rank is located between two known values, it is possible to estimate the desired value using the process of interpolation. Interpolation assumes a regular linear change between the two known values.

9. A stem and leaf display is an alternative procedure for organizing data. Each score is separated into a stem (the first digit or digits) and a leaf (the last digit or digits). The display consists of the stems listed in a column with the leaf for each score written beside its stem. A stem and leaf display combines the characteristics of a table and a graph and produces a concise, well-organized picture of the data.

RESOURCES

If you are using WebTutor along with this book, you will find a WebTutor section corresponding to this chapter. The WebTutor contains a brief summary of Chapter 2, hints for learning the new material and for avoiding common errors, and sample exam items including solutions.

SPSS

The statistics computer package SPSS can be used to perform many of the statistical procedures presented in this chapter. Specifically, SPSS will construct frequency distribution tables including percentages and cumulative percentages, and it will draw histograms and bar graphs showing frequency distributions.

To construct a frequency distribution table, first enter the scores in one column of the data matrix, then click **Analyze** on the tool bar. Next, select **Descriptive Statistics** and select **Frequencies**. Enter the column label for the set of scores into the Variable box. Be sure that the option to Display Frequency Table is selected and then click **OK**. The frequency distribution table will list the score values in a column from smallest to largest, with the percentage and cumulative percentage also listed for each score. Score values that do not occur (zero frequency) are not included in the table, and the program does not group scores into class intervals (all values are listed).

To draw a histogram or bar graph for the scores in a column, click **Analyze** on the tool bar, then select **Descriptive Statistics** and select **Frequencies**. Enter the column label for the set of scores into the Variable box and click **Charts**. Select either Bar Graph or Histogram depending on the scale of measurement for your data, then click **Continue** and click **OK**. After a brief delay, SPSS will display a frequency distribution table and a graph. Note that the graphing program will automatically group scores into class intervals if the range of scores is too wide.

KEY TERMS

frequency distribution	polygon	percentile rank
grouped frequency distribution	relative frequency	cumulative frequency (cf)
range	symmetrical distribution	cumulative percentage ($c\%$)
class interval	positively skewed distribution	interpolation
apparent limits	negatively skewed distribution	stem and leaf display
histogram	tail(s) of a distribution	
bar graph	percentile	

FOCUS ON PROBLEM SOLVING

1. The reason for constructing frequency distributions is to put a disorganized set of raw data into a comprehensible, organized format. Because several different types of frequency distribution tables and graphs are available, one problem is deciding which type should be used. Tables have the advantage of being easier to construct, but graphs generally give a better picture of the data and are easier to understand.

   To help you decide exactly which type of frequency distribution is best, consider the following points:

   a. What is the range of scores? With a wide range, you will need to group the scores into class intervals.

   b. What is the scale of measurement? With an interval or a ratio scale, you can use a polygon or a histogram. With a nominal or an ordinal scale, you must use a bar graph.

2. When using a grouped frequency distribution table, a common mistake is to calculate the interval width by using the highest and lowest values that define each interval. For example, some students are tricked into thinking that an interval identified as 20–24 is only 4 points wide. To determine the correct interval width, you can

   a. Count the individual scores in the interval. For this example, the scores are 20, 21, 22, 23, and 24 for a total of 5 values. Thus, the interval width is 5 points.

   b. Use the real limits to determine the real width of the interval. For example, an interval identified as 20–24 has a lower real limit of 19.5 and an upper real limit of 24.5 (halfway to the next score). Using the real limits, the interval width is

$$24.5 - 19.5 = 5 \text{ points}$$

3. Percentiles and percentile ranks are intended to identify specific locations within a distribution of scores. When solving percentile problems, especially with interpolation, it is helpful to sketch a frequency distribution graph. Use the graph to make a preliminary estimate of the answer before you begin any calculations. For example, to find the 60th percentile, you would want to draw a vertical line through the graph so that slightly more than half (60%) of the distribution is on the left-hand side of the line. Locating this position in your sketch will give you a rough estimate of what the final answer should be. When doing interpolation problems, you should keep several points in mind:

   a. Remember that the cumulative percentage values correspond to the upper real limits of each score or interval.

   b. You should always identify the interval with which you are working. The easiest way to do this is to create a table showing the endpoints on both scales (scores and cumulative percentages). This is illustrated in Example 2.6 on page 56.

   c. The word *interpolation* means *between two poles*. Remember: Your goal is to find an intermediate value between the two ends of the interval. Check your answer to be sure that it is located between the two endpoints. If it is not, then check your calculations.

DEMONSTRATION 2.1

A GROUPED FREQUENCY DISTRIBUTION TABLE

For the following set of $N = 20$ scores, construct a grouped frequency distribution table using an interval width of 5 points. The scores are:

$$14, \quad 8, \quad 27, \quad 16, \quad 10, \quad 22, \quad 9, \quad 13, \quad 16, \quad 12,$$
$$10, \quad 9, \quad 15, \quad 17, \quad 6, \quad 14, \quad 11, \quad 18, \quad 14, \quad 11$$

STEP 1   Set up the class intervals.
The largest score in this distribution is $X = 27$, and the lowest is $X = 6$. Therefore, a frequency distribution table for these data would have 22 rows and would be too large. A grouped frequency distribution table would be better. We have asked specifically for an interval width of 5 points, and the resulting table will have five rows.

X
25–29
20–24
15–19
10–14
5–9

Remember that the interval width is determined by the real limits of the interval. For example, the class interval 25–29 has an upper real limit of 29.5 and a lower real limit of 24.5. The difference between these two values is the width of the interval—namely, 5.

STEP 2   Determine the frequencies for each interval.
Examine the scores, and count how many fall into the class interval of 25–29. Cross out each score that you have already counted. Record the frequency for this class interval. Now repeat this process for the remaining intervals. The result is the following table:

X	f	
25–29	1	(the score $X = 27$)
20–24	1	($X = 22$)
15–19	5	(the scores $X = 16, 16, 15, 17$, and 18)
10–14	9	($X = 14, 10, 13, 12, 10, 14, 11, 14$, and 11)
5–9	4	($X = 8, 9, 9$, and 6)

DEMONSTRATION 2.2

USING INTERPOLATION TO FIND PERCENTILES AND PERCENTILE RANKS

Find the 50th percentile for the set of scores in the grouped frequency distribution table that was constructed in Demonstration 2.1.

STEP 1   Find the cumulative frequency (*cf*) and cumulative percentage values, and add these values to the basic frequency distribution table.

Cumulative frequencies indicate the number of individuals located in or below each category (class interval). To find these frequencies, begin with the bottom interval, and then accumulate the frequencies as you move up the scale. For this example, there are 4 individuals who are in or below the 5–9 interval (*cf* = 4). Moving up the scale, the 10–14 interval contains an additional 9 people, so the cumulative value for this interval is 9 + 4 = 13 (simply add the 9 individuals in the interval to the 4 individuals below). Continue moving up the scale, cumulating frequencies for each interval.

Cumulative percentages are determined from the cumulative frequencies by the relationship

$$c\% = \left(\frac{cf}{N}\right)100\%$$

For example, the *cf* column shows that 4 individuals (out of the total set of $N = 20$) have scores in or below the 5–9 interval. The corresponding cumulative percentage is

$$c\% = \left(\frac{4}{20}\right)100\% = \left(\frac{1}{5}\right)100\% = 20\%$$

The complete set of cumulative frequencies and cumulative percentages is shown in the following table:

X	f	cf	c%
25–29	1	20	100%
20–24	1	19	95%
15–19	5	18	90%
10–14	9	13	65%
5–9	4	4	20%

STEP 2   Locate the interval that contains the value that you want to calculate.

We are looking for the 50th percentile, which is located between the values of 20% and 65% in the table. The scores (upper real limits) corresponding to these two percentages are 9.5 and 14.5, respectively. The interval, measured in terms of scores and percentages, is shown in the following table:

X	C%
14.5	65%
??--------	50%
9.5	20%

STEP 3   Locate the intermediate value as a fraction of the total interval.

Our intermediate value is 50%, which is located in the interval between 65% and 20%. The total width of the interval is 45 points (65 − 20 = 45), and the value of 50% is located 15 points down from the top of the interval. As a fraction, the 50th percentile is located $\frac{15}{45} = \frac{1}{3}$ down from the top of the interval.

S T E P 4   Use the fraction to determine the corresponding location on the other scale.

Our intermediate value, 50%, is located $\frac{1}{3}$ of the way down from the top of the interval. Our goal is to find the score, the X value, that also is located $\frac{1}{3}$ of the way down from the top of the interval.

On the score (X) side of the interval, the top value is 14.5, and the bottom value is 9.5, so the total interval width is 5 points ($14.5 - 9.5 = 5$). The position we are seeking is $\frac{1}{3}$ of the way from the top of the interval. One-third of the total interval is

$$\left(\frac{1}{3}\right)5 = \frac{5}{3} = 1.67 \text{ points}$$

To find this location, begin at the top of the interval, and come down 1.67 points:

$$14.5 - 1.67 = 12.83$$

This is our answer. The 50th percentile is $X = 12.83$.

PROBLEMS

1. Place the following scores in a frequency distribution table. Include columns for proportion (p) and percentage (%) in your table. The scores are:

    9, 10, 7, 7, 5, 7, 10, 9, 8, 7,

    8, 5, 10, 8, 9, 7, 8, 8, 10, 7

2. What information is available about the data in a regular frequency distribution table that is not available in a grouped table?

3. Describe the difference in appearance between a bar graph and a histogram, and describe the circumstances under which each type of graph should be used.

4. If it is appropriate to use a histogram to show the frequency distribution for a set of scores, would it also be appropriate to show the same set of scores in a bar graph? Explain why or why not.

5. Under what circumstances should you use a grouped frequency distribution instead of a regular frequency distribution?

6. Sketch a histogram and a polygon showing the distribution of scores presented in the following table:

X	f
5	3
4	7
3	5
2	3
1	1

7. The following are reading comprehension scores for a third-grade class of 18 students:

    4, 5, 5, 6, 4, 5, 2, 6, 3,

    5, 6, 5, 4, 3, 6, 1, 6, 5

    a. Place the scores in a frequency distribution table.
    b. Sketch a histogram showing the distribution.
    c. Using your graph, answer the following questions:
       (1) What is the shape of the distribution?
       (2) If a score of $X = 4$ is considered typical for third-graders, how would you describe the general reading level of this class?

8. An instructor obtained the following set of scores from a 10-point quiz for a class of 26 students:

    8, 5, 4, 5, 5, 7, 6, 4, 3, 4, 5, 6, 6,

    4, 5, 5, 10, 6, 9, 5, 7, 8, 2, 6, 7, 4

    a. Place the scores in a frequency distribution table.
    b. Sketch a histogram showing the distribution.
    c. Using your graph, answer the following questions:
       (1) What is the shape of the distribution?
       (2) As a whole, how did the class do on the quiz? Were most scores high or low? Was the quiz easy or hard?

9. For the set of scores shown in the following frequency distribution table,
    a. How many scores are in the distribution? ($N = ?$)

b. Find ΣX for this set of scores.

X	f
4	2
3	5
2	3
1	1

10. A set of scores has been organized into the following frequency distribution table. Find each of the following values for the original set of scores:
a. N
b. ΣX

X	f
5	2
4	3
3	5
2	0
1	1

11. Find each value requested for the set of scores summarized in the following table:
a. N
b. ΣX
c. ΣX^2

X	f
5	1
4	2
3	4
2	3
1	1

12. The following table shows the frequency distribution of birth-order position for a sample of $n = 24$ students.
a. What kind of graph would be appropriate for showing this distribution?
b. Sketch the frequency distribution graph.

BIRTH-ORDER POSITION	f
1st born	12
2nd born	5
3rd born	6
4th born	0
5th born	1

13. An instructor at the state college recorded the home state for each student in an introductory psychology class and obtained the following results:

NY NY PA NY NY MA PA NY NH NY
NY VA NY NY NY NY NY OH NY PA
OH NY NY NY MA ME NJ NY NY NY

a. Place the scores in a frequency distribution table.
b. Sketch a bar graph showing the distribution.

14. Place the following set of scores in a frequency distribution table:

1, 3, 1, 1, 4, 1, 4, 5, 6, 2, 1, 1, 5,
1, 3, 2, 1, 6, 2, 4, 5, 2, 3, 2, 3

Compute the proportion and the percentage of individuals with each score. From your frequency distribution table, you should be able to identify the shape of this distribution.

15. For the following set of scores:

5, 6, 2, 3, 6, 5, 6,
4, 1, 5, 6, 3, 4

a. Construct a frequency distribution table.
b. Sketch a polygon showing the distribution.
c. Describe the distribution using the following characteristics:
　　(1) What is the shape of the distribution?
　　(2) What score best identifies the center (average) for the distribution?
　　(3) Are the scores clustered together, or are they spread out across the scale?

16. For the following set of scores:

2, 6, 3, 7, 11, 3, 4, 2, 5,
7, 4, 2, 5, 6, 2, 3, 4,
9, 4, 5, 3, 4, 6, 4, 3

a. Construct a frequency distribution table.
b. Sketch a histogram for these data.
c. Sketch a polygon for these data.
d. What is the shape of the distribution?

17. For the following scores, construct a frequency distribution table using
a. An interval width of 5.
b. An interval width of 10.

64, 75, 50, 67, 86, 66, 62, 64,
71, 47, 57, 74, 63, 67, 56, 65,
70, 87, 48, 50, 41, 66, 73,
60, 63, 45, 78, 68, 53, 75

18. Three sets of data are described by identifying the lowest score and the highest score for each set. Describe how a grouped frequency distribution table should be constructed for each set. That is, give the interval width that you would suggest, and make a list of all class intervals.
  a. 80–127
  b. 21–38
  c. 180–360

19. Complete the cumulative frequency column and the cumulative percentage column for the following table:

X	f	cf	c%
5	7		
4	8		
3	5		
2	3		
1	2		

20. Complete the following frequency distribution table, and find each of the percentiles and percentile ranks requested:

X	f	cf	c%
12	5		
11	7		
10	4		
9	3		
8	0		
7	1		

  a. What is the 20th percentile?
  b. What is the percentile for $X = 10.5$?
  c. What is the 75th percentile?
  d. What is the percentile rank for $X = 11.5$?

21. Find each of the requested values for the distribution of scores shown in the following frequency distribution table:

X	f
14–15	2
12–13	3
10–11	4
8–9	6
6–7	5
4–5	4
2–3	1

  a. What is the 20th percentile?
  b. What is the percentile rank for $X = 11.5$?

c. Use interpolation to find the 25th percentile.
d. Use interpolation to find the percentile rank for $X = 12$.

22. The following frequency distribution presents a set of exam scores for a class of $N = 25$ students.

X	f	cf	c%
90–99	7	25	100%
80–89	8	18	76%
70–79	5	10	40%
60–69	3	5	20%
50–59	1	2	8%
40–49	1	1	4%

  a. Find the 25th percentile.
  b. Find the 49th percentile.
  c. Find the percentile rank for $X = 75$.
  d. Find the percentile rank for $X = 87$.

23. Using the following frequency distribution, find each of the percentiles and percentile ranks requested:

X	f	cf	c%
6	2	20	100%
5	2	18	90%
4	4	16	80%
3	6	12	60%
2	4	6	30%
1	2	2	10%

  a. What is the 80th percentile?
  b. What is the percentile rank for $X = 2.5$?
  c. Use interpolation to find the 75th percentile.
  d. Use interpolation to find the percentile rank for $X = 4$.
  e. What score separates the bottom 25% of the distribution from the rest? That is, find the 25th percentile.

24. A set of scores has been summarized in the following stem and leaf display. For this set of scores:
  a. How many scores are in the 90s?
  b. Identify the individual scores in the 90s.
  c. How many scores are in the 70s?
  d. Identify the individual scores in the 70s.

4	38
5	2
6	461
7	5499
8	2185
9	92407

25. Use a stem and leaf display to organize the following scores:

43, 56, 35, 47, 48, 52, 66, 57, 46,

39, 43, 47, 61, 55, 50, 49, 39, 40

26. Construct a stem and leaf display for the following set of scores:

103, 82, 57, 66, 75, 58, 37, 99,

74, 60, 56, 61, 87, 41, 98, 58,

107, 79, 66, 59, 32, 81, 40,

54, 62, 83, 94, 46, 70, 79

CHAPTER

3

Central Tendency

Tools You Will Need

The following items are considered essential background material for this chapter. If you doubt your knowledge of any of these items, you should review the appropriate chapter or section before proceeding.

- Summation notation (Chapter 1)
- Frequency distributions (Chapter 2)

Preview

In a classic study examining the relationship between heredity and intelligence, Tryon (1940) used a selective breeding program to develop separate strains of "smart" and "dumb" rats. Starting with a large sample of rats, Tryon tested each animal on a maze-learning problem. Based on their error scores for the maze, the brightest rats and the dullest rats were selected from this sample. The brightest males were mated with the brightest females. Similarly, the dullest rats were interbred. This process of testing and selectively breeding was continued for several generations until Tryon had established a line of maze-bright rats and a separate line of maze-dull rats. The results obtained by Tryon are shown in Figure 3.1.

Notice that after seven generations, there is an obvious difference between the two groups. As a rule, the maze-bright animals outperform the maze-dull animals. It is tempting to describe the results of this experiment by saying that the maze-bright rats are better learners than the maze-dull rats. However, notice that there is some overlap between the two groups; not all the dull animals are really dull. In fact, some of the animals bred for brightness are actually poorer learners than some of the animals bred for dullness. What is needed is a simple way of describing the general difference between these two groups while still acknowledging the fact that some individuals may contradict the general trend.

The solution to this problem is to identify the typical or average rat as the representative for each group. Then the experimental results can be described by saying that the typical maze-bright rat is a faster learner than the typical maze-dull rat. On the average, the bright rats really are brighter.

In this chapter, we will introduce the statistical techniques used to identify the typical or average score for a distribution. Although there are several reasons for defining the average score, the primary advantage of an average is that it provides a single number that describes an entire distribution and can be used for comparison with other distributions.

As a footnote, you should know that later research with Tryon's rats revealed that the maze-bright rats were not really more intelligent than the maze-dull rats. Although the bright rats had developed specific abilities that are useful in mazes, the dull rats proved to be just as smart when tested on a variety of other tasks.

Tryon, R. C. (1940). "Genetic differences in maze-learning ability in rats." *The Thirty-ninth Yearbook of the National Society for the Study of Education*, 111–119. Adapted and reprinted with permission of the National Society for the Study of Education.

FIGURE 3.1

Distribution of error scores (a) for the original sample of rats (parents) and (b) for the two separate lines that were selectively bred for either good or poor maze performance (maze-bright and maze-dull).

3.1 OVERVIEW

The general purpose of descriptive statistical methods is to organize and summarize a set of scores. Perhaps the most common method for summarizing and describing a distribution is to find a single value that defines the average score and can serve as a representative for the entire distribution. In statistics, the concept of an average or representative score is called *central tendency*. The goal in measuring central tendency is to describe a distribution of scores by determining a single value that identifies the center of the distribution. Ideally, this central value will be the score that is the best representative value for all of the individuals in the distribution.

DEFINITION   **Central tendency** is a statistical measure to determine a single score that defines the center of a distribution. The goal of central tendency is to find the single score that is most typical or most representative of the entire group.

In everyday language, the goal of central tendency is to identify the "average" or "typical" individual. This average value can then be used to provide a simple description of the entire population or sample. For example, the average family income in Greenwich, Connecticut, is over $225,000. Obviously, not everyone in Greenwich makes this income, but the average should give you a good idea of the neighborhood. Measures of central tendency are also useful for making comparisons between groups of individuals or between sets of figures. For example, weather data indicate that for Seattle, Washington, the average yearly temperature is 53 degrees and the average annual precipitation is 34 inches. By comparison, the average temperature in Phoenix, Arizona, is 71 degrees and the average precipitation is 7.4 inches. The point of these examples is to demonstrate the great advantage of being able to describe a large set of data with a single, representative number. Central tendency characterizes what is typical for a large population and in doing so makes large amounts of data more digestible. Statisticians sometimes use the expression "number crunching" to illustrate this aspect of data description. That is, we take a distribution consisting of many scores and "crunch" them down to a single value that describes them all.

Unfortunately, there is no single, standard procedure for determining central tendency. The problem is that no single measure will always produce a central, representative value in every situation. The three distributions shown in Figure 3.2 should help demonstrate this fact. Before we discuss the three distributions, take a moment to look at the figure and try to identify the "center" or the "most representative score" for each distribution.

1. The first distribution [Figure 3.2(a)] is symmetrical, with the scores forming a distinct pile centered around $X = 5$. For this type of distribution, it is easy to identify the "center," and most people would agree that the value $X = 5$ is an appropriate measure of central tendency.

2. In the second distribution [Figure 3.2(b)], however, problems begin to appear. Now the scores form a negatively skewed distribution, piling up at the high end of the scale around $X = 8$, but tapering off to the left all the way down to $X = 1$. Where is the "center" in this case? Some people might select $X = 8$ as the center because more individuals had this score than any other single value. However, $X = 8$ is clearly not in the middle of the distribution. In fact, the majority of the scores (10 out of 16) have values less than 8, so it seems reasonable that the "center" should be defined by a value that is less than 8.

3. Now consider the third distribution [Figure 3.2(c)]. Again, the distribution is symmetrical, but now there are two distinct piles of scores. Because the distribution is symmetrical with $X = 5$ as the midpoint, you may choose $X = 5$ as the "center." However, none of the scores is located at $X = 5$ (or even close), so this value is not particularly good as a representative score. On the other

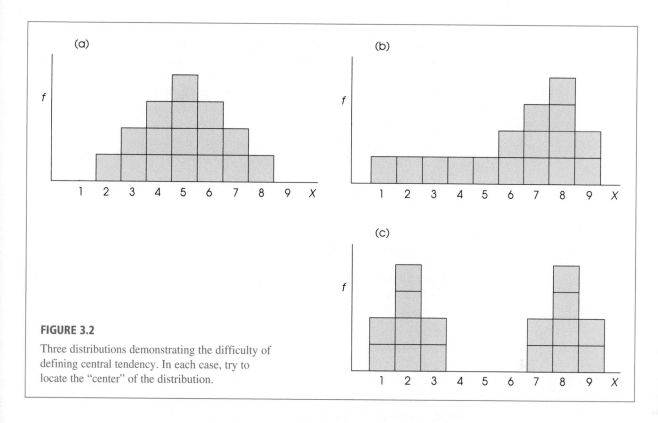

FIGURE 3.2

Three distributions demonstrating the difficulty of defining central tendency. In each case, try to locate the "center" of the distribution.

hand, because there are two separate piles of scores with one group centered at $X = 2$ and the other centered at $X = 8$, it is tempting to say that this distribution has two centers. But can there be two centers?

Clearly, there are problems defining the "center" of a distribution. Occasionally, you will find a nice, neat distribution like the one shown in Figure 3.2(a), where everyone will agree on the center. But you should realize that other distributions are possible and that there may be different opinions concerning the definition of the center. To deal with these problems, statisticians have developed three different methods for measuring central tendency: the mean, the median, and the mode. They are computed differently and have different characteristics. To decide which of the three measures is best for any particular distribution, you should keep in mind that the general purpose of central tendency is to find the single most representative score. Each of the three measures we will present has been developed to work best in a specific situation. We will examine this issue in more detail after we introduce the three measures.

3.2   THE MEAN

The *mean*, commonly known as the arithmetic average, is computed by adding all the scores in the distribution and dividing by the number of scores. The mean for a population will be identified by the Greek letter mu, μ (pronounced "myoo"), and the mean for a sample is identified by M or $\overline{X}$ (read "x-bar").

For years, the convention in statistics textbooks was to use $\overline{X}$ to represent the mean for a sample. However, in manuscripts and in published research reports the letter M is the standard notation for a sample mean. Because you will encounter the letter M when reading research reports and because you should use the letter M when writing research reports, we have decided to use the same notation in this text. Keep in mind that the $\overline{X}$ notation is still appropriate for identifying a sample mean, and you may find it used on occasion, especially in textbooks.

DEFINITION

The **mean** for a distribution is the sum of the scores divided by the number of scores.

The formula for the population mean is

$$\mu = \frac{\Sigma X}{N} \tag{3.1}$$

First, sum all the scores in the population, and then divide by N. For a sample, the computation is done the same way, but the formula uses symbols that signify sample values:

$$\text{sample mean} = M = \frac{\Sigma X}{n} \tag{3.2}$$

In general, we will use Greek letters to identify characteristics of a population and letters of our own alphabet to stand for sample values. If a mean is identified with the symbol M, you should realize that we are dealing with a sample. Also note that n is used as the symbol for the number of scores in the sample.

EXAMPLE 3.1

For a population of $N = 4$ scores,

3,   7,   4,   6

the mean is

$$\mu = \frac{\Sigma X}{N} = \frac{20}{4} = 5$$

ALTERNATIVE DEFINITIONS
FOR THE MEAN

Although the procedure of adding the scores and dividing by the number provides a useful definition of the mean, there are two alternative definitions that may give you a better understanding of this important measure of central tendency.

The first alternative is to think of the mean as the amount each individual would get if the total (ΣX) were divided equally among all the individuals (N) in the distribution. This somewhat socialistic viewpoint is particularly useful in problems where you know the mean and must find the total. Consider the following example.

EXAMPLE 3.2

A group of $n = 6$ boys buys a box of baseball cards at a garage sale and discovers that the box contains a total of 180 cards. If the boys divide the cards equally among themselves, how many cards will each boy get? You should recognize that this problem represents the standard procedure for computing the mean. Specifically, the total (ΣX) is divided by the number (n) to produce the mean, $\frac{180}{6} = 30$ cards for each boy.

You should also recognize that this example demonstrates that it is possible to define the mean as the amount that each individual gets when the total is distributed equally. This new definition can be useful for some problems involving the mean. Consider the following example.

This time we have a group of $n = 4$ boys and we measure the amount of money that each boy has. The data produce a mean of $M = \$5$. Given this information, what is the total amount of money for the whole group? Although you do not know exactly how much money each boy has, the new definition of the mean tells you that if they pool their money together and then distribute the total equally, each boy will get $5. For each of $n = 4$ boys to get $5, the total must be $4(\$5) = \$20$. To check this answer, use the formula for the mean:

$$M = \frac{\Sigma X}{n} = \frac{\$20}{4} = \$5$$

The second alternative definition of the mean describes the mean as a balance point for a distribution. Consider the population consisting of $N = 4$ scores (2, 2, 6, 10). For this population, $\Sigma X = 20$ and $N = 4$, so $\mu = \frac{20}{4} = 5$. Figure 3.3 shows this population drawn as a histogram, with each score represented by a box that is sitting on a seesaw. If the seesaw is positioned so that it pivots at a point equal to the mean, then it will be balanced and will rest level.

FIGURE 3.3

The frequency distribution shown as a seesaw balanced at the mean.

Based on G. H. Weinberg, J. A. Schumaker, & D. Oltman (1981). *Statistics: An Intuitive Approach* (p. 14). Belmont, Calif.: Wadsworth.

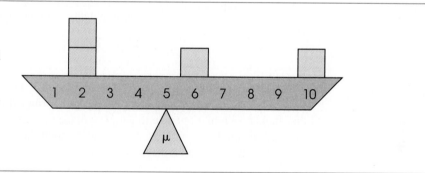

The reason the seesaw is balanced over the mean becomes clear when we measure the distance of each box (score) from the mean:

SCORE	DISTANCE FROM THE MEAN
$X = 2$	3 points below the mean
$X = 2$	3 points below the mean
$X = 6$	1 point above the mean
$X = 10$	5 points above the mean

Notice that the mean balances the distances. That is, the total distance below the mean is the same as the total distance above the mean:

3 points below + 3 points below = 6 points below

1 point above + 5 points above = 6 points above

Because the mean serves as a balance point, the value of the mean will always be located somewhere between the highest score and the lowest score; that is, the mean can never be outside the range of scores. If the lowest score in a distribution is $X = 8$ and the highest is $X = 15$, then the mean *must* be between 8 and 15. If you calculate a value that is outside this range, then you have made an error.

The image of a seesaw with the mean at the balance point is also useful for determining how a distribution is affected if a new score is added or if an existing score is removed. For the distribution in Figure 3.3, for example, what would happen to the mean (balance point) if a new score were added at $X = 10$?

THE WEIGHTED MEAN   Often it is necessary to combine two sets of scores and then find the overall mean for the combined group. Suppose that we begin with two separate samples. The first sample has $n = 12$ scores and $M = 6$. The second sample has $n = 8$ and $M = 7$. If the two samples are combined, what is the mean for the total group?

To calculate the overall mean, we will need two values: the overall sum of the scores for the combined group (ΣX) and the total number of scores in the combined group (n). To determine the overall sum, we will simply find the sum of the scores for the first sample (ΣX_1) and the sum for the second sample (ΣX_2) and then add the two sums together. Similarly, the total number of scores in the combined group can be found by adding the number in the first example (n_1) and the number in the second sample (n_2). With these two values, we can compute the mean using the basic formula

$$\text{overall mean} = M = \frac{\Sigma X \text{ (overall sum for the combined group)}}{n \text{ (total number in the combined group)}}$$

$$= \frac{\Sigma X_1 + \Sigma X_2}{n_1 + n_2}$$

The first sample has $n = 12$ and $M = 6$. Therefore, the total for this sample must be $\Sigma X = 72$. (Remember that when the total is distributed equally, each person gets the mean. For each of 12 people to get $M = 6$, the total must be 72.) In the same way, the second sample has $n = 8$ and $M = 7$, so the total must be $\Sigma X = 56$. Using these values, we obtain an overall mean of

$$\text{overall mean} = M = \frac{\Sigma X_1 + \Sigma X_2}{n_1 + n_2} = \frac{72 + 56}{12 + 8} = \frac{128}{20} = 6.4$$

The following table summarizes the calculations.

FIRST SAMPLE	SECOND SAMPLE	COMBINED SAMPLE
$n = 12$	$n = 8$	$n = 20$ (12 + 8)
$\Sigma X = 72$	$\Sigma X = 56$	$\Sigma X = 128$ (72 + 56)
$M = 6$	$M = 7$	$M = 6.4$

Note that the overall mean is not halfway between the original two sample means. Because the samples are not the same size, one will make a larger contribution to the total group and therefore will carry more weight in determining the overall mean. For this reason, the overall mean we have calculated is called the *weighted mean* (see Box 3.1). In this example, the overall mean of $M = 6.4$ is closer to the value of $M = 6$ (the larger sample) than it is to $M = 7$ (the smaller sample).

In summary, when two samples are combined, the weighted mean is obtained as follows:

STEP 1 Determine the grand total of all the scores in both samples. This sum is obtained by adding the sum of the scores for the first sample (ΣX_1) and the sum of the scores for the second sample (ΣX_2).

STEP 2 Determine the total number of scores in both samples. This value is obtained by adding the number in the first sample (n_1) and the number in the second sample (n_2).

STEP 3 Divide the sum of all the scores (step 1) by the total number of scores (step 2). Expressed as an equation,

$$\text{weighted mean} = \frac{\text{combined sum}}{\text{combined } n} = \frac{\Sigma X_1 + \Sigma X_2}{n_1 + n_2}$$

BOX

3.1   AN ALTERNATIVE PROCEDURE FOR FINDING THE WEIGHTED MEAN

In the text, the weighted mean was obtained by first determining the total number of scores (n) for the two combined samples and then determining the overall sum (ΣX) for the two combined samples. The following example demonstrates how the same result can be obtained using a slightly different conceptual approach.

We begin with the same two samples that were used in the text: One sample has $M = 6$ for $n = 12$ students, and the second sample has $M = 7$ for $n = 8$ students. The goal is to determine the mean for the overall group when the two samples are combined.

Logically, when these two samples are combined, the larger sample (with $n = 12$ scores) will make a greater contribution to the combined group than the smaller sample (with $n = 8$ scores). Thus, the larger sample will carry more weight in determining the mean for the combined group. We will accommodate this fact by assigning a weight to each sample mean so that the weight is determined by the size of the sample. To determine how much weight should be assigned to each sample mean, you simply consider the sample's contribution to the combined group. When the two samples

are combined, the resulting group will have a total of 20 scores ($n = 12$ from the first sample and $n = 8$ from the second). The first sample contributes 12 out of 20 scores and, therefore, is assigned a weight of $\frac{12}{20}$. The second sample contributes 8 out of 20 scores, and its weight is $\frac{8}{20}$. Each sample mean is then multiplied by its weight, and the results are summed to find the weighted mean for the combined sample. For this example,

$$\text{weighted mean} = \left(\frac{12}{20}\right)(6) + \left(\frac{8}{20}\right)(7)$$

$$= \frac{72}{20} + \frac{56}{20}$$

$$= 3.6 + 2.8$$

$$= 6.4$$

Note that this is the same result obtained using the method described in the text.

COMPUTING THE MEAN FROM A FREQUENCY DISTRIBUTION TABLE   When a set of scores has been organized in a frequency distribution table, the calculation of the mean is usually easier if you first remove the individual scores from the table. Table 3.1 shows a distribution of quiz scores organized in a frequency distribution table. To compute the mean for this distribution you must be careful to use both the X values in the first column and the frequencies in the second column. The values in the table show that the distribution consists of one 10, two 9s, four 8s, and one 6, for a total of $n = 8$ scores. Remember that you can determine the number of scores by adding the frequencies, $n = \Sigma f$. To find the sum of the scores, you must be careful to add all eight scores:

$$\Sigma X = 10 + 9 + 9 + 8 + 8 + 8 + 8 + 6 = 66$$

You can also find the sum by computing ΣfX (see p. 38).

Once you have found ΣX and n, you compute the mean as usual. For these data,

$$M = \frac{\Sigma X}{n} = \frac{66}{8} = 8.25$$

TABLE 3.1

Statistics quiz scores for a section of $n = 8$ students.

QUIZ SCORE (X)	f	fX
10	1	10
9	2	18
8	4	32
7	0	0
6	1	6

LEARNING CHECK

1. A sample of $n = 8$ scores has a mean of $M = 6$. What is ΣX for this sample?

2. Sample A has $n = 8$ and $M = 14$. For sample B, $n = 20$ and $M = 6$. If the two samples are combined, then
   a. The mean for the combined sample will be between 14 and 6. (True or false?)
   b. The mean for the combined sample will be closer to 14 than it is to 6. (True or false?)
   c Calculate the mean for the combined sample.

3. Find n, ΣX, and M for the sample of scores shown in the following frequency distribution table:

X	f
4	2
3	4
2	3
1	1

ANSWERS

1. $\Sigma X = 48$

2. a. True   b. False   c. $M = \frac{232}{28} = 8.29$

3. $n = 10$, $\Sigma X = 27$, and $M = \frac{27}{10} = 2.7$

CHARACTERISTICS OF THE MEAN

The mean has many characteristics that will be important in future discussions. In general, these characteristics result from the fact that every score in the distribution contributes to the value of the mean. Specifically, every score must be added into the total in order to compute the mean. We will now discuss four of the more important characteristics.

Changing a score   Changing the value of any score will change the mean. For example, a sample of quiz scores for a psychology lab section consists of 9, 8, 7, 5, and 1. Note that the sample consists of $n = 5$ scores with $\Sigma X = 30$. The mean for this sample is

$$M = \frac{\Sigma X}{n} = \frac{30}{5} = 6.00$$

Now suppose that the score of $X = 1$ is changed to $X = 8$. Note that we have added 7 points to this individual's score, which will also add 7 points to the total (ΣX). After changing the score, the new distribution consists of

9, 8, 7, 5, 8

There are still $n = 5$ scores, but now the total is $\Sigma X = 37$. Thus, the new mean is

$$M = \frac{\Sigma X}{n} = \frac{37}{5} = 7.40$$

Notice that changing a single score in the sample has produced a new mean. You should recognize that changing any score will also change the value of ΣX (the sum of the scores), and thus will always change the value of the mean.

Introducing a new score or removing a score   In general, the mean is determined by two values: ΣX and N (or n). Whenever either of these values is changed, the mean also will be changed. In the preceding example, the value of one score was changed. This produced a change in the total (ΣX) and therefore changed the mean. If you add a new score (or take away a score), you will change both ΣX and n, and you must compute the new mean using the changed values.

Usually, but not always, adding a new score or removing an existing score will change the mean. The exception is when the new score (or the removed score) is exactly equal to the mean. It is easy to visualize the effect of adding or removing a score if you remember that the mean is defined as the balance point for the distribution. Figure 3.4 shows a distribution of scores represented as boxes on a seesaw that is balanced at the mean, $\mu = 7$. Imagine what would happen if we added a new score (a new box) at $X = 10$. Clearly, the seesaw would tip to the right and we would need to slide the pivot point (the mean) higher to restore balance. Thus, adding a new score that is larger than the current mean will cause an increase in the value of the mean.

Again using Figure 3.4, imagine what would happen if we removed the score (the box) that is located at $X = 9$. This time the seesaw would tip to the left and we would have to move the mean lower to restore balance. Thus, removing a score that is greater than the current mean will cause a decrease in the value of the mean.

FIGURE 3.4

A distribution of $N = 5$ scores that is balanced with a mean of $\mu = 7$.

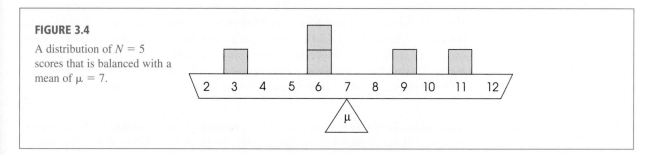

Finally, consider what would happen to the seesaw if we added a new score (or removed an existing score) located exactly at the mean. It should be clear that the seesaw would not tilt in either direction so the mean would stay in exactly the same place.

The following example demonstrates exactly how the new mean is computed when a new score is added to an existing sample.

EXAMPLE 3.3

Remember that the mean is the amount each person gets when the total is divided equally. If each of four people has 7, then the total must be 28.

We begin with a sample of $n = 4$ scores with a mean of $M = 7$. Notice that this sample must have $\Sigma X = 28$. The problem is to determine what happens to the sample mean if a new score of $X = 12$ is added to the sample.

To find the new sample mean, we must first determine the values for n and ΣX. In each case, we begin with the original sample and then consider the impact of adding a new score. The original sample had $n = 4$ scores, so adding one new score brings us to $n = 5$. Similarly, the original sample had $\Sigma X = 28$. We added $X = 12$, so the new total is $\Sigma X = 28 + 12 = 40$. Finally, the new mean is computed using the new values for n and ΣX.

$$M = \frac{\Sigma X}{n} = \frac{40}{5} = 8$$

The entire process can be summarized as follows:

ORIGINAL SAMPLE	NEW SAMPLE ADDING $X = 12$
$n = 4$	$n = 5$
$\Sigma X = 28$	$\Sigma X = 40$
$M = 28/4 = 7$	$M = 40/5 = 8$

Adding or subtracting a constant from each score   If a constant value is added to every score in a distribution, the same constant will be added to the mean. Similarly, if you subtract a constant from every score, the same constant will be subtracted from the mean.

Consider the feeding scores for a sample of $n = 6$ rats shown in Table 3.2. The scores show the amount of food each rat ate during a 24-hour test period. The first column shows food consumption before the rats were injected with a diet drug. Note that the total amount is $\Sigma X = 26$ grams for $n = 6$ rats, so the mean is $M = 4.33$. The following day, the rats are injected with an experimental drug that reduces appetite. Suppose that the effect of the drug is to subtract a constant (2 points) from each rat's feeding score. The resulting scores are shown in the final column in the table. Note that these scores (after the drug) total $\Sigma X = 14$ grams for $n = 6$ rats, so the new mean is $M = 2.33$. Subtracting 2 points from each score has also subtracted 2 points from the mean, from $M = 4.33$ to $M = 2.33$. (It is important to note that experimental effects are practically never as simple as the adding or subtracting of a constant. Nonetheless, the principle of this characteristic of the mean is important and will be addressed in later chapters when we are using statistics to evaluate the effects of experimental manipulations.)

Multiplying or dividing each score by a constant   If every score in a distribution is multiplied by (or divided by) a constant value, the mean will change in the same way.

TABLE 3.2

Amount of food (in grams) consumed before and after diet drug injections

RAT'S IDENTIFICATION	BEFORE DRUG CONSUMPTION	AFTER DRUG CONSUMPTION
A	6	4
B	3	1
C	5	3
D	3	1
E	4	2
F	5	3
	$\Sigma X = 26$	$\Sigma X = 14$
	$M = 4.33$	$M = 2.33$

Multiplying (or dividing) each score by a constant value is a common method for changing the unit of measurement. To change a set of measurements from minutes to seconds, for example, you multiply by 60; to change from inches to feet, you divide by 12. Table 3.3 shows how a sample of $n = 5$ scores originally measured in yards would be transformed into a set of scores measured in feet. The first column shows the original scores with total $\Sigma X = 50$ with $M = 10$ yards. In the second column, each of the original values has been multiplied by 3 (to change from yards to feet), and the resulting values total $\Sigma X = 150$, with $M = 30$ feet. Multiplying each score by 3 has also caused the mean to be multiplied by 3. You should realize, however, that although the numerical values for the individual scores and the sample mean have been multiplied, the actual measurements are not changed.

TABLE 3.3

Measurement of five pieces of wood

ORIGINAL MEASUREMENT IN YARDS	CONVERSION TO FEET (MULTIPLY BY 3)
10	30
9	27
12	36
8	24
11	33
$\Sigma X = 50$	$\Sigma X = 150$
$M = 10$ yards	$M = 30$ feet

LEARNING CHECK

1. **a.** Compute the mean for the following sample of scores:

    6, 1, 8, 0, 5

    b. Add 4 points to each score, and then compute the mean.
    c. Multiply each of the original scores by 5, and then compute the mean.

2. A population has a mean of $\mu = 80$.
    a. If 6 points were added to every score, what would be the value for the new mean?
    b. If every score were multiplied by 2, what would be the value for the new mean?

3. A sample of $n = 5$ scores has a mean of $M = 8$. If you added a new score of $X = 2$ to the sample, what value would you obtain for the new sample mean?

A N S W E R S   **1. a.** $M = \frac{20}{5} = 4$ **b.** $M = \frac{40}{5} = 8$ **c.** $M = \frac{100}{5} = 20$

2. a. The new mean would be 86.   **b.** The new mean would be 160.

3. The original sample has $n = 5$ and $\Sigma X = 40$. With the new score, the sample has $n = 6$ and $\Sigma X = 42$. The new mean is $\frac{42}{6} = 7$.

3.3   THE MEDIAN

The second measure of central tendency we will consider is called the *median*. The median is the score that divides a distribution exactly in half. Exactly one-half of the scores are less than or equal to the median, and exactly one-half are greater than or equal to the median. Because exactly 50% of the scores fall at or below the median, this value is equivalent to the 50th percentile (see Chapter 2).

D E F I N I T I O N

The **median** is the score that divides a distribution exactly in half. Exactly 50% of the individuals in a distribution have scores at or below the median. The median is equivalent to the 50th percentile.

Earlier, when we introduced the mean, specific symbols and notation were used to identify the mean and to differentiate a sample mean and a population mean. For the median, however, there are no symbols or notation. Instead, the median is simply identified by the word *median*. In addition, the definition and the computations for the median are identical for a sample and for a population.

The goal of the median is to determine the precise midpoint of a distribution. The commonsense goal is demonstrated in the following three examples. The three examples are intended to cover all of the different types of data you are likely to encounter.

By midpoint of the distribution we mean that the area in the graph is divided into two equal parts. We are not locating the midpoint between the highest and lowest X values.

METHOD 1: WHEN *N* IS AN ODD NUMBER

With an odd number of scores, you list the scores in order (lowest to highest), and the median is the middle score in the list. Consider the following set of $N = 5$ scores, which have been listed in order:

3,   5,   8,   10,   11

The middle score is $X = 8$, so the median is equal to 8.0. In a graph, the median divides the space or area of the graph in half (Figure 3.5). The amount of area above the median consists of $2\frac{1}{2}$ "boxes," the same as the area below the median (shaded portion).

FIGURE 3.5

The median divides the area in the graph exactly in half.

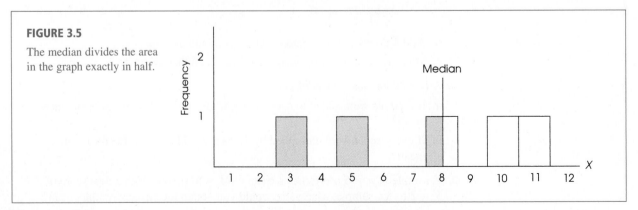

METHOD 2: WHEN *N* IS AN EVEN NUMBER

With an even number of scores in the distribution, you list the scores in order (lowest to highest) and then locate the median by finding the point halfway between the middle two scores. Consider the following population:

$$3, \quad 3, \quad 4, \quad 5, \quad 7, \quad 8$$

Now we select the middle pair of scores (4 and 5), add them together, and divide by 2:

$$\text{median} = \frac{4 + 5}{2} = \frac{9}{2} = 4.5$$

In terms of a graph, we see again that the median divides the area of the distribution exactly in half (Figure 3.6). There are three scores (or boxes) above the median and three below the median.

FIGURE 3.6

The median divides the area in the graph exactly in half.

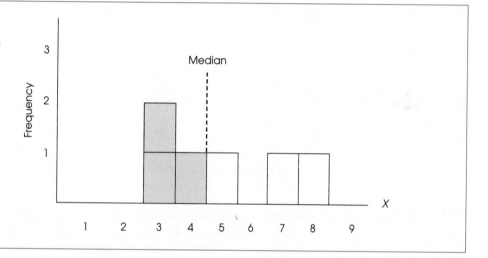

METHOD 3: WHEN THERE ARE SEVERAL SCORES WITH THE SAME VALUE IN THE MIDDLE OF THE DISTRIBUTION

In most cases, one of the two methods already outlined will provide you with a reasonable value for the median. However, when you have more than one individual at the median, these simple procedures may oversimplify the computations. Consider the following set of scores:

$$1, \quad 2, \quad 2, \quad 3, \quad 4, \quad 4, \quad 4, \quad 4, \quad 4, \quad 5$$

There are 10 scores (an even number), so you normally would use method 2 and average the middle pair to determine the median. By this method, the median would be 4.

In many ways, this is a perfectly legitimate value for the median. However, when you look closely at the distribution of scores [see Figure 3.7(a)], you probably get the clear impression that $X = 4$ is not in the middle. The problem comes from the tendency to interpret the score of 4 as meaning exactly 4.00 instead of meaning an interval from 3.5 to 4.5. The simple method of computing the median has determined that the value we want is located in this interval.

To locate the median with greater precision, it will be necessary to split the scores into fractions so that part of each score goes above the median and part goes below the median. In Figure 3.5, for example, the score $X = 8$ was split in half so that $2\frac{1}{2}$ scores

Because the median can be visualized as the midpoint in a graph (dividng the graph into two equal sections), the following graphic demonstration is particularly well suited for finding the median.

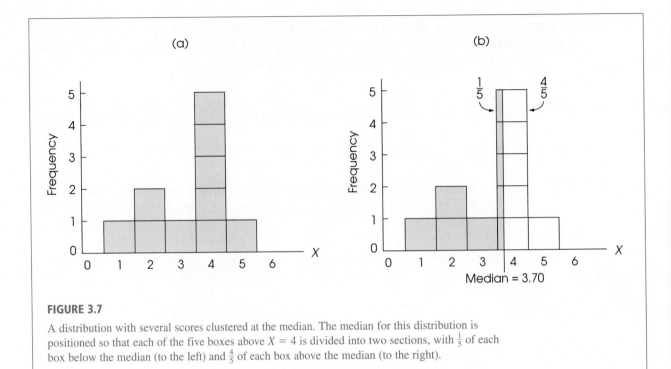

FIGURE 3.7

A distribution with several scores clustered at the median. The median for this distribution is positioned so that each of the five boxes above $X = 4$ is divided into two sections, with $\frac{1}{5}$ of each box below the median (to the left) and $\frac{4}{5}$ of each box above the median (to the right).

are located on each side of the median. The following example demonstrates the process of splitting scores when there are several values located at the position of the median.

EXAMPLE 3.4   For this example, we will use the data shown in Figure 3.7(a). Notice that the figure shows a population of $N = 10$ scores, with each score represented by a box in the histogram. To find the median, we must locate the position of a vertical line that will divide the distribution exactly in half, with 5 boxes on the left-hand side and 5 boxes on the right-hand side.

To begin the process, start at the left-hand side of the distribution, and move up the scale of measurement (the X-axis), counting boxes as you go along. The vertical line corresponding to the median should be drawn at the point where you have counted exactly 5 boxes (50% of the total of 10 boxes). By the time you reach a value of 3.5 on the X-axis, you will have gathered a total of 4 boxes, so that only one more box is needed to give you exactly 50% of the distribution. The problem is that there are 5 boxes in the next interval. The solution is to take only a fraction of each of the 5 boxes so that the fractions combine to give you 1 more box. If you take $\frac{1}{5}$ of each block, the five fifths will combine to make one whole box. This solution is shown in Figure 3.7(b).

Notice that we have drawn a line separating each of the 5 boxes so that $\frac{1}{5}$ is on the left-hand side of the line and $\frac{4}{5}$ are on the right-hand side. Thus, the line should be drawn exactly $\frac{1}{5}$ of the way into the interval containing the 5 boxes. This interval extends from the lower real limit of 3.5 to the upper real limit of 4.5 on the X-axis and has a width of 1.00 point. One-fifth of the interval is 0.20 points ($\frac{1}{5}$ of 1.00). Therefore, the line should be drawn at the point where $X = 3.70$ (the lower real limit of 3.5 + $\frac{1}{5}$ of the interval, or 0.20). This value, $X = 3.70$, is the median and divides the distribution exactly in half.

The process demonstrated in Example 3.4 can be summarized in the following four steps, which can be generalized to any situation where several scores are tied at the median.

STEP 1 Count the number of scores (boxes in the graph) below the tied value.

STEP 2 Find the number of additional scores (boxes) needed to make exactly one-half of the total distribution.

STEP 3 Form a fraction:

$$\frac{\text{number of boxes needed (step 2)}}{\text{number of tied boxes}}$$

STEP 4 Add the fraction (step 3) to the lower real limit of the interval containing the tied scores.
When these steps are incorporated into a formula, we obtain

$$\text{median} = X_{\text{LRL}} + \left(\frac{0.5N - f_{\text{BELOW LRL}}}{f_{\text{TIED}}} \right) \tag{3.3}$$

where X_{LRL} is the lower real limit of the tied values, $f_{\text{BELOW LRL}}$ is the frequency of scores with values below X_{LRL}, and f_{TIED} is the frequency for the tied values. Thus, for Example 3.4,

$$\text{median} = 3.5 + \left(\frac{0.5(10) - 4}{5} \right)$$

$$= 3.5 + \left(\frac{5 - 4}{5} \right)$$

$$= 3.5 + \frac{1}{5}$$

$$= 3.5 + 0.2$$

$$= 3.7$$

You may recognize this four-step procedure as equivalent to the process of interpolation that was used to find percentiles and percentile ranks in Chapter 2. Box 3.2 demonstrates how interpolation can be used to compute the median (the 50th percentile) for the same scores that were used in Example 3.4.

THE MEDIAN, THE MEAN, AND THE MIDDLE

Earlier, we defined the mean as the "balance point" for a distribution because the distances above the mean must have the same total as the distances below the mean. One consequence of this definition is that the mean is always located inside the group of scores; that is, there always will be at least one score above the mean and at least one score below the mean. You should notice, however, that the concept of a balance point

---

BOX
3.2   USING INTERPOLATION TO FIND THE MEDIAN

The same data that are used in Example 3.4 (see Figure 3.7) are organized in the following frequency distribution table:

X	f	cf	c%
5	1	10	100%
4	5	9	90%
3	1	4	40%
2	2	3	30%
1	1	1	10%

The goal is to find the 50th percentile, the median, for this distribution. The score corresponding to 50% cannot be read directly from the table, so it will be necessary to use interpolation.

The intermediate value of 50% is located in the interval between the cumulative percentages of 40% and 90%, which correspond to the real limits of 3.5 and 4.5, respectively.

X	c%
4.5	90%
??	50%
3.5	40%

On the percentage scale, the value of 50% is located 40 percentage points from the top (90%) of the interval, or $\frac{40}{50} = \frac{4}{5}$ of the total distance down in the interval. On the X-scale, the total interval is 1.0 point wide, so $\frac{4}{5}$ of this total distance is

$$\left(\tfrac{4}{5}\right)1.00 = 0.80$$

Starting at the top of the X-scale and moving down 0.80 brings us to a score of $X = 4.50 - 0.80 = 3.70$. This is the score corresponding to the 50th percentile. That is, the score $X = 3.70$ is the median. Notice that this is exactly the same answer we obtained using the graphic method of interpolation in Example 3.4.

---

focuses on distances rather than scores. In particular, it is possible to have a distribution where the vast majority of the scores are located on one side of the mean. Figure 3.8 shows a distribution of $N = 6$ scores where 5 out of 6 scores have values less than the mean. In this figure, the total of the distances above the mean is 8 points and the total of the distances below the mean is 8 points. Although it is reasonable to say that the mean is located in the middle of the distribution if you use the concept of distance to define the term "middle," you should realize that the mean is not necessarily located at the exact center of the group of scores.

The median, on the other hand, defines the middle of the distribution in terms of scores. In particular, the median is always located so that exactly one-half of the scores are on one side and exactly one-half on the other side. For the distribution in Figure 3.8,

FIGURE 3.8

A population of $N = 6$ scores with a mean of $\mu = 4$. Notice that the mean does not necessarily divide the scores into two equal groups. In this example, 5 out of the 6 scores have values less than the mean.

for example, the median is located at $X = 2.5$, with exactly 3 scores above this value and exactly 3 scores below. Thus, it is possible to claim that the median is located in the middle of the distribution, provided that the term "middle" is defined by the number of scores.

In summary, the mean and the median are both methods for defining and measuring central tendency. Although they both define the middle of the distribution, they use different definitions of the term "middle."

LEARNING CHECK

1. Find the median for each distribution of scores:
   a. 3, 10, 8, 4, 10, 7, 6
   b. 13, 8, 10, 11, 12, 10
   c. 3, 4, 3, 2, 1, 3, 2, 4

2. A distribution can have more than one median. (True or false?)

3. If you have a score of 52 on an 80-point exam, then you definitely scored above the median. (True or false?)

4. It is possible for more than 50% of the scores in a distribution to have values greater than the mean. (True or false?)

5. It is possible for more than 50% of the scores in a distribution to have values greater than the median. (True or false?)

ANSWERS

1. a. The median is $X = 7$.   b. The median is $X = 10.5$.
   c. The median is $X = 2.83$ (by interpolation).

2. False

3. False. The value of the median would depend on where the scores are located.

4. True

5. False

3.4 THE MODE

The final measure of central tendency that we will consider is called the *mode*. In its common usage, the word *mode* means "the customary fashion" or "a popular style." The statistical definition is similar in that the mode is the most common observation among a group of scores.

DEFINITION

In a frequency distribution, the **mode** is the score or category that has the greatest frequency.

As with the median, there are no symbols or special notation used to identify the mode or to differentiate between a sample mode and a population mode. In addition, the definition of the mode is the same for a population and for a sample distribution.

The mode is a useful measure of central tendency because it can be used to determine the typical or average value for any scale of measurement, including a nominal scale (see Chapter 1). Consider, for example, the data shown in Table 3.4. These data were obtained by asking a sample of 100 students to name their favorite restaurants in town. For these data, the mode is Luigi's, the restaurant (score) that was named most frequently as a favorite place. Although we can identify a modal response for these data, you should notice that it would be impossible to compute a mean or a median. For example, you cannot add the scores to determine a mean (How much is 5 College Grills plus 42 Luigi's?). Also, the restaurants are listed alphabetically because they do not form any natural order. Thus, it is impossible to list them in order and determine a 50% point (median). In general, the mode is the only measure of central tendency that can be used with data from a nominal scale of measurement.

In a frequency distribution graph, the greatest frequency will appear as the tallest part of the figure. To find the mode, you simply identify the score located directly beneath the highest point in the distribution.

TABLE 3.4

Favorite restaurants named by a sample of $n = 100$ students. *Caution:* The mode is a score or category, not a frequency. For this example, the mode is Luigi's, not $f = 42$.

RESTAURANT	f
College Grill	5
George & Harry's	16
Luigi's	42
Oasis Diner	18
Roxbury Inn	7
Sutter's Mill	12

Although a distribution will have only one mean and only one median, it is possible to have more than one mode. Specifically, it is possible to have two or more scores that have the same highest frequency. In a frequency distribution graph, the different modes will correspond to distinct, equally high peaks. A distribution with two modes is said to be *bimodal*, and a distribution with more than two modes is called *multimodal*. Occasionally, a distribution with several equally high points is said to have no mode.

Technically, the mode is the score with the absolute highest frequency. However, the term *mode* is often used more casually to refer to scores with relatively high frequencies—that is, scores that correspond to peaks in a distribution even though the peaks are not the absolute highest points. For example, Figure 3.9 shows the number of fish caught at various times during the day. There are two distinct peaks in this distribution,

FIGURE 3.9

The relationship between time of day and number of fish caught.

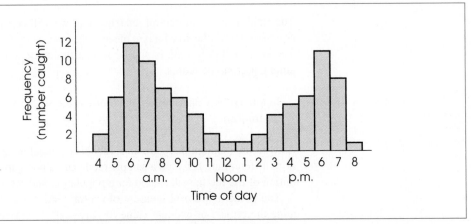

one at 6 A.M. and one at 6 P.M. Each of these values is a mode in the distribution. Note that the two modes do not have identical frequencies. Twelve fish were caught at 6 A.M., and 11 were caught at 6 P.M. Nonetheless, both of these points are called modes. The taller peak is called the *major mode*, and the shorter one is the *minor mode*.

LEARNING CHECK

1. An instructor recorded the number of absences for each student in a class of 20 and obtained the frequency distribution shown in the following table.

NUMBER OF ABSENCES (X)	f
7	1
6	0
5	3
4	2
3	3
2	5
1	4
0	2

   a. Using the mean to define central tendency, what is the average number of absences for this class?

   b. Using the mode to define central tendency, what is the average number of absences for this class?

2. In a recent survey comparing picture quality for three brands of color televisions, 63 people preferred brand A, 29 people preferred brand B, and 58 people preferred brand C. What is the mode for this distribution?

3. It is possible for a distribution to have more than one mode. (True or false?)

ANSWERS

1. a. The mean is 2.65.   b. The mode is $X = 2$ absences.

2. The mode is brand A.

3. True

3.5 SELECTING A MEASURE OF CENTRAL TENDENCY

How do you decide which measure of central tendency to use? The answer to this question depends on several factors. Before we discuss these factors, however, note that with many sets of data it is possible to compute two or even three different measures of central tendency. Often the three measures will produce similar results, but there are situations in which they will be very different (see Section 3.6). Also, note that the mean is most often the preferred measure of central tendency. Because the mean uses every score in the distribution, it usually is a good representative value. Remember that the goal of central tendency is to find the single value that best represents the entire distribution. Besides being a good representative, the mean has the added advantage of being closely related to variance and standard deviation, the most common measures of variability (Chapter 4). This relationship makes the mean a valuable measure for purposes of inferential statistics. For these reasons, and others, the mean generally is considered to be the best of the three measures of central tendency. But there are specific situations in which it is impossible to compute a mean or the mean is not particularly representative. It is in these situations that the mode and the median are used.

WHEN TO USE THE MEDIAN

We will consider four situations in which the median serves as a valuable alternative to the mean. In the first three cases, the data consist of numerical values (interval or ratio scales) where you would normally compute the mean. However, each case also involves a special problem so that it is either impossible to compute the mean, or the calculation of the mean produces a value that is not central or not representative of the distribution. The fourth situation involves measuring central tendency for ordinal data.

Extreme scores or skewed distributions   When a distribution has a few extreme scores, scores that are very different in value from most of the others, then the mean may not be a good representative of the majority of the distribution. The problem comes from the fact that one or two extreme values can have a large influence and cause the mean to be displaced. In this situation, the fact that the mean uses all of the scores equally can be a disadvantage. For example, suppose that a sample of $n = 10$ rats is tested in a T-maze for a food reward. The animals must choose the correct arm of the T (right or left) to find the food in the goal box. The experimenter records the number of errors each rat makes before it solves the maze. Hypothetical data are presented in Figure 3.10.
The mean for this sample is

$$M = \frac{\Sigma X}{n} = \frac{203}{10} = 20.3$$

FIGURE 3.10

Frequency distribution of errors committed before reaching learning criterion.

Notice that the graph shows two *breaks* in the X-axis. Rather than listing all the scores from 0 to 100, the graph jumps directly to the first score, which is $X = 10$, and then jumps directly from $X = 15$ to $X = 100$. The breaks shown in the X-axis are the conventional way of notifying the reader that some values have been omitted.

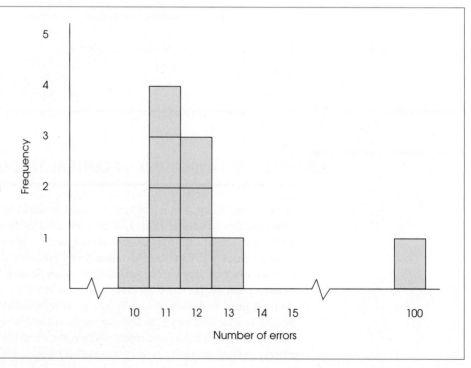

Notice that the mean is not very representative of any score in this distribution. Most of the scores are clustered between 10 and 13. The extreme score of $X = 100$ (a slow learner) inflates the value of ΣX and distorts the mean.

The median, on the other hand, is not easily affected by extreme scores. For this sample, $n = 10$, so there should be five scores on either side of the median. The median is 11.50. Notice that this is a very representative value. Also note that the median would be unchanged even if the slow learner made 1000 errors instead of only 100. The median commonly is used when reporting the average value for a skewed distribution. For example, the distribution of personal incomes is very skewed, with a small segment of the population earning incomes that are astronomical. These extreme values distort the mean, so that it is not very representative of the salaries that most of us earn. As in the previous example, the median is the preferred measure of central tendency when extreme scores exist.

Undetermined values   Occasionally, you will encounter a situation in which an individual has an unknown or undetermined score. In psychology, this often occurs in learning experiments where you are measuring the number of errors (or amount of time) required for an individual to solve a particular problem. For example, suppose that a sample of $n = 6$ people were asked to assemble a wooden puzzle as quickly as possible. The experimenter records how long (in minutes) it takes each individual to arrange all the pieces to complete the puzzle. Table 3.5 presents the outcome of this experiment.

TABLE 3.5

Amount of time to complete puzzle.

PERSON	TIME (MIN.)
1	8
2	11
3	12
4	13
5	17
6	Never finished

Notice that person 6 never completed the puzzle. After an hour, this person still showed no sign of solving the puzzle, so the experimenter stopped him or her. This person has an undetermined score. (There are two important points to be noted. First, the experimenter should not throw out this individual's score. The whole purpose for using a sample is to gain a picture of the population, and this individual tells us that part of the population cannot solve the puzzle. Second, this person should not be given a score of $X = 60$ minutes. Even though the experimenter stopped the individual after 1 hour, the person did not finish the puzzle. The score that is recorded is the amount of time needed to finish. For this individual, we do not know how long this is.)

It is impossible to compute the mean for these data because of the undetermined value. We cannot calculate the ΣX part of the formula for the mean. However, it is possible to compute the median. For these data, the median is 12.5. Three scores are below the median, and three scores (including the undetermined value) are above the median.

Open-ended distributions   A distribution is said to be *open-ended* when there is no upper limit (or lower limit) for one of the categories. The table at the left provides an example of an open-ended distribution, showing the number of children in each family for a sample of $n = 20$ households. The top category in this distribution shows that three of the families have "5 or more" children. This is an open-ended category. Notice

NUMBER OF CHILDREN (X)	f
5 or more	3
4	2
3	2
2	3
1	6
0	4

that it is impossible to compute a mean for these data because you cannot find ΣX (the total number of children for all 20 families). However, you can find the median. For these data, the median is 1.5 (exactly 50% of the families have fewer than 1.5 children).

Ordinal scale   Many researchers believe that it is not appropriate to use the mean to describe central tendency for ordinal data. When scores are measured on an ordinal scale, the median is always appropriate and is usually the preferred measure of central tendency.

You should recall that ordinal measurements allow you to determine direction (greater than or less than) but do not allow you to determine distance. The median is compatible with this type of measurement because it is defined by direction: half the stores are above the median and half are below the median. The mean, on the other hand, defines central tendency in terms of distance. Remember that the mean is the balance point for the distribution, so that the distances above the mean are exactly balanced by the distances below the mean. Because the mean is defined in terms of distances, and because ordinal scales do not measure distance, it is not appropriate to compute a mean for scores from an ordinal scale.

WHEN TO USE THE MODE

We will consider three situations in which the mode is commonly used as an alternative to the mean, or is used in conjunction with the mean to describe central tendency.

Nominal scales   The primary advantage of the mode is that it can be used to measure and describe central tendency for data that are measured on a nominal scale. Recall that the categories that make up a nominal scale are differentiated only by name. Because nominal scales do not measure quantity, it is impossible to compute a mean or a median for data from a nominal scale. Therefore, the mode is the only option for describing central tendency for nominal data.

Discrete variables   Recall that discrete variables are those that exist only in whole, indivisible categories. Often, discrete variables are numerical values, such as the number of children in a family or the number of rooms in a house. When these variables produce numerical scores, it is possible to calculate means. In this situation, the calculated means will usually be fractional values that cannot actually exist. For example, computing means will generate results such as "the average family has 2.4 children and a house with 5.33 rooms." On the other hand, the mode always identifies the most typical case and, therefore, it produces more sensible measures of central tendency. Using the mode, our conclusion would be "the typical, or modal, family has 2 children and a house with 5 rooms." In many situations, especially with discrete variables, people are more comfortable using the realistic, whole-number values produced by the mode.

Describing shape   Because the mode requires little or no calculation, it is often included as a supplementary measure along with the mean or median as a no-cost extra. The value of the mode (or modes) in this situation is that it gives an indication of the shape of the distribution as well as a measure of central tendency. Remember that the mode identifies the location of the peak (or peaks) in the frequency distribution graph. For example, if you are told that a set of exam scores has a mean of 72 and a mode of 80, you should have a better picture of the distribution that would be available from the mean alone (see Section 3.6).

IN THE LITERATURE
REPORTING MEASURES OF CENTRAL TENDENCY

Measures of central tendency are commonly used in the behavioral sciences to summarize and describe the results of a research study. For example, a researcher may report the sample means from two different treatments or the median score for a large sample. These values may be reported in verbal descriptions of the results, in tables, or in graphs.

In reporting the results, many behavioral science journals use guidelines adopted by the American Psychological Association (APA), as outlined in the *Publication Manual of the American Psychological Association* (2001). We will refer to the APA manual from time to time in describing how data and research results are reported in the scientific literature. The APA style typically uses the letter M as the symbol for the sample mean. Thus, a study might state

The treatment group showed fewer errors ($M = 2.56$) on the task than the control group ($M = 11.76$).

When there are many means to report, tables with headings provide an organized and more easily understood presentation. Table 3.6 illustrates this point.

TABLE 3.6

The mean number of errors made on the task for treatment and control groups according to gender.

	TREATMENT	CONTROL
Females	1.45	8.36
Males	3.83	14.77

The median can be reported using the abbreviation *Mdn*, as in "Mdn = 8.5 errors," or it can simply be reported in narrative text, as follows:

The median number of errors for the treatment group was 8.5, compared to a median of 13 for the control group.

There is no special symbol or convention for reporting the mode. If mentioned at all, the mode is usually just reported in narrative text.

The modal hair color was blonde, with 23 out of 50 participants in this category.

Graphs also can be used to report and compare measures of central tendency. Usually graphs are used to display values obtained for sample means, but occasionally you will see sample medians reported in graphs (modes are rarely, if ever, shown in a graph). The value of a graph is that it allows several means (or medians) to be shown simultaneously so it is possible to make quick comparisons between groups or treatment conditions. When preparing a graph, it is customary to list the values for the different groups or treatment conditions on the horizontal axis. Typically, these are the different values that make up the independent variable or the quasi-independent variable. Values for the dependent variable (the scores) are listed on the vertical axis. The means (or medians) are then displayed using a *line graph*, a *histogram*, or a *bar graph*, depending on the scale of measurement used for the independent variable.

Figure 3.11 shows an example of a graph displaying the relationship between drug dose (the independent variable) and food consumption (the dependent variable). In this study, there were five different drug doses (treatment conditions) and they are listed along the horizontal axis. The five means appear as points in the graph. To

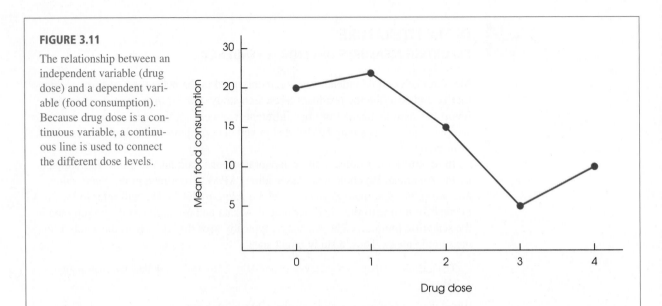

FIGURE 3.11

The relationship between an independent variable (drug dose) and a dependent variable (food consumption). Because drug dose is a continuous variable, a continuous line is used to connect the different dose levels.

construct this graph, a point was placed above each treatment condition so that the vertical position of the point corresponds to the mean score for the treatment condition. The points are then connected with straight lines and the resulting graph is called a *line graph*. A line graph is used when the values on the horizontal axis are measured on an interval or a ratio scale. An alternative to the line graph is a *histogram*. For this example, the histogram would show a bar above each drug dose so that the height of each bar corresponds to the mean food consumption for that group, and with no space between adjacent bars.

Figure 3.12 shows a bar graph displaying the median selling price for single-family homes in different regions of the United States. Bar graphs are used to present means (or medians) when the groups or treatments shown on the *X*-axis are measured on a nominal or an ordinal scale. To construct a bar graph, you simply draw a bar directly above each group or treatment so that the height of the bar corresponds to the mean (or median) for that group or treatment. For a bar graph, a space is left between adjacent bars to indicate that the scale of measurement is nominal or ordinal.

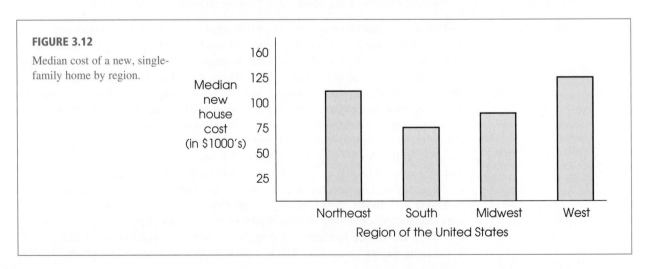

FIGURE 3.12

Median cost of a new, single-family home by region.

When constructing graphs of any type, you should recall the basic rules we introduced in Chapter 2:

1. The height of a graph should be approximately two-thirds to three-quarters of its length.

2. Normally, you start numbering both the X-axis and the Y-axis with zero at the point where the two axes intersect. However, when a value of zero is part of the data, it is common to move the zero point away from the intersection so that the graph does not overlap the axes (see Figure 3.11).

More important, you should remember that the purpose of a graph is to give an accurate representation of the information in a set of data. Box 2.1 in Chapter 2 demonstrates what can happen when these basic principles are ignored.   ❏

3.6 CENTRAL TENDENCY AND THE SHAPE OF THE DISTRIBUTION

We have identified three different measures of central tendency, and often a researcher will calculate all three for a single set of data. Because the mean, the median, and the mode are all trying to measure the same thing (central tendency), it is reasonable to expect that these three values should be related. In fact, there are some consistent and predictable relationships among the three measures of central tendency. In part, the relationships among the mean, median, and mode are determined by the shape of the distribution. We will consider two general types of distributions: symmetrical distributions and skewed distributions.

SYMMETRICAL DISTRIBUTIONS

For a *symmetrical distribution*, the right-hand side of the graph is a mirror image of the left-hand side. By definition, the median will be exactly at the center of a symmetrical distribution because exactly half of the area in the graph will be on either side of the center. The mean also will be exactly at the center of a symmetrical distribution because each individual score in the distribution has a corresponding score on the other side (the mirror image), so that the average of these two values is exactly in the middle. Because all the scores can be paired in this way, the overall average will be exactly at the middle. For any symmetrical distribution, the mean and the median will be the same (Figure 3.13).

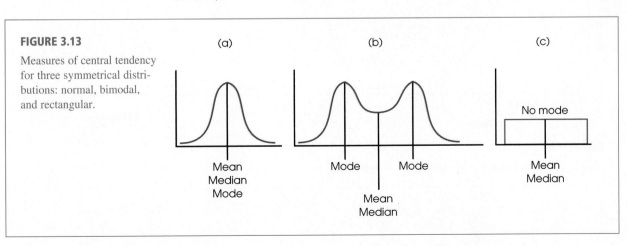

FIGURE 3.13

Measures of central tendency for three symmetrical distributions: normal, bimodal, and rectangular.

If a symmetrical distribution has only one mode, then it must be exactly at the center, so that all three measures of central tendency will have the same value (see Figure 3.13). On the other hand, a bimodal distribution that is symmetrical [Figure 3.13(b)] will have the mean and the median together in the center with the modes on each side. A rectangular distribution [Figure 3.13(c)] has no mode because all X values occur with the same frequency. Still, the mean and the median will be in the center of the distribution and equivalent in value.

SKEWED DISTRIBUTIONS

Distributions are not always symmetrical. In fact, quite often they are lopsided, or *skewed*. For example, Figure 3.14(a) shows a *positively skewed distribution*. In this distribution, the peak (highest frequency) is on the left-hand side. This is the position of the mode. If you examine Figure 3.14(a) carefully, it should be clear that the vertical line drawn at the mode does not divide the distribution into two equal parts. In order to have exactly 50% of the distribution on each side, the median must be located to the right of the mode. Finally, the mean will be located to the right of the median because it is influenced most by extreme scores and will be displaced farthest to the right by the scores in the tail. Therefore, in a positively skewed distribution, the mean will have the largest value, followed by the median and then the mode [see Figure 3.14(a)].

Negatively skewed distributions are lopsided in the opposite direction, with the scores piling up on the right-hand side and the tail tapering off to the left. The grades on an easy exam, for example, will tend to form a negatively skewed distribution [see Figure 3.14(b)]. For a distribution with negative skew, the mode is on the right-hand side (with the peak), while the mean is displaced on the left by the extreme scores in the tail. As before, the median is located between the mean and the mode. In order from highest value to lowest value, the three measures of central tendency will be the mode, the median, and the mean.

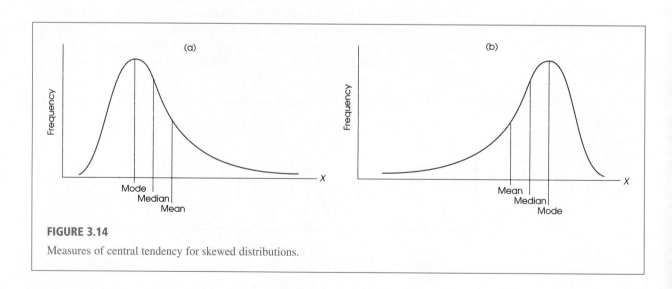

FIGURE 3.14

Measures of central tendency for skewed distributions.

SUMMARY

1. The purpose of central tendency is to determine the single value that identifies the center of the distribution and best represents the entire set of scores. The three standard measures of central tendency are the mode, the median, and the mean.

2. The mean is the arithmetic average. It is computed by summing all the scores and then dividing by the number of scores. Conceptually, the mean is obtained by dividing the total (ΣX) equally among the number of individuals (N or n). The mean can also be defined as the balance point for the distribution. The distances above the mean are exactly balanced by the distances below the mean. Although the calculation is the same for a population or a sample mean, a population mean is identified by the symbol μ, and a sample mean is identified by M.

3. Changing any score in the distribution will cause the mean to be changed. When a constant value is added to (or subtracted from) every score in a distribution, the same constant value is added to (or subtracted from) the mean. If every score is multiplied by a constant, the mean will be multiplied by the same constant. In nearly all circumstances, the mean is the best representative value and is the preferred measure of central tendency.

4. The median is the value that divides a distribution exactly in half. The median is the preferred measure of central tendency when a distribution has a few extreme scores that displace the value of the mean. The median also is used when there are undetermined (infinite) scores that make it impossible to compute a mean.

5. The mode is the most frequently occurring score in a distribution. It is easily located by finding the peak in a frequency distribution graph. For data measured on a nominal scale, the mode is the appropriate measure of central tendency. It is possible for a distribution to have more than one mode.

6. For symmetrical distributions, the mean will equal the median. If there is only one mode, then it will have the same value, too.

7. For skewed distributions, the mode will be located toward the side where the scores pile up, and the mean will be pulled toward the extreme scores in the tail. The median will be located between these two values.

RESOURCES

WebTUTOR

If you are using WebTutor along with this book, you will find a WebTutor section corresponding to this chapter. The WebTutor contains a brief summary of Chapter 3, hints for learning the new material and for avoiding common errors, and sample exam items including solutions.

SPSS

The statistics computer package SPSS can be used to calculate the mean for a set of scores. To calculate the mean, first enter the scores in one column of the data matrix, then click **Analyze** on the tool bar. Then, select **Descriptive Statistics** and select **Descriptives**. Enter the column label for the set of scores into the Variable box and click **OK**. SPSS will produce a summary table listing the maximum score, the minimum score, the mean, and the standard deviation (which is introduced in Chapter 4).

You can also obtain the mean for a set of scores if you use SPSS to display the scores in a frequency distribution histogram (see the SPSS section on page 62 at the end of Chapter 2). The mean and standard deviation are displayed beside the graph.

KEY TERMS

central tendency	mode	bimodal distribution	skewed distribution
mean	major mode	multimodal distribution	positive skew
weighted mean	minor mode	symmetrical distribution	negative skew
median			

FOCUS ON PROBLEM SOLVING

1. Because there are three different measures of central tendency, your first problem is to decide which one is best for your specific set of data. Usually the mean is the preferred measure, but the median may provide a more representative value if you are working with a skewed distribution. With data measured on a nominal scale, you must use the mode.

2. Although the three measures of central tendency appear to be very simple to calculate, there is always a chance for errors. The most common sources of error are listed next.
   a. Many students find it very difficult to compute the mean for data presented in a frequency distribution table. They tend to ignore the frequencies in the table and simply average the score values listed in the X column. You must use the frequencies and the scores! Remember that the number of scores is found by $N = \Sigma f$, and the sum of all N scores is found by ΣfX.

b. The median is the midpoint of the distribution of scores, not the midpoint of the scale of measurement. For a 100-point test, for example, many students incorrectly assume that the median must be $X = 50$. To find the median, you must have the *complete set* of individual scores. The median separates the individuals into two equal-sized groups.

c. The most common error with the mode is for students to report the highest frequency in a distribution rather than the score with the highest frequency. Remember that the purpose of central tendency is to find the most representative score. Therefore, for the following data, the mode is $X = 3$, not $f = 8$.

X	f
4	3
3	8
2	5
1	2

DEMONSTRATION 3.1

COMPUTING MEASURES OF CENTRAL TENDENCY

For the following sample data, find the mean, median, and mode. The scores are:

$$5, \quad 6, \quad 9, \quad 11, \quad 5, \quad 11, \quad 8, \quad 14, \quad 2, \quad 11$$

Compute the mean   Calculating the mean involves two steps:

1. Obtain the sum of the scores, ΣX.

2. Divide the sum by the number of scores, n.

For these data, the sum of the scores is as follows:

$$\Sigma X = 5 + 6 + 9 + 11 + 5 + 11 + 8 + 14 + 2 + 11 = 82$$

We can also observe that $n = 10$. Therefore, the mean of this sample is obtained by

$$M = \frac{\Sigma X}{n} = \frac{82}{10} = 8.2$$

Find the median   The median divides the distribution in half, in that half of the scores are above or equal to the median and half are below or equal to it. In this demonstration, $n = 10$. Thus, the median should be a value that has 5 scores above it and 5 scores below it.

STEP 1   Arrange the scores in order.

$$2, \quad 5, \quad 5, \quad 6, \quad 8, \quad 9, \quad 11, \quad 11, \quad 11, \quad 14$$

STEP 2   With an even number of scores, locate the midpoint between the middle two scores. The middle scores are $X = 8$ and $X = 9$. The median is the midpoint between 8 and 9.

$$\text{median} = \frac{8 + 9}{2} = \frac{17}{2} = 8.5$$

Find the mode   The mode is the X value that has the highest frequency. Looking at the data, we can readily determine that $X = 11$ is the score that occurs most frequently.

DEMONSTRATION 3.2

COMPUTING THE MEAN FROM A FREQUENCY DISTRIBUTION TABLE

Compute the mean for the data in the following table:

X	f
6	1
5	0
4	3
3	3
2	2

To compute the mean from a frequency distribution table, you must use the information in *both* the X and the f columns.

STEP 1   Multiply each X value by its frequency.
You can create a third column labeled fX. For these data,

fX
6
0
12
9
4

STEP 2   Find the sum of the fX values.

$$\Sigma fX = 6 + 0 + 12 + 9 + 4 = 31$$

STEP 3   Find n for these data.
Remember that $n = \Sigma f$.

$$n = \Sigma f = 1 + 0 + 3 + 3 + 2 = 9$$

STEP 4   Divide the sum of fX by n.

$$M = \frac{\Sigma fX}{n} = \frac{31}{9} = 3.44$$

PROBLEMS

1. Explain the general purpose for measuring central tendency.

2. Explain why there is more than one method for measuring central tendency. Why not use a single, standardized method for determining the "average" score?

3. Explain what is meant by each of the following statements:
   a. The mean is the *balance point* of the distribution.
   b. The median is the *midpoint* of the distribution.

4. Identify the circumstances under which the median instead of the mean is the preferred measure of central tendency.

5. Under what circumstances will the mean, the median, and the mode all have the same value?

6. Under what circumstances is the mode the preferred measure of central tendency?

7. Explain why the mean is often not a good measure of central tendency for a skewed distribution.

8. Find the mean, median, and mode for the following sample of scores:

$$1, \quad 4, \quad 3, \quad 5, \quad 4, \quad 2, \quad 4, \quad 2 \quad 5, \quad 3$$

9. Find the mean, median, and mode for the following sample of scores:

$$6, \quad 5, \quad 3, \quad 5, \quad 4, \quad 7, \quad 5, \quad 6, \quad 5, \quad 4$$

10. Find the mean, median, and mode for the set of scores in the following frequency distribution table:

X	f
5	2
4	5
3	3
2	2
1	2

11. Find the mean, median, and mode for the set of scores in the following frequency distribution table:

X	f
8	2
7	3
6	5
5	9
4	1

12. A population of $N = 8$ scores has a mean of $\mu = 12$. What is the value of ΣX for this population?

13. A sample of $n = 15$ scores has a mean of $M = 7$. What is the value of ΣX for this sample?

14. A sample of $n = 6$ scores has a mean of $M = 9$. If one of the scores is changed from $X = 2$ to $X = 14$, what will be the new value for the sample mean?

15. A set of eight scores has a mean of 20. If one of the scores is changed from $X = 42$ to $X = 2$, what will be the new value for the mean?

16. A sample of $n = 10$ scores has a mean of 7. If a new score of $X = 18$ is added to the sample, what will be the new value for the sample mean?

17. A population of $N = 5$ scores has a mean of $\mu = 6$. If a score of $X = 2$ is removed from the population, what will be the new value for the population mean?

18. A sample of $n = 8$ scores has a mean of $M = 10$. One score is removed from the sample and the mean for the remaining scores is $M = 6$. What is the value of the score that was removed? (*Hint:* Find ΣX for the original sample and for the new sample.)

19. A sample of $n = 5$ scores has a mean of $M = 11$. After one new score is added to the sample, the new mean is $M = 12$. What is the value of the new score?

20. A sample of $n = 4$ scores has a mean of $M = 8$. A second sample of $n = 6$ scores has $M = 12$. If the two samples are combined, what value will be obtained for the mean of the combined sample?

21. A sample of $n = 15$ scores has a mean of $M = 30$. A second sample of $n = 5$ scores has a mean of 24. If the two samples are combined, what value will be obtained for the mean of the combined sample?

22. For each of the following situations, identify the measure of central tendency (mean, median, or mode) that would provide the best description of the "average" score:
    a. A researcher asks each individual in a sample of 50 adults to name his/her favorite season (summer, fall, winter, spring).
    b. An insurance company would like to determine how long people remain hospitalized after a routine appendectomy. The data from a large sample indicate that most people are released after 2 or 3 days but a few develop infections and stay in the hospital for weeks.
    c. A teacher measures scores on a standardized reading test for a sample of children from a middle-class, suburban elementary school.

23. The following frequency distribution summarizes the number of absences for each student in a class of $n = 20$:

NUMBER OF ABSENCES (X)	f
5 or more	3
4	4
3	3
2	6
1	3
0	1

a. Find the mode for this distribution.
b. Find the median number of absences for this class.
c. Explain why you cannot compute the mean number of absences using the data provided in the table.

24. On a standardized reading achievement test, the nationwide average for seventh-grade children is $\mu = 7.0$. A seventh-grade teacher is interested in comparing class reading scores with the national average. The scores for the 16 students in this class are as follows:

$$8, \quad 6, \quad 5, \quad 10, \quad 5, \quad 6, \quad 8, \quad 9,$$

$$7, \quad 6, \quad 9, \quad 5, \quad 14, \quad 4, \quad 7, \quad 6$$

a. Find the mean and the median reading scores for this class.
b. If the mean is used to define the class average, how does this class compare with the national norm?
c. If the median is used to define the class average, how does this class compare with the national norm?

25. Research has demonstrated that the amount of sleep needed by the average person changes as a function of age. The following values show the mean hours of sleep needed at each age. Draw a line graph to show the relationship between age and amount of sleep.

1 year, 13 hours

5 years, 11 hours

10 years, 10 hours

15 years, $8\frac{1}{2}$ hours

20 years, 8 hours

26. Does it ever seem to you that the weather is nice during the work week, but lousy on the weekend? Cerveny and Balling (1988) have confirmed that this is not your imagination—pollution accumulating during the work week most likely spoils the weekend weather for people on the Atlantic coast. Consider the following hypothetical data. The amount of rainfall in inches is measured each day during the summer.

SUN-DAY	MON-DAY	TUES-DAY	WEDNES-DAY	THURS-DAY	FRI-DAY	SATUR-DAY
3	1	0	0	1	0	2
0	0	1	2	1	1	3
1	0	0	0	0	0	0
0	1	0	0	0	0	2
3	1	0	1	0	0	0
1	0	0	0	0	0	0
0	0	1	0	0	1	4
5	0	0	1	1	2	3
0	0	0	0	0	0	0
0	0	0	1	0	2	4
1	1	0	0	0	1	1
2	0	1	0	0	0	1

a. For each day, determine the total amount of rainfall.
b. Compute the mean rainfall for each day.
c. Just by looking at the means, what pattern emerges?
d. Use a graph to present the results of this study.

CHAPTER 4

Variability

Tools You Will Need

The following items are considered essential background material for this chapter. If you doubt your knowledge of any of these items, you should review the appropriate chapter or section before proceeding.

- Summation notation (Chapter 1)
- Central tendency (Chapter 3)
  - Mean
  - Median

Preview

It's 10 A.M., and Mary L. is washing her hands for the tenth time today. Before she goes to sleep tonight, she will have washed her hands over 60 times. Is this normal?

To differentiate between normal and abnormal behavior, some psychologists have resorted to using a *statistical model*. For example, we could survey a large sample of individuals and record the number of times each person washes his/her hands during a typical day. Because people are different, the scores should be variable, and the data from this survey should produce a distribution similar to the one shown in Figure 4.1. Notice that most people will have average or moderate scores located in the central part of the distribution. Others, such as Mary L., will deviate from the average. According to the statistical model, those who show substantial deviation are abnormal. Note that this model simply defines abnormal as being unusual or different from normal. The model does not imply that abnormal is necessarily negative or undesirable. Mary L. may have a compulsive personality disorder, or she may be a dentist washing her hands between patients.

The statistical model for abnormality requires two statistical concepts: a measure of the average and a measure of deviation from the average. In Chapter 3, we examined the standard techniques for defining the average score for a distribution. In this chapter, we will examine methods for measuring deviations. The fact that scores deviate from average means that they are variable. Variability is one of the most basic statistical concepts. As an introduction to the concept, this chapter will concentrate on defining and

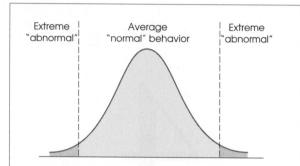

FIGURE 4.1

The statistical model for defining abnormal behavior. The distribution of behavior scores for the entire population is divided into three sections. Those individuals with average scores are defined as normal, and individuals who show extreme deviation from average are defined as abnormal.

measuring variability. In later chapters, we will explore sources of variability and examine how variability affects the interpretation of other statistical measurements.

One additional point should be made before we proceed. You probably have noticed that central tendency and variability are closely related. Whenever one appears, the other usually is close at hand. You should watch for this association throughout the book. The better you understand the relationships between central tendency and variability, the better you will understand statistics.

4.1 OVERVIEW

The term *variability* has much the same meaning in statistics as it has in everyday language; to say that things are variable means that they are not all the same. In statistics, our goal is to measure the amount of variability for a particular set of scores, a distribution. In simple terms, if the scores in a distribution are all the same, then there is no variability. If there are small differences between scores, then the variability is small, and if there are large differences between scores, then the variability is large.

DEFINITION   **Variability** provides a quantitative measure of the degree to which scores in a distribution are spread out or clustered together.

Figure 4.2 shows two distributions of familiar values: Part (a) shows the distribution of adult male heights (in inches), and part (b) shows the distribution of adult male weights (in pounds). Notice that the two distributions differ in terms of central ten-

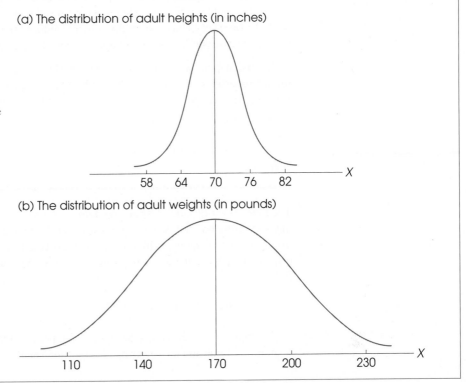

FIGURE 4.2

Population distributions of adult heights and adult weights.

NOTE: For simplicity, we have omitted the vertical axis for these graphs. As always, the height of any point on the curve indicates the relative frequency for that particular score.

(a) The distribution of adult heights (in inches)

(b) The distribution of adult weights (in pounds)

dency. The mean height is 70 inches (5 feet, 10 inches) and the mean weight is 170 pounds. In addition, notice that the distributions differ in terms of variability. For example, most heights are clustered close together, within 5 or 6 inches of the mean. On the other hand, weights are spread over a much wider range. In the weight distribution it is not unusual to find individuals who are located more than 30 pounds away from the mean, and it would not be surprising to find two individuals whose weights differ by more than 30 or 40 pounds. The purpose for measuring variability is to obtain an objective measure of how the scores are spread out in a distribution. In general, a good measure of variability will serve two purposes:

1. Variability describes the distribution. Specifically, it tells whether the scores are clustered close together or are spread out over a large distance. Usually, variability is defined in terms of *distance*. It tells how much distance to expect between one score and another, or how much distance to expect between an individual score and the mean. For example, we know that most adults' heights are clustered close together within 5 or 6 inches of the average. Although more extreme heights exist, they are relatively rare.

2. Variability measures how well an individual score (or group of scores) represents the entire distribution. This aspect of variability is very important for inferential statistics where relatively small samples are used to answer questions about populations. For example, suppose that you selected a sample of one person to represent the entire population. Because most adult heights are within a few inches of the population average (the distances are small), there is a very good chance that you would select someone whose height is within 6 inches of the population mean. On the other hand, the scores are much more spread out

(greater distances) in the distribution of adult weights. In this case, you proba-bly would *not* obtain someone whose weight was within 6 pounds of the popu-lation mean. Thus, variability provides information about how much error to expect if you are using a sample to represent a population.

In this chapter, we will consider three different measures of variability: the range, the interquartile range, and the standard deviation. Of these three, the standard devia-tion (and the related measure of variance) is by far the most important.

4.2   THE RANGE

The range is the distance between the largest score (X_{max}) and the smallest score (X_{min}) in the distribution. In determining this distance, you must also take into account the real limits of the maximum and minimum X values. The range, therefore, is computed as the difference between the upper real limit (URL) for X_{max} and the lower real limit (LRL) for X_{min}.

$$\text{Range} = \text{URL } X_{max} - \text{LRL } X_{min}$$

DEFINITION

The **range** is the difference between the upper real limit of the largest (maxi-mum) X value and the lower real limit of the smallest (minimum) X value.

For example, consider the following data:

$$3, \quad 7, \quad 12, \quad 8, \quad 5, \quad 10$$

For these data, $X_{max} = 12$, with an upper real limit of 12.5, and $X_{min} = 3$, with a lower real limit of 2.5. Thus, the range equals

$$\text{range} = \text{URL } X_{max} - \text{LRL } X_{min}$$
$$= 12.5 - 2.5 = 10$$

When the scores in a distribution are whole numbers, the range also can be obtained as follows:

range = highest X
        − lowest X + 1

The range is perhaps the most obvious way of describing how spread out the scores are—simply find the distance between the maximum and minimum scores. The prob-lem with using the range as a measure of variability is that it is completely determined by the two extreme values and ignores the other scores in the distribution. For exam-ple, the following two distributions have exactly the same range—10 points in each case. However, the scores in the first distribution are clustered together at one end of the range, whereas the scores in the second distribution are spread out over the entire range:

distribution 1:   1,   8,   9,   9,   10,   10

distribution 2:   1,   2,   4,   6,   8,   10

If, for example, these were scores on a 10-point quiz for two different class sections, there are clear differences between the two sections. Nearly all the students in the first section have mastered the material, but there is a wide range of achievement for stu-dents in the second section. A good measure of variability should show this difference.

Because the range does not consider all the scores in the distribution, it often does not give an accurate description of the variability for the entire distribution. For this reason, the range is considered to be a crude and unreliable measure of variability.

4.3 THE INTERQUARTILE RANGE AND SEMI-INTERQUARTILE RANGE

In Chapter 3, we defined the median as the score that divides a distribution exactly in half. In a similar way, a distribution can be divided into four equal parts using *quartiles*. By definition, the first quartile ($Q1$) is the score that separates the lowest 25% of the distribution from the rest. The second quartile ($Q2$) is the score that has exactly two quarters, or 50%, of the distribution below it. Notice that the second quartile and the median are the same. Finally, the third quartile ($Q3$) is the score that divides the bottom three-fourths of the distribution from the top quarter.

You should recognize that the quartiles can also be defined as percentiles (see page 52). The first quartile is identical to the 25th percentile, $Q2$ is the 50th percentile or the median, and $Q3$ is the 75th percentile. Although it is possible to use interpolation to identify the quartiles, a simpler method for finding the values of $Q1$, $Q2$, and $Q3$ is to construct a frequency distribution histogram where each score is represented by a box (see Figure 4.3). Then determine exactly how many boxes are needed to make up exactly one-quarter of the whole set. With a total of $n = 16$ scores (boxes), for example, one-quarter would be exactly 4 boxes. In this case, the first quartile will separate the lowest 4 boxes (25%) from the rest; the second quartile will have exactly 8 boxes (50%) on each side; and the third quartile will separate the lowest 12 boxes (75%) from the highest 4 boxes. Usually you can simply count boxes or fractions of boxes to locate $Q1$, $Q2$, and $Q3$.

FIGURE 4.3

Frequency distribution for a population of $N = 16$ scores. The first quartile is $Q1 = 4.5$. The third quartile is $Q3 = 8.0$. The interquartile range is 3.5 points. Note that the third quartile ($Q3$) divides the two boxes at $X = 8$ exactly in half, so that a total of 4 boxes are above $Q3$ and 12 boxes are below it.

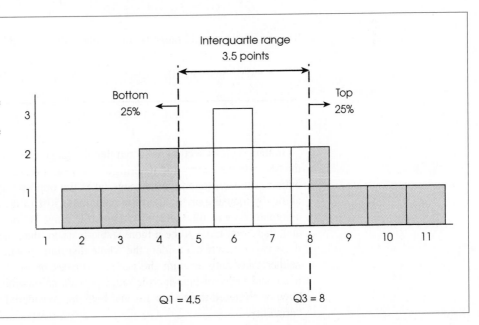

After the quartiles have been determined, the *interquartile range* is defined as the distance between the first quartile and the third quartile. The interquartile range effectively ignores the top 25% and the bottom 25% of the distribution, and measures the range covered by the middle 50% of the distribution.

DEFINITION

The **interquartile range** is the distance between the first quartile and the third quartile:

$$\text{interquartile range} = Q3 - Q1$$

When the interquartile range is used to describe variability, it commonly is transformed into the *semi-interquartile range*. As the name implies, the semi-interquartile range is simply one-half of the interquartile range. Conceptually, the semi-interquartile range measures the distance from the middle of the distribution to the boundaries that define the middle 50%.

DEFINITION

The **semi-interquartile range** is one-half of the interquartile range:

$$\text{semi-interquartile range} = \frac{Q3 - Q1}{2}$$

EXAMPLE 4.1

Figure 4.3 shows a frequency distribution histogram for a set of 16 scores. For this distribution, the first quartile is $Q1 = 4.5$. Exactly 25% of the scores (4 out of 16) are located below $X = 4.5$. Similarly, the third quartile $Q3 = 8.0$. Note that this value separates the bottom 75% of the distribution (12 out of 16 scores) from the top 25%. For this set of scores, the interquartile range is

$$Q3 - Q1 = 8.0 - 4.5 = 3.5$$

The semi-interquartile range is simply one-half of this distance:

$$\text{semi-interquartile range} = \frac{3.5}{2} = 1.75$$

In the previous section we noted that the range measures variability by focusing on the two extreme scores and ignoring all the rest of the scores in the center of the distribution. The semi-interquartile range takes the opposite approach, and measures variability by focusing on the scores in the middle 50% of the distribution and ignoring the extremes. As a result, the semi-interquartile range is not influenced by extreme scores and, therefore, tends to be a more stable measure than the range. However, neither of these two measures considers the whole distribution when measuring variability, and neither takes into account the actual distances between scores. Therefore, both the range and the semi-interquartile range provide an incomplete description of how scattered or clustered the scores are, and both are considered relatively crude measures of variability.

LEARNING CHECK

1. For the following data, find the range and the semi-interquartile range:

$$3, \quad 4, \quad 5, \quad 7, \quad 9, \quad 10, \quad 11, \quad 13$$

2. Consider the distribution of Exercise 1, except replace the score of 13 with a score of 100. What are the new values for the range and the semi-interquartile range? In comparing the answer to the one for the previous exercise, what can you conclude about these measures of variability?

3. Again consider the distribution from Exercise 1 but now add 8 more individuals all with scores of $X = 8$. Now what are the new values for the range and the semi-interquartile range?

ANSWERS

1. Range = URL X_{max} − LRL X_{min} = 13.5 − 2.5 = 11; semi-interquartile range = $(Q3 - Q1)/2$ = (10.5 − 4.5)/2 = 3.

2. Range = 98; semi-interquartile range = 3. The range is greatly affected by extreme scores in the distribution, unlike the semi-interquartile range.

3. The range is still 11 points, but the semi-interquartile range is reduced to 0.5 points.

4.4 STANDARD DEVIATION AND VARIANCE FOR A POPULATION

The standard deviation is the most commonly used and the most important measure of variability. Standard deviation uses the mean of the distribution as a reference point and measures variability by considering the distance between each score and the mean. It determines whether the scores are generally near or far from the mean. That is, are the scores clustered together or scattered? In simple terms, the standard deviation approximates the average distance from the mean.

Although the concept of standard deviation is straightforward, the actual equations will appear complex. Therefore, we will begin by looking at the logic that leads to these equations. If you remember that our goal is to measure the standard, or typical, distance from the mean, then this logic and the equations that follow should be easier to remember.

STEP 1   The first step in finding the standard distance from the mean is to determine the deviation, or distance from the mean, for each individual score. By definition, the deviation for each score is the difference between the score and the mean.

DEFINITION   **Deviation** is distance from the mean:

$$\text{deviation score} = X - \mu$$

For a distribution of scores with $\mu = 50$, if your score is $X = 53$, then your deviation score is

$$X - \mu = 53 - 50 = 3$$

If your score is $X = 45$, then your deviation score is

$$X - \mu = 45 - 50 = -5$$

Notice that there are two parts to a deviation score: the sign ($+$ or $-$) and the number. The sign tells the direction from the mean—that is, whether the score is located above ($+$) or below ($-$) the mean. The number gives the actual distance from the mean. For example, a deviation score of -6 corresponds to a score that is below the mean by 6 points.

STEP 2   Because our goal is to compute a measure of the standard distance from the mean, the obvious next step is to calculate the mean of the deviation scores. To compute this mean, you first add up the deviation scores and then divide by N. This process is demonstrated in the following example.

EXAMPLE 4.2   We start with the following set of $N = 4$ scores. These scores add up to $\Sigma X = 12$, so the mean is $\mu = \frac{12}{4} = 3$. For each score, we have computed the deviation.

X	$X - \mu$
8	$+5$
1	-2
3	0
0	-3
	$0 = \Sigma(X - \mu)$

Notice that the deviation scores add up to zero. This should not be surprising if you remember that the mean serves as a balance point for the distribution. The distances above the mean are equal to the distances below the mean (see page 75). Logically, the deviation scores must *always* add up to zero.

Because the mean deviation is always zero, it is of no value as a measure of variability. It is zero whether the scores are grouped together or all scattered out. The mean deviation score provides no information about variability. (You should note, however, that the constant value of zero can be useful in other ways. Whenever you are working with deviation scores, you can check your calculations by making sure that the deviation scores add up to zero.)

STEP 3   The average of the deviation scores will not work as a measure of variability because it is always zero. Clearly, this problem results from the positive and negative values canceling each other out. The solution is to get rid of the signs ($+$ and $-$). The standard procedure for accomplishing this is to square each deviation score. Using these squared values, you then compute the *mean squared deviation*, which is called *variance*.

DEFINITION   **Population variance** = mean squared deviation. Variance is the mean of the squared deviation scores.

Note that the process of squaring deviation scores does more than simply get rid of plus and minus signs. It results in a measure of variability based on *squared* distances. Although variance is valuable for some of the *inferential* statistical methods covered later, the mean squared distance is not the best *descriptive* measure for variability.

STEP 4   Remember that our goal is to compute a measure of the standard distance from the mean. Variance, the mean squared deviation, is not exactly what we want. The final step simply makes a correction for having squared all the distances. The new measure, the *standard deviation*, is the square root of the variance.

DEFINITION   **Standard deviation** $= \sqrt{\text{variance}}$

Technically, standard deviation is the square root of the mean squared deviation. But, conceptually, standard deviation is easier to understand if you think of it as describing the typical distance of scores from the mean (that is, the typical $X - \mu$). As the name implies, standard deviation measures the standard, or typical, deviation score.

The concept of standard deviation (or variance) is the same for a sample as for a population. However, the details of the calculations differ slightly, depending on whether you have data for a sample or a complete population. We will first consider the formulas for population standard deviation and variance.

SUM OF SQUARED DEVIATIONS (*SS*)   Variance, you should recall, is defined as the mean squared deviation. This mean is computed exactly the same way you compute any mean: First find the sum, then divide by the number of scores:

$$\text{variance} = \text{mean squared deviation} = \frac{\text{sum of squared deviations}}{\text{number of scores}}$$

The value in the numerator of this equation, the sum of the squared deviations, is a basic component of variability, and we will focus on it. To simplify things, it is identified by the notation *SS* (for sum of squared deviations), and it generally is referred to as the *sum of squares*.

DEFINITION   **SS**, or **sum of squares**, is the sum of the squared deviation scores.

There are two formulas you will need to know in order to compute *SS*. These formulas are algebraically equivalent (they always produce the same answer), but they look different and are used in different situations.

The first of these formulas is called the *definitional formula* because the terms in the formula literally define the process of adding up the squared deviations:

$$\text{definitional formula: } SS = \Sigma(X - \mu)^2 \tag{4.1}$$

Note that the formula directs you to square each deviation score, $(X - \mu)^2$, and then add them. The result is the sum of the squared deviations, or *SS*. Following is an example using this formula.

EXAMPLE 4.3   We will compute *SS* for the following set of $N = 4$ scores. These scores have a sum of $\Sigma X = 8$, so the mean is $\mu = \frac{8}{4} = 2$. For each score, we have computed the deviation and the squared deviation. The squared deviations add up to $SS = 22$. These steps are illustrated in the following computational table:

Caution: The definition formula requires that you first square the deviations and then add them. Note that we have done these computations in a table. Do not confuse this table with a frequency distribution table.

X	$X - \mu$	$(X - \mu)^2$	
1	-1	1	$\Sigma X = 8$
0	-2	4	$N = 4$
6	$+4$	16	$\mu = 2$
1	-1	$\underline{1}$	
		$22 = \Sigma(X - \mu)^2$	

The second formula for SS is called the *computational formula* (or the machine formula) because it works directly with the scores (X values) and, therefore, is generally easier to use for calculations, especially with an electronic calculator:

$$\text{computational formula: } SS = \Sigma X^2 - \frac{(\Sigma X)^2}{N} \tag{4.2}$$

The first part of this formula, ΣX^2, directs you to square each score and then sum the squared values. The second part requires you to add up the scores, ΣX, square this total, and then divide the result by N. The use of this formula is shown in Example 4.4 with the same set of scores we used for the definitional formula.

EXAMPLE 4.4

The computational formula is used to calculate SS for the same set of $N = 4$ scores we used in Example 4.3. First, compute ΣX. Then square each score, and compute ΣX^2. These calculations are shown in the following table. The two sums are used in the formula to compute SS.

X	X^2
1	1
0	0
6	36
1	1
$\Sigma X = 8$	$\Sigma X^2 = 38$

$$SS = \Sigma X^2 - \frac{(\Sigma X)^2}{N}$$

$$= 38 - \frac{(8)^2}{4}$$

$$= 38 - \frac{64}{4}$$

$$= 38 - 16$$

$$= 22$$

Notice that the two formulas produce exactly the same value for SS. Although the formulas look different, they are in fact equivalent.

The definitional formula should be very easy to learn if you simply remember that SS stands for the sum of the squared deviations. If you use notation to write out "the sum of" (Σ) "squared deviations" $(X - \mu)^2$, then you have the definitional formula. Unfortunately, the terms in the computational formula do not translate directly into "sum of squared deviations," so you need to memorize this formula.

The definitional formula for *SS* is the most direct way of calculating the sum of squares, but it can be awkward to use for most sets of data. In particular, if the mean is not a whole number, then the deviation scores will have fractions or decimals, and the calculations become difficult. In addition, calculations with decimals or fractions introduce the opportunity for rounding error, which makes the results less accurate. For these reasons, the computational formula is used most of the time. If you have a small group of scores and the mean is a whole number, then the definitional formula is fine; otherwise, use the computational formula.

FORMULAS FOR POPULATION STANDARD DEVIATION AND VARIANCE

With the definition and calculation of *SS* behind you, the equations for variance and standard deviation become relatively simple. Remember that variance is defined as the mean squared deviation. The mean is the sum divided by *N*, so the equation for variance is

$$\text{variance} = \frac{SS}{N}$$

In the same way that sum of squares, or *SS*, is used to refer to the sum of squared deviations, the term *mean square*, or *MS*, is often used to refer to variance, which is the mean squared deviation.

Standard deviation is the square root of variance, so the equation for standard deviation is

$$\text{standard deviation} = \sqrt{\frac{SS}{N}}$$

There is one final bit of notation before we work completely through an example computing *SS*, variance, and standard deviation. Like the mean (μ), variance and standard deviation are parameters of a population and will be identified by Greek letters. To identify the standard deviation, we use the Greek letter sigma (the Greek letter *s*, standing for standard deviation). The capital letter sigma (Σ) has been used already, so we now use the lowercase sigma, σ:

$$\text{population standard deviation} = \sigma = \sqrt{\frac{SS}{N}} \tag{4.3}$$

The symbol for population variance should help you remember the relationship between standard deviation and variance. If you square the standard deviation, you will get the variance. The symbol for variance is sigma squared, σ^2:

$$\text{population variance} = \sigma^2 = \frac{SS}{N} \tag{4.4}$$

SUMMARY OF COMPUTATION FOR VARIANCE AND STANDARD DEVIATION

The following example demonstrates the complete process of calculating the variance and the standard deviation for a population of scores. Before we begin the example, however, we will briefly review the basic steps in the calculation.

STEP 1   Find the distance from the mean for each individual.

STEP 2   Square each distance.

STEP 3   Find the sum of the squared distances. This value is called *SS* or *sum of squares*. (*Note:* SS can also be obtained using the computational formula instead of steps 1–3.)

STEP 4   Find the mean of the squared distances. This value is called *variance* and measures the average squared distance from the mean.

STEP 5   Take the square root of the variance. This value is called *standard deviation* and provides a measure of the standard distance from the mean.

EXAMPLE 4.5   The following population of scores will be used to demonstrate the calculation of *SS*, variance, and standard deviation:

1, 9, 5, 8, 7

These five scores add up to $\Sigma X = 30$, so the mean is $\frac{30}{5} = 6$. Before we do any other calculations, remember that the purpose of standard deviation is to measure the standard distance from the mean. The scores we are working with are shown in a histogram in Figure 4.4 so you can see the variability more easily. Note that the scores closest to the mean are only 1 point away. Also, the score farthest from the mean is 5 points away. For this distribution, the largest distance from the mean is 5 points and the smallest distance is 1 point. Thus, the standard distance should be somewhere between 1 and 5. By looking at a distribution in this way, you should be able to make a rough estimate of the standard deviation. In this case, the standard deviation should be between 1 and 5, probably around 3 points.

FIGURE 4.4

A frequency distribution histogram for a population of $N = 5$ scores. The mean for this population is $\mu = 6$. The smallest distance from the mean is 1 point, and the largest distance is 5 points. The standard distance (or standard deviation) should be between 1 and 5 points.

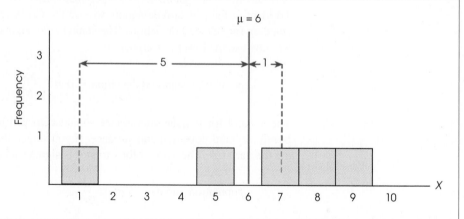

Making a preliminary judgment of standard deviation can help you avoid errors in calculation. If, for example, you worked through the formulas and ended up with a value of $\sigma = 12$, you should realize immediately that you have made an error. (If the largest deviation is only 5 points, then it is impossible for the standard deviation to be 12.)

Now we will start the calculations. The first step is to find *SS* for this set of scores. Because the mean is a whole number ($\mu = 6$), we can use the definitional formula for

SS. Therefore, we first find each deviation, then square each deviation, then sum the squared deviations. This process is shown in the following table.

X	$X - \mu$	$(X - \mu)^2$	
1	-5	25	$\Sigma X = 30$
9	$+3$	9	$N = 5$
5	-1	1	$\mu = 6$
8	$+2$	4	
7	$+1$	1	
		$40 = \Sigma(X - \mu)^2 = SS$	

$$\sigma^2 = \frac{SS}{N}$$

$$= \frac{40}{5} = 8$$

$$\sigma = \sqrt{8} = 2.83$$

For this set of scores, the variance is $\sigma^2 = 8$, and the standard deviation is $\sigma = \sqrt{8} = 2.83$. Note that the value for the standard deviation is in excellent agreement with our preliminary estimate of the standard distance from the mean.

LEARNING CHECK

1. Write brief definitions of variance and standard deviation.

2. Find SS, variance, and standard deviation for the following population of scores: 10, 10, 10, 10, 10. (*Note:* You should be able to answer this equation without doing any calculations.)

3. **a.** Sketch a frequency distribution histogram for the following population of scores: 1, 3, 3, 9. Using this histogram, make an estimate of the standard deviation (i.e., the standard distance from the mean).

   b. Calculate SS, variance, and standard deviation for these scores. How well does your estimate from part a compare with the real standard deviation?

ANSWERS

1. Variance is the mean squared distance from the mean. Standard deviation is the square root of variance and provides a measure of the standard distance from the mean.

2. Because there is no variability in the population, SS, variance, and standard deviation are all equal to zero.

3. **a.** Your sketch should show a mean of $\mu = 4$. The score closest to the mean is $X = 3$, and the farthest score is $X = 9$. The standard deviation should be somewhere between 1 point and 5 points.

   b. For this population, $SS = 36$; the variance is $\frac{36}{4} = 9$; the standard deviation is $\sqrt{9} = 3$.

GRAPHIC REPRESENTATION OF THE MEAN AND STANDARD DEVIATION

In frequency distribution graphs, we will identify the position of the mean by drawing a vertical line and labeling it with μ or M (see Figure 4.5). Because the standard deviation measures distance from the mean, it will be represented by an arrow drawn from the mean outward for a distance equal to the standard deviation (see Figure 4.5). For rough sketches, you can identify the mean with a vertical line in the middle of the distribution. The arrow for the standard deviation should extend approximately halfway from the mean to the most extreme score.

FIGURE 4.5

The graphic representation of a population with a mean of $\mu = 40$ and a standard deviation of $\sigma = 4$.

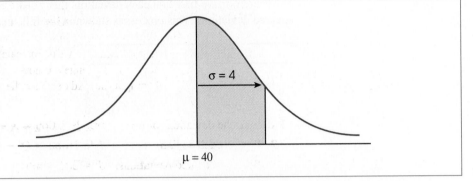

$\sigma = 4$

$\mu = 40$

4.5   STANDARD DEVIATION AND VARIANCE FOR SAMPLES

The goal of inferential statistics is to use the limited information from samples to draw general conclusions about populations. The basic assumption of this process is that samples should be representative of the populations from which they come. This assumption poses a special problem for variability because samples consistently tend to be less variable than their populations. An example of this general tendency is shown in Figure 4.6. The fact that a sample tends to be less variable than its population means that sample variability gives a *biased* estimate of population variability. This bias is in the direction of underestimating the population value rather than being right on the mark. To correct for the bias, it is necessary to make an adjustment in the calculation of variability when you are working with sample data. The intent of the adjustment is to make the resulting value for sample variability a more accurate estimate of the population variability.

A sample statistic is said to be *biased* if, on the average, it consistently overestimates or underestimates the corresponding population parameter.

FIGURE 4.6

The population of adult heights forms a normal distribution. If you select a sample from this population, you are most likely to obtain individuals who are near average in height. As a result, the scores in the sample will be less variable (spread out) than the scores in the population.

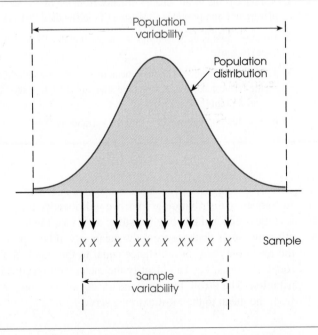

Population variability

Population distribution

X X   X   X X   X   X X   X   X   Sample

Sample variability

The calculation of variance and standard deviation for a sample follows the same steps that were used to find population variance and standard deviation. Except for minor changes in notation, the first three steps in this process are exactly the same for a sample as they were for a population. That is, the calculation of SS is the same for a sample as it is for a population. The changes in notation involve using M for the sample mean instead of μ, and using n (instead of N) for the number of scores. Thus, SS for a sample is found by

1. Find the deviation for each score: deviation $= X - M$
2. Square each deviation: squared deviation $= (X - M)^2$
3. Sum the squared deviations: $SS = \Sigma(X - M)^2$

These three steps can be summarized in a definitional formula for SS:

$$\text{Definitional formula: } SS = \Sigma(X - M)^2 \tag{4.5}$$

The value of SS can also be obtained using the computational formula. Using sample notation, this formula is:

$$\text{Computational formula: } SS = \Sigma X^2 - \frac{(\Sigma X)^2}{n} \tag{4.6}$$

After you compute SS, however, it becomes critical to differentiate between samples and populations. To correct for the bias in sample variability, it is necessary to make an adjustment in the formulas for sample variance and standard deviation. With this in mind, sample variance (identified by the symbol s^2) is defined as

$$\text{sample variance} = s^2 = \frac{SS}{n - 1} \tag{4.7}$$

Sample standard deviation (identified by the symbol s) is simply the square root of the variance.

$$\text{sample standard deviation} = s = \sqrt{\frac{SS}{n - 1}} \tag{4.8}$$

Remember, sample variability tends to underestimate population variability unless some correction is made.

Notice that these sample formulas use $n - 1$ instead of n. This is the adjustment that is necessary to correct for the bias in sample variability. The effect of the adjustment is to increase the value you will obtain. Dividing by a smaller number ($n - 1$ instead of n) produces a larger result and makes sample variance an accurate, or unbiased, estimator of population variance.

A complete example showing the calculation of sample variance and standard deviation will now be worked out.

EXAMPLE 4.6   We have selected a sample of $n = 7$ scores from a population. The scores are 1, 6, 4, 3, 8, 7, 6. The frequency distribution histogram for this sample is shown in Figure 4.7. Before we begin any calculations, you should be able to look at the sample distribution and make a preliminary estimate of the outcome. Remember that standard deviation measures the standard distance from the mean. For this sample the mean is $M = 5\left(\frac{35}{7} = 5\right)$. The scores closest to the mean are $X = 4$ and $X = 6$, both of which are exactly 1 point away. The score farthest from the mean is $X = 1$, which is 4 points away. With the smallest distance from the mean equal to 1 and the largest

FIGURE 4.7

The frequency distribution histogram for a sample of $n = 7$ scores. The sample mean is $M = 5$. The smallest distance from the mean is 1 point, and the largest distance from the mean is 4 points. The standard distance (standard deviation) should be between 1 and 4 points or about 2.5.

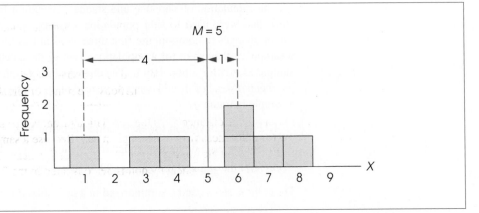

distance equal to 4, we should obtain a standard distance somewhere around 2.5 (between 1 and 4).

Now let's begin the calculations. First, we will find SS for this sample. Because there are only a few scores and the mean is a whole number, the definitional formula will be easy to use. You should try this formula for practice. Meanwhile, we will work with the computational formula.

The following computational table illustrates the work:

Caution: The calculation of SS is the same for a sample or a population. Do not use $n - 1$ in the formula for SS.

X	X²
1	1
6	36
4	16
3	9
8	64
7	49
6	36

$\Sigma X = 35$
$\Sigma X^2 = 211$
$n = 7$

$$SS = \Sigma X^2 - \frac{(\Sigma X)^2}{n}$$

$$= 211 - \frac{(35)^2}{7}$$

$$= 211 - \frac{1225}{7}$$

$$= 211 - 175 = 36$$

SS for this sample is 36. You should obtain exactly the same answer using the definitional formula. Continuing the calculations,

$$\text{sample variance} = s^2 = \frac{SS}{n - 1} = \frac{36}{7 - 1} = 6$$

Finally, the standard deviation is

$$s = \sqrt{s^2} = \sqrt{6} = 2.45$$

Note that the value we obtained is in excellent agreement with our preliminary prediction (Figure 4.7).

Remember that the formulas for sample variance and standard deviation were constructed so that the sample variability would provide a good estimate of population variability. For this reason, the sample variance is often called *estimated population variance*, and the sample standard deviation is called *estimated population standard de-*

viation. When you have only a sample to work with, the sample variance and standard deviation provide the best possible estimates of the population variability.

SAMPLE VARIABILITY AND DEGREES OF FREEDOM

Although the concept of a deviation score and the calculation of *SS* are almost exactly the same for samples and populations, the minor differences in notation are really very important. Specifically, with a population, you find the deviation for each score by measuring its distance from the population mean. With a sample, on the other hand, you measure distances from the sample mean. Because a sample mean is usually not identical to the population mean, the two methods for measuring distance will produce different results. This phenomenon is demonstrated in the following example.

EXAMPLE 4.7

We will begin with a population with a mean of $\mu = 4$ and select a sample of $n = 3$ scores (see Figure 4.8). The three scores are $X = 8$, $X = 3$, and $X = 4$; they sum to $\Sigma X = 15$, so the sample mean is $M = \frac{15}{3} = 5$. Note that the sample mean, $M = 5$, is not exactly the same as the population mean, $\mu = 4$. The difference is an example of sampling error.

Next we will use two different methods to compute the value of *SS* for the sample: First, we will calculate deviations and squared deviations using the population mean, $\mu = 4$. Then, we will repeat the calculations using the sample mean, $M = 5$. Both calculations are shown in the following table.

Sampling error refers to the fact that a sample statistic should be close to the corresponding population parameter, but usually there will be some difference or error.

WITH $\mu = 4$			WITH $M = 5$		
X	$X - \mu$	$(X - \mu)^2$	X	$X - M$	$(X - M)^2$
8	+4	16	8	+3	9
3	−1	1	3	−2	4
4	0	0	4	−1	1
		$SS = 17$			$SS = 14$

Using $\mu = 4$, the sum of the squared deviations is $SS = 17$.   Using $M = 5$, the sum of the squared deviations is $SS = 14$.

There are two points to be made from the previous example. First, notice that the

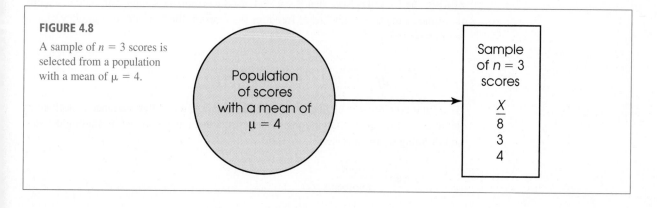

FIGURE 4.8

A sample of $n = 3$ scores is selected from a population with a mean of $\mu = 4$.

sum of the squared deviations (*SS*) computed with the sample mean is *smaller* than the value of *SS* computed with the population mean. This is another example of the general point that we made earlier. Sample variability tends to underestimate population variability (see page 116). Although we will not go into the mathematical proof, it is a general fact that *SS* computed from a sample mean will *always* be less than or equal to the value of *SS* computed with the real population mean.

Second, the reason that sample variability underestimates population variability comes from the fact that whenever you use the sample mean to compute deviations, there is a restriction on the variability of the scores in the sample. The restriction comes from the fact that you must know the value of the sample mean *before* you can compute deviations and *SS*. In Example 4.7, we had a sample of $n = 3$ scores with a mean of $M = 5$. Given this information, as soon as you know that the first two scores are $X = 8$ and $X = 3$, you also know that the third score must be $X = 4$. (In order to have a mean of $M = 5$, the three scores must total 15. Because the first two scores add up to 11, the third score has to be $X = 4$.) Thus, the final score in the sample is not free to vary, but rather is restricted by the sample mean.

In general, with a sample of n scores, the first $n - 1$ scores are completely independent and are free to vary. The final score, however, is dependent on the earlier scores and is not free to vary. Specifically, the final score must produce the correct ΣX to result in the correct sample mean. In Example 4.7 we had a sample of $n = 3$ scores with a mean of $M = 5$. The first score in the sample was $X = 8$. Notice that the second score could be any value. We could, for example, have selected a second score of $X = 1$. The final score in the sample, however, is dependent on the first two. If the first two scores are $X = 8$ and $X = 1$, then the final score must be $X = 6$ so that the sample totals $\Sigma X = 15$ and has a mean of $M = 5$.

Whenever you compute *SS* or variance for a sample, the sample is said to have $n - 1$ *degrees of freedom*. That is, the first $n - 1$ scores are independent and are free to vary, but the final score is determined by the earlier scores and the sample mean.

DEFINITION

Degrees of freedom or *df*, for a sample is defined as the number of independent scores. To compute sample variance, you must first know the sample mean. As a result, the first $n - 1$ scores in the sample are independent and free to vary, but the final score is restricted. Thus, the sample variance has degrees of freedom equal to $df = n - 1$, where n is the number of scores in the sample.

Because sample variability tends to underestimate population variability, we made an adjustment in the formula for sample variance. Specifically, we use $n - 1$ (instead of n) in the formula. Note that the $n - 1$ used to compute sample variance is exactly the same as the $n - 1$ degrees of freedom. As a result, the formula for sample variance is often written as

$$s^2 = \frac{SS}{df}$$

As a final note on the calculation of sample variance, recall that variance is defined as the mean of the squared deviations. As always, a mean is computed by finding the sum and dividing by the number:

$$\text{mean} = \frac{\text{sum}}{\text{number}}$$

Thus, the variance for a population is expressed as

$$\text{mean squared deviation} = \frac{\text{sum of squared deviations}}{\text{number}} = \frac{SS}{N}$$

However, to calculate sample variance, we find the sum of the squared deviations (SS) and divide by the number of scores that are free to vary. This number is $n - 1 = df$.

$$s^2 = \frac{\text{sum of squared deviations}}{\text{number of scores free to vary}} = \frac{SS}{n - 1} = \frac{SS}{df}$$

Later in this book, we will use the concept of degrees of freedom for inferential statistics. For now, remember that whenever you compute sample variance, only $n - 1$ of the scores are free to vary; $df = n - 1$.

BIASED AND UNBIASED STATISTICS

The formula for sample variance has been developed so that the value obtained for a sample variance will provide an *unbiased* estimate of the corresponding population variance. This does not mean that each individual sample variance will be exactly equal to its population variance. In fact, some sample variances will overestimate the population value and some will underestimate it. However, the average of all the sample variances will produce an accurate estimate of the population variance. This is the idea behind the concept of an unbiased statistic.

D E F I N I T I O N S

A sample statistic is **unbiased** if the average value of the sample statistic, obtained over many different samples, is equal to the population parameter. On the other hand, if the average value for a sample statistic consistently underestimates or overestimates the corresponding population parameter, then the statistic is **biased**.

The following example demonstrates the concept of biased and unbiased statistics.

E X A M P L E 4 . 8

We begin with a population that consists of exactly $N = 6$ scores: 0, 0, 3, 3, 9, 9. With a few simple calculations you should be able to verify that this population has a mean of $\mu = 4$ and a variance of $\sigma^2 = 14$.

Next, we will select samples of $n = 2$ scores from this population. In fact, we will obtain every single possible sample with $n = 2$. The complete set of samples is listed in Table 4.1. Notice that the samples are listed systematically to ensure that every possible sample is included. We begin by listing all the samples that have $X = 0$ as the first score, then all the samples with $X = 3$ as the first score, and so on. Notice that the table shows a total of 9 samples.

Finally, we have computed the mean and the variance for each sample. Note that we have computed the sample variance two different ways. First, we examine what would happen if there was no correction for bias and the sample variance was computed by simply dividing SS by n. Second, we examine the real sample variances where SS is divided by $n - 1$ to produce an unbiased measure of variance. You should verify our calculations by computing one or two of the values for yourself. The complete set of sample means and sample variances is presented in Table 4.1.

Now, direct your attention to the column of sample means. For this example, the original population has a mean of $\mu = 4$. Although none of the samples has a mean exactly equal to 4, if you consider the complete set of sample means, you will find that the 9 sample means add up to a total of 36, so the average of the sample means is $\frac{36}{9} = 4$. Note that the average of the sample means is exactly equal to the population

Technical note: We have structured this example to mimic "sampling with replacement," which will be covered in Chapter 6.

TABLE 4.1

The set of all the possible samples for $n = 2$ selected from the population described in Example 4.8. The mean is computed for each sample, and the variance is computed two different ways: (1) dividing by n, which is incorrect and produces a biased statistic; and (2) dividing by $n - 1$, which is correct and produces an unbiased statistic.

SAMPLE	FIRST SCORE	SECOND SCORE	SAMPLE STATISTICS MEAN M	BIASED VARIANCE (USING n)	UNBIASED VARIANCE (USING $n - 1$)
1	0	0	0.00	0.00	0.00
2	0	3	1.50	2.25	4.50
3	0	9	4.50	20.25	40.50
4	3	0	1.50	2.25	4.50
5	3	3	3.00	0.00	0.00
6	3	9	6.00	9.00	18.00
7	9	0	4.50	20.25	40.50
8	9	3	6.00	9.00	18.00
9	9	9	9.00	0.00	0.00
		Totals	36.00	63.00	126.00

mean. This is what is meant by the concept of an unbiased statistic. On average, the sample values provide an accurate representation of the population. In this example, the average of the 9 sample means is exactly equal to the population mean.

Next, consider the column of biased sample variances where we divided by n. For the original population the variance is $\sigma^2 = 14$. The 9 sample variances, however, add up to a total of 63, which produces an average variance of $\frac{63}{9} = 7$. Note that the average of these sample variances is *not* equal to the population variance. If the sample variance is computed by dividing by n, the resulting values will not produce an accurate estimate of the population variance. On average, these sample variances underestimate the population variance and, therefore, are biased statistics.

Finally, consider the column of sample variances that are computed using $n - 1$. Although the population has a variance of $\sigma^2 = 14$, you should notice that none of the samples has a variance exactly equal to 14. However, if you consider the complete set of sample variances, you will find that the 9 values add up to a total of 126, which produces an average variance of $\frac{126}{9} = 14.00$. Thus, the average of the sample variances is exactly equal to the population variance. On average, the sample variance (computed using $n - 1$) produces an accurate, unbiased estimate of the population variance.

In summary, both the sample mean and the sample variance (using $n - 1$) are examples of unbiased statistics. This fact makes the sample mean and sample variance extremely valuable for use as inferential statistics. Although no individual sample is likely to have a mean and variance exactly equal to the population values, both the sample mean and the sample variance, on average, do provide accurate estimates of the corresponding population values.

IN THE LITERATURE
REPORTING THE STANDARD DEVIATION

In reporting the results of a study, the researcher often provides descriptive information for both central tendency and variability. The dependent variables in psychology research frequently involve measures taken on interval or ratio scales. Thus, the mean (central tendency) and the standard deviation (variability) are commonly reported together. In many journals, especially those following APA style, the symbol *SD* is used for the sample standard deviation. For example, the results might state:

Children who viewed the violent cartoon displayed more aggressive responses ($M = 12.45$, $SD = 3.7$) than those who viewed the control cartoon ($M = 4.22$, $SD = 1.04$).

When reporting the descriptive measures for several groups, the findings may be summarized in a table. Table 4.2 illustrates the results of hypothetical data.

TABLE 4.2

The number of aggressive responses in male and female children after viewing cartoons.

	TYPE OF CARTOON	
	Violent	Control
Males	$M = 15.72$	$M = 6.94$
	$SD = 4.43$	$SD = 2.26$
Females	$M = 3.47$	$M = 2.61$
	$SD = 1.12$	$SD = 0.98$

Sometimes the table will also indicate the sample size, *n*, for each group. You should remember that the purpose of the table is to present the data in an organized, concise, and accurate manner.   ❏

LEARNING CHECK

1. **a.** Sketch a frequency distribution histogram for the following sample of scores: 1, 1, 9, 1. Using your histogram, make an estimate of the standard deviation for this sample.

    b. Calculate *SS*, variance, and standard deviation for this sample. How well does your estimate from part a compare with the real standard deviation?

2. If the scores in the previous exercise were for a population, what value would you obtain for *SS*?

3. Explain why the formulas for sample variance and standard deviation use $n - 1$ instead of *n*.

ANSWERS

1. **a.** Your graph should show a sample mean of $M = 3$. The score farthest from the mean is $X = 9$, and the closest score is $X = 1$. You should estimate the standard deviation to be between 2 points and 6 points.

    b. For this sample, $SS = 48$; the sample variance is $\frac{48}{3} = 16$; the sample standard deviation is $\sqrt{16} = 4$.

2. $SS = 48$ whether the data are from a sample or a population.

3. Without some correction, sample variability is biased because the scores in a sample tend to be less variable than the scores in the population. To correct for this bias, the formulas divide by $n - 1$ instead of *n*.

4.6   MORE ABOUT VARIANCE AND STANDARD DEVIATION

Variance and standard deviation are two of the most important elements in all of statistics. These two statistical measures are used extensively in both descriptive methods and inferential methods. Although variance and standard deviation can both be defined by mathematical formulas, it is much better to understand the concept behind these two measures and to understand how they are used. In the following sections, we will examine some of the uses and characteristics of the variance and the standard deviation.

TRANSFORMATIONS OF SCALE

Occasionally, it is convenient to transform a set of scores by adding a constant to each score or by multiplying each score by a constant value. This is done, for example, when you want to "curve" a set of exam scores by adding a fixed amount to each individual's grade or when you want to change the unit of measurement (to convert from minutes to seconds, multiply each X by 60). What happens to the standard deviation when the scores are transformed in this manner?

The easiest way to determine the effect of a transformation is to remember that the standard deviation is a measure of distance. If you select any two scores and see what happens to the distance between them, you also will find out what happens to the standard deviation.

1. **Adding a constant to each score will not change the standard deviation** If you begin with a distribution that has $\mu = 40$ and $\sigma = 10$, what happens to σ if you add 5 points to every score? Consider any two scores in this distribution: Suppose, for example, that these are exam scores and that you had $X = 41$ and your friend had $X = 43$. The distance between these two scores is $43 - 41 = 2$ points. After adding the constant, 5 points, to each score, your score would be $X = 46$, and your friend would have $X = 48$. The distance between scores is still 2 points. Adding a constant to every score will not affect any of the distances and, therefore, will not change the standard deviation. This fact can be seen clearly if you imagine a frequency distribution graph. If, for example, you add 10 points to each score, then every score in the graph will be moved 10 points to the right. The result is that the entire distribution is shifted to a new position 10 points up the scale. Note that the mean moves along with the scores and is increased by 10 points. However, the variability does not change because each of the deviation scores $(X - \mu)$ does not change.

2. **Multiplying each score by a constant causes the standard deviation to be multiplied by the same constant** Consider the same distribution of exam scores we looked at earlier. If $\mu = 40$ and $\sigma = 10$, what would happen to σ if each score were multiplied by 2? Again, we will look at two scores, $X = 41$ and $X = 43$, with a distance between them equal to 2 points. After all the scores have been multiplied by 2, these scores become $X = 82$ and $X = 86$. Now the distance between scores is 4 points, twice the original distance. Multiplying each score causes each distance to be multiplied, so the standard deviation also is multiplied by the same amount.

STANDARD DEVIATION AND DESCRIPTIVE STATISTICS

Because standard deviation requires extensive calculations, there is a tendency to get lost in the arithmetic and forget what standard deviation is and why it is important. Standard deviation is primarily a descriptive measure; it describes how variable, or how spread out, the scores are in a distribution. Behavioral scientists must deal with the variability that comes from studying people and animals. People are not all the same; they have different attitudes, opinions, talents, IQs, and personalities. Although we can calculate the average value for any of these variables, it is equally important to describe the variability. Standard deviation describes variability by measuring *distance from the mean*. In any distribution, some individuals will be close to the mean, and others will be relatively far from the mean. Standard deviation provides a measure of the typical, or standard, distance from the mean.

In addition to describing an entire distribution, standard deviation also allows us to interpret individual scores. We know, for example, that a person with an IQ of 110 is above average but not exceptional; with a standard distance of $\sigma = 15$ points, a score that is 10 points above average is not extreme. If we are comparing heights, on the other hand, a person who is 10 inches taller than average is exceptional; with a small standard deviation of only 4 or 5 inches, a 10-point difference is extreme.

The mean and the standard deviation are the most common values used to describe a set of data. A research report, for example, typically will not list all of the individual scores but rather will summarize the data by reporting the mean and standard deviation. When you are given these two descriptive statistics, you should be able to visualize the entire set of data. For example, consider a sample with a mean of $M = 36$ and a standard deviation of $s = 4$. Although there are several different ways to picture the data, one simple technique is to imagine (or sketch) a histogram where each score is represented by a box in the graph. For this sample, the data can be pictured as a pile of boxes (scores) with the center of the pile located at a value of $M = 36$. The individual scores or boxes are scattered on both sides of the mean with some of the boxes relatively close to the mean and some farther away. In this example, the standard distance from the mean is $s = 4$ points. Your image of the distribution should have most of the boxes within 4 points of the mean and only a few boxes located more than 4 points away from $M = 36$. One possibility for the resulting image is shown in Figure 4.9.

Notice that Figure 4.9 not only shows the mean and the standard deviation, but also uses these two values to reconstruct the underlying scale of measurement (the X values along the horizontal line). The scale of measurement helps complete the picture of the entire distribution and helps to relate each individual score to the rest of the group. In this example, you should realize that a score of $X = 34$ is located near the center of the distribution, only slightly below the mean. On the other hand, a score of $X = 44$ is an extremely high score, located far out in the right-hand tail of the distribution.

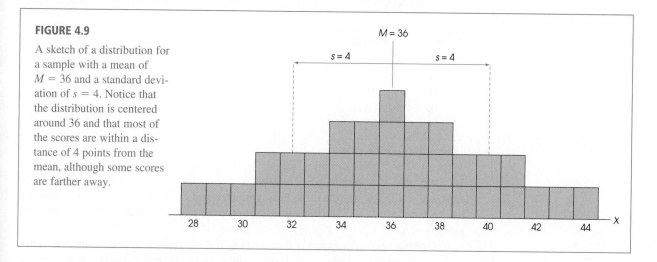

FIGURE 4.9

A sketch of a distribution for a sample with a mean of $M = 36$ and a standard deviation of $s = 4$. Notice that the distribution is centered around 36 and that most of the scores are within a distance of 4 points from the mean, although some scores are farther away.

The general point of this discussion is that the mean and standard deviation are not simply abstract concepts or mathematical equations. Instead, these two values should be concrete and meaningful, especially in the context of a set of scores. The mean and standard deviation are central concepts for most of the statistics that will be presented in the following chapters. A good understanding of these two statistics will help you with the more complex procedures that follow. (See Box 4.1.)

BOX 4.1   AN ANALOGY FOR THE MEAN AND THE STANDARD DEVIATION

Although the basic concepts of the mean and the standard deviation are not overly complex, the following analogy often helps students gain a more complete understanding of these two statistical measures.

In our local community, the site for a new high school was selected because it provides a central location. An alternative site on the western edge of the community was considered, but this site was rejected because it would require extensive busing for students living on the east side. In this example, the location of the high school is analogous to the concept of the mean; just as the high school is located in the center of the community, the mean is located in the center of the distribution of scores.

For each student in the community, it is possible to measure the distance between home and the new high school. Some students live only a few blocks from the new school and others live as much as 3 miles away. The average distance that a student must travel to school was calculated to be 0.80 miles. The average distance from the school is analogous to the concept of the standard deviation; that is, the standard deviation measures the standard distance from an individual score to the mean.

VARIANCE AND INFERENTIAL STATISTICS

In very general terms, the goal of inferential statistics is to detect meaningful and significant patterns in research results. The basic question is whether the sample data reflect patterns that exist in the population, or do the sample data simply show random fluctuations that occur by chance? Variability plays an important role in the inferential process because the variability in the data influences how easy it is to see patterns. In general, low variability means that existing patterns can be seen clearly, whereas high variability tends to obscure any patterns that might exist. The following two samples provide a simple demonstration of how variance can influence the perception of patterns. Your task is to examine each sample briefly and then estimate the sample mean.

SAMPLE 1 X	SAMPLE 2 X
34	26
35	10
36	64
35	40

After a few seconds examining sample 1, you should realize that the sample mean is $M = 35$, and you should be very confident that this is an accurate estimate. With sample 2, on the other hand, the task is much more difficult. Although both samples have a mean of $M = 35$, it is much easier to see the mean in sample 1 than it is in sample 2.

The difference between the two samples is variability. In the first sample the scores are all clustered close together and variability is small. In this situation it is easy to see the sample mean. However, the scores in the second sample are spread out over a wide range and variability is large. With high variability, it is not easy to identify the location of the mean. In general, high variability makes it difficult to see any patterns in the data.

The preceding example demonstrates how variability can affect the ability to identify the mean for a single sample. In most research studies, however, the goal is to compare means for two (or more) sets of data. For example:

Is the mean level of depression lower after therapy than it was before therapy?

Is the mean attitude score for men different from the mean score for women?

Is the mean reading achievement score higher for students in a special program than for students in regular classrooms?

In each of these situations, the goal is to find a clear difference between two means that would demonstrate a significant, meaningful pattern in the results. Once again variability plays an important role in determining whether or not a clear pattern exists. Consider the following data representing hypothetical results from two experiments, each comparing two separate samples. In each case, your task is to determine whether or not there appears to be any consistent difference between the two samples.

EXPERIMENT A		EXPERIMENT B	
SAMPLE 1	SAMPLE 2	SAMPLE 1	SAMPLE 2
35	39	31	46
34	40	15	21
36	41	57	61
35	40	37	32

For each experiment, the data have been constructed so that there is a 5-point difference between the two samples: On average, the scores in sample 2 are 5 points higher than the scores in sample 1. The 5-point difference is relatively easy to see in Experiment A where the variability is low, but the same 5-point difference is relatively difficult to see in Experiment B where the variability is large. Again, high variability tends to obscure any patterns in the data. This general fact is perhaps even more convincing when the data are presented in a graph. Figure 4.10 shows the two sets of data from Experiments A and B that we just considered. Notice that the data from

FIGURE 4.10

Graphs showing the results from two experiments. In Experiment A, the variability within samples is small and it is easy to see the 5-point mean difference between the two samples. In Experiment B, however, the 5-point mean difference between samples is obscured by the large variability within samples.

Experiment A show a clear difference between the two samples [Figure 4.10(a)]. On the other hand, the two samples from Experiment B [Figure 4.10(b)] seem to be mixed together randomly with no clear difference between samples. If the two samples represented two different experimental treatments, then the results from Experiment A (low variability) would suggest that there is a real difference between the treatments. However, the results from Experiment B tend to indicate that there is no difference between treatments.

In the context of inferential statistics, the variance that exists in a set of sample data is often classified as *error variance*. This term is used to indicate that the sample variance represents unexplained and uncontrolled differences between scores. As the error variance increases, it becomes more difficult to see any systematic differences or patterns that might exist in the data. An analogy is to think of variance as the static that exists on a radio or the "snow" on a television screen. In general, variance makes it difficult to get a clear signal from the data. High variance can make it difficult or impossible to see the mean for a set of scores, or to see the mean difference between two sets of data, or to see any meaningful patterns in the results of a research study.

4.7   COMPARING MEASURES OF VARIABILITY

By far the most commonly used measure of variability is standard deviation (together with the related measure of variance). Nonetheless, there are situations for which the range or the semi-interquartile range may be preferred. The advantages and disadvantages of each of these three measures will be discussed next.

In simple terms, two considerations determine the value of any statistical measurement:

1. The measure should provide a stable and reliable description of the scores. Specifically, it should not be greatly affected by minor details in the set of data.

2. The measure should have a consistent and predictable relationship with other statistical measurements.

We will examine each of these considerations separately.

FACTORS THAT AFFECT VARIABILITY

1. **Extreme scores**   Of the three measures of variability, the range is most affected by extreme scores. A single extreme value will have a large influence on the range. In fact, the range is determined exclusively by the two extremes of the distribution. Standard deviation and variance also are influenced by extreme scores. Because these measures are based on squared deviations, a single extreme value can have a disproportionate effect. For example, a score that is 10 points away from the mean will contribute $10^2 = 100$ points to the SS. For this reason, standard deviation and variance should be interpreted carefully in distributions with one or two extreme values. Because the semi-interquartile range focuses on the middle of the distribution, it is least affected by extreme values. For this reason, the semi-interquartile range often provides the best measure of variability for distributions that are very skewed or that have a few extreme scores.

2. **Sample size**   As you increase the number of scores in a sample, you also tend to increase the range because each additional score has the potential to replace the current highest or lowest value in the set. Thus, the range is directly related to sample size. This relationship between sample size and variability is unacceptable. A researcher should

not be able to influence variability by manipulating sample size. Standard deviation, variance, and the semi-interquartile range are relatively unaffected by sample size and, therefore, provide better measures.

3. Stability under sampling If you take several different samples from the same population, you should expect the samples to be similar. Specifically, if you compute variability for each of the separate samples, you should expect to obtain similar values. Because all of the samples come from the same source, it is reasonable that there should be some "family resemblance." When standard deviation and variance are used to measure variability, the samples tend to have similar variability. For this reason, standard deviation and variance are said to be stable under sampling. The semi-interquartile range also provides a reasonably stable measure of variability. The range, however, will change unpredictably from sample to sample and is said to be unstable under sampling.

4. Open-ended distributions When a distribution does not have any specific boundary for the highest score or the lowest score, it is open-ended. This can occur when you have infinite or undetermined scores. For example, a subject who cannot solve a problem has taken an undetermined or infinite amount of time to reach the solution. In an open-ended distribution, you cannot compute the range, or the standard deviation, or the variance. In this situation, the only available measure of variability is the semi-interquartile range.

RELATIONSHIP WITH OTHER STATISTICAL MEASURES

As noted earlier, variance and standard deviation are computed from squared deviation scores. Because they are based on squared distances, these measures fit into a coherent system of mathematical relationships that underlies many of the statistical techniques we will examine in this book. Although we generally will not present the underlying mathematics, you will notice that variance and standard deviation appear repeatedly. This is because they are valuable measures of variability. Also, you should notice that variance and standard deviation have a direct relationship to the mean (they are based on deviations from the mean). Therefore, the mean and standard deviation tend to be reported together. Because the mean is the most common measure of central tendency, the standard deviation will be the most common measure of variability.

Because the median and the semi-interquartile range are both based on percentiles, they share a common foundation and tend to be associated. When the median is used to report central tendency, the semi-interquartile range is commonly used to report variability.

The range has no direct relationship to any other statistical measure. For this reason, it is rarely used in conjunction with other statistical techniques.

SUMMARY

1. The purpose of variability is to determine how spread out the scores are in a distribution. There are four basic measures of variability: the range, the semi-interquartile range, the variance, and the standard deviation.

The range is the distance between the upper real limit of the largest X and the lower real limit of the smallest X in the distribution. The semi-interquartile range is one-half the distance between the first quartile and the third quartile. Variance is defined as the mean squared deviation. Standard deviation is the square root of the variance.

2. Standard deviation and variance are the most commonly used measures of variability. Both of these measures are based on the observation that each individual score can be described in terms of its deviation or distance from the mean. The standard deviation is a measure of the standard distance from the mean. The variance is equal to the mean of the squared deviations—that is, the mean squared distance from the mean.

3. To calculate variance or standard deviation, you first need to find the sum of the squared deviations, SS. There are two methods for calculating SS:
   I. By definition, you can find SS using the following steps:
      a. Find the deviation $(X - \mu)$ for each score.
      b. Square each deviation.
      c. Sum the squared deviations.
   This process can be summarized in a formula as follows:

   Definitional formula: $SS = \Sigma(X - \mu)^2$

   II. The sum of the squared deviations can also be found using a computational formula, which is especially useful when the mean is not a whole number:

   Computational formula: $SS = \Sigma X^2 - \dfrac{(\Sigma X)^2}{N}$

4. Variance is the mean squared deviation and is obtained by finding the sum of the squared deviations and then dividing by the number. For a population, variance is

$$\sigma^2 = \frac{SS}{N}$$

For a sample, only $n - 1$ of the scores are free to vary (degrees of freedom or $df = n - 1$), so sample variance is

$$s^2 = \frac{SS}{n - 1} = \frac{SS}{df}$$

Using $n - 1$ in the sample formula makes the sample variance an accurate and unbiased estimate of the population variance.

5. Standard deviation is the square root of the variance. For a population, this is

$$\sigma = \sqrt{\frac{SS}{N}}$$

Sample standard deviation is

$$s = \sqrt{\frac{SS}{n - 1}} = \sqrt{\frac{SS}{df}}$$

6. Adding a constant value to every score in a distribution will not change the standard deviation. Multiplying every score by a constant, however, will cause the standard deviation to be multiplied by the same constant.

RESOURCES

WebTUTOR

If you are using WebTutor along with this book, there is a WebTutor section corresponding to this chapter. The WebTutor contains a brief summary of Chapter 4, hints for learning the concepts and the formulas for variability, cautions about common errors, and sample exam items including solutions.

The statistics computer package SPSS can be used to calculate the standard deviation for a sample of scores. First, enter the scores in one column of the data matrix and then click **Analyze** on the tool bar. Then, select **Descriptive Statistics** and select **Descriptives**. Enter the column label for the set of scores into the Variable box and click **OK**. SPSS will produce a summary table listing the maximum score, the minimum score, the mean, and the standard deviation. Note that SPSS computes the *sample* standard deviation using $n - 1$. If your scores are intended to be a population, you must multiply the sample standard deviation by the square root of $(n - 1)/n$ to obtain the population standard deviation.

You can also obtain the standard deviation for a sample if you use SPSS to display the scores in a frequency distribution histogram (see the SPSS section on page 62 at the end of Chapter 2). The mean and standard deviation are displayed beside the graph.

KEY TERMS

variability	semi-interquartile range	variance	sum of squares (SS)
range	deviation score	standard deviation	degrees of freedom (df)
interquartile range	biased statistic	unbiased statistic	

FOCUS ON PROBLEM SOLVING

1. The purpose of variability is to provide a measure of how spread out the scores are in a distribution. Usually this is described by the standard deviation. Because the calculations are relatively complicated, it is wise to make a preliminary estimate of the standard deviation before you begin. Remember that standard deviation provides a measure of the typical, or standard, distance from the mean. Therefore, the standard deviation must have a value somewhere between the largest and the smallest deviation scores. As a rule of thumb, the standard deviation should be about one-fourth of the range.

2. Rather than trying to memorize all the formulas for SS, variance, and standard deviation, you should focus on the definitions of these values and the logic that relates them to each other:

    Deviation is distance from the mean.

    SS is the sum of the squared deviations.

    Variance is the mean of the squared deviations.

    Standard deviation is the square root of variance.

    The only formula you should need to memorize is the computational formula for SS.

3. If you heed the warnings in the following list, you may avoid some of the more common mistakes in solving variability problems.
    a. Because the calculation of standard deviation requires several steps of calculation, students often get lost in the arithmetic and forget what they are trying to compute. It helps to examine the data before you begin and make a rough estimate of the mean and the standard deviation.
    b. For SS, a computational table (page 118) helps organize your work. However, do *not* attempt to compute SS from a frequency distribution table.
    c. The variance formulas for populations and samples are slightly different. Be sure that you know whether the data come from a sample or a population before you begin calculations.
    d. A common error is to use $n - 1$ in the computational formula for SS when you have scores from a sample. Remember that the SS formula always uses n (or N). After you compute SS for a sample, you must correct for the sample bias by using $n - 1$ in the formulas for variance and standard deviation.

DEMONSTRATION 4.1

COMPUTING MEASURES OF VARIABILITY

For the following sample data, compute the variance and standard deviation. The scores are:

$$10, \quad 7, \quad 6, \quad 10, \quad 6, \quad 15$$

Compute the sum of squares   For SS, we will use the definitional formula:

$$SS = \Sigma(X - M)^2$$

STEP 1   Calculate the sample mean for these data.

$$M = \frac{\Sigma X}{n} = \frac{54}{6} = 9$$

STEP 2   Compute the deviation scores, $(X - M)$, for every X value. This is facilitated by making a computational table listing all the X values in one column and all the deviation scores in another column. Remember: Do not use a frequency distribution table.

X	X − M
10	10 − 9 = +1
7	7 − 9 = −2
6	6 − 9 = −3
10	10 − 9 = +1
6	6 − 9 = −3
15	15 − 9 = +6

STEP 3   Square the deviation scores. This is shown in a new column labeled $(X - M)^2$.

X	X − M	(X − M)²
10	+1	1
7	−2	4
6	−3	9
10	+1	1
6	−3	9
15	+6	36

STEP 4   Sum the squared deviation scores to obtain the value for SS.

$$SS = \Sigma(X - M)^2 = 1 + 4 + 9 + 1 + 9 + 36 = 60$$

Compute the sample variance   For sample variance, we divide SS by $n - 1$ (also known as degrees of freedom).

STEP 1   Compute degrees of freedom, $n - 1$.

$$\text{degrees of freedom} = df = n - 1 = 6 - 1 = 5$$

STEP 2   Divide SS by df.

$$s^2 = \frac{SS}{n - 1} = \frac{60}{5} = 12$$

Compute the sample standard deviation   The sample standard deviation is simply the square root of the sample variance.

$$s = \sqrt{\frac{SS}{n - 1}} = \sqrt{\frac{60}{5}} = \sqrt{12} = 3.46$$

PROBLEMS

1. In words, explain what is measured by each of the following:
  a. SS
  b. Variance
  c. Standard deviation

2. A population has $\mu = 100$ and $\sigma = 20$. If you select a single score from this population, on the average, how close would it be to the population mean? Explain your answer.

3. Can SS ever have a value less than zero? Explain your answer.

4. In general, what does it mean for a sample to have a standard deviation of zero? Describe the scores in such a sample.

5. Explain what it means to say that the sample variance provides an *unbiased* estimate of the population variance.

6. a. A sample of $n = 6$ scores has $SS = 60$. What is the variance for this sample?
  b. A population of $N = 6$ scores has $SS = 60$. What is the variance for this population?

7. Calculate SS, variance, and standard deviation for the following sample:

  4, 7, 3, 1, 5

8. For the following scores:

  1, 0, 4, 1

  a. Calculate the mean. (Note that the value of the mean does not depend on whether the set of scores is considered to be a sample or a population.)
  b. Find the deviation for each score, and check that the deviations sum to zero.
  c. Square each deviation, and compute SS. (Again, note that the value of SS is independent of whether the set of scores is a sample or a population.)

9. There are two different formulas or methods that can be used to calculate SS.
  a. Under what circumstances is the definitional formula easy to use?
  b. Under what circumstances is the computational formula preferred?

10. Calculate SS, variance, and standard deviation for the following sample of $n = 4$ scores: 0, 3, 0, 3. (*Note:* The computational formula for SS works best with these scores).

11. Calculate SS, variance, and standard deviation for the following sample of $n = 9$ scores: 2, 0, 0, 0, 0, 2, 0, 2, 0. (*Note:* The computational formula for SS works best with these scores.)

12. Calculate SS, variance, and standard deviation for the following population of $N = 6$ scores: 5, 0, 9, 3, 8, 5. (*Note:* The definitional formula for SS works well with these scores.)

13. Calculate SS, variance, and standard deviation for the following population of $N = 4$ scores: 6, 8, 0, 6. (*Note:* The definitional formula for SS works well with these scores.)

14. A distribution of scores has a mean of $M = 42$.
   a. If the standard deviation is $s = 12$, would a score of $X = 48$ be considered an extreme value? Explain your answer.
   b. If the standard deviation is $s = 2$, would a score of $X = 48$ be considered an extreme value? Explain your answer.

15. On an exam with $\mu = 75$, you obtain a score of $X = 80$.
   a. Would you prefer that the exam distribution had $\sigma = 2$ or $\sigma = 10$? (Sketch each distribution, and locate the position of $X = 80$ in each one.)
   b. If your score is $X = 70$, would you prefer $\sigma = 2$ or $\sigma = 10$? (Again, sketch each distribution to determine how the value of σ affects your position relative to the rest of the class.)

16. For the following sample of $n = 5$ scores:

$$10, \quad 0, \quad 6, \quad 2, \quad 2$$

   a. Sketch a histogram showing the sample distribution.
   b. Locate the value of the sample mean in your sketch, and make an estimate of the sample standard deviation (as done in Example 4.6).
   c. Compute SS, variance, and standard deviation for the sample. (How well does your estimate compare with the actual value of s?)

17. For the following sample of $n = 8$ scores:

$$6, \quad 8, \quad 4, \quad 3, \quad 5, \quad 7, \quad 4, \quad 3$$

   a. Sketch a histogram showing the sample distribution.
   b. Locate the value of the sample mean in your sketch, and make an estimate of the sample standard deviation (as done in Example 4.6).
   c. Compute SS, variance, and standard deviation for the sample. (How well does your estimate compare with the actual value of s?)

18. For the following population of $N = 5$ scores:

$$11, \quad 2, \quad 0, \quad 8, \quad 4$$

   a. Sketch a histogram showing the population distribution.
   b. Locate the value of the population mean in your sketch, and make an estimate of the standard deviation (as done in Example 4.5).
   c. Compute SS, variance, and standard deviation for the population. (How well does your estimate compare with the actual value of σ?)

19. For the following population of $N = 12$ scores:

$$6, \quad 10, \quad 4, \quad 4, \quad 6, \quad 7, \quad 11, \quad 7, \quad 3, \quad 11, \quad 6, \quad 12$$

   a. Sketch a histogram showing the population distribution.
   b. Locate the value of the population mean in your sketch, and make an estimate of the standard deviation (as done in Example 4.5).
   c. Compute SS, variance, and standard deviation for the population. (How well does your estimate compare with the actual value of σ?)

20. Sketch a normal distribution (see Figure 4.5, page 116) with $\mu = 50$ and $\sigma = 20$.
   a. Locate each of the following scores in your sketch, and indicate whether you consider each score to be an extreme value (high or low) or a central value:

$$65, \quad 55, \quad 40, \quad 47$$

   b. Make another sketch showing a distribution with $\mu = 50$, but this time with $\sigma = 2$. Now locate each of the four scores in the new distribution, and indicate whether they are extreme or central. (*Note:* The value of the standard deviation can have a dramatic effect on the location of a score within a distribution.)

21. For the following population of $N = 4$ scores:

$$2, \quad 0, \quad 8, \quad 2$$

   a. Use the definitional formula to compute SS; then find the population variance and the standard deviation.
   b. Add 3 points to each score; then compute SS, variance, and standard deviation for the new population.
   c. Multiply each of the original scores by 2; then compute SS, variance, and standard deviation for the new population.
   d. When a constant is added to each score, what happens to the deviation scores? What happens to the standard deviation?
   e. When each score is multiplied by a constant, what happens to the deviation scores? What happens to the standard deviation?

22. For the following population of scores:

$$1, \quad 6, \quad 9, \quad 0, \quad 4$$

   a. Find the mean for the population, and compute the deviation score for each individual.
   b. Show that the deviation scores sum to zero.

c. Square each deviation, and find the sum of squared deviations (*SS*).

d. Now assume that the set of scores is a sample instead of a population, and repeat parts a, b, and c. How does the distinction between a sample and a population affect the calculation of *SS*?

23. A study examines the relationship between level of arousal and problem solving. Three samples are used, consisting of subjects with low, moderate, or high levels of arousal. The researcher measures the number of problems successfully completed during a problem-solving task. The data are as follows:

LEVEL OF AROUSAL		
LOW	MODERATE	HIGH
2	20	9
6	17	8
5	12	8
7	16	6
5	18	7
5	19	10
4	17	6
4	16	8
7	18	7
6	17	8

a. Compute the mean and standard deviation for each sample. Present these values in a table, using APA style.

b. Looking at the data, how would you describe the relationship between level of arousal and task performance?

24. When the sun cannot be seen (overcast day), homing pigeons find their way back to their roosts by using magnetic cues from the Earth (Walcott, 1972).

Consider the following study. One sample of pigeons has a magnet fastened on their heads to interfere with their ability to detect the Earth's magnetic field. A second sample has a nonmagnetic bar of equal size and weight placed on their heads. The pigeons are driven far from their roosts and let go. The researcher measures the error of the pigeons' heading; that is, how many degrees there are between the pigeon's course and the true course to the home roost. Hypothetical data are as follows:

MAGNETIC BAR	NONMAGNETIC BAR
40	8
36	10
50	17
33	5
27	6

a. Identify the independent and dependent variables.

b. Compute the means for the treatment groups.

c. Compute the variance and standard deviation.

d. Looking at the descriptive statistics, do magnetic cues seem to help homing pigeons? Explain your answer.

25. For the data in the following sample:

1, 4, 3, 6, 2, 7, 18, 3, 7, 2, 4, 3

a. Sketch a frequency distribution histogram.

b. Compute the mean and standard deviation.

c. Find the median and the semi-interquartile range.

d. Which measures of central tendency and variability provide a better description of the sample? Explain your answer.

P A R T
II

Foundations of Inferential Statistics

Recall from Chapter 1 that statistical methods are classified into two general categories: *descriptive statistics* that attempt to organize and summarize data, and *inferential statistics* that use the limited information from samples to answer general questions about populations. In most research situations, both kinds of statistics are used to gain a complete understanding of the research results. In the introduction to Part I we presented a hypothetical research situation examining the relationship between television violence and the violent behavior exhibited by preschool children. One group of children watches a violent TV program immediately before play time each day and a second group watches a nonviolent program during the same time. The researcher then observes the children while they play and records the number of violent acts for each child. Suppose the results show that the children watching violent TV displayed more violent behavior than the children watching nonviolent TV. Can the researcher use this sample of 50 children to make a general conclusion that TV violence has an effect on children's behavior? That is, can the researcher generalize the results from this specific sample to the general population? Using samples to answer questions about populations is the purpose of inferential statistics.

Before we begin to look at inferential statistics, however, it is necessary to present some additional information about samples. We know that it is possible to obtain hundreds or even thousands of different samples from the same population. We need to determine how all the different samples are related to each other and how individual samples are related to the population from which they were obtained. Finally, we need a system for designating which samples are representative of their populations and which are not.

In the next three chapters we will develop the concepts and skills that form the foundation for inferential statistics. In general, these three chapters will establish formal, quantitative relationships between samples and populations. After we have developed the relationships between samples and populations, we will be prepared to begin inferential statistics. That is, we can begin to use sample data as the basis for drawing conclusions about populations.

C H A P T E R

5

z-Scores: Location of Scores and Standardized Distributions

Tools You Will Need

The following items are considered essential background material for this chapter. If you doubt your knowledge of any of these items, you should review the appropriate chapter and section before proceeding.

- The mean (Chapter 3)
- The standard deviation (Chapter 4)
- Basic algebra (math review, Appendix A)

Preview

While on vacation recently, 1000 miles from home, I visited a shopping mall. Although I had never been to this particular mall before, I began to realize after a few minutes that it was essentially identical to the old familiar mall where I usually shop at home. The specialty shops were all the same, the food court looked the same, and although some of the department stores had different names, they were basically the same as the stores back home. As the similarities began to sink in, my first response was disappointment. I had traveled 1000 miles to a different city in a different state, only to find that nothing was really different.

On second thought, however, I decided that there are definite advantages to having shopping malls standardized all across the country. First, the new mall was immediately familiar and easily understandable. I did not need to spend much time getting oriented so that I could determine which stores I needed to visit in order to find the items I wanted. Second, because I knew that the items and sizes were standardized, I could be confident that a "large" T-shirt or a pair of size $9\frac{1}{2}$ shoes would fit just as well as the same items purchased back home.

In the same way that shopping malls and shoe sizes are standardized to make them familiar and understandable, researchers often will standardize a set of scores to make them more meaningful and easier to understand. The scores obtained from IQ tests are probably the most familiar example of standardized scores. Specifically, your IQ score is *not* simply a report of the number of questions you answered correctly. In fact, several tests are used to measure IQ (the Stanford-Binet, the WAIS-R, and the WISC-R, for example), some with many questions and some with relatively few. In addition, each test often contains a number of subtests that measure different aspects of IQ. Thus, your actual performance on an IQ test might be reported as follows:

Vocabulary subtest: 8 correct

Arithmetic subtest: 9 correct

General Comprehension subtest: 11 correct

Picture Arrangement subtest: 4 correct

Object Assembly subtest: 4 correct

etc.

Very few people could make much sense of this report. Fortunately, however, IQ scores are standardized, so that each individual's test result can be reported as a single meaningful score. Specifically, the scores are *standardized*, so that they form a distribution with a mean of 100 and a standard deviation of 15. As a result, nearly everyone understands that an IQ of 98 is slightly below average or that an IQ of 150 is substantially above average.

In the preceding chapters, we concentrated on methods for describing entire distributions using the basic parameters of central tendency and variability. In this chapter, we will examine a procedure for *standardizing distributions* and for describing specific locations within a distribution. In particular, we will convert each individual score into a new, *standardized score*, so that the standardized value provides a meaningful description of its exact location within the distribution. We will use the mean as a reference point to determine whether the individual is above or below average. The standard deviation will serve as a yardstick for measuring how much an individual differs from the group average. A good conceptual understanding of the mean and the standard deviation will make this chapter much easier.

5.1   INTRODUCTION TO z-SCORES

In the previous two chapters, we introduced the concepts of the mean and the standard deviation as methods for describing an entire distribution of scores. Now we will shift attention to the individual scores within a distribution. In this chapter, we introduce a statistical technique that uses the mean and the standard deviation to transform each score (*X* value) into a *z-score* or a *standard score*. The purpose of z-scores, or standard scores, is to identify and describe the exact location of every score in a distribution.

The following example demonstrates why z-scores are useful and introduces the general concept of transforming *X* values into z-scores.

EXAMPLE 5.1   Suppose you received a score of $X = 76$ on a statistics exam. How did you do? It should be clear that you need more information to predict your grade. Your score of $X = 76$ could be one of the best scores in the class, or it might be the lowest score in the distribution. To find the location of your score, you must have information about the other scores in the distribution. It would be useful, for example, to know the mean for the class. If the mean were $\mu = 70$, you would be in a much better position than if the mean were $\mu = 85$. Obviously, your position relative to the rest of the class depends on the mean. However, the mean by itself is not sufficient to tell you the exact location of your score. Suppose you know that the mean for the statistics exam is $\mu = 70$ and your score is $X = 76$. At this point, you know that your score is above the mean, but you still do not know exactly where it is located. You may have the highest score in the class, or you may be only slightly above average. Figure 5.1 shows two possible distributions of exam scores. Both distributions have a mean of $\mu = 70$, but for one distribution, the standard deviation is $\sigma = 3$, and for the other, $\sigma = 12$. Notice that the relative location of $X = 76$ is very different for these two distributions. When the standard deviation is $\sigma = 3$, your score of $X = 76$ is in the extreme right-hand tail, one of the highest scores in the distribution. However, in the other distribution, where $\sigma = 12$, your score is only slightly above average. Thus, your relative location within the distribution depends on the mean and the standard deviation as well as on your actual score.

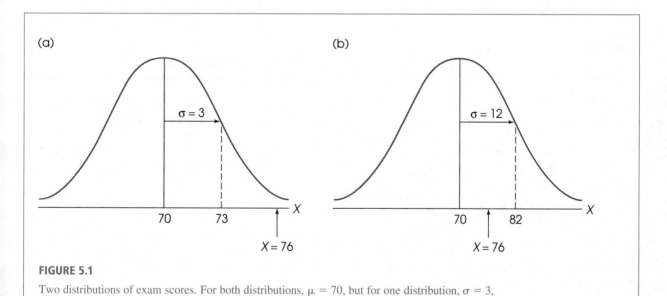

FIGURE 5.1

Two distributions of exam scores. For both distributions, $\mu = 70$, but for one distribution, $\sigma = 3$, and for the other, $\sigma = 12$. The position of $X = 76$ is very different for these two distributions.

The purpose of the preceding example is to demonstrate that a score *by itself* does not necessarily provide much information about its position within a distribution. These original, unchanged scores that are the direct result of measurement are often called *raw scores*. To make raw scores more meaningful, they are often transformed into new values that contain more information. This transformation is one purpose for *z*-scores. In particular, we will transform X values into z-scores so that the resulting z-scores tell exactly where the original scores are located.

A second purpose for z-scores is to *standardize* an entire distribution. A common example of a standardized distribution is the distribution of IQ scores. Although there are several different tests for measuring IQ, the tests usually are standardized so that they have a mean of 100 and a standard deviation of 15. Because all the different tests are standardized, it is possible to understand and compare IQ scores even though they come from different tests. For example, we all understand that an IQ score of 95 is a little below average, *no matter which IQ test was used*. Similarly, an IQ of 145 is extremely high, *no matter which IQ test was used*. In general terms, the process of standardizing takes different distributions and makes them equivalent. The advantage of this process is that it is possible to compare distributions even though they may have been quite different before standardization.

In summary, the process of transforming X values into z-scores serves two useful purposes:

1. Each z-score will tell the exact location of the original X value within the distribution.

2. The z-scores will form a standardized distribution that can be directly compared to other distributions that also have been transformed into z-scores.

Each of these purposes will be discussed in the following sections.

5.2   *z*-SCORES AND LOCATION IN A DISTRIBUTION

One of the primary purposes of a z-score is to describe the exact location of a score within a distribution. The z-score accomplishes this goal by transforming each X value into a signed number ($+$ or $-$) so that

1. The *sign* tells whether the score is located above ($+$) or below ($-$) the mean, and

2. The *number* tells the distance between the score and the mean in terms of the number of standard deviations.

Thus, in a distribution of standardized IQ scores with $\mu = 100$ and $\sigma = 15$, a score of $X = 130$ would be transformed into $z = +2.00$. The z value indicates that the score is located above the mean ($+$) by a distance of 2 standard deviations (30 points).

DEFINITION

A **z-score** specifies the precise location of each X value within a distribution. The sign of the z-score ($+$ or $-$) signifies whether the score is above the mean (positive) or below the mean (negative). The numerical value of the z-score specifies the distance from the mean by counting the number of standard deviations between X and μ.

Notice that a z-score always consists of two parts: a sign ($+$ or $-$) and a magnitude. Both parts are necessary to describe completely where a raw score is located within a distribution.

Whenever you are working with z-scores, you should imagine or draw a picture similar to Figure 5.2. Although you should realize that not all distributions are normal, we will use the normal shape as an example when showing z-scores.

Figure 5.2 shows a population distribution with various positions identified by their z-score values. Notice that all z-scores above the mean are positive and all z-scores below the mean are negative. The sign of a z-score tells you immediately whether the score is located above or below the mean. Also, note that a z-score of $z = +1.00$ corresponds to a position exactly 1 standard deviation above the mean. A z-score of $z = +2.00$ is always located exactly 2 standard deviations above the mean. The numerical value of the z-score tells you the number of standard deviations from the mean. Finally, you should notice that Figure 5.2 does not give any specific values for the population mean or the standard deviation. The locations identified by z-scores are the same for *all distributions*, no matter what mean or standard deviation the distributions may have.

FIGURE 5.2

The relationship between z-score values and locations in a population distribution.

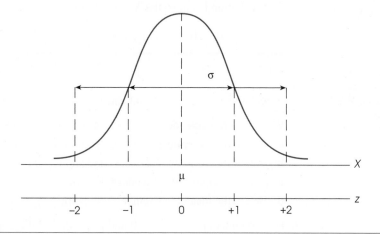

Now we can return to the two distributions shown in Figure 5.1 and use a z-score to describe the position of $X = 76$ within each distribution as follows:

In Figure 5.1(a), the score $X = 76$ corresponds to a z-score of $z = +2.00$. That is, the score is located above the mean by exactly 2 standard deviations.

In Figure 5.1(b), the score $X = 76$ corresponds to a z-score of $z = +0.50$. In this distribution, the score is located above the mean by exactly $\frac{1}{2}$ standard deviation.

The definition of a z-score indicates that each X value has a corresponding z-score. The following examples demonstrate the relationship between X values and z-scores within a distribution.

EXAMPLE 5.2

A distribution of scores has a mean of $\mu = 50$ and a standard deviation of $\sigma = 8$.

a. For this distribution, what is the z-score corresponding to $X = 58$? Because 58 is *above* the mean, the z-score has a positive sign. The score is 8 points greater than the mean. This distance is exactly 1 standard deviation (because $\sigma = 8$), so the z-score is

$$z = +1$$

This z-score indicates that the raw score is located 1 standard deviation above the mean.

    b. What is the z-score corresponding to $X = 46$? The z-score will be negative because 46 is *below* the mean. The X value is 4 points away from the mean. This distance is exactly $\frac{1}{2}$ standard deviation; therefore, the z-score is

$$z = -\tfrac{1}{2}$$

This z-score tells us that the X value is $\frac{1}{2}$ standard deviation below the mean.

    c. For this distribution, what raw score corresponds to a z-score of $+2$? This z-score indicates that the X value is 2 standard deviations above the mean. One standard deviation is 8 points, so 2 standard deviations are 16 points. Therefore, the score we are looking for is 16 points above the mean. The mean for the distribution is 50, so the X value is

$$X = 50 + 16 = 66$$

LEARNING CHECK

1. A negative z-score always indicates a location below the mean. (True or false?)

2. What z-score value identifies each of the following locations in a distribution?
   a. Above the mean by 2 standard deviations
   b. Below the mean by $\frac{1}{2}$ standard deviation
   c. Above the mean by $\frac{1}{4}$ standard deviation
   d. Below the mean by 3 standard deviations

3. For a population with $\mu = 50$ and $\sigma = 10$, find the z-score for each of the following scores:
   a. $X = 55$ b. $X = 40$ c. $X = 30$

4. For a population with $\mu = 50$ and $\sigma = 10$, find the X value corresponding to each of the following z-scores:
   a. $z = +1.00$ b. $z = -0.50$ c. $z = +2.00$

ANSWERS

1. True

2. a. $z = +2$ b. $z = -\frac{1}{2}$ or -0.50 c. $z = +\frac{1}{4}$ or $+0.25$ d. $z = -3$

3. a. $z = +0.50$ b. $z = -1.00$ c. $z = -2.00$

4. a. $X = 60$ b. $X = 45$ c. $X = 70$

THE z-SCORE FORMULA

The z-score definition is adequate for transforming back and forth from X values to z-scores as long as the arithmetic is easy to do in your head. For more complicated values, it is best to have an equation to help structure the calculations. Fortunately, the relationship between X values and z-scores can be easily expressed in a formula. The formula for transforming scores into z-scores is

$$z = \frac{X - \mu}{\sigma} \tag{5.1}$$

The numerator of the equation, $X - \mu$, is a *deviation score* (Chapter 4, page 109); it measures the distance in points between X and μ and indicates whether X is located above or below the mean. We divide this difference by σ because we want the z-score to measure distance in terms of standard deviation units. The formula performs exactly the same arithmetic that is used with the z-score definition, and it provides a structured equation to organize the calculations when the numbers are more difficult. The following examples demonstrate the use of the z-score formula.

EXAMPLE 5.3   A distribution of scores has a mean of $\mu = 100$ and a standard deviation of $\sigma = 10$. What z-score corresponds to a score of $X = 120$ in this distribution?

According to the definition, the z-score will have a value of $+2$ because the score is located above the mean by exactly 2 standard deviations. Using the z-score formula, we obtain

$$z = \frac{X - \mu}{\sigma} = \frac{120 - 100}{10} = \frac{20}{10} = 2.00$$

The formula produces exactly the same result that is obtained using the z-score definition.

EXAMPLE 5.4   A distribution of scores has a mean of $\mu = 86$ and a standard deviation of $\sigma = 7$. What z-score corresponds to a score of $X = 95$ in this distribution?

Note that this problem is not particularly easy, especially if you try to use the z-score definition and perform the calculations in your head. However, the z-score formula organizes the numbers and allows you to finish the final arithmetic with your calculator. Using the formula, we obtain

$$z = \frac{X - \mu}{\sigma} = \frac{95 - 86}{7} = \frac{9}{7} = 1.29$$

According to the formula, a score of $X = 95$ corresponds to $z = 1.29$. The z-score indicates a location that is above the mean (positive) by slightly more than $1\frac{1}{4}$ standard deviations.

DETERMINING A RAW SCORE (X) FROM A z-SCORE   Although the z-score equation (Formula 5.1) works well for transforming X values into z-scores, it can be awkward when you are trying to work in the opposite direction and change z-scores back into X values. Therefore, we will create a different version of the formula that is easier to use when you need to transform z-scores into X values. To develop the new formula, we will begin with a sample problem.

> For a distribution with a mean of $\mu = 60$ and a standard deviation of $\sigma = 5$, what X value corresponds to a z-score of $z = -2.00$?

To solve this problem, we will use the z-score definition and carefully monitor the step-by-step process. The value of the z-score indicates that X is located 2 standard deviations below the mean. Thus, the first step in the calculation is to determine the distance corresponding to 2 standard deviations. This distance is obtained by multiplying the z-score value (-2) by the standard deviation value (5). In symbols,

$$z\sigma = (-2)(5) = -10 \text{ points}$$

(Our score is located below the mean by 10 points.)

The next step is to start at the mean and go down by 10 points to find the value of X. In symbols,

$$X = \mu - 10 = 60 - 10 = 50$$

The two steps can be combined to form a single formula:

$$X = \mu + z\sigma \tag{5.2}$$

Note that the value of $z\sigma$ is the *deviation* of X. In the example, this value was -10, or 10 points below the mean. Formula 5.2 simply combines the mean and the deviation from the mean to determine the exact value of X.

Finally, you should realize that Formula 5.1 and Formula 5.2 are actually two different versions of the same equation. If you begin with either formula and use algebra to shuffle the terms around, you will soon end up with the other formula. We will leave this as an exercise for those who want to try it.

LEARNING CHECK

1. A distribution of scores has a mean of $\mu = 80$ and a standard deviation of $\sigma = 12$. Find the z-score value corresponding to each of the following scores. *Note:* You may use the z-score definition or Formula 5.1 to determine these values.

   a. $X = 98$ **b.** $X = 86$ **c.** $X = 77$ **d.** $X = 75$

2. A distribution of scores has a mean of $\mu = 80$ and a standard deviation of $\sigma = 12$. Find the X value corresponding to each of the following z-scores. *Note:* You may use the z-score definition or Formula 5.2 to determine these values.

   a. $z = -1.25$ **b.** $z = 2.50$ **c.** $z = -0.75$ **d.** $z = 1.00$

ANSWERS

1. **a.** $z = 1.50$ **b.** $z = 0.50$ **c.** $z = -0.25$ **d.** $z = -0.42$

2. **a.** $X = 65$ **b.** $X = 110$ **c.** $X = 71$ **d.** $X = 92$

5.3 USING z-SCORES TO STANDARDIZE A DISTRIBUTION

It is possible to transform every X value in a distribution into a corresponding z-score. The result of this process is that the entire distribution of X values is transformed into a distribution of z-scores (see Figure 5.3). The new distribution of z-scores has characteristics that make the *z-score transformation* a very useful tool. Specifically, if every X value is transformed into a z-score, then the distribution of z-scores will have the following properties:

1. Shape. The shape of the z-score distribution will be the same as the original distribution of raw scores. If the original distribution is negatively skewed, for example, then the z-score distribution will also be negatively skewed. If the original distri-

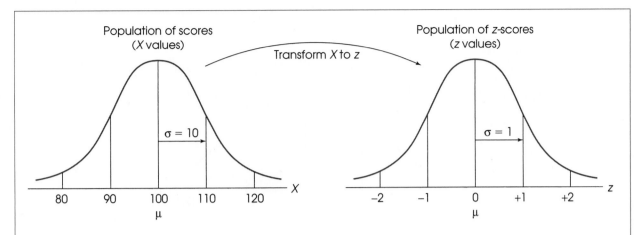

FIGURE 5.3

An entire population of scores is transformed into z-scores. The transformation does not change the shape of the population but the mean is transformed into a value of 0 and the standard deviation is transformed to a value of 1.

bution is normal, the distribution of z-scores will also be normal. Transforming raw scores into z-scores does not change anyone's position in the distribution. For example, any raw score that is above the mean by 1 standard deviation will be transformed to a z-score of +1.00, which is still above the mean by 1 standard deviation. Transforming a distribution from X values to z values does not move scores from one position to another; the procedure simply relabels each score (see Figure 5.3). Because each individual score stays in its same position within the distribution, the overall shape of the distribution does not change.

2. The Mean. The z-score distribution will *always* have a mean of zero. In Figure 5.3, the original distribution of X values has a mean of $\mu = 100$. When this value, $X = 100$, is transformed into a z-score, the result is

$$z = \frac{X - \mu}{\sigma} = \frac{100 - 100}{10} = 0$$

Thus, the original population mean is transformed into a value of zero in the z-score distribution. The fact that the z-score distribution has a mean of zero makes it easy to identify locations. Recall from the definition of z-scores that all positive values are above the mean and all negative values are below the mean. A value of zero makes the mean a convenient reference point.

3. The Standard Deviation. The distribution of z-scores will always have a standard deviation of 1. In Figure 5.3, the original distribution of X values has $\mu = 100$ and $\sigma = 10$. In this distribution, a value of $X = 110$ is above the mean by exactly 10 points or 1 standard deviation. When $X = 110$ is transformed, it becomes $z = +1.00$, which is above the mean by exactly 1 point in the z-score distribution. Thus, the standard deviation corresponds to a 10-point distance in the X distribution and is transformed into a 1-point distance in the z-score distribution. The advantage of having a standard deviation of 1 is that the numerical value of a z-score is exactly the same as the number of standard deviations from the mean. For example, a z-score value of 2 is exactly 2 standard deviations from the mean.

FIGURE 5.4

Following a *z*-score transformation, the *X*-axis is relabeled in *z*-score units. The distance that is equivalent to 1 standard deviation on the *X*-axis ($\sigma = 10$ points in this example) corresponds to 1 point on the *z*-score scale.

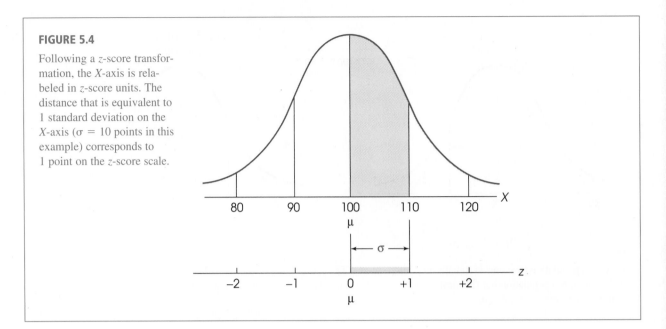

In Figure 5.3, we showed the *z*-score transformation as a process that changed a distribution of *X* values into a new distribution of *z*-scores. In fact, there is no need to create a whole new distribution. Instead, you can think of the *z*-score transformation as simply *relabeling* the values along the *X*-axis. That is, after a *z*-score transformation, you still have the same distribution, but now each individual is labeled with a *z*-score instead of an X value. Figure 5.4 demonstrates this concept with a single distribution that has two sets of labels: the *X* values along one line and the corresponding *z*-scores along another line. Notice that the mean for the distribution of *z*-scores is zero and the standard deviation is 1.

When *any* distribution (with any mean or standard deviation) is transformed into *z*-scores, the resulting distribution will always have a mean of $\mu = 0$ and a standard deviation of $\sigma = 1$. Because all *z*-score distributions have the same mean and the same standard deviation, the *z*-score distribution is called a *standardized distribution*.

DEFINITION   A **standardized distribution** is composed of scores that have been transformed to create predetermined values for μ and σ. Standardized distributions are used to make dissimilar distributions comparable.

A *z*-score distribution is an example of a standardized distribution with $\mu = 0$ and $\sigma = 1$. That is, when any distribution (with any mean or standard deviation) is transformed into *z*-scores, the transformed distribution will always have $\mu = 0$ and $\sigma = 1$. The advantage of standardizing is that it makes it possible to compare different scores or different individuals even though they may come from completely different distributions.

DEMONSTRATION OF A z-SCORE TRANSFORMATION

Although the basic characteristics of a *z*-score distribution have been explained logically, the following example provides a concrete demonstration that a *z*-score transformation will create a new distribution with a mean of zero, a standard deviation of 1, and the same shape as the original population.

EXAMPLE 5.5   We begin with a population of $N = 6$ scores consisting of the following values: 0, 6, 5, 2, 3, 2. This population has a mean of $\mu = \frac{18}{6} = 3$ and a standard deviation of $\sigma = 2$ (check the calculations for yourself).

Each of the X values in the original population is then transformed into a z-score as summarized in the following table.

$X = 0$	Below the mean by $1\frac{1}{2}$ standard deviation	$z = -1.50$
$X = 6$	Above the mean by $1\frac{1}{2}$ standard deviations	$z = +1.50$
$X = 5$	Above the mean by 1 standard deviation	$z = +1.00$
$X = 2$	Below the mean by $\frac{1}{2}$ standard deviation	$z = -0.50$
$X = 3$	Exactly equal to the mean—no deviation	$z = 0$
$X = 2$	Below the mean by $\frac{1}{2}$ standard deviation	$z = -0.50$

The frequency distribution for the original population of X values is shown in Figure 5.5(a) and the corresponding distribution for the z-scores is shown in Figure 5.5(b). A simple comparison of the two distributions demonstrates the results of a z-score transformation.

1. The two distributions have exactly the same shape. Each individual has exactly the same relative position in the X distribution and in the z-score distribution.

2. After the transformation to z-scores, the mean of the distribution becomes $\mu = 0$. The individual with a score of $X = 3$ is located exactly at the mean in the X distribution and this individual is transformed into $z = 0$, exactly at the mean in the z-score distribution.

3. After the transformation, the standard deviation becomes $\sigma = 1$. For example, in the original distribution of scores, the individual with $X = 5$ is located above

FIGURE 5.5

Transforming a distribution of raw scores (top) into z-scores (bottom) will not change the shape of the distribution.

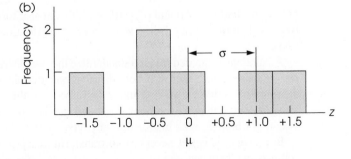

the mean by 2 points, which is a distance of exactly 1 standard deviation. After transformation, this same individual is located above the mean by 1 point, which is exactly 1 standard deviation in the z-score distribution.

Finally, this example provides an opportunity to point out some other characteristics of z-scores. First, you should realize that each z-score is also a deviation score. For example, $z = -1.50$ is below the mean by 1.50 points. The fact that each z-score is its own deviation score has two interesting consequences:

1. For a z-score distribution, $\Sigma z = 0$.
2. For a z-score distribution, $\Sigma z^2 = SS = N$.

The first point is relatively easy to explain because the sum of the deviations is always zero for any distribution. The second point begins with the fact that SS is the sum of the squared deviations, hence $SS = \Sigma z^2$. The fact that this sum is equal to N requires a few logical steps. First, for a distribution of z-scores, the standard deviation is 1, which means that the variance is also 1. Next, we take the formula for variance and substitute the equality $\Sigma z^2 = SS$. The result is

$$\sigma^2 = 1 = \frac{SS}{N} = \frac{\Sigma z^2}{N}$$

For the final fraction to equal 1, it is necessary that $\Sigma z^2 = N$.

For the simple distribution of z-scores in Example 5.5 (see Figure 5.5), there are $N = 6$ scores, and the sum of the squared z-scores is

$$\Sigma z^2 = (-1.50)^2 + (1.50)^2 + (1.00)^2 + (-0.50)^2 + 0^2 + (-0.50)^2$$

$$= 2.25 + 2.25 + 1.00 + 0.25 + 0 + 0.25$$

$$= 6.00$$

USING z-SCORES FOR MAKING COMPARISONS

When two scores come from different distributions, it is impossible to make any direct comparison between them. Suppose, for example, Bob received a score of $X = 60$ on a psychology exam and a score of $X = 56$ on a biology test. For which course should Bob expect the better grade?

Because the scores come from two different distributions, you cannot make any direct comparison. Without additional information, it is even impossible to determine whether Bob is above or below the mean in either distribution. Before you can begin to make comparisons, you must know the values for the mean and standard deviation for each distribution. Suppose the biology scores had $\mu = 48$ and $\sigma = 4$, and the psychology scores had $\mu = 50$ and $\sigma = 10$. With this new information, you could sketch the two distributions, locate Bob's score in each distribution, and compare the two locations.

An alternative procedure is to standardize the two distributions by transforming both sets of scores into z-scores. If both distributions are transformed to z-scores, then both distributions will have $\mu = 0$ and $\sigma = 1$. Because the two distributions are now the same (at least they have the same mean and standard deviation) we can compare Bob's z-score for biology with his z-score for psychology.

In practice, it is not necessary to transform every score in a distribution to make comparisons between two scores. We need to transform only the two scores in question. In Bob's case, we must find the z-scores for his psychology and biology scores. For psychology, Bob's z-score is

Be sure to use the μ and σ values for the distribution to which X belongs.

$$z = \frac{X - \mu}{\sigma} = \frac{60 - 50}{10} = \frac{10}{10} = +1.0$$

For biology, Bob's z-score is

$$z = \frac{56 - 48}{4} = \frac{8}{4} = +2.0$$

Note that Bob's z-score for biology is $+2.0$, which means that his test score is 2 standard deviations above the class mean. On the other hand, his z-score is $+1.0$ for psychology, or 1 standard deviation above the mean. In terms of relative class standing, Bob is doing much better in the biology class. Notice that it is meaningful to make a direct comparison of the two z-scores. A z-score of $+2.00$ *always* indicates a higher position than a z-score of $+1.00$ because all z-score values are based on a standardized distribution with $\mu = 0$ and $\sigma = 1$.

LEARNING CHECK

1. A normal-shaped distribution with $\mu = 50$ and $\sigma = 8$ is transformed into z-scores. Describe the shape, mean, and standard deviation for the z-score distribution.

2. Why is it possible to compare scores from different distributions after the distributions have been transformed into z-scores?

3. A set of mathematics exam scores has $\mu = 70$ and $\sigma = 8$. A set of English exam scores has $\mu = 74$ and $\sigma = 16$. For which exam would a score of $X = 78$ have a higher standing?

4. What is the advantage of having $\mu = 0$ for a distribution of z-scores?

ANSWERS

1. The z-score distribution would be normal (same shape) with $\mu = 0$ and $\sigma = 1$.

2. Comparison is possible because the two distributions will have the same mean ($\mu = 0$) and the same standard deviation ($\sigma = 1$) after transformation.

3. $z = 1.00$ for the mathematics exam, which is a higher standing than $z = 0.25$ for the English exam.

4. With $\mu = 0$, you know immediately that any positive value is above the mean and any negative value is below the mean.

5.4 OTHER STANDARDIZED DISTRIBUTIONS BASED ON z-SCORES

TRANSFORMING z-SCORES TO A DISTRIBUTION WITH A PREDETERMINED μ AND σ

Although z-score distributions have distinct advantages, many people find them cumbersome because they contain negative values and decimals. For this reason, it is common to standardize a distribution by transforming z-scores to a distribution with a predetermined mean and standard deviation that are whole round numbers. The goal is to create a new (standardized) distribution that has "simple" values for the mean and standard deviation but does not change any individual's location within the distribution. Standardized scores of this type are frequently used in psychological or educational testing. For example, raw scores of the Scholastic Aptitude Test (SAT) are transformed to a standardized distribution that has $\mu = 500$ and $\sigma = 100$. For intelligence tests, raw scores are frequently converted to standard scores that have a mean of 100 and a standard deviation of 15. Because most IQ tests are standardized so that they have the same mean and standard deviation, it is possible to compare IQ scores even though they may come from different tests.

The procedure for standardizing a distribution to create new values for μ and σ involves two-steps:

1. The original raw scores are transformed into *z*-scores.

2. The *z*-scores are then transformed into new *X* values so that the specific μ and σ are attained.

This procedure ensures that each individual has exactly the same *z*-score location in the new distribution as in the original distribution. The following example demonstrates the standardization procedure.

EXAMPLE 5.6   An instructor gives an exam to a psychology class. For this exam, the distribution of raw scores has a mean of $\mu = 57$ with $\sigma = 14$. The instructor would like to simplify the distribution by transforming all scores into a new, standardized distribution with $\mu = 50$ and $\sigma = 10$. To demonstrate this process, we will consider what happens to two specific students: Maria, who has a raw score of $X = 64$ in the original distribution; and Joe, whose original raw score is $X = 43$.

STEP 1   Transform each of the original raw scores into *z*-scores. For Maria, $X = 64$, so her *z*-score is

$$z = \frac{X - \mu}{\sigma} = \frac{64 - 57}{14} = +0.5$$

Remember: The values of μ and σ are for the distribution from which *X* was taken.

For Joe, $X = 43$, and his *z*-score is

$$z = \frac{X - \mu}{\sigma} = \frac{43 - 57}{14} = -1.0$$

STEP 2   Change the *z*-scores into standardized scores so that the new distribution has a mean of $\mu = 50$ and a standard deviation of $\sigma = 10$.

Maria's *z*-score, $z = +0.50$, indicates that she is located above the mean by $\frac{1}{2}$ standard deviation. In the new, standardized distribution, this location corresponds to $X = 55$ (above the mean by 5 points).

Joe's *z*-score, $z = -1.00$, indicates that he is located below the mean by exactly 1 standard deviation. In the new distribution, this location corresponds to $X = 40$ (below the mean by 10 points).

The results of this two-step transformation process are summarized in Table 5.1. Note that Joe, for example, has exactly the same *z*-score ($z = -1.00$) in both the original distribution and the new standardized distribution. This means that Joe's position relative to the other students in the class has not changed.

TABLE 5.1

A demonstration of how two individual scores are changed when a distribution is standardized. See Example 5.6

	ORIGINAL SCORES $\mu = 57$ AND $\sigma = 14$	Z-SCORE LOCATION	STANDARDIZED SCORES $\mu = 50$ AND $\sigma = 10$
Maria	$X = 64 \longrightarrow$	$z = +0.50 \longrightarrow$	$X = 55$
Joe	$X = 43 \longrightarrow$	$z = -1.00 \longrightarrow$	$X = 40$

Figure 5.6 provides another demonstration of the concept that standardizing a distribution does not change the individual positions within the distribution. The figure shows the original exam scores from Example 5.6, with a mean of $\mu = 57$ and a standard deviation of $\sigma = 14$. In the original distribution, Joe is located at a score of $X = 43$. In addition to the original scores, we have included a second scale showing the z-score value for each location in the distribution. In terms of z-scores, Joe is located at a value of $z = -1.00$. Finally, we have added a third scale showing the standardized scores where the mean is $\mu = 50$ and the standard deviation is $\sigma = 10$. For the standardized scores, Joe is located at $X = 40$. Note that Joe is always in the same place in the distribution. The only thing that changes is the number that is assigned to Joe: For the original scores, Joe is at 43; for the z-scores, Joe is at -1.00; and for the standardized scores, Joe is at 40.

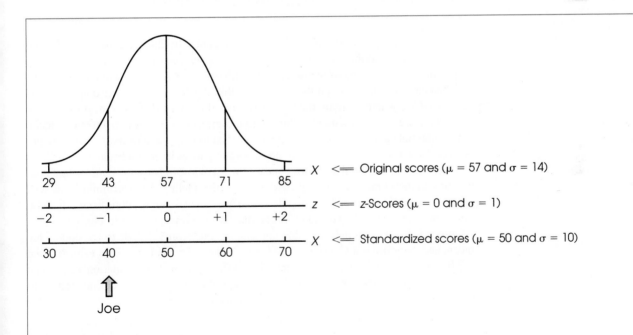

FIGURE 5.6

The distribution of exam scores from Example 5.6. The original distribution was standardized to produce a new distribution with $\mu = 50$ and $\sigma = 10$. Note that each individual is identified by an original score, a z-score, and a new, standardized score. For example, Joe has an original score of 43, a z-score of -1.00, and a standardized score of 40.

LEARNING CHECK

1. A population has $\mu = 37$ and $\sigma = 2$. If this distribution is transformed into a new distribution with $\mu = 100$ and $\sigma = 20$, what new values will be obtained for each of the following scores: 35, 36, 37, 38, 39?

2. For the following population, $\mu = 7$ and $\sigma = 4$. The scores are 2, 4, 6, 10, 13.
   a. Transform this distribution so $\mu = 50$ and $\sigma = 20$.
   b. Compute μ and σ for the new population. (You should obtain $\mu = 50$) and $\sigma = 20$.)

ANSWERS

1. The five scores 35, 36, 37, 38, and 39 are transformed into 80, 90, 100, 110, and 120, respectively.

2. a. The original scores, 2, 4, 6, 10, and 13, are transformed into 25, 35, 45, 65, and 80, respectively.
   b. The new scores add up to $\Sigma X = 250$, so the mean is $\frac{250}{5} = 50$. The SS for the transformed scores is 2000, the variance is 400, and the new standard deviation is 20.

5.5   LOOKING AHEAD TO INFERENTIAL STATISTICS

The preceding chapters focused on descriptive statistics. In Chapter 2, for example, you learned how to describe an entire distribution by presenting the scores in a table or a graph. In Chapters 3 and 4, we described distributions of scores by computing means and standard deviations. Even the z-scores in this chapter were presented as a method for *describing* where an individual score is located within a distribution. As we move forward, however, the focus will shift from descriptive to inferential statistics. Recall that inferential statistics involve using sample data to help make decisions about populations. As you will see, z-scores are an important part of inferential statistics.

Once again, inferential statistics concern the relationship between samples and populations. In this context, z-scores provide an objective method for determining how well a specific sample represents its population. If we select a single score from a population, a sample of $n = 1$, we can calculate the z-score corresponding to our X value. A z-score value near zero indicates that our sample (our score) is close to the population mean, and therefore is a good representative. On the other hand, a z-score value beyond $+2.00$ or -2.00 is an extreme value, very different from the population mean, and therefore not a representative value. The concept of using z-scores to define "representative" and "nonrepresentative" samples is shown in Figure 5.7.

USING z-SCORES IN A RESEARCH STUDY

Usually we will use inferential statistics to help interpret the results from a research study. A researcher typically begins with a general question about how a treatment will affect the individuals in a population. A sample is obtained, and the sample is used in a research study to evaluate the effect of the treatment. Figure 5.8 shows the general research situation we have described.

To decide whether or not the treatment has an effect, the researchers will compare the sample (after treatment) with the original population. If the treated sample is noticeably different from the original population, then the researcher has evidence that the

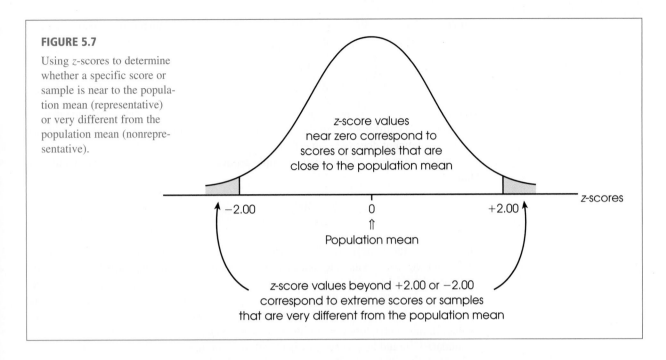

FIGURE 5.7

Using *z*-scores to determine whether a specific score or sample is near to the population mean (representative) or very different from the population mean (nonrepresentative).

z-score values near zero correspond to scores or samples that are close to the population mean

−2.00   0   +2.00   z-scores

⇑

Population mean

z-score values beyond +2.00 or −2.00 correspond to extreme scores or samples that are very different from the population mean

treatment does have an effect. On the other hand, if the sample is still similar to the original population, then the researcher must conclude that there is no clear evidence that the treatment has an effect. One issue in this process is deciding whether the sample is "still similar" to the population or is "noticeably different" from the population. We can help to clarify this issue with the help of *z*-scores. Specifically, we can compute the *z*-score corresponding to the sample, and then use the *z*-score value to make a decision. For example, in Figure 5.8, the original population has a mean of $\mu = 400$

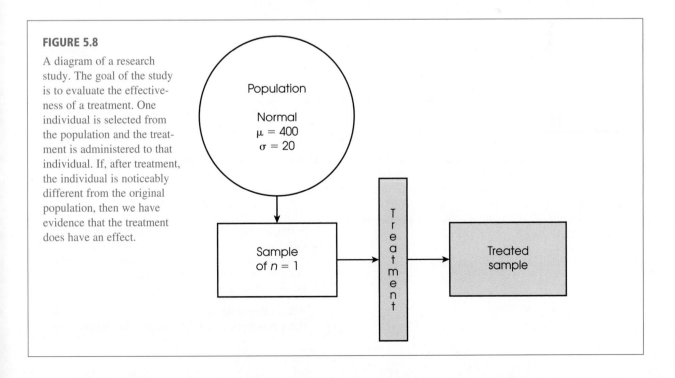

FIGURE 5.8

A diagram of a research study. The goal of the study is to evaluate the effectiveness of a treatment. One individual is selected from the population and the treatment is administered to that individual. If, after treatment, the individual is noticeably different from the original population, then we have evidence that the treatment does have an effect.

Population

Normal
$\mu = 400$
$\sigma = 20$

Sample
of $n = 1$

Treatment

Treated
sample

and a standard deviation of $\sigma = 20$. If our sample (after treatment) produced a score of $X = 450$, then we would compute

$$z = \frac{X - \mu}{\sigma} = \frac{450 - 400}{20} = \frac{50}{20} = 2.50$$

Because this is an extreme value for a *z*-score (see Figure 5.7), we can conclude that the sample is very different from the original population and therefore provides evidence that the treatment does have an effect.

On the other hand, if our treated sample produced a value of $X = 405$, we would obtain

$$z = \frac{X - \mu}{\sigma} = \frac{405 - 400}{20} = \frac{5}{20} = 0.25$$

In this case, the sample is still very close to the original population mean (see Figure 5.7), and we would conclude that there is no evidence that the treatment has an effect.

For this example, we have used *z*-score values of ± 2.00 to set the boundaries between samples that are still similar and samples that are noticeably different from the population. This choice was purely arbitrary; that is, we could have selected any *z*-score value. In future chapters we will introduce a rationale for determining exactly where the boundaries should be placed. In Chapter 6, for example, we will introduce the concept of probability. This will allow us to determine that a sample with a *z*-score beyond 2.00 is not only an extreme value, it also is extremely unlikely.

Finally, you should realize that the research study shown in Figure 5.8 is somewhat oversimplified. Among other things, the study uses a sample of only $n = 1$ individual. In a real study, a researcher would probably use a group of $n = 25$ or $n = 100$ individuals. In Chapter 7 we will develop statistical techniques that allow for any size sample. Specifically, we will be able to compute *z*-scores and find probabilities for any sample, no matter how many individuals are involved.

SUMMARY

1. Each X value can be transformed into a *z*-score that specifies the exact location of X within the distribution. The sign of the *z*-score indicates whether the location is above (positive) or below (negative) the mean. The numerical value of the *z*-score specifies the number of standard deviations between X and μ.

2. The *z*-score formula is used to transform X values into *z*-scores:

$$z = \frac{X - \mu}{\sigma}$$

3. To transform *z*-scores back into X values, solve the *z*-score equation for X:

$$X = \mu + z\sigma$$

4. When an entire distribution of X values is transformed into *z*-scores, the result is a distribution of *z*-scores. The *z*-score distribution will have the same shape as the distribution of raw scores, and it always will have a mean of 0 and a standard deviation of 1.

5. When comparing raw scores from different distributions, it is necessary to standardize the distributions

with a z-score transformation. The distributions will then be comparable because they will have the same parameters ($\mu = 0$, $\sigma = 1$). In practice, it is necessary to transform only those raw scores that are being compared.

6. In certain situations, such as psychological testing, a distribution may be standardized by converting the original X values into z-scores and then converting the z-scores into a new distribution of scores with predetermined values for the mean and the standard deviation.

7. In inferential statistics, z-scores provide an objective method for determining how well a specific score (a sample of $n = 1$) represents its population. A z-score near zero indicates that the score is close to the population mean, and therefore is representative. A z-score beyond ± 2.00 indicates that the score is very different from the population mean, and therefore is not representative.

RESOURCES

WebTUTOR

If you are using WebTutor along with this book, there is a WebTutor section corresponding to this chapter. The WebTutor contains a brief summary of Chapter 5, hints for learning about z-scores, cautions about common errors, and sample exam items including solutions.

SPSS

The statistics computer package SPSS can be used to calculate the z-scores corresponding to a set of X values. First, enter the scores (the X values) in one column of the data matrix and then click **Analyze** on the tool bar. Then, select **Descriptive Statistics** and select **Descriptives**. Enter the column label for the set of scores into the Variable box. Select the option to **Save standardized values as variables**, and click **OK**. SPSS will produce a summary table listing the maximum score, the minimum score, the mean, and the standard deviation. In addition, if you go back to the SPSS Data Editor, you will find a new column in the data matrix showing the z-score corresponding to each X value.

Caution: The SPSS program computes the mean and standard deviation assuming that the set of scores is a sample. Thus, the z-scores are computed using the sample standard deviation instead of the population standard deviation. For most applications, these "sample z-scores" serve the same purpose as regular z-scores; that is, they standardize the distribution so that the sample mean is zero and the sample standard deviation is one, and they identify specific locations within the sample distribution. However, if your set of scores is intended to be a population, SPSS will not produce the correct z-score values. You can fix the SPSS values by multiplying each z-score by the square root of $n/(n - 1)$.

KEY TERMS

raw score	deviation score	standardized distribution
z-score	z-score transformation	standard score

FOCUS ON PROBLEM SOLVING

1. When you are converting an X value to a z-score (or vice versa), do not rely entirely on the formula. You can avoid careless mistakes if you use the definition of a z-score (sign and numerical value) to make a preliminary estimate of the answer before you begin computations. For example, a z-score of $z = -0.85$ identifies a score located *below* the mean by almost 1 standard deviation. When computing the X value for this z-score, be sure that your answer is smaller than the mean, and check that the distance between X and μ is slightly less than the standard deviation.

   A common mistake when computing z-scores is to forget to include the sign of the z-score. The sign is determined by the deviation score ($X - \mu$) and should be carried through all steps of the computation. If, for example, the correct z-score is $z = -2.0$, then an answer of $z = 2.0$ is wrong. In the first case, the raw score is 2 standard deviations *below* the mean. But the second (and incorrect) answer indicates that the X value is 2 standard deviations *above* the mean. These are clearly different answers, and only one can be correct. What is the best advice on how to avoid careless errors? Sketch the distribution, showing the mean and the raw score (or z-score) in question. This way you will have a concrete frame of reference for each problem.

2. When comparing scores from distributions that have different standard deviations, it is important to be sure that you use the correct value for σ in the z-score formula. Use the σ value for the distribution from which the raw score in question was taken.

3. Remember that a z-score specifies a relative position within the context of a specific distribution. A z-score is a relative value, not an absolute value. For example, a z-score of $z = -2.0$ does not necessarily suggest a very low raw score—it simply means that the raw score is among the lowest within that specific group.

DEMONSTRATION 5.1

TRANSFORMING *X* VALUES INTO *z*-SCORES

A distribution of scores has a mean of $\mu = 60$ with $\sigma = 12$. Find the z-score for $X = 75$.

STEP 1 Determine the sign of the z-score.

First, determine whether X is above or below the mean. This will determine the sign of the z-score. For this demonstration, X is larger than (above) μ, so the z-score will be positive.

STEP 2 Find the distance between X and μ.

The distance is obtained by computing a deviation score.

$$\text{deviation score} = X - \mu = 75 - 60 = 15$$

Thus, the score, $X = 75$, is 15 points above μ.

STEP 3 Convert the distance to standard deviation units.

Converting the distance from step 2 to σ units is accomplished by dividing the distance by σ. For this demonstration,

$$\frac{15}{12} = 1.25$$

Thus, $X = 75$ is 1.25 standard deviations from the mean.

STEP 4   Combine the sign from step 1 with the number from step 2.

The raw score is above the mean, so the z-score must be positive (step 1). For these data,

$$z = +1.25$$

In using the z-score formula, the sign of the z-score will be determined by the sign of the deviation score, $X - \mu$. If X is larger than μ, then the deviation score will be positive. However, if X is smaller than μ, then the deviation score will be negative. For this demonstration, formula 5.1 is used as follows:

$$z = \frac{X - \mu}{\sigma} = \frac{75 - 60}{12} = \frac{+15}{12} = +1.25$$

DEMONSTRATION 5.2

CONVERTING z-SCORES TO X VALUES

For a population with $\mu = 60$ and $\sigma = 12$, what is the X value corresponding to $z = -0.50$?

Notice that in this situation we know the z-score and must find X.

STEP 1   Locate X in relation to the mean.

The sign of the z-score is negative. This tells us that the X value we are looking for is below μ.

STEP 2   Determine the distance from the mean (deviation score).

The magnitude of the z-score tells us how many standard deviations there are between X and μ. In this case, X is $\frac{1}{2}$ standard deviation from the mean. In this distribution, 1 standard deviation is 12 points ($\sigma = 12$). Therefore, X is one-half of 12 points from the mean, or

$$(0.5)(12) = 6 \text{ points}$$

STEP 3   Find the X value.

Starting with the value of the mean, use the direction (step 1) and the distance (step 2) to determine the X value. For this demonstration, we want to find the score that is 6 points below $\mu = 60$. Therefore,

$$X = 60 - 6 = 54$$

Equation 5.2 is used to convert a z-score to an X value. For this demonstration, we obtain the following, using the formula:

$$X = \mu + z\sigma$$
$$= 60 + (-0.50)(12)$$
$$= 60 + (-6) = 60 - 6$$
$$= 54$$

Notice that the sign of the z-score determines whether the deviation score is added to or subtracted from the mean.

PROBLEMS

1. Describe exactly what information is provided by a z-score.

2. A positively skewed distribution has $\mu = 80$ and $\sigma = 12$. If this entire distribution is transformed into z-scores, describe the shape, mean, and standard deviation for the resulting distribution of z-scores.

3. A distribution of scores has a mean of $\mu = 70$. In this distribution, a score of $X = 80$ is located 10 points above the mean.
   a. Assume that the standard deviation is $\sigma = 5$. Sketch the distribution, and locate the position of $X = 80$. What z-score corresponds to $X = 80$?
   b. Now assume that the standard deviation is $\sigma = 20$. Again, sketch the distribution, and locate the position of $X = 80$. Now what z-score corresponds to $X = 80$?

4. For a distribution of raw scores, $\mu = 45$. The z-score for $X = 55$ is computed, and a value of $z = -2.00$ is obtained. Regardless of the value for the standard deviation, why must this z-score be incorrect?

5. For a population with $\mu = 100$ and $\sigma = 20$,
   a. Find the z-score corresponding to each of the following X values:

   $X = 110$ $X = 140$ $X = 80$

   $X = 60$ $X = 105$ $X = 130$

   b. Find the X value corresponding to each of the following z-scores:

   $z = 2.50$ $z = -0.50$ $z = -1.50$

   $z = 0.25$ $z = -1.00$ $z = 0.75$

6. For a population with $\mu = 30$ and $\sigma = 4$,
   a. Find the z-score corresponding to each of the following X values:

   $X = 34$ $X = 36$ $X = 38$

   $X = 28$ $X = 22$ $X = 24$

   b. Find the X value corresponding to each of the following z-scores:

   $z = 0.75$ $z = 1.50$ $z = 2.50$

   $z = -0.25$ $z = -1.00$ $z = -1.25$

7. A population of scores has $\mu = 60$ and $\sigma = 12$. Find the z-score corresponding to each of the following X values from this population:

   $X = 66$ $X = 48$ $X = 84$

   $X = 55$ $X = 70$ $X = 75$

8. A population of scores has $\mu = 85$ and $\sigma = 20$. Find the raw score (X value) corresponding to each of the following z-scores in this population:

   $z = 1.25$ $z = -2.30$ $z = -1.50$

   $z = 0.60$ $z = -0.40$ $z = 2.10$

9. For a population with $\mu = 80$ and $\sigma = 20$,
   a. Find the z-score for each of the following X values. (*Note:* You should be able to find these values using the definition of a z-score. You should not need to use a formula or do any serious calculations.)

   $X = 75$ $X = 90$ $X = 110$

   $X = 95$ $X = 60$ $X = 40$

   b. Find the score (X value) that corresponds to each of the following z-scores. (*Note:* You should be able to find these values using the definition of a z-score. You should not need to use a formula or do any serious calculations.)

   $z = 2.50$ $z = -0.50$ $z = -1.50$

   $z = 0.25$ $z = -0.75$ $z = 1.00$

10. For a population with $\mu = 45$ and $\sigma = 7$, find the z-score for each of the following X values. (*Note:* You probably will need to use the formula and a calculator to find these values.)

    $X = 47$ $X = 35$ $X = 40$

    $X = 60$ $X = 55$ $X = 42$

11. A distribution of scores has a standard deviation of $\sigma = 10$. Find the z-score corresponding to each of the following values.
    a. A score that is 20 points above the mean.
    b. A score 10 points below the mean.
    c. A score 15 points above the mean.
    d. A score 30 points below the mean.

12. A score that is 12 points above the mean corresponds to a z-score of $z = +2.00$. What is the standard deviation for this population?

13. For a population with $\mu = 90$, a raw score of $X = 93$ corresponds to $z = +0.50$. What is the standard deviation for this distribution?

14. For a population with $\mu = 60$, a score of $X = 52$ corresponds to $z = -2.00$. What is the standard deviation for this distribution?

15. For a population with $\sigma = 12$, a score of $X = 87$ corresponds to $z = -0.25$. What is the mean for this distribution?

16. For a population with $\sigma = 5$, a score of $X = 55$ corresponds to $z = 2.00$. What is the mean for this distribution?

17. For a population of exam scores, a score of $X = 58$ corresponds to $z = +0.50$ and a score of $X = 46$ corresponds to $z = -1.00$. Find the mean and the standard deviation for the population. (*Hint:* Sketch the distribution and locate the two scores in your sketch.)

18. In a distribution of scores, a raw score of $X = 43$ corresponds to $z = 1.00$ and a score of $X = 49$ corresponds to $z = 2.00$. Find the mean and the standard deviation for the distribution of scores.

19. A population has $\mu = 80$.
   a. If the standard deviation is $\sigma = 2$, would a score of $X = 86$ be described as a central score or an extremely high score?
   b. If the standard deviation is $\sigma = 12$, would a score of $X = 86$ be described as a central score or an extremely high score?

20. Suppose that you have a score of $X = 55$ on an exam with $\mu = 50$. Which standard deviation would give you the better grade: $\sigma = 5$ or $\sigma = 10$?

21. Answer the question as in Problem 20, but this time assume that the mean for the exam is $\mu = 60$.

22. On Tuesday afternoon, Bill earned a score of $X = 73$ on an English test with $\mu = 65$ and $\sigma = 8$. The same day, John earned a score of $X = 63$ on a math test with $\mu = 57$ and $\sigma = 3$. Who should expect the better grade, Bill or John? Explain your answer.

23. Mary had two exams last week. She got a score of $X = 48$ on a statistics exam where the class averaged $\mu = 50$ with $\sigma = 8$, and she got a score of $X = 68$ on a biology exam where the class average was $\mu = 74$ with $\sigma = 3$. For which exam should Mary expect the better grade? Explain your answer.

24. A distribution with $\mu = 65$ and $\sigma = 6$ is being standardized to produce a new mean of $\mu = 50$ and a new standard deviation of $\sigma = 10$. Find the new standardized value for each of the following scores from the original distribution: 68, 59, 77, and 56.

25. A distribution with $\mu = 38$ and $\sigma = 4$ is being standardized to produce a new mean of $\mu = 100$ and a new standard deviation of $\sigma = 20$. Find the new standardized value for each of the following scores from the original distribution: 34, 36, 37, 41, and 46.

26. A population consists of the following scores:

$$12, \quad 1, \quad 10, \quad 3, \quad 7, \quad 3$$

   a. Compute μ and σ for the population.
   b. Find the z-score for each raw score in the population.
   c. Transform each score into a new standardized value so that the standardized distribution has a mean of $\mu = 100$ and a standard deviation of $\sigma = 20$.

27. A population consists of the following $N = 6$ scores: 0, 4, 6, 1, 3, and 4.
   a. Compute μ and σ for the population.
   b. Find the z-score for each score in the population.
   c. Transform the original population into a new population of $N = 6$ scores with a mean of $\mu = 50$ and a standard deviation of $\sigma = 10$.

28. A population consists of the following $N = 5$ scores: 0, 6, 4, 3, and 12.
   a. Compute μ and σ for the population.
   b. Find the z-score for each score in the population.
   c. Transform the original population into a new population of $N = 5$ scores with a mean of $\mu = 60$ and a standard deviation of $\sigma = 8$.

C H A P T E R

6

Probability

Tools You Will Need

The following items are considered essential background material for this chapter. If you doubt your knowledge of any of these items, you should review the appropriate chapter or section before proceeding.

- Proportions (math review, Appendix A)
  - Fractions
  - Decimals
  - Percentages
- Basic algebra (math review, Appendix A)
- Upper and lower real limits (Chapters 1 and 2)
- Percentiles and percentile ranks (Chapter 2)
- Quartiles and semi-interquartile range (Chapter 4)
- z-Scores (Chapter 5)

Preview

If you were to read a novel or a newspaper (or this entire textbook), which of the following would you be more likely to encounter:

1. A word beginning with the letter *K*?
2. A word with a *K* as its third letter?

If you think about this question and answer honestly, you probably will decide that words beginning with a *K* are more probable.

A similar question was asked a group of participants in an experiment reported by Tversky and Kahneman (1973). Their participants estimated that words beginning with *K* are twice as likely as words with a *K* as the third letter. In truth, the relationship is just the opposite. There are more than twice as many words with a *K* in the third position as there are words beginning with a *K*. How can people be so wrong? Do they completely misunderstand probability?

When you were deciding which type of *K* words are more likely, you probably searched your memory and tried to estimate which words are more common. How many words can you think of that start with the letter *K*? How many words can you think of that have a *K* as the third letter? Because you have had years of practice alphabetiz-

ing words according to their first letter, you should find it much easier to search your memory for words beginning with a *K* than to search for words with a *K* in the third position. Consequently, you would conclude that first-letter *K* words are more common and are, therefore, more likely to occur in a book.

Notice that when you use the strategy of counting words during your search, you are estimating their frequencies. From these frequencies, you estimate the proportions of these words in the population of words. Most people think of probability in this way—as a proportion based on how often an outcome occurs. As you will see in this chapter, this idea is a perfectly reasonable approach to probability. In fact, you will see that probability and proportion often are interchangeable concepts.

As for the Tversky and Kahneman study, your error in judging the relative probabilities of *K* words was not due to a misunderstanding of probability. Instead, you simply were misled by the availability of the two types of words in your memory. If you had actually searched through the words in this text (instead of those in your memory), you probably would have found more third-letter *K* words, and you would have concluded (correctly) that these words are more likely.

6.1 OVERVIEW

In Chapter 1, we introduced the idea that research studies begin with a general question about an entire population, but the actual research is conducted using a sample. In this situation, the role of inferential statistics is to use the sample data as the basis for answering questions about the population. To accomplish this goal, inferential procedures are typically built around the concept of probability. Specifically, the relationships between samples and populations are usually defined in terms of probability. Suppose, for example, you are selecting a single marble from a jar the contains 50 black and 50 white marbles. (In this example, the jar of marbles is the *population* and the single marble to be selected is the *sample*.) Although you cannot guarantee the exact outcome of your sample, it is possible to talk about the potential outcomes in terms of probabilities. In this case, you have a fifty-fifty chance of getting either color. Now consider another jar (population) that has 90 black and only 10 white marbles. Again, you cannot specify the exact outcome of a sample, but now you know that the sample probably will be a black marble. By knowing the makeup of a population, we can determine the probability of obtaining specific samples. In this way, probability gives us a connection between populations and samples, and this connection will be the foundation for the inferential statistics to be presented in the chapters that follow.

You may have noticed that the preceding examples begin with a population and then use probability to describe the samples that could be obtained. This is exactly backward from what we want to do with inferential statistics. Remember: The goal of inferential statistics is to begin with a sample and then answer general questions about the population. We will reach this goal in a two-stage process. In the first stage, we develop probability as a bridge from population to samples. This stage involves identifying the types of samples that probably would be obtained from a specific population. Once this bridge is established, we simply reverse the probability rules to allow us to move from samples to populations (see Figure 6.1). The process of reversing the probability relationship can be demonstrated by considering again the two jars of marbles we looked at earlier. (One jar has 50 black and 50 white marbles; the other jar has 90 black and only 10 white marbles.) This time, suppose that you are blindfolded when the sample is selected and that your task is to use the sample to help you to decide which jar was used. If you select a sample of $n = 4$ marbles and all are black, where did the sample come from? It should be clear that it would be relatively unlikely (low probability) to obtain this sample from jar 1; in four draws, you almost certainly would get at least 1 white marble. On the other hand, this sample would have a high probability of coming from jar 2, where nearly all the marbles are black. Your decision, therefore, is that the sample probably came from jar 2. Notice that you now are using the sample to make an inference about the population.

FIGURE 6.1

The role of probability in inferential statistics. Probability is used to predict what kind of samples are likely to be obtained from a population. Thus, probability establishes a connection between samples and populations. Inferential statistics rely on this connection when they use sample data as the basis for making conclusions about populations.

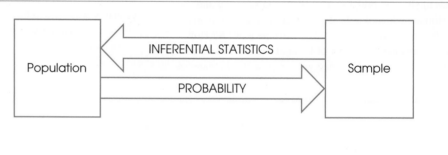

6.2   INTRODUCTION TO PROBABILITY

Probability is a huge topic that extends far beyond the limits of introductory statistics, and we will not attempt to examine it all here. Instead, we will concentrate on the few concepts and definitions that are needed for an introduction to inferential statistics. We begin with a relatively simple definition of probability.

DEFINITION   For a situation in which several different outcomes are possible, we define the **probability** for any particular outcome as a fraction or proportion. If the possible outcomes are identified as A, B, C, D, and so on, then

$$\text{probability of } A = \frac{\text{number of outcomes classified as } A}{\text{total number of possible outcomes}}$$

For example, if you are selecting a card from a complete deck, there are 52 possible outcomes. The probability of selecting the king of hearts is $p = \frac{1}{52}$. The probability of selecting an ace is $p = \frac{4}{52}$ because there are 4 aces in the deck.

To simplify the discussion of probability, we will use a notation system that eliminates a lot of the words. The probability of a specific outcome will be expressed with a lowercase p (for probability) followed by the specific outcome in parentheses. For example, the probability of selecting a king from a deck of cards will be written as p(king). The probability of obtaining heads for a coin toss will be written as p(heads).

You should note that probability is defined as a proportion. This definition makes it possible to restate any probability problem as a proportion problem. For example, the probability problem "What is the probability of obtaining a king from a deck of cards?" can be restated as "Out of the whole deck, what proportion are kings?" In each case, the answer is $\frac{4}{52}$, or "4 out of 52." This translation from probability to proportion may seem trivial now, but it will be a great aid when the probability problems become more complex. In most situations, we are concerned with the probability of obtaining a particular sample from a population. The terminology of *sample* and *population* will not change the basic definition of probability. For example, the whole deck of cards can be considered as a population, and the single card we select is the sample.

The definition we are using identifies probability as a fraction or a proportion. If you work directly from this definition, the probability values you obtain will be expressed as fractions. For example, if you are selecting a card at random,

$$p(\text{spade}) = \frac{13}{52} = \frac{1}{4}$$

Of if you are tossing a coin,

$$p(\text{heads}) = \frac{1}{2}$$

You should be aware that these fractions can be expressed equally well as either decimals or percentages:

If you are unsure how to convert from fractions to decimals or percentages, you should review the section on proportions in the math review, Appendix A.

$$p = \frac{1}{4} = 0.25 = 25\%$$

$$p = \frac{1}{2} = 0.50 = 50\%$$

By convention, probability values most often are expressed as decimal values. But you should realize that any of these three forms is acceptable.

You also should note that all the possible probability values are contained in a limited range. At one extreme, when an event never occurs, the probability is zero or 0% (see Box 6.1). At the other extreme, when an event always occurs, the probability is 1, or 100%. For example, suppose that you have a jar containing 10 white marbles. The probability of randomly selecting a black marble is

$$p(\text{black}) = \frac{0}{10} = 0$$

The probability of selecting a white marble is

$$p(\text{white}) = \frac{10}{10} = 1$$

BOX	
6.1	**ZERO PROBABILITY**

An event that never occurs has a probability of zero. However, the opposite of this statement is not always true: A probability of zero does not mean that the event is guaranteed never to occur. Whenever there are an extremely large number of possible events, the probability of any specific event is assigned the value zero. This is done because the probability value tends toward zero as the number of possible events get large. Consider, for example, the series

$$\frac{1}{10} \quad \frac{1}{100} \quad \frac{1}{1000} \quad \frac{1}{10,000} \quad \frac{1}{100,000}$$

Note that the value of the fraction is getting smaller and smaller, headed toward zero. At the far extreme, when the number of possible events is so large that it cannot be specified, the probability of a single, specific event is said to be zero.

$$\frac{1}{\text{infinite number}} = 0$$

Consider, for example, the fish in the ocean. If there were only 10 fish, then the probability of selecting any particular one would be $p = \frac{1}{10}$. Note that if you add up the probabilities for all 10 fish, you get a total of 1.00. In reality, the number of fish in the ocean is essentially infinite, and, for all practical purposes, the probability of catching any specific fish is $p = 0$. However, this does not mean that you are doomed to fail whenever you go fishing. The zero probability simply means that you cannot predict in advance which fish you will catch. Note that each individual fish has a probability near or, practically speaking, equal to zero. However, there are so many fish that when you add up all the "zeros," you still get a total of 1.00. In probability for large populations, a value of zero does not mean never. But, practically speaking, it does mean very, very close to never.

Finally, you can determine a probability from a frequency distribution table by computing the proportion for the X value in question. In Chapter 2, we computed this proportion as follows:

$$p = \frac{f}{N}$$

X	F
9	1
8	3
7	4
6	2

Once again, probability and proportion are equivalent. Also note that this formula is perfectly consistent with the definition of probability. That is, the "number of outcomes classified as A" is the frequency (f) for that particular score, and the "total number of possible outcomes" is the number of scores in the entire distribution, N. Consider the distribution of scores shown to the left, which has been summarized in a frequency distribution table.

For this distribution of scores, what is the probability of selecting a score of $X = 8$? Stated as a proportion, we may ask, What proportion of all scores in the distribution has a value of 8? There are 10 scores ($N = \Sigma f$) in the distribution, and 3 of them are $X = 8$. Therefore,

$$p(X = 8) = \frac{f}{N} = \frac{3}{10} = 0.30$$

RANDOM SAMPLING   For the preceding definition of probability to be accurate, it is necessary that the outcomes be obtained by a process called random sampling.

DEFINITION   A **random sample** must satisfy two requirements:

1. Each individual in the population has an *equal chance* of being selected.
2. If more than one individual is to be selected for the sample, there must be *constant probability* for each and every selection.

Each of the two requirements for random sampling has some interesting consequences. The first assures that there is no bias in the selection process. For a population with N individuals, each individual must have the same probability, $p = 1/N$, of being selected. This means, for example, that you would not get a random sample of people in your city by selecting names from a yacht club membership list. Similarly, you would not get a random sample of college students by selecting individuals from your psychology classes. You also should note that the first requirement of random sampling prohibits you from applying the definition of probability to situations where the possible outcomes are not equally likely. Consider, for example, the question of whether or not there is life on Mars. There are only two possible alternatives.

1. There is life on Mars.
2. There is no life on Mars.

However, you cannot conclude that the probability of life on Mars is $p = \frac{1}{2}$.

The second requirement also is more interesting than may be apparent at first glance. Consider, for example, the selection of $n = 2$ cards from a complete deck. For the first draw, the probability of obtaining the jack of diamonds is

$$p(\text{jack of diamonds}) = \frac{1}{52}$$

Now, for the second draw, what is the probability of obtaining the jack of diamonds? Assuming that you still are holding the first card, there are two possibilities:

$$p(\text{jack of diamonds}) = \frac{1}{51} \text{ if the first card was not the jack of diamonds}$$

or

$$p(\text{jack of diamonds}) = 0 \text{ if the first card was the jack of diamonds}$$

In either case, the probability is different from its value for the first draw. This contradicts the requirement for random sampling, which says that the probability must stay constant. To keep the probabilities from changing from one selection to the next, it is necessary to replace each sample before you make the next selection. This is called *sampling with replacement*. The second requirement for random samples (constant probability) demands that you sample with replacement.

(*Note:* The definition that we are using identifies one type of random sampling, often called a *simple random sample* or an *independent random sample*. This kind of sampling is important for the mathematical foundation of many of the statistics we will encounter later. However, you should realize that other definitions exist for the concept of random sampling. In particular, it is very common to define random sampling without the requirement of constant probabilities—that is, without replacement. In addition, there are many different sampling techniques that are used when researchers are selecting individuals to participate in research studies.)

PROBABILITY AND FREQUENCY DISTRIBUTIONS

The situations in which we are concerned with probability usually will involve a population of scores that can be displayed in a frequency distribution graph. If you think of the graph as representing the entire population, then different portions of the graph will represent different portions of the population. Because probability and proportion are equivalent, a particular proportion of the graph corresponds to a particular probability in the population. Thus, whenever a population is presented in a frequency distribution graph, it will be possible to represent probabilities as proportions of the graph. The relationship between graphs and probabilities is demonstrated in the following example.

EXAMPLE 6.1

We will use a very simple population that contains only $N = 10$ scores with values 1, 1, 2, 3, 3, 4, 4, 4, 5, 6. This population is shown in the frequency distribution graph in Figure 6.2. If you are taking a random sample of $n = 1$ score from this population, what is the probability of obtaining a score greater than 4? In probability notation,

$$p(X > 4) = ?$$

Using the definition of probability, there are 2 scores that meet this criterion out of the total group of $N = 10$ scores, so the answer would be $p = \frac{2}{10}$. This answer can be obtained directly from the frequency distribution graph if you recall that probability and proportion measure the same thing. Looking at the graph (Figure 6.2), what proportion of the population consists of scores greater than 4? The answer is the shaded part of the distribution—that is, 2 squares out of the total of 10 squares in the distribution. Notice that we now are defining probability as proportion of *area* in the frequency distribution graph. This provides a very concrete and graphic way of representing probability.

Using the same population once again, what is the probability of selecting a score less than 5? In symbols,

$$p(X < 5) = ?$$

Going directly to the distribution in Figure 6.2, we now want to know what part of the graph is not shaded. The unshaded portion consists of 8 out of the 10 blocks $\left(\frac{8}{10} \text{ of the area of the graph}\right)$, so the answer is $p = \frac{8}{10}$.

FIGURE 6.2

A frequency distribution histogram for a population that consists of $N = 10$ scores. The shaded part of the figure indicates the portion of the whole population that corresponds to scores greater than $X = 4$. The shaded portion is two-tenths $\left(p = \frac{2}{10}\right)$ of the whole distribution.

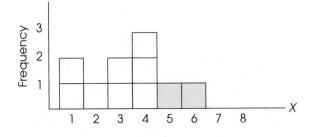

LEARNING CHECK

1. The animal colony in the psychology department contains 20 male rats and 30 female rats. Of the 20 males, 15 are white and 5 spotted. Of the 30 females, 15 are white, and 15 are spotted. Suppose that you randomly select 1 rat from this colony.
   a. What is the probability of obtaining a female?
   b. What is the probability of obtaining a white male?
   c. Which selection is more likely, a spotted male or a spotted female?

2. A jar contains 10 red marbles and 30 blue marbles.
   a. If you randomly select 1 marble from the jar, what is the probability of obtaining a red marble?
   b. If you take a *random sample* of $n = 3$ marbles from the jar and the first two marbles are both blue, what is the probability that the third marble will be red?

3. Suppose that you are going to select a random sample of $n = 1$ score from the distribution in Figure 6.2. Find the following probabilities:
   a. $p(X > 2)$ b. $p(X > 5)$ c. $p(X < 3)$

ANSWERS

1. a. $p = \frac{30}{50} = 0.60$ b. $p = \frac{15}{50} = 0.30$

   c. A spotted female ($p = 0.30$) is more likely than a spotted male ($p = 0.10$).

2. a. $p = \frac{10}{40} = 0.25$

   b. $p = \frac{10}{40} = 0.25$ Remember that random sampling requires sampling with replacement.

3. a. $p = \frac{7}{10} = 0.70$ b. $p = \frac{1}{10} = 0.10$ c. $p = \frac{3}{10} = 0.30$

6.3 PROBABILITY AND THE NORMAL DISTRIBUTION

The normal distribution was first introduced in Chapter 2 as an example of a commonly occurring shape for population distributions. An example of a normal distribution is shown in Figure 6.3.

Note that the normal distribution is symmetrical, with the highest frequency in the middle and frequencies tapering off as you move toward either extreme. Although the exact shape for the normal distribution is defined by an equation (see Figure 6.3), the normal shape can also be described by the proportions of area contained in each section of the distribution. Statisticians often identify sections of a normal distribution by using z-scores. Figure 6.4 shows a normal distribution with several sections marked in z-score units. You should recall that z-scores measure positions in a distribution in terms of standard deviations from the mean. (Thus, $z = +1$ is 1 standard deviation above the mean, $z = +2$ is 2 standard deviations above the mean, and so on.) The graph shows the percentage of scores that fall in each of these sections. For example, the section between the mean ($z = 0$) and the point that is 1 standard deviation above the mean ($z = 1$) makes up 34.13% of the scores. Similarly, 13.59% of the scores fall between 1 and 2 standard deviations from the mean. In this way it is possible to define a normal distribution in terms of its proportions; that is, a distribution is normal if and only if it has all the right proportions.

FIGURE 6.3

The normal distribution. The exact shape of the normal distribution is specified by an equation relating each X value (score) with each Y value (frequency). The equation is

$$Y = \frac{1}{\sqrt{2\pi\sigma^2}} e^{-(X-\mu)^2/2\sigma^2}$$

(π and e are mathematical constants.) In simpler terms, the normal distribution is symmetrical with a single mode in the middle. The frequency tapers off as you move farther from the middle in either direction.

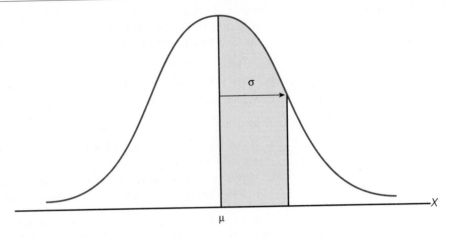

There are two additional points to be made about the distribution shown in Figure 6.4. First, you should realize that the sections on the left side of the distribution have exactly the same proportions as the corresponding sections on the right side because the normal distribution is symmetrical. Second, because the locations in the distribution are identified by z-scores, the proportions shown in the figure apply to *any normal distribution* regardless of the values for the mean and the standard deviation. Remember: When any distribution is transformed into z-scores, the mean becomes zero and the standard deviation becomes one.

FIGURE 6.4

The normal distribution following a z-score transformation.

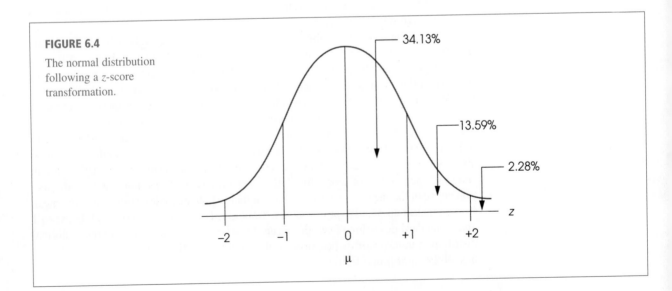

Because the normal distribution is a good model for many naturally occurring distributions and because this shape is guaranteed in some circumstances (as you will see in Chapter 7), we will devote considerable attention to this particular distribution.

The process of answering probability questions about a normal distribution is introduced in the following example.

EXAMPLE 6.2   Adult heights form a normal distribution with a mean of 68 inches and a standard deviation of 6 inches. Given this information about the population and the known proportions for a normal distribution (Figure 6.4), we can determine the probability associated with specific samples. For example, what is the probability of randomly selecting an individual from this population who is taller than 6 feet 8 inches ($X = 80$ inches)?

Restating this question in probability notation, we get

$$p(X > 80) = ?$$

We will follow a step-by-step process to find the answer to this question.

1. First, the probability question is translated into a proportion question: Out of all possible adult heights, what proportion is greater than 80?

2. We know that "all possible adult heights" is simply the population distribution. This population is shown in Figure 6.5(a). The mean is $\mu = 68$, so the score $X = 80$ is to the right of the mean. Because we are interested in all heights greater than 80, we shade in the area to the right of 80. This area represents the proportion we are trying to determine.

3. Identify the exact position of $X = 80$ by computing a z-score. For this example,

$$z = \frac{X - \mu}{\sigma} = \frac{80 - 68}{6} = \frac{12}{6} = 2.00$$

That is, a height of $X = 80$ inches is exactly 2 standard deviations above the mean and corresponds to a z-score of $z = +2.00$ [see Figure 6.5(b)].

4. The proportion we are trying to determine may now be expressed in terms of its z-score:

$$p(z > 2.00) = ?$$

FIGURE 6.5

The distribution for Example 6.2.

(a)

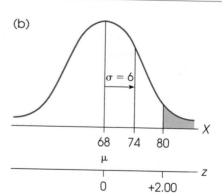

(b)

According to the proportions shown in Figure 6.4, all normal distributions, regardless of the values for μ and σ, will have 2.28% of the scores in the tail beyond $z = +2.00$. Thus, for the population of adult heights,

$$p(X > 80) = p(z > +2.00) = 2.28\%$$

THE UNIT NORMAL TABLE

Before we attempt any more probability questions, we must introduce a more useful tool than the graph of the normal distribution shown in Figure 6.4. The graph shows proportions for only a few selected z-score values. A more complete listing of z-scores and proportions is provided in the *unit normal table*. This table lists proportions of the normal distribution for a full range of possible z-score values.

The complete unit normal table is provided in Appendix B Table B.1, and a portion of the table is reproduced in Figure 6.6. Notice that the table is structured in a four-column format. The first column (A) lists z-score values corresponding to different positions in a normal distribution. If you imagine a vertical line drawn through a normal distribution, then the exact location of the line can be described by one of the z-score values listed in column A. You should also realize that a vertical line will separate the distribution into two sections: a larger section called the *body* and a smaller section called the *tail*. Columns B and C in the table identify the proportion of the distribution in each of the two sections. Column B presents the proportion in the body (the larger portion), and column C presents the proportion in the tail. Finally, we have added a fourth column, column D, that identifies the proportion of the distribution *between* the mean and the z-score. Using the portion of the table shown in Figure 6.6, you should

FIGURE 6.6

A portion of the unit normal table. This table lists proportions of the normal distribution corresponding to each z-score value. Column A of the table lists z-scores. Column B lists the proportion in the body of the normal distribution up to the z-score value. Column C lists the proportion of the normal distribution that is located in the tail of the distribution beyond the z-score value. Column D lists the proportion between the mean and the z-score value.

(A) z	(B) Proportion in Body	(C) Proportion in Tail	(D) Proportion Between Mean and z
0.00	.5000	.5000	.0000
0.01	.5040	.4960	.0040
0.02	.5080	.4920	.0080
0.03	.5120	.4880	.0120
0.21	.5832	.4168	.0832
0.22	.5871	.4129	.0871
0.23	.5910	.4090	.0910
0.24	.5948	.4052	.0948
0.25	.5987	.4013	.0987
0.26	.6026	.3974	.1026
0.27	.6064	.3936	.1064
0.28	.6103	.3897	.1103
0.29	.6141	.3859	.1141
0.30	.6179	.3821	.1179
0.31	.6217	.3783	.1217
0.32	.6255	.3745	.1255
0.33	.6293	.3707	.1293
0.34	.6331	.3669	.1331

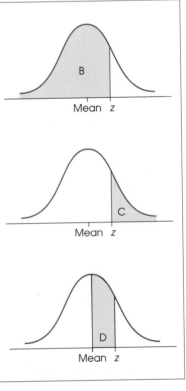

verify that a vertical line drawn through a normal distribution at $z = +0.25$ will separate the distribution into two sections with the larger section containing 0.5987 (59.87%) of the distribution and the smaller section containing 0.4013 (40.13%) of the distribution. Also, there is exactly 0.0987 (9.87%) of the distribution between the mean and $z = +0.25$.

To make full use of the unit normal table, there are a few facts to keep in mind:

1. The *body* always corresponds to the larger part of the distribution whether it is on the right-hand side or the left-hand side. Similarly, the *tail* is always the smaller section whether it is on the right or the left.

2. Because the normal distribution is symmetrical, the proportions on the right-hand side are exactly the same as the corresponding proportions on the left-hand side. For example, the proportion in the right-hand tail beyond $z = +1.00$ is exactly the same as the proportion in the left-hand tail beyond $z = -1.00$.

3. Although the z-score values will change signs (+ and −) from one side to the other, the proportions will always be positive. Thus, column C in the table always lists the proportion in the tail whether it is the right-hand tail or the left-hand tail.

PROBABILITIES, PROPORTIONS, AND z-SCORES

The unit normal table lists relationships between z-score locations and proportions in a normal distribution. For any z-score location, you can use the table to look up the corresponding proportions. Similarly, if you know the proportions, you can use the table to look up the specific z-score location. Because we have defined probability as equivalent to proportion, you can also use the unit normal table to look up probabilities for normal distributions. The following examples demonstrate a variety of different ways that the unit normal table can be used.

Finding proportions/probabilities for specific z-score values   For each of the following examples, we begin with a specific z-score value and then use the unit normal table to find probabilities or proportions associated with the z-score.

EXAMPLE 6.3A   What proportion of the normal distribution corresponds to z-score values greater than $z = 1.00$? First, you should sketch the distribution and shade in the area you are trying to determine. This is shown in Figure 6.7(a). In this case, the shaded portion is the

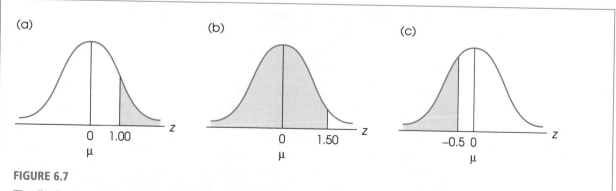

FIGURE 6.7

The distributions for Example 6.3A–6.3C.

tail of the distribution beyond $z = 1.00$. To find this shaded area, you simply look up $z = 1.00$ in column A of the unit normal table. Then you read column C (tail) for the proportion. Using the table in Appendix B, you should find that the answer is 0.1587.

You also should notice that this same problem could have been phrased as a probability question. Specifically, we could have asked, "For a normal distribution, what is the probability of selecting a z-score value greater than $z = +1.00$?" Because probability is the same as proportion, the answer is $p(z > 1.00) = 0.1587$ (or 15.87%).

EXAMPLE 6.3B   For a normal distribution, what is the probability of selecting a z-score less than $z = 1.50$? In symbols, $p(z < 1.50) = ?$ Because probability is equivalent to proportion, our goal is to determine what proportion of the z-scores have values less than 1.50. A normal distribution is shown in Figure 6.7(b) and $z = 1.50$ is located in the distribution. Note that we have shaded all the values to the left of (less than) $z = 1.50$. This is the portion we are trying to find. Clearly the shaded portion is more than 50% so it corresponds to the body of the distribution. Therefore, we find $z = 1.50$ in the unit normal table and read the proportion from column B. The answer is $p(z < 1.50) = 0.9332$ (or 93.32%).

EXAMPLE 6.3C   Many problems will require that you find proportions for negative z-scores. For example, what proportion of the normal distribution corresponds to the tail beyond $z = -0.50$? That is, $p(z < -0.50)$. This portion has been shaded in Figure 6.7(c). To answer questions with negative z-scores, simply remember that the normal distribution is symmetrical with a z-score of zero at the mean, positive values to the right, and negative values to the left. The proportion in the left tail beyond $z = -0.50$ is identical to the proportion in the right tail beyond $z = +0.50$. To find this proportion, look up $z = 0.50$ in column A, and find the proportion in column C (tail). You should get an answer of 0.3085 (30.85%).

Moving to the left on the X-axis results in smaller X values and smaller z-scores. Thus, a z-score of -3.00 reflects a smaller value than a z-score of -1.

Finding the z-score location that corresponds to specific proportions   The preceding examples all involved using a z-score value in column A to look up a proportion in column B or C. You should realize, however, that the table also allows you to begin with a known proportion and then look up the corresponding z-score.
In general, the unit normal table can be used for two purposes:

1. If you know a specific location (z-score) in a normal distribution, you can use the table to look up the corresponding proportions.

2. If you know a specific proportion (or proportions), you can use the table to look up the exact z-score location in the distribution.

The following examples demonstrate how the table can be used to find specific z-scores if you begin with known proportions.

EXAMPLE 6.4A   For a normal distribution, what z-score separates the top 10% from the remainder of the distribution? To answer this question, we have sketched a normal distribution [Figure 6.8(a)] and drawn a vertical line that separates the highest 10% (approximately) from the rest. The problem is to locate the exact position of this line.

FIGURE 6.8

The distributions for
Examples 6.4A and 6.4B.

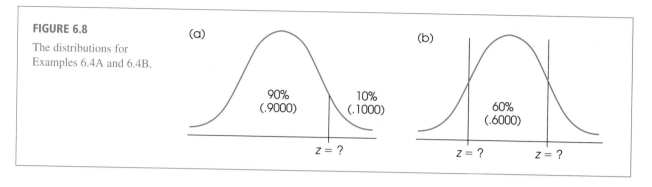

For this distribution, we know that the tail contains 0.1000 (10%) and the body contains 0.9000 (90%). To find the z-score value, you simply locate 0.1000 in column C
or 0.9000 in column B of the unit normal table. Note that you probably will not find
the exact proportion, but you can use the closest value listed in the table. For example, you will not find 0.1000 listed in column C but you can use 0.1003, which is
listed. Once you have found the correct proportion in the table, simply read the corresponding z-score. For this example, the z-score that separates the extreme 10% in the
tail is $z = 1.28$. At this point you must be careful because the table does not differentiate between the right-hand tail and the left-hand tail of the distribution. Specifically,
the final answer could be either $z = +1.28$, which separates 10% in the right-hand
tail, or $z = -1.28$, which separates 10% in the left-hand tail. For this problem we
want the right-hand tail (the highest 10%), so the z-score value is $z = +1.28$.

EXAMPLE 6.4B   For a normal distribution, what z-score values form the boundaries that separate the
middle 60% of the distribution from the rest of the scores?

   Again, we have sketched a normal distribution [Figure 6.8(b)] and drawn vertical
lines in the approximate locations. For this example, we want slightly more than half
of the distribution in the central section, with the remainder split equally between the
two tails. The problem is to find the z-score values that define the exact locations for
the lines. To find the z-score values, we begin with the known proportions: 0.6000 in
the center and 0.4000 divided equally between the two tails. Although these proportions can be used in several different ways, this example provides an opportunity to
demonstrate how column D in the table can be used to solve problems. For this problem, the 0.6000 in the center can be divided in half with exactly 0.3000 to the right of
the mean and exactly 0.3000 to the left. Each of these sections corresponds to the
proportion listed in column D. Looking in column D for a value of 0.3000, you will
discover that this exact proportion is not in the table, but the closest value is 0.2995.
Reading across the row to column A, you should find a z-score value of $z = 0.84$.
Looking again at the sketch [Figure 6.8(b)], you can see that the right-hand line is
located at $z = +0.84$ and the left-hand line is located at $z = -0.84$.

   You may have noticed that we have sketched distributions for each of the preceding
problems. As a general rule, you should always sketch a distribution, locate the mean
with a vertical line, and shade in the portion you are trying to determine. Look at your
sketch. It will indicate which columns to use in the unit normal table. If you make a
habit of drawing sketches, you will avoid careless errors in using the table.

1. Find the proportion of the normal distribution that is associated with the following sections of a graph:

   a. $z < +1.00$ **b.** $z > +0.80$ **c.** $z < -2.00$

   d. $z > -0.33$ **e.** $z > -0.50$

2. For a normal distribution, find the z-score location that separates the distribution as follows:

   a. Separate the highest 30% from the rest of the distribution.

   b. Separate the lowest 40% from the rest of the distribution.

   c. Separate the highest 75% from the rest of the distribution.

1. a. 0.8413   **b.** 0.2119   **c.** 0.0228   **d.** 0.6293   **e.** 0.6915

   2. a. $z = 0.52$ **b.** $z = -0.25$ **c.** $z = -0.67$

6.4   PROBABILITIES AND PROPORTIONS FOR SCORES FROM A NORMAL DISTRIBUTION

In the preceding section, we used the unit normal table to find probabilities and proportions corresponding to specific z-score values. In most situations, however, it will be necessary to find probabilities for specific X values. Consider the following example:

> It is known that SAT scores form a normal distribution with a mean of $\mu = 500$ and a standard deviation of $\sigma = 100$. Given this information, what is the probability of randomly selecting an individual with an SAT score greater than 650?

This problem is asking for a specific probability or proportion of a normal distribution. However, before we can look up the answer in the unit normal table, we must first transform the SAT scores (X values) into z-scores. Thus, to solve this new kind of probability problem, we must add one new step to the process. Specifically, in order to answer probability questions about scores (X values) from a normal distribution, you must use the following two-step procedure:

1. Transform the X-values into z-scores.

2. Use the unit normal table to look up the proportions corresponding to the z-score values.

This process is demonstrated in the following examples. Once again, we suggest that you sketch the distribution and shade the portion you are trying to find in order to avoid careless mistakes.

EXAMPLE 6.5   We will now answer the probability question about SAT scores that was presented at the beginning of this section. Specifically, what is the probability of randomly selecting a person with an SAT score greater than 650?

$$p(X > 650) = ?$$

Restated in terms of proportions, we want to find the proportion of the SAT distribution that corresponds to scores greater than 650. The distribution is drawn in Figure 6.9, and the portion we want has been shaded.

FIGURE 6.9

The distribution of SAT scores. The problem is to find the probability or proportion of the distribution corresponding to scores greater than 650.

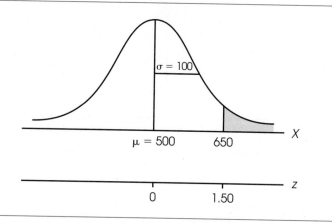

The first step is to change the X values into z-scores. In particular, the score of $X = 650$ is changed to

$$z = \frac{X - \mu}{\sigma} = \frac{650 - 500}{100} = \frac{150}{100} = 1.50$$

Next, look up the z-score value in the unit normal table. Because we want the proportion in the tail of the distribution (Figure 6.9), the answer will be found in column C. Consulting the table, we see that a z-score of 1.50 corresponds to a proportion of 0.0668. The probability of randomly selecting an individual with an SAT score greater than 650 is 0.0668.

$$p(X > 650) = 0.0668 \text{ (or 6.68\%)}$$

Finding proportions/probabilities located between two scores   The next example demonstrates the process of finding the probability of selecting a score that is located *between* two specific values. Although these problems can be solved using the proportions of columns B and C (body and tail), they are often easier to solve with the proportions listed in column D. We will consider two versions of this type of problem: (1) the two scores are on opposite sides of the mean, and (2) the two scores are on the same side of the mean. As you will see, to solve the first type of problem we *add* the column D proportions, and to solve the second type, we *subtract*.

EXAMPLE 6.6   The highway department conducted a study measuring driving speeds on a local section of interstate highway. They found an average speed of $\mu = 58$ miles per hour with a standard deviation of $\sigma = 10$. The distribution was approximately normal. Given this information, what proportion of the cars are traveling between 55 and 65 miles per hour? Using probability notation, we can express the problem as

$$p(55 < X < 65) = ?$$

FIGURE 6.10

The distribution for
Example 6.6.

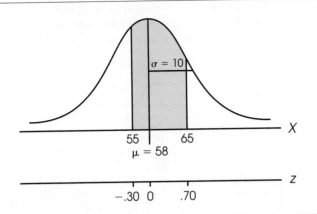

The distribution of driving speeds is shown in Figure 6.10 with the appropriate area shaded. The first step is to determine the z-score corresponding to the X value at each end of the interval.

$$\text{For } X = 55: \quad z = \frac{X - \mu}{\sigma} = \frac{55 - 58}{10} = \frac{-3}{10} = -0.30$$

$$\text{For } X = 65: \quad z = \frac{X - \mu}{\sigma} = \frac{65 - 58}{10} = \frac{7}{10} = 0.70$$

Looking again at Figure 6.10, we see that the proportion we are seeking can be divided into two sections: (1) the area left of the mean, and (2) the area right of the mean. The first area is the proportion between the mean and $z = -0.30$ and the second is the proportion between the mean and $z = +0.70$. Using column D of the unit normal table, these two proportions are 0.1179 and 0.2580. The total proportion is obtained by adding these two sections:

$$p(55 < X < 65) = p(-0.30 < z < +0.70) = 0.1179 + 0.2580 = 0.3759$$

EXAMPLE 6.7   Using the same distribution of driving speeds from the previous example, what proportion of cars are traveling between 65 and 75 miles per hour?

$$p(65 < X < 75) = ?$$

The distribution is shown in Figure 6.11 with the appropriate area shaded. Again, we start by determining the z-score corresponding to each end of the interval.

$$\text{For } X = 65: \quad z = \frac{X - \mu}{\sigma} = \frac{65 - 58}{10} = \frac{7}{10} = 0.70$$

$$\text{For } X = 75: \quad z = \frac{X - \mu}{\sigma} = \frac{75 - 58}{10} = \frac{17}{10} = 1.70$$

According to column D in the unit normal table, the proportion between the mean and $z = 1.70$ is $p = 0.4554$. Note that this proportion includes the section that we want, but it also includes an extra, unwanted section between the mean and $z = 0.70$.

FIGURE 6.11

The distribution for
Example 6.7.

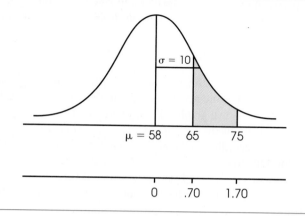

Again, using column D, we see that the unwanted section is $p = 0.2580$. To obtain
the correct answer, we subtract the unwanted portion from the total proportion between the mean and $z = 1.70$.

$$p(65 < X < 75) = p(0.70 < z < 1.70) = 0.4554 - 0.2580 = 0.1974$$

Finding scores corresponding to specific proportions or probabilities   In the previous two examples, the problem was to find the proportion or probability corresponding to specific X values. The two-step process for finding these proportions is shown
in Figure 6.12. Thus far, we have only considered examples that move in a clockwise
direction around the triangle shown in the figure; that is, we start with an X value that
is transformed into a z-score, and then we use the unit normal table to look up the appropriate proportion. You should realize, however, that it is possible to reverse this two-
step process so that we move backwards, or counterclockwise, around the triangle. This
reverse process will allow us to find the score (X value) corresponding to a specific
proportion in the distribution. Following the lines in Figure 6.12, we will begin with a

FIGURE 6.12

Determining probabilities or
proportions for a normal
distribution is shown as a
two-step process with
z-scores as an intermediate
stop along the way. Note that
you cannot move directly
along the dashed line be-
tween X values and probabi-
ities or proportions. Instead,
you must follow the solid
lines around the corner.

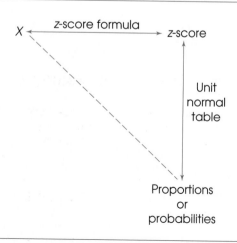

specific proportion, use the unit normal table to look up the corresponding z-score, and then transform the z-score into an X value. The following example demonstrates this process.

EXAMPLE 6.8   Scores on the SAT form a normal distribution with $\mu = 500$ and $\sigma = 100$. What is the minimum score necessary to be in the top 15% of the SAT distribution? This problem is shown graphically in Figure 6.13.

In this problem, we begin with a proportion (15% = 0.15), and we are looking for a score. According to the map in Figure 6.12, we can move from p (proportion) to X (score) by going via z-scores. The first step is to use the unit normal table to find the z-score that corresponds to a proportion of 0.15. Because the proportion is located beyond z in the tail of the distribution, we will look in column C for a proportion of 0.1500. Note that you may not find 0.1500 exactly, but locate the closest value possible. In this case, the closest value in the table is 0.1492, and the z-score that corresponds to this proportion is $z = 1.04$.

The next step is to determine whether the z-score is positive or negative. Remember that the table does not specify the sign of the z-score. Looking at the graph in Figure 6.13, you should realize that the score we want is above the mean, so the z-score is positive, $z = +1.04$.

Now you are ready for the last stage of the solution—that is, changing the z-score into an X value. Using z-score equation 5.2 (page 144) and the known values of μ, σ, and z, we obtain

$$X = \mu + z\sigma$$
$$= 500 + 1.04(100)$$
$$= 500 + 104$$
$$= 604$$

The conclusion for this example is that you must have an SAT score of at least 604 to be in the top 15% of the distribution.

FIGURE 6.13

The distribution of SAT scores. The problem is to locate the score that separates the top 15% from the rest of the distribution. A line is drawn to divide the distribution roughly into 15% and 85% sections.

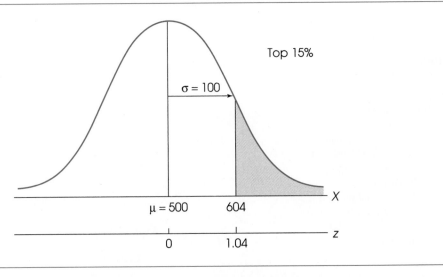

EXAMPLE 6.9   This time the problem is to find the range of values that defines the middle 80% of the distribution of SAT scores. The distribution is shown in Figure 6.14 with the appropriate section shaded.

The 80% (0.8000) in the middle of the distribution can be split in half with 0.4000 to the right of the mean and 0.4000 to the left. Looking up 0.4000 in column D of the unit normal table, you will discover that the exact proportion is not listed but the closest value is 0.3997. Reading across to column A, the corresponding z-score is $z = 1.28$. Thus, the z-score at the right-hand boundary is $z = +1.28$ and the z-score at the left boundary is $z = -1.28$. In either case, a z-score of 1.28 indicates a location that is 1.28 standard deviations away from the mean. In this case,

$$1.28\sigma = 1.28(100) = 128 \text{ points}$$

Therefore, the right-hand boundary ($z = +1.28$) is $X = 500 + 128 = 628$. The left-hand boundary ($z = -1.28$) is $X = 500 - 128 = 372$. The middle 80% of the distribution corresponds to X values between 372 and 628.

FIGURE 6.14

The distribution of SAT scores. The problem is to find the scores that determine the middle 80%.

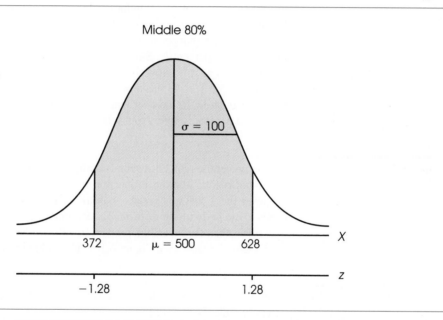

LEARNING CHECK   **1.** For a normal distribution with a mean of 80 and a standard deviation of 10, find each probability value requested.

    a. $p(X > 85) = ?$ **b.** $p(X < 95) = ?$
    c. $p(X > 70) = ?$ **d.** $p(75 < X < 100) = ?$

2. For a normal distribution with a mean of 100 and a standard deviation of 20, find each value requested.

    a. What score separates the top 40% from the bottom 60% of the distribution?

    b. What is the minimum score needed to be in the top 5% of this distribution?

    c. What scores form the boundaries for the middle 60% of this distribution?

3. What is the probability of selecting a score greater than 45 from a positively skewed distribution with $\mu = 40$ and $\sigma = 10$? (Be careful.)

ANSWERS
1. **a.** $p = 0.3085$ (30.85%)   **b.** $p = 0.9332$ (93.32%)
    c. $p = 0.8413$ (84.13%)   **d.** $p = 0.6687$ (66.87%)

2. **a.** $z = +0.25$; $X = 105$ **b.** $z = +1.64$; $X = 132.8$
    c. $z = \pm0.84$; boundaries are 83.2 and 116.8

3. You cannot obtain the answer. The unit normal table cannot be used to answer this question because the distribution is not normal.

6.5 PERCENTILES AND PERCENTILE RANKS

In Chapter 2, we introduced percentiles and percentile ranks as techniques for describing an individual's relative standing within a distribution. Recall that the percentile rank of a particular score is defined as the percentage of individuals in the distribution with scores at or below that particular score. The particular score associated with a percentile rank is called a percentile. Suppose, for example, that you have a score of $X = 43$ on an exam and that you know that exactly 60% of the class had scores of 43 or lower. Then your score $X = 43$ has a percentile rank of 60%, and your score would be called the 60th percentile. Remember that percentile rank refers to a percentage of the distribution, and percentile refers to a score.

FINDING PERCENTILE RANKS

Because percentile rank corresponds to the percentage *below* a particular score, we will always be looking for the proportion of the distribution to the *left* of the X value. Whenever the X value is greater than the mean, the proportion to the left of X corresponds to the body of the distribution, column B in the table (see Figure 6.15). If the X value is less than the mean, the proportion will correspond to the tail of the distribution, column C in the table (see Figure 6.16). The following examples illustrate this point.

EXAMPLE 6.10A

A population is normally distributed with $\mu = 100$ and $\sigma = 10$. What is the percentile rank for $X = 114$?

Because a percentile rank indicates one's standing relative to all lower scores, we must focus on the area of the distribution to the *left* of $X = 114$. The distribution is shown in Figure 6.15, and the area of the curve containing all scores below $X = 114$ is shaded. The proportion for this shaded area will give us the percentile rank.

Because the distribution is normal, we can use the unit normal table to find this proportion. The first step is to compute the z-score for the X value we are considering.

$$z = \frac{X - \mu}{\sigma} = \frac{114 - 100}{10} = \frac{14}{10} = 1.40$$

The next step is to consult the unit normal table. Note that the shaded area in Figure 6.15 makes up the body of the graph. The proportion for this area is presented in column B. For $z = 1.40$, column B indicates a proportion of 0.9192. The percentile rank for $X = 114$ is 91.92%.

FIGURE 6.15

The distribution for Example 6.10A. The proportion for the shaded area provides the percentile rank for $X = 114$.

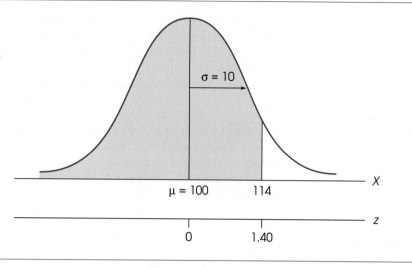

EXAMPLE 6.10B   For the distribution in Example 6.10A, what is the percentile rank for $X = 92$?

This example is diagrammed in Figure 6.16. The score $X = 92$ is placed in the left side of the distribution because it is below the mean. Again, percentile ranks deal with the area of the distribution below the score in question. Therefore, we have shaded the area to the left of $X = 92$.

First, the X value is transformed to a z-score:

$$z = \frac{X - \mu}{\sigma} = \frac{92 - 100}{10} = \frac{-8}{10} = -0.80$$

FIGURE 6.16

The distribution for Example 6.10B. The proportion for the shaded area provides the percentile rank for $X = 92$.

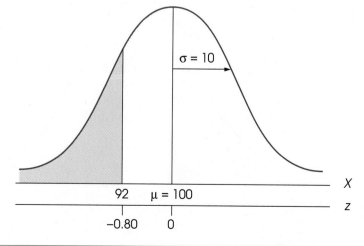

Remember that the normal distribution is symmetrical. Therefore, the proportion in the right-hand tail beyond $z = 0.80$ is identical to that in the left-hand tail beyond $z = -0.80$.

Now the unit normal table can be consulted. The proportion in the left-hand tail beyond $z = -0.80$ can be found in column C. According to the unit normal table, for $z = 0.80$, the proportion in the tail is $p = 0.2119$. This also is the area beyond $z = -0.80$. Thus, the percentile rank for $X = 92$ is 21.19%. That is, a score of 92 is greater than 21.19% of the scores in the distribution.

FINDING PERCENTILES

The process of finding a particular percentile is identical to the process used in Example 6.8. You are given a percentage (this time a percentile rank), and you must find the corresponding X value (the percentile). You should recall that finding an X value from a percentage requires the intermediate step of determining the z-score for that proportion of the distribution. The following example demonstrates this process for percentiles.

EXAMPLE 6.11

A population is normally distributed with $\mu = 60$ and $\sigma = 5$. For this population, what is the 34th percentile?

In this example, we are looking for an X value (percentile) that has 34% (or $p = 0.3400$) of the distribution below it. This problem is illustrated in Figure 6.17. Note that 34% is roughly equal to one-third of the distribution, so the corresponding shaded area in Figure 6.17 is located entirely on the left-hand side of the mean. In this problem, we begin with a proportion (34% = 0.3400), and we are looking for a score (the percentile). The first step in moving from a proportion to a score is to find the z-score (Figure 6.12). You must look in the unit normal table to find the z-score that corresponds to a proportion of 0.3400. Because the proportion is in the tail beyond z, you must look in column C for a proportion of 0.3400. Although the exact value of 0.3400 is not listed in the table, you may use the closest value possible, which is 0.3409. The z-score corresponding to this value is $z = -0.41$. Notice that we have added a negative sign to the z-score value because the position we want is located below the mean. Thus, the X value for which we are looking has a z-score of $z = -0.41$.

Be careful: The table does not differentiate positive and negative z-scores. You must look at your sketch of the distribution to determine the sign.

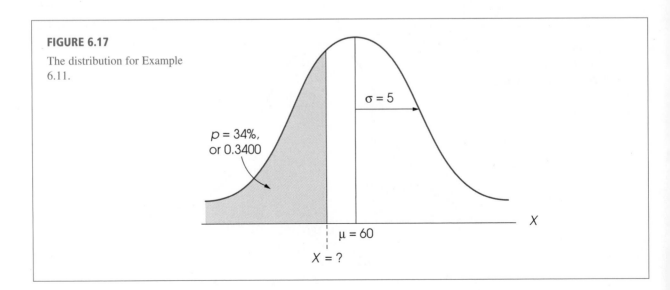

FIGURE 6.17

The distribution for Example 6.11.

$\sigma = 5$

$p = 34\%$, or 0.3400

$\mu = 60$

$X = ?$

X

The next step is to convert the z-score to an X value. Using the z-score formula solved for X, we obtain

$$X = \mu + z\sigma$$
$$= 60 + (-0.41)(5)$$
$$= 60 - 2.05$$
$$= 57.95$$

The 34th percentile for this distribution is $X = 57.95$. This answer makes sense because the 50th percentile for this example is 60 (the mean and median). Therefore, the 34th percentile has to be a value less than 60.

QUARTILES   Percentiles divide the distribution into 100 equal parts, each corresponding to 1% of the distribution. The area in a distribution can also be divided into four equal parts called quartiles, each corresponding to 25%. We first looked at quartiles in Chapter 4 (page 107) in considering the semi-interquartile range. The first quartile ($Q1$) is the score that separates the lowest 25% of the distribution from the rest. Thus, the first quartile is the same as the 25th percentile. Similarly, the second quartile ($Q2$) is the score that has 50% (two quarters) of the distribution below it. You should recognize that $Q2$ is the median or 50th percentile of the distribution. Finally, the third quartile ($Q3$) is the X value that has 75% (three quarters) of the distribution below it. The $Q3$ for a distribution is also the 75th percentile.

For a normal distribution, the first quartile always corresponds to $z = -0.67$. You should verify this value by checking the unit normal table to determine the z-score value corresponding to 25% (0.2500) in the tail of the distribution. Similarly, the third quartile corresponds to $z = +0.67$. Finally, the second quartile, the median, corresponds to $z = 0$ because the normal distribution is symmetrical and the mean equals the median (see Figure 6.18). Thus, for any normal distribution, the first quartile is located below the mean by 0.67 standard deviations; the third quartile is above the mean by 0.67 standard deviations, and the second quartile is exactly at the mean. The following example demonstrates the calculation of quartiles for a normal distribution.

FIGURE 6.18

The z-scores corresponding to the first, second, and third quartiles in a normal distribution.

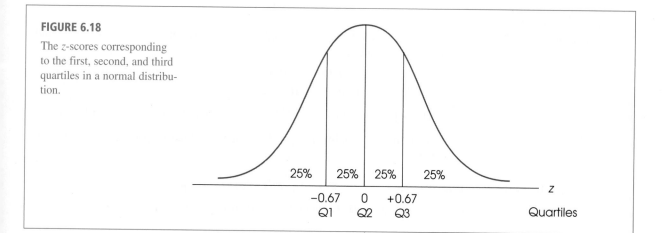

EXAMPLE 6.12   Reading achievement scores for third-grade children form a normal distribution with a mean of $\mu = 50$ and a standard deviation of $\sigma = 10$. For this distribution, the first quartile is located below the mean by $0.67\sigma = 0.67(10) = 6.7$ points. Thus, $Q1 = 43.3$ $(50 - 6.7)$. Similarly, the third quartile is 6.7 points above the mean, so $Q3 = 56.7$ $(50 + 6.7)$. The second quartile is equal to the mean, $Q2 = 50$.

In Chapter 4, the semi-interquartile range was defined as one-half the distance between the first and third quartile, or

$$\text{semi-interquartile range} = \frac{(Q3 - Q1)}{2}$$

For this example, the semi-interquartile range is

$$\text{semi-interquartile range} = \frac{(Q3 - Q1)}{2}$$

$$= \frac{(56.7 - 43.3)}{2}$$

$$= \frac{13.4}{2}$$

$$= 6.7$$

Notice that $Q1$ and $Q3$ are the same distance from the mean (6.7 points in the previous example). $Q1$ and $Q3$ will always be equidistant from the mean for normal distributions because normal distributions are symmetrical. Therefore, one-half of the distance between $Q1$ and $Q3$ (the semi-interquartile range by definition) will also equal the distance of $Q3$ from the mean (see Figure 6.18). The distance of $Q3$ from the mean can be obtained simply by multiplying the z-score for $Q3$ ($z = 0.67$) times the standard deviation. Using this shortcut greatly simplifies the computation. For a normal distribution,

Remember: $z\sigma$ is a deviation score, or distance from the mean (Chapter 5).

$$\text{semi-interquartile range} = 0.67\sigma \qquad (6.1)$$

Remember that this simplified formula is used *only* for normal distributions.

LEARNING CHECK   1. A population is normally distributed and has a mean of $\mu = 90$ with $\sigma = 8$. Find the following values:

    a. The percentile rank for $X = 100$ **b.** The percentile rank for $X = 88$

    c. The 85th percentile   **d.** The 10th percentile

2. For the population in Exercise 1, find $Q1$, $Q3$, and the semi-interquartile range.

3. For a normal distribution, any positive z-score will have a percentile rank greater than 50%. (True or false?)

ANSWERS   **1. a.** 89.44%   **b.** 40.13%   **c.** $X = 98.32$ **d.** $X = 79.76$

2. $Q1 = 84.64$, $Q3 = 95.36$, semi-interquartile range $= 5.36$

3. True

6.6 PROBABILITY AND THE BINOMIAL DISTRIBUTION

When a variable is measured on a scale consisting of exactly two categories, the resulting data are called binomial. The term *binomial* can be loosely translated as "two names," referring to the two categories on the measurement scale.

Binomial data can occur when a variable naturally exists with only two categories. For example, people can be classified as male or female, and a coin toss results in either heads or tails. It also is common for a researcher to simplify data by collapsing the scores into two categories. For example, a psychologist may use personality scores to classify people as either high or low in aggression.

In binomial situations, the researcher often knows the probabilities associated with each of the two categories. With a balanced coin, for example, $p(\text{heads}) = p(\text{tails}) = \frac{1}{2}$. The question of interest is the number of times each category occurs in a series of trials or in a sample of individuals. For example:

What is the probability of obtaining 15 heads in 20 tosses of a balanced coin?

What is the probability of obtaining more than 40 introverts in a sampling of 50 college freshmen?

As we shall see, the normal distribution serves as an excellent model for computing probabilities with binomial data.

THE BINOMIAL DISTRIBUTION

To answer probability questions about binomial data, we need to examine the binomial distribution. To define and describe this distribution, we first introduce some notation.

1. The two categories are identified as A and B.
2. The probabilities (or proportions) associated with each category are identified as

   $p = p(A) = $ the probability of A

   $q = p(B) = $ the probability of B

   Notice that $p + q = 1.00$ because A and B are the only two possible outcomes.
3. The number of individuals or observations in the sample is identified by n.
4. The variable X refers to the number of times category A occurs in the sample.

Notice that X can have any value from 0 (none of the sample is in category A) to n (all the sample is in category A).

DEFINITION

Using the notation presented here, the **binomial distribution** shows the probability associated with each value of X from $X = 0$ to $X = n$.

A simple example of a binomial distribution is presented next.

EXAMPLE 6.13

Figure 6.19 shows the binomial distribution for the number of heads obtained in 2 tosses of a balanced coin. This distribution shows that it is possible to obtain as many as 2 heads or as few as 0 heads in 2 tosses. The most likely outcome (highest probability) is to obtain exactly 1 head in 2 tosses. The construction of this binomial distribution is discussed in detail next.

For this example, the event we are considering is a coin toss. There are two possible outcomes, heads and tails. We assume the coin is balanced, so

$$p = p(\text{heads}) = \frac{1}{2}$$

$$q = p(\text{tails}) = \frac{1}{2}$$

We are looking at a sample of $n = 2$ tosses, and the variable of interest is

$$X = \text{the number of heads}$$

To construct the binomial distribution, we will look at all the possible outcomes from tossing a coin 2 times. The complete set of 4 outcomes is listed below.

1ST TOSS	2ND TOSS	
Heads	Heads	(both heads)
Heads	Tails	(each sequence has
Tails	Heads	exactly 1 head)
Tails	Tails	(no heads)

Notice that there are 4 possible outcomes when you toss a coin 2 times. Only 1 of the 4 outcomes has 2 heads, so the probability of obtaining 2 heads is $p = \frac{1}{4}$. Similarly, 2 of the 4 outcomes have exactly 1 head, so the probability of 1 head is $p = \frac{2}{4} = \frac{1}{2}$. Finally, the probability of no heads ($X = 0$) is $p = \frac{1}{4}$. These are the probabilities shown in Figure 6.19.

Note that this binomial distribution can be used to answer probability questions. For example, what is the probability of obtaining at least 1 head in 2 tosses? According to the distribution shown in Figure 6.19, the answer is $\frac{3}{4}$.

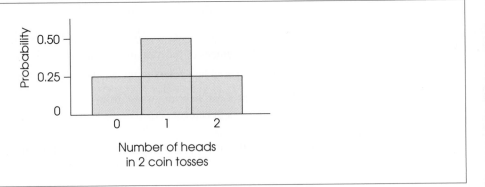

FIGURE 6.19

The binomial distribution showing the probability for the number of heads in 2 tosses of a balanced coin.

Similar binomial distributions have been constructed for the number of heads in 4 tosses of a balanced coin and in 6 tosses of a coin (Figure 6.20). It should be obvious from the binomial distributions shown in Figures 6.19 and 6.20 that the binomial distribution tends toward a normal shape, especially when the sample size (*n*) is relatively large.

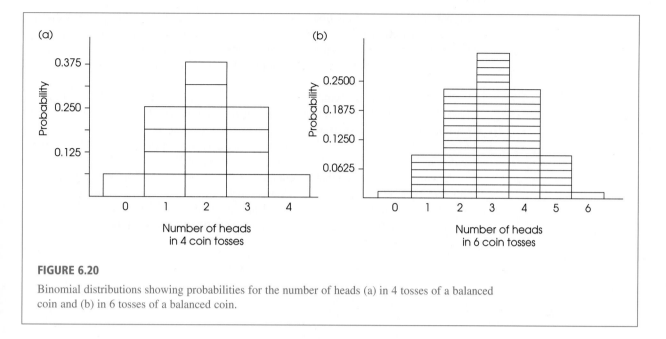

FIGURE 6.20

Binomial distributions showing probabilities for the number of heads (a) in 4 tosses of a balanced coin and (b) in 6 tosses of a balanced coin.

It should not be surprising that the binomial distribution tends to be normal. With $n = 10$ coin tosses, for example, the most likely outcome would be to obtain around $X = 5$ heads. On the other hand, values far from 5 would be very unlikely—you would not expect to get all 10 heads or all 10 tails (0 heads) in 10 tosses. Notice that we have described a normal-shaped distribution: The probabilities are highest in the middle (around $X = 5$), and they taper off as you move toward either extreme.

THE NORMAL APPROXIMATION TO THE BINOMIAL DISTRIBUTION

We have stated that the binomial distribution tends to approximate a normal distribution, particularly when n is large. To be more specific, the binomial distribution will be a nearly perfect normal distribution when pn and qn are both equal to or greater than 10. Under these circumstances, the binomial distribution will approximate a normal distribution with the following parameters:

The value of 10 for pn or qn is a general guide, not an absolute cutoff. Values slightly less than 10 still provide a good approximation. However, with smaller values the normal approximation becomes less accurate as a substitute for the binomial distribution.

Mean: $\mu = pn$ (6.2)

standard deviation: $\sigma = \sqrt{npq}$ (6.3)

Within this normal distribution, each value of X has a corresponding z-score,

$$z = \frac{X - \mu}{\sigma} = \frac{X - pn}{\sqrt{npq}}$$ (6.4)

Coin tosses produce discrete events. In a series of coin tosses, you may observe 1 head, 2 heads, 3 heads, and so on, but no values between them are possible (p. 18).

The fact that the binomial distribution tends to be normal in shape means that we can compute probability values directly from z-scores and the unit normal table.

It is important to remember that the normal distribution is only an approximation of a true binomial distribution. Binomial values, such as the number of heads in a series of coin tosses, are *discrete*. The normal distribution is *continuous*. However, the normal approximation provides an extremely accurate model for computing binomial probabilities in many situations. Figure 6.21 shows the difference between the discrete binomial distribution (histogram) and the normal distribution (smooth curve). Although the two distributions are slightly different, the area under the distributions is nearly equivalent. *Remember: It is the area under the distribution that is used to find probabilities.*

To gain maximum accuracy when using the normal approximation, you must remember that each X value in the binomial distribution is actually an interval, not a point on the scale. In the histogram, this interval is represented by the width of the bar. In Figure 6.21, for example, $X = 6$ is actually an interval bounded by the real limits of 5.5 and 6.5. To find the probability of obtaining a score of $X = 6$, you should find the area of the normal distribution that is contained between the two real limits. If you are using the normal approximation to find the probability of obtaining a score greater than $X = 6$, you should use the area beyond the real limit boundary of 6.5. The following two examples demonstrate how the normal approximation to the binomial distribution is used to compute probability values.

FIGURE 6.21

The relationship between the binomial distribution and the normal distribution. The binomial distribution is always a discrete histogram, and the normal distribution is a continuous, smooth curve. Each X value is represented by a bar in the histogram or a section of the normal distribution.

EXAMPLE 6.14A

In the psychology department, three-fourths of the students are female and only one-fourth are male. If a random sample of 48 students is selected, what is the probability that the sample will have exactly 14 males?

For this example, the probability of selecting a male is $p = \frac{1}{4}$ and the probability of selecting a female is $q = \frac{3}{4}$. With a sample of $n = 48$, the example meets the criteria for using the normal approximation to the binomial distribution:

$$pn = \frac{1}{4}(48) = 12 \qquad qn = \frac{3}{4}(48) = 36 \qquad \text{Both are greater than 10.}$$

Therefore, the distribution for the number of males in a sample of $n = 48$ will be normal with a mean of $\mu = pn = 12$ and a standard deviation of $\sigma = \sqrt{npq} = \sqrt{\left(\frac{1}{4}\right)\left(\frac{3}{4}\right)(48)} = \sqrt{9} = 3$.

Remember to use the real limits when determining probabilities from the normal approximation to the binomial distribution.

This distribution is shown in Figure 6.22. We want the proportion of this distribution that corresponds to $X = 14$—that is, the portion bounded by $X = 13.5$ and $X = 14.5$. This portion is shaded in Figure 6.22. To find the shaded portion, we first convert each X value into a z-score. For $X = 13.5$,

$$z = \frac{X - \mu}{\sigma} = \frac{13.5 - 12}{3} = 0.50$$

For $X = 14.5$,

$$z = \frac{X - \mu}{\sigma} = \frac{14.5 - 12}{3} = 0.83$$

Looking up these values in the unit normal table, we find

a. The area beyond $z = 0.50$ is 0.3085.

b. The area beyond $z = 0.83$ is 0.2033.

Therefore, the area between the two z-scores is

$$0.3085 - 0.2033 = 0.1052$$

You should remember that this value is an approximation to the exact probability value. However, it is a reasonably accurate approximation. To illustrate the accuracy of the normal approximation, the exact probability of getting $X = 14$ males from the binomial distribution is 0.1015. Notice that the normal approximation value of 0.1052 is very close to the exact probability of 0.1015. As we mentioned earlier, the normal distribution provides an excellent approximation for finding binomial probabilities.

FIGURE 6.22

The binomial distribution (normal approximation) for the number of males in a sample of $n = 48$ psychology majors. The shaded area corresponds to the probability of obtaining exactly 14 males when the probability of selecting a male is $p = \frac{1}{4}$. Note that the score $X = 14$ is defined by its real limits.

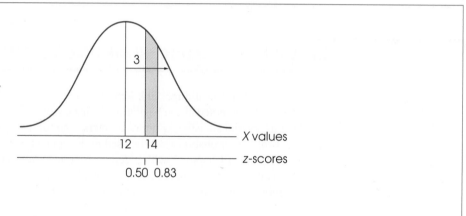

EXAMPLE 6.14B

Caution: If the question had asked for the probability of 14 or more males, you would have used the area beyond 13.5. Read the question carefully.

Once again, suppose that you are selecting a random sample of $n = 48$ psychology majors. If the probability of selecting a male is $p = \frac{1}{4}$, what is the probability that the sample will have more than 14 males?

The normal approximation to this binomial distribution is shown in Figure 6.22. Because we want the probability of obtaining *more than* 14 males, we must find the area located in the tail of the distribution beyond $X = 14.5$. (Remember: A score of 14 correct corresponds to the interval from 13.5 to 14.5. We want the area beyond this interval.) The first step is to find the z-score corresponding to the boundary $X = 14.5$.

$$z = \frac{X - \mu}{\sigma} = \frac{14.5 - 12}{3} = 0.83$$

Now we can look up the probability in the unit normal table. In this case, we want the proportion in the tail of the distribution beyond $z = 0.83$.

The value from the unit normal table is 0.2033. This is the answer we want. If we select a random sample of 48 psychology majors, the probability of getting more than 14 males is $p = 0.2033$, or 20.33%.

LEARNING CHECK

1. Under what circumstances is the normal distribution an accurate approximation to the binomial distribution?

2. A multiple-choice test consists of 48 questions with 4 possible answers for each question. What is the probability that you would get more than 20 questions correct just by guessing?

3. If you toss a balanced coin 36 times, you would expect, on the average, to get 18 heads and 18 tails. What is the probability of obtaining exactly 18 heads in 36 tosses?

ANSWERS

1. When pn and qn are both greater than 10

2. $p = \frac{1}{4}$, $q = \frac{3}{4}$; $p(X > 20.5) = p(z > 2.83) = 0.0023$

3. $z = \pm 0.17$, $p = 0.1350$

6.7   LOOKING AHEAD TO INFERENTIAL STATISTICS

Probability forms a direct link between samples and the populations from which they come. As we noted at the beginning of this chapter, this link will be the foundation for the inferential statistics in future chapters. The following example provides a brief preview of how probability will be used in the context of inferential statistics.

We ended Chapter 5 with a demonstration of how inferential statistics are used to help interpret the results of a research study. A general research situation was shown in Figure 5.8 and is repeated here in Figure 6.23. Note that the research begins with a pop-

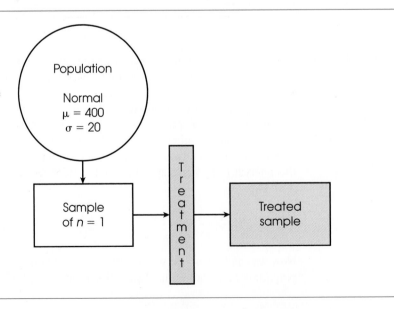

FIGURE 6.23

A diagram of a research study. One individual is selected from the population and receives a treatment. The goal is to determine whether or not the treatment has an effect.

ulation that forms a normal distribution with a mean of $\mu = 400$ and a standard deviation of $\sigma = 20$. One individual is selected from the population and a treatment is administered to that individual. The goal for the study is to evaluate the effect of the treatment.

To determine whether or not the treatment has an effect, the researcher will simply compare the treated individual with the original population. If the individual still has a score around 400 (the original population mean), then we must conclude that the treatment appears to have no effect. On the other hand, if the treated individual has a score that is noticeably different from 400, then the researcher has evidence that the treatment does have an effect. Notice that the study is using a sample (the individual) to help answer a question about a population; this is the essence of inferential statistics.

The problem for the researcher is determining exactly what is meant by "noticeably different" from 400. If the treated individual has a score of $X = 415$, is that enough to say that the treatment has an effect? What about $X = 420$ or $X = 450$? In Chapter 5, we suggested that z-scores provide one method for solving this problem. Specifically, we suggested that a z-score value beyond $z = 2.00$ (or -2.00) was an extreme value and therefore noticeably different. However, the choice of $z = \pm2.00$ was purely arbitrary. Now, we have another tool, *probability*, to help us decide exactly where to set the boundaries.

Figure 6.24 shows the original population from our hypothetical research study. Note that most of the scores are located close to $\mu = 400$. Also note that we have added boundaries separating the middle 95% of the distribution from the extreme 5% in the tails. The boundaries are located at $z = +1.96$ and $z = -1.96$. We obtained these boundaries from the unit normal table using the following procedure:

1. The 95% in the middle of the distribution is split in half, with 47.5% to the right of the mean and 47.5% to the left.

2. We located 47.5% (0.4750) in column D of the unit normal table and found the corresponding z-score to be $z = 1.96$.

FIGURE 6.24

Using probability to evaluate a treatment effect. Values that are extremely unlikely to be obtained from the original population are viewed as evidence of a treatment effect.

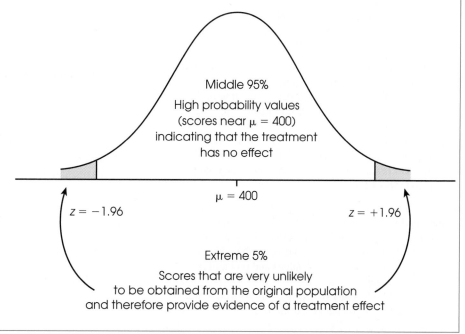

Middle 95%
High probability values
(scores near $\mu = 400$)
indicating that the treatment
has no effect

$\mu = 400$

$z = -1.96$ $z = +1.96$

Extreme 5%
Scores that are very unlikely
to be obtained from the original population
and therefore provide evidence of a treatment effect

The boundaries set at $z = \pm 1.96$ provide objective criteria for deciding whether or not our sample is noticeably different from the original population. Specifically, a sample that falls beyond the $z = \pm 1.96$ boundaries is not only extreme, it also is extremely unlikely, where "extremely unlikely" is defined as a probability that is 5% or less. If our treated individual has a score that is located beyond the $z = \pm 1.96$ boundaries, we can reach a logical conclusion as follows:

a. It is very unlikely that this score comes directly from the original population.

b. Therefore, this individual is noticeably different from the original population.

c. Therefore, it appears that the treatment has an effect.

Note that probabilities allow us to separate a distribution into those scores that are likely to be obtained (high probability) and those scores that are extremely unlikely (low probability). The extremely unlikely values provide us with criteria for determining whether or not a treatment has caused individuals to be different.

SUMMARY

1. The probability of a particular event A is defined as a fraction or proportion:

$$p(A) = \frac{\text{number of outcomes classified as } A}{\text{total number of possible outcomes}}$$

2. This definition is accurate only for a random sample. There are two requirements that must be satisfied for a random sample:
   a. Every individual in the population has an equal chance of being selected.
   b. When more than one individual is being selected, the probabilities must stay constant. This means there must be sampling with replacement.

3. All probability problems can be restated as proportion problems. The "probability of selecting a king from a deck of cards" is equivalent to the "proportion of the deck that consists of kings." For frequency distributions, probability questions can be answered by determining proportions of area. The "probability of selecting an individual with an IQ greater than 108" is equivalent to the "proportion of the whole population that consists of IQs above 108."

4. For normal distributions, probabilities (proportions) can be found in the unit normal table. The table provides a listing of the proportions of a normal distribution that correspond to each z-score value. With the table, it is possible to move between X values and probabilities using a two-step procedure:
   a. The z-score formula (Chapter 5) allows you to transform X to z or to change z back to X.

   b. The unit normal table allows you to look up the probability (proportion) corresponding to each z-score or the z-score corresponding to each probability.

5. Percentiles and percentile ranks measure the relative standing of a score within a distribution. Percentile rank is the percentage of individuals with scores at or below a particular X value. A percentile is an X value that is identified by its rank. The percentile rank always corresponds to the proportion to the left of the score in question.

6. The binomial distribution is used whenever the measurement procedure simply classifies individuals into exactly two categories. The two categories are identified as A and B, with probabilities of

$$p(A) = p \quad \text{and} \quad p(B) = q$$

7. The binomial distribution gives the probability for each value of X, where X equals the number of occurrences of category A in a series of n events. For example, X equals the number of heads in $n = 10$ tosses of a coin.

8. When pn and qn are both at least 10, the binomial distribution is closely approximated by a normal distribution with

$$\mu = pn$$
$$\sigma = \sqrt{npq}$$

9. In the normal approximation to the binomial distribution, each value of X has a corresponding z-score:

$$z = \frac{X - \mu}{\sigma} = \frac{X - pn}{\sqrt{npq}}$$

With the z-score and the unit normal table, you can find probability values associated with any value of X. For maximum accuracy, you should use the appropriate real limits for X when computing z-scores and probabilities.

RESOURCES

WebTUTOR

If you are using WebTutor along with this book, there is a WebTutor section corresponding to this chapter. The WebTutor contains a brief summary of Chapter 6, hints for learning the new material and for avoiding common errors, and sample exam items including solutions.

SPSS

The statistics computer package SPSS is not structured to compute probabilities. However, the program does report probability values as part of the inferential statistics that we will examine later in this book. In the context of inferential statistics, the probabilities are called *significance levels*, and they warn researchers about the probability of misinterpreting their research results.

KEY TERMS

probability	percentile rank	percentile
random sample	unit normal table	normal approximation (binomial)
sampling with replacement	binomial distribution	

FOCUS ON PROBLEM SOLVING

1. We have defined probability as being equivalent to a proportion, which means that you can restate every probability problem as a proportion problem. This definition is particularly useful when you are working with frequency distribution graphs where the population is represented by the whole graph and probabilities (proportions) are represented by portions of the graph. When working problems with the normal distribution, you always should start with a sketch of the distribution. You should shade the portion of the graph that reflects the proportion you are looking for.

2. When using the unit normal table, you must remember that the proportions in the table (columns B, C, and D) correspond to specific portions of the distribution. This is

important because you will often need to translate the proportions from a problem into specific proportions provided in the table. Also, remember that the table allows you to move back and forth between z-scores and proportions. You can look up a given z to find a proportion, or you can look up a given proportion to find the corresponding z-score.

3. Remember that the unit normal table shows only positive z-scores in column A. However, since the normal distribution is symmetrical, the proportions in the table apply to both positive and negative z-score values.

4. A common error for students is to use negative values for proportions on the left-hand side of the normal distribution. Proportions (or probabilities) are always positive: 10% is 10% whether it is in the left or right tail of the distribution.

5. The proportions in the unit normal table are accurate only for normal distributions. If a distribution is not normal, you cannot use the table.

6. When determining percentiles or percentile ranks, it helps to sketch the distribution first. Remember that the percentile rank always corresponds to the proportion to the left of the X value.

7. For maximum accuracy when using the normal approximation to the binomial distribution, you must remember that each X value is an interval bounded by real limits. For example, to find the probability of obtaining an X value greater than 10, you should use the real limit 10.5 in the z-score formula. Similarly, to find the probability of obtaining an X value less than 10, you should use the real limit 9.5.

DEMONSTRATION 6.1

FINDING PROBABILITY FROM THE UNIT NORMAL TABLE

A population is normally distributed with a mean of $\mu = 45$ and a standard deviation of $\sigma = 4$. What is the probability of randomly selecting a score that is greater than 43? In other words, what proportion of the distribution consists of scores greater than 43?

STEP 1   Sketch the distribution.
    You should always start by sketching the distribution, identifying the mean and standard deviation ($\mu = 45$ and $\sigma = 4$ in this example). You should also find the approximate location of the specified score and draw a vertical line through the distribution at that score. The score of $X = 43$ is lower than the mean, and, therefore, it should be placed somewhere to the left of the mean. Figure 6.25(a) shows the preliminary sketch.

STEP 2   Shade in the distribution.
    Read the problem again to determine whether you want the proportion greater than the score (to the right of your vertical line) or less than the score (to the left of the line). Then shade in the appropriate portion of the distribution. In this demonstration, we are considering scores greater than 43. Thus, we shade in the distribution to the right of this score [Figure 6.25(b)]. Notice that if the shaded area covers more than one-half of the distribution, then the probability should be greater than 0.5000.

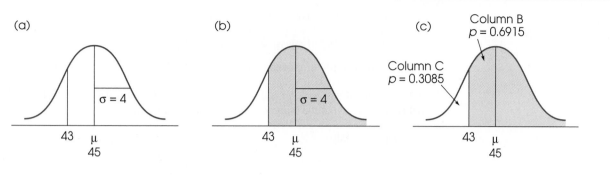

FIGURE 6.25

Sketches of the distribution for Demonstration 6.1.

S T E P   3   Transform the X value to a z-score.

Remember: To get a probability from the unit normal table, we must first convert the X value to a z-score.

$$z = \frac{X - \mu}{\sigma} = \frac{43 - 45}{4} = \frac{-2}{4} = -0.5$$

S T E P   4   Consult the unit normal table.

Look up the z-score (ignoring the sign) in the unit normal table. Find the proportions in the table that are associated with the z-score, and write the proportions in the appropriate regions on the figure. Remember that column B gives the proportion of area in the body of the distribution (more than 50%), and column C gives the area in the tail beyond z. Figure 6.25(c) shows the proportions for different regions of the normal distribution.

S T E P   5   Determine the probability.

If one of the proportions in step 4 corresponds exactly to the entire shaded area, then that proportion is your answer. Otherwise, you will need to do some additional arithmetic. In this example, the proportion in the shaded area is given in column B:

$$p(X > 43) = 0.6915$$

DEMONSTRATION 6.2

PROBABILITY AND THE BINOMIAL DISTRIBUTION

Suppose that you forgot to study for a quiz and now must guess on every question. It is a true-false quiz with $n = 40$ questions. What is the probability that you will get at least 26 questions correct just by chance? Stated in symbols,

$$p(X \geq 26) = ?$$

STEP 1   Identify p and q.
This problem is a binomial situation, in which

$$p = \text{probability of guessing correctly} = 0.50$$

$$q = \text{probability of guessing incorrectly} = 0.50$$

With $n = 40$ quiz items, both pn and qn are greater than 10. Thus, the criteria for the normal approximation to the binomial distribution are satisfied:

$$pn = 0.50(40) = 20$$

$$qn = 0.50(40) = 20$$

STEP 2   Identify the parameters, and sketch the distribution.
The normal approximation will have a mean and a standard deviation as follows:

$$\mu = pn = 0.5(40) = 20$$

$$\sigma = \sqrt{npq} = \sqrt{10} = 3.16$$

Figure 6.26 shows the distribution. We are looking for the probability of getting $X = 26$ or more questions correct. Remember that when determining probabilities from the binomial distribution, we must use real limits. Because we are interested in scores equal to or greater than 26, we will use the lower real limit for 26 (25.5). By using the lower real limit, we include the entire interval (25.5 to 26.5) that corresponds to $X = 26$. In Figure 6.26, everything to the right of $X = 25.5$ is shaded.

STEP 3   Compute the z-score, and find the probability.
The z-score for $X = 25.5$ is calculated as follows:

$$z = \frac{X - pn}{\sqrt{npq}} = \frac{25.5 - 20}{3.16} = +1.74$$

The shaded area in Figure 6.26 corresponds to column C of the unit normal table. For the z-score of 1.74, the proportion of the shaded area is $P = 0.0409$. Thus, the probability of getting at least 26 questions right just by guessing is

$$p(X \geq 26) = 0.0409 \text{ (or 4.09\%)}$$

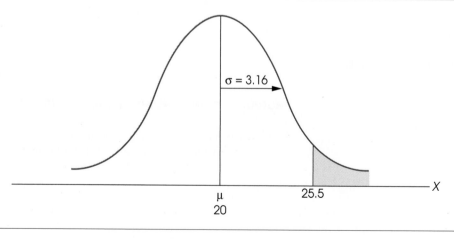

FIGURE 6.26

The normal approximation to a binomial distribution with $\mu = 20$ and $\sigma = 3.16$. The proportion of all scores equal to or greater than 26 is shaded. Notice that the real lower limit (25.5) for $X = 26$ is used.

PROBLEMS

1. In a psychology class of 90 students, there are 30 males and 60 females. Of the 30 men, only 5 are freshmen. Of the 60 women, 10 are freshmen. If you randomly sample an individual from this class,
   a. What is the probability of obtaining a female?
   b. What is the probability of obtaining a freshman?
   c. What is the probability of obtaining a male freshman?

2. A jar contains 10 black marbles and 40 white marbles.
   a. If you randomly select a marble from the jar, what is the probability that you will get a white marble?
   b. If you are selecting a random sample of $n = 3$ marbles and the first 2 marbles are both white, what is the probability that the third marble will be black?

3. What requirements must be satisfied to have a *random sample*?

4. What is sampling with replacement, and why is it used?

5. For each of the following z-score values, sketch a normal distribution and draw a vertical line at the location of the z-score. Then, determine whether the tail is to the right or the left of the z-score and find the proportion of the distribution located in the tail.
   a. $z = 1.00$ b. $z = -2.00$
   c. $z = 1.50$ d. $z = -0.75$

6. For each of the following z-score values, sketch a normal distribution and draw a vertical line at the location of the z-score. Then, determine whether the body is to the right or the left of the z-score and find the proportion of the distribution located in the body.
   a. $z = 0.25$ b. $z = -1.50$
   c. $z = 1.75$ d. $z = -0.40$

7. Find each of the following probabilities for a normal distribution that has been transformed into z-scores.
   a. $p(z > 1.50)$ b. $p(z > -2.00)$
   c. $p(z < 0.50)$ d. $p(z < -0.75)$

8. For a normal distribution, find the probability of randomly selecting a z-score
   a. Between $z = +0.50$ and $z = -0.50$
   b. Between $z = +1.00$ and $z = -1.00$
   c. Between $z = +1.50$ and $z = -1.50$
   d. Between $z = +2.00$ and $z = -2.00$

9. For a normal distribution, identify the z-score location that separates the distribution into the two sections described in each of the following:
   a. What z-score separates the top 10% of the distribution (right side) from the bottom 90% (left side)?

   b. What z-score separates the top 40% from the bottom 60%?
   c. What z-score separates the top 80% from the bottom 20%?

10. For a normal distribution, identify the z-score location that would separate the distribution into sections as described in each of the following:
    a. What z-score would separate the distribution so that there is 20% in the tail on the right-hand side?
    b. What z-score would separate the distribution so that there is 25% in the tail on the right-hand side?
    c. What z-score would separate the distribution so that there is 15% in the tail on the left-hand side?
    d. What z-score would separate the distribution so that there is 30% in the tail on the left-hand side?

11. For each of the following pairs of z-scores, find the proportion of a normal distribution that is located between the two z-score values.
    a. Between $z = 1.50$ and $z = 2.00$
    b. Between $z = -1.00$ and $z = -0.50$
    c. Between $z = 0$ and $z = 2.00$
    d. Between $z = -1.50$ and $z = 0.25$

12. For a normal distribution, find the z-score values that divide the distribution as described in each of the following.
    a. What z-scores separate the middle 80% from the extreme 20% in the tails?
    b. What z-scores separate the middle 85% from the extreme 15% in the tails?
    c. What z-scores separate the middle 90% from the extreme 10% in the tails?
    d. What z-scores separate the middle 95% from the extreme 5% in the tails?

13. Answer each of the following for a positively skewed distribution. (*Caution:* This is a trick question.)
    a. Find the proportion of the distribution located in the tail beyond $z = 2.00$.
    b. What z-score value separates the top 40% of the distribution from the rest?

14. A normal distribution has $\mu = 80$ and $\sigma = 20$. For each of the following scores, indicate whether the tail of the distribution is located to the right or the left of the score, and find the proportion of the distribution located in the tail.
    a. $X = 85$ b. $X = 92$
    c. $X = 70$ d. $X = 64$

15. The distribution of IQ scores is normal with $\mu = 100$ and $\sigma = 15$.
    a. What proportion of the population have IQ scores above 130?

b. What proportion of the population have IQ scores below 90?

c. What proportion of the population have IQ scores above 110?

16. The distribution of scores on the SAT is approximately normal with $\mu = 500$ and $\sigma = 100$.

    a. What proportion of the population have SAT scores above 650?

    b. What proportion of the population have SAT scores below 540?

    c. What is the minimum SAT score needed to be in the highest 20% of the population?

    d. What SAT score separates the top 60% from the rest of the distribution?

17. The distribution of SAT scores is normal with $\mu = 500$ and $\sigma = 100$.

    a. What SAT score, X value, separates the top 10% from the rest of the distribution?

    b. What SAT score, X value, separates the top 30% from the rest of the distribution?

    c. What SAT score, X value, separates the top 40% from the rest of the distribution?

18. A population forms a normal distribution with $\mu = 68$ and $\sigma = 6$. Find the following probabilities:

    a. The probability of selecting a score with a value less than 60

    b. The probability of selecting a score with a value greater than 58

    c. The probability of selecting a score with a value greater than 74

19. A normal distribution has $\mu = 225$ with $\sigma = 15$. Find the following probabilities:

    a. $p(X > 227)$ **b.** $p(X > 212)$

    c. $p(X < 230)$ **d.** $p(X < 220)$

20. A normal distribution has a mean of 80 and a standard deviation of 10. For this distribution, find each of the following probability values:

    a. $p(X > 75) = ?$ **b.** $p(X < 65) = ?$

    c. $p(X < 100) = ?$ **d.** $P(65 < X < 95) = ?$

    e. $P(84 < X < 90) = ?$

21. The distribution of SAT scores is normal with $\mu = 500$ and $\sigma = 100$.

    a. What SAT scores, X values, separate the middle 60% from the rest of the distribution?

    b. What SAT scores, X values, separate the middle 80% from the rest of the distribution?

    c. What SAT scores, X values, separate the middle 95% from the rest of the distribution?

22. A normal distribution has a mean of $\mu = 120$ with $\sigma = 15$. Find the following values:

    a. The 15th percentile

    b. The 88th percentile

    c. The percentile rank for $X = 142$

    d. The percentile rank for $X = 102$

    e. The percentile rank for $X = 120$

    f. The semi-interquartile range

23. For a normal distribution with $\mu = 200$ and $\sigma = 50$,

    a. Find the 90th percentile.

    b. Find the 30th percentile.

    c. Find the 75th percentile.

24. For a normal distribution with $\mu = 500$ and $\sigma = 100$,

    a. What is the percentile rank for $X = 550$?

    b. What is the percentile rank for $X = 650$?

    c. What is the percentile rank for $X = 450$?

    d. What is the percentile rank for $X = 350$?

25. One common test for extrasensory perception (ESP) requires participants to predict the suit of a card that is randomly selected from a complete deck. If a participant has no ESP and is just guessing, find each of the probabilities requested.

    a. What is the probability of a correct prediction on each trial?

    b. What is the probability of correctly predicting more than 18 out of 48 cards?

    c. What is the probability of correctly predicting at least 18 out of 48 cards?

26. A trick coin has been weighted so that the probability of heads is 0.8 and the probability of tails is 0.2.

    a. If you toss the coin 100 times, how many heads would you expect on the average?

    b. What is the probability of obtaining more than 95 heads in 100 tosses?

    c. What is the probability of obtaining fewer than 95 heads in 100 tosses?

    d. What is the probability of obtaining exactly 95 heads in 100 tosses?

27. For a balanced coin:

    a. What is the probability of getting more than 30 heads in 50 tosses?

    b. What is the probability of getting more than 60 heads in 100 tosses?

    c. Parts a and b of this question both asked for the probability of getting more than 60% heads in a series of coin tosses $\left(\frac{30}{50} = \frac{60}{100} = 60\%\right)$. Explain why the two probabilities are different.

28. A true-false test has 36 questions. If a passing grade on this test is $X = 24$ (or more) correct, what is the probability of obtaining a passing grade just by guessing?

C H A P T E R

7

Probability and Samples: The Distribution of Sample Means

Tools You Will Need

The following items are considered essential background material for this chapter. If you doubt your knowledge of any of these items, you should review the appropriate chapter and section before proceeding.

- Random sampling (Chapter 6)
- Probability and the normal distribution (Chapter 6)
- *z*-Scores (Chapter 5)

Preview

Now that you have some understanding of probability, consider the following problem.

> Imagine an urn filled with balls. Two-thirds of the balls are one color, and the remaining one-third are a second color. One individual selects 5 balls from the urn and finds that 4 are red and 1 is white. Another individual selects 20 balls and finds that 12 are red and 8 are white. Which of these two individuals should feel more confident that the urn contains two-thirds red balls and one-third white balls, rather than the opposite?*

When Tversky and Kahneman (1974) presented this problem to a group of experimental participants, they found that most people felt that the first sample (4 out of 5) provided much stronger evidence and, therefore, should give more confidence. At first glance, it may appear that this is the correct decision. After all, the first sample contained $\frac{4}{5}$ = 80% red balls, and the second sample contained only $\frac{12}{20}$ = 60% red balls. However, you should also notice that the two samples differ in another important respect: the sample size. One sample contains only $n = 5$, and the other sample contains $n = 20$. The correct answer to the problem is that the larger sample (12 out of 20) gives a much stronger justification for concluding that the balls in the urn are predominantly red. It appears that most people tend to focus on the sample proportion and pay very little attention to the sample size.

The importance of sample size may be easier to appreciate if you approach the urn problem from a different perspective. Suppose that you are the individual assigned responsibility for selecting a sample and then deciding which color is in the majority. Before you select your sample, you are offered a choice of selecting either a sample of 5 balls or a sample of 20 balls. Which would you prefer? It should be clear that the larger sample would be better. With a small number, you risk obtaining an unrepresentative sample. By chance, you could end up with 3 white balls and 2 red balls even though the reds outnumber the whites by 2 to 1. The larger sample is much more likely to provide an accurate representation of the population. This is an example of the *law of large numbers*, which states that large samples will be representative of the population from which they are selected. One final example should help demonstrate this law. If you were tossing a coin, you probably would not be greatly surprised to obtain 3 heads in a row. However, if you obtained a series of 20 heads in a row, you almost certainly would suspect a trick coin. The large sample has more authority.

In this chapter, we will examine the relationship between samples and populations. More specifically, we will consider the relationship between sample means and the population mean. As you will see, sample size is one of the primary considerations in determining how well a sample mean represents the population mean.

---

*Adapted from Tversky, A., and Kahneman, D. (1974). Judgments under uncertainty: Heuristics and biases. *Science, 185,* 1124–1131. Copyright 1974 by the AAAS.

7.1 OVERVIEW

The preceding two chapters presented the topics of *z*-scores and probability. Whenever a score is selected from a population, you should be able to compute a *z*-score that describes exactly where the score is located in the distribution. And, if the population is normal, you should be able to determine the probability value for obtaining any individual score. In a normal distribution, for example, a *z*-score of +2.00 corresponds to an extreme score out in the tail of the distribution, and a score at least this large has a probability of only $p = .0228$.

However, the *z*-scores and probabilities that we have considered so far are limited to situations in which the sample consists of a single score. Most research studies use much larger samples such as $n = 25$ preschool children or $n = 100$ laboratory rats. In this chapter we will extend the concepts of *z*-scores and probability to cover situations with larger samples. In particular, we will introduce a procedure for transforming a sample mean into a *z*-score. Thus, a researcher will be able to compute a *z*-score that describes an entire sample. As always, a z-score value near zero indicates a central, representative sample; a z-value beyond +2.00 or −2.00 indicates an extreme sample. Thus, it will be possible to describe how any specific sample is related to all the other possible samples. In addition, we will be able to use the z-score values to look up probabilities for obtaining certain samples, no matter how many scores the sample contains.

In general, the difficulty of working with samples is that a sample provides an incomplete picture of the population. Suppose, for example, a researcher selects a sample of $n = 25$ students from the State College. Although the sample should be representative of the entire student population, there will almost certainly be some segments of the population that are not included in the sample. In addition, any statistics that are computed for the sample will not be identical to the corresponding parameters for the entire population. For example, the average IQ for the sample of 25 students will not be the same as the overall mean IQ for the entire population. This difference, or *error* between sample statistics and the corresponding population parameters, is called *sampling error*.

DEFINITION

Sampling error is the discrepancy, or amount of error, between a sample statistic and its corresponding population parameter.

The concept of sampling error is demonstrated in Figure 1.2 on page 8.

Furthermore, samples are variable; they are not all the same. If you take two separate samples from the same population, the samples will be different. They will contain different individuals, they will have different scores, and they will have different sample means. How can you tell which sample gives the best description of the population? Can you even predict how well a sample will describe its population? What is the probability of selecting a sample that has a certain sample mean? These questions can be answered once we establish the set of rules that relate samples to populations.

7.2 THE DISTRIBUTION OF SAMPLE MEANS

As noted, two separate samples probably will be different even though they are taken from the same population. The samples will have different individuals, different scores, different means, and the like. In most cases, it is possible to obtain thousands of different samples from one population. With all these different samples coming from the same population, it may seem hopeless to try to establish some simple rules for the relationships between samples and populations. Fortunately, however, the huge set of possible samples forms a relatively simple, orderly, and predictable pattern that makes it possible to predict the characteristics of a sample with some accuracy. The ability to predict sample characteristics is based on the *distribution of sample means*.

DEFINITION

The **distribution of sample means** is the collection of sample means for all the possible random samples of a particular size (n) that can be obtained from a population.

Notice that the distribution of sample means contains *all the possible samples*. It is necessary to have all the possible values in order to compute probabilities. For example, if the entire set contains exactly 100 samples, then the probability of obtaining any specific sample is 1 out of 100: $p = \frac{1}{100}$.

Also, you should notice that the distribution of sample means is different from distributions we have considered before. Until now we always have discussed distributions of scores; now the values in the distribution are not scores, but statistics (sample means). Because statistics are obtained from samples, a distribution of statistics is referred to as a sampling distribution.

DEFINITION

A **sampling distribution** is a distribution of statistics obtained by selecting all the possible samples of a specific size from a population.

Thus, the distribution of sample means is an example of a sampling distribution. In fact, it often is called the sampling distribution of *M*.

If you actually wanted to construct the distribution of sample means, you would first select a random sample of a specific size (*n*) from a population, calculate the sample mean, and add the sample mean to a frequency distribution. Then you would select another random sample. Again, you calculate the sample mean and add it to the frequency distribution. This process is shown in Figure 7.1. You continue selecting samples and calculating means, over and over. Remember that all the samples have the same number of scores (*n*). Eventually, you would have the complete set of all the possible random samples, and your frequency distribution would show the distribution of sample means.

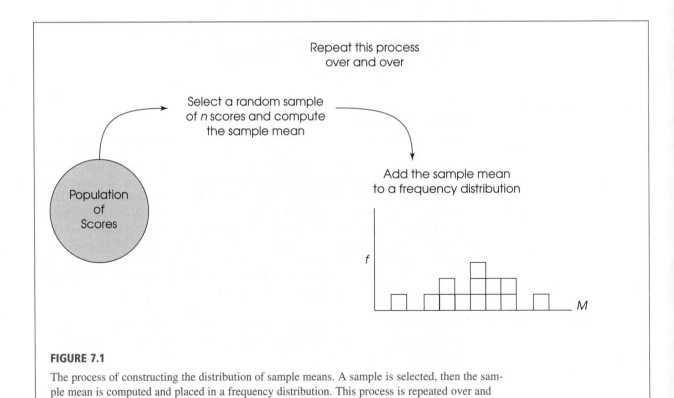

FIGURE 7.1

The process of constructing the distribution of sample means. A sample is selected, then the sample mean is computed and placed in a frequency distribution. This process is repeated over and over until all the possible random samples are obtained and the complete set of sample means is in the distribution.

We will demonstrate the process of constructing a distribution of sample means in Example 7.1, but first we use common sense and a little logic to predict the general characteristics of the distribution.

1. The sample means should pile up around the population mean. Samples are not expected to be perfect but they are representative of the population. As a result, most of the sample means should be relatively close to the population mean.

2. The pile of sample means should tend to form a normal-shaped distribution. Logically, most of the samples should have means close to μ, and it should be relatively rare to find sample means that are substantially different from μ. As a result, the sample means should pile up in the center of the distribution (around μ) and the frequencies should taper off as the distance between M and μ increases. This describes a normal-shaped distribution.

3. In general, the larger the sample size, the closer the sample means should be to the population mean, μ. Logically, a large sample should be a better representative than a small sample. Thus, the sample means obtained with a large sample size should cluster relatively close to the population mean; the means obtained from small samples should be more widely scattered.

As you will see, each of these three commonsense characteristics is an accurate description of the distribution of sample means. The following example demonstrates the process of constructing the distribution of sample means by repeatedly selecting samples from a population.

EXAMPLE 7.1   Consider a population that consists of only 4 scores: 2, 4, 6, 8. This population is pictured in the frequency distribution histogram in Figure 7.2.

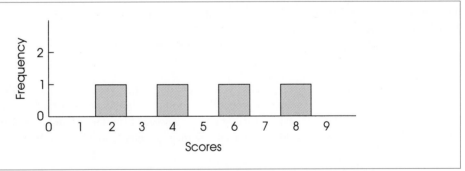

FIGURE 7.2

Frequency distribution histogram for a population of 4 scores: 2, 4, 6, 8.

We are going to use this population as the basis for constructing the distribution of sample means for $n = 2$. Remember: This distribution is the collection of sample means from all the possible random samples of $n = 2$ from this population. We begin by looking at all the possible samples. Each of the 16 different samples is listed in Table 7.1.

Next, we compute the mean, M, for each of the 16 samples (see the last column of Table 7.1). The 16 means are then placed in a frequency distribution histogram in Figure 7.3. This is the distribution of sample means. Note that the distribution in Figure 7.3 demonstrates two of the characteristics that we predicted for the distribution of sample means.

Remember that random sampling requires sampling with replacement.

1. The sample means tend to pile up around the population mean. For this example, the population mean is $\mu = 5$, and the sample means are clustered around a value of 5. It should not surprise you that the sample means tend to approximate the population mean. After all, samples are supposed to be representative of the population.

2. The distribution of sample means is approximately normal in shape. This is a characteristic that will be discussed in detail later and will be extremely useful because we already know a great deal about probabilities and the normal distribution (Chapter 6).

TABLE 7.1

All the possible samples of $n = 2$ scores that can be obtained from the population presented in Figure 7.2. Notice that the table lists *random samples*. This requires sampling with replacement, so it is possible to select the same score twice. Also note that samples are listed systematically. The first four examples are all the possible samples that have $X = 2$ as the first score; the next four samples all have $X = 4$ as the first score; etc. This way we are sure to have all the possible samples listed, although the samples probably would not be selected in this order.

| | SCORES | | SAMPLE MEAN |
SAMPLE	FIRST	SECOND	(M)
1	2	2	2
2	2	4	3
3	2	6	4
4	2	8	5
5	4	2	3
6	4	4	4
7	4	6	5
8	4	8	6
9	6	2	4
10	6	4	5
11	6	6	6
12	6	8	7
13	8	2	5
14	8	4	6
15	8	6	7
16	8	8	8

Remember that our goal in this chapter is to answer probability questions about samples with $n > 1$.

Finally, you should notice that we can use the distribution of sample means to answer probability questions about sample means (see Box 7.1). For example, if you take a sample of $n = 2$ scores from the original population, what is the probability of obtaining a sample mean greater than 7? In symbols,

$$p(M > 7) = ?$$

Because probability is equivalent to proportion, the probability question can be restated as follows: Of all the possible sample means, what proportion has values greater than 7? In this form, the question is easily answered by looking at the distribution of sample means. All the possible sample means are pictured (Figure 7.3), and only 1 out of the 16 means has a value greater than 7. The answer, therefore, is 1 out of 16, or $p = \frac{1}{16}$.

THE CENTRAL LIMIT THEOREM

Example 7.1 demonstrates the construction of the distribution of sample means for a relatively simple, specific situation. In most cases, however, it will not be possible to list all the samples and compute all the possible sample means. Therefore, it is neces-

FIGURE 7.3

The distribution of sample means for $n = 2$. The distribution shows the 16 sample means from Table 7.1.

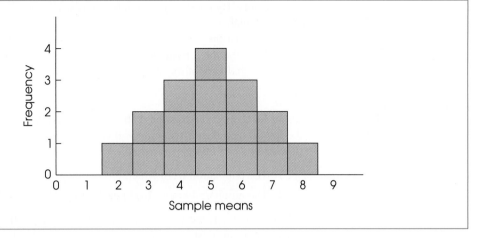

BOX

7.1   PROBABILITY AND THE DISTRIBUTION OF SAMPLE MEANS

I have a bad habit of losing playing cards. This habit is compounded by the fact that I always save the old deck in the hope that someday I will find the missing cards. As a result, I have a drawer filled with partial decks of playing cards. Suppose that I take one of these almost-complete decks, shuffle the cards carefully, and then randomly select one card. What is the probability that I will draw a king?

You should realize that it is impossible to answer this probability question. To find the probability of selecting a king, you must know how many cards are in the deck and exactly which cards are missing. (It is crucial that you know whether or not any kings are missing.) The point of this simple example is that any probability question requires that you have complete

information about the population from which the sample is being selected. In this case, you must know all the possible cards in the deck before you can find the probability for selecting any specific card.

In this chapter, we are examining probability and sample means. In order to find the probability for any specific sample mean, you first must know *all the possible sample means*. Therefore, we begin by defining and describing the set of all possible sample means that can be obtained from a particular population. Once we have specified the complete set of all possible sample means (i.e., the distribution of sample means), we will be able to find the probability of selecting any specific sample mean.

sary to develop the general characteristics of the distribution of sample means that can be applied in any situation. Fortunately, these characteristics are specified in a mathematical proposition known as the *central limit theorem*. This important and useful theorem serves as a cornerstone for much of inferential statistics. Following is the essence of the theorem:

> *Central Limit Theorem:* For any population with mean μ and standard deviation σ, the distribution of sample means for sample size n will have a mean of μ and a standard deviation of $\sigma/\sqrt{n}$, and will approach a normal distribution as n approaches infinity.

The value of this theorem comes from two simple facts. First, it describes the distribution of sample means for *any population*, no matter what shape, or mean, or standard deviation. Second, the distribution of sample means "approaches" a normal distribution very rapidly. By the time the sample size reaches $n = 30$, the distribution is almost perfectly normal.

Notice that the central limit theorem describes the distribution of sample means by identifying the three basic characteristics that describe any distribution: shape, central tendency, and variability. We will examine each of these.

THE SHAPE OF THE DISTRIBUTION OF SAMPLE MEANS

It has been observed that the distribution of sample means tends to be a normal distribution. In fact, this distribution will be almost perfectly normal if either one of the following two conditions is satisfied:

1. The population from which the samples are selected is a normal distribution.

2. The number of scores (n) in each sample is relatively large, around 30 or more.

(As n gets larger, the distribution of sample means will closely approximate a normal distribution. In most situations for which $n > 30$, the distribution is almost normal regardless of the shape of the original population.)

The fact that the distribution of sample means tends to be normal should not be surprising. Whenever you take a sample from a population, you expect the sample mean to be near to the population mean. When you take lots of different samples, you expect the sample means to "pile up" around μ, resulting in a normal-shaped distribution. You can see this tendency emerging (although it is not yet normal) in Figure 7.3.

THE MEAN OF THE DISTRIBUTION OF SAMPLE MEANS: THE EXPECTED VALUE OF M

The expected value of M is often identified by the symbol μ_M, signifying the "mean of the sample means." However, μ_M is always equal to μ, so we will continue to use the symbol μ to refer to the mean for the population of scores and the mean for the distribution of sample means.

You probably noticed in Example 7.1 that the distribution of sample means is centered around the mean of the population from which the samples were obtained. In fact, the average value of all the sample means is exactly equal to the value of the population mean. This fact should be intuitively reasonable; the sample means are expected to be close to the population mean, and they do tend to pile up around μ. The formal statement of this phenomenon is that the mean of the distribution of sample means always will be identical to the population mean. This mean value is called the *expected value of M*.

In commonsense terms, a sample mean is "expected" to be near its population mean. When all of the possible sample means are obtained, the average value will be identical to μ.

The fact that the average value of M is equal to μ was first introduced in Chapter 4 (page 122) in the context of *biased* versus *unbiased* statistics. The sample mean is an example of an unbiased statistic, which means that on average the sample statistic produces a value that is exactly equal to the corresponding population parameter. In this case, the average value of all the sample means is exactly equal to μ.

DEFINITION

The mean of the distribution of sample means is equal to μ (the population mean) and is called the **expected value of M**.

THE STANDARD ERROR OF M

So far, we have considered the shape and the central tendency of the distribution of sample means. To completely describe this distribution, we need one more characteristic, variability. The value we will be working with is the standard deviation for the distribution of sample means, and it is called the *standard error of M*. Remember that a sample is not expected to provide a perfectly accurate reflection of its population.

Although a sample mean should be representative of the population mean, there typically will be some error between the sample and the population. The standard error measures exactly how much difference should be expected on average between M and μ.

DEFINITION

The standard deviation of the distribution of sample means is called the **standard error of** M. The standard error measures the standard amount of difference between M and μ that is reasonable to expect simply by chance.

$$\text{standard error} = \sigma_M = \text{standard distance between } M \text{ and } \mu$$

The notation that is used to identify the standard error is σ_M. The σ indicates that we are measuring a standard deviation or a standard distance from the mean. The subscript M indicates that we are measuring the standard deviation for a distribution of sample means. The standard error is an extremely valuable measure because it specifies precisely how well a sample mean estimates its population mean—that is, how much error you should expect, on the average, between M and μ. Remember that one basic reason for taking samples is to use the sample data to answer questions about the population. Specifically, we can use the sample mean as an estimate of the population mean. Although we do not expect a sample mean to be exactly the same as the population mean, it should provide a good estimate. The standard error tells how good the estimate will be (see Box 7.2).

The magnitude of the standard error is determined by two factors: (1) the size of the sample and (2) the standard deviation of the population from which the sample is selected. We will examine each of these factors.

The sample size   Earlier we predicted, based on common sense, that the distribution of sample means would be tightly clustered around μ for large samples and more widely scattered for smaller samples. It should be intuitively reasonable that the size of a sample should influence how accurately the sample represents its population. Specifically, a large sample should be more accurate than a small sample. In general, as the sample size increases, the error between the sample mean and the population mean should decrease. This rule is also known as the *law of large numbers*.

BOX
7.2   **THE ERROR BETWEEN M AND μ**

Consider a population consisting of $N = 3$ scores: 1, 8, 9. The mean for this population is $\mu = 6$. Using this population, try to find a random sample of $n = 2$ scores with a sample mean (M) exactly equal to the population mean. (Try selecting a few samples; record the scores and the sample mean for each of your samples.)

You may have guessed that we constructed this problem so that it is impossible to obtain a sample mean that is identical to μ. The point of this exercise is to emphasize the notion of sampling error. Samples are not identical to their populations, and a sample mean generally will not provide a perfect estimate of the population mean. The purpose of standard error is to provide a quantitative measure of the difference (or error) between sample means and the population mean. Standard error is the standard distance between M and μ.

D E F I N I T I O N   The **law of large numbers** states that the larger the sample size (n), the more probable it is that the sample mean will be close to the population mean.

The population standard deviation   As we noted earlier, as the size of the sample decreases, the standard error increases. At the extreme, the smallest possible sample and the largest possible error occur when the sample consists of $n = 1$ score. At this extreme, the sample is a single score, X, and the sample mean is also X. In this case, the standard error measures the standard distance between the score, X, and the population mean, μ. However, we already know the standard distance between X and μ is the standard deviation. By definition, σ is the standard distance between X and μ. Then, when $n = 1$, the standard error and the standard deviation are identical.

$$\text{When } n = 1, \text{ standard error} = \sigma_M = \sigma = \text{standard deviation}$$

You can think of the standard deviation as the "starting point" for standard error. When $n = 1$, σ and σ_M are the same. As the sample size increases, the standard error decreases in relation to n.

The formula for standard error incorporates both the standard deviation and the sample size.

$$\text{standard error} = \sigma_M = \frac{\sigma}{\sqrt{n}} \tag{7.1}$$

Note that the formula satisfies all the requirements for the concept of standard error. Specifically, as the sample size increases, the error decreases. As n decreases, the error increases. At the extreme, when $n = 1$, the standard error is equal to the standard deviation.

In Equation 7.1 and in most of the preceding discussion, we have defined standard error in terms of the population standard deviation. However, the population standard deviation (σ) and the population variance (σ^2) are directly related, and it is easy to substitute variance into the equation for standard error. Using the simple equality $\sigma = \sqrt{\sigma^2}$, the equation for standard error can be rewritten as follows:

$$\text{standard error} = \sigma_M = \frac{\sigma}{\sqrt{n}} = \frac{\sqrt{\sigma^2}}{\sqrt{n}} = \sqrt{\frac{\sigma^2}{n}} \tag{7.2}$$

Throughout the rest of this chapter (and in Chapter 8), we will continue to define standard error in terms of the standard deviation (Equation 7.1). However, in later chapters (starting in Chapter 9) the formula based on variance (Equation 7.2) will become more useful.

Table 7.2 demonstrates the general relationship between standard error and sample size. Notice that standard error decreases regularly as the sample size increases. Technically, the magnitude of the standard error decreases in proportion to the square root of n. The square root relationship can be seen directly when standard error is calculated in terms of the standard deviation (second column in Table 7.2). The more general relationship that error decreases as n increases is apparent when standard error is expressed in terms of variance (third column). In this case, the variance is directly divided by the sample size. In either case, the simple, commonsense rule applies: The bigger the sample, the more accurately the sample mean represents the population mean.

TABLE 7.2

The relationship between
standard error and sample size

SAMPLE SIZE	STANDARD ERROR IN TERMS OF σ	STANDARD ERROR IN TERMS OF σ^2	STANDARD ERROR FOR $\sigma = 10$
$n = 1$	$\sigma/\sqrt{1}$	$\sqrt{\sigma^2/1}$	10.00
$n = 2$	$\sigma/\sqrt{2}$	$\sqrt{\sigma^2/2}$	7.07
$n = 3$	$\sigma/\sqrt{3}$	$\sqrt{\sigma^2/3}$	5.77
$n = 4$	$\sigma/\sqrt{4}$	$\sqrt{\sigma^2/4}$	5.00
$n = 5$	$\sigma/\sqrt{5}$	$\sqrt{\sigma^2/5}$	4.47
$n = 6$	$\sigma/\sqrt{6}$	$\sqrt{\sigma^2/6}$	4.08
$n = 7$	$\sigma/\sqrt{7}$	$\sqrt{\sigma^2/7}$	3.78
$n = 8$	$\sigma/\sqrt{8}$	$\sqrt{\sigma^2/8}$	3.54

LEARNING CHECK

1. A population of scores is normal with $\mu = 80$ and $\sigma = 20$.

   a. Describe the distribution of sample means for samples of size $n = 16$ selected from this population. (Describe shape, central tendency, and variability for the distribution.)

   b. How would the distribution of sample means be changed if the sample size were $n = 100$ instead of $n = 16$?

2. As sample size increases, the value of the standard error also increases. (True or false?)

3. Under what circumstances will the distribution of sample means be a normal-shaped distribution?

ANSWERS

1. a. The distribution of sample means is normal with a mean (expected value) of $\mu = 80$ and a standard error of $\sigma_M = 20/\sqrt{16} = 5$.

   b. The distribution would still be normal with a mean of 80 but the standard error would be reduced to $\sigma_M = 20/\sqrt{100} = 2$.

2. False. Standard error decreases as n increases.

3. The distribution of sample means will be normal if the sample size is relatively large (around $n = 30$ or more) or if the population of scores is normal.

7.3 PROBABILITY AND THE DISTRIBUTION OF SAMPLE MEANS

The primary use of the distribution of sample means is to find the probability associated with any specific sample. Recall that probability is equivalent to proportion. Because the distribution of sample means presents the entire set of all possible M's, we can use proportions of this distribution to determine probabilities. The following example demonstrates this process.

EXAMPLE 7.2

The population of scores on the SAT forms a normal distribution with $\mu = 500$ and $\sigma = 100$. If you take a random sample of $n = 25$ students, what is the probability that the sample mean will be greater than $M = 540$?

Caution: Whenever you have a probability question about a sample mean, you must use the distribution of sample means.

First, you can restate this probability question as a proportion question: Out of all the possible sample means, what proportion has values greater than 540? You know about "all the possible sample means"; this is simply the distribution of sample means. The problem is to find a specific portion of this distribution. The parameters of this distribution are the following:

a. The distribution is normal because the population of SAT scores is normal.

b. The distribution has a mean of 500 because the population mean is $\mu = 500$.

c. The distribution has a standard error of $\sigma_M = 20$:

$$\sigma_M = \frac{\sigma}{\sqrt{n}} = \frac{100}{\sqrt{25}} = \frac{100}{5} = 20$$

This distribution of sample means is shown in Figure 7.4.

We are interested in sample means greater than 540 (the shaded area in Figure 7.4), so the next step is to use a z-score to locate the exact position of $M = 540$ in the distribution. The value 540 is located above the mean by 40 points, which is exactly 2 standard deviations (in this case, exactly 2 standard errors). Thus, the z-score for $M = 540$ is $z = +2.00$.

Because this distribution of sample means is normal, you can use the unit normal table to find the probability associated with $z = +2.00$. The table indicates that 0.0228 of the distribution is located in the tail of the distribution beyond $z = +2.00$. Our conclusion is that it is very unlikely, $p = 0.0228$ (2.28%), to obtain a random sample of $n = 25$ students with an average SAT score greater than 540.

FIGURE 7.4

The distribution of sample means for $n = 25$. Samples were selected from a normal population with $\mu = 500$ and $\sigma = 100$.

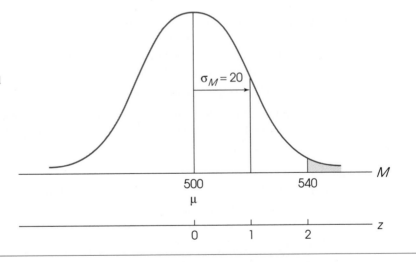

A z-SCORE FOR SAMPLE MEANS

As demonstrated in Example 7.2, it is possible to use a z-score to describe the position of any specific sample within the distribution of sample means. The z-score tells exactly where a specific sample is located in relation to all the other possible samples that could have been obtained. A z-score of $z = +2.00$, for example, indicates that the sample mean is much larger than usually would be expected: It is greater than the expected value of M by twice the standard distance. The z-score for each sample mean can be computed by using the standard z-score formula with a few minor changes. First, the value we are locating is a sample mean rather than a score, so the formula uses M in place of X. Second, the standard deviation for this distribution is measured by the standard error, so the formula uses σ_M in place of σ (see Box 7.3). The resulting formula, giving the z-score value corresponding to any sample mean, is

Caution: When computing z for a single score, use the standard deviation, σ. When computing z for a sample mean, you must use the standard error, σ_M.

$$z = \frac{M - \mu}{\sigma_M}$$

(7.3)

Every sample mean has a z-score that describes its position in the distribution of sample means. Using z-scores and the unit normal table, it is possible to find the probability associated with any specific sample mean (as in Example 7.2). Example 7.3 demonstrates that it also is possible to make quantitative predictions about the kinds of samples that should be obtained from any population.

BOX

7.3   THE DIFFERENCE BETWEEN STANDARD DEVIATION AND STANDARD ERROR

A constant source of confusion for many students is the difference between standard deviation and standard error. Remember that standard deviation measures the standard distance between a *score* and the population mean, $X - \mu$. Whenever you are working with a distribution of scores, the standard deviation is the appropriate measure of variability. Standard error, on the other hand, measures the standard distance between a *sample mean* and the population mean, $M - \mu$. Whenever you have a question concerning a sample, the standard error is the appropriate measure of variability.

If you still find the distinction confusing, there is a simple solution. Namely, if you always use standard error, you always will be right. Consider the formula for standard error:

$$\text{standard error} = \sigma_M = \frac{\sigma}{\sqrt{n}}$$

If you are working with a single score, then $n = 1$, and the standard error becomes

$$\text{standard error} = \sigma_M = \frac{\sigma}{\sqrt{n}} = \frac{\sigma}{\sqrt{1}}$$

$$= \sigma = \text{standard deviation}$$

Thus, standard error always measures the standard distance from the population mean, whether you have a sample of $n = 1$ or $n = 100$.

EXAMPLE 7.3   Once again, the distribution of SAT scores forms a normal distribution with a mean of $\mu = 500$ and a standard deviation of $\sigma = 100$. For this example, we are going to determine what kind of sample mean is likely to be obtained as the average SAT score for a random sample of $n = 25$ students.

First, we select a level of probability; for this example we will use 80%. Then we will determine the range of values that is expected for a sample mean 80% of the time. We begin with the distribution of sample means. Remember that this distribution is the collection of all the possible sample means, and it will show which samples are likely to be obtained and which are not.

As demonstrated in Example 7.2, the distribution of sample means for $n = 25$ will be normal with an expected value of $\mu = 500$ and a standard error of $\sigma_M = 20$. The

FIGURE 7.5

The middle 80% of the distribution of sample means for $n = 25$. Samples were selected from a normal population with $\mu = 500$ and $\sigma = 100$.

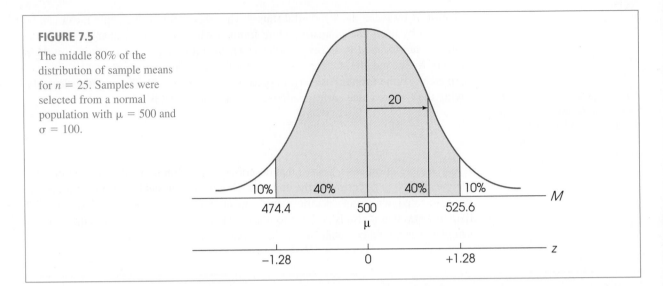

distribution of sample means is shown in Figure 7.5. Looking at the distribution, it should be clear that a sample of $n = 25$ students should have a mean SAT score around 500. To be more precise, we can determine the range of values that is expected 80% of the time by locating the middle 80% of the distribution. Because the distribution is normal, we can use the unit normal table. The 80% in the middle is split in half with 40% (0.4000) to the right of the mean and 40% (0.4000) to the left. Looking in column D of the table (proportion between the mean and z) for 0.4000, we find a corresponding z-score value of $z = 1.28$. Thus, the z-score boundaries for the middle 80% are $z = +1.28$ and $z = -1.28$. By definition, a z-score of 1.28 represents a location that is 1.28 standard deviations (or standard errors) from the mean. For this example, the distance from the mean is 1.28(20) = 25.6 points. The mean is $\mu = 500$, so a distance of 25.6 in both directions produces a range of values from 474.4 to 525.6.

Thus, 80% of all the possible sample means are contained in a range between 474.4 and 525.6. If we select a sample of $n = 25$ students, we can be 80% confident that the mean SAT score for the sample will be in this range.

The point of Example 7.3 is that the distribution of sample means makes it possible to predict the value that ought to be obtained for a sample mean. We know, for example, that a sample of $n = 25$ students ought to have a mean SAT score around 500. More specifically, we are 80% confident that the value of the sample mean will be between 474.4 and 525.6. The ability to predict sample means in this way will be a valuable tool for the inferential statistics that follow.

LEARNING CHECK   **1.** A population has $\mu = 80$ and $\sigma = 12$. Find the z-score corresponding to each of the following sample means:

    a. $M = 84$ for a sample of $n = 9$ scores

    b. $M = 74$ for a sample of $n = 16$ scores

    c. $M = 81$ for a sample of $n = 36$ scores

2. A random sample of $n = 25$ scores is selected from a normal population with $\mu = 90$ and $\sigma = 10$.

    a. What is the probability that the sample mean will have a value greater than 94?

    b. What is the probability that the sample mean will have a value less than 91?

3. A positively skewed distribution has $\mu = 60$ and $\sigma = 8$.

    a. What is the probability of obtaining a sample mean greater than $M = 62$ for a sample of $n = 4$ scores? (Be careful. This is a trick question.)

    b. What is the probability of obtaining a sample mean greater than $M = 62$ for a sample of $n = 64$ scores?

A N S W E R S   **1. a.** The standard error is $12/\sqrt{9} = 4$. The z-score is $z = +1.00$.

    b. The standard error is $12/\sqrt{16} = 3$. The z-score is $z = -2.00$.

    c. The standard error is $12/\sqrt{36} = 2$. The z-score is $z = +0.50$.

2. a. The standard error is $10/\sqrt{25} = 2$. A mean of $M = 94$ corresponds to $z = +2.00$, and $p = 0.0228$.

    b. A mean of $M = 91$ corresponds to $z = +0.50$, and $p = 0.6915$.

3. a. The distribution of sample means does not satisfy either of the criteria for being normal. Therefore, you cannot use the unit normal table, and it is impossible to find the probability.

    b. With $n = 64$, the distribution of sample means is nearly normal. The standard error is $8/\sqrt{64} = 1$, the z-score is $+2.00$, and the probability is 0.0228.

7.4 MORE ABOUT STANDARD ERROR

At the beginning of this chapter, we introduced the idea that it is possible to obtain thousands of different samples from a single population. Each sample will have its own individuals, its own scores, and its own sample mean. The distribution of sample means provides a method for organizing all of the different sample means into a single picture that shows how the sample means are related to each other and how they are related to the overall population mean. Figure 7.6 shows a prototypical distribution of sample means. To emphasize the fact that the distribution contains many different samples, we have constructed this figure so that the distribution is made up of hundreds of small boxes, each box representing a single sample mean. Also notice that the sample means tend to pile up around the population mean (μ), forming a normal-shaped distribution as predicted by the central limit theorem.

The distribution shown in Figure 7.6 provides a concrete example for reviewing the general concepts of sampling error and standard error. Although the following points may seem obvious, they are intended to provide you with a better understanding of these two statistical concepts.

1. Sampling Error. The general concept of sampling error is that a sample typically will not provide a perfectly accurate representation of its population. More specifically, there typically will be some discrepancy (or error) between a statistic computed for a sample and the corresponding parameter for the population. As you look at Figure 7.6,

FIGURE 7.6

An example of a typical distribution of sample means. Each of the small boxes represents the mean obtained for one sample.

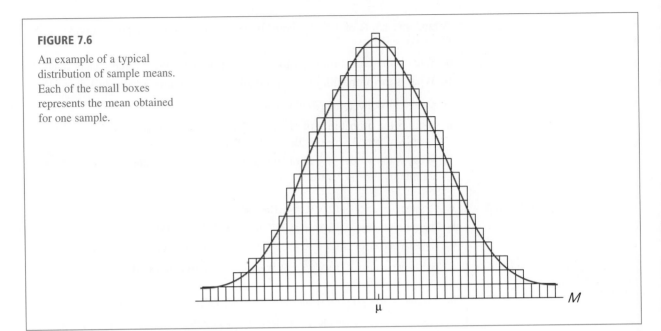

notice that the individual sample means are not exactly equal to the population mean. In fact, 50% of the samples will have means that are smaller than μ (the entire left-hand side of the distribution). Similarly, 50% of the samples will produce means that over-estimate the true population mean. In general, there will be some discrepancy, or *sampling error*, between the mean for a sample and the mean for the population from which the sample was obtained.

2. Standard Error. Again, looking at Figure 7.6, notice that most of the sample means are relatively close to the population mean (those in the center of the distribution). These samples provide a fairly accurate representation of the population. On the other hand, some samples will produce means that are out in the tails of the distribution, relatively far way from the population mean. These extreme sample means do not accurately represent the population. For each individual sample, you can measure the error (or distance) between the sample mean and the population mean. For some samples, the error will be relatively small, but for other samples, the error will be relatively large. The *standard error* provides a way to measure the "average" or standard distance between a sample mean and the population mean.

Thus, the standard error provides a method for defining and measuring sampling error. Knowing the standard error gives researchers a good indication of how accurately their sample data represent the populations they are studying. In most research situations, for example, the population mean is unknown, and the researcher selects a sample to help obtain information about the unknown population. Specifically, the sample mean provides information about the value of the unknown population mean. The sample mean is not expected to give a perfectly accurate representation of the population mean; there will be some error, and the standard error tells *exactly how much error*, on the average, should exist between the sample mean and the unknown population mean.

The following example demonstrates the use of the standard error and provides additional information about the relationship between standard error and standard deviation.

EXAMPLE 7.4   We will begin with a population that is normally distributed with a mean of $\mu = 80$ and a standard deviation of $\sigma = 20$. Next, we will take a sample from this population and examine how accurately the sample mean represents the population mean. More specifically, we will examine how sample size affects accuracy by considering three different samples: one with $n = 1$ score, one with $n = 4$ scores, and one with $n = 100$ scores.

Figure 7.7 shows the distributions of sample means based on samples of $n = 1$, $n = 4$, and $n = 100$. Each distribution shows the collection of all possible sample means that could be obtained for that particular sample size. Notice that all three sampling distributions are normal (because the original population is normal), and all three have the same mean, $\mu = 80$, which is the expected value of M. However, the three distributions differ greatly with respect to variability. We will consider each one separately.

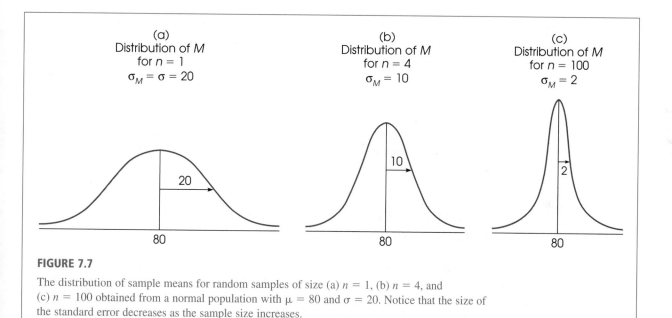

FIGURE 7.7

The distribution of sample means for random samples of size (a) $n = 1$, (b) $n = 4$, and (c) $n = 100$ obtained from a normal population with $\mu = 80$ and $\sigma = 20$. Notice that the size of the standard error decreases as the sample size increases.

The smallest sample size is $n = 1$. When a sample consists of a single score, the mean for the sample will equal the value of that score, $M = X$. Thus, when $n = 1$, the distribution of sample means is identical to the original population of scores. In this case, the standard error for the distribution of sample means is equal to the standard deviation for the original population. Equation 7.1 confirms this observation.

$$\sigma_M = \frac{\sigma}{\sqrt{n}} = \frac{20}{\sqrt{1}} = 20$$

In one sense, the population standard deviation is the "starting point" for the standard error. With the smallest possible sample, $n = 1$, the standard error is equal to the standard deviation [see Figure 7.7(a)].

As the sample size increases, however, the standard error gets smaller. For $n = 4$, the standard error is

$$\sigma_M = \frac{\sigma}{\sqrt{n}} = \frac{20}{\sqrt{4}} = \frac{20}{2} = 10$$

That is, the typical (or standard) distance between M and μ is 10 points. Figure 7.7(b) illustrates this distribution. Notice that the sample means in this distribution approximate the population mean more closely than in the previous distribution where $n = 1$.

With a sample of $n = 100$, the standard error is still smaller.

$$\sigma_M = \frac{\sigma}{\sqrt{n}} = \frac{20}{\sqrt{100}} = \frac{20}{10} = 2$$

A sample of $n = 100$ scores should provide a sample mean that is a much better estimate of μ than you would obtain with a sample of $n = 4$ or $n = 1$. As shown in Figure 7.7(c), there is very little error between M and μ (you would expect only a 2-point error, on the average). The sample means pile up very close to μ.

In summary, this example illustrates that with the smallest possible sample ($n = 1$), the standard error and the population standard deviation are the same. When sample size is increased, the standard error gets smaller, and the sample means tend to approximate μ more closely. Thus, standard error defines the relationship between sample size and the accuracy with which M represents μ.

IN THE LITERATURE
REPORTING STANDARD ERROR

As we will see later, the standard error plays a very important role in inferential statistics. Because of its crucial role, the standard error for a sample mean, rather than the sample standard deviation, is often reported in scientific papers. Scientific journals vary in how they refer to the standard error, but frequently the symbols *SE* and *SEM* (for standard error of the mean) are used. The standard error is reported in two ways. Much like the standard deviation, it may be reported in a table along with the sample means (see Table 7.3). Alternatively, the standard error may be reported in graphs.

TABLE 7.3

The mean self-consciousness scores for participants who were working in front of a video camera and those who were not (controls)

	n	MEAN	SE
Control	17	32.23	2.31
Camera	15	45.17	2.78

Figure 7.8 illustrates the use of a bar graph to display information about the sample mean and the standard error. In this experiment, two samples (groups A and B) are given different treatments, and then the subjects' scores on a dependent variable are recorded. The mean for group A is $M = 15$, and for group B, it is $M = 30$. For both samples, the standard error of M is $\sigma_M = 4$. Note that the mean is represented by the height of the bar, and the standard error is depicted on the graph by brackets at the top of each bar. Each bracket extends 1 standard error above and 1 standard error below the sample mean. Thus, the graph illustrates the mean for each group plus or minus 1 standard error ($M \pm SE$). When you glance at Figure 7.8, not only do you get a "picture" of the sample means, but also you get an idea of how much error you should expect for those means.

FIGURE 7.8

The mean ($\pm SE$) score for treatment groups A and B.

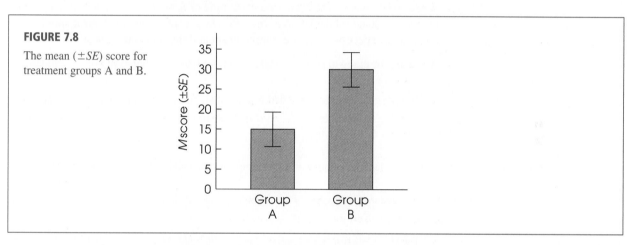

Figure 7.9 shows how sample means and standard error are displayed on a line graph. In this study, two samples (groups A and B) receive different treatments and then are tested on a task for four trials. The number of mistakes committed on each trial is recorded for all subjects. The graph shows the mean (M) number of mistakes committed for each group on each trial. The brackets show the size of the standard error for each sample mean. Again, the brackets extend 1 standard error above and below the value of the mean.

FIGURE 7.9

The mean ($\pm SE$) number of mistakes made for groups A and B on each trial.

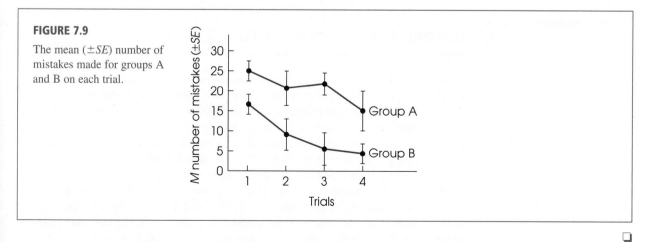

LEARNING CHECK

1. A population has a standard deviation of $\sigma = 20$.

   a. If a single score is selected from this population, how close, on average, would you expect the score to be to the population mean?

   b. If a sample of $n = 4$ scores is selected from the population, how close, on average, would you expect the sample mean to be to the population mean?

   c. If a sample of $n = 25$ scores is selected from the population, how close, on average, would you expect the sample mean to be to the population mean?

2. Can the magnitude of the standard error ever be larger than the population standard deviation? Explain your answer.

3. A researcher plans to select a random sample of $n = 36$ individuals from a population with a standard deviation of $\sigma = 18$. On average, how much error should the researcher expect between the sample mean and the population mean?

4. A researcher plans to select a random sample from a population with a standard deviation of $\sigma = 20$.

   a. How large a sample is needed to have a standard error of 10 points or less?

   b. How large a sample is needed if the researcher wants the standard error to be 2 points or less?

ANSWERS

1. a. The standard deviation, $\sigma = 20$, measures the standard distance between a score and the mean.

   b. The standard error is $20/\sqrt{4} = 10$ points.

   c. The standard error is $20/\sqrt{25} = 4$ points.

2. No. The standard error is always less than or equal to the standard deviation. The two values are equal only when $n = 1$.

3. The standard error is $18/\sqrt{36} = 3$ points.

4. a. A sample of $n = 4$ or larger.

   b. A sample of $n = 100$ or larger.

7.5 LOOKING AHEAD TO INFERENTIAL STATISTICS

Inferential statistics are methods that use sample data as the basis for drawing general conclusions about populations. However, we have noted that a sample is not expected to give a perfectly accurate reflection of its population. In particular, there will be some error or discrepancy between a sample statistic and the corresponding population parameter. In this chapter, we have observed that a sample mean will not be exactly equal to the population mean. The standard error of M specifies how much difference is expected on average between the mean for a sample and the mean for the population.

The natural differences that exist between samples and populations introduce a certain amount of uncertainty and error into all inferential processes. For example, there is always a margin of error that must be considered whenever a researcher uses a sample mean as the basis for drawing a conclusion about a population mean. Remember that the same mean is not perfect. In the next eight chapters we will introduce a variety of statistical methods that all use sample means to draw inferences about population means.

In each case, the distribution of sample means and the standard error will be critical elements in the inferential process. Before we begin this series of chapters, we will pause briefly to demonstrate how the distribution of sample means, along with z-scores and probability, can help us use sample means to draw inferences about population means.

EXAMPLE 7.5   Suppose that a psychologist is planning a research study to evaluate the effect of a new growth hormone. It is known that regular, adult rats (with no hormone) weigh an average of $\mu = 400$ grams. Of course, not all rats are the same size, and the distribution of their weights is normal with $\sigma = 20$. The psychologist plans to select a sample of $n = 25$ newborn rats, inject them with the hormone, and then measure their weights when they become adults. The structure of this research study is shown in Figure 7.10.

The psychologist will make a decision about the effect of the hormone by comparing the sample of treated rats with the regular untreated rats in the original population. If the treated rats in the sample are noticeably different from untreated rats, then the researcher has evidence that the hormone has an effect. The problem is to determine exactly how much difference is necessary before we can say that the sample is *noticeably different*.

The distribution of sample means and the standard error can help researchers make this decision. In particular, the distribution of sample means can be used to show exactly what would be expected for a sample of rats who do not receive any hormone injections. This allows researchers to make a simple comparison between

a. The sample of treated rats (from the research study)

b. Samples of untreated rats (from the distribution of sample means)

FIGURE 7.10

The structure of the research study described in Example 7.5. The purpose of the study is to determine whether or not the treatment (a growth hormone) has an effect on weight for rats.

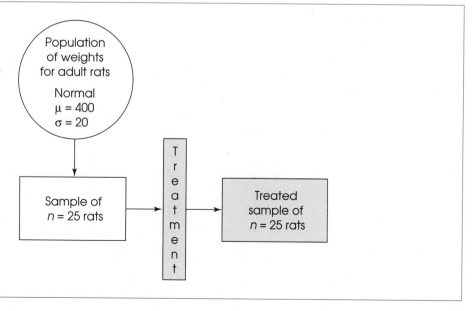

According to the central limit theorem, the distribution of sample means for $n = 25$ untreated rats will have the following characteristics:

1. It will be a normal distribution, because the population of rat weights is normal.

2. It will have an expected value of 400, because the population mean is $\mu = 400$.

3. It will have a standard error of $\sigma_M = 20/\sqrt{25} = 20/5 = 4$, because the population standard deviation is $\sigma = 20$ and the sample size is $n = 25$.

The distribution of sample means is shown in Figure 7.11. Notice that a sample of $n = 25$ untreated rats (without the hormone) should have a mean weight around 400 grams. To be more precise, we can use z-scores to determine the middle 95% of all the possible sample means. As demonstrated in Chapter 6 (page 191), the middle 95% of a normal distribution is located between z-score boundaries of $z = +1.96$ and $z = -1.96$ (check the unit normal table). These z-score boundaries are shown in Figure 7.11, and correspond to sample means of 392.16 and 407.84.

Thus, a sample of untreated rats is almost guaranteed (95% probability) to have a sample mean between 392.16 and 407.84. If our sample has a mean within this range, then we must conclude that our sample of treated rats does not look any different than a sample of untreated rats. In this case, we conclude that the hormone does not appear to have any effect.

On the other hand, if the mean for the treated sample is outside the 95% range, then we can conclude that the sample of treated rats is not the same as a sample of untreated rats. In this case, the research results do provide clear evidence that the hormone has an effect.

FIGURE 7.11

The distribution of sample means for samples of $n = 25$ untreated rats (from Example 7.5).

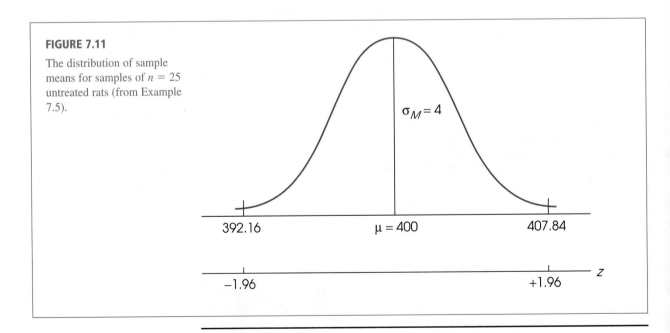

In Example 7.5 we used the distribution of sample means, together with z-scores and probability, to provide a description of what is reasonable to expect for an untreated sample. Then, we evaluated the effect of a treatment by determining whether or not the treated sample was noticeably different from an untreated sample. This procedure forms the foundation for the inferential technique known as *hypothesis testing* that is introduced in the following chapter.

STANDARD ERROR AS A MEASURE OF RELIABILITY

The research situation shown in Figure 7.10 introduces one final issue concerning sample means and standard error. In Figure 7.10, as in most research studies, the researcher must rely on a *single* sample to provide an accurate representation of the population being investigated. As we have noted, however, if you take two different samples from the same population, you will get different individuals with different scores and different sample means. Thus, every researcher must face the nagging question, "If I had taken a different sample, would I have obtained different results?"

The importance of this question is directly related to the degree of similarity among all the different samples. For example, if there is a high level of consistency from one sample to another, then a researcher can be reasonably confident that the specific sample being studied will provide a good measurement of the population. That is, when all the samples are similar, then it does not matter which one you have selected. On the other hand, if there are big differences from one sample to another, then the researcher is left with some doubts about the accuracy of his/her specific sample. In this case, a different sample could have produced vastly different results.

In this context, the standard error can be viewed as a measure of the *reliability* of a sample mean. The term *reliability* refers to the consistency of different measurements of the same thing. More specifically, a measurement procedure is said to be reliable if you make two different measurements of the same thing and obtain identical (or nearly identical) values. If you view a sample as a "measurement" of a population, then a sample mean is a "measurement" of a population mean. If the means from different samples are all nearly identical, then the sample mean provides a reliable measure of the population. On the other hand, if there are big differences from one sample to another, then the sample mean is an unreliable measure of the population mean.

Earlier, in Figure 7.7, we showed how the size of the standard error is related to sample size. The same figure can now be used to demonstrate how standard error provides a measure of reliability for a sample mean.

Look back at Figure 7.7(b), and see that it shows a distribution of sample means based on $n = 4$ scores. In this distribution, the standard error is $\sigma_M = 10$, and it is relatively easy to find two samples with means that differ by 10 or 20 points. In this situation, a researcher cannot be confident that the mean for any individual sample is reliable; that is, a different sample could provide a very different mean.

Now consider what happens when the sample size is increased. Figure 7.7(c) shows the distribution of sample means based on $n = 100$. For this distribution, the standard error is $\sigma_M = 2$. With a larger sample and a smaller standard error, all of the sample means are clustered close together. With $n = 100$, a researcher can be much more confident that the mean for any individual sample is reliable; that is, if a second sample were taken, it would probably produce a sample mean that is essentially equivalent to the mean for the first sample.

LEARNING CHECK

1. A population forms a normal distribution with a mean of $\mu = 50$ and a standard deviation of $\sigma = 10$.

    a. A sample of $n = 4$ scores from this population has a mean of $M = 55$. Would you describe this as a relatively typical sample or is the mean an extreme value? Explain your answer.

    b. If the sample from part a had $n = 25$ scores, would it be considered typical or extreme?

2. An automobile manufacturer claims that a newly introduced model will average $\mu = 45$ miles per gallon with $\sigma = 2$. A sample of $n = 4$ cars is tested and averages only $M = 42$ miles per gallon. Is this sample mean likely to occur if the manufacturer's claim is true? Specifically, is the sample mean within the range of values that would be expected 95% of the time? (Assume that the distribution of mileage scores is normal.)

ANSWERS

1. a. With $n = 4$ the standard error is 5, and the sample mean corresponds to $z = 1.00$. This is a relatively typical value.

    b. With $n = 25$ the standard error is 2, and the sample mean corresponds to $z = 2.50$. This is an extreme value.

2. With $n = 4$, the standard error is $\sigma_M = 1$. If the real mean is $\mu = 45$, then 95% of all sample means should be within $1.96(1) = 1.96$ points of $\mu = 45$. This is a range of values from 43.04 to 46.96. Our sample mean is outside this range, so it is not the kind of sample that ought to be obtained if the manufacturer's claim is true.

SUMMARY

1. The distribution of sample means is defined as the set of M's for all the possible random samples for a specific sample size (n) that can be obtained from a given population. According to the *central limit theorem*, the parameters of the distribution of sample means are as follows:

    a. *Shape*. The distribution of sample means will be normal if either one of the following two conditions is satisfied:

    (1) The population from which the samples are selected is normal.

    (2) The size of the samples is relatively large (around $n = 30$ or more).

    b. *Central Tendency*. The mean of the distribution of sample means will be identical to the mean of the population from which the samples are selected. The mean of the distribution of sample means is called the expected value of M.

    c. *Variability*. The standard deviation of the distribution of sample means is called the standard error of M and is defined by the formula

$$\sigma_M = \frac{\sigma}{\sqrt{n}} \quad \text{or} \quad \sigma_M = \sqrt{\frac{\sigma^2}{n}}$$

Standard error measures the standard distance between a sample mean (M) and the population mean (μ).

2. One of the most important concepts in this chapter is the standard error. The standard error is the standard deviation of the distribution of sample means. It measures the standard distance between a sample mean (M) and the population mean (μ). The standard error tells how much error to expect if you are using a sample mean to estimate a population mean.

3. The location of each M in the distribution of sample means can be specified by a z-score:

$$z = \frac{M - \mu}{\sigma_M}$$

Because the distribution of sample means tends to be normal, we can use these z-scores and the unit normal table to find probabilities for specific sample means. In particular, we can identify which sample means are likely and which are very unlikely to be obtained from any given population. This ability to find probabilities for samples is the basis for the inferential statistics in the chapters ahead.

4. In general terms, the standard error measures how much discrepancy you should expect, due to chance, between a sample statistic and a population parameter. Statistical inference involves using sample statistics to make a general conclusion about a population parameter. Thus, standard error plays a crucial role in inferential statistics.

RESOURCES

WebTUTOR

If you are using WebTutor along with this book, there is a WebTutor section corresponding to this chapter. The WebTutor contains a brief summary of Chapter 7, hints for learning about the distribution of sample means, cautions about common errors, and sample exam items including solutions.

In this chapter we defined the standard error in terms of the population standard deviation, σ, and the sample size, n. Specifically,

$$\text{standard error} = \frac{\sigma}{\sqrt{n}}$$

In later chapters, however, we will switch to an *estimated standard error* that is defined in terms of the sample standard deviation, s, and the sample size:

$$\text{estimated standard error} = \frac{s}{\sqrt{n}}$$

The estimated standard error serves the same purpose as the "real" standard error; that is, it measures how much difference, on average, there is between a sample mean and the population mean. The statistics computer package SPSS can be used to calculate the estimated standard error for a sample mean. First, enter the scores in one column of the data matrix and then click **Analyze** on the tool bar. Then, select **Descriptive Statistics** and select **Descriptives**. Enter the column label for the set of scores into the Variable box. Click **Options** and select **S.E. mean**, then click **Continue** and click **OK**. SPSS will produce a summary table listing the maximum score, the minimum score, the mean and the estimated standard error of the mean, and the standard deviation.

The estimated standard error will also appear in many of the inferential statistics that we will examine in later chapters.

KEY TERMS

sampling error

distribution of sample means

sampling distribution

central limit theorem

expected value of M

standard error of M

law of large numbers

FOCUS ON PROBLEM SOLVING

1. Whenever you are working probability questions about sample means, you must use the distribution of sample means. Remember that every probability question can be restated as a proportion question. Probabilities for sample means are equivalent to proportions of the distribution of sample means.

2. When computing probabilities for sample means, the most common error is to use standard deviation (σ) instead of standard error (σ_M) in the z-score formula. Standard deviation measures the typical deviation (or "error") for a single score. Standard error measures the typical deviation (or error) for a sample. Remember: The larger the sample is, the more accurately the sample represents the population; that is, the larger the sample, the smaller the error.

$$\text{Standard error} = \sigma_M = \frac{\sigma}{\sqrt{n}}$$

3. Although the distribution of sample means is often normal, it is not always a normal distribution. Check the criteria to be certain the distribution is normal before you use the unit normal table to find probabilities (see item 1a of the Summary). Remember that all probability problems with the normal distribution are easier if you sketch the distribution and shade in the area of interest.

DEMONSTRATION 7.1

PROBABILITY AND THE DISTRIBUTION OF SAMPLE MEANS

For a normally distributed population with $\mu = 60$ and $\sigma = 12$, what is the probability of selecting a random sample of $n = 36$ scores with a sample mean greater than 64?

In symbols, for $n = 36$

$$p(M > 64) = ?$$

Notice that we may rephrase the probability question as a proportion question. Out of all the possible sample means for $n = 36$, what proportion has values greater than 64?

STEP 1   Sketch the distribution.

We are looking for a specific proportion of *all possible sample means*. Therefore, we will have to sketch the distribution of sample means. We should include the expected value, the standard error, and the specified sample mean.

The expected value for this demonstration is $\mu = 60$. The standard error is

$$\sigma_M = \frac{\sigma}{\sqrt{n}} = \frac{12}{\sqrt{36}} = \frac{12}{6} = 2$$

Remember that we must use the standard error, *not* the standard deviation, because we are dealing with the distribution of sample means.

Find the approximate location of the sample mean, and place a vertical line through the distribution. In this demonstration, the sample mean is 64. It is larger than the expected value of $\mu = 60$ and, therefore, is placed on the right side of the distribution.

Figure 7.12(a) depicts the preliminary sketch.

STEP 2   Shade the appropriate area of the distribution.

Determine whether the problem asks for a proportion greater than ($>$) or less than ($<$) the specified sample mean. Then shade the appropriate area of the distribution. In this demonstration, we are looking for the area greater than $M = 64$, so we shade the area on the right-hand side of the vertical line [Figure 7.12(b)].

STEP 3   Compute the z-score for the sample mean.

Use the z-score formula for sample means. Remember that it uses standard error in the denominator.

$$z = \frac{M - \mu}{\sigma_M} = \frac{64 - 60}{2} = \frac{4}{2} = 2.00$$

STEP 4   Consult the unit normal table.

Look up the z-score in the unit normal table, and note the two proportions in columns B and C. Jot them down in the appropriate areas of the distribution [Figure 7.12(c)]. For this demonstration, the value in column C (the tail beyond z) corresponds exactly to the proportion we want (the shaded area). Thus, for $n = 36$,

$$p(M > 64) = p(z > +2.00) = 0.0228$$

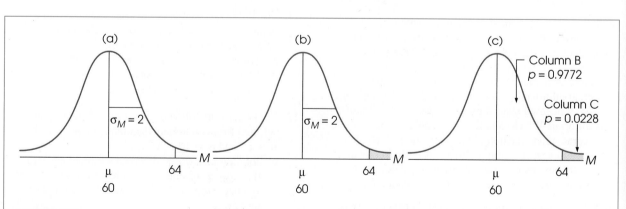

FIGURE 7.12

Sketches of the distributions for Demonstration 7.1.

PROBLEMS

1. Briefly define each of the following:
   a. Distribution of sample means
   b. Expected value of M
   c. Standard error of M

2. For each of the following, assume that the sample was selected from a population with $\mu = 75$ and $\sigma = 20$.
   a. What is the expected value of M for a sample of $n = 4$ scores?
   b. What is the standard error of M for a sample of $n = 4$ scores?
   c. What is the expected value of M for a sample of $n = 25$ scores?
   d. What is the standard error of M for a sample of $n = 25$ scores?

3. Describe the distribution of sample means (shape, expected value, and standard error) for samples of size $n = 36$ selected from a population with $\mu = 65$ and $\sigma = 12$.

4. The distribution of sample means is not always a normal distribution. Under what circumstances will the distribution of sample means *not* be normal?

5. The distribution of SAT scores is normal with $\mu = 500$ and $\sigma = 100$.
   a. If you selected a random sample of $n = 4$ scores from this population, how much error would you expect between the sample mean and the population mean?
   b. If you selected a random sample of $n = 25$ scores, how much error would you expect between the sample mean and the population mean?
   c. How much error would you expect for a sample of $n = 100$ scores?

6. IQ scores form normal distributions with $\sigma = 15$. However, the mean IQ varies from one population to another. For example, the mean IQ for registered voters is different from the mean for nonregistered voters. A researcher would like to use a sample to obtain information about the mean IQ for the population of licensed clinical psychologists in the state of California. Which sample mean should provide a more accurate representation of the population: a sample of $n = 9$ or a sample of $n = 25$? In each case, compute exactly how much error the researcher should expect between the sample mean and the population mean.

7. Standard error measures the standard distance between a sample mean and the population mean. For a population with $\sigma = 30$,
   a. How large a sample would be needed to have a standard error of 10 points or less?

   b. How large a sample would be needed to have a standard error of 5 points or less?
   c. Can the standard error ever be larger than the population standard deviation? Explain your answer.

8. For a population with a standard deviation of $\sigma = 20$:
   a. How large a sample would be needed to have a standard error less than 10 points?
   b. How large a sample would be needed to have a standard error less than 4 points?
   c. How large a sample would be needed to have a standard error less than 2 points?

9. A distribution of scores has $\sigma = 6$, but the value of the mean is unknown. A researcher plans to select a sample from the population in order to learn more about the unknown mean.
   a. If the sample consists of a single score ($n = 1$), how accurately should the score represent the population mean? That is, how much error, on average, should the researcher expect between the score and the population mean?
   b. If the sample consists of $n = 9$ scores, how accurately should the sample mean represent the population mean?
   c. If the sample consists of $n = 36$ scores, how much error, on average, would be expected between the sample mean and the population mean?

10. A population has $\mu = 60$ and $\sigma = 10$. Find the z-score corresponding to each of the following sample means:
    a. A sample of $n = 4$ with $M = 55$
    b. A sample of $n = 25$ with $M = 64$
    c. A sample of $n = 100$ with $M = 62$

11. The following samples were each selected from a population with $\mu = 35$ and $\sigma = 8$. Calculate the z-score corresponding to each sample mean.
    a. A sample of $n = 4$ scores with $M = 37$
    b. A sample of $n = 16$ scores with $M = 37$
    c. A sample of $n = 64$ scores with $M = 37$

12. For each of the following, assume that the sample was selected from a normally distributed population with $\mu = 65$ and $\sigma = 10$.
    a. For a sample of $n = 4$, would a sample mean of $M = 70$ be considered an extreme value compared to other samples of $n = 4$ scores? Explain your answer.
    b. For a sample of $n = 25$, would a sample mean of $M = 70$ be considered an extreme value compared to other samples of $n = 25$ scores? Explain your answer.

13. A normally distributed population has a mean of $\mu = 80$ and a standard deviation of $\sigma = 20$. Calculate the z-score for each of the following samples. Based on the z-score value, determine whether the sample mean is a typical, representative value or an extreme value for samples of this size.
   a. A sample of $n = 4$ scores with a mean of $M = 90$
   b. A sample of $n = 25$ scores with a mean of $M = 84$
   c. A sample of $n = 100$ scores with a mean of $M = 84$

14. If you are taking a random sample from a normally distributed population with $\mu = 100$ and $\sigma = 16$, which of the following outcomes is more likely? Explain your answer. (*Hint:* Calculate the z-score for each sample mean.)
   a. A sample mean greater than 106 for a sample of $n = 4$
   b. A sample mean greater than 103 for a sample of $n = 36$

15. A random sample of $n = 16$ scores is selected from a normal distribution with a mean of $\mu = 50$ and a standard deviation of $\sigma = 10$.
   a. What is the probability that the sample mean will have a value between 45 and 55?
   b. What is the probability that the sample mean will have a value between 48 and 52?
   c. What range of values has a 95% probability of containing the sample mean?

16. A normal distribution has $\mu = 100$ and $\sigma = 20$.
   a. Sketch the distribution of sample means for random samples of $n = 25$.
   b. Using z-scores, find the boundaries that separate the middle 95% of the sample means from the extreme 5% in the tails of the distribution.
   c. A sample mean of $M = 106$ is computed for a sample of $n = 25$ scores. Is this sample mean in the extreme 5%?

17. The distribution of SAT scores is normal with a mean of $\mu = 500$ and a standard deviation of $\sigma = 100$. If the mean SAT score is computed for a sample of $n = 100$ students,
   a. What range of values should contain the sample mean 90% of the time? (That is, find the middle 90% of the distribution of sample means.)
   b. What range of values should contain the sample mean 95% of the time?
   c. What range of values should contain the sample mean 99% of the time?

18. A normal population has $\mu = 70$ and $\sigma = 12$.
   a. Sketch the population distribution. What proportion of the scores have values greater than $X = 73$?

   b. Sketch the distribution of sample means for samples of size $n = 16$. What proportion of the sample means have values greater than 73?

19. For a normal population with $\mu = 70$ and $\sigma = 20$, what is the probability of obtaining a sample mean greater than 75
   a. For a random sample of $n = 4$ scores?
   b. For a random sample of $n = 16$ scores?
   c. For a random sample of $n = 100$ scores?

20. A normal population has $\mu = 80$ and $\sigma = 12$.
   a. Using this population, sketch the distribution of sample means for $n = 4$. Mark the location of the mean (expected value), and show the standard error in your sketch.
   b. Using your sketch from part (a), what proportion of samples of $n = 4$ will have sample means greater than 86? (Locate the position of $M = 86$, find the corresponding z-score, and use the unit normal table to determine the proportion.)
   c. Sketch the distribution of sample means based on $n = 16$.
   d. Using your sketch from part (c), what proportion of samples of $n = 16$ will have sample means greater than 86?

21. A normal population has $\mu = 55$ and $\sigma = 8$.
   a. Sketch the distribution of sample means for samples of size $n = 16$ selected from this population.
   b. What proportion of samples based on $n = 16$ will have sample means greater than 59?
   c. What proportion of samples with $n = 16$ will have sample means less than 56?
   d. What proportion of samples with $n = 16$ will have sample means between 53 and 57?

22. A normal distribution has $\mu = 35$ and $\sigma = 8$.
   a. Sketch the distribution of sample means for samples of $n = 4$ selected from this population. Show the expected value of M and the standard error in your sketch.
   b. For a sample of $n = 4$, what is the probability of selecting a random sample with a mean greater than $M = 32$?
   c. Sketch the distribution of sample means for samples of $n = 16$ selected from this population. Show the expected value of M and the standard error in your sketch.
   d. For a sample of $n = 16$, what is the probability of selecting a random sample with a mean greater than $M = 32$?

23. A normally distributed population has $\mu = 80$ and $\sigma = 20$.
   a. Sketch the distribution of sample means for samples of $n = 25$ selected from this population. Show

the expected value of M and the standard error in your sketch.

b. What is the probability of obtaining a sample mean greater than $M = 85$ for a random sample of $n = 25$ scores?

c. Sketch the distribution of sample means for samples of $n = 100$ selected from this population. Show the expected value of M and the standard error in your sketch.

d. What is the probability of obtaining a sample mean greater than $M = 85$ for a random sample of $n = 100$ scores?

24. The average age for licensed drivers in the county is $\mu = 42.6$ years with a standard deviation of $\sigma = 12$. The distribution is approximately normal.

a. A researcher obtained a sample of $n = 25$ drivers who received parking tickets. The average age for these drivers was $M = 40.5$. Is this a reasonable outcome for a sample of $n = 25$, or is the sample mean very different from what would normally be expected? (*Hint:* Compute the z-score for the sample mean.)

b. The same researcher also obtained a sample of $n = 16$ drivers who received speeding tickets. The average age for this sample was $M = 34.4$. Is this a reasonable outcome for a sample of $n = 16$, or is the sample mean very different from what would usually be expected? (Again, compute the z-score for the sample mean.)

25. The average age for registered voters in the county is $\mu = 39.7$ years with $\sigma = 11.8$. The distribution of ages is approximately normal. During a recent jury trial in the county courthouse, a statistician noted that the average age for the 12 jurors was $M = 50.4$ years.

a. How likely is it to obtain a jury this old or older by chance?

b. Is it reasonable to conclude that this jury is not a random sample of registered voters?

26. Boxes of sugar are filled by machine with considerable accuracy. The distribution of box weights is normal and has a mean of 32 ounces with a standard deviation of only 2 ounces. A quality control inspector takes a sample of $n = 16$ boxes and finds the sample contains, on the average, $M = 31$ ounces of sugar. If the machine is working properly, what is the probability of obtaining a sample of 16 boxes that averages 31 ounces or less? Should the inspector suspect that the filling machinery needs repair?

27. A researcher evaluated the effectiveness of relaxation training in reducing anxiety. One sample of anxiety-ridden people received relaxation training, while a second sample did not. Then anxiety scores were measured for all participants, using a standardized test. Use the information that is summarized in the following table to complete this exercise.

THE EFFECT OF RELAXATION TRAINING ON ANXIETY SCORES

GROUP	MEAN ANXIETY SCORE	SE
Control group	36	7
Relaxation training	18	5

a. Construct a bar graph that incorporates all of the information in the table.

b. Looking at your graph, do you think the relaxation training really worked? Explain your answer.

28. Research examining sleep behavior demonstrates a relationship between age and average amount of sleep. The following data show the mean amount of sleep and the standard error for five samples representing five different ages. Sketch a line graph showing these data including the standard error for each sample mean.

1 year: $M = 13$ hours with $SE = 2$
5 years: $M = 11$ hours with $SE = 1\frac{1}{2}$
10 years: $M = 10$ hours with $SE = 1\frac{1}{2}$
15 years: $M = 8\frac{1}{2}$ hours with $SE = 1$
20 years: $M = 8$ hours with $SE = 1$

In the next eight chapters we will examine a variety of statistical methods, all of which use sample means as the basis for drawing inferences about population means. The primary application of these inferential methods is to help researchers interpret the outcome of their research studies. In a typical study, the goal is to demonstrate a difference between two (or more) treatment conditions. For example, a researcher hopes to demonstrate that a group of children who are exposed to violent TV programs will behave more aggressively than children who are shown nonviolent TV programs. In this situation, the data consist of one sample mean representing the scores in one treatment condition and another sample mean representing the scores from a different treatment. The researcher hopes to find a difference between the two sample means, and would like to generalize the mean difference to the entire population.

The problem is that two sample means can be different even when there is no difference whatsoever in the population. As you saw in Chapter 1 (see Figure 1.2), two samples can have different means even when they are selected from the same population. Thus, even though a researcher may obtain a sample mean difference in a research study, it does not necessarily indicate that there is a mean difference in the population.

In this part we will examine the statistical methods that help researchers decide when it is reasonable to generalize a sample mean difference to the rest of the population.

CHAPTER

8

Introduction to Hypothesis Testing

Tools You Will Need

The following items are considered essential background material for this chapter. If you doubt your knowledge of any of these items, you should review the appropriate chapter or section before proceeding.

- *z*-Scores (Chapter 5)
- Distribution of sample means (Chapter 7)
  - Expected value
  - Standard error
  - Probability of sample means

Preview

In the spring of 1998 a group of parents in Rochester, New York, filed a $185-million lawsuit against a major corporation claiming that environmental pollutants from the company had caused a cancer cluster. A "cluster" is defined as a disproportionate number of similar illnesses that could have a common cause, occurring in a defined geographic area. The parents had gathered data showing an unusually large number of childhood cancers concentrated in a relatively small area of the city. The cancers, which are fairly rare, affect the brain and nervous system of children and are typically fatal. The fact that a large number of cancers were concentrated in a single neighborhood seemed to indicate that there must be some factor that was responsible for causing the cluster.

One part of the company's defense involved challenging the existence of a real cluster. They argued that the number of cases and the concentration of cases were not unusual, but were well within the boundaries of what would be normal to expect by chance. For an analogy, imagine that a huge map of the city is pasted on one wall and that you are standing across the room, blindfolded, with a handful of darts. After throwing 20 or 30 darts blindly at the map, you remove your blindfold and observe the pattern that you have created. What is almost guaranteed to happen is that the darts will *not* be spaced evenly across the map. Instead, there will be large sections with no darts at all, and other sections where darts are concentrated in clusters. In this example, the clusters are not caused by any external factor; they are simply the result of chance.

The point of this story is that whenever there appear to be patterns in a set of data, you should not automatically assume that some factor is causing the patterns to appear. Instead, you should realize that chance, by itself, can produce what appear to be meaningful patterns in the data. The problem is to determine what kinds of outcomes are reasonable to expect just by chance, and what outcomes go beyond chance. This is one of the fundamental problems for inferential statistics.

In this chapter we will introduce one of the basic inferential statistical procedures that researchers use to interpret their results. The procedure is called *hypothesis testing*, and it is intended to help researchers differentiate between real and random patterns in the data. In the context of a research study, the goal is to decide whether the results indicate a real relationship between two variables or whether the results simply show the random fluctuation that would be expected just by chance.

As we develop the hypothesis-testing procedure, you will realize that we are using all of the basic elements that were developed in the previous three chapters. In particular, a hypothesis test is a relatively simple logical process that uses z-scores, probability, and standard error to help researchers make reasonable conclusions about the significance of their research results. A good understanding of the material presented in Chapters 5, 6, and 7 will make this new topic much easier to learn.

8.1 THE LOGIC OF HYPOTHESIS TESTING

It usually is impossible or impractical for a researcher to observe every individual in a population. Therefore, researchers usually collect data from a sample and then use the sample data to help answer questions about the population. Hypothesis testing is a statistical procedure that allows researchers to use sample data to draw inferences about the population of interest.

Hypothesis testing is one of the most commonly used inferential procedures. In fact, most of the remainder of this book will examine hypothesis testing in a variety of different situations and applications. Although the details of a hypothesis test will change from one situation to another, the general process will remain constant. In this chapter, we will introduce the general procedure for a hypothesis test. You should notice that we will use the statistical techniques that have been developed in the preceding three chapters: That is, we will combine the concepts of z-scores, probability, and the distribution of sample means to create a new statistical procedure known as a *hypothesis test*.

DEFINITION   A **hypothesis test** is a statistical method that uses sample data to evaluate a hypothesis about a population parameter.

In very simple terms, the logic underlying the hypothesis-testing procedure is as follows:

1. First, we state a hypothesis about a population. Usually the hypothesis concerns the value of a population parameter. For example, we might hypothesize that the mean IQ for registered voters in the United States is $\mu = 110$.

2. Before we actually select a sample, we use the hypothesis to predict the characteristics that the sample should have. For example, if we hypothesize that the population mean IQ is $\mu = 110$, then we would predict that our sample should have a mean *around* 110. Remember: The sample should be similar to the population, but you should always expect a certain amount of error.

3. Next, we obtain a random sample from the population. For example, we might select a random sample of $n = 200$ registered voters and compute the mean IQ for the sample.

4. Finally, we compare the obtained sample data with the prediction that was made from the hypothesis. If the sample mean is consistent with the prediction, we will conclude that the hypothesis is reasonable. But if there is a big discrepancy between the data and the prediction, we will decide that the hypothesis is wrong.

A hypothesis test is typically used in the context of a research study. That is, a researcher completes a research study and then uses a hypothesis test to evaluate the results. Depending on the type of research and the type of data, the details of the hypothesis test will change from one research situation to another. In later chapters, we will examine different versions of hypothesis testing that are used for different kinds of research. For now, however, we will focus on the basic elements that are common to all hypothesis tests. To accomplish this general goal, we will examine a hypothesis test as it applies to the simplest possible research study. In particular, we will look at the situation in which a researcher is using one sample to examine one unknown population.

The unknown population   Figure 8.1 shows the general research situation that we will use to introduce the process of hypothesis testing. Notice that the researcher begins with a known population. This is the set of individuals as they exist *before treatment*. For this example, we are assuming that the original set of scores forms a normal distri-

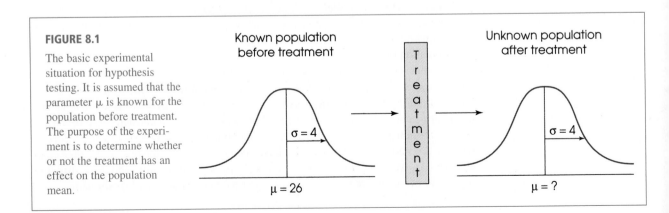

FIGURE 8.1

The basic experimental situation for hypothesis testing. It is assumed that the parameter μ is known for the population before treatment. The purpose of the experiment is to determine whether or not the treatment has an effect on the population mean.

Known population before treatment

$\sigma = 4$

$\mu = 26$

Treatment

Unknown population after treatment

$\sigma = 4$

$\mu = ?$

bution with $\mu = 26$ and $\sigma = 4$. The purpose of the research is to determine the effect of the treatment on the individuals in the population. That is, the goal is to determine what happens to the population *after the treatment is administered.*

To simplify the hypothesis-testing situation, one basic assumption is made about the effect of the treatment: If the treatment has any effect, it is simply to add a constant amount to (or subtract a constant amount from) each individual's score. You should recall from Chapters 3 and 4 that adding (or subtracting) a constant will not change the shape of the population, nor will it change the standard deviation. Thus, we will assume that the population after treatment has the same shape as the original population and the same standard deviation as the original population. This assumption is incorporated into the situation shown in Figure 8.1.

Note that the unknown population, after treatment, is the focus of the research question. Specifically, the purpose of the research is to determine what would happen if the treatment were administered to every person in the population.

The sample in the research study The goal of the hypothesis test is to determine whether or not the treatment has any effect on the individuals in the population (see Figure 8.1). Usually, however, we cannot administer the treatment to the entire population so the actual research study is conducted using a sample. Figure 8.2 shows the structure of the research study from two different perspectives.

1. The top half of the figure (part a) shows the study as it actually would be conducted. That is, a sample is selected from the original population, the treatment is administered to the sample, and the sample is measured to evaluate the effect of the treatment.

2. The bottom half of the figure (part b) shows the study from the point of view of the hypothesis test. In this case, the treatment is administered to the entire population, producing the *unknown* population that we saw in Figure 8.1. Then, a sample is selected from the treated population. From this perspective, we have a sample that comes from an unknown population. The hypothesis test will use the sample to evaluate a hypothesis about the unknown population mean.

Note that the unknown population in Figure 8.2(b) is actually *theoretical* (the treatment is never administered to the entire population). However, we do have a *real* sample that represents the population, and we can use the sample to test a hypothesis about the unknown population.

A hypothesis test is a formalized procedure that follows a standard series of operations. In this way, researchers have a standardized method for evaluating the results of their research studies. Other researchers will recognize and understand exactly how the data were evaluated and how conclusions were reached. To emphasize the formal structure of a hypothesis test, we will present hypothesis tests as a four-step process that will be used throughout the rest of the book. The following example will be used to provide a concrete foundation for introducing the hypothesis-testing procedure.

EXAMPLE 8.1   Psychologists have noted that stimulation during infancy can have profound effects on the development of infant rats. It has been demonstrated that stimulation (for example, increased handling) results in eyes opening sooner, more rapid brain maturation, faster growth, and larger body weight (see Levine, 1960). Based on the data, one might theorize that increased stimulation early in life can be beneficial. Suppose a researcher who is interested in this developmental theory would like to determine whether or not stimulation during infancy has an effect on human development.

FIGURE 8.2

Two views of the general research situation. From both perspectives a treatment is administered to a sample, and the treated sample can be compared with the original, untreated population. The hypothesis test is concerned with the unknown population that would exist if the treatment were administered to the entire population.

(a) The actual research situation

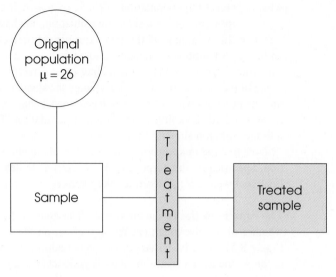

(b) The research situation from the point of view of the hypothesis test

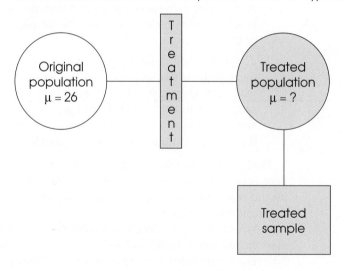

It is known from national health statistics that the mean weight for 2-year-old children is $\mu = 26$ pounds. The distribution of weights is normal with $\sigma = 4$ pounds. The researcher's plan is to obtain a sample of $n = 16$ newborn infants and give their parents detailed instructions for giving their children increased handling and stimulation. At age 2, each of the 16 infants will be weighed, and the mean weight for the sample will be computed. If the mean weight for the sample is noticeably different from the mean for the general population, then the researcher can conclude that the increased handling does appear to have an effect on development. On the other hand, if the mean weight for the sample is around 26 pounds (the same as the general population mean), then the researcher must conclude that increased handling does not appear to have any effect.

Figure 8.2 depicts the research situation that was described in the preceding example. Notice that the population after treatment is unknown. Specifically, we do not know what will happen to the mean weight for 2-year-old children if the whole population receives special handling. However, we do have a sample of $n = 16$ children who have received special handling, and we can use this sample to draw inferences about the unknown population. The following four steps outline the hypothesis-testing procedure that allows us to use the sample data to answer questions about the unknown population.

STEP 1: STATE THE HYPOTHESES

As the name implies, the process of hypothesis testing begins by stating a hypothesis about the unknown population. Actually, we state two opposing hypotheses. Notice that both hypotheses are stated in terms of population parameters.

The first and most important of the two hypotheses is called the *null hypothesis*. The null hypothesis states that the treatment has no effect. In general, the null hypothesis states that there is no change, no effect, no difference—nothing happened, hence the name *null*. The null hypothesis is identified by the symbol H_0. (The *H* stands for *hypothesis*, and the zero subscript indicates that this is the *zero-effect*, null hypothesis.) For the study in Example 8.1, the null hypothesis states that increased handling during infancy has *no effect* on body weight for the population of infants. In symbols, this hypothesis is

$$H_0: \quad \mu_{\text{infants handled}} = 26 \text{ pounds} \qquad \text{(Even with extra handling, the mean weight at 2 years is still 26 pounds.)}$$

> The goal of inferential statistics is to make general statements about the population by using sample data. Therefore, when testing hypotheses, we make our predictions about the population parameters.

DEFINITION

The **null hypothesis** (H_0) states that in the general population there is no change, no difference, or no relationship. In the context of an experiment, H_0 predicts that the independent variable (treatment) will have *no effect* on the dependent variable for the population.

The second hypothesis is simply the opposite of the null hypothesis, and it is called the *scientific* or *alternative hypothesis* (H_1). This hypothesis states that the treatment has an effect on the dependent variable.

DEFINITION

The **alternative hypothesis** (H_1) states that there is a change, a difference, or a relationship for the general population. In the context of an experiment, H_1 predicts that the independent variable (treatment) *will have an effect* on the dependent variable.

> The null hypothesis and the alternative hypothesis are mutually exclusive and exhaustive. They cannot both be true. The data will determine which one should be rejected.

For this example, the alternative hypothesis predicts that handling does alter growth for the population. In symbols, H_1 is represented as

$$H_1: \quad \mu_{\text{infants handled}} \neq 26 \qquad \text{(With handling, the mean will be different from 26 pounds.)}$$

Notice that the alternative hypothesis simply states that there will be some type of change. It does not specify whether the effect will be increased or decreased growth. In some circumstances, it is appropriate to specify the direction of the effect in H_1. For example, the researcher might hypothesize that increased handling will increase growth ($\mu > 26$ pounds). This type of hypothesis results in a directional hypothesis test, which will be examined in detail later in this chapter. For now we will concentrate on nondirectional tests, where the hypotheses simply state that the treatment has some effect

(H_1) or has no effect (H_0). For this example, we are examining whether handling in infancy does alter growth in some way (H_1) or has no effect (H_0). You should also note that both hypotheses refer to a population whose mean is unknown—namely, the population of infants who receive extra handling early in life.

STEP 2: SET THE CRITERIA FOR A DECISION

The researcher will eventually use the data from the sample to evaluate the credibility of the null hypothesis. The data will either provide support for the null hypothesis or tend to refute the null hypothesis. In particular, if there is a big discrepancy between the data and the hypothesis, we will conclude that the hypothesis is wrong.

To formalize the decision process, we will use the null hypothesis to predict the kind of sample mean that ought to be obtained. Specifically, we will determine exactly what sample means are consistent with the null hypothesis and what sample means are at odds with the null hypothesis. To accomplish this, we begin by examining all the possible sample means that could be obtained if the null hypothesis is true. For our example, this is the distribution of sample means for $n = 16$. According to the null hypothesis, the population mean is still $\mu = 26$ pounds, so the distribution of sample means should also be centered at $\mu = 26$. The distribution of sample means is then divided into two sections:

1. Sample means that are likely to be obtained if H_0 is true; that is, sample means that are close to the null hypothesis

2. Sample means that are very unlikely to be obtained if H_0 is true; that is, sample means that are very different from the null hypothesis

The distribution of sample means divided into these two sections is shown in Figure 8.3. Notice that the high-probability samples are located in the center of the distri-

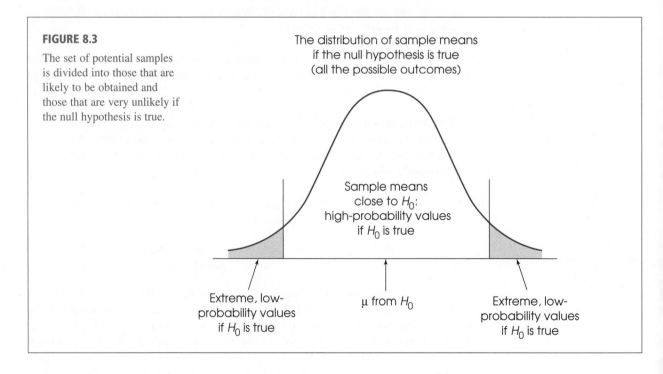

FIGURE 8.3

The set of potential samples is divided into those that are likely to be obtained and those that are very unlikely if the null hypothesis is true.

The distribution of sample means if the null hypothesis is true (all the possible outcomes)

Sample means close to H_0: high-probability values if H_0 is true

Extreme, low-probability values if H_0 is true

μ from H_0

Extreme, low-probability values if H_0 is true

bution and have sample means close to the value specified in the null hypothesis. On the other hand, the low-probability samples are located in the extreme tails of the distribution. After the distribution has been divided in this way, we can compare our sample data with the values in the distribution. Specifically, we can determine whether our sample mean is consistent with the null hypothesis (like the values in the center of the distribution) or whether our sample mean is very different from the null hypothesis (like the values in the extreme tails).

The alpha level   To find the boundaries that separate the high-probability samples from the low-probability samples, we must define exactly what is meant by "low" probability and "high" probability. This is accomplished by selecting a specific probability value, which is known as the *level of significance* or the *alpha level* for the hypothesis test. The alpha (α) value is a small probability that is used to identify the low-probability samples. By convention, commonly used alpha levels are $\alpha = .05$ (5%), $\alpha = .01$ (1%), and $\alpha = .001$ (0.1%). For example, with $\alpha = .05$, we will separate the most unlikely 5% of the sample means (the extreme values) from the most likely 95% of the sample means (the central values).

With rare exceptions, an alpha level is never larger than .05.

The extremely unlikely values, as defined by the alpha level, make up what is called the *critical region*. These extreme values in the tails of the distribution define outcomes that are inconsistent with the null hypothesis; that is, they are very unlikely to occur if the null hypothesis is true. Whenever the data from a research study produce a sample mean that is located in the critical region, we will conclude that the data are inconsistent with the null hypothesis, and we will reject the null hypothesis.

DEFINITIONS

The **alpha level** or the **level of significance** is a probability value that is used to define the very unlikely sample outcomes if the null hypothesis is true.

The **critical region** is composed of extreme sample values that are very unlikely to be obtained if the null hypothesis is true. The boundaries for the critical region are determined by the alpha level. If sample data fall in the critical region, the null hypothesis is rejected.

The boundaries for the critical region   To determine the exact location for the boundaries that define the critical region, we will use the alpha-level probability and the unit normal table. In most cases, the distribution of sample means is normal, and the unit normal table will provide the precise z-score location for the critical region boundaries. With $\alpha = .05$, for example, the boundaries separate the extreme 5% from the middle 95%. Because the extreme 5% is split between two tails of the distribution, there is exactly 2.5% (or 0.0250) in each tail. In the unit normal table, you can look up a proportion of 0.0250 in column C (the tail) and find that the z-score boundary is $z = 1.96$. Thus, for any normal distribution, the extreme 5% is in the tails of the distribution beyond $z = 1.96$ and $z = -1.96$. These values define the boundaries of the critical region for a hypothesis test using $\alpha = .05$ (see Figure 8.4).

Similarly, an alpha level of $\alpha = .01$ means that 1% or .0100 is split between the two tails. In this case, the proportion in each tail is .0050, and the corresponding z-score boundaries are $z = \pm2.58$ (±2.57 is equally good). For $\alpha = .001$, the boundaries are located at $z = \pm3.30$. You should verify these values in the unit normal table and be sure that you understand exactly how they are obtained.

FIGURE 8.4

The critical region (very unlikely outcomes) for $\alpha = .05$.

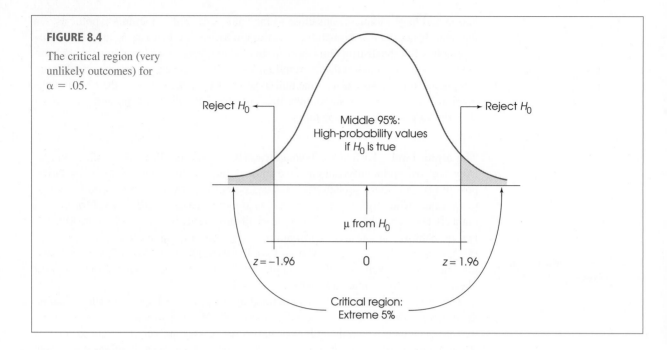

Reject H_0 ← → Reject H_0

Middle 95%:
High-probability values
if H_0 is true

μ from H_0

$z = -1.96$ 0 $z = 1.96$

Critical region:
Extreme 5%

STEP 3: COLLECT DATA AND COMPUTE SAMPLE STATISTICS

The next step in hypothesis testing is to obtain the sample data. At this time, a random sample of infants would be selected, and their parents would be trained to provide the additional daily handling that constitutes the treatment for this study. Then the body weight for each infant would be measured at 2 years of age. Notice that the data are collected *after* the researcher has stated the hypotheses and established the criteria for a decision. This sequence of events helps ensure that a researcher makes an honest, objective evaluation of the data and does not tamper with the decision criteria after the experimental outcome is known.

Next, the raw data from the sample are summarized with the appropriate statistics: For this example, the researcher would compute the sample mean. Now it is possible for the researcher to compare the sample mean (the data) with the null hypothesis. This is the heart of the hypothesis test: comparing the data with the hypothesis.

The comparison is accomplished by computing a z-score that describes exactly where the sample mean is located relative to the hypothesized population mean from H_0. In step 2, we constructed the distribution of sample means that would be expected if the null hypothesis were true—that is, the entire set of sample means that could be obtained if the treatment has no effect (see Figure 8.4). Now we will calculate a z-score that identifies where our sample mean is located in this hypothesized distribution. The z-score formula for a sample mean is

$$z = \frac{M - \mu}{\sigma_M}$$

In the formula, the value of the sample mean (M) is obtained from the sample data, and the value of μ is obtained from the null hypothesis. Thus, the z-score formula can be expressed in words as follows:

$$z = \frac{\text{sample mean} - \text{hypothesized population mean}}{\text{standard error between } M \text{ and } \mu}$$

Notice that the top of the z-score formula measures how much difference there is between the data and the hypothesis. The bottom of the formula measures the standard distance that ought to exist between the sample mean and the population mean.

STEP 4: MAKE A DECISION   In the final step, the researcher uses the z-score value obtained in step 3 to make a decision about the null hypothesis according to the criteria established in step 2. There are two possible decisions, and both are stated in terms of the null hypothesis (see Box 8.1).

One possible decision is to *reject the null hypothesis*. This decision is made whenever the sample data fall in the critical region. By definition, a sample value in the critical region indicates that there is a big discrepancy between the sample and the null hypothesis (the sample is in the extreme tail of the distribution). This kind of outcome is very unlikely to occur if the null hypothesis is true, so our conclusion is to reject H_0. In this case, the data provide convincing evidence that the null hypothesis is wrong, and we conclude that the treatment really did have an effect on the individuals in the sample. For the example we have been considering, suppose the sample of $n = 16$ infants produced a sample mean of $M = 30$ pounds at age 2. With $n = 16$ and $\sigma = 4$, the standard error for the sample mean is

$$\sigma_M = \frac{\sigma}{\sqrt{n}} = \frac{4}{\sqrt{16}} = \frac{4}{4} = 1$$

The z-score for the sample is

$$z = \frac{M - \mu}{\sigma_M} = \frac{30 - 26}{1} = \frac{4}{1} = 4.00$$

BOX 8.1   REJECTING THE NULL HYPOTHESIS VERSUS PROVING THE ALTERNATIVE HYPOTHESIS

It may seem awkward to pay so much attention to the null hypothesis. After all, the purpose of most experiments is to show that a treatment does have an effect, and the null hypothesis states that there is no effect. The reason for focusing on the null hypothesis, rather than the alternative hypothesis, comes from the limitations of inferential logic. Remember that we want to use the sample data to draw conclusions, or inferences, about a population. Logically, it is much easier to demonstrate that a universal (population) hypothesis is false than to demonstrate that it is true. This principle is shown more clearly in a simple example. Suppose you make the universal statement "all dogs have four legs" and you intend to test this hypothesis by using a sample of one dog. If the dog in your sample does have four legs, have you proved the statement? It should be clear that one four-legged dog does not prove the general statement to be true. On the other hand, suppose the dog in your sample has only three legs. In this case, you have proved the statement to be false. Again, it is much easier to show that something is false than to prove that it is true.

Hypothesis testing uses this logical principle to achieve its goals. It would be difficult to state "the treatment has an effect" as the hypothesis and then try to prove that this is true. Therefore, we state the null hypothesis, "the treatment has no effect," and try to show that it is false. The end result still is to demonstrate that the treatment does have an effect. That is, we find support for the alternative hypothesis by disproving (rejecting) the null hypothesis.

With an alpha level of $\alpha = .05$, this z-score is far beyond the boundary of 1.96. Because the sample z-score is in the critical region, we reject the null hypothesis and conclude that the special handling did have an effect on the infants' weights.

The second possibility occurs when the sample data are not in the critical region. In this case, the data are reasonably close to the null hypothesis (in the center of the distribution). Because the data do not provide strong evidence that the null hypothesis is wrong, our conclusion is to *fail to reject the null hypothesis*. This conclusion means that the treatment does not appear to have an effect. For the research study examining extra handling of infants, suppose our sample of $n = 16$ produced a mean weight of $M = 27$ pounds. As before, the standard error for the sample mean is $\sigma_M = 1$, and the null hypothesis states that $\mu = 26$. These values produce a z-score of

$$z = \frac{M - \mu}{\sigma_M} = \frac{27 - 26}{1} = \frac{1}{1} = 1.00$$

The z-score of 1.00 is not in the critical region. Therefore, we fail to reject the null hypothesis and conclude that the special handling does not appear to have an effect on weight.

The two possible decisions may be easier to understand if you think of a research study as an attempt to gather evidence to prove that a treatment works. From this perspective, the research study has two possible outcomes:

1. You gather enough evidence to demonstrate convincingly that the treatment really works. That is, you reject the null hypothesis and conclude that the treatment does have an effect.

2. The evidence you gather from the research study is not convincing. In this case, all you can do is conclude that there is not enough evidence. The research study has failed to demonstrate that the treatment has an effect, and the statistical decision is to fail to reject the null hypothesis.

A CLOSER LOOK AT THE
z-SCORE STATISTIC

The z-score statistic that is used in the hypothesis test is the first specific example of what is called a *test statistic*. The term *test statistic* simply indicates that the sample data are converted into a single, specific statistic that is used to test the hypotheses. In the chapters that follow, we will introduce several other test statistics that are used in a variety of different research situations. However, each of the new test statistics will have the same basic structure and will serve the same purpose as the z-score. We have already described the z-score equation as a formal method for comparing the sample data and the population hypothesis. In this section, we will discuss the z-score from two other perspectives that may give you a better understanding of hypothesis testing and the role that z-scores play in this inferential technique. In each case, keep in mind that the z-score will serve as a general model for other test statistics that will come in future chapters.

The z-score formula as a recipe   The z-score formula, like any formula, can be viewed as a recipe. If you follow instructions and use all the right ingredients, the formula will produce a z-score. In the hypothesis-testing situation, however, you do not have all the necessary ingredients. Specifically, you do not know the value for the population mean (μ), which is one component or ingredient in the formula.

This situation is similar to trying to follow a cake recipe where one of the ingredients is not clearly listed. For example, there may be a grease stain that obscures one part of your recipe. In this situation, what do you do? One possibility is to make a hypothesis about the missing ingredient. You might, for example, hypothesize that the missing ingredient is 2 cups of flour. To test the hypothesis, you simply add the rest of the ingredients along with the hypothesized flour and see what happens. If the cake turns out to be good, then you can be reasonably confident that your hypothesis was correct. But if the cake is terrible, then you conclude that your hypothesis was wrong.

In a hypothesis test with z-scores, you do essentially the same thing. You do not know the value for the population mean, μ. Therefore, you make a hypothesis (the null hypothesis) about the unknown value μ and plug it into the formula along with the rest of the ingredients. To evaluate your hypothesis, you simply look at the result. If the formula produces a z-score near zero (which is where z-scores are supposed to be), then you conclude that your hypothesis was correct. On the other hand, if the formula produces an unlikely result (an extremely large or small value), then you conclude that your hypothesis was wrong.

The z-score formula as a ratio   In the context of a hypothesis test, the z-score formula has the following structure:

$$z = \frac{M - \mu}{\sigma_M} = \frac{\text{sample mean} - \text{hypothesized population mean}}{\text{standard error between } M \text{ and } \mu}$$

Notice that the numerator of the formula involves a direct comparison between the sample data and the null hypothesis. In particular, the numerator measures the obtained difference between the sample mean and the hypothesized population mean. The standard error in the denominator of the formula measures how much difference ought to exist between the sample mean and the population mean just by chance. Thus, the z-score formula (and most other test statistics) forms a ratio

$$z = \frac{\text{obtained difference}}{\text{difference due to chance}}$$

Thus, for example, a z-score of $z = 3.00$ means that the obtained difference between the sample and the hypothesis is 3 times bigger than would be expected by chance.

In general, the purpose of a test statistic is to determine whether the result of the research study (the obtained difference) is more than would be expected by chance alone. Whenever the test statistic has a value greater than 1.00, it means that the obtained result (numerator) is greater than chance (denominator). However, you should realize that most researchers are not satisfied with results that are simply more than chance. Instead, conventional research standards require results that are substantially more than chance. Specifically, hypothesis tests require that the test statistic ratio is large enough to satisfy the criterion set by the alpha level. With alpha levels of .05 or .01, this usually means that the ratio must have a value around 2 or 3; that is, the obtained difference must be 2 or 3 times bigger than chance before the null hypothesis can be rejected.

LEARNING CHECK

1. What does the null hypothesis predict about a population or a treatment effect?

2. Define the *critical region* for a hypothesis test.

3. As the alpha level gets smaller, the size of the critical region also gets smaller. (True or false?)

4. A small value (near zero) for the z-score statistic is evidence that the null hypothesis should be rejected. (True or false?)

5. A decision to reject the null hypothesis means you have demonstrated that the treatment has no effect. (True or false?)

1. The null hypothesis predicts that the treatment has no effect and the population is unchanged.

2. The critical region consists of sample outcomes that are very unlikely to be obtained if the null hypothesis is true.

3. True. As alpha gets smaller, the critical region is moved farther out into the tails of the distribution.

4. False. A z-score near zero indicates that the data support the null hypothesis.

5. False. Rejecting the null hypothesis means you have concluded that the treatment does have an effect.

8.2   UNCERTAINTY AND ERRORS IN HYPOTHESIS TESTING

Hypothesis testing is an *inferential process*, which means that it uses limited information as the basis for reaching a general conclusion. Specifically, a sample provides only limited or incomplete information about the whole population, and yet a hypothesis test uses a sample to draw a conclusion about the population. In this situation, there is always the possibility that an incorrect conclusion will be made. Although sample data are usually representative of the population, there is always a chance that the sample is misleading and will cause a researcher to make the wrong decision about the research results. In a hypothesis test, there are two different kinds of errors that can be made.

TYPE I ERRORS

It is possible that the data will lead you to reject the null hypothesis when in fact the treatment has no effect. Remember: Samples are not expected to be identical to their populations, and some extreme samples can be very different from the populations they are supposed to represent. If a researcher selects one of these extreme samples just by chance, then the data from the sample may give the appearance of a strong treatment effect, even though there is no real effect. In the previous section, for example, we discussed a research study examining how increased handling during infancy affects body weight. Suppose the researcher selected a sample of $n = 16$ infants who were genetically destined to be bigger than average even without any special treatment. When these infants are measured at 2 years of age, they will look bigger than average even though the special handling (the treatment) may have had no effect. In this case, the researcher is likely to conclude that the treatment has had an effect, when in fact it really did not. This is an example of what is called a *Type I error*.

DEFINITION

A **Type I error** occurs when a researcher rejects a null hypothesis that is actually true. In a typical research situation, a Type I error means the researcher concludes that a treatment does have an effect when in fact it has no effect.

You should realize that a Type I error is not a stupid mistake in the sense that a researcher is overlooking something that should be perfectly obvious. On the contrary, the researcher is looking at sample data that appear to show a clear treatment effect. The researcher then makes a careful decision based on the available information. The problem is that the information from the sample is misleading.

In most research situations, the consequences of a Type I error can be very serious. Because the researcher has rejected the null hypothesis and believes that the treatment has a real effect, it is likely that the researcher will report or even publish the research results. A Type I error, however, means that this is a false report. Thus, Type I errors lead to false reports in the scientific literature. Other researchers may try to build theories or develop other experiments based on the false results. A lot of precious time and resources may be wasted.

Fortunately, the hypothesis test is structured to control and minimize the risk of committing a Type I error. Although extreme and misleading samples are possible, they are relatively unlikely. For an extreme sample to produce a Type I error, the sample mean must be in the critical region. However, the critical region is structured so that it is *very unlikely* to obtain a sample mean in the critical region when H_0 is true. Specifically, the alpha level determines the probability of obtaining a sample mean in the critical region when H_0 is true. Thus, the alpha level defines the probability or risk of a Type I error. This fact leads to an alternative definition of the alpha level.

DEFINITION   The **alpha level** for a hypothesis test is the probability that the test will lead to a Type I error. That is, the alpha level determines the probability of obtaining sample data in the critical region even though the null hypothesis is true.

In summary, whenever the sample data are in the critical region, the appropriate decision for a hypothesis test is to reject the null hypothesis. Normally this is the correct decision because the treatment has caused the sample to be different from the original population; that is, the treatment effect has literally pushed the sample mean into the critical region. In this case, the hypothesis test has correctly identified a real treatment effect. Occasionally, however, sample data will be in the critical region just by chance, without any treatment effect. When this occurs, the researcher will make a Type I error; that is, the researcher will conclude that a treatment effect exists, when in fact it does not. Fortunately, the risk of a Type I error is small and is under the control of the researcher. Specifically, the probability of a Type I error is equal to the alpha level.

TYPE II ERRORS   Whenever a researcher rejects the null hypothesis, there is a risk of a Type I error. Similarly, whenever a researcher fails to reject the null hypothesis, there is a risk of a *Type II error*. By definition, a Type II error is the failure to reject a false null hypothesis. In more straightforward English, a Type II error means that a treatment effect really exists, but the hypothesis test fails to detect it.

DEFINITION   A **Type II error** occurs when a researcher fails to reject a null hypothesis that is really false. In a typical research situation, a Type II error means that the hypothesis test has failed to detect a real treatment effect.

A Type II error occurs when the sample mean is not in the critical region even though the treatment has had an effect on the sample. Often this happens when the effect of the treatment is relatively small. In this case, the treatment does influence the sample, but the magnitude of the effect is not big enough to move the sample mean into the critical region. Because the sample is not substantially different from the original population (it is not in the critical region), the statistical decision is to fail to reject the null hypothesis and to conclude that there is not enough evidence to say there is a treatment effect.

The consequences of a Type II error are usually not as serious as those from a Type I error. In general terms, a Type II error means that the research data do not show the results that the researcher had hoped to obtain. The researcher can accept this outcome and conclude that the treatment either has no effect or has only a small effect that is not worth pursuing. Or the researcher can repeat the experiment (usually with some improvements) and try to demonstrate that the treatment really does work.

Unlike a Type I error, it is impossible to determine a single, exact probability value for a Type II error. Instead, the probability of a Type II error depends on a variety of factors and therefore is a function, rather than a specific number. Nonetheless, the probability of a Type II error is represented by the symbol β, the Greek letter beta.

In summary, a hypothesis test always leads to one of two decisions:

1. The sample data provide sufficient evidence to reject the null hypothesis and conclude that the treatment has an effect.

2. The sample data do not provide enough evidence to reject the null hypothesis. In this case, you fail to reject H_0 and conclude that the treatment does not appear to have an effect.

In either case, there is a chance that the data are misleading and the decision is wrong. The complete set of decisions and outcomes is shown in Table 8.1. The risk of an error is especially important in the case of a Type I error, which can lead to a false report. Fortunately, the probability of a Type I error is determined by the alpha level, which is completely under the control of the researcher. At the beginning of a hypothesis test, the researcher states the hypotheses and selects the alpha level, which immediately determines the risk that a Type I error will be made.

TABLE 8.1

Possible outcomes of a statistical decision

		ACTUAL SITUATION	
		NO EFFECT, H_0 TRUE	EFFECT EXISTS, H_0 FALSE
EXPERIMENTER'S DECISION	Reject H_0	Type I error	Decision correct
	Retain H_0	Decision correct	Type II error

SELECTING AN ALPHA LEVEL   As you have seen, the alpha level for a hypothesis test serves two very important functions. First, alpha helps determine the boundaries for the critical region by defining the concept of "very unlikely" outcomes. More importantly, alpha determines the probability of a Type I error. When you select a value for alpha at the beginning of a hypothesis test, your decision influences both of these functions.

The primary concern when selecting an alpha level is to minimize the risk of a Type I error. Thus, alpha levels tend to be very small probability values. By convention, the largest permissible value is $\alpha = .05$. An alpha level of .05 means that there is a 5% risk, or a 1-in-20 probability, of committing a Type I error. For many research situations, the .05 level of significance is viewed as an unacceptably large risk. Consider, for example, a scientific journal containing 20 research articles, each evaluated with an alpha level of .05. With a 5% risk of a Type I error, 1 out of the 20 articles is probably a false report. For this reason, most scientific publications require a more stringent alpha level, such as .01 or .001. (For more information on the origins of the .05 level of significance, see the excellent short article by Cowles and Davis, 1982.)

At this point, it may appear that the best strategy for selecting an alpha level is to choose the smallest possible value in order to minimize the risk of a Type I error. However, there is a different kind of risk that develops as the alpha level is lowered. Specifically, a lower alpha level means less risk of a Type I error, but it also means that the hypothesis test demands more evidence from the research results.

The trade-off between the risk of a Type I error and the demands of the test is controlled by the boundaries of the critical region. For the hypothesis test to conclude that the treatment does have an effect, the sample data must be in the critical region. If the treatment really has an effect, it should cause the sample to be different from the original population; essentially, the treatment should push the sample into the critical region. However, as the alpha level is lowered, the boundaries for the critical region move farther out and become more difficult to reach. Figure 8.5 shows how the boundaries for the critical region move farther away as the alpha level decreases. Notice that lower alpha levels produce more extreme boundaries for the critical region. Thus, an extremely small alpha level, such as .000001 (one in a million), would mean almost no risk of a Type I error but would push the critical region so far out that it would become essentially impossible to ever reject the null hypothesis; that is, it would require an enormous treatment effect before the sample data would reach the critical boundaries.

In general, researchers try to maintain a balance between the risk of a Type I error and the demands of the hypothesis test. Alpha levels of .05, .01, and .001 are considered reasonably good values because they provide a relatively low risk of error without placing excessive demands on the research results.

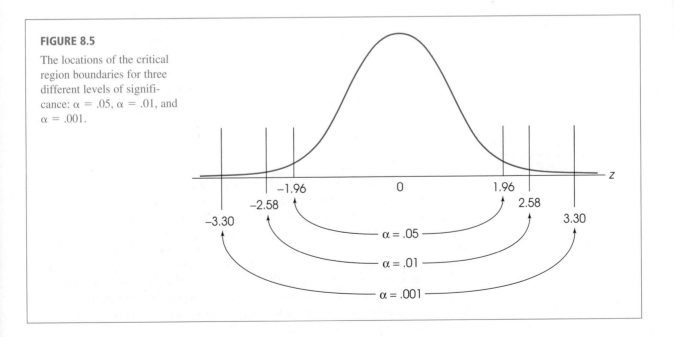

FIGURE 8.5

The locations of the critical region boundaries for three different levels of significance: $\alpha = .05$, $\alpha = .01$, and $\alpha = .001$.

1. What is a Type I error?

2. Why is the consequence of a Type I error considered serious?

3. If the alpha level is changed from $\alpha = .05$ to $\alpha = .01$, the probability of a Type I error increases. (True or false?)

4. Define a Type II error.

5. What research situation is likely to lead to a Type II error?

1. A Type I error is rejecting a true null hypothesis—that is, saying that the treatment has an effect when in fact it does not.

2. A Type I error often results in a false report. A researcher reports or publishes a treatment effect that does not exist.

3. False. The probability of a Type I error is α.

4. A Type II error is the failure to reject a false null hypothesis. In terms of a research study, a Type II error occurs when a study fails to detect a treatment effect that really exists.

5. A Type II error is likely to occur when the treatment effect is very small. In this case, a research study is more likely to fail to detect the effect.

8.3   AN EXAMPLE OF A HYPOTHESIS TEST

At this time, we have introduced all the elements of a hypothesis test. In this section, we will present a complete example of the hypothesis-testing process and discuss how the results from a hypothesis test are presented in a research report. For purposes of demonstration, the following scenario will be used to provide a concrete background for the hypothesis-testing process.

EXAMPLE 8.2   Alcohol appears to be involved in a variety of birth defects, including low birth weight and retarded growth. A researcher would like to investigate the effect of pre-natal alcohol on birth weight. A random sample of $n = 16$ pregnant rats is obtained. The mother rats are given daily doses of alcohol. At birth, one pup is selected from each litter to produce a sample of $n = 16$ newborn rats. The average weight for the sample is $M = 15$ grams. The researcher would like to compare the sample with the general population of rats. It is known that regular newborn rats (not exposed to alcohol) have an average weight of $\mu = 18$ grams. The distribution of weights is normal with $\sigma = 4$. Figure 8.6 shows the overall research situation. Notice that the researcher's question concerns the unknown population that is exposed to alcohol, and the data consist of a sample representing this unknown population.

FIGURE 8.6

The structure of a research study to determine whether prenatal alcohol affects birth weight.

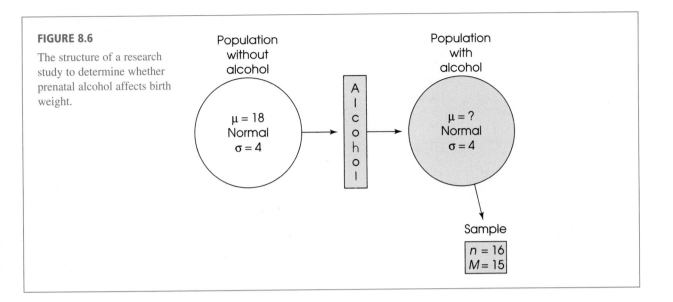

The following steps outline the hypothesis test that evaluates the effect of alcohol exposure on birth weight.

STEP 1 *State the hypotheses, and select the alpha level.* Both hypotheses concern the population that is exposed to alcohol where the population mean is unknown (the population on the right-hand side of Figure 8.6). The null hypothesis states that exposure to alcohol has no effect on birth weight. Thus, the population of rats with alcohol exposure should have the same mean birth weight as the regular, unexposed rats. In symbols,

$$H_0: \quad \mu_{\text{alcohol exposure}} = 18 \qquad \text{(Even with alcohol exposure, the rats still average 18 grams at birth.)}$$

The alternative hypothesis states that alcohol exposure does affect birth weight, so the exposed population should be different from the regular rats. In symbols,

$$H_1: \quad \mu_{\text{alcohol exposure}} \neq 18 \qquad \text{(Alcohol exposure will change birth weight.)}$$

Notice that both hypotheses concern the unknown population. For this test, we will use an alpha level of $\alpha = .05$. That is, we are taking a 5% risk of committing a Type I error.

STEP 2 *Set the decision criteria by locating the critical region.* By definition, the critical region consists of outcomes that are very unlikely if the null hypothesis is true. To locate the critical region we will go through a three-stage process that is portrayed in Figure 8.7. We begin with the population of birth weight scores for rats that have been exposed to alcohol. According to the null hypothesis, the mean for this population is $\mu = 18$ grams. In addition, it is known that the population of birth weight scores is normal with $\sigma = 4$ (see Figure 8.6). Next, we use the hypothesized population to determine the complete set of sample outcomes that could be obtained if the null hypothesis is true. Because the researcher plans to use a sample of $n = 16$ rats, we construct the distribution of sample means for $n = 16$. This distribution of sample means is often called the *null distribution* because it specifies exactly what kinds of

FIGURE 8.7

Locating the critical region as a three-step process. You begin with the population of scores that is predicted by the null hypothesis. Then, you construct the distribution of sample means for the sample size that is being used. The distribution of sample means corresponds to all the possible outcomes that could be obtained if H_0 is true. Finally, you use z-scores to separate the extreme outcomes (as defined by the alpha level) from the high-probability outcomes. The extreme values determine the critical region.

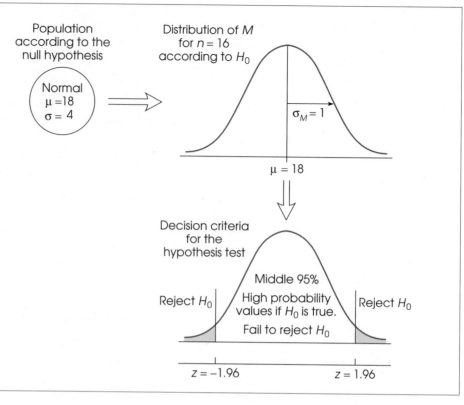

research outcomes should be obtained according to the null hypothesis. For this example, the null distribution is centered at $\mu = 18$ (from H_0) and has a standard error of $\sigma_M = \sigma/\sqrt{n} = 4/\sqrt{16} = 1$. Finally, we use the null distribution to identify the critical region, which consists of those outcomes that are very unlikely if the null hypothesis is true. With $\alpha = .05$, the critical region consists of the extreme 5% of the distribution. As we saw earlier, for any normal distribution, z-scores of $z = \pm 1.96$ separate the middle 95% from the extreme 5% (a proportion of 0.0250 in each tail). Thus, we have identified the sample means that according to the null hypothesis have a high probability of occurrence, and the sample means that are very unlikely to occur. It is the unlikely sample means, those with z-score values beyond ± 1.96, that form the critical region for the test.

STEP 3   *Collect the data, and compute the test statistic.* At this point, we would select our sample of $n = 16$ pups whose mothers had received alcohol during pregnancy. The birth weight would be recorded for each pup and the sample mean computed. As noted, we obtained a sample mean of $M = 15$ grams. The sample mean is then converted to a z-score, which is our test statistic.

$$z = \frac{M - \mu}{\sigma_M} = \frac{15 - 18}{1} = \frac{-3}{1} = -3.00$$

STEP 4   *Make a decision.* The z-score computed in step 3 has a value of -3.00, which is beyond the boundary of -1.96. Therefore, the sample mean is located in the critical region. This is a very unlikely outcome if the null hypothesis is true, so our decision is to reject the null hypothesis. In addition to this statistical decision concerning the null hypothesis, it is customary to state a conclusion about the results of the research study. For this example, we conclude that prenatal exposure to alcohol does have a significant effect on birth weight.

IN THE LITERATURE
REPORTING THE RESULTS OF THE STATISTICAL TEST

A special jargon and notational system are used in published reports of hypothesis tests. When you are reading a scientific journal, for example, you typically will not be told explicitly that the researcher evaluated the data using a z-score as a test statistic with an alpha level of .05. Nor will you be told that "the null hypothesis is rejected." Instead, you will see a statement such as

> The treatment with medication had a significant effect on people's depression scores, $z = 2.45, p < .05$.

Let us examine this statement piece by piece. First, what is meant by the term *significant*? In statistical tests, this word indicates that the result is different from what would be expected due to chance. A significant result means that the null hypothesis has been rejected. (Note that the null hypothesis for this study would state that the medication *has no effect* on depression scores for the population.) That is, the data are significant because the sample mean falls in the critical region and is not what we would have expected to obtain if H_0 were true.

DEFINITION   Findings are said to be **significant** or **statistically significant** when the null hypothesis has been rejected. Thus, if results achieve statistical significance, the researcher concludes that a treatment effect occurred.

Next, what is the meaning of $z = 2.45$? The z indicates that a z-score was used as the test statistic to evaluate the sample data and that its value is 2.45. Finally, what is meant by $p < .05$? This part of the statement is a conventional way of specifying the alpha level that was used for the hypothesis test. More specifically, we are being told that an outcome as extreme as the result of the experiment would occur by chance with a probability (p) that is less than .05 (alpha).

In circumstances where the statistical decision is to *fail to reject H_0*, the report might state that

The APA style does not use a leading zero in a probability value that refers to a level of significance.

> There was no evidence that the medication had an effect on depression scores, $z = 1.30$, $p > .05$.

In this case, we are saying that the obtained result, $z = 1.30$, is not unusual (not in the critical region) and is relatively likely to occur by chance (the probability is greater than .05). Thus, H_0 was not rejected.

Sometimes students become confused trying to differentiate between $p < .05$ and $p > .05$. Remember that you reject the null hypothesis with extreme, low-probability values, located in the critical region in the tails of the distribution. Thus, a significant result that rejects the null hypothesis corresponds to $p < 0.5$. See Figure 8.8.

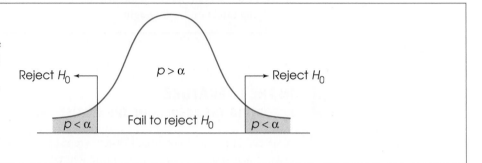

FIGURE 8.8

Sample means that fall in the critical region (shaded areas) have a probability *less than* alpha ($p < \alpha$). H_0 should be rejected. Sample means that do not fall in the critical region have a probability *greater than* alpha ($p > \alpha$).

When a hypothesis test is conducted using a computer program, the printout will often include not only a z-score value but also an exact value for p, the probability that the result occurred by chance alone. In this case, researchers are encouraged to report the exact p value instead of using the less-than or greater-than notation. For example, a research report might state that the treatment effect was significant, with $z = 2.45$, $p = .0142$. When using exact values for p, however, you must still satisfy the traditional criterion for significance; specifically, the p value must be smaller than .05 to be considered statistically significant. Remember: The p value is the probability that the result would occur simply by chance (without any treatment effect), which is also the probability of a Type I error. It is essential that this probability is very small.

Finally, you should notice that in scientific reports, the researcher does not actually state that "the null hypothesis was rejected." Instead, the researcher reports that the effect of the treatment was statistically significant. Likewise, when H_0 is not rejected, one simply states that the treatment effect was not statistically significant or that there was no evidence for a treatment effect. In fact, when you read scientific reports, you will note that the terms *null hypothesis* and *alternative hypothesis* are rarely mentioned. Nevertheless, H_0 and H_1 are part of the logic of hypothesis testing even if they are not formally stated in a scientific report. Because of their central role in the process of hypothesis testing, you should be able to identify and state these hypotheses. ❏

ASSUMPTIONS FOR HYPOTHESIS TESTS WITH z-SCORES

The mathematics that are used for a hypothesis test are based on a set of assumptions. When these assumptions are satisfied, you can be confident that the test produces a justified conclusion. However, if the assumptions are not satisfied, the hypothesis test may be compromised. In practice, researchers are not overly concerned with the assumptions underlying a hypothesis test because the tests usually work well even when the assumptions are violated. However, you should be aware of the fundamental conditions that are associated with each type of statistical test to ensure that the test is being used appropriately. The assumptions for hypothesis tests with z-scores are summarized below.

Random sampling   It is assumed that the subjects used to obtain the sample data were selected randomly. Remember, we wish to generalize our findings from the sample to the population. We accomplish this task when we use sample data to test a hypothesis about the population. Therefore, the sample must be representative of the population from which it has been drawn. Random sampling helps to ensure that it is representative.

Independent observations The values in the sample must consist of *independent* observations. In everyday terms, two observations are independent if there is no consistent, predictable relationship between the first observation and the second. More precisely, two events (or observations) are independent if the occurrence of the first event has no effect on the probability of the second event. Specific examples of independence and nonindependence are examined in Box 8.2. Usually, this assumption is satisfied by using a *random* sample, which also helps ensure that the sample is representative of the population and that the results can be generalized to the population.

The value of σ is unchanged by the treatment A critical part of the z-score formula in a hypothesis is the standard error, σ_M. To compute the value for the standard error,

---

BOX

8.2   INDEPENDENT OBSERVATIONS

Independent observations are a basic requirement for nearly all hypothesis tests. The critical concern is that each observation or measurement is not influenced by any other observation or measurement. An example of independent observations is the set of outcomes obtained in a series of coin tosses. Assuming that the coin is balanced, each toss has a 50–50 chance of coming up either heads or tails. More important, each toss is *independent* of the tosses that came before. On the fifth toss, for example, there is a 50% chance of heads no matter what happened on the previous four tosses; the coin does not remember what happened earlier and is not influenced by the past. (*Note:* Many people fail to believe in the independence of events. For example, after a series of four tails in a row, it is tempting to think that the probability of heads must increase because the coin is overdue to come up heads. This is a mistake, called the "gambler's fallacy." Remember that the coin does not know what happened on the preceding tosses and cannot be influenced by previous outcomes.)

In most research situations, the requirement for independent observations is typically satisfied by using a random sample of separate, unrelated individuals. Thus, the measurement obtained for each individual is not influenced by other subjects in the sample. The following two situations demonstrate circumstances in which the observations are *not* independent.

1. A researcher is interested in examining television preferences for children. To obtain a sample of $n = 20$ children, the researcher selects 4 children from family A, 3 children from family B, 5 children from family C, 2 children from family D, and 6 children from family E.

   It should be obvious that the researcher does *not* have 20 independent observations. Within each family, the children probably share television preference (at least they watch the same shows). Thus, the response for each child is likely to be related to the responses of his or her siblings.

2. A researcher is interested in people's ability to judge distances. A sample of 20 people is obtained. All 20 participants are gathered together in the same room. The researcher asks the first person to estimate the distance between New York and Miami. The second person is asked the same question, then the third person, the fourth person, and so on.

   Again, the observations that the researcher obtains are not independent: The response of each participant is probably influenced by the responses of the previous participants. For example, consistently low estimates by the first few participants could create pressure to conform and thereby increase the probability that the following participants would also produce low estimates.

we must know the sample size (n) and the population standard deviation (σ). In a hypothesis test, however, the sample comes from an *unknown* population (see Figures 8.2 and 8.6). If the population is really unknown, it would suggest that we do not know the standard deviation and, therefore, we cannot calculate the standard error. To solve this dilemma, we must make an assumption. Specifically, we assume that the standard deviation for the unknown population (after treatment) is the same as it was for the population before treatment.

Actually, this assumption is the consequence of a more general assumption that is part of many statistical procedures. This general assumption states that the effect of the treatment is to add a constant amount to (or subtract a constant amount from) every score in the population. You should recall that adding (or subtracting) a constant will change the mean but will have no effect on the standard deviation. You also should note that this assumption is a theoretical ideal. In actual experiments, a treatment generally will not show a perfect and consistent additive effect.

Normal sampling distribution   To evaluate hypotheses with z-scores, we have used the unit normal table to identify the critical region. This table can be used only if the distribution of sample means is normal.

LEARNING CHECK

1. A researcher would like to know whether room temperature will affect eating behavior in rats. The lab temperature is usually kept at 72 degrees, and under these conditions, the rats eat an average of $\mu = 10$ grams of food each day. The amount of food varies from one rat to another, forming a normal distribution with $\sigma = 2$. The researcher selects a sample of $n = 4$ rats and places them in a room where the temperature is kept at 65 degrees. The daily food consumption for these rats averaged $M = 13$ grams. Do these data indicate that temperature has a significant effect on eating? Test at the .05 level of significance.

2. In a research report, the term *significant* is used when the null hypothesis is rejected. (True or false?)

3. In a research report, the results of a hypothesis test include the phrase "$p < .01$." This means that the test failed to reject the null hypothesis. (True or false?)

ANSWERS

1. The null hypothesis states that temperature has no effect, or $\mu = 10$ grams even at the lower temperature. With $\alpha = .05$, the critical region consists of z-scores beyond $z = \pm 1.96$. The sample data produce $z = 3.00$. The decision is to reject H_0 and conclude that temperature has a significant effect on eating behavior.

2. True

3. False. The probability is *less than* .01, which means it is very unlikely that the result occurred by chance. In this case, the data are in the critical region, and H_0 is rejected.

8.4   DIRECTIONAL (ONE-TAILED) HYPOTHESIS TESTS

The hypothesis-testing procedure presented in Section 8.2 was the standard, or *two-tailed*, test format. The term *two-tailed* comes from the fact that the critical region is located in both tails of the distribution. This format is by far the most widely accepted procedure for hypothesis testing. Nonetheless, there is an alternative that will be discussed in this section.

Usually a researcher begins an experiment with a specific prediction about the direction of the treatment effect. For example, a special training program is expected to *increase* student performance, or alcohol consumption is expected to *slow* reaction times. In these situations, it is possible to state the statistical hypotheses in a manner that incorporates the directional prediction into the statement of H_0 and H_1. The result is a directional test, or what commonly is called a *one-tailed test*.

DEFINITION

In a directional hypothesis test, or a one-tailed test, the statistical hypotheses (H_0 and H_1) specify either an increase or a decrease in the population mean score. That is, they make a statement about the direction of the effect.

Suppose, for example, a researcher is using a sample of $n = 16$ laboratory rats to examine the effect of a new diet drug. It is known that under regular circumstances these rats eat an average of 10 grams of food each day. The distribution of food consumption is normal with $\sigma = 4$. The expected effect of the drug is to reduce food consumption. The purpose of the experiment is to determine whether or not the drug really works.

THE HYPOTHESES FOR A DIRECTIONAL TEST

Because a specific direction is expected for the treatment effect, it is possible for the researcher to perform a directional test. The first step (and the most critical step) is to state the statistical hypotheses. Remember that the null hypothesis states that there is no treatment effect and that the alternative hypothesis says that there is an effect. For this example, the predicted effect is that the drug will reduce food consumption. Thus, the two hypotheses would state:

H_0:   Food consumption is not reduced

H_1:   Food consumption is reduced

To express these directional hypotheses in symbols, it usually is easier to begin with the alternative hypothesis. Again, we know that regular rats eat an average of $\mu = 10$ grams of food, and H_1 says that food consumption will be reduced by the diet drug. Therefore, expressed in symbols, H_1 states

H_1: $\mu < 10$ (With the drug, food consumption is less than 10 grams per day.)

The null hypothesis states that the drug does not reduce food consumption:

H_0: $\mu \geq 10$ (With the drug, food consumption is at least 10 grams per day.)

THE CRITICAL REGION FOR DIRECTIONAL TESTS

The critical region is determined by sample values that are very unlikely if the null hypothesis is true. In other words, the critical region is defined by sample values that refute the null hypothesis because they provide evidence that the treatment really does work. In this example, the treatment is intended to reduce food consumption. Therefore, only sample values that are substantially less than $\mu = 10$ would indicate that the treatment worked and thereby lead to rejecting H_0. Thus, the critical region is located entirely in one tail of the distribution (see Figure 8.9). This is why the directional test commonly is called *one-tailed*.

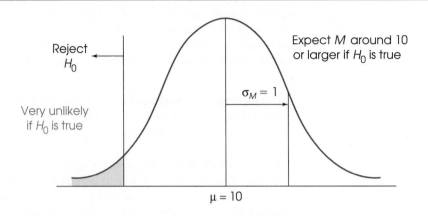

FIGURE 8.9

The distribution of sample means for $n = 16$ if H_0 is true. The null hypothesis states that the diet drug has no effect, so the population mean will be 10 or larger.

Reject H_0

Expect M around 10 or larger if H_0 is true

$\sigma_M = 1$

Very unlikely if H_0 is true

$\mu = 10$

Notice that a directional (one-tailed) test requires two changes in the step-by-step hypothesis-testing procedure.

1. In the first step of the hypothesis test, the directional prediction is incorporated into the statement of the hypotheses.

2. In the second step of the process, the critical region is located entirely in one tail of the distribution.

After these two changes, the remainder of a one-tailed test proceeds exactly the same as a regular two-tailed test. Specifically, you calculate the z-score statistic and then make a decision about H_0 depending on whether or not the z-score is in the critical region.

Next we present a complete example of a one-tailed test. Once again, we will use the experiment examining the effect of extra handling on the physical growth of infants to demonstrate the hypothesis-testing procedure.

EXAMPLE 8.3   It is known that under regular circumstances the population of 2-year-old children has an average weight of $\mu = 26$ pounds. The distribution of weights is normal with $\sigma = 4$. The researcher selects a random sample of $n = 4$ newborn infants and instructs the parents to provide each child with extra handling. The researcher predicts that the extra handling will stimulate the infants' growth and produce an *increase* in their weight at age 2. The researcher records the weight of each child at age 2 and obtains a sample mean of $M = 29.5$ pounds.

STEP 1   *State the hypotheses.* Because the researcher is predicting that extra handling will produce an increase in weight, it is possible to do a directional test. The null hypothesis states that the treatment does not have the predicted effect. In symbols,

Simply because you can do a directional test does not mean that you must use a directional test. A two-tailed test is always acceptable.

H_0: $\mu \leq 26$ (There is no increase in weight.)

The alternative hypothesis states that extra handling will increase physical growth. In symbols,

H_1: $\mu > 26$ (There is an increase in weight.)

STEP 2 *Locate the critical region.* To find the critical region, we look at all the possible sample means for $n = 4$ that could be obtained if H_0 were true. This is the distribution of sample means. It will be normal (because the population is normal), it will have a standard error of $\sigma_M = 4/\sqrt{4} = 2$, and it will have a mean of $\mu = 26$ if the null hypothesis is true. The distribution is shown in Figure 8.10.

For a one-tailed test with $\alpha = .05$, the entire 5% is in one tail of the distribution. Check the unit normal table with .05 in the tail, and you should find that $z = 1.65$.

If the extra handling does increase growth, as predicted, then the sample should have an average weight substantially greater than 26 pounds. That is, a large value for the sample mean would indicate that H_0 is wrong and should be rejected. Therefore, it is only large values that comprise the critical region. With $\alpha = .05$, the most likely 95% of the distribution is separated from the most unlikely 5% by a z-score of $z = +1.65$ (see Figure 8.10).

STEP 3 *Obtain the sample data.* The mean for the sample is $M = 29.5$. This value corresponds to a z-score of

$$z = \frac{M - \mu}{\sigma_M} = \frac{29.5 - 26}{2} = \frac{3.5}{2} = 1.75$$

STEP 4 *Make a statistical decision.* A z-score of $z = +1.75$ indicates that our sample mean is in the critical region. This is a very unlikely outcome if H_0 is true, so the statistical decision is to reject H_0. The conclusion is that extra handling does result in increased growth for infants.

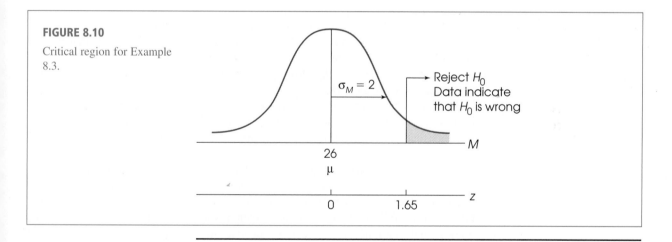

FIGURE 8.10

Critical region for Example 8.3.

COMPARISON OF ONE-TAILED VERSUS TWO-TAILED TESTS

The general goal of hypothesis testing is to determine whether or not a particular treatment has any effect on a population. The test is performed by selecting a sample, administering the treatment to the sample, and then comparing the result with the original population. If the treated sample is noticeably different from the original population, then we conclude that the treatment has an effect, and we reject H_0. On the other hand, if the treated sample is still similar to the original population, then we conclude that there is no evidence for a treatment effect, and we fail to reject H_0. The critical factor in this decision is the *size of the difference* between the treated sample and the original population. A large difference is evidence that the treatment worked; a small difference is not sufficient to say that the treatment has any effect.

The major distinction between one-tailed and two-tailed tests is in the criteria they use for rejecting H_0. A one-tailed test allows you to reject the null hypothesis when the difference between the sample and the population is relatively small, provided the difference is in the specified direction. A two-tailed test, on the other hand, requires a relatively large difference independent of direction. This point is illustrated in the following example.

EXAMPLE 8.4   Consider again the experiment examining extra handling and infant growth (see Example 8.3). If we had used a standard two-tailed test, then the hypotheses would have been

H_0: $\mu = 26$ pounds   (There is no treatment effect.)

H_1: $\mu \neq 26$ pounds   (Handling does affect growth.)

With $\alpha = .05$, the critical region would consist of any z-score beyond the $z = \pm 1.96$ boundaries.

If we obtained the same sample data, $M = 29.5$ pounds, which corresponds to a z-score of $z = 1.75$, our statistical decision would be "fail to reject H_0."

Note that with the two-tailed test (Example 8.4), the difference between the data ($M = 29.5$) and the hypotheses ($\mu = 26$) is not big enough to conclude that the hypothesis is wrong. In this case, we are saying that the data do not provide sufficient evidence to justify rejecting H_0. However, with the one-tailed test (Example 8.3), the same data led us to reject H_0.

All researchers agree that one-tailed tests are different from two-tailed tests. However, there are several ways to interpret the difference. One group of researchers contends that a two-tailed test is more rigorous and, therefore, more convincing than a one-tailed test. Remember that the two-tailed test demands more evidence to reject H_0 and thus provides a stronger demonstration that a treatment effect has occurred.

Other researchers feel that one-tailed tests are preferred because they are more sensitive. That is, a relatively small treatment effect may be significant with a one-tailed test but fail to reach significance with a two-tailed test. Also, there is the argument that one-tailed tests are more precise because they test hypotheses about a specific directional effect instead of an indefinite hypothesis about a general effect.

In general, two-tailed tests should be used in research situations where there is no strong directional expectation or where there are two competing predictions. For example, a two-tailed test would be appropriate for a study in which one theory predicts an increase in scores but another theory predicts a decrease. On the other hand, one-tailed tests should be used only in situations where the directional prediction is made before the research is conducted and there is a strong justification for making the directional prediction. In particular, you should never use a one-tailed test as a second attempt to salvage a significant result from a research study where the first analysis with a two-tailed test failed to reach significance.

LEARNING CHECK

1. A researcher predicts that a treatment will lower scores. If this researcher uses a one-tailed test, will the critical region be in the right- or left-hand tail of the distribution?

2. A psychologist is examining the effects of early sensory deprivation on the development of perceptual discrimination. A sample of $n = 9$ newborn kittens is obtained. These kittens are raised in a completely dark environment for 4 weeks, after which they receive normal visual stimulation. At age 6 months, the kittens are tested on a visual discrimination task. The average score for this sample is $M = 32$. It is known that under normal circumstances cats score an average of $\mu = 40$ on this task. The distribution of scores is normal with $\sigma = 12$. The researcher is predicting that the early sensory deprivation will reduce the kittens' performance on the discrimination task. Use a one-tailed test with $\alpha = .01$ to test this hypothesis.

ANSWERS

1. The left-hand tail

2. The hypotheses are H_0: $\mu \geq 40$ and H_1: $\mu < 40$. The critical region is determined by z-scores less than -2.33. The z-score for these sample data is $z = -2.00$. Fail to reject H_0.

8.5 THE GENERAL ELEMENTS OF HYPOTHESIS TESTING: A REVIEW

The z-score hypothesis test presented in this chapter is one specific example of the general process of hypothesis testing. In later chapters, we will examine many other hypothesis tests. Although the details of the hypothesis test will vary from one situation to another, all of the different tests use the same basic logic and consist of the same elements. In this section, we will present a generic hypothesis test that outlines the basic elements and logic common to all tests. The better you understand the general process of a hypothesis test, the better you will understand inferential statistics.

Hypothesis tests consist of five basic elements organized in a logical pattern. The elements and their relationships to each other are as follows:

1. Hypothesized Population Parameter. The first step in a hypothesis test is to state a null hypothesis. You will notice that the null hypothesis typically provides a specific value for an unknown population parameter. The specific value predicted by the null hypothesis is the first element of hypothesis testing.

2. Sample Statistic. The sample data are used to calculate a sample statistic that corresponds to the hypothesized population parameter. Thus far we have considered only situations where the null hypothesis specifies a value for the population mean, and the corresponding sample statistic is the sample mean.

3. Estimate of Error. To evaluate our findings, we must know what types of sample data are likely to occur simply due to chance. Typically, a measure of standard error is used to provide this information. The general purpose of standard error is to provide a measure of how much difference it is reasonable to expect between a sample statistic and the corresponding population parameter. Remember that samples (statistics) are not expected to provide a perfectly accurate picture of populations (parameters). There always will be some error or discrepancy between a statistic and a parameter; this is the concept of sampling error. The standard error tells how much error is expected just by chance.

In Chapter 7, we introduced the standard error of M. This is the first specific example of the general concept of standard error, and it provides a measure of how much error is expected between a sample mean (statistic) and the corresponding population mean (parameter). In later chapters, you will encounter other examples of standard error, each of which measures the standard distance (expected by chance) between a specific statistic and its corresponding parameter. Although the exact formula for standard error will change from one situation to another, you should recall that standard error is basically determined by two factors:

a. The variability of the scores. Standard error is directly related to the variability of the scores: The larger the variability, the larger the standard error.

b. The size of the sample. Standard error is inversely related to sample size: The larger the sample, the smaller the standard error.

4. The Test Statistic. In this chapter, the test statistic is a z-score:

$$z = \frac{M - \mu}{\sigma_M}$$

This z-score statistic provides a model for other test statistics that will follow. In particular, a test statistic typically forms a *ratio*. The numerator of the ratio is simply the obtained difference between the sample statistic and the hypothesized parameter. The denominator of the ratio is the standard error measuring how much difference is expected by chance. Thus, the generic form of a test statistic is

$$\text{test statistic} = \frac{\text{obtained difference}}{\text{difference expected by chance}}$$

In general, the purpose of a test statistic is to determine whether or not the result of an experiment (the obtained difference) is more than expected by chance alone. Thus, a test statistic that has a value greater than 1.00 indicates that the obtained result (numerator) is greater than chance (denominator). In addition, the test statistic provides a measure of *how much* greater than chance. A value of 3.00, for example, indicates that the obtained difference is three times greater than would be expected by chance.

5. The Alpha Level. The final element in a hypothesis test is the alpha level, or level of significance. The alpha level provides a criterion for interpreting the test statistic. As we noted, a test statistic greater than 1.00 indicates that the obtained result is greater than expected by chance. However, researchers demand that the likelihood of a result be not simply "more than chance," but *significantly more than chance*. The alpha level provides a criterion for "significance."

Earlier in this chapter (page 249), we noted that when a significant result is reported in the literature, it always is accompanied by a p value; for example, $p < .05$ means that there is less than a 5% probability that the result occurred by chance. In general, a *significant* result means that the researcher is very confident that the result is not simply due to chance. A result that is significant at the .05 (5%) level, for example, means that the researcher is 95% confident that the obtained difference is greater than what one would expect by chance alone. Similarly, the .01 (1%) level indicates 99% confidence, and the .001 (.1%) level indicates 99.9% confidence.

To gain more confidence (rejecting H_0 with a smaller alpha), the results must demonstrate a larger difference (treatment effect). As you know, the alpha level is used to obtain a critical value for the test statistic. With the z-score test, for example, $\alpha = .05$ produces a critical value of ± 1.96. In general, the alpha level determines how big the test statistic ratio must be before you can claim a *significant* result. The following values are *not* intended to be precise, but rather they should provide you with a *rough idea* of how the alpha level is related to the test statistic:

a. For $\alpha = .05$, the criterion for significance is a test statistic ratio of at least 2.00. That is, the obtained difference (numerator) must be at least *twice as big* as chance (denominator). *Note:* The exact value for the z-score test is 1.96.

b. For $\alpha = .01$, the criterion for significance is a test statistic ratio of at least 2.50. That is, the obtained difference must be at least *two and one-half times bigger* than chance. *Note:* The exact value for the z-score test is 2.58.

c. For $\alpha = .001$, the criterion for significance is a test statistic ratio of at least 3.00. That is, the obtained difference must be at least *three times bigger* than chance. *Note:* The exact value for the z-score test is 3.30.

Again, these values are *approximate* and provide a rough "rule of thumb." You always will need to consult a statistical table (for example, the unit normal table) to determine the exact critical values for your hypothesis test. Finally, we remind you to watch for the five basic elements described in this section as you encounter different examples of hypothesis testing throughout the rest of the book.

8.6 CONCERNS ABOUT HYPOTHESIS TESTING: MEASURING EFFECT SIZE

Although hypothesis testing is the most commonly used technique for evaluating and interpreting research data, a number of scientists have expressed a variety of concerns about the hypothesis-testing procedure (for example, see Loftus, 1996, and Hunter, 1997). The following are some of the questions and criticisms that have been presented:

1. One common criticism is that a hypothesis test produces an absolute, all-or-none decision: Either you reject the null hypothesis, or you don't. Suppose, for example, a hypothesis test for one treatment produces a z-score of $z = 2.00$ and a test for a second treatment produces a z-score of 1.90. Assuming both tests are done with $\alpha = .05$, then both tests have the same critical boundary of $z = 1.96$. According to this standard, the first test would reject H_0 and conclude that the treatment has a significant effect. The second test would fail to reject H_0 and conclude that the treatment does not have a significant effect. Although the hypothesis test makes completely different decisions about the two treatments, it could be argued that they are really almost identical: How much difference is there between $z = 1.90$ and $z = 2.00$? The counterargument is that you have to draw a line somewhere. If you decided to reject H_0 with $z = 1.90$ just because it is close to the criterion, what about $z = 1.80$, or 1.70, and so on? Although the .05 level of significance may be arbitrary, some criterion is essential.

2. A second criticism is that the whole idea of a null hypothesis is artificial. Remember: The null hypothesis states that the treatment has no effect, meaning absolutely none. The criticism is based on the real-world observation that *every* treatment must have *some* effect. Anytime a researcher administers any treatment, it has to have some effect, however small, on the subjects. If this is true, it means that the null hypothesis is always wrong. If you follow this line of reasoning, it means that you can never make a Type I error: If H_0 cannot be true, you can never reject a true H_0. The counterargument to this criticism is that the null hypothesis, even if it is flawed, is a logical necessity. Because we are using an *inferential* process, it is impossible to prove that a hypothesis is true; however, it is possible to show that a hypothesis is false (or at least very unlikely). See Box 8.1 for more discussion of this issue.

3. A much more serious criticism of a hypothesis test focuses on what the test actually does and does not do. Specifically, when a hypothesis test rejects the null hypothesis and concludes that a treatment has a significant effect, it does not mean that the treatment has a "substantial" effect. Instead, a significant result simply means that the hypothesis test has ruled out *chance* (sampling error) as a plausible explanation for the results. That is, the hypothesis test has determined that the results obtained in the research study are very unlikely to have occurred if there really is no treatment effect. The hypothesis test reaches this conclusion by (1) calculating the standard error, which measures how much difference is reasonable to expect just by chance, and (2) demonstrating that the obtained mean difference is substantially bigger than the standard error.

   Notice that the test is making a *relative* comparison: the size of the treatment effect is being evaluated relative to the difference expected by chance. If the standard error (chance) is very small, then the treatment effect can also be very small and still be bigger than chance. Thus, a significant effect does not necessarily mean a big effect.

The idea that a hypothesis test evaluates the relative size of a treatment effect, rather than the absolute size, is illustrated in the following example.

EXAMPLE 8.5 We begin with a population of scores that forms a normal distribution with $\mu = 80$ and $\sigma = 10$. A sample is selected from the population and a treatment is administered to the sample. After treatment, the sample mean is found to be $M = 81$. Does this sample provide evidence of a statistically significant treatment effect?

Although there is only a 1-point difference between the sample mean and the orginal population mean, the difference may be enough to be significant. In particular, the outcome of the hypothesis test depends on the sample size.

For example, with a sample of $n = 25$ the standard error is

$$\sigma_M = \frac{\sigma}{\sqrt{n}} = \frac{10}{\sqrt{25}} = \frac{10}{5} = 2.00$$

and the z-score for $M = 81$ is

$$z = \frac{M - \mu}{\sigma_M} = \frac{81 - 80}{2} = \frac{1}{2} = 0.50$$

For an alpha level of .05, the critical region would begin at $z = 1.96$. Our z-score fails to reach the critical region, so we fail to reject the null hypothesis. In this case, the 1-point difference between M and μ is not significant because it is being evaluated relative to a standard error of 2 points.

Now consider the outcome with a sample of $n = 400$. The standard error is

$$\sigma_M = \frac{\sigma}{\sqrt{n}} = \frac{10}{\sqrt{400}} = \frac{10}{20} = 0.50$$

and the z-score for $M = 81$ is

$$z = \frac{M - \mu}{\sigma_M} = \frac{81 - 80}{0.5} = \frac{1}{0.5} = 2.00$$

Now the z-score is beyond the 1.96 boundary, so we reject the null hypothesis and conclude that there is a signficant effect. In this case, the 1-point difference between M and μ is considered statistically significant because it is being evaluated relative to a standard error of only 0.5 points.

The point of Example 8.5 is that a small treatment effect can still be statistically significant. If the sample size is large enough, any treatment effect, no matter how small, can be enough for us to reject the null hypothesis.

MEASURING EFFECT SIZE

As noted in the previous section (point #3), one concern with hypothesis testing is that a hypothesis test does not really evaluate the absolute size of a treatment effect. To correct this problem, it is recommended that whenever researchers report a statistically significant effect, they also provide a report of the effect size (see the guidelines presented by the APA Task Force on Statistical Inference; Wilkinson, 1999). Therefore, as we present different hypothesis tests we will also present different options for measuring and reporting effect size.

One of the simplest and most direct methods for measuring effect size is Cohen's d. Cohen (1988) recommended that effect size can be standardized by measuring the mean difference in terms of the standard deviation. The resulting measure of effect size is computed as

$$\text{Cohen's } d = \frac{\text{mean difference}}{\text{standard deviation}} \tag{8.1}$$

The standard deviation is included in the calculation to standardize the size of the mean difference in much the same way that z-scores standardize locations in a distribution. The standardization process is demonstrated in Figure 8.11. The top portion of the figure (part a) shows the results of a treatment that produces a 15-point mean difference in SAT scores; before treatment, the average SAT score is 500, and after treatment the average is 515. Notice that the standard deviation for SAT scores is $\sigma = 100$, so the 15-point difference appears to be small. For this example, Cohen's d is

$$\text{Cohen's } d = \frac{\text{mean difference}}{\text{standard deviation}} = \frac{15}{100} = 0.15$$

See Box 8.3.

Now consider the treatment effect shown in Figure 8.11(b). This time, the treatment produces a 15-point mean difference in IQ scores; before treatment the average IQ is 100, and after treatment the average is 115. Because IQ scores have a standard devia-

FIGURE 8.11

The appearance of a 15-point treatment effect in two different situations. In part (a), the standard deviation is σ = 100 and the 15-point effect is relatively small. In part (b), the standard deviation is σ = 15 and the 15-point effect is relatively large. Cohen's *d* uses the standard deviation to help measure effect size.

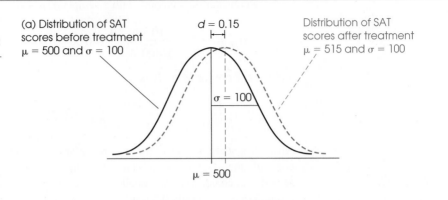

(a) Distribution of SAT scores before treatment μ = 500 and σ = 100

$d = 0.15$

Distribution of SAT scores after treatment μ = 515 and σ = 100

σ = 100

μ = 500

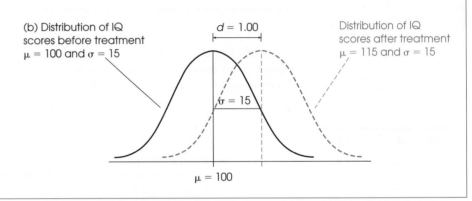

(b) Distribution of IQ scores before treatment μ = 100 and σ = 15

$d = 1.00$

Distribution of IQ scores after treatment μ = 115 and σ = 15

σ = 15

μ = 100

tion of σ = 15, the 15-point mean difference now appears to be large. For this example, Cohen's *d* is

$$\text{Cohen's } d = \frac{\text{mean difference}}{\text{standard deviation}} = \frac{15}{15} = 1.00$$

Cohen (1988) also suggested criteria for evaluating the size of a treatment effect as shown in Table 8.2.

TABLE 8.2

Evaluating effect size with Cohen's *d*

MAGNITUDE OF D	EVALUATION OF EFFECT SIZE
$0 < d < 0.2$	Small effect (mean difference less than 0.2 standard deviation)
$0.2 < d < 0.8$	Medium effect (mean difference around 0.5 standard deviation)
$d > 0.8$	Large effect (mean difference greater than 0.8 standard deviation)

As one final demonstration of Cohen's *d*, consider the two hypothesis tests in Example 8.5. For each test, the original population had a mean of μ = 80 with a standard deviation of σ = 10. For each test, the mean for the treated sample was *M* = 81. Although one test used a sample of *n* = 25 and the other test used a sample of *n* = 400, the sample size is not considered when computing Cohen's *d*. Therefore, both of the hypothesis tests would produce the same value:

$$\text{Cohen's } d = \frac{\text{mean difference}}{\text{standard deviation}} = \frac{1}{10} = 0.10$$

BOX
8.3   OVERLAPPING DISTRIBUTIONS

Figure 8.11(b) shows the results of a treatment with a Cohen's d of 1.00; that is, the effect of the treatment is to increase the mean by one full standard deviation. According to the guidelines in Table 8.2, a value of $d = 1.00$ is considered a large treatment effect. However, looking at the figure, you may get the impression that there really isn't that much difference between the distribution before treatment and the distribution after treatment. In particular, there is substantial overlap between the two distributions, so that many of the individuals who receive the treatment are not any different from the individuals who do not receive the treatment.

The overlap between distributions is a basic fact of life in most research situations; it is extremely rare for the scores after treatment to be *completely different* (no overlap) from the scores before treatment. Consider, for example, children's heights at different ages. Everyone

knows that 8-year-old children are taller than 6-year-old children; on average, the difference is 3 or 4 inches. However, this does not mean that all 8-year-old children are taller than all 6-year-old children. In fact, there is considerable overlap between the two distributions, so that the tallest among the 6-year-old children are actually taller than most 8-year-old children. In fact, the height distributions for the two age groups would look a lot like the two distributions in Figure 8.11(b). Although there is a clear *mean difference* between the two distributions, there still can be substantial overlap.

Cohen's d measures the degree of separation between two distributions, and a separation of one standard deviation ($d = 1.00$) represents a large difference. Eight-year-old children really are bigger than 6-year-old children.

LEARNING CHECK

1. Briefly explain why a "statistically significant" effect is not necessarily a "substantial" effect.

2. Compute Cohen's d for the hypothesis test presented in Example 8.2.

ANSWERS

1. "Statistically significant" means that the result is greater than would be expected by chance. However, chance is measured by the standard error and, when the sample size is large, the standard error can be very small. In this case, a small treatment effect can still be significant.

2. For Example 8.2, the mean difference is 3 points and the standard deviation is $\sigma = 4$. Cohen's d is 0.75.

8.7   STATISTICAL POWER

The purpose of a hypothesis test is to determine whether or not a particular treatment has an effect. The null hypothesis states that there is no effect, and the researcher is hoping that the sample data will provide evidence to reject this hypothesis. In previous sections, we saw that the hypothesis-testing procedure always involves some risk of reaching a wrong conclusion. Specifically, a researcher may commit a Type I error by rejecting a true null hypothesis. This risk is minimized by selecting an alpha level (usually .05 or .01) that determines the maximum probability of committing a Type I error. You should also recall that every hypothesis test has the potential for resulting in a Type II error, the failure to reject a false null hypothesis. In simple terms, a Type II error means that the treatment really does have an effect, but the hypothesis test failed to discover it.

In this section, we reverse our perspective on hypothesis testing. Rather than examine the potential for making an error, we examine the probability of reaching the correct decision. Remember that the researcher's goal is to demonstrate that the experimental treatment actually does have an effect. This is the purpose of conducting the experiment in the first place. If the researcher is correct and the treatment really does have an effect, then what is the probability that the hypothesis test will correctly identify it? This is a question concerning the *power* of a statistical test.

DEFINITION
> The **power** of a statistical test is the probability that the test will correctly reject a false null hypothesis. That is, power is the probability that the test will correctly detect a real treatment effect.

It should be clear that the concepts of power and Type II error are closely related. *When a treatment effect exists*, the hypothesis test will have one of two results:

1. It can fail to discover the existing treatment effect (a Type II error).

2. It can correctly detect the presence of a treatment effect (rejecting a false null hypothesis).

Remember that the probabilities for all events must add up to 1.00 (100%).

We already have noted that the probability of a Type II error is identified by the symbol beta, β. Because there are only two possible outcomes when a treatment effect exists, and one has a probability of β, the other outcome must have a probability of $1 - \beta$. Thus, the probability of correctly rejecting a false null hypothesis is $1 - \beta$. If, for example, a hypothesis test has a probability of .20 (20% chance) of failing to detect a treatment effect (Type II error), then it must have a probability of .80 (80% chance) of successfully detecting it. Thus, the power of a statistical test is determined by

$$\text{power} = p(\text{reject a false } H_0) = 1 - \beta \tag{8.2}$$

POWER AND THE SIZE OF THE TREATMENT EFFECT

Although we have defined power, we have not attempted to compute a value for this probability. The difficulty in specifying this value comes from the fact that power depends on the size of the treatment effect. Therefore, it does not have a single value for every hypothesis test. When a treatment has a large effect, it will be easy to detect this effect, and power will be high. On the other hand, when the treatment effect is very small, it will be difficult to detect, and power will be low. Thus, rather than talking about power as a single value, we must examine the different values of power associated with different magnitudes of treatment effect. We can illustrate this in a set of graphs that represent different treatment effects, but, first, one more reminder is necessary. Remember: The null hypothesis is rejected whenever sample data are in the critical region. With this in mind, we can restate the definition of power as follows:

DEFINITION
> **Power** is the probability of obtaining sample data in the critical region when the null hypothesis is false.
>
> *Note:* Because "data in the critical region" is equivalent to "rejecting the null hypothesis," this definition is equivalent to our original definition of power.

The following hypothesis-testing scenario will be used to help demonstrate the relationship between power and the size of the treatment effect.

EXAMPLE 8.6   A researcher is interested in evaluating the effects of a cold medicine on reaction time. It is known that reaction times typically form a normal distribution with $\mu = 200$ milliseconds and $\sigma = 40$. The researcher selects a sample of $n = 16$ individuals. Each person is given a standard dose of the cold medicine, and 30 minutes later each person's reaction time is measured. The goal is to use the data from the sample to determine whether or not the medicine has an effect on reaction time. We will begin the hypothesis test using an alpha level of .05.

The first step in the hypothesis test is to state the hypotheses. For this example, the null hypothesis states that the medication has no effect on reaction time. In symbols, H_0: $\mu = 200$.

Next, we locate the critical region. For this example, we consider all the possible sample means that could be obtained from a population with $\mu = 200$, and define the critical region as the extreme, unlikely values in the tails of this distribution. With $\alpha = .05$, the boundaries for the critical region are $z = 1.96$ and $z = -1.96$. This distribution, which we will call the *null distribution*, is shown in Figure 8.12. Notice that the null distribution is centered at $\mu = 200$ (which is the value that is expected for our sample mean if the null hypothesis is true) and has a standard error of $\sigma_M = 40/\sqrt{16} = 10$. Also notice that we have indicated the critical region by using a grey shading for all of the area beyond $z = 1.96$ and $z = -1.96$.

FIGURE 8.12

The distribution of sample means for a population with a hypothesized mean of $\mu = 200$. Notice that the critical region, determined by $\alpha = .05$, is shown by shading the entire area beyond $z = 1.96$ and $z = -1.96$.

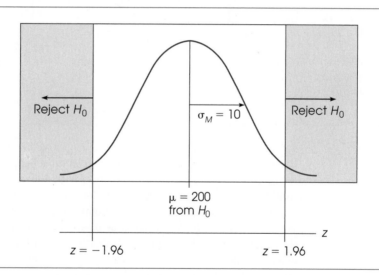

Example 8.6 presented a typical hypothesis test. Now, however, we will return to the topic of statistical power. Specifically, we will consider two possibilities for how much effect the cold medicine actually has on reaction time. First, we will examine the outcome if the medicine produces a 20-point increase in reaction time, and then we will examine what happens if the medicine causes a 40-point increase in reaction time.

Power for a 20-point effect   If the effect of the medicine is to increase reaction time by 20 points, then the population mean for people taking the medicine increases to $\mu = 220$. A sample of individuals who had taken the medication would be expected to have a sample mean around 220, and the set of all the possible sample means should pile up around 220. In Figure 8.13 we show the original null distribution and critical region from the hypothesis test, and a new distribution that contains all the sample means that could be obtained if the cold medicine has a 20-point effect. Notice that if the medicine has a 20-point effect, roughly 50% of the possible sample means are located in the critical region. Thus, there is a 50% probability that we would obtain a sample mean in the critical region and reject the null hypothesis. You should note that the value of 50% is the power of the test; that is, there is a 50% probability that the test would correctly reject the null hypothesis if the treatment has a 20-point effect.

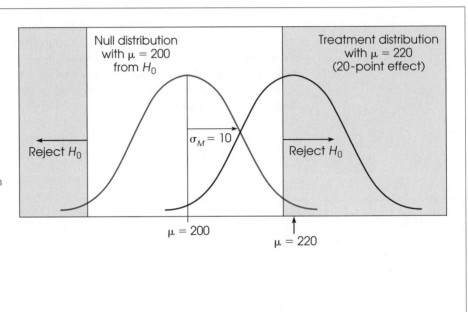

FIGURE 8.13

The null distribution and critical region from Figure 8.12, and a treatment distribution showing the set of sample means that would be obtained with a 20-point treatment effect. Notice that the null distribution is centered at $\mu = 200$ (no effect) and the treatment distribution is centered at $\mu = 220$ (showing a 20-point effect). Also notice that with a 20-point effect, more than 50% of the sample means would be in the critical region and would lead to rejecting the null hypothesis.

Power for a 40-point effect   Now consider what happens if the effect of the medicine is to increase reaction time by 40 points. With a 40-point effect, the population mean would increase from $\mu = 200$ to $\mu = 240$. A sample of individuals who had taken the cold medicine should produce a sample mean around 240, and the entire distribution of sample means would be centered at $\mu = 240$. This distribution of sample means is shown along with the original null distribution in Figure 8.14. Notice that when the treatment has a 40-point effect, nearly all of the possible sample means (around 95%) are located in the critical region. This means that there is a 95% probability of obtaining a sample mean in the critical region and rejecting the null hypothesis if the medicine causes a 40-point increase in reaction time. In this case, the power of the hypothesis test is roughly 95%.

In general you should realize that the power of a hypothesis test is directly related to the size of the treatment effect. When a treatment has a large effect, there is a relatively high probability that the hypothesis test will detect it; when a treatment effect is small, there is a relatively low probability that it will be detected. As a secondary point, you should realize that it is possible to compute a precise value for the power of a hypothesis test. In the preceding examples, we simply estimated power by looking at the graphs. However, if you locate the exact position of the critical region boundary, it is possible to look up the exact proportion of the sample means that fall into the critical region. Although we will not demonstrate the details of the calculations, those of you

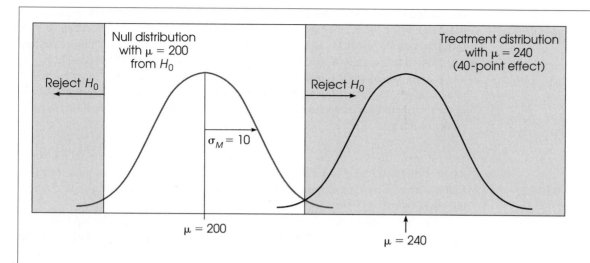

FIGURE 8.14

The null distribution and critical region from Figure 8.12, and a treatment distribution showing the set of sample means that would be obtained with a 40-point treatment effect. Notice that the null distribution is centered at $\mu = 200$ (no effect) and the treatment distribution is centered at $\mu = 240$ (showing a 40-point effect). Also notice that with a 40-point effect, nearly all of the possible sample means would be in the critical region and would lead to rejecting the null hypothesis.

who want to try should find that power equals 51.60% for the 20-point effect, and power is 97.93% for the 40-point effect.

OTHER FACTORS THAT AFFECT POWER

In addition to the size of the treatment effect, there are a variety of other factors that influence the power of a hypothesis test. In this section we discuss three factors that are under the direct control of the researcher.

Alpha level   Reducing the alpha level for a hypothesis test will also reduce the power of the test. For example, lowering α from .05 to .01 will lower the power of the hypothesis test. The effect of reducing the alpha level can be seen by referring again to Figure 8.14. In this figure, the null distribution is on the left and the treatment distribution (for a 40-point effect) is shown on the right. The power of the test is determined by the proportion of the treatment distribution that is in the critical region. Remember: Power is defined as the probability that the sample data will be in the critical region and reject H_0 when there is a real treatment effect. Thus, any factor that causes more of the treatment distribution to be in the critical region will increase power, and any factor that causes less of the treatment distribution to be in the critical region will decrease power. In Figure 8.14, the boundaries for the critical region are drawn using $\alpha = .05$. Specifically, the critical region on the right-hand side begins at $z = 1.96$. If α were changed to .01, the boundary would be moved farther to the right, out to $z = 2.58$. It should be clear that moving the critical boundary to the right means that less of the treatment distribution (the right-hand distribution) would be in the critical region. Thus, there would be a lower probability of rejecting the null hypothesis and a lower value for the power of the test.

One-tailed versus two-tailed tests   Changing from a regular two-tailed test to a one-tailed test will increase the power of the hypothesis test. Again, this effect can be seen by referring to Figure 8.14. The figure shows the boundaries for the critical region using a two-tailed test with $\alpha = .05$, so that the critical region on the right-hand side begins at $z = 1.96$. Changing to a one-tailed test would move the critical boundary to the left to a value of $z = 1.65$. Moving the boundary to the left will cause a larger proportion of the treatment distribution to be in the critical region and, therefore, will increase the power of the test.

Sample size   In general, the larger the sample is, the better it will represent the population. If there actually is a treatment effect in the population, you are more likely to find it with a large sample than with a small sample. Thus, the power of a test can be increased by using larger samples.

Figure 8.15 demonstrates how the sample size influences power. The top half of the figure reproduces Figure 8.13 and shows a null distribution (on the left) and a 20-point

FIGURE 8.15

The relationship between sample size and power. The top figure (a) shows a null distribution and a 20-point treatment distribution based on samples of $n = 16$ and a standard error of 10 points. Notice that the right-hand critical boundary is located in the middle of the treatment distribution so that roughly 50% of the treated samples fall in the critical region.

In the bottom figure (b) the distributions are based on samples of $n = 100$ and the standard error is reduced to 4 points. In this case, essentially all of the treated samples fall in the critical region and the hypothesis test has power of nearly 100%.

(a) With $n = 16$ the standard error is $\sigma_M = 10$ points.

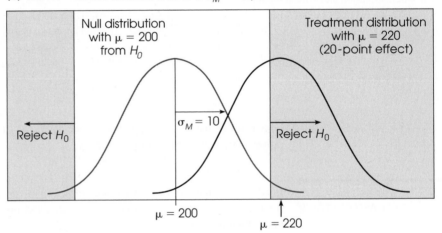

(b) With $n = 100$ the standard error is $\sigma_M = 4$ points.

treatment effect distribution (on the right). Both distributions display the distribution of sample means based on samples of $n = 16$ scores selected from a population with $\sigma = 40$. Thus, both distributions have a standard error of $\sigma_M = 40/\sqrt{16} = 10$. Notice that roughly 50% of the sample means in the treatment distribution are located in the critical region, which means that this particular test situation has power equal to 50%.

Now consider the two distributions in the bottom half of Figure 8.15. Again, we show a null distribution on the left and a 20-point treatment distribution on the right. Now, however, the two distributions represent the distribution of sample means that would be obtained with $n = 100$ scores in each sample. Notice that with $n = 100$ the standard error is reduced to $\sigma_M = 40/\sqrt{100} = 4$. With a standard error of only 4 points, nearly all of the treatment distribution is located in the critical region, producing a value of roughly 99% for the power of the hypothesis test. Again, the basic relationship is that larger samples tend to produce greater power for a hypothesis test.

LEARNING CHECK

1. For a particular hypothesis test, $1 - \beta$ equals .50 for a 5-point treatment effect. Will the power be greater or less for a 10-point treatment effect?

2. As the power of a test increases, what happens to the probability of a Type II error?

3. If a researcher uses $\alpha = .01$ rather than $\alpha = .05$, then power will increase. (True or false?)

4. What is the effect of increasing sample size on power?

ANSWERS

1. The hypothesis test is more likely to detect a 10-point effect, so power will be greater.

2. As power increases, the probability of a Type II error decreases.

3. False.   4. Power increases as sample size gets larger.

SUMMARY

1. Hypothesis testing is an inferential procedure that uses the data from a sample to draw a general conclusion about a population. The procedure begins with a hypothesis about an unknown population. Then a sample is selected, and the sample data provide evidence that either supports or refutes the hypothesis.

2. In this chapter, we introduced hypothesis testing using the simple situation in which a sample mean is used to test a hypothesis about an unknown population mean. We begin with an unknown population, generally a population that has received a treatment. The question is to determine whether or not the treatment has had an effect on the population mean (see Figure 8.1).

3. Hypothesis testing is structured as a four-step process that will be used throughout the remainder of the book.

a. State the null hypothesis (H_0), and select an alpha level. The null hypothesis states that there is no effect or no change. In this case, H_0 states that the mean for the treated population is the same as the mean before treatment. The alpha level, usually $\alpha = .05$ or $\alpha = .01$, provides a definition of the term *very unlikely* and determines the risk of a Type I error. An alternative hypothesis (H_1), which is the exact opposite of the null hypothesis, is also stated.

b. Locate the critical region. The critical region is defined as sample outcomes that would be very unlikely to occur if the null hypothesis is true. The alpha level defines "very unlikely." For example, with $\alpha = .05$, the critical region is defined as sample means in the extreme 5% of the distribution of sample means. When the distribution is normal, the extreme 5% corresponds to z-scores beyond $z = \pm 1.96$.

c. Collect the data, and compute the test statistic. The sample mean is transformed into a z-score by the formula

$$z = \frac{M - \mu}{\sigma_M}$$

The value of μ is obtained from the null hypothesis. The z-score test statistic identifies the location of the sample mean in the distribution of sample means. Expressed in words, the z-score formula is

$$z = \frac{\text{sample mean} - \begin{array}{c}\text{hypothesized}\\\text{population mean}\end{array}}{\text{standard error}}$$

d. Make a decision. If the obtained z-score is in the critical region, we reject H_0 because it is very unlikely that these data would be obtained if H_0 were true. In this case, we conclude that the treatment has changed the population mean. If the z-score is not in the critical region, we fail to reject H_0 because the data are not significantly different from the null hypothesis. In this case, the data do not provide sufficient evidence to indicate that the treatment has had an effect.

4. Whatever decision is reached in a hypothesis test, there is always a risk of making the incorrect decision. There are two types of errors that can be committed.

A Type I error is defined as rejecting a true H_0. This is a serious error because it results in falsely reporting a treatment effect. The risk of a Type I error is determined by the alpha level and therefore is under the experimenter's control.

A Type II error is defined as the failure to reject a false H_0. In this case, the experiment fails to report an effect that actually occurred. The probability of a Type II error cannot be specified as a single value and depends in part on the size of the treatment effect. It is identified by the symbol β (beta).

5. When a researcher expects that a treatment will change scores in a particular direction (increase or decrease), it is possible to do a directional or one-tailed test. The first step in this procedure is to incorporate the directional prediction into the hypotheses. For example, if the prediction is that a treatment will increase scores, the null hypothesis says that there is no increase and the alternative hypothesis states that there is an increase. To locate the critical region, you must determine what kind of data would refute the null hypothesis by demonstrating that the treatment worked as predicted. These outcomes will be located entirely in one tail of the distribution, so the entire critical region (5% or 1% depending on α) will be in one tail.

6. A one-tailed test is used when there is prior justification for making a directional prediction. These a priori reasons may be previous reports and findings or theoretical considerations. In the absence of the a priori basis, a two-tailed test is appropriate. In this situation, one might be unclear regarding what to expect in the study, or one might be testing competing theories.

7. In addition to using a hypothesis test to evaluate the *significance* of a treatment effect, it is recommended that you also measure and report the *effect size*. One measure of effect size is Cohen's d, which is a standardized measure of the mean difference. Cohen's d is computed as

$$\text{Cohen's } d = \frac{\text{mean difference}}{\text{standard deviation}}$$

8. The power of a hypothesis test is defined as the probability that the test will correctly reject the null hypothesis. Power is identified as

$$\text{power} = 1 - \beta$$

where β is the probability of a Type II error.

9. To illustrate the power for a hypothesis test, you must first identify the treatment and null distributions. Also, you must specify the magnitude of the treatment effect. Next, you locate the critical region in the null distribution. The power of the hypothesis test is the portion of the treatment distribution that is located beyond the boundary (critical value) of the critical region.

10. As the size of the treatment effect increases, statistical power increases. Also, power is influenced by several factors that can be controlled by the experimenter:
   a. Increasing the alpha level will increase power.
   b. A one-tailed test will have greater power than a two-tailed test.
   c. A large sample will result in more power than a small sample.

RESOURCES

If you are using WebTutor along with this book, there is a WebTutor section corresponding to this chapter. The WebTutor contains a brief summary of Chapter 8, hints for learning the new material and for avoiding common errors, and a review of the new vocabulary
that accompanies hypothesis testing. As usual, WebTutor also presents some sample exam
items including solutions.

SPSS

The statistics computer package SPSS is not structured to perform hypothesis tests using
z-scores. In truth, the z-score test presented in Chapter 8 is rarely, if ever, used in actual
research situations. The problem with the z-score hypothesis is that it requires that you
know the value of the population standard deviation, and this is usually an unrealistic
assumption. That is, researchers rarely have detailed information about the populations
they wish to study. Instead, they must obtain information entirely from samples. In the
following chapters, beginning with Chapter 9, we will introduce a new hypothesis-testing
statistic that is based entirely on sample data. The new statistic, called a t statistic, is included in SPSS and will be introduced in the next chapter.

KEY TERMS

hypothesis testing	Type II error	test statistic	effect size
null hypothesis	level of significance	beta	Cohen's d
alternative hypothesis	alpha level	directional test	power
Type I error	critical region	one-tailed test	

FOCUS ON PROBLEM SOLVING

1. Hypothesis testing involves a set of logical procedures and rules that enable us to make
   general statements about a population when all we have are sample data. This logic is
   reflected in the four steps that have been used throughout this chapter. Hypothesis-
   testing problems will become easier to tackle when you learn to follow the steps.

STEP 1   State the hypotheses, and set the alpha level.

STEP 2   Locate the critical region.

S T E P 3   Compute the z-score for the sample statistic.

S T E P 4   Make a decision about H_0 based on the result of step 3.

A nice benefit of mastering these steps is that all hypothesis tests that will follow use the same basic logic outlined in this chapter.

2. Students often ask, "What alpha level should I use?" Or a student may ask, "Why is an alpha of .05 used?" as opposed to something else. There is no single correct answer to either of these questions. Keep in mind the aim of setting an alpha level in the first place: *to reduce the risk of committing a Type I error*. Therefore, you would not want to set α to something like .20. In that case, you would be taking a 20% risk of committing a Type I error—reporting an effect when one actually does not exist. Most researchers would find this level of risk unacceptable. Instead, researchers generally agree to the convention that $\alpha = .05$ is the greatest risk one should take in making a Type I error. Thus, the .05 level of significance is frequently used and has become the "standard" alpha level. However, some researchers prefer to take even less risk and use alpha levels of .01 and smaller.

3. Take time to consider the implications of your decision about the null hypothesis. The null hypothesis states that there is no effect. Therefore, if your decision is to reject H_0, you should conclude that the sample data provide evidence for a treatment effect. However, it is an entirely different matter if your decision is to fail to reject H_0. Remember that when you fail to reject the null hypothesis, the results are inconclusive. It is impossible to *prove* that H_0 is correct; therefore, you cannot state with certainty that "there is no effect" when H_0 is not rejected. At best, all you can state is that "there is insufficient evidence for an effect" (see Box 8.1).

4. It is very important that you understand the structure of the z-score formula (page 241). It will help you understand many of the other hypothesis tests that will be covered later.

5. When you are doing a directional hypothesis test, read the problem carefully, and watch for key words (such as increase or decrease, raise or lower, and more or less) that tell you which direction the researcher is predicting. The predicted direction will determine the alternative hypothesis (H_1) and the critical region. For example, if a treatment is expected to *increase* scores, H_1 would contain a *greater than* symbol, and the critical region would be in the tail associated with high scores.

DEMONSTRATION 8.1

HYPOTHESIS TEST WITH z

A researcher begins with a known population—in this case, scores on a standardized test that are normally distributed with $\mu = 65$ and $\sigma = 15$. The researcher suspects that special training in reading skills will produce a change in the scores for the individuals in the population. Because it is not feasible to administer the treatment (the special training) to everyone in the population, a sample of $n = 25$ individuals is selected, and the treatment is given to this sample. Following treatment, the average score for this sample is $M = 70$. Is there evidence that the training has an effect on test scores?

S T E P 1   State the hypothesis and select an alpha level.

Remember that the goal of hypothesis testing is to use sample data to make general conclusions about a population. The hypothesis always concerns an unknown population. For this demonstration, the researcher does not know what would happen if the entire population were given the treatment. Nevertheless, it is possible to make hypotheses about the treated population.

Specifically, the null hypothesis says that the treatment has no effect. According to H_0, the unknown population (after treatment) is identical to the original population (before treatment). In symbols,

H_0: $\mu = 65$ (After special training, the mean is still 65.)

The alternative hypothesis states that the treatment does have an effect that causes a change in the population mean. In symbols,

H_1: $\mu \neq 65$ (After special training, the mean is different from 65.)

At this time, you also select the alpha level. Traditionally, α is set at .05 or .01. If there is particular concern about a Type I error, or if a researcher desires to present overwhelming evidence for a treatment effect, a smaller alpha level can be used (such as $\alpha = .001$). For this demonstration, we will set alpha to .05. Thus, we are taking a 5% risk of committing a Type I error.

STEP 2   Locate the critical region.

You should recall that the critical region is defined as the set of outcomes that is very unlikely to be obtained if the null hypothesis is true. Therefore, obtaining sample data from this region would lead us to reject the null hypothesis and conclude there is an effect. Remember that the critical region is a region of rejection.

We begin by looking at all possible outcomes that could be obtained and then use the alpha level to determine the outcomes that are unlikely. For this demonstration, we look at the distribution of sample means for samples of $n = 25$—that is, all possible sample means that could be obtained if H_0 were true. The distribution of sample means will be normal because the original population is normal. It will have an expected value of $\mu = 65$ and a standard error of

$$\sigma_M = \frac{\sigma}{\sqrt{n}} = \frac{15}{\sqrt{25}} = \frac{15}{5} = 3$$

With $\alpha = .05$, we want to identify the most unlikely 5% of this distribution. The most unlikely part of a normal distribution is in the tails. Therefore, we divide our alpha level evenly between the two tails, 2.5% or $p = .0250$ per tail. In column C of the unit normal table, find $p = .0250$. Then find its corresponding z-score in column A. The entry is $z = 1.96$. The boundaries for the critical region are -1.96 (on the left side) and $+1.96$ (on the right side). The distribution with its critical region is shown in Figure 8.16.

FIGURE 8.16

The critical region for Demonstration 8.1 consists of the extreme tails with boundaries of $z = -1.96$ and $z = +1.96$.

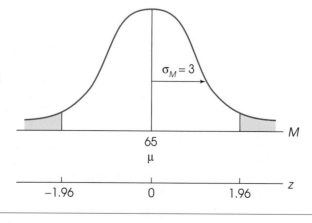

S T E P 3   Obtain the sample data, and compute the test statistic.

For this demonstration, the researcher obtained a sample mean of $M = 70$. This sample mean corresponds to a z-score of

$$z = \frac{M - \mu}{\sigma_M} = \frac{70 - 65}{3} = \frac{5}{3} = +1.67$$

S T E P 4   Make a decision about H_0, and state the conclusion.

The z-score we obtained is not in the critical region. This indicates that our sample mean of $M = 70$ is not an extreme or unusual value to be obtained from a population with $\mu = 65$. Therefore, our statistical decision is to *fail to reject* H_0. Our conclusion for the study is that the data do not provide sufficient evidence that the special training changes test scores.

DEMONSTRATION 8.2

EFFECT SIZE USING COHEN'S *d*

We will compute Cohen's d using the research situation and the data from Demonstration 8.1. Again, the original population mean was $\mu = 65$ and, after treatment (special training), the sample mean was $M = 70$. Thus, there is a 5-point mean difference. Using the population standard deviation, $\sigma = 15$, we obtain an effect size of

$$\text{Cohen's } d = \frac{\text{mean difference}}{\text{standard deviation}} = \frac{5}{15} = 0.33$$

According to Cohen's evaluation standards (Table 8.2), this is a medium treatment effect.

PROBLEMS

1. In the z-score formula as it is used in a hypothesis test,
   a. Explain what is measured by $M - \mu$ in the numerator.
   b. Explain what is measured by the standard error in the denominator.

2. The value that is obtained for the z-score in a hypothesis test is determined by several factors. For each of the following, indicate whether the effect would be to produce a larger z-score value (farther from zero) or a smaller z-score value (closer to zero). In each case, assume that all the other components of the z-score remain constant.
   a. Increase the sample size.
   b. Increase the population standard deviation.
   c. Increase the difference between the sample mean and the value of μ specified in the null hypothesis.

3. What happens to the boundaries for the critical region when the alpha level is lowered—for example, from .05 to .01? Also, what happens to the probability of a Type I error when the alpha level is lowered?

4. Briefly explain the advantage of using an alpha level of .01 versus a level of .05. In general, what is the disadvantage of using a smaller alpha level?

5. Discuss the errors that can be made in hypothesis testing.
   a. What is a Type I error? Why might it occur?
   b. What is a Type II error? How does it happen?

6. The term *error* is used two different ways in the context of a hypothesis test. First, there is the concept of standard error, and, second, there is the concept of a Type I error.
   a. What factor can a researcher control that will reduce the risk of a Type I error?
   b. What factor can a researcher control that will reduce the standard error?

7. A researcher did a one-tailed hypothesis test using an alpha level of .01. For this test, H_0 was rejected. A colleague analyzed the same data but used a two-tailed test with $\alpha = .05$. In this test, H_0 was *not* rejected. Can both analyses be correct? Explain your answer.

8. A researcher would like to test the effectiveness of a newly developed growth hormone. The researcher knows that under normal circumstances laboratory rats reach an average weight of $\mu = 950$ grams at 10 weeks of age. The distribution of weights is normal with $\sigma = 30$. A random sample of $n = 25$ newborn rats is obtained, and the hormone is given to each rat. When the rats in the sample reach 10 weeks old, each rat is weighed. The mean weight for this sample is $M = 974$.
   a. Identify the independent and the dependent variables for this study.
   b. Assuming a two-tailed test, state the null hypothesis in a sentence that includes the independent variable and the dependent variable.
   c. Using symbols, state the hypotheses (H_0 and H_1) for the two-tailed test.
   d. Sketch the appropriate distribution, and locate the critical region for $\alpha = .05$.
   e. Calculate the test statistic (z-score) for the sample.
   f. What decision should be made about the null hypothesis, and what decision should be made about the effect of the hormone?

9. A researcher is evaluating the effectiveness of a new physical education program for elementary school children. The program is designed to reduce competition and increase individual self-esteem. A sample of $n = 16$ children is selected and the children are placed in the new program. After 3 months, each child is given a standardized self-esteem test. For the general population of elementary school children, the scores on the self-esteem test form a normal distribution with $\mu = 40$ and $\sigma = 8$.
   a. If the researcher obtains a sample mean of $M = 42$, is this enough evidence to conclude that the program has a significant effect? Assume a two-tailed test with $\alpha = .05$.
   b. If the sample mean is $M = 44$, is this enough to demonstrate a significant effect? Again, assume a two-tailed test with $\alpha = .05$.
   c. Briefly explain why you reach different conclusions for part (a) and part (b).

10. State College is evaluating a new English composition course for freshmen. A random sample of $n = 25$ freshmen is obtained and the students are placed in the course during their first semester. One year later, a writing sample is obtained for each student and the writing samples are graded using a standardized evaluation technique. The average score for the sample is

$M = 76$. For the general population of college students, writing scores form a normal distribution with a mean of $\mu = 70$.
   a. If the writing scores for the population have a standard deviation of $\sigma = 20$, does the sample provide enough evidence to conclude that the new composition course has a significant effect? Assume a two-tailed test with $\alpha = .05$.
   b. If the population standard deviation is $\sigma = 10$, is the sample sufficient to demonstrate a significant effect? Again, assume a two-tailed test with $\alpha = .05$.
   c. Briefly explain why you reach different conclusions for part (a) and part (b).

11. Suppose that scores on the Scholastic Achievement Test (SAT) form a normal distribution with $\mu = 500$ and $\sigma = 100$. A high school counselor has developed a special course designed to boost SAT scores. A random sample of $n = 16$ students is selected to take the course and then the SAT. The sample has an average score of $M = 554$. Does the course have an effect on SAT scores?
   a. What are the dependent and independent variables for this experiment?
   b. Perform the hypothesis test using the four steps outlined in the chapter. Use a two-tailed test with $\alpha = .05$.
   c. If $\alpha = .01$ were used instead, what z-score values would be associated with the critical region?
   d. For part c, what decision should be made regarding H_0? Compare to part (b), and explain the difference.

12. A mood questionnaire has been standardized so that the scores form a normal distribution with $\mu = 50$ and $\sigma = 15$. A psychologist would like to use this test to examine how the environment affects mood. A sample of $n = 25$ individuals is obtained, and the individuals are given the mood test in a darkly painted, dimly lit room with plain metal desks and no windows. The average score for the sample is $M = 43$.
   a. Do the sample data provide sufficient evidence to conclude that the environment has a significant effect on mood? Use a two-tailed test at the .05 level of significance.
   b. Repeat the hypothesis test using the .01 level of significance.
   c. Compare the results from part (a) and part (b). How does the level of significance influence the outcome of a hypothesis test?

13. A normal population has a mean of $\mu = 100$. A sample of $n = 36$ is selected from the population, and a treatment is administered to the sample. After treatment, the sample mean is computed to be $M = 106$.
   a. Assuming that the population standard deviation is $\sigma = 12$, use the data to test whether or not the

treatment has a significant effect. Use a two-tailed test with $\alpha = .05$.

b. Repeat the hypothesis test, but this time assume that the population standard deviation is $\sigma = 30$.

c. Compare the results from part (a) and part (b). How does the population standard deviation influence the outcome of a hypothesis test?

14. A sample of $n = 4$ individuals is selected from a normal population with $\mu = 70$ and $\sigma = 10$. A treatment is administered to the individuals in the sample, and after the treatment, the sample mean is found to be $M = 75$.

a. On the basis of the sample data, can you conclude that the treatment has a significant effect? Use a two-tailed test with $\alpha = .05$.

b. Suppose that the sample consisted of $n = 25$ individuals and produced a mean of $M = 75$. Repeat the hypothesis test at the .05 level of significance.

c. Compare the results from part (a) and part (b). How does the sample size influence the outcome of a hypothesis test?

15. A sample of $n = 9$ scores is obtained from a normal population with $\sigma = 12$. The sample mean is $M = 60$.

a. With a two-tailed test and $\alpha = .05$, use the sample data to test the hypothesis that the population mean is $\mu = 65$.

b. With a two-tailed test and $\alpha = .05$, use the sample data to test the hypothesis that the population mean is $\mu = 55$.

c. In parts (a) and (b) of this problem, you should find that $\mu = 65$ and $\mu = 55$ are both acceptable hypotheses. Explain how two different values can both be acceptable.

16. A psychologist has developed a standardized test for measuring the vocabulary skills of 4-year-old children. The scores on the test form a normal distribution with $\mu = 60$ and $\sigma = 10$. A researcher would like to use this test to investigate the hypothesis that children who grow up as single children develop vocabulary skills at a different rate than children in large families. A sample of $n = 25$ single children is obtained, and the mean test score for this sample is $M = 63$.

a. On the basis of this sample, can the researcher conclude that vocabulary skills for single children are significantly different from those of the general population? Use a two-tailed test at the .05 level of significance.

b. Perform the same test assuming that the researcher had used a sample of $n = 100$ single children and obtained the same sample mean, $M = 63$.

c. You should find that the larger sample (part b) produces a different conclusion than the smaller sample (part a). Explain how the sample size influences the outcome of the hypothesis test.

17. In 1975, a nationwide survey revealed that U.S. grade-school children spent an average of $\mu = 8.4$ hours per week doing homework. The distribution of homework times was normal with $\sigma = 3.2$. Last year, a sample of $n = 100$ grade-school students was given the same survey. For this sample, the mean number of home-work hours was $M = 7.1$.

a. Has there been a significant change in the home-work habits of grade-school children? Use a two-tailed test with $\alpha = .05$.

b. Compute Cohen's d to measure the effect size (that is, the amount of change).

18. A sample of $n = 16$ individuals is selected from a population that forms a normal distribution with $\mu = 40$. A treatment is administered to the sample and, after treatment, the sample is measured to evaluate the effect of the treatment.

a. Assuming that the population standard deviation is $\sigma = 8$, compute Cohen's d to measure the effect size for a sample mean of $M = 42$.

b. Assuming $\sigma = 2$, compute Cohen's d to measure the effect size for a sample mean of $M = 42$.

c. Assuming $\sigma = 8$, compute Cohen's d to measure the effect size for a sample mean of $M = 48$.

d. Assuming $\sigma = 2$, compute Cohen's d to measure the effect size for a sample mean of $M = 48$.

19. Reading achievement test scores for fifth-grade students form a normal distribution with a mean of $\mu = 60$ and a standard deviation of $\sigma = 10$. A researcher would like to evaluate a new reading program by administering the program to a sample of fifth-grade students and then measuring their reading achievement.

a. Suppose that the researcher uses a sample of $n = 16$ students and obtains a mean test score of $M = 62$. Use a hypothesis test with $\alpha = .05$ to determine whether the program has a significant effect. Then compute Cohen's d to measure the effect size.

b. Now suppose that the researcher uses a sample of $n = 100$ and obtains a mean of $M = 62$. Again, use a hypothesis test to evaluate the significance of the program effect and compute Cohen's d to measure the effect size.

c. Comparing your answers from (a) and (b), explain how the hypothesis test and Cohen's d are influenced by sample size.

20. A researcher would like to determine whether there is any relationship between students' grades and where they choose to sit in the classroom. Specifically, the researcher suspects that the better students choose to sit in the front of the room. To test this hypothesis, the researcher asks her colleagues to help identify a sample of $n = 100$ students who all sit in the front row in

at least one class. At the end of the semester, the grades are obtained for these students and the average grade point average is $M = 3.25$. For the same semester, the average grade point average for the entire college is $\mu = 2.95$ with $\sigma = 1.10$.

a. Use a two-tailed test with $\alpha = .01$ to determine whether students who sit in the front of the classroom have significantly different grade point averages than other students.

b. Use a one-tailed test with $\alpha = .01$ to determine whether students who sit in the front of the classroom have significantly higher grade point averages than other students.

21. A developmental psychologist has prepared a training program that, according to a psychological theory, should improve problem-solving ability. For the population of six-year-olds, the average score on a standardized problem-solving test is known to be $\mu = 80$ with $\sigma = 10$. To test the effectiveness of the training program, a random sample of $n = 18$ six-year-old children is selected. After training, the average score for this sample is $M = 84.44$. Can the experimenter conclude that the program has an effect?

a. Use a two-tailed test with $\alpha = .05$.

b. What decision would be made using a one-tailed test with $\alpha = .05$?

22. On a standardized anagram task (anagrams are sets of scrambled letters that must be arranged to form words), people successfully complete an average of $\mu = 26$ anagrams with $\sigma = 4$. This distribution is normal. A researcher would like to demonstrate that the arousal from anxiety is distracting and will decrease task performance. A sample of $n = 14$ anxiety-ridden participants is tested on the task. The average number of anagrams solved is $M = 23.36$.

a. Do the anxiety-ridden participants show a decrease in task performance? Test with alpha set at .01 for one tail.

b. If a two-tailed test with $\alpha = .01$ was used, what conclusion should be drawn?

23. A psychologist examined the effect of chronic alcohol abuse on memory. In this experiment, a standardized memory test was used. Scores on this test for the general population form a normal distribution with $\mu = 50$ and $\sigma = 6$. A sample of $n = 22$ alcohol abusers has a mean score of $M = 47$. Is there evidence for memory impairment among alcoholics? Use $\alpha = .01$ for a one-tailed test.

24. A psychologist would like to determine whether there is any relationship between people's personality traits and the colors they choose when buying a new car. Based on records from the Department of Motor Vehicles, the psychologist obtains a sample of $n = 25$ people who recently purchased a new red car. These people were all given a brief questionnaire measuring introversion/extroversion. The average score for this sample was $M = 54.5$. The personality questionnaire is standardized so that the population forms a normal distribution with a mean of $\mu = 50$ and $\sigma = 10$ (higher scores indicate more extroversion).

a. Using a two-tailed test with $\alpha = .05$, does the sample provide sufficient evidence to conclude that people who select red cars have personality traits that are significantly different from those of the general population?

b. What conclusion is obtained for a two-tailed test with $\alpha = .01$?

c. If the researcher used a one-tailed test, predicting that people with red cars are more extroverted, are the results significant at the .01 level?

25. Researchers have often noted increases in violent crimes when it is very hot. In fact, Reifman, Larrick, and Fein (1991) noted that this relationship even extends to baseball. That is, there is a much greater chance of a batter being hit by a pitch when the temperature increases. Consider the following hypothetical data. Suppose over the past 30 years, during any given week of the major league season, an average of $\mu = 12$ players are hit by wild pitches. Assume the distribution is nearly normal with $\sigma = 3$. For a sample of $n = 4$ weeks in which the daily temperature was extremely hot, the weekly average of hit-by-pitch players was $M = 15.5$. Are players more likely to get hit by pitches during hot weeks? Set alpha to .05 for a one-tailed test.

26. A psychologist develops a new inventory to measure depression. Using a very large standardization group of "normal" individuals, the mean score on this test is $\mu = 55$ with $\sigma = 12$, and the scores are normally distributed. To determine whether the test is sensitive in detecting those individuals that are severely depressed, a random sample of patients who are described as depressed by a therapist is selected and given the test. Presumably, the higher the score on the inventory, the more depressed the patient is. The data are as follows: 59, 60, 60, 67, 65, 90, 89, 73, 74, 81, 71, 71, 83, 83, 88, 83, 84, 86, 85, 78, 79. Do patients score significantly differently on the test? Test with the .01 level of significance for two tails.

27. What happens to power as the treatment distribution gets farther from the null distribution?

28. Suppose that a researcher normally uses an alpha level of .01 for hypothesis tests, but this time uses an alpha level of .05. What does this change in alpha level do to the amount of power? What does it do to the risk of a Type I error?

29. Explain why power for a hypothesis test can never be less than the alpha level for the test. (*Hint:* Sketch a null distribution, and then add several different treatment distributions to your sketch. What happens to power as the treatment distribution gets very close to the null distribution?)

30. A researcher wants a statistical test to be powerful, yet would also like to avoid a Type I error. Which of the following approaches would achieve these goals? Explain your answer.
    a. Increase the alpha level (for example, from .05 to .10).
    b. Use a small alpha level, but increase the sample size.
    c. Use a one-tailed test.

CHAPTER

9

Introduction to the *t* Statistic

Tools You Will Need

The following items are considered essential background material for this chapter. If you doubt your knowledge of any of these items, you should review the appropriate chapter or section before proceeding.

- Sample standard deviation (Chapter 4)
- Degrees of freedom (Chapter 4)
- Hypothesis testing (Chapter 8)

Preview

Numerous accounts suggest that for many animals, including humans, direct stare from another animal is aversive (e.g., Cook, 1977). Try it out for yourself. Make direct eye contact with a stranger in a cafeteria. Chances are the person will display avoidance by averting his or her gaze or turning away from you. Some insects, such as moths, have even developed eye-spot patterns on the wings or body to ward off predators (mostly birds) who may have a natural fear of eyes (Blest, 1957). Suppose that a comparative psychologist is interested in determining whether or not the birds that feed on these insects show an avoidance of eye-spot patterns.

Using methods similar to those of Scaife (1976), the researcher performed the following experiment. A sample of $n = 16$ moth-eating birds is selected. The animals are tested in an apparatus that consists of a two-chambered box. The birds are free to roam from one side of the box to the other through a doorway in the partition that separates the two chambers. In one chamber, there are two eye-spot patterns painted on the wall. The other side of the box has plain walls. One at a time, the researcher tests each bird by placing it in the doorway between the chambers. Each subject is left in the apparatus for 60 minutes, and the amount of time spent in the plain chamber is recorded.

What kind of data should the experimenter expect if the null hypothesis is true? If the animals show no aversion to the eye spots, they should spend, on the average, half of the time in the plain side. Therefore, the null hypothesis would predict that

$$H_0: \mu_{\text{plain side}} = 30 \text{ minutes}$$

Suppose the researcher collected data for the sample and found that the amount of time spent on the plain side was $M = 35$ minutes. The researcher now has two of the needed pieces of information to compute a z-score test statistic—the sample data (M) and the hypothesized mean (according to H_0). All that is needed is the population standard deviation so that the standard error can be computed. Note that the investigator does not have this information about the population. All that is known is that the population of animals should, on the average, spend half of the 60-minute period in the plain chamber if the eye spots have no effect on behavior. Without complete information about the population, how can the researcher test the hypothesis with a z-score? The answer is simple—z cannot be used for the test statistic because the standard error cannot be computed. However, it is possible to estimate the standard error using the sample data and to compute a test statistic that is similar in structure to the z-score. This new test statistic is called the t statistic.

9.1 OVERVIEW

In the previous chapter, we presented the statistical procedures that permit researchers to use a sample mean to test hypotheses about an unknown population. These statistical procedures were based on a few basic concepts, which we summarize as follows:

Remember that the expected value of the distribution of sample means is μ, the population mean.

1. A sample mean (M) is expected more or less to approximate its population mean (μ). This permits us to use the sample mean to test a hypothesis about the population mean.

2. The standard error provides a measure of how well a sample mean approximates the population mean. Specifically, the standard error determines how much difference between M and μ is reasonable to expect just by chance.

$$\sigma_M = \frac{\sigma}{\sqrt{n}} \quad \text{or} \quad \sigma_M = \sqrt{\frac{\sigma^2}{n}}$$

3. To quantify our inferences about the population, we compare the obtained sample mean (M) with the hypothesized population mean (μ) by computing a z-score test statistic.

$$z = \frac{M - \mu}{\sigma_M} = \frac{\text{obtained difference between data and hypothesis}}{\text{standard distance expected by chance}}$$

The goal of the hypothesis test is to determine whether or not the obtained result is significantly greater than would be expected by chance. When the z-scores form a normal distribution, we are able to use the unit normal table (Appendix B) to find the critical region for the hypothesis test.

THE PROBLEM WITH z-SCORES

The shortcoming of using a z-score as an inferential statistic is that the z-score formula requires more information than is usually available. Specifically, a z-score requires that we know the value of the population standard deviation (or variance), which is needed to compute the standard error. In most situations, however, the standard deviation for the population is not known. In this case, we cannot compute the standard error and we cannot compute a z-score for the hypothesis test. In this chapter we introduce a new procedure, called a t statistic, that is used instead of a z-score for hypothesis testing when the population standard deviation is unknown.

9.2 THE *t* STATISTIC—AN ALTERNATIVE TO *z*

As previously noted, the limitation of z-scores in hypothesis testing is that the population standard deviation (or variance) must be known. More often than not, however, the variability of the scores in the population is not known. In fact, the whole reason for conducting a hypothesis test is to gain knowledge about an *unknown* population. This situation appears to create a paradox: You want to use a z-score to find out about an unknown population, but you must know about the population before you can compute a z-score. Fortunately, there is a relatively simple solution to this problem. When the variability for the population is not known, we use the sample variability in its place.

In Chapter 4, the sample variance was developed specifically to provide an unbiased estimate of the corresponding population variance. You should recall the formulas for sample variance and sample standard deviation as follows:

The concept of degrees of freedom, $df = n - 1$, was introduced in Chapter 4 (page 120) and is discussed later in this chapter (page 283).

$$\text{sample variance} = s^2 = \frac{SS}{n-1} = \frac{SS}{df}$$

$$\text{sample standard deviation} = s = \sqrt{\frac{SS}{n-1}} = \sqrt{\frac{SS}{df}}$$

Using the sample values, we can now *estimate* the standard error. Recall from Chapters 7 and 8 that the value of the standard error can be computed using either standard deviation or variance:

$$\text{standard error} = \sigma_M = \frac{\sigma}{\sqrt{n}} \qquad \text{or} \qquad \sigma_M = \sqrt{\frac{\sigma^2}{n}}$$

Now we will estimate the standard error by simply substituting the sample variance or standard deviation in place of the unknown population value:

$$\text{estimated standard error} = s_M = \frac{s}{\sqrt{n}} \qquad \text{or} \qquad s_M = \sqrt{\frac{s^2}{n}} \qquad (9.1)$$

Notice that the symbol for the estimated standard error is s_M instead of σ_M, indicating that the estimated value is computed from sample data rather than from the actual population parameter.

DEFINITION   The **estimated standard error (s_M)** is used as an estimate of the real standard error σ_M when the value of σ is unknown. It is computed from the sample variance or sample standard deviation and provides an estimate of the standard distance between a sample mean M and the population mean μ.

Finally, you should recognize that we have shown formulas for standard error (actual or estimated) using both the standard deviation and the variance. In the past (Chapters 7 and 8), we concentrated on the formula using the standard deviation. At this point, however, we will shift our focus to the formula based on variance. Thus, throughout the remainder of this chapter, and in following chapters, the *estimated standard error of M* typically will be presented and computed using

$$s_M = \sqrt{\frac{s^2}{n}}$$

There are two reasons for making this shift from standard deviation to variance:

1. In Chapter 4 (page 121) we saw that the sample variance is an *unbiased* statistic; on average, the sample variance (s^2) will provide an accurate and unbiased estimate of the population variance (σ^2). Therefore, the most accurate way to estimate the standard error is to use the sample variance to estimate the population variance.

2. In future chapters we will encounter other versions of the *t* statistic that require variance (instead of standard deviation) in the formulas for estimated standard error. To maximize the similarity from one version to another, we will use variance in the formula for *all* of the different *t* statistics. Thus, whenever we present a *t* statistic, the estimated standard error will be computed as

$$\text{estimated standard error} = \sqrt{\frac{\text{sample variance}}{\text{sample size}}}$$

Now we can substitute the estimated standard error in the denominator of the *z*-score formula. The result is a new test statistic called a *t statistic*:

$$t = \frac{M - \mu}{s_M} \qquad (9.2)$$

DEFINITION   The **_t_ statistic** is used to test hypotheses about an unknown population mean μ when the value of σ is unknown. The formula for the *t* statistic has the same structure as the *z*-score formula, except that the *t* statistic uses the estimated standard error in the denominator.

The only difference between the *t* formula and the *z*-score formula is that the *z*-score uses the actual population variance, σ^2 (or the standard deviation), and the *t* formula uses the corresponding sample variance (or standard deviation) when the population value is not known.

$$z = \frac{M - \mu}{\sigma_M} = \frac{M - \mu}{\sqrt{\sigma^2/n}} \qquad t = \frac{M - \mu}{s_M} = \frac{M - \mu}{\sqrt{s^2/n}}$$

DEGREES OF FREEDOM AND THE *t* STATISTIC

In this chapter, we have introduced the *t* statistic as a substitute for a *z*-score. The basic difference between these two is that the *t* statistic uses sample variance (s^2) and the *z*-score uses the population variance (σ^2). To determine how well a *t* statistic approximates a *z*-score, we must determine how well the sample variance approximates the population variance.

In Chapter 4, we introduced the concept of degrees of freedom (page 120). Reviewing briefly, you must know the sample mean before you can compute sample variance. This places a restriction on sample variability such that only $n - 1$ scores in a sample are free to vary. The value $n - 1$ is called the *degrees of freedom* (or *df*) for the sample variance.

$$\text{degrees of freedom} = df = n - 1 \tag{9.3}$$

DEFINITION

Degrees of freedom describe the number of scores in a sample that are free to vary. Because the sample mean places a restriction on the value of one score in the sample, there are $n - 1$ degrees of freedom for the sample (see Chapter 4).

The greater the value of *df* is for a sample, the better s^2 represents σ^2, and the better the *t* statistic approximates the *z*-score. This should make sense because the larger the sample (*n*) is, the better the sample represents its population. Thus, the degrees of freedom associated with s^2 also describe how well *t* represents *z*.

THE *t* DISTRIBUTION

Every sample from a population can be used to compute a *z*-score or a *t* statistic. If you select all the possible samples of a particular size (*n*), then the entire set of resulting *z*-scores will form a *z*-score distribution. In the same way, the set of all possible *t* statistics will form a *t* distribution. As we saw in Chapter 7, the distribution of *z*-scores computed from sample means tends to be a normal distribution. For this reason, we consulted the unit normal table to find the critical region when using *z*-scores to test hypotheses about a population. The *t* distribution will approximate a normal distribution in the same way that a *t* statistic approximates a *z*-score. How well a *t* distribution approximates a normal distribution is determined by degrees of freedom. In general, the greater the sample size (*n*) is, the larger the degrees of freedom ($n - 1$) are, and the better the *t* distribution approximates the normal distribution (see Figure 9.1).

THE SHAPE OF THE *t* DISTRIBUTION

The exact shape of a *t* distribution changes with degrees of freedom. In fact, statisticians speak of a "family" of *t* distributions. That is, there is a different sampling distribution of *t* (a distribution of all possible sample *t* values) for each possible number of degrees of freedom. As *df* gets very large, the *t* distribution gets closer in shape to a normal *z*-score distribution (see Box 9.1). A quick glance at Figure 9.1 reveals that distributions of *t* are bell-shaped and symmetrical and have a mean of zero. However, the

FIGURE 9.1

Distributions of the *t* statistic for different values of degrees of freedom are compared to a normal *z*-score distribution. Like the normal distribution, *t* distributions are bell-shaped and symmetrical and have a mean of zero. However, *t* distributions have more variability, indicated by the flatter and more spread-out shape. The larger the value of *df* is, the more closely the *t* distribution approximates a normal distribution.

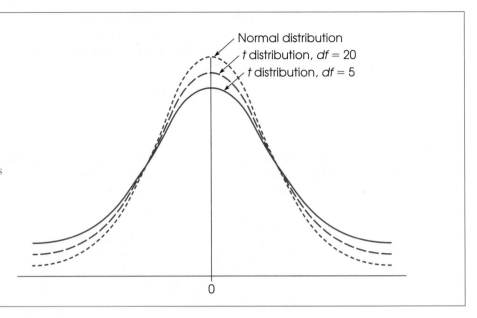

B O X	**THE CRITICAL VALUES IN A NORMAL *z* DISTRIBUTION AND A *t* DISTRIBUTION:**
9.1	**THE INFLUENCE OF DEGREES OF FREEDOM**

As previously noted, a *t* distribution approximates a normal *z* distribution. How well it approximates a normal distribution depends on the value of *df*. This can be seen by simply comparing critical values for *z* with critical values of *t* at various degrees of freedom. For example, if we test a hypothesis with a *z*-score using a two-tailed test and $\alpha = .05$, the critical *z*-scores will be $+1.96$ and -1.96. However, suppose the population standard deviation was not known and we conducted this hypothesis test with a *t* statistic. For a sample of $n = 4$, the *t* statistic would have $df = 3$. For 3 degrees of freedom, the critical values of *t* would be $+3.182$ and -3.182 (with alpha still set at .05). When *df* is small, the *t* distribution is flatter and more spread out. Consequently, the tails have a greater area for the

t distribution compared to a normal distribution (see Figure 9.1). The extreme 5% of the distribution will be farther from the mean and have a larger critical value in the *t* distribution. If, however, we use a sample of $n = 31$, then $df = 30$, and the critical *t* values will be $+2.042$ and -2.042. These values are very close to the critical *z*-score values (± 1.96). If the sample is made even larger—say, $n = 121$—the critical values get even closer to the *z* values. For 120 degrees of freedom, the critical values of *t* are $+1.980$ and -1.980 when $\alpha = .05$. Thus, the difference between a *t* distribution and normal *z*-distribution becomes negligible when a very large sample is used. (*Note:* Determining critical values for *t* is discussed on pages 285–286.)

---

t distribution has more variability than a normal *z* distribution, especially when *df* values are small (Figure 9.1). The *t* distribution tends to be flatter and more spread out, whereas the normal *z* distribution has more of a central peak.

The reason that the *t* distribution is flatter and more variable than the normal *z*-score distribution becomes clear if you look at the structure of the formulas for *z* and *t*. For a particular population, the top of the *z*-score formula, $M - \mu$, can take on different values because *M* will vary from one sample to another. However, the value of the bottom of the z-score formula, σ_M, is constant. The standard error will not vary from sample to sample because it is derived from the population variance. The implication is that samples that have the same value for M should also have the same z-score.

On the other hand, the standard error in the t formula is not a constant because it is estimated. That is, s_M is based on the sample variance, which will vary in value from sample to sample. The result is that samples can have the same value for M, yet different values of t because the estimated error will vary from one sample to another. Therefore, a t distribution will have more variability than the normal z distribution. It will look flatter and more spread out. When the value of df increases, the variability in the t distribution decreases, and it more closely resembles the normal distribution because with greater df, s_M will more closely estimate σ_M, and when df is very large, they are nearly the same.

DETERMINING PROPORTIONS AND PROBABILITIES FOR *t* DISTRIBUTIONS

Just as we used the unit normal table to locate proportions associated with z-scores, we will use a t distribution table to find proportions for t statistics. The complete t distribution table is presented in Appendix B, page 691, and a portion of this table is reproduced in Table 9.1. The two rows at the top of the table show proportions of the t distribution contained in either one or two tails, depending on which row is used. The first column of the table lists degrees of freedom for the t statistic. Finally, the numbers in the body of the table are the t values that mark the boundary between the tails and the rest of the t distribution.

TABLE 9.1

A portion of the t-distribution table. The numbers in the table are the values of t that separate the tail from the main body of the distribution. Proportions for one or two tails are listed at the top of the table, and df values for t are listed in the first column.

	PROPORTION IN ONE TAIL					
	0.25	0.10	0.05	0.025	0.01	0.005
	PROPORTION IN TWO TAILS COMBINED					
df	0.50	0.20	0.10	0.05	0.02	0.01
1	1.000	3.078	6.314	12.706	31.821	63.657
2	0.816	1.886	2.920	4.303	6.965	9.925
3	0.765	1.638	2.353	3.182	4.541	5.841
4	0.741	1.533	2.132	2.776	3.747	4.604
5	0.727	1.476	2.015	2.571	3.365	4.032
6	0.718	1.440	1.943	2.447	3.143	3.707

For example, with $df = 3$, exactly 5% of the t distribution is located in the tail beyond $t = 2.353$ (see Figure 9.2). To find this value, you locate $df = 3$ in the first column and locate 0.05 (5%) in the one-tail proportion row. When you line up these two values in the table, you should find $t = 2.353$. Similarly, 5% of the t distribution is located in the tail beyond $t = -2.353$ (see Figure 9.2). Finally, notice that a total of 10% is contained in the two tails beyond $t = \pm 2.353$ (check the proportion value in the "two-tails combined" row at the top of the table).

A close inspection of the t distribution table in Appendix B will demonstrate a point we made earlier: As the value for df increases, the t distribution becomes more and more similar to a normal distribution. For example, examine the column containing t values for a 0.05 proportion in two tails. You will find that when $df = 1$, the t values that separate the extreme 5% (0.05) from the rest of the distribution are $t = \pm 12.706$.

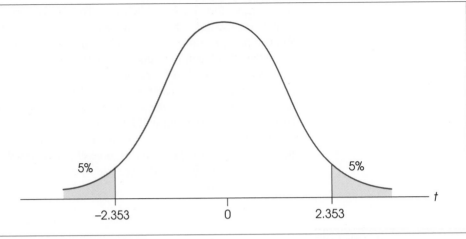

FIGURE 9.2

The *t* distribution with *df* = 3. Note that 5% of the distribution is located in the tail beyond *t* = 2.353. Also, 5% is in the tail beyond *t* = −2.353. Thus, a total proportion of 10% (0.10) is in the two tails beyond *t* = ±2.353.

5%   5%

−2.353   0   2.353   *t*

As you read down the column, however, you should find that the critical *t* values become smaller and smaller, ultimately reaching ±1.96. You should recognize ±1.96 as the *z*-score values that separate the extreme 5% in a normal distribution. Thus, as *df* increases, the proportions in a *t* distribution become more like the proportions in a normal distribution. When the sample size (and degrees of freedom) is sufficiently large, the difference between a *t* distribution and the normal distribution becomes negligible.

Caution: The *t* distribution table printed in this book has been abridged and does not include entries for every possible *df* value. For example, the table lists *t* values for *df* = 40 and for *df* = 60, but does not list any entries for *df* values between 40 and 60. Occasionally, you will encounter a situation in which your *t* statistic has a *df* value that is not listed in the table. In these situations, you should look up the critical *t* for both of the surrounding *df* values listed and then use the *larger* value for *t*. If, for example, you have *df* = 53 (not listed), look up the critical *t* value for both *df* = 40 and *df* = 60 and *then use the larger t value.* If your sample *t* statistic is greater than the larger value listed, you can be certain that the data are in the critical region, and you can confidently reject the null hypothesis.

LEARNING CHECK

1. Under what circumstances is a *t* statistic used instead of a *z*-score for a hypothesis test?

2. To conduct a hypothesis test with a *z*-score statistic, you must know the variance or the standard deviation for the population of scores. (True or false?)

3. In general, a distribution of *t* statistics is flatter and more spread out than the standard normal distribution. (True or false?)

4. A sample of *n* = 15 scores would produce a *t* statistic with *df* = 16. (True or false?)

5. For *df* = 15, find the value(s) of *t* associated with each of the following:
   a. The top 5% of the distribution.
   b. The middle 95% versus the extreme 5% of the distribution.
   c. The middle 99% versus the extreme 1% of the distribution.

A N S W E R S   1. A *t* statistic is used instead of a *z*-score when the population standard deviation or variance is not known.

2. True

3. True

4. False. With $n = 15$, the *df* value would be 14.

5. **a.** $t = +1.753$ **b.** $t = \pm2.131$ **c.** $t = \pm2.947$

9.3 HYPOTHESIS TESTS WITH THE *t* STATISTIC

In the hypothesis-testing situation, we begin with a population with an unknown mean and an unknown variance, often a population that has received some treatment (Figure 9.3). The goal is to use a sample from the treated population (a treated sample) as the basis for determining whether or not the treatment has any effect.

FIGURE 9.3

The basic experimental situation for using the *t* statistic or the *z*-score is presented. It is assumed that the parameter μ is known for the population before treatment. The purpose of the experiment is to determine whether or not the treatment has an effect. We ask, Is the population mean after treatment the same as or different from the mean before treatment? A sample is selected from the treated population to help answer this question.

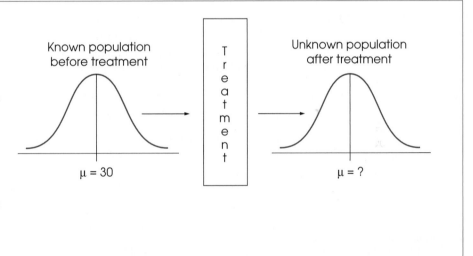

Known population before treatment

$\mu = 30$

Treatment

Unknown population after treatment

$\mu = ?$

As always, the null hypothesis states that the treatment has no effect; specifically, H_0 states that the population mean is unchanged. Thus, the null hypothesis provides a specific value for the unknown population mean. The sample data provide a value for the sample mean. Finally, the variance and estimated standard error are computed from the sample data. When these values are used in the *t* formula, the result becomes

$$t = \frac{\text{sample mean (from the data)} - \text{population mean (hypothesized from } H_0)}{\text{estimated standard error (computed from the sample data)}}$$

As with the z-score formula, the t statistic forms a ratio. The numerator measures the actual difference between the sample data (M) and the population hypothesis (μ). The denominator measures how much difference is expected just by chance (error). When the obtained difference between the data and the hypothesis (numerator) is much greater than chance (denominator), we will obtain a large value for t (either large positive or large negative). In this case, we conclude that the data are not consistent with the hypothesis, and our decision is to "reject H_0." On the other hand, when the difference between the data and the hypothesis is small relative to the standard error, we will obtain a t statistic near zero, and our decision will be "fail to reject H_0."

We will now review the basic steps of the hypothesis-testing procedure.

STEPS AND PROCEDURES

For hypothesis tests with a t statistic, we use the same steps that we used with z-scores (Chapter 8). The major difference is that we are now required to estimate standard error because σ^2 is unknown. Consequently, we compute a t statistic rather than a z-score and consult the t distribution table rather than the unit normal table to find the critical region.

STEP 1   State the hypotheses, and set the alpha level. The null hypothesis states that the treatment has no effect; there is no change in the population mean. The alternative hypothesis says that there is an effect. Both hypotheses are stated in terms of the unknown population parameter, μ.

STEP 2   Locate the critical region. The exact shape of the t distribution and, therefore, the critical t values vary with degrees of freedom. Thus, to find a critical region in a t distribution, it is necessary to determine the value for df. Then the critical region can be located by consulting the t distribution table (Appendix B).

STEP 3   Collect the sample data, and compute the test statistic. When σ^2 is unknown, the test statistic is a t statistic (Equation 9.2).

STEP 4   Evaluate the null hypothesis. If the t statistic we obtained in step 3 falls within the critical region (exceeds the value of a critical t), then H_0 is rejected. We can conclude that a treatment effect exists. However, if the obtained t value does not lie in the critical region, then we fail to reject H_0, and we conclude that we failed to observe evidence for an effect in our study.

HYPOTHESIS-TESTING EXAMPLE

Let us return to the research problem posed in the Preview to demonstrate the steps and procedures of hypothesis testing. Recall that direct eye contact is avoided by many animals. Some animals, such as moths, have evolved large eye-spot patterns, presumably to ward off predators that have a natural aversion to direct gaze. The experiment will assess the effect of exposure to eye-spot patterns on the behavior of moth-eating birds, using procedures similar to Scaife's (1976) study.

EXAMPLE 9.1

To test the effectiveness of eye-spot patterns in deterring predation, a sample of $n = 16$ insectivorous birds is selected. The animals are tested in a box that has two separate chambers (see Figure 9.4). The birds are free to roam from one chamber to another through a doorway in a partition. On the wall of one chamber, two large eye-spot patterns have been painted. The other chamber has plain walls. The birds are tested one at a time by placing them in the doorway in the center of the apparatus. Each animal is left in the box for 60 minutes, and the amount of time spent in the

FIGURE 9.4

Apparatus used in Example 9.1.

plain chamber is recorded. Suppose that the sample of $n = 16$ birds spent an average of $M = 39$ minutes in the plain side, with $SS = 540$. Can we conclude that eye-spot patterns have an effect on behavior? Note that although it is possible to predict a value for μ, we have no information about the population variance.

STEP 1 State the hypotheses, and select an alpha level. If the null hypothesis were true, then the eye-spot patterns would have no effect on behavior. The animals should show no preference for either side of the box. That is, they should spend half of the 60-minute test period in the plain chamber. In symbols, the null hypothesis would state that

$$H_0: \quad \mu_{\text{plain side}} = 30 \text{ minutes}$$

The alternative hypothesis states that the eye patterns have an effect on behavior, and the birds will either avoid the painted walls or be attracted to the painted eye spots. (Note that a directional, one-tailed test would specify one of the two alternatives.) The nondirectional alternative hypothesis would be expressed as follows:

$$H_1: \quad \mu_{\text{plain side}} \neq 30 \text{ minutes}$$

We will set the level of significance at $\alpha = .05$ for two tails.

STEP 2 Locate the critical region. The test statistic is a *t* statistic because the population variance is not known. The exact shape of the *t* distribution and, therefore, the proportions under the *t* distribution depend on the number of degrees of freedom associated with the sample. To find the critical region, *df* must be computed:

$$df = n - 1 = 16 - 1 = 15$$

For a two-tailed test at the .05 level of significance and with 15 degrees of freedom, the critical region consists of *t* values greater than $+2.131$ or less than -2.131. Figure 9.5 depicts the critical region in this *t* distribution.

FIGURE 9.5

The critical region in the *t* distribution for $\alpha = .05$ and $df = 15$.

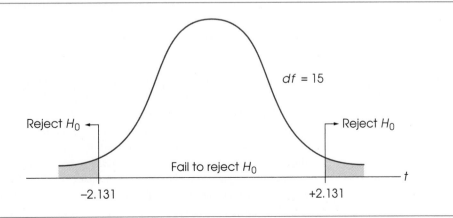

STEP 3 Calculate the test statistic. The *t* statistic typically requires much more computation than is necessary for a *z*-score. Therefore, we recommend that you divide the calculations into a three-stage process as follows.

    a. First, calculate the sample variance. Remember that the population variance is unknown, and you must use the sample value in its place. (This is why we are using a *t* statistic instead of a *z*-score.)

$$s^2 = \frac{SS}{n-1} = \frac{SS}{df}$$

$$= \frac{540}{15}$$

$$= 36$$

    b. Next, use, the sample variance (s^2) to compute the estimated standard error. This value will be the denominator of the *t* statistic and measures how much error is reasonable to expect by chance between a sample mean and the corresponding population mean.

$$s_M = \sqrt{\frac{s^2}{n}}$$

$$= \sqrt{\frac{36}{16}}$$

$$= \sqrt{2.25}$$

$$= 1.50$$

    c. Finally, compute the *t* statistic for the sample data.

$$t = \frac{M - \mu}{s_M}$$

$$= \frac{39 - 30}{1.50} = \frac{9}{1.50} = 6.00$$

STEP 4 Make a decision regarding H_0. The obtained *t* statistic of 6.00 falls into the critical region on the right-hand side of the *t* distribution (Figure 9.5). Our statistical decision is to reject H_0 and conclude that the presence of eye-spot patterns does influence behavior. Specifically, the average amount of time that the birds spent on the plain side is *significantly* different from the 30 minutes that would be expected merely by chance. As can be seen from the sample mean, there is a tendency for the birds to avoid the eye spots and spend more time on the plain side of the box.

DESCRIPTIVE STATISTICS AND HYPOTHESIS TESTS

Example 9.1 presented a hypothesis test using the *t* statistic. You should recognize that the hypothesis test is an example of inferential statistics; that is, the hypothesis test is a procedure that uses sample data to draw inferences about a population. Although we have focused attention on inferential statistics in the past two or three chapters, do not forget that descriptive statistics are also an important part of data analysis. Typically, the first step in the data analysis process is to compute the descriptive statistics that summarize and describe the results. For the data from Example 9.1 the fundamental descriptive statistics are the sample mean and the sample standard deviation. These two values should give you a good picture of the data, and they should help you to understand the outcome of the hypothesis test that followed.

In Example 9.1 the mean amount of time that the birds spent on the plain side of the box was reported as $M = 39$ minutes. The standard deviation can be computed from the values $n = 16$ and $SS = 540$:

$$\text{sample standard deviation} = s = \sqrt{\frac{SS}{n-1}} = \sqrt{\frac{540}{15}} = \sqrt{36} = 6$$

These two values, M and s, should allow you to visualize the complete set of data. Specifically, the scores in the sample pile up around $M = 39$ and are scattered with a standard distance of about 6 points in either direction from the mean. To help you visualize the sample, we created a set of $n = 16$ scores with a mean of $M = 39$ and a standard deviation of $s = 6$. The set of scores is shown in Figure 9.6.

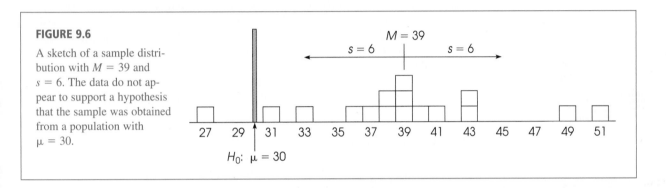

FIGURE 9.6

A sketch of a sample distribution with $M = 39$ and $s = 6$. The data do not appear to support a hypothesis that the sample was obtained from a population with $\mu = 30$.

You should also realize that the sample distribution presented in Figure 9.6 can be used to help understand the outcome of the hypothesis test in Example 9.1. In the hypothesis test, the null hypothesis stated that the sample was obtained from a population with $\mu = 30$. To help visualize the relationship between the sample data and the null hypothesis, we have drawn a vertical line in Figure 9.6 so that the line is located at the position specified by H_0. Looking at Figure 9.6, do the scores in the sample appear to be clustered around $\mu = 30$? Does it seem reasonable that this sample came from a population with a mean of $\mu = 30$? It should be fairly clear that the answer to both is no. You should also realize that the same conclusion was reached in the hypothesis test. Specifically, the hypothesis test rejected the null hypothesis and concluded that the population mean was not equal to 30.

In simple terms, a hypothesis test is a formal procedure that does exactly the same thing that you did while looking at Figure 9.6. That is, the hypothesis test looks at the data in relation to the null hypothesis, and then decides whether or not the data seem to fit the hypothesis.

MEASURING EFFECT SIZE FOR THE *t* STATISTIC

In Chapter 8 we noted that one criticism of a hypothesis test is that it does not really evaluate the size of the treatment effect. Instead, a hypothesis test simply determines whether the treatment effect is greater than chance, where "chance" is measured by the standard error. To correct for this prob'em, it is recommended that the results from a hypothesis test be accompanied by a report of effect size such as Cohen's *d*.

When Cohen's *d* was originally introduced (page 261), the formula was presented as:

$$\text{Cohen's } d = \frac{\text{mean difference}}{\text{standard deviation}}$$

For hypothesis tests with the *t* statistic, the population standard deviation is unknown, so we will simply use the sample standard deviation in its place. Thus, for a *t* statistic

$$\text{Cohen's } d = \frac{\text{mean difference}}{\text{sample standard deviation}} \qquad (9.4)$$

The following example demonstrates the use of Cohen's *d* to measure effect size for a hypothesis test using a *t* statistic.

EXAMPLE 9.2   For the eye-spot study in Example 9.1, the birds averaged $M = 39$ minutes on the plain side of the box. If the eye spots had no effect (as stated by the null hypothesis), the population mean would be $\mu = 30$ minutes. Thus, the results demonstrate that the eye spots had a 9-minute effect on the birds ($39 - 30$). For this study, the sample standard deviation is

$$\text{sample standard deviation} = s = \sqrt{\frac{SS}{df}} = \sqrt{\frac{540}{15}} = \sqrt{36} = 6$$

Thus, Cohen's *d* for this example is

$$\text{Cohen's } d = \frac{\text{mean difference}}{\text{sample standard deviation}} = \frac{9}{6} = 1.50$$

According to the standards suggested by Cohen (Table 8.2, page 262), this is a large treatment effect.

Measuring the percentage of variance explained, r^2   An alternative method for measuring effect size is to determine how much of the variability in the scores is explained by the treatment effect. The concept behind this measure is that the treatment causes the scores to increase (or decrease), which means that the treatment is causing the scores to vary. If we can measure how much of the variability is explained by the treatment, we will obtain a measure of the size of the treatment effect.

To demonstrate this concept we will use the data from the hypothesis test in Example 9.1. Recall that the null hypothesis stated that the treatment (eye-spot pattern) has no effect on the birds' behavior. According to the null hypothesis, the birds should show no preference between the two sides of the box, and therefore should spend an average of $\mu = 30$ minutes on the plain side. However, if you look at the data in Figure 9.6, the scores do not appear to be clustered around $\mu = 30$. Instead, the scores are shifted to the right so that they are centered around the sample mean, $M = 39$. This shift is the treatment effect. To measure the size of the treatment effect we will calculate deviations from the mean and the sum of squared deviations, *SS*, two different ways.

Figure 9.7(a) shows the original set of scores. For each score, the deviation from $\mu = 30$ is shown as a colored line. Recall that $\mu = 30$ comes from the null hypothesis and represents the mean that should be obtained if the treatment has no effect. Note that most of the scores are located on the right-hand side of $\mu = 30$. This shift to the right

(a) Original scores, including the treatment effect

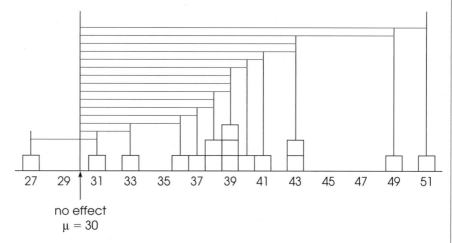

no effect
μ = 30

(b) Adjusted scores with the treatment effect removed

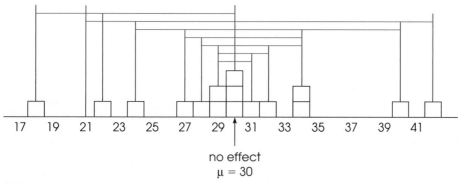

no effect
μ = 30

FIGURE 9.7

Deviations from μ = 30 for the scores from Example 9.1. The colored lines in part (a) show the deviations from the mean for the original scores, including the treatment effect. In part (b), the colored lines show the deviations from the mean for the adjusted scores, after the treatment effect is removed.

is the treatment effect. Specifically, the treatment caused the birds to spend more time on the plain side of the box, which means that their scores are generally greater than 30. Thus, the treatment effect is contributing to the size of the deviations.

Next we will eliminate the treatment effect and see how the deviations are affected. In this example, the treatment has increased the scores so that they are centered at $M = 39$, instead of μ = 30. To remove the treatment effect, we will simply subtract 9 points from each score. The adjusted scores are shown in Figure 9.7(b) and, once again, the deviations from μ = 30 are shown as colored lines. First, notice that the adjusted scores are centered at μ = 30, indicating that there is no treatment effect. Also notice that the deviations, the colored lines, are noticeably smaller when the treatment effect is removed.

To measure how much the variability is reduced when the treatment effect is removed, we will compute the sum of squared deviations, *SS*, for each set of scores. The calculations for the original scores [Figure 9.7(a)] are shown in the left-hand columns of Table 9.2, and the calculations for the adjusted scores [Figure 9.7(b)] are shown in the right-hand columns. Note that the total variability, including the treatment effect, is *SS* = 1836. However, when the treatment effect is removed, the variability is reduced to *SS* = 540. The difference between these two values, 1296 points, is the amount of variability that is accounted for by the treatment effect. This value is usually reported as a proportion or percentage of the total variability:

$$\frac{\text{variability accounted for}}{\text{total variability}} = \frac{1296}{1836} = 0.706 = 70.6\% \text{ of the total variability}$$

The letter *r* is the traditional symbol used for a correlation. The concept of r^2 will be discussed further when we consider correlations in Chapter 16.

Thus, removing the treatment effect reduces the variability by 70.6%. This value is called the *percentage of variance accounted for by the treatment* and is identified as r^2.

Rather than computing r^2 directly by comparing two different calculations for *SS*, the value can be found from a single equation based on the outcome of the *t* test.

$$r^2 = \frac{t^2}{t^2 + df} \tag{9.5}$$

For the hypothesis test in Example 9.1, we obtained *t* = 6.00 with *df* = 15. These values produce

$$r^2 = \frac{6^2}{6^2 + 15} = \frac{36}{51} = 0.706 \text{ or } 70.6\%$$

Note that this is exactly the same value we obtained with the direct calculation of the percentage of variability accounted for by the treatment.

TABLE 9.2

Calculation of *SS*, the sum of squared deviations, for the data in Figure 9.7. The three columns on the left show the calculations for the original scores, including the treatment effect. The three columns on the right show the calculations for the adjusted scores after the treatment effect has been removed.

	CALCULATION OF *SS* INCLUDING THE TREATMENT EFFECT			CALCULATION OF *SS* AFTER THE TREATMENT EFFECT IS REMOVED	
SCORE	DEVIATION FROM $\mu = 30$	SQUARED DEVIATION	ADJUSTED SCORE	DEVIATION FROM $\mu = 30$	SQUARED DEVIATION
27	−3	9	27 − 9 = 18	−12	144
31	1	1	31 − 9 = 22	−8	64
33	3	9	33 − 9 = 24	−6	36
36	6	36	36 − 9 = 26	−3	9
37	7	49	37 − 9 = 28	−2	4
38	8	64	38 − 9 = 29	−1	1
38	8	64	38 − 9 = 29	−1	1
39	9	81	39 − 9 = 30	0	0
39	9	81	39 − 9 = 30	0	0
39	9	81	39 − 9 = 30	0	0
40	10	100	40 − 9 = 31	1	1
41	11	121	41 − 9 = 32	2	4
43	13	169	43 − 9 = 34	4	16
43	13	169	43 − 9 = 34	4	16
49	19	361	49 − 9 = 40	10	100
51	21	441	51 − 9 = 42	12	144
		1836 = *SS*			540 = *SS*

Interpreting r^2   In addition to developing the Cohen's *d* measure of effect size, Cohen (1988) also proposed criteria for evaluating the size of a treatment effect that is measured by r^2. The criteria were actually suggested for evaluating the size of a correlation, *r*, but are easily extended to apply to r^2. Cohen's standards for interpreting r^2 are shown in Table 9.3.

TABLE 9.3

Criteria for interpreting the value of r^2 as proposed by Cohen (1988)

PERCENTAGE OF VARIANCE EXPLAINED, r^2	
$0.01 < r^2 < 0.09$	Small effect
$0.09 < r^2 < 0.25$	Medium effect
$r^2 > 0.25$	Large effect

Sample Variance and Effect Size   Examples 9.1 and 9.2 provide an opportunity to demonstrate a point that was made back in Chapter 4 (page 126): In the context of inferential statistics, variance is bad. High variance means that there are big differences from one score to another, which makes it difficult to see any consistent trends or patterns in the data. Using more literary terms, in the world of inferential statistics, variance is the villain.

In Example 9.1, the hypothesis test concluded that the eye-spot patterns had a *significant* effect on the bird's behavior, and in Example 9.2 we concluded that the size of the effect was *large*. Both conclusions were based on a sample of $n = 16$ birds with a mean of $M = 39$ minutes on the plain side with a standard deviation of $s = 6$. Now consider what happens to our conclusions if we *increase the sample variance*. Specifically, we will repeat the hypothesis test and recalculate Cohen's *d* using a sample standard deviation of $s = 36$ instead of the original $s = 6$.

EXAMPLE 9.3   First, we will repeat the hypothesis test from Example 9.1, but now we will use a sample standard deviation of $s = 36$. With $s = 36$, the estimated standard error is

$$\text{estimated standard error} = s_M = \sqrt{\frac{s^2}{n}} = \sqrt{\frac{36^2}{16}} = \sqrt{\frac{1296}{16}} = \sqrt{81} = 9$$

The *t* statistic is

$$t = \frac{M - \mu}{s_M} = \frac{39 - 30}{9} = \frac{9}{9} = 1.00$$

As shown in Figure 9.5, the critical region for this test begins at $t = \pm 2.131$. Our result, $t = 1.00$, is not in the critical region. We fail to reject the null hypothesis and conclude that the eye-spot patterns do not have a significant effect on behavior. Notice that the increased variability has eliminated what was once a significant effect.

Now consider what happens to Cohen's *d* when the standard deviation is increased to $s = 36$. With a mean difference of 9 minutes ($M = 39$ versus $\mu = 30$ from H_0) and $s = 36$,

$$\text{Cohen's } d = \frac{\text{mean difference}}{\text{sample standard deviation}} = \frac{9}{36} = 0.25$$

According to the standards in Table 8.2, $d = 0.25$ represents a small to moderate effect size. In this case, the increased variability has eliminated what was originally a large treatment effect.

Finally, we will examine the effect of sample variance on r^2, the percentage of variance accounted for by the treatment. For the original data, with $s = 6$, we obtained $r^2 = 0.706$. However, when the sample standard deviation was increased to $s = 36$, we obtained a *t* statistic of $t = 1.00$ with $df = 15$, and the value of r^2 becomes

$$r^2 = \frac{t^2}{t^2 + df} = \frac{1}{1 + 15} = \frac{1}{16} = 0.0625$$

Once again, the increased variability has dramatically lowered the effect size.

In general, high variability reduces the ability to detect any patterns in the sample data. In particular, high variability reduces the likelihood of finding a significant effect in a hypothesis test, and it reduces the effect size as measured by Cohen's *d* or r^2.

IN THE LITERATURE
REPORTING THE RESULTS OF A *t* TEST

In Chapter 8, we noted the conventional style for reporting the results of a hypothesis test, according to APA format. First, recall that a scientific report typically uses the term *significant* to indicate that the null hypothesis has been rejected and the term *not significant* to indicate failure to reject H_0. Additionally, there is a prescribed format for reporting the calculated value of the test statistic, degrees of freedom, and alpha level for a *t* test. This format parallels the style introduced in Chapter 8 (page 249). For Example 9.1, we calculated a *t* statistic of $+6.00$ with $df = 15$ and decided to reject H_0 with alpha set at .05. In a scientific report, this information is conveyed in a concise statement, as follows:

> The subjects averaged $M = 39$ minutes on the plain side of the apparatus with $SD = 6$. Statistical analysis indicates that the time spent on the side without eye spots was significantly more than would be expected by chance, $t(15) = 6.00$, $p < .05$, and that the effect size was large, $d = 1.50$.

In the first statement, the mean ($M = 39$) and the standard deviation ($SD = 6$) are reported as previously noted (Chapter 4, page 123). The next statement provides the results of the statistical analysis. Note that the degrees of freedom are reported in parentheses immediately after the symbol *t*. The value for the obtained *t* statistic follows (6.00), and next is the probability of committing a Type I error (less than 5%).

The statement p < .05 was explained in Chapter 8, page 249.

Finally, the effect size is reported, Cohen's $d = 1.50$.

Often, researchers will use a computer to perform a hypothesis test like the one in Example 9.1. In addition to calculating the mean, standard deviation, and the *t* statistic for the data, the computer usually calculates and reports the *exact probability* associated with the *t* value. In Example 9.1 we determined that any *t* value beyond ± 2.131 has a probability of less than .05 (see Figure 9.5). Thus, the obtained *t* value, $t = 6.00$, is reported as being very unlikely, $p < .05$. A computer printout, however, would have included an exact probability for our specific *t* value.

Whenever a specific probability value is available, you are encouraged to use it in a research report. For example, if the computer reports $t = 3.18$ and $p = 0.002$, your research report should state "$p = .002$" instead of the less specific "$p < .05$." As one final caution, we should note that occasionally a *t* value is so extreme that the computer reports $p = 0.000$. The zero value does not mean that the probability is literally zero; instead, it means that the computer has rounded off the probability value to three decimal places and obtained a result of 0.000. In this situation, you do not know the exact probability value, but you can report $p < .001$.   ❏

LEARNING CHECK	**1.** In Example 9.1, the sample of $n = 16$ birds spent an average of $M = 39$ minutes on the plain side with a sample standard deviation of $s = 6$. The data were sufficient to conclude that the eye spots had a significant effect on behavior (we rejected H_0) and produced an effect size of $d = 1.50$. If the sample size were changed to $n = 4$, how would the hypothesis test and the measure of effect size be affected? Using the same sample mean and sample standard deviation as before, repeat the hypothesis test and the calculation of Cohen's *d* with a sample size of $n = 4$ birds instead of $n = 16$. In general, how does sample size influence the hypothesis test? How does sample size influence Cohen's *d*?
ANSWERS	**1.** With $n = 4$, the standard error is $s_M = 3$ and the hypothesis test produces $t = 3.00$. With $df = 3$, this result is not enough to conclude that there is a significant effect. The results for the hypothesis test change because the standard error depends on the size of the sample. With $n = 4$, Cohen's *d* is still 1.50. The value of Cohen's *d* does not depend on sample size.

DIRECTIONAL HYPOTHESES AND ONE-TAILED TESTS

As noted in Chapter 8, the nondirectional (two-tailed) test is more commonly used than the directional (one-tailed) alternative. On the other hand, a directional test may be used in some research situations, such as exploratory investigations or pilot studies or where there is a priori justification (for example, a theory or previous findings). The following example demonstrates a directional hypothesis test with a *t* statistic, using the same experimental situation that was presented in Example 9.1.

EXAMPLE 9.4   The research question is whether eye-spot patterns will affect the behavior of birds placed in a special testing box. The researcher is expecting the birds to avoid the eye-spot patterns. Therefore, the researcher predicts that the birds will spend most of the hour on the plain side of the box. For this example we will use the same sample data that were used in the original hypothesis test in Example 9.1. Specifically, the researcher tested a sample of $n = 16$ birds and obtained a mean of $M = 39$ minutes spent on the plain side of the box with $SS = 540$.

STEP 1   State the hypotheses, and select an alpha level. With most directional tests, it is usually easier to state the hypothesis in words, including the directional prediction, and then convert the words into symbols. For this example, the researcher is predicting that the eye-spot patterns will cause the birds to *increase* the amount of time that they spend in the plain side of the box; that is, more than half of the hour should be spent on the plain side. In general, the null hypothesis states that the predicted effect will not happen. For this study, the null hypothesis states that the birds will *not* spend more than half of the hour on the plain side. In symbols,

H_0: $\mu_{\text{plain side}} \leq 30$ minutes   (Not more than half of the hour on the plain side)

Similarly, the alternative states that the treatment will work. In this case, H_1 states that the birds will spend more than half of the hour on the plain side. In symbols,

H_1: $\mu_{plain\ side} > 30$ minutes   (More than half of the hour on the plain side)

We will set the level of significance at $\alpha = .05$.

STEP 2   Locate the critical region. In this example, the researcher is predicting that the sample mean (M) will be greater than 30. The null hypothesis states that the population mean is $\mu = 30$ (or less). If you examine the structure of the *t* statistic formula, it should be clear that a positive *t* statistic would support the researcher's prediction and refute the null hypothesis.

$$t = \frac{M - \mu}{s_M}$$

The problem is to determine how large a positive value is necessary to reject H_0. To find the critical value, you must look in the *t* distribution table using the one-tail proportions. With a sample of $n = 16$, the *t* statistic will have $df = 15$; using $\alpha = .05$, you should find a critical *t* value of 1.753. Figure 9.8 depicts the critical region in the *t* distribution.

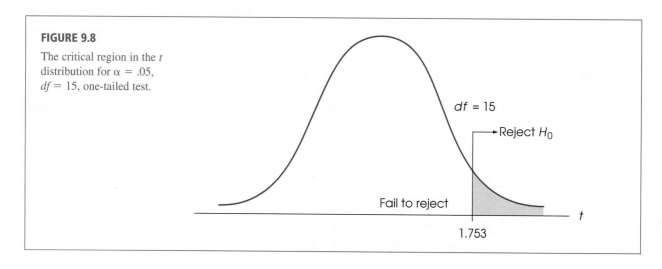

FIGURE 9.8

The critical region in the *t* distribution for $\alpha = .05$, $df = 15$, one-tailed test.

STEP 3   Calculate the test statistic. The computation of the *t* statistic is the same for either a one-tailed or a two-tailed test. Earlier (in Example 9.1), we found that the data for this experiment produce a test statistic of $t = 6.00$.

STEP 4   Make a decision. The test statistic is in the critical region, so we reject H_0. In terms of the experimental variables, we have decided that the birds spent significantly more time on the plain side of the box than on the side with eye-spot patterns.

ASSUMPTIONS OF THE *t* TEST   There are two basic assumptions necessary for hypothesis tests with the *t* statistic.

   1. The values in the sample must consist of *independent* observations.
       In everyday terms, two observations are independent if there is no consistent, predictable relationship between the first observation and the second. More precisely, two events (or observations) are independent if the occurrence of the

first event has no effect on the probability of the second event. Specific examples of independence and nonindependence were presented in Box 8.2. Usually this assumption is satisfied by using a *random* sample, which also helps ensure that the sample is representative of the population and that the results can be generalized to the population.

2. The population sampled must be normal.

This assumption is a necessary part of the mathematics underlying the development of the *t* statistic and the *t* distribution table. However, violating this assumption has little practical effect on the results obtained for a *t* statistic, especially when the sample size is relatively large. With very small samples, a normal population is important. With larger samples, this assumption can be violated without affecting the validity of the hypothesis test. If you have reason to suspect that the population is not normal, use a large sample to be safe.

THE VERSATILITY OF THE *t* TEST

The obvious advantage of hypothesis testing with the *t* statistic (as compared to *z*-scores) is that you do not need to know the value of the population variance. This means that we still can do a hypothesis test even though we have little or no information about the population. One result of this extra versatility is that it is possible to do *t* tests in circumstances where hypothesis tests with *z*-scores would be impossible.

Both the *t*-statistic and the *z*-score tests were introduced as methods for determining whether or not a treatment has any effect on a dependent variable. For both tests, we assumed that you start with a *known* population before treatment, and the research question (hypotheses) concerned the *unknown* population mean after treatment (see Figures 8.1 and 9.3). With the *t*-statistic, however, we no longer need a known population before treatment. All we need to compute a *t* statistic is a sample and a sensible null hypothesis. For example, the hypothesis tests evaluating the effects of eye-spot patterns did not involve a population before treatment and a population after treatment. Instead, we simply had one unknown population, one sample, and a logical null hypothesis. Some similar examples follow. Note in each case that the hypothesis is not dependent on knowing the actual population mean before treatment and that the rest of the *t* statistic can be computed entirely from the obtained sample data.

1. A researcher would like to examine the accuracy of people's judgment of time when they are distracted. Individuals are placed in a waiting room where many distracting events occur for a period of 12 minutes. The researcher then asks each person to judge how much time has passed. The goal is to determine whether the distractions cause people to overestimate or underestimate the amount of time. The null hypothesis would state that distraction has no effect and that time judgments are accurate. That is,

$$H_0: \quad \mu = 12 \text{ minutes}$$

2. A researcher has developed a 7-point rating scale to measure people's opinions concerning different issues. The research presents an issue to each participant and asks for an opinion on the scale where a rating of 1 indicates "strongly disapprove" and a rating of 7 indicates "strongly approve." A rating of 4 (the center of the scale) indicates "neutral or no opinion." For any given issue, the null hypothesis would state that there is no consistently positive or consistently negative opinion in the population. Instead, the null hypothesis states that the overall population opinion is neutral.

$$H_0: \quad \mu = 4.00$$

LEARNING CHECK

1. A researcher would like to evaluate the effect of a new cold medication on reaction time. It is known that under regular circumstances the distribution of reaction times is normal with $\mu = 200$. A sample of $n = 4$ participants is obtained. Each person is given the new cold medication, and 1 hour later reaction time is measured for each individual. The average reaction time for this sample is $M = 215$ with $SS = 300$. On the basis of these data, can the researcher conclude that the cold medication has an effect on reaction time? Test at the .05 level of significance.

   a. State the hypotheses.

   b. Locate the critical region.

   c. Compute the t statistic.

   d. Make a decision.

2. Suppose the researcher in the previous problem predicted that the medication would increase reaction time and decided to use a one-tailed test to evaluate this prediction.

   a. State the hypotheses for the one-tailed test.

   b. Locate the one-tailed critical region for $\alpha = .05$.

ANSWERS

1. a. $H_0: \mu = 200$ (The medication has no effect on reaction time.)
   $H_1: \mu \neq 200$ (Reaction time is changed.)

   b. With $df = 3$ and $\alpha = .05$, the critical region consists of t values beyond ± 3.182.

   c. The sample variance is $s^2 = 100$, the standard error is 5, and $t = 3.00$.

   d. The t statistic is not in the critical region. Fail to reject the null hypothesis, and conclude that the data do not provide enough evidence to say that the medication affects reaction time.

2. a. $H_0: \mu \leq 200$ (Reaction time is not increased.)
   $H_1: \mu > 200$ (Reaction time is increased.)

   b. With $df = 3$ and $\alpha = .05$, the one-tailed critical region consists of values beyond $t = +2.353$.

SUMMARY

1. The t statistic is used instead of a z-score for hypothesis testing when the population standard deviation (or variance) is unknown.

2. To compute the t statistic, you must first calculate the sample variance (or standard deviation) as a substitute for the unknown population value.

$$s^2 = \frac{SS}{df} \qquad s = \sqrt{\frac{SS}{df}}$$

   Next, the standard error is *estimated* by substituting s^2 or s in the formula for standard error. The estimated standard error is calculated in the following manner:

$$s_M = \sqrt{\frac{s^2}{n}} \qquad \text{or} \qquad s_M = \frac{s}{\sqrt{n}}$$

Finally, a t statistic is computed using the estimated standard error. The t statistic is used as a substitute for a z-score that cannot be computed when the population variance or standard deviation is unknown.

$$t = \frac{M - \mu}{s_M}$$

3. The structure of the t formula is similar to that of the z-score in that

$$z \text{ or } t = \frac{\text{sample mean} - \text{population mean}}{\text{(estimated) standard error}}$$

For a hypothesis test, you hypothesize a value for the unknown population mean and plug the hypothesized value into the equation along with the sample mean and the estimated standard error, which are computed from the sample data. If the hypothesized mean produces an extreme value for t, you conclude that the hypothesis was wrong.

4. The t distribution is an approximation of the normal z distribution. To evaluate a t statistic that is obtained for a sample mean, the critical region must be located in a t distribution. There is a family of t distributions, with the exact shape of a particular distribution of t values depending on degrees of freedom $(n - 1)$. Therefore, the critical t values will depend on the value for df associated with the t test. As df increases, the shape of the t distribution approaches a normal distribution.

5. When a t statistic is used for a hypothesis test, Cohen's d can be computed to measure effect size. In this situation, the sample standard deviation is used in the formula:

$$\text{Cohen's } d = \frac{\text{mean difference}}{\text{standard deviation}} = \frac{M - \mu}{s}$$

6. A second measure of effect size is r^2, which measures the percentage of the variability that is accounted for by the treatment effect. This value is computed as follows:

$$r^2 = \frac{t^2}{t^2 + df}$$

RESOURCES

WebTUTOR

If you are using WebTutor along with this book, there is a WebTutor section corresponding to this chapter. The WebTutor contains a brief summary of Chapter 9, hints for learning the concepts and the formulas for the t score hypothesis test, cautions about common errors, and sample exam items including solutions.

SPSS

The statistics computer package SPSS can be used to conduct a hypothesis test using the t statistic. First, the scores from the sample are entered in one column of the data matrix. Then, click **Analyze** on the tool bar, select **Compare Means**, and click on **One-Sample T Test**. Enter the column label for the set of scores into the Test Variable box, enter the hypothesized value for the population mean (from the null hypothesis) into the Test Value box, and click **OK.** SPSS will produce a summary table listing the number of scores, the mean, the standard deviation, and the standard error for the sample. The results of the hypothesis test are summarized by presenting the calculated t value, the degrees of freedom, the level of significance (probability of a Type I error), and the size of the mean difference. In addition, the output includes a 95% confidence interval that provides a range of values estimating how much difference the treatment has produced. (Confidence intervals will be presented in Chapter 12.)

KEY TERMS

| estimated standard error | t statistic | degrees of freedom | t distribution | percentage of variance accounted for (r^2) |

FOCUS ON PROBLEM SOLVING

1. The first problem we confront in analyzing data is determining the appropriate statistical test. Remember that you can use a z-score for the test statistic only when the value for σ is known. If the value for σ is not provided, then you must use the t statistic.

2. For a t test, students sometimes use the unit normal table to locate the critical region. This, of course, is a mistake. The critical region for a t test is obtained by consulting the t distribution table. Notice that to use this table you first must compute the value for degrees of freedom (df).

3. For the t test, the sample variance is used to find the value for estimated standard error. Remember that when computing the sample variance, use $n - 1$ in the denominator (see Chapter 4). When computing estimated standard error, use n in the denominator.

DEMONSTRATION 9.1

A HYPOTHESIS TEST WITH THE *t* STATISTIC

A psychologist has prepared an "Optimism Test" that is administered yearly to graduating college seniors. The test measures how each graduating class feels about its future—the higher the score, the more optimistic the class. Last year's class had a mean score of $\mu = 15$. A sample of $n = 9$ seniors from this year's class was selected and tested. The scores for these seniors are as follows:

$$7 \quad 12 \quad 11 \quad 15 \quad 7 \quad 8 \quad 15 \quad 9 \quad 6$$

On the basis of this sample, can the psychologist conclude that this year's class has a different level of optimism than last year's class?

Note that this hypothesis test will use a t statistic because the population variance (σ^2) is not known.

STEP 1   State the hypotheses, and select an alpha level.

The statements for the null hypothesis and the alternative hypothesis follow the same form for the t statistic and the z-score test.

$$H_0: \quad \mu = 15 \qquad \text{(There is no change.)}$$

$$H_1: \quad \mu \neq 15 \qquad \text{(This year's mean is different.)}$$

For this demonstration, we will use $\alpha = .05$, two tails.

STEP 2   Locate the critical region.

Remember that for hypothesis tests with the t statistic, we must consult the t distribution table to find the critical t values. With a sample of $n = 9$ students, the t statistic will have degrees of freedom equal to

$$df = n - 1 = 9 - 1 = 8$$

For a two-tailed test with $\alpha = .05$ and $df = 8$, the critical t values are $t = \pm 2.306$. These critical t values define the boundaries of the critical region. The obtained t value must be more extreme than either of these critical values to reject H_0.

STEP 3   Obtain the sample data, and compute the test statistic.

For the t formula, we need to determine the values for the following:

1. The sample mean, M

2. The estimated standard error, s_M

We will also have to compute sums of squares (SS) and the variance (s^2) for the sample in order to get the estimated standard error.

The sample mean. For these data, the sum of the scores is

$$\Sigma X = 7 + 12 + 11 + 15 + 7 + 8 + 15 + 9 + 6 = 90$$

Therefore, the sample mean is

$$M = \frac{\Sigma X}{n} = \frac{90}{9} = 10$$

Sum of squares. We will use the definitional formula for sum of squares,

$$SS = \Sigma(X - M)^2$$

The following table summarizes the steps in the computation of SS:

X	$X - M$	$(X - M)^2$
7	$7 - 10 = -3$	9
12	$12 - 10 = +2$	4
11	$11 - 10 = +1$	1
15	$15 - 10 = +5$	25
7	$7 - 10 = -3$	9
8	$8 - 10 = -2$	4
15	$15 - 10 = +5$	25
9	$9 - 10 = -1$	1
6	$6 - 10 = -4$	16

For this demonstration problem, the sum of squares value is

$$SS = \Sigma(X - M)^2 = 9 + 4 + 1 + 25 + 9 + 4 + 25 + 1 + 16$$
$$= 94$$

Sample variance.

$$s^2 = \frac{SS}{n - 1} = \frac{94}{8} = 11.75$$

Estimated standard error. The estimated standard error for these data is

$$s_M = \sqrt{\frac{s^2}{n}} = \sqrt{\frac{11.75}{9}} = 1.14$$

The t statistic. Now that we have the estimated standard error and the sample mean, we can compute the t statistic. For this demonstration,

$$t = \frac{M - \mu}{s_M} = \frac{10 - 15}{1.14} = \frac{-5}{1.14} = -4.39$$

STEP 4   Make a decision about H_0, and state a conclusion.
The t statistic we obtained ($t = -4.39$) is in the critical region. Thus, our sample data are unusual enough to reject the null hypothesis at the .05 level of significance. We can conclude that there is a significant difference in level of optimism between this year's and last year's graduating classes, $t(8) = -4.39, p < .05$, two-tailed.

DEMONSTRATION 9.2

EFFECT SIZE WITH COHEN'S d AND r^2

We will compute Cohen's d for the same data used for the hypothesis test in Demonstration 9.1. The mean optimism score for the sample from this year's class was 5 points lower than the mean from last year ($M = 10$ versus $\mu = 15$). In Demonstration 9.1 we computed a sample variance of $s^2 = 11.75$, so the standard deviation is $\sqrt{11.75} = 3.43$. With these values,

$$\text{Cohen's } d = \frac{\text{mean difference}}{\text{standard deviation}} = \frac{5}{3.43} = 1.46$$

To calculate the percentage of variance explained by the treatment effect, r^2, we need the value of t and the df value from the hypothesis test. In Demonstration 9.1 we obtained $t = -4.39$ with $df = 8$. Using these values in Equation 9.4, we obtain

$$r^2 = \frac{t^2}{t^2 + df} = \frac{(-4.39)^2}{(-4.39)^2 + 8} = \frac{19.27}{27.27} = 0.71$$

PROBLEMS

1. What factor determines whether you should use a z-score or a t statistic for a hypothesis test?

2. A sample of $n = 25$ scores has a mean of $M = 42$ and a sample variance of $s^2 = 100$.
   a. Compute the sample standard deviation (s) and the estimated standard error for the sample mean (s_M).
   b. Briefly describe what is measured by the sample standard deviation and by the estimated standard error.

3. Several different factors influence the value obtained for a t statistic. For each of the following, describe how the value of t is affected. In each case, assume that all other factors are held constant.
   a. What happens to the value of t when the variability of the scores in the sample increases?
   b. What happens to the value of t when the number of scores in the sample increases?
   c. What happens to the value of t when the difference between the sample mean and the hypothesized population mean increases?

4. Why is a t distribution generally more variable than a normal distribution?

5. What is the relationship between the value for degrees of freedom and the shape of the t distribution? What happens to the critical value of t for a particular alpha level when df increases in value?

6. A sample has a mean of $M = 63$ and a variance of $s^2 = 36$.
   a. If the sample consists of $n = 4$ scores, then compute the sample standard deviation (s) and the estimated standard error for the sample mean (s_M).
   b. If the sample consists of $n = 16$ scores, then compute the sample standard deviation (s) and the estimated standard error for the sample mean (s_M).

7. The following sample was obtained from a population with unknown parameters.

   Scores: 8, 0, 8, 8

a. Compute the sample mean and standard deviation. (Note that these are descriptive values that summarize the sample data.)

b. Compute the estimated standard error for *M*. (Note that this is an inferential value that describes how accurately the sample mean represents the unknown population mean.)

8. The following sample was obtained from a population with unknown parameters.

Scores: 9,   1,   13,   1

a. Compute the sample mean and standard deviation. (Note that these are descriptive values that summarize the sample data.)

b. Compute the estimated standard error for *M*. (Note that this is an inferential value that describes how accurately the sample mean represents the unknown population mean.)

9. The following sample was obtained from an unknown population.

Scores: 3,   11,   3,   3

a. Compute the sample mean and the sample standard deviation. (Note that these values are descriptive statistics that summarize the sample data.)

b. How accurately does the sample mean represent the unknown population mean? Compute the estimated standard error for *M*. (Note that this value is an inferential statistic that allows you to use the sample data to answer questions about the population.)

10. One characteristic of *t* statistics is that you must first compute the sample variance or sample standard deviation before it is possible to calculate the value for *t*. For each of the following samples:

Sample 1:   2,   10,   2,   2

Sample 2:   1,   5,   1,   1

a. Calculate the sample variance and the sample standard deviation. (Note that the standard deviation is a descriptive statistic that describes how the scores are scattered around the mean.)

b. Calculate the estimated standard error. (Note that the standard error is an inferential statistic that measures how accurately the sample mean represents the population mean.)

c. Although the two samples are the same size, you should find that one provides a more accurate representation of the population than the other (one has less error). In general, explain how sample variance is related to the magnitude of the estimated standard error.

11. For each of the following samples:

Sample 1:   1,   7,   1,   1

Sample 2:   8,   6,   3,   5,   5,   13,   7,   6,   10

a. Calculate the sample variance and the standard deviation. (Note that the standard deviation is a descriptive statistic that describes how the scores are scattered around the mean.)

b. Calculate the estimated standard error. (Note that the standard error is an inferential statistic that measures how accurately the sample mean represents the population mean.)

c. Although the two samples have exactly the same variance, you should find that one provides a more accurate representation of the population than the other (one has less error). In general, explain how sample size is related to the magnitude of the estimated standard error.

12. A random sample of $n = 9$ individuals is obtained from a population with a mean of $\mu = 80$. A treatment is administered to each individual in the sample and, after treatment, each individual is measured. The average score for the treated sample is $M = 86$ with $SS = 288$.

a. How much difference is there between the mean for the treated sample and the mean for the original population? (*Note:* In a hypothesis test, this value forms the numerator of the *t* statistic.)

b. How much difference is expected just by chance between the sample mean and its population mean? That is, find the standard error for *M*. (*Note:* In a hypothesis test, this value is the denominator of the *t* statistic.)

c. Based on the sample data, does the treatment have a significant effect? Use a two-tailed test with $\alpha = .05$.

13. A standardized reading achievement test for fifth-grade students has a nationwide mean of $\mu = 70$. A teacher would like to know whether the students in her fifth-grade class are significantly different from the national average. The test is given to the class of $n = 25$ students and the average score is $M = 75$ with $SS = 2400$.

a. Using a two-tailed test with $\alpha = .05$, is this class significantly different from the national average?

b. Using a two-tailed test with $\alpha = .01$, is the class significantly different from the national average?

c. Explain why you reach a different conclusion when $\alpha = .01$ instead of .05.

14. A newspaper article reported that the typical American family spent an average of $\mu = \$81$ for Halloween candy and costumes last year. A sample of $n = 16$

families this year produced a mean of $M = \$85$ with $SS = 6000$. Do these data indicate a significant change in holiday spending? Use a two-tailed test with $\alpha = .05$.

15. A researcher is evaluating the effect of a new treatment. A sample is selected from a population with $\mu = 60$ and the treatment is administered to the sample. After treatment, the individuals in the sample produce a mean score of $M = 66$ with a standard deviation of $s = 10$.

    a. Assuming that the sample consists of $n = 25$ individuals, compute the estimated standard error for the sample mean and find the boundaries for the critical region for a two-tailed hypothesis test with $\alpha = .05$. Is the 6-point difference between the sample mean and the original population enough to conclude that the treatment has a significant effect?

    b. Now assume that the sample has only $n = 4$ individuals. Again, compute the estimated standard error for the sample mean and find the boundaries for the critical region for a two-tailed hypothesis test with $\alpha = .05$. With $n = 4$, is the 6-point difference between the sample mean and the original population enough to conclude that the treatment has a significant effect?

    c. Based on your answers to parts (a) and (b), how does sample size influence the size of the standard error and the outcome of a hypothesis test?

    d. Use Cohen's *d* to measure effect size for the treatment, first for a sample of $n = 25$ and then for a sample of $n = 4$. How does sample size influence Cohen's *d*?

16. A sample of $n = 9$ individuals is selected from a population with a mean of $\mu = 100$. A treatment is administered to the individuals and the sample mean is then found to be $M = 105$.

    a. Assuming that the sample standard deviation is $s = 12$, use a two-tailed test with $\alpha = .05$ to determine whether the treatment had a significant effect.

    b. Now assume that the sample standard deviation is $s = 3$. Once again, use a two-tailed test with $\alpha = .05$ to determine whether the treatment had a significant effect.

    c. Based on your answers to parts (a) and (b), how does the sample standard deviation influence the size of the standard error and the outcome of a hypothesis test?

    d. Use Cohen's *d* to measure the effect size for the sample in part (a), where $s = 12$, and for the sample in part (b), where $s = 3$. How does sample standard deviation affect Cohen's *d*?

17. When adults are given a set of items to hold in memory, they automatically rehearse the items to prevent forgetting. Young children, however, do not sponta-

neously use rehearsal. As a result, when given a small set of items to remember, 2-year-old children on average can recall only $\mu = 2$ items. A researcher would like to determine whether memory can be improved by teaching 2-year-old children to use rehearsal. A sample of $n = 16$ children is obtained and the children are trained to use rehearsal during a memory task. After training, the children in the sample show an average recall of $M = 4.3$ items, with $SS = 60$.

    a. Use a one-tailed test with $\alpha = .05$ to determine whether the data are sufficient to conclude that the rehearsal training produces a significant increase in memory.

    b. Compute Cohen's *d* and r^2 for these data to measure the effect size.

18. A researcher suspects that there is a direct relationship between hand–eye coordination and academic success in mathematics. A sample of $n = 16$ students who have demonstrated above average performance on a second-grade mathematics achievement test is selected. These students are given a standardized hand–eye coordination task where the average score for this age group is $\mu = 55$. The 16 students in the sample produced a mean of $M = 61$ with $SS = 540$.

    a. Are the data sufficient to conclude that the high math achievement students have hand–eye coordination scores that are significantly *different from* the general population? Use a two-tailed test with $\alpha = .01$.

    b. Are the data sufficient to conclude that the high math achievement students have hand–eye coordination scores that are significantly *better than* the general population? Use a one-tailed test with $\alpha = .01$.

19. A major corporation in the Northeast noted that last year its employees averaged $\mu = 5.8$ absences during the winter season (December to February). In an attempt to reduce absences, the company offered free flu shots to all employees this year. For a sample of $n = 100$ people who took the shots, the average number of absences this winter was $M = 3.6$ with $SS = 396$.

    a. Do these data indicate a significant decrease in the number of absences? Use a one-tailed test with $\alpha = .05$.

    b. Compute Cohen's *d* and r^2 to measure effect size for this study.

20. A social psychologist suspects that the increased availability of e-mail may have an effect on the methods that people use to communicate. The researchers obtained records from 1995 indicating that the average American spent $\mu = 2.1$ hours per month on long distance phone calls. A random sample of $n = 36$ people was obtained and their phone records were

monitored for all of last year. The people in the sample averaged $M = 1.9$ hours of long distance calls per month with $SS = 12.6$. Do these data indicate a significant change in the use of long distance telephone? Use a two-tailed test with $\alpha = .05$.

21. A random sample has a mean of $M = 81$ with $s = 10$. Use this sample to test the null hypothesis that $\mu = 85$ for each of the following situations.
   a. Assume that the sample size is $n = 25$, and use $\alpha = .05$ for the two-tailed hypothesis test.
   b. Assume that the sample size is $n = 400$, and use $\alpha = .05$ for the two-tailed hypothesis test.
   c. In general, how does sample size contribute to the outcome of a hypothesis test?

22. A random sample of $n = 16$ scores has a mean of $M = 48$. Use this sample to test the null hypothesis that the population mean is $\mu = 45$ for each of the following situations:
   a. Assume that the sample has a variance of $s^2 = 4$ and use a two-tailed test with $\alpha = .05$.
   b. Assume the sample has a variance of $s^2 = 400$ and use a two-tailed test with $\alpha = .05$.
   c. In general, how does the magnitude of the sample variance contribute to the outcome of a hypothesis test?

23. A researcher would like to examine the effects of humidity on eating behavior. It is known that laboratory rats normally eat an average of $\mu = 21$ grams of food each day. The researcher selects a random sample of $n = 100$ rats and places them in a controlled atmosphere room where the relative humidity is maintained at 90%. The daily food consumption for the sample averages $M = 18.7$ with $SS = 2475$. On the basis of this sample, can the researcher conclude that humidity affects eating behavior? Use a two-tailed test at the .05 level.

24. A psychologist would like to determine whether there is a relation between depression and aging. It is known that the general population averages $\mu = 40$ on a standardized depression test. The psychologist obtains a sample of $n = 36$ individuals who are all over the age of 70. The average depression score for this sample is $M = 44.5$ with $SS = 5040$. On the basis of this sample, can the psychologist conclude that depression for elderly people is significantly different from depression in the general population? Use a two-tailed test at the .05 level of significance.

25. One of the original tests of ESP involves using Zener cards. Each card shows 1 of 5 different symbols (square, circle, star, wavy lines, cross). One person randomly picks a card and concentrates on the symbol it shows. A second person, in a different room, attempts to identify the symbol that was selected. Chance performance on this task (just guessing) should lead to correct identification of 1 out of 5 cards. A psychologist used the Zener cards to evaluate a sample of $n = 9$ participants who claimed to have ESP. Each was tested on a series of 100 cards, and the number correct for each individual is as follows: 18, 23, 24, 22, 19, 28, 15, 26, 25.
   Chance performance on a series of 100 cards would be $\mu = 20$ correct. Did this sample perform significantly better than chance? Use a one-tailed test at the .05 level of significance.

26. Several years ago, a survey in a school district revealed that the average age at which students first tried an alcoholic drink was $\mu = 14$ years. To determine whether anything has changed, a random sample of students was asked questions about alcohol use. The age at which drinking behavior first began was reported by members of the sample as follows: 11, 13, 14, 12, 10. Has there been a change in the mean age at which drinking began? Use $\alpha = .05$ for two tails.

27. On a standardized spatial skills task, normative data reveal that people typically get $\mu = 15$ correct solutions. A psychologist tests $n = 7$ individuals who have brain injuries in the right cerebral hemisphere. For the following data, determine whether or not right-hemisphere damage results in significantly reduced performance on the spatial skills task. Test with alpha set at .05 with one tail. The data are as follows: 12, 16, 9, 8, 10, 17, 10.

CHAPTER

10

The *t* Test for Two Independent Samples

Tools You Will Need

The following items are considered essential background material for this chapter. If you doubt your knowledge of any of these items, you should review the appropriate chapter or section before proceeding.

- Sample variance (Chapter 4)
- Standard error formulas (Chapter 7)
- The *t* statistic (Chapter 9)
  - Distribution of *t* values
  - *df* for the *t* statistic
  - Estimated standard error

Preview

In a classic study in the area of problem solving, Katona (1940) compared the effectiveness of two methods of instruction. One group of participants was shown the exact, step-by-step procedure for solving a problem, and then these participants were required to memorize the solution. This method was called *learning by memorization* (later called the *expository* method). Participants in a second group were encouraged to study the problem and find the solution on their own. Although these participants were given helpful hints and clues, the exact solution was never explained. This method was called *learning by understanding* (later called the *discovery* method).

Katona's experiment included the problem shown in Figure 10.1. This figure shows a pattern of five squares made of matchsticks. The problem is to change the pattern into exactly four squares by moving only three matches. (All matches must be used, none can be removed, and all the squares must be the same size.) Two groups of participants learned the solution to this problem. One group learned by understanding, and the other group learned by memorization. After 3 weeks, both groups returned to be tested again. The two groups did equally well on the matchstick problem they had learned earlier. But when they were given two new problems (similar to the matchstick problem), the *understanding* group performed much better than the *memorization* group.

The outcome of Katona's experiment probably is no surprise. However, you should realize that the experimental design is different from any we have considered before.

This experiment involved *two separate samples*. Previously, we examined statistical techniques for evaluating the data from only one sample. In this chapter, we will present the statistical procedures that allow a researcher to use two samples to evaluate the difference between two experimental treatment conditions.

Incidentally, if you still have not discovered the solution to the matchstick problem, keep trying. According to Katona's results, it would be very poor teaching strategy for us to give you the answer.

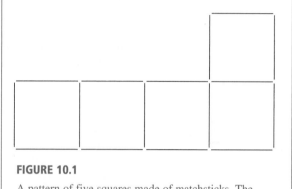

FIGURE 10.1

A pattern of five squares made of matchsticks. The problem is to change the pattern into exactly four squares by moving only three matchsticks.

Katona, G. (1940). *Organizing and Memorizing*. New York: Colombia University Press, Reprinted by permission.

10.1 OVERVIEW

Until this point, all the inferential statistics we have considered involve using one sample as the basis for drawing conclusions about one population. Although these *single-sample* techniques are used occasionally in real research, most research studies require the comparison of two (or more) sets of data. For example, a social psychologist may want to compare men and women in terms of their attitudes toward abortion, an educational psychologist may want to compare two methods for teaching mathematics, or a clinical psychologist may want to evaluate a therapy technique by comparing depression scores for patients before therapy with their scores after therapy. In each case, the research question concerns a mean difference between two sets of data.

There are two general research strategies that can be used to obtain the two sets of data to be compared:

1. The two sets of data could come from two completely separate samples. For example, the study could involve a sample of men compared with a sample of

women. Or the study could compare one sample of students taught by method A and a second sample taught by method B.

2. The two sets of data could both come from the same sample. For example, the researcher could obtain one set of scores by measuring depression for a sample of patients before they begin therapy and then obtain a second set of data by measuring the same individuals after 6 weeks of therapy.

The first research strategy, using completely separate samples, is called an *independent-measures* research design or a *between-subjects* design. These terms emphasize the fact that the design involves separate and independent samples and makes a comparison between two groups of individuals. In this chapter, we will examine the statistical techniques used to evaluate the data from an independent-measures design. More precisely, we will introduce the hypothesis test that allows researchers to use the data from two separate samples to evaluate the mean difference between two populations or between two treatment conditions.

The second research strategy, where the two sets of data are obtained from the same sample, is called a *repeated-measures* research design or a *within-subjects* design. The statistical analysis for repeated measures will be introduced in Chapter 11. Also, at the end of Chapter 11, we will discuss some of the advantages and disadvantages of independent-measures versus repeated-measures research designs.

The typical structure of an independent-measures research study is shown in Figure 10.2. Notice that the research study is using two separate samples to answer a question about two populations.

DEFINITION   A research design that uses a separate sample for each treatment condition (or for each population) is called an **independent-measures** research design or a **between-subjects** design.

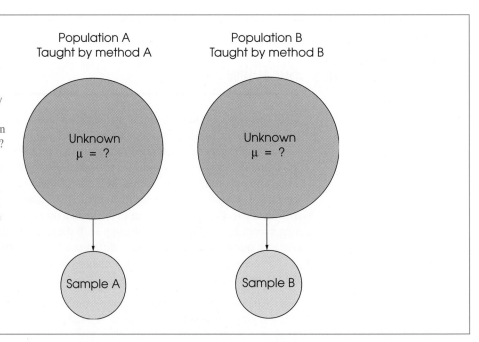

FIGURE 10.2

Do the achievement scores for children taught by method A differ from the scores for children taught by method B? In statistical terms, are the two population means the same or different? Because neither of the two population means is known, it will be necessary to take two samples, one from each population. The first sample will provide information about the mean for the first population, and the second sample will provide information about the second population.

Population A
Taught by method A

Unknown
$\mu = ?$

Sample A

Population B
Taught by method B

Unknown
$\mu = ?$

Sample B

10.2 THE *t* STATISTIC FOR AN INDEPENDENT-MEASURES RESEARCH DESIGN

Because an independent-measures study involves two separate samples, we will need some special notation to help specify which data go with which sample. This notation involves the use of subscripts, which are small numbers written beside a sample statistic. For example, the number of scores in the first sample would be identified by n_1; for the second sample, the number of scores would be n_2. The sample means would be identified by M_1 and M_2. The sums of squares would be SS_1 and SS_2.

THE HYPOTHESES FOR AN INDEPENDENT-MEASURES TEST

The goal of an independent-measures research study is to evaluate the mean difference between two populations (or between two treatment conditions). Using subscripts to differentiate the two populations, the mean for the first population is μ_1, and the second population mean is μ_2. The difference between means is simply $\mu_1 - \mu_2$. As always, the null hypothesis states that there is no change, no effect, or, in this case, no difference. Thus, in symbols, the null hypothesis for the independent-measures test is

$$H_0: \quad \mu_1 - \mu_2 = 0 \qquad \text{(No difference between the population means)}$$

You should notice that the null hypothesis could also be stated as $\mu_1 = \mu_2$. However, the first version of H_0 produces a specific numerical value (zero) that will be used in the calculation of the *t* statistic. Therefore, we prefer to phrase the null hypothesis in terms of the difference between the two population means.

The alternative hypothesis states that there is a mean difference between the two populations,

$$H_1: \quad \mu_1 - \mu_2 \neq 0 \qquad \text{(There is a mean difference.)}$$

Equivalently, the alternative hypothesis can simply state that the two population means are not equal: $\mu_1 \neq \mu_2$.

THE FORMULAS FOR AN INDEPENDENT-MEASURES HYPOTHESIS TEST

The independent-measures hypothesis test will be based on another *t* statistic. The formula for this new *t* statistic will have the same general structure as the *t* statistic formula that was introduced in Chapter 9. To help distinguish between the two *t* formulas, we will refer to the original formula (Chapter 9) as the *single-sample t statistic* and we will refer to the new formula as the *independent-measures t statistic*. Because the new independent-measures *t* will include data from two separate samples and hypotheses about two populations, the formulas may appear to be a bit overpowering. However, the new formulas will be easier to understand if you view them in relation to the single-sample *t* formulas from Chapter 9. In particular, there are two points to remember:

1. The basic structure of the *t* statistic is the same for both the independent-measures and the single-sample hypothesis tests. In both cases,

$$t = \frac{\text{sample statistic} - \text{hypothesized population parameter}}{\text{estimated standard error}}$$

2. The independent-measures *t* is basically a *two-sample t that doubles all the elements of the single-sample t formulas.*

To demonstrate the second point, we will examine the two *t* formulas piece by piece.

The overall *t* formula   The single-sample *t* uses one sample mean to test a hypothesis about one population mean. The sample mean and the population mean appear in the numerator of the *t* formula, which measures how much difference there is between the sample data and the population hypothesis.

$$t = \frac{\text{sample mean} - \text{population mean}}{\text{estimated standard error}} = \frac{M - \mu}{s_M}$$

The independent-measures *t* uses the difference between *two* sample means to evaluate the difference between *two* population means. Thus, the independent-measures *t* formula is

$$t = \frac{\text{sample mean difference} - \text{population mean difference}}{\text{estimated standard error}} = \frac{(M_1 - M_2) - (\mu_1 - \mu_2)}{s_{(M_1 - M_2)}}$$

In this formula, the value of $M_1 - M_2$ is obtained from the sample data and the value for $\mu_1 - \mu_2$ comes from the null hypothesis. Finally, note that the null hypothesis sets the population mean difference equal to zero, so the independent measures *t* formula can be simplified further,

$$t = \frac{\text{sample mean difference}}{\text{estimated standard error}}$$

In this form, the *t* statistic is a simple ratio comparing the actual mean difference (numerator) with the difference that is expected by chance (denominator).

The standard error   In each of the *t*-score formulas, the standard error in the denominator measures how accurately the sample statistic represents the population parameter. In the single-sample *t* formula, the standard error measures the amount of error expected for a sample mean and is represented by the symbol s_M. For the independent-measures *t* formula, the standard error measures the amount of error that is expected when you use a sample mean difference $(M_1 - M_2)$ to represent a population mean difference $(\mu_1 - \mu_2)$. The standard error for the sample mean difference is represented by the symbol $s_{(M_1 - M_2)}$.

Caution: Do not let the notation for standard error confuse you. In general, standard error measures how accurately a statistic represents a parameter. The symbol for standard error takes the form $s_{\text{statistic}}$. When the statistic is a sample mean, M, the symbol for standard error is s_M. For the independent-measures test, the statistic is a sample mean difference $(M_1 - M_2)$, and the symbol for standard error is $s_{(M_1 - M_2)}$. In each case, the standard error tells how much discrepancy is reasonable to expect just by chance between the statistic and the corresponding population parameter.

To develop the formula for $s_{(M_1 - M_2)}$ we will consider the following three points:

1. Each of the two sample means represents it own population mean, but in each case there is some error.

   M_1 approximates μ_1 with some error

   M_2 approximates μ_2 with some error

   Thus, there are two sources of error.

2. The amount of error associated with each sample mean can be measured by computing the standard error of *M*. In Chapter 9 (Equation 9.1), we calculated the standard error for a single sample mean as

$$s_M = \sqrt{\frac{s^2}{n}}$$

3. For the independent-measures *t* statistic, we want to know the total amount of error involved in using *two* sample means to approximate *two* population means. To do this, we will find the error from each sample separately and then add the two errors together. The resulting formula for standard error is

$$s_{(M_1 - M_2)} = \sqrt{\frac{s_1^2}{n_1} + \frac{s_2^2}{n_2}} \tag{10.1}$$

Because the independent-measures *t* statistic uses two sample means, the formula for the estimated standard error simply combines the error for the first sample mean and the error for the second sample mean (see Box 10.1 on page 314).

POOLED VARIANCE Although Equation 10.1 accurately presents the concept of standard error for the independent-measures *t* statistic, this formula is limited to situations where the two samples are exactly the same size (that is, $n_1 = n_2$). In situations where the two sample sizes are different, the formula is *biased* and, therefore, inappropriate. The bias comes from the fact that Equation 10.1 treats the two sample variances equally. However, when the sample sizes are different, the two sample variances are not equally good and should not be treated equally. In Chapter 7, we introduced the law of large numbers, which states that statistics obtained from large samples tend to be better (more accurate) estimates of population parameters than statistics obtained from small samples. This same fact holds for sample variances: The variance obtained from a large sample will be a more accurate estimate of σ^2 than the variance obtained from a small sample.

To correct for the bias in the sample variances, the independent-measures *t* statistic will combine the two sample variances into a single value called the *pooled variance*. The pooled variance is obtained by averaging or "pooling" the two sample variances using a procedure that allows the bigger sample to carry more weight in determining the final value.

You should recall that when there is only one sample, the sample variance is computed as

$$s^2 = \frac{SS}{df}$$

For the independent-measures *t* statistic, there are two *SS* values and two *df* values (one from each sample). The values from the two samples are combined to compute what is called the *pooled variance*. The pooled variance is identified by the symbol s_p^2 and is computed as

$$s_p^2 = \frac{SS_1 + SS_2}{df_1 + df_2} \tag{10.2}$$

BOX

10.1   THE VARIABILITY OF DIFFERENCE SCORES

It may seem odd that the independent-measures *t* statistic *adds* together the two sample errors when it *subtracts* to find the difference between the two sample means. The logic behind this apparently unusual procedure is demonstrated here.

We begin with two populations, I and II (see Figure 10.3). The scores in population I range from a high of 70 to a low of 50. The scores in population II range from 30 to 20. We will use the range as a measure of how spread out (variable) each population is:

For population I, the scores cover a range of 20 points.

For population II, the scores cover a range of 10 points.

If we randomly select a score from population I and a score from population II and compute the difference between these two scores ($X_1 - X_2$), what range of values is possible for these differences? To answer this question, we need to find the biggest possible difference

and the smallest possible difference. Look at Figure 10.3; the biggest difference occurs when $X_1 = 70$ and $X_2 = 20$. This is a difference of $X_1 - X_2 = 50$ points. The smallest difference occurs when $X_1 = 50$ and $X_2 = 30$. This is a difference of $X_1 - X_2 = 20$ points. Notice that the differences go from a high of 50 to a low of 20. This is a range of 30 points:

range for population I (X_1 scores) = 20 points

range for population II (X_2 scores) = 10 points

range for the differences ($X_1 - X_2$) = 30 points

The variability for the difference scores is found by *adding* together the variabilities for the two populations.

In the independent-measures *t* statistics, we are computing the variability (standard error) for a sample mean difference. To compute this value, we add together the variability for each of the two sample means.

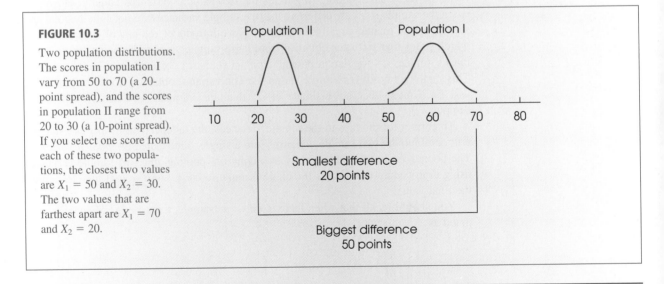

FIGURE 10.3

Two population distributions. The scores in population I vary from 50 to 70 (a 20-point spread), and the scores in population II range from 20 to 30 (a 10-point spread). If you select one score from each of these two populations, the closest two values are $X_1 = 50$ and $X_2 = 30$. The two values that are farthest apart are $X_1 = 70$ and $X_2 = 20$.

With one sample, variance is computed as *SS* divided by *df*. With two samples, the two *SS*s are divided by the two *df*s to compute pooled variance. The pooled variance is actually an average of the two sample variances, and the value of the pooled variance will always be located between the two sample variances. A more detailed discussion of the pooled variance is presented in Box 10.2.

> **BOX**
>
> **10.2   A CLOSER LOOK AT THE POOLED VARIANCE**

The pooled variance is actually an average of the two sample variances, but the formula is structured so that each sample is weighted by the magnitude of its *df* value. Thus, the larger sample carries more weight in determining the final value. The following examples demonstrate this point.

Equal Sample Size   We begin with two samples that are exactly the same size. The first sample has $n = 6$ scores with $SS = 50$, and the second sample has $n = 6$ scores with $SS = 30$. Individually, the two sample variances are

$$\text{Variance for sample 1:} \quad s^2 = \frac{SS}{df} = \frac{50}{5} = 10$$

$$\text{Variance for sample 2:} \quad s^2 = \frac{SS}{df} = \frac{30}{5} = 6$$

The pooled variance for these two samples is

$$s_p^2 = \frac{SS_1 + SS_2}{df_1 + df_2} = \frac{50 + 30}{5 + 5} = \frac{80}{10} = 8.00$$

Note that the pooled variance is exactly halfway between the two sample variances. Because the two samples are exactly the same size, the pooled variance is simply the average of the two individual sample variances.

Unequal Samples Sizes   Now consider what happens when the samples are not the same size. This time the first sample has $n = 3$ scores with $SS = 20$, and the second sample has $n = 9$ scores with $SS = 48$. Individually, the two sample variances are

$$\text{Variance for sample 1:} \quad s^2 = \frac{SS}{df} = \frac{20}{2} = 10$$

$$\text{Variance for sample 2:} \quad s^2 = \frac{SS}{df} = \frac{48}{8} = 6$$

The pooled variance for these two samples is

$$s_p^2 = \frac{SS_1 + SS_2}{df_1 + df_2} = \frac{20 + 48}{2 + 8} = \frac{68}{10} = 6.80$$

This time the pooled variance is not located halfway between the two sample variances. Instead, the pooled value is closer to the variance for the large sample ($s^2 = 6$) than to the variance for the small sample ($s^2 = 10$). The larger sample carries more weight when the pooled variance is computed.

Using the pooled variance in place of the individual sample variances, we can now obtain an unbiased measure of the standard error for a sample mean difference. The resulting formula for the independent-measures standard error is

$$\text{two-sample standard error} = s_{(M_1 - M_2)} = \sqrt{\frac{s_p^2}{n_1} + \frac{s_p^2}{n_2}} \tag{10.3}$$

Conceptually, this standard error measures how accurately the difference between two sample means represents the difference between the two population means. The formula combines the error for the first sample mean with the error for the second sample mean. Notice that the pooled variance from the two samples is used to compute the standard error for the two samples.

THE FINAL FORMULA AND DEGREES OF FREEDOM

The complete formula for the independent-measures t statistic is as follows:

$$t = \frac{(M_1 - M_2) - (\mu_1 - \mu_2)}{s_{(M_1 - M_2)}}$$

$$= \frac{\text{sample mean difference} - \text{population mean difference}}{\text{estimated standard error}} \quad (10.4)$$

In the formula, the estimated standard error in the denominator is calculated using Equation 10.3, and requires calculation of the pooled variance using Equation 10.2.

The degrees of freedom for the independent-measures t statistic are determined by the df values for the two separate samples:

$$df \text{ for the } t \text{ statistic} = df \text{ for the first sample} + df \text{ for the second sample}$$

$$= df_1 + df_2$$

$$= (n_1 - 1) + (n_2 - 1) \quad (10.5)$$

The independent-measures t statistic will be used for hypothesis testing. Specifically, we will use the difference between sample means $(M_1 - M_2)$ as the basis for testing hypotheses about the difference between population means $(\mu_1 - \mu_2)$. In this context, the overall structure of the t statistic can be reduced to the following:

$$t = \frac{\text{data} - \text{hypothesis}}{\text{error}}$$

This same structure is used for both the single-sample t from Chapter 9 and the new independent-measures t that was introduced in the preceding pages. Table 10.1 identifies each component of these two t statistics and should help reinforce the point that we made earlier in the chapter; that is, the independent-measures t statistic simply doubles each aspect of the single-sample t statistic.

TABLE 10.1

The basic elements of a t statistic for the single-sample t and the independent-measures t

	SAMPLE DATA	HYPOTHESIZED POPULATION PARAMETER	ESTIMATED STANDARD ERROR	SAMPLE VARIANCE
Single-sample t statistic	M	μ	$\sqrt{\dfrac{s^2}{n}}$	$s^2 = \dfrac{SS}{df}$
Independent-measures t statistic	$(M_1 - M_2)$	$(\mu_1 - \mu_2)$	$\sqrt{\dfrac{s_p^2}{n_1} + \dfrac{s_p^2}{n_2}}$	$s_p^2 = \dfrac{SS_1 + SS_2}{df_1 + df_2}$

LEARNING CHECK

1. Describe the general characteristics of an independent-measures research study.

2. Identify the two sources of error that are reflected in the standard error for the independent-measures t statistic.

3. An independent-measures t statistic is used to evaluate the mean difference between two treatments using a sample of $n = 10$ in one treatment and a separate sample of $n = 15$ in the other treatment. What is the df value for the t test?

4. The first sample from an independent-measures research study has $n = 5$ scores with $SS = 40$. The second sample has $n = 5$ scores with $SS = 32$.

  a. Compute the variance for each of the two samples.

  b. Compute the pooled variance for the two samples. (Because the two samples are exactly the same size, you should find that the pooled variance is equal to the average of the two sample variances.)

5. The first sample from an independent-measures research study has $n = 5$ scores with $SS = 40$. The second sample has $n = 7$ scores with $SS = 48$.

  a. Compute the variance for each of the two samples.

  b. Compute the pooled variance for the two samples. (Because the two samples are not the same size, you should find that the pooled variance is not halfway between the two sample variances. Because the larger sample carries more weight, the pooled variance is closer to the variance from the larger sample.)

ANSWERS   **1.** An independent-measures study uses a separate sample to represent each of the treament conditions or populations being compared.

2. The two sources of error come from the fact that two sample means are being used to represent two population means. Thus, M_1 represents its population mean with some error and M_2 represents its population mean with some error. The two errors combine in the standard error for the independent-measures *t* statistic.

3. The *t* statistic has $df = df_1 + df_2 = 9 + 14 = 23$.

4. a. The first sample has a variance of 10 and the second sample has a variance of 8.

  b. The pooled variance is 9 (exactly halfway between 10 and 8).

5. a. The first sample has a variance of 10 and the second sample has a variance of 8.

  b. The pooled variance is 8.8, which is weighted toward the variance for the larger sample.

10.3 HYPOTHESIS TESTS WITH THE INDEPENDENT-MEASURES *t* STATISTIC

The independent-measures *t* statistic uses the data from two separate samples to help decide whether or not there is a significant mean difference between two populations or between two treatment conditions. A complete example of a hypothesis test with two independent samples follows.

EXAMPLE 10.1

Remember that an independent-measures design means there are separate samples for each treatment condition.

In recent years, psychologists have demonstrated repeatedly that using mental images can greatly improve memory. Here we present a hypothetical experiment designed to examine this phenomenon.

The psychologist first prepares a list of 40 pairs of nouns (for example, dog/bicycle, grass/door, lamp/piano). Next, two groups of participants are obtained (two separate samples). Participants in one group are given the list for 5 minutes and instructed to memorize the 40 noun pairs. Participants in another group receive the same list of words, but in addition to the regular instructions, they are told to form a mental image for each pair of nouns (imagine a dog riding a bicycle, for example). Note that the two samples are identical except that one group is using mental images to help learn the list.

Later each group is given a memory test in which they are given the first word from each pair and asked to recall the second word. The psychologist records the number of words correctly recalled for each individual. The data from this experiment are as follows. On the basis of these data, can the psychologist conclude that mental images affected memory?

DATA (NUMBER OF WORDS RECALLED)			
GROUP 1 (IMAGES)		GROUP 2 (NO IMAGES)	
18	31	24	13
19	29	23	17
23	26	16	20
29	21	17	15
30	24	19	26
$n = 10$		$n = 10$	
$M = 25$		$M = 19$	
$SS = 200$		$SS = 160$	

STEP 1 State the hypotheses, and select the alpha level.

Directional hypotheses could be used and would specify whether imagery should increase or decrease recall scores.

H_0: $\mu_1 - \mu_2 = 0$ (No difference; imagery has no effect.)

H_1: $\mu_1 - \mu_2 \neq 0$ (Imagery produces a difference.)

We will set $\alpha = .05$.

STEP 2 This is an independent-measures design. The *t* statistic for these data will have degrees of freedom determined by

$$df = df_1 + df_2$$
$$= (n_1 - 1) + (n_2 - 1)$$
$$= 9 + 9$$
$$= 18$$

The *t* distribution for $df = 18$ is presented in Figure 10.4. For $\alpha = .05$, the critical region consists of the extreme 5% of the distribution and has boundaries of $t = +2.101$ and $t = -2.101$.

FIGURE 10.4

The *t* distribution with $df = 18$. The critical region for $\alpha = .05$ is shown.

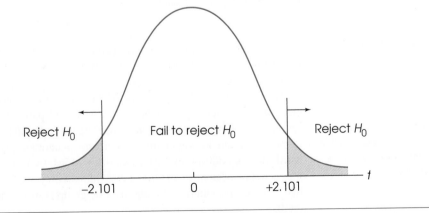

Reject H_0 Fail to reject H_0 Reject H_0

−2.101   0   +2.101   *t*

STEP 3   Obtain the data, and compute the test statistic. The data are as given, so all that remains is to compute the t statistic. Because the independent-measures t formula is relatively complex, the calculations can be simplified by dividing the process into three parts.

Caution: The pooled variance combines the two samples to obtain a single estimate of variance. In the formula, the two samples are combined in a single fraction.

First, find the pooled variance for the two samples:

$$s_p^2 = \frac{SS_1 + SS_2}{df_1 + df_2}$$

$$= \frac{200 + 160}{9 + 9}$$

$$= \frac{360}{18}$$

$$= 20$$

Caution: The standard error adds the errors from two separate samples. In the formula, these two errors are added as two separate fractions. In this case, the two errors are equal because the sample sizes are the same.

Second, use the pooled variance to compute the estimated standard error:

$$s_{(M_1 - M_2)} = \sqrt{\frac{s_p^2}{n_1} + \frac{s_p^2}{n_2}} = \sqrt{\frac{20}{10} + \frac{20}{10}}$$

$$= \sqrt{2 + 2}$$

$$= \sqrt{4}$$

$$= 2$$

Third, compute the t statistic:

$$t = \frac{(M_1 - M_2) - (\mu_1 - \mu_2)}{s_{(M_1 - M_2)}} = \frac{(25 - 19) - 0}{2}$$

$$= \frac{6}{2}$$

$$= 3.00$$

STEP 4   Make a decision. The obtained value ($t = 3.00$) is in the critical region. In this example, the obtained sample mean difference is three times greater than would be expected by chance (the standard error). This result is very unlikely if H_0 is true. Therefore, we reject H_0 and conclude that using mental images produced a significant difference in memory performance. More specifically, the group using images recalled significantly more words than the group with no images.

DESCRIPTIVE STATISTICS AND THE HYPOTHESIS TEST

In addition to understanding the hypothesis test in Example 10.1, you should not lose sight of the descriptive statistics that give you a picture of the data by summarizing the original scores. If you can visualize the sample data, you will often have a better understanding of the results from the hypothesis test. In the following section we will demonstrate the relationship between descriptive statistics and the hypothesis test by reexamining the data from Example 10.1.

In Example 10.1, two separate samples, each with $n = 10$ scores, were used to evaluate two treatment conditions (imagery versus no-imagery). The first sample had a mean of $M = 25$ with $SS = 200$. For this sample the standard deviation is

$$s = \sqrt{\frac{SS}{n-1}} = \sqrt{\frac{200}{9}} = 4.71$$

Using the values for the mean and the standard deviation, you should be able to visualize the sample distribution. For this sample, the scores pile up around $M = 25$ and are scattered about 4 or 5 points in both directions ($s = 4.71$). Thus, most of the scores have values between 21 and 29.

The second sample has a mean of $M = 19$ and $SS = 160$. The sample standard deviation is

$$s = \sqrt{\frac{SS}{n-1}} = \sqrt{\frac{160}{9}} = 4.22$$

For this sample, the distribution is centered around $M = 19$, with most of the scores located within 4 or 5 points of the mean. Figure 10.5 shows the two sample distributions. Notice that the actual distributions match the descriptions we created from the sample means and standard deviations.

If you compare the two sample distributions in Figure 10.5, you should realize that the two samples form two separate and distinct piles of scores: The fact that the two samples form distinct distributions suggests that the samples come from two distinct populations. Recall that the null hypothesis for the independent-measures test states that the two populations are the same, $\mu_1 = \mu_2$. If this hypothesis is true, you would expect

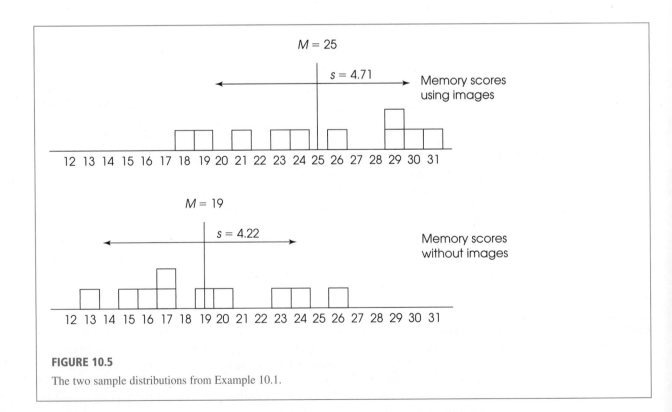

FIGURE 10.5

The two sample distributions from Example 10.1.

the two samples to be mixed together, forming a single pile centered around the common population mean. The alternative hypothesis, on the other hand, states that the population means are different, so you should expect to obtain two different and distinct samples. When the hypothesis test was conducted for these data (Example 10.1) the decision was to reject the null hypothesis and conclude that the two population means are different. This is exactly the same decision that you would make from simply looking at the sample data. Specifically, the two samples look different and appear to come from different populations.

MEASURING EFFECT SIZE FOR THE INDEPENDENT-MEASURES t

As noted in Chapters 8 and 9, a hypothesis test is usually accompanied by a report of effect size to provide an indication of the absolute magnitude of the treatment effect. One technique for measuring effect size is Cohen's d, which produces a standardized measure of mean difference. In its general form, Cohen's d is defined as

$$d = \frac{\text{mean difference}}{\text{standard deviation}}$$

In the context of an independent-measures research study, the mean difference is simply the difference between sample means ($M_1 - M_2$), and the standard deviation (or pooled standard deviation) is obtained by taking the square root of the pooled variance. Thus,

$$d = \frac{M_1 - M_2}{\sqrt{s_p^2}} \tag{10.6}$$

For the data from Example 10.1, the two sample means are 25 and 19, and the pooled variance is 20. Cohen's d for these data is

$$d = \frac{M_1 - M_2}{\sqrt{s_p^2}} = \frac{25 - 19}{\sqrt{20}} = \frac{6}{4.47} = 1.34$$

This value is perfectly consistent with the picture of the data in Figure 10.5. Notice that the distance between the two sample means is slightly more than 1 standard deviation, indicating that Cohen's d should be slightly larger than 1.00.

The independent-measures t hypothesis test also allows for measuring effect size by computing the percentage of variance accounted for, r^2. As we saw in Chapter 9, r^2 measures how much of the variability in the scores can be explained by the treatment effects. For example, the participants in the imagery study (Example 10.1) have different memory scores. In statistical terms, their scores are variable. However, some of the variability can be explained by knowing which group the scores are in; people in the imagery condition tended to have higher scores than people in the no-imagery condition. By measuring exactly how much of the variability can be explained, we can obtain a measure of how big the treatment effect actually is. The calculation of r^2 for the independent-measures t is exactly the same as it was for the single-sample t in Chapter 9:

$$r^2 = \frac{t^2}{t^2 + df} \tag{10.7}$$

For the data in Example 10.1 (the imagery study), we obtained $t = 3.00$ with $df = 18$. These values produce an r^2 of

$$r^2 = \frac{3^2}{3^2 + 18} = \frac{9}{9 + 18} = \frac{9}{27} = 0.333$$

For this example, 0.333 (or 33.3%) of the variance in the memory scores is determined by the imagery versus no-imagery conditions. That is, the memory scores obtained in the research study are partially determined by the two treatment conditions (images versus no images), and the part that is determined amounts to 33.3% (1/3) of the variance in the scores.

Although the value of r^2 is usually obtained by using Equation 10.7, it is possible to determine the percentage of variability directly by computing *SS* values for the set of scores. The following example demonstrates this process using the data from the imagery study in Example 10.1 (see Figure 10.5).

EXAMPLE 10.2

The memory study described in Example 10.1 compared memory scores for two groups of participants; one group that was instructed to use images and another group that received no imagery instructions. If we assume that the null hypothesis is true and there is no difference between the two treatments (imagery and no-imagery), then there should be no treatment effect to cause a mean difference between the two samples. In this case, the two samples can be combined to form a single set of $n = 20$ scores with an overall mean of $M = 22$. The two samples are shown as a single distribution in Figure 10.6(a).

FIGURE 10.6

The two groups of scores from Example 10.1 (Figure 10.5) combined into a single distribution. The original scores, including the treatment effect, are shown in part (a). Part (b) shows the adjusted scores, after the treatment effect has been removed.

For this example, however, there is a treatment effect that causes the imagery group to have higher memory scores than the no-imagery group. Specifically, the imagery group has a mean score of $M = 25$, which is three points higher than the overall mean of $M = 22$. Similarly, the no-imagery group has a mean score of $M = 19$, which is three points lower than the overall mean. Thus, the treatment effect causes the imagery scores to move away from $M = 22$ toward the right, and the no-imagery scores to move away from $M = 22$ toward the left. Notice that the imagery scores are clustered on the right side of the distribution in Figure 10.6(a), and the no-imagery scores are clustered on the left. Thus, the treatment effect has shifted the scores away from the mean and therefore increased the variability.

To determine how much the treatment effect contributes to the variability, we will remove the treatment effect and examine the resulting scores. To eliminate the treatment effect, we adjust the scores by adding 3 points to each score in the no-imagery group and subtracting 3 points from each score in the imagery group. After this adjustment, both groups have a mean of $M = 22$, and there is no mean difference between the two treatments. The adjusted scores are shown in Figure 10.6(b).

It should be clear that the adjusted scores in Figure 10.6(b) are less variable (more closely clustered) than the original scores in Figure 10.6(a). That is, removing the treatment effect reduces the variability. To determine exactly how much the treatment influences variability, we will compute *SS*, the sum of squared deviations, for each set of scores. The calculations are shown in Table 10.2. For the original set of scores, including the treatment effect, the total variability is $SS = 540$. When the treatment effect is removed, the variability is reduced to $SS = 360$. The difference between

TABLE 10.2

Calculation of *SS* for the two sets of scores shown in Figure 10.6. The calculations in the left-hand columns are for the original scores, including the treatment effect. The values in the right-hand columns are for the adjusted scores, after the treatment effect is removed.

	SCORE	CALCULATION OF *SS* INCLUDING THE TREATMENT EFFECT — DEVIATION FROM $M = 22$	SQUARED DEVIATION	ADJUSTED SCORE	CALCULATION OF *SS* AFTER THE TREATMENT EFFECT IS REMOVED — DEVIATION FROM $M = 22$	SQUARED DEVIATION
	13	−9	81	13 + 3 = 16	−6	36
	15	−7	49	15 + 3 = 18	−4	16
	16	−6	36	16 + 3 = 19	−3	9
No	17	−5	25	17 + 3 = 20	−2	4
Imagery	17	−5	25	17 + 3 = 20	−2	4
Scores	19	−3	9	19 + 3 = 22	0	0
	20	−2	4	20 + 3 = 23	1	1
	23	1	1	23 + 3 = 26	4	16
	24	2	4	24 + 3 = 27	5	25
	26	4	16	26 + 3 = 29	7	49
	18	−4	16	18 − 3 = 15	−7	49
	19	−3	9	19 − 3 = 16	−6	36
	21	−1	1	21 − 3 = 18	−4	16
	23	1	1	23 − 3 = 20	−2	4
Imagery	24	2	4	24 − 3 = 21	−1	1
Scores	26	4	16	26 − 3 = 23	1	1
	29	7	49	29 − 3 = 26	4	16
	29	7	49	29 − 3 = 26	4	16
	30	8	64	30 − 3 = 27	5	25
	31	9	81	31 − 3 = 28	6	36
			$SS = 540$			$SS = 360$

these two values, 180 points, is the amount of variability that is explained by the treatment effect. When expressed as a proportion of the total variability, we obtain

$$\frac{\text{variability explained by the treatment}}{\text{total variability}} = \frac{180}{540} = 0.333 = 33.3\%$$

You should recognize that this is exactly the same value we obtained for r^2 using Equation 10.7.

IN THE LITERATURE:
REPORTING THE RESULTS OF AN INDEPENDENT-MEASURES *t* TEST

In Chapter 4 (page 123), we demonstrated how the mean and the standard deviation are reported in APA format. In Chapter 9 (page 296), we illustrated the APA style for reporting the results of a *t* test. Now we will use the APA format to report the results of Example 10.1, an independent-measures *t* test. A concise statement might read as follows:

> The group using mental images recalled more words ($M = 25$, $SD = 4.71$) than the group that did not use mental images ($M = 19$, $SD = 4.22$). This difference was significant, $t(18) = 3.00$, $p < .05$, $d = 1.34$.

You should note that standard deviation is not a step in the computations for the independent-measures *t* test, yet it is useful when providing descriptive statistics for each treatment group. It is easily computed when doing the *t* test because you need *SS* and *df* for both groups to determine the pooled variance. Note that the format for reporting *t* is exactly the same as that described in Chapter 9 (page 296) and that the measure of effect size is reported immediately after the results of the hypothesis test.

Also, as we noted in Chapter 9, if an exact probability is available from a computer analysis, it should be reported. For the data in Example 10.1, the computer analysis reports a probability value of $p = 0.008$ for $t = 3.00$ with $df = 18$. In the research report, this value would be included as follows:

> The difference was significant, $t(18) = 3.00$, $p = .008$, $d = 1.34$.   ❏

DIRECTIONAL HYPOTHESES AND ONE-TAILED TESTS

When planning an independent-measures experiment, a researcher usually has some expectation or specific prediction for the outcome. For the memory experiment described in Example 10.1, the psychologist clearly expects the group using images to have higher memory scores than the group without images. This kind of directional prediction can be incorporated into the statement of the hypotheses, resulting in a directional, or one-tailed, test. Recall from Chapter 8 that one-tailed tests can be less conservative than two-tailed tests, and should be used when clearly justified by theory or previous findings. The following example demonstrates the procedure for stating hypotheses and locating the critical region for a one-tailed test using the independent-measures *t* statistic.

EXAMPLE 10.3

We will use the same experimental situation that was described in Example 10.1. The researcher is using an independent-measures design to examine the effect of mental images on memory. The prediction is that the imagery group will have higher memory scores.

STEP 1   State the hypotheses and select the alpha level. As always, the null hypothesis says that there is no effect, and the alternative hypothesis says that there is an effect. For this example, the *predicted* effect is that images will produce higher scores. Thus, the two hypotheses are:

$$H_0: \quad \mu_{images} \leq \mu_{no\ images} \qquad \text{(Imagery scores are not higher.)}$$

$$H_1: \quad \mu_{images} > \mu_{no\ images} \qquad \text{(Imagery scores are higher, as predicted.)}$$

Note that it is usually easier to state the hypotheses in words before you try to write them in symbols. Also, it usually is easier to begin with the alternative hypothesis (H_1), which states that the treatment works as predicted. Then the null hypothesis is just the opposite of H_1. Also note that the equal sign goes in the null hypothesis, indicating *no difference* between the two treatment conditions. The idea of zero difference is the essence of the null hypothesis, and the numerical value of zero will be used for $(\mu_1 - \mu_2)$ during the calculation of the t statistic.

STEP 2   Locate the critical region. For a directional test, the critical region will be located entirely in one tail of the distribution. Rather than trying to determine which tail, positive or negative, is the correct location, we suggest you identify the criteria for the critical region in a two-step process as follows. First, look at the data and determine whether or not the sample mean difference is in the direction that was predicted. If the answer is no, then the data obviously do not support the predicted treatment effect, and you can stop the analysis. On the other hand, if the difference is in the predicted direction, then the second step is to determine whether the difference is large enough to be significant. To test for significance, simply find the one-tailed critical value in the t distribution table. If the sample t statistic is more extreme (either positive or negative) than the critical value, then the difference is significant.

For this example, the imagery group scored 6 points higher, as predicted. With $df = 18$, the one-tailed critical value for $\alpha = .05$ is $t = 1.734$.

STEP 3   Collect the data and calculate the test statistic. The details of the calculations were shown in Example 10.1. The data produce a t statistic of $t = 3.00$.

STEP 4   Make a decision. The t statistic of $t = 3.00$ is well beyond the critical boundary of $t = 1.734$. Therefore, we reject the null hypothesis and conclude that scores in the imagery condition are significantly higher than scores in the no-imagery condition, as predicted.

LEARNING CHECK   1. A researcher report states that there is a significant difference between treatments for an independent-measures design with $t(28) = 2.27$.

  a. How many individuals participated in the research study? (*Hint:* Start with the df value.)

  b. Should the report state that $p > .05$ or $p < .05$?

2. A developmental psychologist would like to examine the difference in verbal skills for 6-year-old girls compared with 6-year-old boys. A sample of $n = 5$ girls and $n = 5$ boys is obtained, and each child is given a standardized verbal abilities test.

The data are as follows:

GIRLS	BOYS
$M = 83$	$M = 71$
$SS = 200$	$SS = 120$

Are these data sufficient to indicate a significant difference in verbal skills for 6-year-old girls and 6-year-old boys? Test at the .05 level of significance.

3. Suppose that the psychologist in the previous question predicted that girls have higher verbal scores at age 5. What value of *t* would be necessary to conclude that girls score significantly higher using a one-tailed test with $\alpha = .01$?

4. Calculate the effect size for the data in Question 2. Compute Cohen's *d* and r^2.

ANSWERS
1. **a.** The $df = 28$, so the total number of participants is 30.
   b. A significant result is indicated by $p < .05$.
2. Pooled variance $= 40$, the standard error $= 4$, and $t = 3.00$. With $df = 8$, the decision is to reject the null hypothesis.
3. With $df = 8$ and $\alpha = .01$ the critical value is $t = 2.896$.
4. For these data, $d = 1.90$ (a very large effect size), and $r^2 = 0.53$ (or 53%).

10.4 FACTORS INFLUENCING THE INDEPENDENT-MEASURES *t* AND EFFECT SIZE

The independent-measures (or between-subjects) research design uses two separate samples to evaluate the mean difference between two populations or between two treatment conditions. The sample data can be analyzed with a hypothesis test that evaluates the *significance* of the mean difference—that is, whether the mean difference between the two samples is greater than would be expected simply by chance. In addition, we can measure the *size* of the treatment effect using either Cohen's *d* or r^2. In each case, the outcome of the statistical analysis is influenced by three factors:

1. The size of the mean difference between the two samples
2. The magnitude of the sample variance
3. The number of scores in each sample

We will use the data in Figure 10.7 to examine each of the three factors. Figure 10.7 shows the results from a research study comparing two treatments. Notice that the study uses two separate samples, each with $n = 9$, and there is a 5-point mean difference between the two samples: $M = 8$ for treatment 1 and $M = 13$ for treatment 2. Also notice that there is no overlap between the two distributions; all of the scores for treatment 2 are higher than all of the scores for treatment 1. In this situation, there is a very clear and obvious difference between the two treatments.

For the hypothesis test, the data produce a pooled variance of 1.50, an estimated standard error of 0.58. The t statistic is

$$t = \frac{\text{mean difference}}{\text{estimated standard error}} = \frac{5}{0.58} = 8.62$$

With $df = 16$, this value is far into the critical region (for $\alpha = .05$ or $\alpha = .01$), so we reject the null hypothesis and conclude that there is a significant difference between the two treatments.

To measure effect size, these data produce a Cohen's d of $5/1.22 = 4.10$, which is a very large effect, and a percentage of variance accounted for equal to

$$r^2 = \frac{t^2}{t^2 + df} = \frac{(8.62)^2}{(8.62)^2 + 16} = \frac{74.30}{90.30} = 0.82 \quad \text{(or 82\%)}$$

By all measures these data show a very large and significant effect.

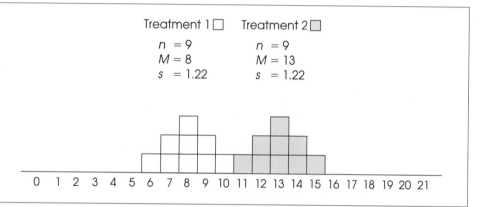

FIGURE 10.7

Two sample distributions representing two different treatments. These data show a significant difference between treatments, $t(16) = 8.62$, $p < .01$, and both measures of effect size indicate a very large treatment effect, $d = 4.10$ and $r^2 = 0.82$.

THE EFFECT OF THE MEAN DIFFERENCE BETWEEN SAMPLES

The size of the sample mean difference influences the hypothesis test and the measures of effect size in a very predictable way. In particular, a big difference between the two samples increases the likelihood of a significant effect and increases the likelihood of a large effect size. The following example demonstrates the role of sample mean difference.

EXAMPLE 10.4

Subtracting a constant does not affect the variance or standard deviation.

We will begin with the data in Figure 10.7 and consider what happens if the mean difference between samples is reduced. For example, suppose that the distribution for treatment 2 is moved to the left until its center is located at $M = 9$. The move produces considerable overlap between the two distributions and reduces the mean difference between samples to only 1 point. Notice that this change does not affect the sample variance or the standard error.

After moving the distributions closer together, the hypothesis test produces a t statistic of

$$t = \frac{\text{mean difference}}{\text{estimated standard error}} = \frac{1}{0.58} = 1.72$$

With $df = 16$, this value is not in the critical region for $\alpha = .05$, so we fail to reject the null hypothesis and conclude that there is no signficant difference between the treatments. With a mean difference of only 1 point, Cohen's *d* becomes $1/1.22 = 0.82$, which is substantially reduced from its former value. Finally, the measure of percentage of variance accounted for becomes

$$r^2 = \frac{t^2}{t^2 + df} = \frac{(1.72)^2}{(1.72)^2 + 16} = \frac{2.96}{18.96} = 0.16 \quad \text{(or 16\%)}$$

In general, the size of the sample mean difference has exactly the effect that would be expected. A smaller mean difference decreases the likelihood of finding significance in a hypothesis test, and reduces the measures of effect size.

THE EFFECT OF SAMPLE VARIANCE   In Chapters 4 and 9, we made the point that large values for sample variance can make it difficult to see any significant patterns in the results from a research study. The same holds true for the independent-measures *t*. The following example demonstrates how large variance can lower the likelihood of statistical significance and reduce effect size.

E X A M P L E   1 0 . 5   Figure 10.8 shows another set of data from a research study comparing two treatments. To create these scores, we simply increased the variability for the two samples shown in Figure 10.7. Notice that there are still $n = 9$ scores in each sample, and the two sample means are still $M = 8$ and $M = 13$. However, the sample variances have been greatly increased: Each sample now has $s^2 = 44.25$ as compared with $s^2 = 1.5$ for the data in Figure 10.7. Notice that the increased variance means that the scores are now spread out over a wider range, with the result that the two samples are mixed together without any clear distinction between them.

The absence of a clear difference between the two samples is supported by the hypothesis test. The pooled variance is 44.25, the estimated standard error is 3.14, and the independent-measures *t* statistic is

$$t = \frac{\text{mean difference}}{\text{estimated standard error}} = \frac{5}{3.14} = 1.59$$

With $df = 16$ and $= .05$, this value is not in the critical region. Therefore, we fail to reject the null hypothesis and conclude that there is no significant difference between the two treatments. Although there is still a 5-point difference between sample means (as in Figure 10.7), the 5-point difference is not significant with the increased variance.

The increased variance also affects measures of effect size. For these data, the pooled standard deviation (square root of pooled variance) is $\sqrt{44.25} = 6.65$. With a mean difference of 5 points, Cohen's $d = 5/6.65$ and $a = 0.75$. This is still a large effect, but substantially less than the value we obtained with smaller variance in Example 10.7. Finally, the measure of r^2 produces

$$r^2 = \frac{t^2}{t^2 + df} = \frac{(1.59)^2}{(1.59)^2 + 16} = \frac{2.53}{18.53} = 0.14 \quad \text{(or 14\%)}$$

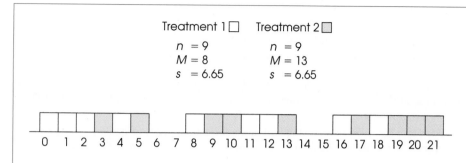

FIGURE 10.8

Two sample distributions representing two different treatments. These data show exactly the same mean difference as the scores in Figure 10.7; however, the variance has been greatly increased. With the increased variance, there is no longer a significant difference between treatments, $t(16) = 1.59$, $p > .05$, and both measures of effect size are substantially reduced, $d = 0.75$ and $r^2 = 0.14$.

Although the data in Figure 10.8 show a 5-point mean difference between treatment (just as in Figure 10.7), the 5-point difference is

a. not easy to see in the figure

b. not significant as determined by the hypothesis test

c. a much smaller effect as measured by Cohen's *d* or by r^2

In general, increasing sample variance decreases the likelihood of obtaining a significant difference and reduces measures of effect size.

THE EFFECT OF SAMPLE SIZE

Although hypothesis tests and measures of effect size are both statistical techniques for evaluating treatment effects, they serve very different purposes. The hypothesis test focuses on the significance of the effect, determining whether or not the effect is more than can be reasonably explained by chance. Measures of effect size, on the other hand, are concerned with the size of the treatment effect. This difference is demonstrated when you consider how sample size influences these two statistical methods. In simple terms, a treatment effect with a large sample is more convincing than the same effect with a small sample. Thus, sample size has a large influence on statistical significance. However, a 5-point treatment effect is still only 5 points whether it is demonstrated with a large sample or a small sample. Thus, sample size has little or no influence on measures of effect size. The influence of sample size is demonstrated in the following example.

EXAMPLE 10.6

Once again, we consider the data that were used in Example 10.5 (Figure 10.8). These data do not produce a significant difference ($t = 1.59$, $p > .05$), and the effect size was measured as $d = 0.75$ and $r^2 = 0.14$. In Example 10.5, however, each of the samples had only $n = 9$ scores. Suppose the same data were obtained with $n = 100$ in each sample. That is, suppose that the sample means are still $M = 8$ and $M = 13$, and the two sample variances are still $s^2 = 44.25$. With these values and $n = 100$ for each sample, the pooled variance is 44.25, and the estimated standard error is

$$\sqrt{\frac{44.25}{100} + \frac{44.25}{100}} = \sqrt{0.44 + 0.44} = \sqrt{0.88} = 0.94$$

The *t* statistic is

$$t = \frac{\text{mean difference}}{\text{estimated standard error}} = \frac{5}{0.94} = 5.32$$

With $df = 16$, this value is far into the critical region (for $\alpha = .05$ or $\alpha = .01$), so we reject the null hypothesis and conclude that there is a significant difference between the two treatments.

For these data, the pooled standard deviation (square root of pooled variance is $\sqrt{44.25} = 6.65$. With a mean difference of 5 points, Cohen's $d = 5/6.65 = 0.75$. Finally, the measure of r^2 produces

$$r^2 = \frac{t^2}{t^2 + df} = \frac{(5.32)^2}{(5.32)^2 + 198} = \frac{28.30}{226.30} = 0.125 \quad \text{(or 12.5\%)}$$

Increasing the sample size from $n = 9$ to $n = 100$ changed the outcome of the hypothesis test from not significant (in Example 10.5) to significant. However, the increase in sample size had little or no effect on the measures of effect size. The value of r^2 changes only slightly from 14% (in Example 10.5) to 12.5%. As we have noted in earlier examples, Cohen's d does not consider sample size, so the value of d is the same whether $n = 9$ or $n = 100$.

In general, a mean difference obtained with large samples is more convincing than a mean difference obtained with small samples. Look once again at Figure 10.8; the 5-point mean difference is not easy to see and is not statistically significant. However, the samples in the figure contain only $n = 9$ scores. If the same 5-point difference is maintained as the sample size grows to $n = 100$, then it becomes a more convincing difference and ultimately a significant difference as we found in Example 10.6. Thus, tests of statistical significance are influenced by sample size, but measures of effect size (such as r^2 and Cohen's d) are not.

10.5 ASSUMPTIONS UNDERLYING THE INDEPENDENT-MEASURES *t* FORMULA

There are three assumptions that should be satisfied before you use the independent-measures *t* formula for hypothesis testing:

1. The observations within each sample must be independent (see page 251).

2. The two populations from which the samples are selected must be normal.

3. The two populations from which the samples are selected must have equal variances.

The first two assumptions should be familiar from the single-sample *t* hypothesis test presented in Chapter 9. As before, the normality assumption is the less important of the two, especially with large samples. When there is reason to suspect that the populations are far from normal, you should compensate by ensuring that the samples are relatively large.

The third assumption is referred to as *homogeneity of variance* and states that the two populations being compared must have the same variance. You may recall a similar assumption for both the *z*-score and the single-sample *t*. For those tests, we assumed

Remember: Adding a constant to (or subtracting a constant from) each score does not change the standard deviation.

that the effect of the treatment was to add a constant amount to (or subtract a constant amount from) each individual score. As a result, the population standard deviation after treatment was the same as it had been before treatment. We now are making essentially the same assumption but phrasing it in terms of variances.

Recall that the pooled variance in the *t*-statistic formula is obtained by averaging together the two sample variances. It makes sense to average these two values only if they both are estimating the same population variance—that is, if the homogeneity of variance assumption is satisfied. If the two sample variances represent different population variances, then the average is meaningless. (*Note:* There is no meaning to the value obtained by averaging two unrelated numbers. For example, what is the significance of the number obtained by averaging your shoe size and the last two digits of your social security number?)

The importance of the homogeneity assumption increases when there is a large discrepancy between the sample sizes. With equal (or nearly equal) sample sizes, this assumption is less critical, but still important.

The homogeneity of variance assumption is quite important because violating this assumption can negate any meaningful interpretation of the data from an independent-measures experiment. Specifically, when you compute the *t* statistic in a hypothesis test, all the numbers in the formula come from the data except for the population mean difference, which you get from H_0. Thus, you are sure of all the numbers in the formula except one. If you obtain an extreme result for the *t* statistic (a value in the critical region), you conclude that the hypothesized value was wrong. But consider what happens when you violate the homogeneity of variance assumption. In this case, you have two questionable values in the formula (the hypothesized population value and the meaningless average of the two variances). Now if you obtain an extreme *t* statistic, you do not know which of these two values is responsible. Specifically, you cannot reject the hypothesis because it may have been the pooled variance that produced the extreme *t* statistic. Without satisfying the homogeneity of variance requirement, you cannot accurately interpret a *t* statistic, and the hypothesis test becomes meaningless.

How do you know whether or not the homogeneity of variance assumption is satisfied? One simple test involves just looking at the two sample variances. Logically, if the two population variances are equal, then the two sample variances should be very similar. When the two sample variances are reasonably close, you can be reasonably confident that the homogeneity assumption has been satisfied and proceed with the test. However, if one sample variance is more than three or four times larger than the other, then there is reason for concern. A more objective procedure involves a statistical test to evaluate the homogeneity assumption. Although there are many different statistical methods for determining whether or not the homogeneity of variance assumption has been satisfied, Hartley's *F*-max test is one of the simplest to compute and to understand. An additional advantage is that this test can also be used to check homogeneity of variance with more than two independent samples. Later, in Chapter 13, we will examine statistical methods for comparing several different samples, and Hartley's test will be useful again.

The following example demonstrates the *F*-max test for two independent samples.

EXAMPLE 10.7

The *F*-max test is based on the principle that a sample variance provides an unbiased estimate of the population variance. Therefore, if the population variances are the same, the sample variances should be very similar. The procedure for using the *F*-max test is as follows:

1. Compute the sample variance, $s^2 = SS/df$, for each of the separate samples.

2. Select the largest and the smallest of these sample variances and compute

$$F\text{-max} = \frac{s^2(\text{largest})}{s^2(\text{smallest})}$$

A relatively large value for *F*-max indicates a large difference between the sample variances. In this case, the data suggest that the population variances are different and that the homogeneity assumption has been violated. On the other hand, a small value of *F*-max (near 1.00) indicates that the sample variances are similar and that the homogeneity assumption is reasonable.

3. The *F*-max value computed for the sample data is compared with the critical value found in Table B.3 (Appendix B). If the sample value is larger than the table value, then you conclude that the variances are different and that the homogeneity assumption is not valid.

To locate the critical value in the table, you need to know

   a. k = number of separate samples. (For the independent-measures *t* test, $k = 2$.)

   b. $df = n - 1$ for each sample variance. The Hartley test assumes that all samples are the same size.

   c. The alpha level. The table provides critical values for $\alpha = .05$ and $\alpha = .01$. Generally a test for homogeneity would use the larger alpha level.

Example: Two independent samples each have $n = 10$. The sample variances are 12.34 and 9.15. For these data,

$$F\text{-max} = \frac{s^2(\text{largest})}{s^2(\text{smallest})} = \frac{12.34}{9.15} = 1.35$$

With $\alpha = .05$, $k = 2$, and $df = n - 1 = 9$, the critical value from the table is 4.03. Because the obtained *F*-max is smaller than this critical value, you conclude that the data do not provide evidence that the homogeneity-of-variance assumption has been violated.

SUMMARY

1. The independent-measures *t* statistic uses the data from two separate samples to draw inferences about the mean difference between two populations or between two different treatment conditions.

2. The formula for the independent-measures *t* statistic has the same structure as the original *z*-score or the single-sample *t*:

$$t = \frac{\text{sample statistic} - \text{population parameter}}{\text{estimated standard error}}$$

For the independent-measures *t*, the sample statistic is the sample mean difference ($M_1 - M_2$). The population parameter is the population mean difference, ($\mu_1 - \mu_2$). The estimated standard error for the sample mean difference is computed by combining the errors for the two sample means. The resulting formula is

$$t = \frac{(M_1 - M_2) - (\mu_1 - \mu_2)}{s_{(M_1 - M_2)}}$$

where the estimated standard error is

$$s_{(M_1 - M_2)} = \sqrt{\frac{s_p^2}{n_1} + \frac{s_p^2}{n_2}}$$

The pooled variance in the formula, s_p^2, is the weighted mean of the two sample variances:

$$s_p^2 = \frac{SS_1 + SS_2}{df_1 + df_2}$$

This *t* statistic has degrees of freedom determined by the sum of the *df* values for the two samples:

$$df = df_1 + df_2$$
$$= (n_1 - 1) + (n_2 - 1)$$

3. For hypothesis testing, the null hypothesis normally states that there is no difference between the two population means:

$$H_0\text{: } \mu_1 = \mu_2 \quad \text{or} \quad \mu_1 - \mu_2 = 0$$

4. When a hypothesis test with an independent-measures t statistic indicates a significant difference, it is recommended that you also compute a measure of the effect size. One measure of effect size is Cohen's d, which is a standardized measured of the mean difference. For the independent-measures t statistic,

$$d = \frac{M_1 - M_2}{\sqrt{s_p^2}}$$

A second common measure of effect size is the percentage of variance accounted for by the treatment effect. This measure is identified by r^2 and is computed as

$$r^2 = \frac{t^2}{t^2 + df}$$

5. Appropriate use and interpretation of the t statistic require that the data satisfy the homogeneity of variance assumption. This assumption stipulates that the two populations have equal variances. An informal test of this assumption can be made by simply comparing the two sample variances: If the two sample variances are approximately equal, the t test is justified. Hartley's F-max test provides a statistical technique for determining whether or not the data satisfy the homogeneity assumption.

RESOURCES

WebTUTOR

If you are using WebTutor along with this book, there is a WebTutor section corresponding to this chapter. The WebTutor contains a brief summary of Chapter 10, hints for learning the concepts and the formulas for the independent-measures t test, cautions about common errors, and sample exam items including solutions.

The statistics computer package SPSS can be used to conduct an independent-measures t hypothesis test. First, the data are entered in what is called a *stacked* format, which means that all the scores from *both samples* are entered in one column of the data matrix. Values are then entered into a second column to designate the sample or treatment condition corresponding to each of the scores. For example, enter all of the scores in the first column. In the second column, enter a 1 beside each score from sample #1 and enter a 2 beside each score from sample #2. Then, click **Analyze** on the tool bar, select **Compare Means**, and click on **Independent-Samples T Test**. Enter the label for the column containing the scores into the Test Variable box. Then click in the Group Variable box and enter the label for the column that contains the group number for each score. Next, click on **Define Groups**. Assuming that you used a 1 to identify group #1 and a 2 to identify group #2, enter the values 1 and 2 into the appropriate group boxes, then click **Continue** and click **OK**. SPSS will produce a summary table showing the number of scores, the mean, the standard deviation, and the standard error for each of the two samples. SPSS also conducts a test for homogeneity of variance, using Levene's test instead of the F-max test. Remember that this test should *not* be significant so you want the reported Sig. value to be greater than .05. Next, the results of the hypothesis test are presented using two different assumptions; we will focus on the top row where equal variances are assumed. The test results include the calculated *t* value, the degrees of freedom, the level of significance (probability of a Type I error), and the size of the mean difference. Finally, the output includes a report of the standard error for the mean difference and a 95% confidence interval that provides a range of values estimating how much difference exists between the two treatment conditions. (Confidence intervals will be presented in Chapter 12.)

KEY TERMS

independent-measures
   research design

between-subjects research
   design

pooled variance

homogeneity of variance

FOCUS ON PROBLEM SOLVING

1. As you learn more about different statistical methods, one basic problem will be deciding which method is appropriate for a particular set of data. Fortunately, it is easy to identify situations in which the independent-measures *t* statistic is used. First, the data will always consist of two separate samples (two *ns*, two *Ms*, two *SS*s, and so on). Second, this *t* statistic is always used to answer questions about a mean difference: On the average, is one group different (better, faster, smarter) than the other group? If you examine the data and identify the type of question that a researcher is asking, you should be able to decide whether or not an independent-measures *t* is appropriate.

2. When computing an independent-measures *t* statistic from sample data, we suggest that you routinely divide the formula into separate stages rather than trying to do all the calculations at once. First, find the pooled variance. Second, compute the standard error. Third, compute the *t* statistic.

3. One of the most common errors for students involves confusing the formulas for pooled variance and standard error. When computing pooled variance, you are "pooling" the two samples together into a single variance. This variance is computed as a *single fraction*, with two *SS* values in the numerator and two *df* values in the denominator. When computing the standard error, you are adding the error from the first sample and the error from the second sample. These two separate errors add as *two separate fractions* under the square root symbol.

DEMONSTRATION 10.1

THE INDEPENDENT-MEASURES *t* TEST

In a study of jury behavior, two samples of participants were provided details about a trial in which the defendant was obviously guilty. Although group 2 received the same details as group 1, the second group was also told that some evidence had been withheld from the jury by the judge. Later the participants were asked to recommend a jail sentence. The length of term suggested by each participant is presented here. Is there a significant difference between the two groups in their responses?

Group 1 scores: 4   4   3   2   5   1   1   4

Group 2 scores: 3   7   8   5   4   7   6   8

There are two separate samples in this study. Therefore, the analysis will use the independent-measures t test.

STEP 1   State the hypothesis, and select an alpha level.

H_0: $\mu_1 - \mu_2 = 0$ (For the population, knowing evidence has been withheld has no effect on the suggested sentence.)

H_1: $\mu_1 - \mu_2 \neq 0$ (For the population, knowledge of withheld evidence has an effect on the jury's response.)

We will set the level of significance to $\alpha = .05$, two tails.

STEP 2   Identify the critical region.

For the independent-measures t statistic, degrees of freedom are determined by

$$df = n_1 + n_2 - 2$$
$$= 8 + 8 - 2$$
$$= 14$$

The t distribution table is consulted, for a two-tailed test with $\alpha = .05$ and $df = 14$. The critical t values are $+2.145$ and -2.145.

STEP 3   Compute the test statistic.

We are computing an independent-measures t statistic. To do this, we will need the mean and SS for each sample, pooled variance, and estimated standard error.

Sample means and sums of squares   The means (M) and sums of squares (SS) for the samples are computed as follows:

SAMPLE 1		SAMPLE 2	
X	X^2	X	X^2
4	16	3	9
4	16	7	49
3	9	8	64
2	4	5	25
5	25	4	16
1	1	7	49
1	1	6	36
4	16	8	64
$\Sigma X = 24$	$\Sigma X^2 = 88$	$\Sigma X = 48$	$\Sigma X^2 = 312$

$$n_1 = 8 \qquad\qquad n_2 = 8$$

$$M_1 = \frac{\Sigma X}{n} = \frac{24}{8} = 3 \qquad M_2 = \frac{\Sigma X}{n} = \frac{48}{8} = 6$$

$$SS_1 = \Sigma X^2 - \frac{(\Sigma X)^2}{n} \qquad SS_2 = \Sigma X^2 - \frac{(\Sigma X)^2}{n}$$

$$= 88 - \frac{(24)^2}{8} \qquad\qquad = 312 - \frac{(48)^2}{8}$$

$$= 88 - \frac{576}{8} \qquad\qquad = 312 - \frac{2304}{8}$$

$$= 88 - 72 \qquad\qquad = 312 - 288$$

$$SS_1 = 16 \qquad\qquad SS_2 = 24$$

Pooled variance For these data, the pooled variance equals

$$s_p^2 = \frac{SS_1 + SS_2}{df_1 + df_2} = \frac{16 + 24}{7 + 7} = \frac{40}{14} = 2.86$$

Estimated standard error Now we can calculate the estimated standard error for mean differences.

$$s_{(M_1 - M_2)} = \sqrt{\frac{s_p^2}{n_1} + \frac{s_p^2}{n_2}} = \sqrt{\frac{2.86}{8} + \frac{2.86}{8}} = \sqrt{0.358 + 0.358}$$

$$= \sqrt{0.716} = 0.85$$

The t statistic Finally, the *t* statistic can be computed.

$$t = \frac{(M_1 - M_2) - (\mu_1 - \mu_2)}{s_{(M_1 - M_2)}} = \frac{(3 - 6) - 0}{0.85} = \frac{-3}{0.85}$$

$$= -3.53$$

STEP 4   Make a decision about H_0, and state a conclusion.

The obtained *t* value of -3.53 falls in the critical region of the left tail (critical $t = \pm2.145$). Therefore, the null hypothesis is rejected. The participants that were informed about the withheld evidence gave significantly longer sentences, $t(14) = -3.53$, $p < .05$, two tails.

DEMONSTRATION 10.2

EFFECT SIZE FOR THE INDEPENDENT-MEASURES *t*

We will compute Cohen's *d* and r^2 for the jury decision data in Demonstration 10.1. For these data, the two sample means are $M_1 = 3$ and $M_2 = 6$, and the pooled variance is 2.86. Therefore, Cohen's *d* is

$$d = \frac{M_1 - M_2}{\sqrt{s_p^2}} = \frac{3 - 6}{\sqrt{2.86}} = \frac{3}{1.69} = 1.78$$

With a t value of $t = 3.53$ and $df = 14$, the percentage of variance accounted for is

$$r^2 = \frac{t^2}{t^2 + df} = \frac{(3.53)^2}{(3.53)^2 + 14} = \frac{12.46}{26.46} = 0.47 \quad \text{(or 47\%)}$$

PROBLEMS

1. Describe the general characteristics of a research study for which an independent-measures t would be the appropriate test statistic.

2. What is measured by the estimated standard error that is used for the independent-measures t statistic?

3. Describe the homogeneity of variance assumption, and explain why it is important for the independent-measures hypothesis test.

4. What happens to the value of the independent-measures t statistic as the difference between the two sample means increases? What happens to the t value as the variability of the scores in the two samples increases?

5. One sample has $n = 8$ scores with $SS = 56$, and a second sample has $n = 10$ scores with $SS = 108$.
   a. Find the variance for each sample.
   b. Find the pooled variance for the two samples.

6. One sample has $SS = 63$, and a second sample has $SS = 45$.
   a. Assuming that $n = 10$ for both samples, find each of the sample variances, and calculate the pooled variance. You should find that the pooled variance is exactly halfway between the two sample variances.
   b. Now assume that $n = 10$ for the first sample and $n = 16$ for the second. Find each of the sample variances, and calculate the pooled variance. You should find that the pooled variance is closer to the variance for the larger sample ($n = 16$).

7. One sample has $SS = 32$ and a second sample has $SS = 48$.
   a. Calculate the variance for each sample and the pooled variance, assuming that $n = 5$ for both samples. You should find that the pooled variance is halfway between the two sample variances.
   b. Now assume that the first sample has $n = 9$ scores and the second has $n = 4$. Calculate the variance for each sample and the pooled variance. This time you should find that the pooled variance is closer to the value for the larger sample.

8. In the independent-measures hypothesis test, the difference between two sample means is used to test a hypothesis about the difference between two population means. The sample mean difference is not expected to be exactly equal to the population mean difference, and the estimated standard error measures how much discrepancy is reasonable to expect on average. For each of the following situations, compute the estimated standard error for the sample mean difference.
   a. First sample has $n = 8$ and $SS = 416$; second sample has $n = 8$ and $SS = 480$.
   b. First sample has $n = 8$ and $SS = 170$; second sample has $n = 4$ and $SS = 70$.

9. One sample has $n = 15$ with $SS = 1660$, and a second sample has $n = 15$ with $SS = 1700$.
   a. Find the pooled variance for the two samples.
   b. Find the standard error for the sample mean difference.
   c. If the sample mean difference is 8 points, is this enough to reject the null hypothesis and conclude that there is a significant difference at the .05 level?
   d. If the sample mean difference is 12 points, is this enough to indicate a significant difference at the .05 level?
   e. Calculate the percentage of variance accounted for (r^2) to measure the effect size for an 8-point mean difference and for a 12-point mean difference.

10. One sample has $n = 6$ scores with $SS = 500$ and a second sample has $n = 9$ with $SS = 670$.
   a. Calculate the pooled variance for the two samples.
   b. Calculate the estimated standard error for the sample mean difference.
   c. If the sample mean difference is 10 points, is this enough to reject the null hypothesis for an independent-measures hypothesis test using $\alpha = .05$? Assume a two-tailed test.
   d. If the sample mean difference is 15 points, is this enough to reject the null hypothesis for an independent-measures hypothesis test using $\alpha = .05$? Assume a two-tailed test.
   e. Calculate Cohen's d to measure the effect size for a 10-point mean difference and for a 15-point mean difference.

11. A researcher is interested in determining whether difficulties experienced during birth have any influence on the eventual development of mother–child relationships. By examining hospital records from 15 years ago, the researcher identifies 12 women who had relatively easy deliveries and another 12 women who had relatively difficult deliveries. All 24 women gave birth to daughters. The researcher then locates the 24 children (now 15 years old) and gives each one a questionnaire measuring the quality of the relationship between mother and daughter. The data for the two samples are as follows:

EASY DELIVERY	DIFFICULT DELIVERY
$M = 47$	$M = 52$
$SS = 946$	$SS = 1166$

Do these data provide evidence for a significant connection between mother–daughter relationships and difficulty during childbirth? Use a two-tailed test with a .05 level of significance.

12. A researcher using an independent-measures design to compare two treatments obtained the following data:

TREATMENT 1	TREATMENT 2
$n = 4$	$n = 4$
$M = 58$	$M = 52$
$SS = 84$	$SS = 108$

a. Calculate the variance for each of the two samples and then compute the pooled variance. (You should find that the pooled variance is exactly halfway between the two sample variances.)

b. Do these data indicate a significant difference between the two treatments? Use a two-tailed test with $\alpha = .05$.

13. The researcher from the previous problem attempted to repeat the experiment using larger samples. The data for the second attempt are as follows:

TREATMENT 1	TREATMENT 2
$n = 16$	$n = 16$
$M = 58$	$M = 52$
$SS = 420$	$SS = 540$

a. Calculate the variance for each of the two samples and then compute the pooled variance.

b. Do these data indicate a significant difference between the two treatments? Use a two-tailed test with $\alpha = .05$.

c. You should find that these two samples have exactly the same means and variances as the two samples in Problem 12, and the pooled variance is the same for both sets of data. However, in this problem the sample data were sufficient to reject the null hypothesis, but in the previous problem the decision was to fail to reject H_0. Explain why the two sets of data lead to different decisions.

14. The data in Problems 12 and 13 produce the same pooled variance (32) and the same sample mean difference (6 points). However, the two problems use different-sized samples and result in different values for t ($t = 1.50$ with $df = 6$ for Problem 12, and $t = 3.00$ with $df = 30$ for Problem 13).

a. With this information, compute the effect size using Cohen's d (standardized mean difference) for both problems.

b. Compute effect size using r^2 (percentage of variance accounted for) for both problems.

c. You should find that one measure of effect size is influenced by sample size and the other is not. Explain how the sample size is included in one formula but not in the other.

15. The animal learning course in the psychology department requires that each student train a rat to perform certain behaviors. The student's grade is partially determined by the rat's performance. The instructor for this course has noticed that some students are very comfortable working with the rats and seem to be very successful training their rats. The instructor suspects that these students may have previous experience with pets that gives them an advantage in the class. To test this hypothesis, the instructor gives the entire class a questionnaire at the beginning of the course. One question determines whether or not each student currently has a pet of any type at home. Based on the responses to this question, the instructor divides the class into two groups and compares the rats' learning scores for the two groups. The data are as follows:

RATS' SCORES FOR STUDENTS WITH PETS	RATS' SCORES FOR STUDENTS WITHOUT PETS
$n = 10$	$n = 15$
$M = 78$	$M = 66$
$SS = 1420$	$SS = 2030$

a. Are these data sufficient to indicate a significant difference between groups? Use a two-tailed test with $\alpha = .05$.

b. Use Cohen's d to measure the effect size for this study.

16. Friedman and Rosenman (1974) have classified people into two categories: Type A personalities and Type B personalities. Type As are hard-driving, competitive, and ambitious. Type Bs are more relaxed, easy-going people. One factor that differentiates these two groups is the chronically high level of frustration experienced by Type As. To demonstrate this phenomenon, separate samples of Type As and Type Bs are obtained, with $n = 8$ in each sample. The individual participants are all given a frustration inventory measuring level of frustration. The average score for the Type As is $M = 84$ with $SS = 740$, and the Type Bs average $M = 71$ with $SS = 660$. Do these data indicate a significant difference between the two groups? Test at the .01 level of significance.

17. In a classic study of problem solving, Duncker (1945) asked participants to mount a candle on a wall in an upright position so that it would burn normally. One group was given a candle, a book of matches, and a box of tacks. A second group was given the same items, except the tacks and the box were presented separately as two distinct items. The solution to this problem involves using the tacks to mount the box on the wall, which creates a shelf for the candle. Duncker reasoned that the first group of participants would have trouble seeing a "new" function for the box (a shelf) because it was already serving a function (holding tacks). For each participant, the amount of time to solve the problem was recorded. Data similar to Duncker's are as follows:

TIME TO SOLVE PROBLEM (IN SEC.)	
BOX OF TACKS	TACKS AND BOX SEPARATE
128	42
160	24
113	68
101	35
94	47

Do these data indicate a significant difference between the two conditions? Test at the .01 level of significance.

18. Siegel (1990) found that elderly people who owned dogs were less likely to pay visits to their doctors after upsetting events than were those who did not own pets. Similarly, consider the following hypothetical data. A sample of elderly dog owners is compared to a similar group (in terms of age and health) who do not own dogs. The researcher records the number of visits to the doctor during the past year for each person. The data are as follows:

CONTROL GROUP	DOG OWNERS
12	8
10	5
6	9
9	4
15	6
12	
14	

a. Is there a significant difference in the number of doctor visits between dog owners and control subjects? Use a two-tailed test with $\alpha = .05$.
b. Calculate the value of r^2 (percentage of variance accounted for) for these data.

19. A researcher reports an independent-measures t statistic of $t = 2.53$ with $df = 24$.
a. How many individuals participated in the researcher's experiment?
b. Can the researcher conclude that there is a significant difference between the two samples with $\alpha = .05$, two tails?
c. Can the researcher conclude that the mean difference is significant at the .01 level, two tails?

20. Consider the following data from an independent measures study:

TREATMENT 1	TREATMENT 2
3	12
5	10
7	8
1	14

a. Compute the mean for each sample.
b. Compute SS for each sample, pooled variance, and estimated standard error.
c. Is there evidence for a treatment effect? Test with $\alpha = .05$, two tails.

21. Consider the following data from an independent measures study:

TREATMENT 1	TREATMENT 2
0	12
4	10
0	4
12	18

a. Compute the mean for each sample and compare them to the means for treatments 1 and 2 of Problem 20.

b. Compute *SS* for each sample, pooled variance, and estimated standard error. Compare these results to part (b) of Problem 20.

c. Is there evidence for a treamtent effect? Test with $\alpha = .05$, two tails.

d. Compare part (c) of this problem with part (c) of the previous problem. Why is there a different outcome?

22. Consider the following data from an independent measures study:

TREATMENT 1	TREATMENT 2
3	7
5	5
7	3
1	9

a. Compute the mean for each sample. Compare these results to those of Problem 20.

b. Compute *SS* for each sample, pooled variance, and estimated standard error. Compare these results to those of Problem 20.

c. Is there a statistically significant treatment effect? Test with $\alpha = .05$, two tails.

d. Compare the results of part (c) to those of Problem 20. Why do the outcomes of the studies differ?

23. A researcher studied the effect of feedback on estimation of length. Two samples of participants are given practice estimating the lengths of lines drawn by the researcher on a chalkboard. One group receives no feedback about the accuracy of the estimates. The second group receives feedback ("too long," "too short") for accuracy. Then everyone is tested for accuracy of length estimation. The amount of error, in inches, is measured for all participants. The following table gives the data.

NO FEEDBACK	FEEDBACK
6	3
7	3
4	1
7	5

a. Does feedback have a significant effect on accuracy? Use a two-tailed test with $\alpha = .05$.

b. Calculate Cohen's *d* and r^2 (percentage of variance accounted for) to measure the effect size for this study.

The *t* Test for Two Related Samples

Tools You Will Need

The following items are considered essential background material for this chapter. If you doubt your knowledge of any of these items, you should review the appropriate chapter or section before proceeding.

- Introduction to the *t* statistic (Chapter 9)
  - Estimated standard error
  - Degrees of freedom
  - *t* distribution
  - Hypothesis tests with the *t* statistic
- Independent-measures design (Chapter 10)

Preview

At the Olympic level of competition, even the smallest factors can make the difference between winning and losing. Pelton (1983) reports on one study demonstrating this point. The study recorded the scores obtained by Olympic marksmen when firing *during* heartbeats compared with the scores for shots fired *between* heartbeats. The results show the marksmen had significantly higher scores when they shot between heartbeats. Apparently the small vibrations caused by a heartbeat were enough to disturb the marksmen's aim.

To conduct this study, the researchers monitored each individual's heart during a shooting session (see Figure 11.1). The shots fired during heartbeats were then separated from the shots fired between heartbeats, and two separate scores were computed for each marksman. Thus, the study involved only one sample of participants, but produced two sets of scores. The goal of the study is to look for a mean difference between the two sets of scores.

In the previous chapter, we introduced a statistical procedure for evaluating the mean difference between two sets of data (the independent-measures t statistic). However, the independent-measures t statistic is intended for research situations involving two separate and independent samples. You should realize that the two sets of shooting scores are not independent samples. In fact, for each participant, the score for shots fired during a heartbeat is *directly related*, one to one, to the score for shots fired between heartbeats because both scores come from the same individual. This kind of study is called a repeated-measures (or related-samples) design because the data are obtained by literally repeating measurements, under different conditions, for the same sample. Because the two sets of measurements are related, rather than independent, a different statistical analysis will be required. In this chapter, we will introduce the statistical methods used to evaluate and interpret mean differences for two related samples.

A repeated-measures study (like that of the Olympic shooters) provides an excellent example of a distinction we made in Chapter 1. Specifically, we noted that the term

sample is used in two different ways. First, the word is used to refer to a set of individuals selected to participate in a research study. For this example, the study involved one sample of Olympic marksmen. However, the term *sample* is also used to refer to a set of scores. By this definition, the marksmen produced two samples of scores. To help clarify the distinction between individual subjects and their scores, the generic term *sample* is often modified, so that the group of individuals is called a *subject sample* and the set of scores is called a *statistical sample*. Using this terminology, a repeated-measures study involves one subject sample and produces two statistical samples.

Finally, we should note that independent measures and repeated measures are two basic research strategies that are available whenever a researcher is planning a study that will compare treatment conditions. That is, a researcher may choose to use two separate samples, one for each of the treatments, or a researcher may choose to use one sample and measure each individual in both of the treatment conditions. Later in this chapter, we will take a closer look at the differences between independent-measures and repeated-measures designs and discuss the advantages and disadvantages of each.

FIGURE 11.1

Illustration of the timing of shots taken between or during a heartbeat.

11.1 OVERVIEW

In the previous chapter, we introduced the independent-measures research design as one strategy for comparing two treatment conditions or two populations. The independent-measures design is characterized by the fact that two separate samples are used to obtain the two sets of data that are to be compared. In this chapter, we will examine an alternative research strategy known as a *repeated-measures* research design. With a repeated-measures design, two sets of data are obtained from the same sample of individuals. For example, a group of patients could be measured before therapy and then measured again after therapy. Or a group of individuals could be tested in a quiet, comfortable environment and then tested again in a noisy, stressful environment to examine the effects of stress on performance. In each case, notice that the same variable is being measured twice for the same set of individuals; that is, we are literally repeating measurements on the same sample.

DEFINITION   A **repeated-measures** study is one in which a single sample of individuals is measured more than once on the same dependent variable. The same subjects are used in all of the treatment conditions.

The main advantage of a repeated-measures study is that it uses exactly the same individuals in all treatment conditions. Thus, there is no risk that the participants in one treatment are substantially different from the participants in another. With an independent-measures design, on the other hand, there is always a risk that the results are biased because the individuals in one sample are much different (smarter, faster, more extroverted, and so on) than the individuals in the other sample. At the end of this chapter we will present a more detailed comparison of repeated-measures studies and independent-measures studies, considering the advantages and disadvantages of both types of research.

Occasionally, researchers will try to approximate the advantages of a repeated-measures design by using a technique known as *matched subjects*. A matched-subjects design involves two separate samples, but each individual in one sample is matched one-to-one with an individual in the other sample. Typically, the individuals are matched on one or more variables that are considered to be especially important for the study. For example, a researcher studying verbal learning might want to be certain that the two samples are matched in terms of IQ and gender. In this case, a male participant with an IQ of 120 in one sample would be matched with another male with an IQ of 120 in the other sample. Although the subjects in one sample are not *identical* to the subjects in the other sample, the matched-subjects design at least ensures that the two samples are equivalent (or matched) with respect to some specific variables.

DEFINITION   In a **matched-subjects** study, each individual in one sample is matched with a subject in the other sample. The matching is done so that the two individuals are equivalent (or nearly equivalent) with respect to a specific variable that the researcher would like to control.

Of course, it is possible to match subjects on more than one variable. For example, a researcher could match pairs of subjects on age, gender, race, and IQ. In this case, for example, a 22-year-old white female with an IQ of 115 who was in one sample would be matched with another 22-year-old white female with an IQ of 115 in the second sample. The more variables that are used, however, the more difficult it is to find matching subjects. The goal of the matching process is to simulate a repeated-measures design as closely as possible. In a repeated-measures design, the matching is perfect because the same individual is used in both conditions. In a matched-subjects design, however, the best you can get is a degree of match that is limited to the variable(s) that are used for the matching process.

In a repeated-measures design, or a matched-subjects design, the data consist of two sets of scores (two samples) with the scores in one sample directly related, one-to-one,

A matched-subjects study occasionally is called a *matched-samples design*. But the subjects in the samples must be matched one-to-one before you can use the statistical techniques in this chapter.

with the scores in the second sample. For this reason, the two research designs are statistically equivalent and are grouped together under the common name *related-samples* designs (or correlated-samples designs). In this chapter, we will focus our discussion on repeated-measures designs because they are overwhelmingly the more common example of related-groups designs. However, you should realize that the statistical techniques used for repeated-measures studies can be applied directly to data from a matched-subjects study.

Now we will examine the statistical techniques that allow a researcher to use the sample data from a repeated-measures study to draw inferences about the general population.

11.2   THE *t* STATISTIC FOR RELATED SAMPLES

The *t* statistic for related samples is structurally similar to the other *t* statistics we have examined. As we shall see, it is essentially the same as the single-sample *t* statistic covered in Chapter 9. One major distinction of the related-samples *t* is that it is based on difference scores rather than raw scores (*X* values). In this section, we examine difference scores and develop the *t* statistic for related samples.

DIFFERENCE SCORES: THE DATA FOR A RELATED-SAMPLES STUDY

Table 11.1 presents hypothetical data from a research study examining how reaction time is affected by a common, over-the-counter cold medication. The first score for each person (X_1) is a measurement of reaction time before the medication was administered. The second score (X_2) is a measure of reaction time one half hour after taking the medication. Because we are interested in how the medication affects reaction time, we have computed the difference between the first score and the second score for each individual. The *difference scores*, or *D* values, are shown in the last column of the table. Typically, the difference scores are obtained by subtracting the first score (before treatment) from the second score (after treatment) for each person:

$$\text{difference score} = D = X_2 - X_1 \tag{11.1}$$

Note that the sign of each *D* score tells you the direction of the change. Person A, for example, shows a decrease in reaction time after taking the medication (a negative change), but person B shows an increase (a positive change).

TABLE 11.1

Reaction time measurements taken before and after taking an over-the-counter cold medication.

PERSON	BEFORE MEDICATION (X_1)	AFTER MEDICATION (X_2)	DIFFERENCE D
A	215	210	−5
B	221	242	21
C	196	219	23
D	203	228	25

$$\Sigma D = 64$$

$$M_D = \frac{\Sigma D}{n} = \frac{64}{4} = 16$$

The sample of difference scores (D values) will serve as the sample data for the hypothesis test. To compute the t statistic, we will use the number of D scores (n) as well as the sample mean (M_D) and the value of SS for the sample of D scores.

THE HYPOTHESES FOR A RELATED-SAMPLES TEST

The researcher's goal is to use the sample of difference scores to answer questions about the general population. In particular, the researcher would like to know whether there is any difference between the two treatment conditions for the general population. Note that we are interested in a population of *difference scores*. That is, we would like to know what would happen if every individual in the population was measured in two treatment conditions (X_1 and X_2) and a difference score (D) was computed for everyone. Specifically, we are interested in the mean for the population of difference scores. We will identify this population mean difference with the symbol μ_D (using the subscript letter D to indicate that we are dealing with D values rather than X scores).

As always, the null hypothesis states that for the general population there is no effect, no change, or no difference. For a repeated-measures study, the null hypothesis states that the mean difference for the general population is zero. In symbols,

$$H_0: \quad \mu_D = 0$$

Again, this hypothesis refers to the mean for the entire population of difference scores. According to this hypothesis, it is possible that some individuals will show positive difference scores and some will show negative scores, but the differences are random and unsystematic and will balance out to zero.

The alternative hypothesis states that there is a treatment effect that causes the scores in one treatment condition to be systematically higher (or lower) than the scores in the other condition. In symbols,

$$H_1: \quad \mu_D \neq 0$$

According to H_1, the difference scores for the individuals in the population tend to be consistently positive (or negative), indicating a consistent, predictable difference between the two treatments. See Box 11.1 for further discussion of H_0 and H_1.

THE t STATISTIC FOR RELATED SAMPLES

Figure 11.2 shows the general situation that exists for a repeated-measures hypothesis test. You may recognize that we are facing essentially the same situation that we encountered in Chapter 9. In particular, we have a population for which the mean and the standard deviation are unknown, and we have a sample that will be used to test a hypothesis about the unknown population. In Chapter 9, we introduced a t statistic that allowed us to use the sample mean as a basis for testing hypotheses about the population mean. This t-statistic formula will be used again here to develop the repeated-measures t test. To refresh your memory, the single-sample t statistic (Chapter 9) is defined by the formula

$$t = \frac{M - \mu}{s_M}$$

In this formula, the sample mean, M, is calculated from the data, and the value for the population mean, μ, is obtained from the null hypothesis. The estimated standard error, s_M, is also calculated from the data and provides a measure of how much difference it is reasonable to expect between the sample mean and the population mean.

BOX

11.1   ANALOGIES FOR H_0 AND H_1 IN THE REPEATED-MEASURES TEST

An Analogy for H_0: Intelligence is a fairly stable characteristic; that is, you do not get noticeably smarter or dumber from one day to the next. However, if we gave you an IQ test every day for a week, we probably would get seven different numbers. The day-to-day changes in your IQ scores would probably be small and would be random. Some days your IQ score is slightly higher, and some days it is slightly lower. On average, the day-to-day changes in IQ should balance out to zero. This is the situation that is predicted by the null hypothesis for a repeated-measures test. According to H_0, any changes that occur either for an individual or for a sample are just due to chance, and in the long run, they will average out to zero.

An Analogy for H_1: On the other hand, suppose we evaluate the effects of a fitness training program by

measuring the strength of the muscles in your right arm. We will measure your grip strength every day over a 4-week period. We probably will find small differences in your scores from one day to the next, just as we did with the IQ scores. However, the day-to-day changes in grip strength will not be random. Although your grip strength may decrease occasionally, there should be a general trend toward increased strength as you go through the training program. Thus, most of the day-to-day changes should show an increase. This is the situation that is predicted by the alternative hypothesis for the repeated-measures test. According to H_1, the changes that occur are systematic and predictable and will not average out to zero.

FIGURE 11.2

A sample of $n = 4$ people is selected from the population. Each individual is measured twice, once in treatment I and once in treatment II, and a difference score, D, is computed for each individual. This sample of difference scores is intended to represent the population. Note that we are using a sample of difference scores to represent a population of difference scores. Specifically, we are interested in the mean difference for the general population. The null hypothesis states that for the general population there is no consistent or systematic difference between the two treatments, so the population mean difference is $\mu_D = 0$.

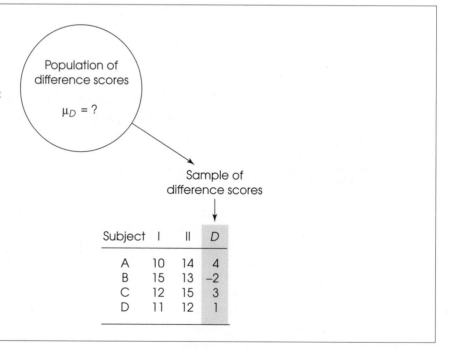

For the repeated-measures design, the sample data are difference scores and are identified by the letter D, rather than X. Therefore, we will use Ds in the formula to emphasize that we are dealing with difference scores instead of X values. Also, the population mean that is of interest to us is the population mean difference (the mean amount of change for the entire population), and we identify this parameter with the symbol

μ_D. With these simple changes, the t formula for the repeated-measures design becomes

$$t = \frac{M_D - \mu_D}{s_{M_D}}$$

(11.2)

In this formula, the estimated standard error, s_{M_D}, is computed in exactly the same way as it is computed for the single-sample t statistic. To calculate the estimated standard error, the first step is to compute the variance (or the standard deviation) for the sample of D scores.

$$s^2 = \frac{SS}{n-1} = \frac{SS}{df} \qquad \text{or} \qquad s = \sqrt{\frac{SS}{df}}$$

The estimated standard error is then computed using the sample variance (or sample standard deviation) and the sample size, n.

$$s_{M_D} = \sqrt{\frac{s^2}{n}} \qquad \text{or} \qquad s_{M_D} = \frac{s}{\sqrt{n}}$$

Notice that all of the calculations are done using the difference scores (the D scores) and that there is only one D score for each subject. With a sample of n subjects, there will be exactly n D scores, and the t statistic will have $df = n - 1$. Remember that n refers to the number of D scores, not the number of X scores in the original data.

You should also note that the repeated-measures t statistic is conceptually similar to the t statistics we have previously examined:

$$t = \frac{\text{sample statistic} - \text{population parameter}}{\text{estimated standard error}}$$

In this case, the sample data are represented by the sample mean of the difference scores (M_D), the population hypothesis is the value predicted by H_0 for μ_D, and the amount of sampling error is measured by the standard error of the sample mean (s_{M_D}).

LEARNING CHECK

1. What characteristics differentiate a repeated-measures study from an independent-measures research study?

2. How are the difference scores (D values) obtained for a repeated-measures hypothesis test?

3. Stated in words and stated in symbols, what is the null hypothesis for a repeated-measures hypothesis test?

ANSWERS

1. A repeated-measures study obtains two sets of data from one sample by measuring each individual twice. An independent-measures study uses two separate samples to obtain two sets of data.

2. The difference score for each subject is obtained by subtracting the subject's first score from the second score. In symbols, $D = X_2 - X_1$.

3. The null hypothesis states that the mean difference for the entire population is zero. That is, if every individual is measured in both treatments and a difference score is computed for each individual, then the average difference score for the entire population will be equal to zero. In symbols, $\mu_D = 0$.

11.3 HYPOTHESIS TESTS FOR THE REPEATED-MEASURES DESIGN

In a repeated-measures study, we are interested in whether or not there is a systematic difference between the scores in the first treatment condition and the scores in the second treatment condition. The hypothesis test will use the difference scores obtained from a sample to evaluate the overall mean difference, μ_D, for the entire population. According to the null hypothesis, there is no consistent or systematic difference between treatments and $\mu_D = 0$. The alternative hypothesis, on the other hand, says that there is a systematic difference and $\mu_D \neq 0$. The purpose of the hypothesis test is to decide between these two options.

The hypothesis test is based on the repeated-measures *t* formula,

$$t = \frac{M_D - \mu_D}{s_{M_D}}$$

The numerator measures the actual difference between the data (M_D) and the hypothesis (μ_D), and the denominator measures the standard difference that is expected by chance. A large value for the *t* statistic (either positive or negative) indicates that the obtained difference is greater than would be expected by chance.

The hypothesis test with the repeated-measures *t* statistic follows the same four-step process that we have used for other tests. The complete hypothesis-testing procedure is demonstrated in Example 11.1.

EXAMPLE 11.1 A researcher in behavioral medicine believes that stress often makes asthma symptoms worse for people who suffer from this respiratory disorder. Because of the suspected role of stress, the investigator decides to examine the effect of relaxation training on the severity of asthma symptoms. A sample of 5 patients is selected for the study. During the week before treatment, the investigator records the severity of their symptoms by measuring how many doses of medication are needed for asthma attacks. Then the patients receive relaxation training. For the week following training, the researcher once again records the number of doses required by each patient. Table 11.2 shows the data and summarizes the findings. Do these data indicate that relaxation training alters the severity of symptoms?

TABLE 11.2

The number of doses of medication needed for asthma attacks before and after relaxation training.

PATIENT	WEEK BEFORE TRAINING	WEEK AFTER TRAINING	D	D²
A	9	4	−5	25
B	4	1	−3	9
C	5	5	0	0
D	4	0	−4	16
E	5	1	−4	16

$$\Sigma D = -16 \qquad \Sigma D^2 = 66$$

$$M_D = \frac{\Sigma D}{n} = \frac{-16}{5} = -3.2$$

$$SS = \Sigma D^2 - \frac{(\Sigma D)^2}{n} = 66 - \frac{(-16)^2}{5}$$

$$= 66 - 51.2 = 14.8$$

Because the data consist of difference scores (*D* values) instead of *X* scores, the formulas for the mean and *SS* also use *D* in place of *X*. However, you should recognize the computational formula for *SS* that was introduced in Chapter 4.

STEP 1   State the hypotheses, and select the alpha level.

$$H_0: \quad \mu_D = 0 \quad \text{(There is no change in symptoms.)}$$

$$H_1: \quad \mu_D \neq 0 \quad \text{(There is a change.)}$$

The level of significance is set at $\alpha = .05$ for a two-tailed test.

STEP 2   Locate the critical region. For this example, $n = 5$, so the t statistic will have $df = n - 1 = 4$. From the t-distribution table, you should find that the critical values are $+2.776$ and -2.776. These values are shown in Figure 11.3.

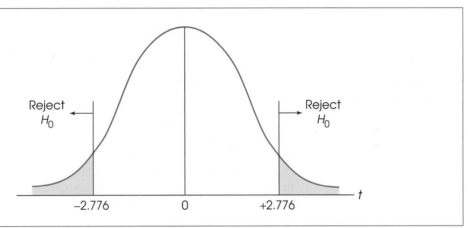

FIGURE 11.3

The critical regions with $\alpha = .05$ and $df = 4$ begin at $+2.776$ and -2.776 in the t distribution. Obtained values of t that are more extreme than these values will lie in a critical region. In that case, the null hypothesis would be rejected.

STEP 3   Calculate the t statistic. Table 11.2 shows the sample data and the calculations for $M_D = -3.2$ and $SS = 14.8$. As we have done with the other t statistics, we will present the computation of the t statistic as a three-step process.

First, compute the variance for the sample. Remember that the population variance (σ^2) is unknown, and we must use the sample value in its place.

$$s^2 = \frac{SS}{n-1} = \frac{14.8}{4} = 3.7$$

Next, use the sample variance to compute the estimated standard error.

$$s_{M_D} = \sqrt{\frac{s^2}{n}} = \sqrt{\frac{3.7}{5}} = \sqrt{0.74} = 0.86$$

Finally, use the sample mean (M_D) and the hypothesized population mean (μ_D) along with the estimated standard error to compute the value for the t statistic.

$$t = \frac{M_D - \mu_D}{s_{M_D}} = \frac{-3.2 - 0}{.86} = -3.72$$

STEP 4   Make a decision. The t value we obtained falls in the critical region (see Figure 11.3). The investigator rejects the null hypothesis and concludes that relaxation training does affect the amount of medication needed to control the asthma symptoms.

MEASURING EFFECT SIZE FOR THE REPEATED-MEASURES t

As we noted with other hypothesis tests, whenever a treatment effect is found to be statistically significant, it is recommended that you also report a measure of the absolute magnitude of the effect. The two commonly used measures of effect size are Cohen's d and r^2, the percentage of variance accounted for. Using the data from Example 11.1, we will demonstrate how these two measures are used to calculate effect size.

Cohen's d provides a standardized measure of the mean difference between treatments. The standardized process simply divides the actual mean difference by the standard deviation. For the repeated-measures t, the mean difference is simply M_D, and the standard deviation is simply the square root of the sample variance. Thus, Cohen's d is computed as follows:

$$\text{Cohen's } d = \frac{\text{mean difference}}{\text{standard deviation}} = \frac{M_D}{s} \qquad (11.3)$$

Because we are measuring the size of the effect and not the direction, we have ignored the minus sign for the sample mean difference.

For the repeated-measures study presented in Example 11.1, the data produce

$$d = \frac{M_D}{s} = \frac{3.20}{\sqrt{3.7}} = \frac{3.20}{1.92} = 1.67$$

You should recall that any value greater than 0.80 is considered to be a large effect and these data are clearly in that category.

Percentage of variance is computed using the obtained t value and the df value from the hypothesis test, exactly as was done for the independent-measures t (see page 321). For the data in Example 11.1, we obtain

$$r^2 = \frac{t^2}{t^2 + df} = \frac{(-3.72)^2}{(-3.72)^2 + 4} = \frac{13.84}{13.84 + 4} = 0.776$$

For these data, nearly 78% of the variance in the medication scores is accounted for by the difference after training compared with before training.

DESCRIPTIVE STATISTICS AND THE HYPOTHESIS TEST

Often, a close look at the sample data from a research study will make it easier to see the size of the treatment effect and to understand the outcome of the hypothesis test. In Example 11.1, we obtained a sample of $n = 5$ individuals who produce a mean difference of $M_D = -3.20$ and a variance of $s^2 = 3.7$. The standard deviation for the sample is

$$s = \sqrt{s^2} = \sqrt{3.7} = 1.92$$

The sample mean and standard deviation describe a set of scores centered at $M_D = -3.20$ with most of the scores located in a range from -2 to -5. Although you should be able to visualize the sample distribution based on the values for M_D and s, the actual set of difference scores is shown in Figure 11.4. In addition to showing the sample scores, we have highlighted the position of 0, which is the value specified in the null hypothesis and the value indicating no treatment effect. Notice that the actual sample is displaced away from zero, which is why we rejected the null hypothesis; the data are not consistent with a population mean of $\mu_D = 0$. In addition, note that the sample mean is more than 1 standard deviation below zero, which is consistent with Cohen's $d = 1.67$. For these data, the picture of the sample distribution (Figure 11.4) should help you to understand the measure of effect size and the outcome of the hypothesis test.

FIGURE 11.4

The sample of difference scores from Example 11.1. The mean is $M_D = -3.20$ and the standard deviation is $s = 1.92$. It does not appear that this sample was obtained from a population with a mean of $\mu_D = 0$.

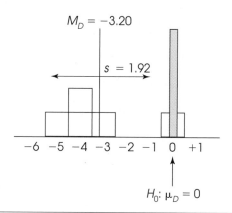

$M_D = -3.20$

$s = 1.92$

−6 −5 −4 −3 −2 −1   0   +1

$H_0: \mu_D = 0$

IN THE LITERATURE:
REPORTING THE RESULTS OF A REPEATED-MEASURES *t* TEST

As we have seen in Chapters 9 and 10, the APA format for reporting the results of *t* tests consists of a concise statement that incorporates the *t* value, degrees of freedom, and alpha level. One typically includes values for means and standard deviations, either in a statement or a table (Chapter 4). For Example 11.1, we observed a mean difference of $M_D = -3.2$ with $s = 1.92$. We obtained a *t* statistic of $t = -3.72$ with $df = 4$, and our decision was to reject the null hypothesis at the .05 level of significance. Also, Cohen's *d* was found to be $d = 1.67$. A published report of this study might summarize the conclusion as follows:

> Relaxation training resulted in a decrease ($M = 3.2$, $SD = 1.92$) in the number of doses of medication needed to control asthma symptoms. This reduction was statistically significant, $t(4) = -3.72, p < .05, d = 1.67$.

In reporting the mean change, we did not bother to include the negative sign because we explicitly noted that the change was a decrease.

If the hypothesis test is conducted on a computer and an exact probability value is included in the printout, then the exact *p* value should be reported. For the data in Example 11.1, the computer analysis reports $t = -3.72$ and $p = 0.020$. With the exact value, the research report would state:

> The reduction was statistically significant, $t(4) = -3.72, p = .020, d = 1.67$.   ❏

DIRECTIONAL HYPOTHESES AND ONE-TAILED TESTS

In many repeated-measures and matched-subjects studies, the researcher has a specific prediction concerning the direction of the treatment effect. For example, in the study described in Example 11.1, the researcher expects relaxation training to reduce the severity of asthma symptoms and therefore to reduce the amount of medication needed for asthma attacks. This kind of directional prediction can be incorporated into the statement of hypotheses, resulting in a directional, or one-tailed, hypothesis test. The following example demonstrates how the hypotheses and critical region are determined in a directional test.

EXAMPLE 11.2   We will reexamine the experiment presented in Example 11.1. The researcher is using a repeated-measures design to investigate the effect of relaxation training on the severity of asthma symptoms. The researcher predicts that people will need less medication after training than before, which will produce negative difference scores.

$$D = X_2 - X_1 = \text{after} - \text{before}$$

STEP 1   State the hypotheses, and select the alpha level. As always, the null hypothesis states that there is no effect. For this example, the researcher predicts that relaxation training will reduce the need for medication. The null hypothesis says it will not. In symbols,

$$H_0: \quad \mu_D \geq 0 \qquad \text{(Medication is not reduced after training.)}$$

The alternative hypothesis says that the treatment does work. For this example, H_1 states that the training does reduce medication. In symbols,

$$H_1: \quad \mu_D < 0 \qquad \text{(Medication is reduced after training.)}$$

It often is easier to start by stating H_1, which says that the treatment works as predicted. Then H_0 is simply the opposite of H_1.

STEP 2   Locate the critical region. The researcher is predicting negative difference scores if the treatment works. Hence, a negative t statistic would tend to support the experimental prediction and refute H_0. With a sample of $n = 5$ participants, the t statistic will have $df = 4$. Looking in the t-distribution table for $df = 4$ and $\alpha = .05$ for a one-tailed test, we find a critical value of 2.132. Thus, a t statistic that is more extreme than -2.132 will be sufficient to reject H_0. The t distribution with the one-tailed critical region is shown in Figure 11.5.

FIGURE 11.5

The one-tailed critical region for $\alpha = .05$ in the t distribution with $df = 4$.

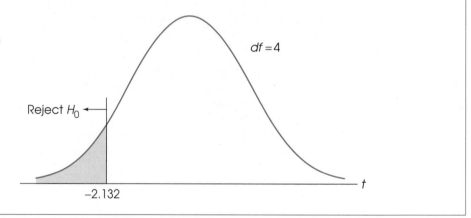

STEP 3   Compute the t statistic. We calculated the t statistic in Example 11.1, where we obtained $t = -3.72$.

STEP 4   Make a decision. The obtained t statistic is well beyond the critical boundary. Therefore, we reject H_0 and conclude that the relaxation training did significantly reduce the amount of medication needed for asthma attacks.

LEARNING CHECK

1. A researcher would like to examine the effect of hypnosis on cigarette smoking. The researcher selects a sample of smokers ($n = 4$) for the study, and records the number of cigarettes smoked on the day prior to treatment. The participants are then hypnotized and given the posthypnotic suggestion that each time they light a cigarette, they will experience a horrible taste and feel nauseous. The data are as follows:

THE NUMBER OF CIGARETTES SMOKED BEFORE AND AFTER HYPNOSIS		
PARTICIPANT	BEFORE TREATMENT	AFTER TREATMENT
1	19	12
2	35	36
3	20	13
4	31	24

   a. Find the difference scores (D values) for this sample.
   b. The difference scores have $M_D = -5$ and $SS = 48$. Do these data indicate that hypnosis has a significant effect on cigarette smoking? Test with $\alpha = .05$.
   c. Are the data sufficient to conclude that hypnosis significantly *reduces* cigarette smoking? Use a one-tailed test with $\alpha = .05$.

2. Compute the effect size using both Cohen's d and r^2 for the hyponosis treatment in Question 1.

3. A computer printout for a repeated-measures t test reports a p value of $p = .021$.
   a. Can the researcher claim a significant effect with $\alpha = .01$?
   b. Is the effect significant with $\alpha = .05$?

ANSWERS

1. a. The difference scores are -7, 1, -7, -7.
   b. For these data, $s^2 = 16$, $s_{M_D} = 2$, and $t = -2.50$. With $\alpha = .05$, we fail to reject H_0; these data do not provide sufficient evidence to conclude that hypnosis has a significant effect on cigarette smoking.
   c. For the directional test, the critical value is $t = -2.353$. The sample data produce a t statistic of $t = -2.50$, which is in the critical region. Reject H_0 and conclude that the hypnosis significantly reduced smoking.

2. For these data, $d = -5/4 = -1.25$ and $r^2 = 0.68$.

3. a. The exact p value, $p = .021$, is not less than $\alpha = .01$. Therefore, the effect is not significant for $\alpha = .01$ ($p > .01$).
   b. The p value is less than .05, so the effect is significant with $\alpha = .05$.

11.4 HYPOTHESIS TESTING WITH A MATCHED-SUBJECTS DESIGN

As noted in Section 11.1, matched-subjects studies involve two related samples. Each participant in the second sample has been matched to a participant in the first sample on some variable that the researcher wishes to control. In this way, the researcher attempts to make the groups equivalent at the start of the study, so that any differences observed can be attributed to the treatment (independent variable). As noted earlier, data from the matched-subjects design may be analyzed using the related-samples *t* test. The following example illustrates this analysis.

EXAMPLE 11.3 A psychologist studies the effectiveness of a newly developed reading program. One sample of students takes the new reading course. A second sample serves as a control group and takes the regular coursework. The groups are later tested in reading comprehension to assess the effectiveness of the new program. However, the researcher is concerned that the children in one group might simply be better readers than the children in the other group. A difference in reading ability could cause one group to score higher than the other whether or not the reading program has any effect. To avoid this problem, each participant in the first group is matched one-to-one with a participant in the second group in terms of their reading achievement scores. For example, if a child in the first group has a reading achievement score of 75, then the researcher finds another child with a score of 75 to place in the second group. Thus, the study consists of *matched pairs* of children, with one child from each pair in the control group and the other child in the reading program. In this way, the researcher knows that the two groups are identical in terms of reading ability at the beginning of the study. If there are differences between groups after the treatment, the researcher can be more confident that the treatment has caused the differences. The data for this study are summarized in Table 11.3. Is there evidence that the reading program has a significant effect?

TABLE 11.3

Reading comprehension scores for children in a matched-subjects study of a new reading program.

	CONTROL	READING PROGRAM	D	D^2
Matched pair A	6	15	+9	81
Matched pair B	5	15	+10	100
Matched pair C	11	17	+6	36
Matched pair D	6	13	+7	49

$$\Sigma D = +32 \quad \Sigma D^2 = 266$$

$$M_D = \frac{\Sigma D}{n} = \frac{+32}{4} = +8$$

$$SS = \Sigma D^2 - \frac{(\Sigma D)^2}{n} = 266 - \frac{(32)^2}{4}$$

$$= 266 - \frac{1024}{4}$$

$$= 266 - 256 = 10$$

STEP 1 State the hypotheses, and select an alpha level.

H_0: $\mu_D = 0$ (There is no effect on reading comprehension.)

H_1: $\mu_D \neq 0$ (There is an effect.)

We will set the alpha level to .01 for a two-tailed test.

STEP 2   Locate the critical region. For this example, $df = n - 1 = 3$. The *t* distribution table indicates the critical values are $+5.841$ and -5.841.

STEP 3   Calculate the *t* statistic. This study used a matched-subjects design; thus, a related-samples *t* statistic is appropriate. The values for *SS* and M_D were calculated in Table 11.3.

Calculation of the *t* statistic proceeds in a three-step process: First, compute the variance for the sample of *D* scores.

$$s^2 = \frac{SS}{n - 1} = \frac{10}{3} = 3.33$$

Next, use the sample variance to compute the estimated standard error.

$$s_{M_D} = \sqrt{\frac{s^2}{n}} = \sqrt{\frac{3.33}{4}} = \sqrt{0.83} = 0.91$$

Finally, compute the *t* statistic.

$$t = \frac{M_D - \mu_D}{s_{M_D}} = \frac{8 - 0}{0.91} = 8.79$$

STEP 4   Make a decision. The obtained *t* value of $t = 8.79$ is more extreme than the critical value of $+5.841$. Therefore, it falls within the critical region on the right side of the *t* distribution. The null hypothesis can be rejected, and it can be concluded that the new reading program has a significant effect on reading comprehension, $t(3) = 8.79$, $p < .01$, two tails.

11.5   USES AND ASSUMPTIONS FOR RELATED-SAMPLES *t* TESTS

COMPARING REPEATED-MEASURES AND INDEPENDENT-MEASURES DESIGNS

In many research situations, it is possible to use either a repeated-measures study or an independent-measures study to compare two treatment conditions. The independent-measures design would use two separate samples (one in each treatment condition) and the repeated-measures design would use only one sample with the same individuals in both treatments. The decision about which design to use is often made by considering the advantages and disadvantages of the two designs. In general, the repeated-measures design has most of the advantages.

Number of subjects   A repeated-measures design typically requires fewer subjects than an independent-measures design. The repeated-measures design uses the subjects more efficiently because each individual is measured in both of the treatment conditions. This can be especially important when there are relatively few subjects available (for example, when you are studying a rare species or individuals in a rare profession).

Study changes over time   The repeated-measures design is especially well suited for studying learning, development, or other changes that take place over time. Remember that this design involves measuring individuals at one time and then returning to measure the same individuals at a later time. In this way, a researcher can observe behaviors that change or develop over time.

Individual differences   The primary advantage of a repeated-measures design is that it reduces or eliminates problems caused by individual differences. *Individual differences* are characteristics such as age, IQ, gender, and personality that vary from one individual to another. These individual differences can influence the scores obtained in a research study, and they can affect the outcome of a hypothesis test. Consider the data in Table 11.4 . The first set of data represents the results from a typical independent-measures study and the second set represents a repeated-measures study. Note that we have identified each participant by name to help demonstrate the effects of individual differences.

TABLE 11.4

Hypothetical data showing the results from an independent-measures study and a repeated-measures study. The two sets of data use exactly the same numerical scores and they both show the same 5-point mean difference between treatments.

INDEPENDENT-MEASURES STUDY (2 SEPARATE SAMPLES)		REPEATED-MEASURES STUDY (SAME SAMPLE IN BOTH TREATMENTS)		
TREATMENT 1	TREATMENT 2	TREATMENT 1	TREATMENT 2	D
(John) $X = 18$	(Sue) $X = 15$	(John) $X = 18$	(John) $X = 15$	-3
(Mary) $X = 27$	(Tom) $X = 20$	(Mary) $X = 27$	(Mary) $X = 20$	-7
(Bill) $X = 33$	(Dave) $X = 28$	(Bill) $X = 33$	(Bill) $X = 28$	-5
$M = 26$	$M = 21$			$M_D = -5$
$SS = 114$	$SS = 86$			$SS = 8$

For the independent-measures data, note that every score represents a different person. For the repeated-measures study, on the other hand, the same participants are measured in both of the treatment conditions. This difference between the two designs has some important consequences.

1. We have constructed the data so that both research studies have exactly the same scores and they both show the same 5-point mean difference between treatments. In each case, the researcher would like to conclude that the 5-point difference was caused by the treatments. However, with the independent-measures design, there is always the possibility that the participants in treatment 1 have different characteristics than those in treatment 2. For example, the three participants in treatment 2 may be more intelligent than those in treatment 1 and their higher intelligence caused them to have higher scores. Note that this problem disappears with the repeated-measures design. Specifically, with repeated measures there is no possibility that the participants in one treatment are different from those in another treatment because the same participants are used in all the treatments.

2. Although the two sets of data contain exactly the same scores and have exactly the same 5-point mean difference, you should realize that they are very different in terms of the variance used to compute standard error. For the independent-measures study, you calculate the SS or variance for the scores in each of the two separate samples. Note that in each sample there are big differences between subjects. In treatment 1, for example, Bill has a score of 33 and John's score is only 18. These individual differences produce a relatively large sample variance and a large standard error. For the independent-measures study, the standard error is 5.77, which produces a *t* statistic of $t = 0.87$. For these data,

the hypothesis test concludes that there is no significant difference between treatments.

In the repeated-measures study, the *SS* and variance are computed for the difference scores. If you examine the repeated-measures data in Table 11.4, you will see that the big differences between John and Bill that exist in treatment 1 and in treatment 2 are eliminated when you get to the difference scores. Because the individual differences are eliminated, the variance and standard error are dramatically reduced. For the repeated-measures study, the standard error is 1.15 and the *t* statistic is $t = -4.35$. With the repeated-measures *t*, the data show a significant difference between treatments. Thus, one big advantage of a repeated-measures study is that it reduces variance by removing individual differences, which increases the chances of finding a significant result.

CARRYOVER EFFECTS AND PROGRESSIVE ERROR

The general goal of a repeated-measures *t* test is to demonstrate that two different treatment conditions result in two different means. However, the outcome of a repeated-measures study may be contaminated by other factors that can cause the two means to be different when there actually is no difference between the two treatments. Two such factors, specifically associated with repeated-measures designs, are *carryover effects* and *progressive error*.

A carryover effect occurs when a subject's response in the second treatment is altered by lingering aftereffects from the first treatment. Some examples of carryover effects follow:

1. Imagine that a researcher is comparing the effectiveness of two drugs by testing both drugs, one after the other, on the same group of subjects. If the second drug is tested too soon after the first, there may be some of the first drug still in the subject's system, and this residual could exaggerate or minimize the effects of the second drug.

2. Imagine a researcher comparing performance on two tasks that vary in difficulty. If subjects are given a very difficult task first, their poor performance may cause them to lose motivation, so that performance suffers when they get to the second task.

In each of these examples, the researcher will observe a difference between the two treatment means. However, the difference in performance is not caused by the treatments; instead, it is caused by carryover effects.

Progressive error occurs when a subject's performance or response changes consistently over time. For example, a subject's performance may decline over time as a result of fatigue. Or a subject's performance may improve over time as a result of practice. In either case, a researcher would observe a mean difference in performance between the first treatment and the second treatment, but the change in performance is not due to the treatments; instead, it is caused by progressive error.

To help distinguish between carryover effects and progressive error, remember that carryover effects are related directly to the first treatment. Progressive error, on the other hand, occurs as a function of time, independent of which treatment condition is presented first or second. Both of these factors, however, can influence the outcome of a repeated-measures study, and they can make it difficult for a researcher to interpret the results: Is the mean difference caused by the treatments, or is it caused by other factors?

One way to deal with carryover effects or progressive error is to *counterbalance* the order of presentation of treatments. That is, the subjects are randomly divided into two groups, with one group receiving treatment 1 followed by treatment 2 and the other group receiving treatment 2 followed by treatment 1. When there is reason to expect strong carryover effects or large progressive error, your best strategy is not to use a repeated-measures design. Instead, use independent measures with a separate sample for each treatment condition, or use a matched-subjects design so that each individual participates in only one treatment condition.

ASSUMPTIONS OF THE RELATED-SAMPLES *t* TEST

The related-samples *t* statistic requires two basic assumptions:

1. The observations within each treatment condition must be independent (see page 251). Notice that the assumption of independence refers to the scores *within* each treatment. Inside each treatment, the scores are obtained from different individuals and should be independent of one another.

2. The population distribution of difference scores (*D* values) must be normal.

As before, the normality assumption is not a cause for concern unless the sample size is relatively small. In the case of severe departures from normality, the validity of the *t* test may be compromised with small samples. However, with relatively large samples ($n > 30$), this assumption can be ignored.

LEARNING CHECK

1. What assumptions must be satisfied for repeated-measures *t* tests to be valid?

2. Describe some situations for which a repeated-measures design is well suited.

3. How is a matched-subjects design similar to a repeated-measures design? How do they differ?

4. The data from a research study consist of 10 scores in each of two different treatment conditions. How many individual subjects would be needed to produce these data

   a. For an independent-measures design?

   b. For a repeated-measures design?

   c. For a matched-subjects design?

ANSWERS

1. The observations within a treatment are independent. The population distribution of *D* scores is assumed to be normal.

2. The repeated-measures design is suited to situations in which a particular type of subject is not readily available for study. This design is helpful because it uses fewer subjects (only one sample is needed). Certain questions are addressed more adequately by a repeated-measures design—for example, anytime one would like to study changes across time in the same individuals. Also, when individual differences are large, a repeated-measures design is helpful because it reduces the amount of this type of error in the statistical analysis.

3. They are similar in that the role of individual differences in the experiment is reduced. They differ in that there are two samples in a matched-subjects design and only one in a repeated-measures study.

4. a. The independent-measures design would require 20 subjects (two separate samples with $n = 10$ in each).

   b. The repeated-measures design would require 10 subjects (the same 10 individuals are measured in both treatments).

   c. The matched-subjects design would require 20 subjects (10 matched pairs).

SUMMARY

1. In a related-samples research study, the individuals in one treatment condition are directly related, one-to-one, with the individuals in the other treatment condition(s). The most common related-samples study is a repeated-measures design, where the same sample of individuals is tested in all of the treatment conditions. This design literally repeats measurements on the same subjects. An alternative is a matched-subjects design, where the individuals in one sample are matched one-to-one with individuals in another sample. The matching is based on a variable relevant to the study.

2. The repeated-measures t test begins by computing a difference between the first and second measurements for each subject (or the difference for each matched pair). The difference scores, or D scores, are obtained by

$$D = X_2 - X_1$$

The sample mean, M_D, and sample variance, s^2, are used to summarize and describe the set of difference scores.

3. The formula for the repeated-measures t statistic is

$$t = \frac{M_D - \mu_D}{s_{M_D}}$$

In the formula, the null hypothesis specifies $\mu_D = 0$, and the estimated standard error is computed by

$$s_{M_D} = \sqrt{\frac{s^2}{n}}$$

4. A repeated-measures design may be preferred to an independent-measures study when one wants to observe changes in behavior in the same subjects, as in learning or developmental studies. An important advantage of the repeated-measures design is that it removes or reduces individual differences, which in turn lowers sample variability and tends to increase the chances for obtaining a significant result.

5. For a repeated-measures design, effect size can be measured using either r^2 (the percentage of variance accounted for) or Cohen's d (the standardized mean difference). The value of r^2 is computed the same for both independent- and repeated-measures designs,

$$r^2 = \frac{t^2}{t^2 + df}$$

Cohen's d is defined as the sample mean difference divided by standard deviation for both repeated- and independent-measures designs. For repeated-measures studies,

$$d = \frac{M_D}{s}$$

RESOURCES

WebTUTOR

If you are using WebTutor along with this book, there is a WebTutor section corresponding to this chapter. The WebTutor contains a brief summary of Chapter 11, hints for learning the concepts and the formulas for the repeated-measures t test, cautions about common errors, and sample exam items including solutions.

The statistics computer package SPSS can be used to conduct a repeated-measures *t* hypothesis test. First, the data are entered into two columns in the data matrix with the first score for each participant in the first column and the second score in the second column. Then, click **Analyze** on the tool bar, select **Compare Means**, and click on **Paired-Samples T Test**. Next you must enter the labels for the two data columns into the Paired Variables box. (Highlight the first-score column as Variable 1 and highlight the second-score column as Variable 2, then click the arrow to move your selections into the box.) Then click **OK**. SPSS will produce a summary table showing descriptive statistics for each of the two sets of scores, and a table showing the correlation between the first and second score (correlations are presented in Chapter 17). Finally, SPSS conducts the *t* test for the difference scores. The output shows the mean difference, the standard deviation and the standard error for the difference scores, as well as the value for *t*, the value for *df*, and the level of significance (*p* value). The output also includes a 95% confidence interval that provides a range of values estimating how much difference exists between the two treatment conditions. (Confidence intervals are discussed in Chapter 12.)

Note: If your data consist of difference scores (instead of pairs of scores) you can enter the difference scores in one column and use the **One-Sample T Test** option, entering a value of zero in the Test Value box (see the SPSS section in Chapter 9).

KEY TERMS

repeated-measures design	difference scores	estimated standard error for M_D	carryover effects
within-subjects design	repeated-measures	individual differences	progressive error
matched-subjects design	*t* statistic		

FOCUS ON PROBLEM SOLVING

1. Once data have been collected, we must then select the appropriate statistical analysis. How can you tell whether the data call for a repeated-measures *t* test? Look at the experiment carefully. Is there only one sample of subjects? Are the same subjects tested a second time? If your answers are yes to both of these questions, then a repeated-measures *t* test should be done. There is only one situation in which the repeated-measures *t* can be used for data from two samples, and that is for *matched-subjects* studies (page 343).

2. The repeated-measures *t* test is based on difference scores. In finding difference scores, be sure you are consistent with your method. That is, you may use either $X_2 - X_1$ or $X_1 - X_2$ to find *D* scores, but you must use the same method for all subjects.

DEMONSTRATION 11.1

A REPEATED-MEASURES *t* TEST

A major oil company would like to improve its tarnished image following a large oil spill. Its marketing department develops a short television commercial and tests it on a sample of $n = 7$ participants. People's attitudes about the company are measured with a short questionnaire, both before and after viewing the commercial. The data are as follows:

PERSON	X_1 (BEFORE)	X_2 (AFTER)
A	15	15
B	11	13
C	10	18
D	11	12
E	14	16
F	10	10
G	11	19

Was there a significant change? Note that participants are being tested twice—once before and once after viewing the commercial. Therefore, we have a repeated-measures experiment.

STEP 1   State the hypotheses, and select an alpha level.
   The null hypothesis states that the commercial has no effect on people's attitude, or in symbols,

$$H_0: \quad \mu_D = 0 \quad \text{(The mean difference is zero.)}$$

The alternative hypothesis states that the commercial does alter attitudes about the company, or

$$H_1: \quad \mu_D \neq 0 \quad \text{(There is a mean change in attitudes.)}$$

For this demonstration, we will use an alpha level of .05 for a two-tailed test.

STEP 2   Locate the critical region.
   Degrees of freedom for the repeated-measures t test are obtained by the formula

$$df = n - 1$$

For these data, degrees of freedom equal

$$df = 7 - 1 = 6$$

The t distribution table is consulted for a two-tailed test with $\alpha = .05$ for $df = 6$. The critical t values for the critical region are $t = \pm 2.447$.

STEP 3   Obtain the sample data, and compute the test statistic.
   The first step in computing the repeated-measures t statistic is to calculate basic descriptive statistics for the sample data. This involves finding the difference scores (D values) and then computing the sample mean and SS for the D scores.

The difference scores   The following table illustrates the computations of the D values for our sample data. Remember that $D = X_2 - X_1$.

X_1	X_2	D
15	15	$15 - 15 = 0$
11	13	$13 - 11 = +2$
10	18	$18 - 10 = +8$
11	12	$12 - 11 = +1$
14	16	$16 - 14 = +2$
10	10	$10 - 10 = 0$
11	19	$19 - 11 = +8$

The sample mean of the D values   The sample mean for the difference scores is equal to the sum of the D values divided by n. For these data,

$$\Sigma D = 0 + 2 + 8 + 1 + 2 + 0 + 8 = 21$$

$$M_D = \frac{\Sigma D}{n} = \frac{21}{7} = 3$$

Sum of squares for D scores   We will use the computational formula for SS. The following table summarizes the calculations:

D	D^2	
0	0	$\Sigma D = 21$
2	4	$\Sigma D^2 = 0 + 4 + 64 + 1 + 4 + 0 + 64 = 137$
8	64	
1	1	$SS = \Sigma D^2 - \dfrac{(\Sigma D)^2}{n} = 137 - \dfrac{(21)^2}{7}$
2	4	
0	0	$SS = 137 - \dfrac{441}{7} = 137 - 63 = 74$
8	64	

Variance for the D scores   The variance for the sample of D scores is

$$s^2 = \frac{SS}{n-1} = \frac{74}{6} = 12.33$$

Estimated standard error for M_D   The estimated standard error for the sample mean difference is computed as follows:

$$s_{M_D} = \sqrt{\frac{s^2}{n}} = \sqrt{\frac{12.33}{7}} = \sqrt{1.76} = 1.33$$

The repeated-measures t statistic   Now we have the information required to calculate the t statistic.

$$t = \frac{M_D - \mu_D}{s_{M_D}} = \frac{3 - 0}{1.33} = 2.26$$

S T E P 4   Make a decision about H_0, and state the conclusion.

The obtained t value is not extreme enough to fall in the critical region. Therefore, we fail to reject the null hypothesis. We conclude that there is no evidence that the commercial will change people's attitudes, $t(6) = 2.26$, $p > .05$, two-tailed. (Note that we state that p is *greater than* .05 because we failed to reject H_0.)

DEMONSTRATION 11.2

EFFECT SIZE FOR THE REPEATED-MEASURES t

We will compute Cohen's d and r^2 for the data in Demonstration 11.1. The data produced a sample mean difference of $M_D = 3.00$ with a sample variance of $s^2 = 12.33$. Based on these values, Cohen's d is

$$\text{Cohen's } d = \frac{\text{mean difference}}{\text{standard deviation}} = \frac{M_D}{s} = \frac{3.00}{\sqrt{12.33}} = \frac{3.00}{3.51} = 0.86$$

The hypothesis test produced $t = 2.26$ with $df = 6$. Based on these values,

$$r^2 = \frac{t^2}{t^2 + df} = \frac{(2.26)^2}{(2.26)^2 + 6} = \frac{5.11}{11.11} = 0.46 \quad (\text{or } 46\%)$$

PROBLEMS

1. For the following studies, indicate whether or not a repeated-measures t test is the appropriate analysis. Explain your answers.
  a. A researcher is examining the effectiveness of an SAT training program for high school seniors. One sample of seniors takes the 3-week training program and another sample serves as a control group (no training). Both groups then take the SAT, and the researcher compares the mean scores for the two groups.
  b. Another researcher does a similar study. A sample of seniors is obtained and each student takes the SAT. Then the seniors participate in the 3-week training program, after which they take the SAT again. The researcher measures how much the mean changes from the first test to the second.

2. What is the primary advantage of a repeated-measures design over an independent-measures design?

3. Explain the difference between a matched-subjects design and a repeated-measures design.

4. A researcher conducts an experiment comparing two treatment conditions and obtains data with 10 scores for each treatment condition.
  a. If the researcher used an independent-measures design, how many subjects participated in the experiment?
  b. If the researcher used a repeated-measures design, how many subjects participated in the experiment?
  c. If the researcher used a matched-subjects design, how many subjects participated in the experiment?

5. Each of the following describes a sample of difference scores (D scores) from a repeated-measures study:

  Sample 1: $n = 10$, $M_D = +4.00$, $s = 10$

  Sample 2: $n = 10$, $M_D = +12.00$, $s = 10$

  a. For each sample, use the mean and standard deviation to sketch the distribution (or simply describe the distribution). Identify where a value of zero is located relative to the pile of scores.
  b. Which of the two samples would be more likely to reject a null hypothesis stating that the population mean difference is zero ($\mu_D = 0$)? Explain your answer.

6. Each of the following describes a sample of difference scores (D scores) from a repeated-measures study:

  Sample 1: $n = 10$, $M_D = +4.00$, $s = 10$

  Sample 2: $n = 10$, $M_D = +4.00$, $s = 2$

  a. For each sample, use the mean and standard deviation to sketch the distribution (or simply describe the distribution). Identify where a value of zero is located relative to the pile of scores.
  b. Which of the two samples would be more likely to reject a null hypothesis stating that the population mean difference is zero ($\mu_D = 0$)? Explain your answer.

7. A sample of $n = 16$ difference scores (D values) from a repeated-measures experiment has $SS = 440$.
  a. Compute the standard error of the mean for this sample.
  b. Would a sample mean difference of $M_D = 3$ be large enough to be significant with $\alpha = .05$? Assume a two-tailed test.

8. A sample of difference scores (*D* values) from a repeated-measures experiment has a mean of $M_D = 5.00$
with a variance of $s^2 = 16$.
  a. If $n = 4$, is this sample sufficient to reject the null
  hypothesis using a two-tailed test with $\alpha = .05$?
  b. Would you reject H_0 if $n = 16$? Again, assume a
  two-tailed test with $\alpha = .05$.

9. A researcher uses a repeated-measures study to evaluate the effectiveness of magnetic therapy for treating
chronic pain. A sample of $n = 9$ patients is obtained,
each of whom has been treated for severe elbow pain
for at least 1 year. Each patient is given a questionnaire to evaluate the current level of pain. The patients
are then instructed to wear a magnetic band around
their elbows for 2 weeks. After the 2-week period, the
patients again complete the questionnaire evaluating
pain. For this sample, the reported level of pain decreased by an average of $M_D = 12.5$ points with
$SS = 1152$. Do these data indicate a significant change
in the level of pain? Use a two-tailed test with
$\alpha = .01$.

10. A researcher suspects that increasing the level of lighting during the winter months will have a positive
effect on people's moods. To test this hypothesis, the
researcher identifies a sample of $n = 36$ college students living on one floor of a dormitory. Each student
is given a mood questionnaire at the beginning of
February, and then all the lights on the dormitory floor
are changed from 75-watt to 100-watt bulbs. After
4 weeks, the researcher again measures each student's
mood and records the amount of difference between
the two measurements. For this sample, the students'
mood scores increased by an average of $M_D = +8.6$
points with $SS = 2835$.
  a. Do these data show a significant change in mood
  following the increase in lighting? Use a two-tailed
  test with $\alpha = .05$.
  b. Calculate Cohen's *d*, the standardized mean difference, to measure effect size for this study.

11. Many over-the-counter cold medications come with a
warning that the medicine may cause drowsiness. To
evaluate this effect, a researcher measures reaction
time for a sample of $n = 36$ participants. Each participant is then given a dose of a popular cold medicine,
and reaction time is measured again. For this sample,
reaction time increases after the medication by an
average of $M_D = 24$ milliseconds with $s = 8$.
  a. Do these data indicate that the cold medicine has a
  significant effect on reaction time? Use a two-tailed
  test with $\alpha = .05$.
  b. Calculate r^2, the percentage of variance in the
  reaction time scores that is accounted for by the
  medication.

12. One of the benefits of aerobic exercise is the release of
endorphins, which are natural chemicals in the brain
that produce a feeling of general well-being. A sample
of $n = 16$ participants is obtained, and each person's
tolerance for pain is tested before and after a
50-minute session of aerobic exercise. On average, the
pain tolerance for this sample is $M_D = 10.5$ points
higher after exercise than it was before. The *SS* for the
sample of difference scores is $SS = 960$.
  a. Do these data indicate that exercise produces a
  significant increase in pain tolerance? Use a one-
  tailed test with $\alpha = .01$.
  b. Calculate r^2, the percentage of variance in the
  pain tolerance scores that is accounted for by the
  exercise.

13. A psychologist is testing a new drug for its painkilling
effects. The pain threshold is measured for a sample of
$n = 9$ participants by determining the intensity of an
electric shock that causes discomfort. After the initial
baseline is established, each participant receives the
drug, and the pain threshold is measured once again.
For this sample, the pain threshold has increased by an
average of $M_D = 1.6$ milliamperes with $s = 0.4$ after
receiving the drug.
  a. Do these data indicate that the drug produces a
  significant increase in pain tolerance? Use a one-
  tailed test with $\alpha = .05$.
  b. Calculate r^2, the percentage of variance accounted
  for by the new drug.

14. The level of lymphocytes (white blood cells) in the
blood is associated with susceptibility to disease—
lower levels indicate greater susceptibility. In a study
of the effects of stress on physical well-being,
Schleifer and colleagues (1983) recorded the lymphocyte counts for men who were going through a period
of extreme emotional stress (the death of a spouse).
For a sample of $n = 20$ men, the lymphocyte counts
dropped an average of $M_D = 0.09$ with $s = 0.08$. Do
these data indicate a significant change? Test at the
.05 level of significance.

15. The following data are from a research study comparing two treatment conditions.
  a. Assume that the data are from an independent-
  measures study using two separate samples. Use a
  two-tailed test with $\alpha = .05$ to determine whether
  there is a significant difference between the two
  treatments.
  b. Now assume that the data are from a repeated-
  measures study using the same sample of $n = 4$
  participants in both treatments. Use a two-tailed test
  with $\alpha = .05$ to determine whether there is a significant difference between the two treatments.

c. Whether the data are from an independent-measures or a repeated-measures study, the sample mean difference is 4 points. However, the variance and the estimated standard error are very different depending on which study is used. Explain why the variance and standard error are lower in the repeated-measures test.

TREATMENT	
I	II
7	10
9	12
10	17
6	9
$M = 8$	$M = 12$
$SS = 10$	$SS = 38$

16. The following data are exactly the same as the scores shown in Problem 15 except that we have added 40 points to the first score in each treatment condition. The result is that each condition now contains one score that is very different from the others. In a research study, an odd score can occur when one individual is very different from the other participants in the study.

a. Assume that the data are from an independent-measures study using two separate samples. Use a two-tailed test with $\alpha = .05$ to determine whether there is a significant difference between the two treatments. How does the outcome compare with the results from Problem 15?

b. Now assume that the data are from a repeated-measures study using the same sample of $n = 4$ participants in both treatments. Use a two-tailed test with $\alpha = .05$ to determine whether there is a significant difference between the two treatments. How does the outcome compare with the results from Problem 15?

c. Changing the first score in each treatment produced a huge increase in the individual differences (the differences from one score to the next). How do large individual differences affect the independent-measures t? How do large individual differences affect the repeated-measures t?

TREATMENT	
I	II
47	50
9	12
10	17
6	9
$M = 18$	$M = 22$
$SS = 1130$	$SS = 1078$

17. A researcher studies the effect of cognitive psychotherapy on positive self-regard. The number of positive statements made about oneself is recorded for each participant during the first meeting. After 8 weekly therapy sessions, the measures are repeated for each person. For the following data,
a. Calculate the difference scores and M_D.
b. Compute SS, sample variance, and estimated standard error.
c. Is there a significant treatment effect? Use $\alpha = .05$, two tails.

PARTICIPANT	BEFORE TREATMENT	AFTER TREATMENT
A	3	12
B	5	10
C	7	8
D	1	14

18. Another researcher did the exact same study as described in Problem 17, and the data are presented here.
a. Calculate the difference scores and M_D. Compare the results to part (a) of the previous problem.
b. Compute SS, sample variance, and estimated standard error. Compare theses results to those of the previous problem.
c. Is there a significant treatment effect? Use $\alpha = .05$, two tails.
d. Compare the outcome of part (c) to that of the previous problem. Explain why the outcomes differ.

PARTICIPANT	BEFORE TREATMENT	AFTER TREATMENT
A	0	12
B	4	10
C	0	4
D	12	18

19. The previous two problems both have a sample mean difference of $M_D = 7$ with $n = 4$ scores. However, for Problem 17 the sample variance is $s^2 = 26.67$, which results in $t = 2.71$ (not significant), and in Problem 18, $s^2 = 12$, which produces $t = 4.05$ (significant).
a. Compute Cohen's d and r^2 to measure effect size for both sets of data.
b. Explain how effect size is related to sample variance.

20. The following data are from an experiment comparing two treatments:

TREATMENT 1	TREATMENT 2
10	11
2	5
1	2
15	18
7	9
$M = 7$	$M = 9$
$SS = 134$	$SS = 150$

a. Assume that the data are from an independent-measures experiment using two separate samples, each with $n = 5$ subjects. Do the data indicate a significant difference between the two treatments? Use a two-tailed test with $\alpha = .05$.

b. Now assume that the data are from a repeated-measures experiment using the same sample of $n = 5$ subjects in both treatments. Calculate the difference score (D) for each subject and use the sample of difference scores to determine whether there is a significant difference between the two treatments. Again, use a two-tailed test with $\alpha = .05$.

c. You should find that the repeated-measures test in part (b) reaches a different conclusion than the independent-measures test in part (a). How do you explain the different outcomes? (*Hint:* Look at the variances or SS values for the two different tests. This is an example of the general fact that repeated-measures designs reduce variance by removing individual differences.)

21. A researcher for a cereal company wanted to demonstrate the health benefits of eating oatmeal. A sample of 9 volunteers was obtained and each participant ate a fixed diet without any oatmeal for 30 days. At the end of the 30-day period, cholesterol was measured for each individual. Then the participants began a second 30-day period in which they repeated exactly the same diet except for 2 cups of oatmeal each day. After the second 30-day period, cholesterol levels were measured again. The scores for the 9 participants are as follows:

PARTICIPANT	A	B	C	D	E	F	G	H	I
1st score	145	187	130	155	152	112	120	208	167
2nd score	145	185	119	140	140	115	111	199	159
Difference	0	−2	−11	−15	−12	+3	−9	−9	−8

Are the data sufficient to demonstrate a significant change in cholesterol level after 30 days with an oatmeal diet? Use a two-tailed test with $\alpha = .05$.

22. A variety of research results suggest that visual images interfere with visual perception. In one study, Segal and Fusella (1970) had participants watch a screen, looking for brief presentations of a small blue arrow. On some trials, the participants were also asked to form a mental image (for example, imagine a volcano). The results show that participants made more errors while forming images than while not forming images. Data similar to the Segal and Fusella results are as follows. Do these data indicate a significant difference between the two conditions? Use a two-tailed test with $\alpha = 05$.

PARTICIPANT	ERRORS WITH IMAGE	ERRORS WITHOUT IMAGE
A	13	4
B	9	2
C	12	10
D	7	8
E	10	6
F	8	6
G	9	4

23. People with agoraphobia are so filled with anxiety about being in public places that they seldom leave their homes. Knowing this is a difficult disorder to treat, a researcher tries a long-term treatment. A sample of individuals report how often they have ventured out of the house in the past month. Then they receive relaxation training and are introduced to trips away from the house at gradually increasing durations. After 2 months of treatment, participants report the number of trips out of the house they made in the last 30 days. The data are as follows:

PARTICIPANT	BEFORE TREATMENT	AFTER TREATMENT
A	0	4
B	0	0
C	3	14
D	3	23
E	2	9
F	0	8
G	0	6

Does the treatment have a significant effect on the number of trips a person takes? Test with $\alpha = .05$, two tails.

24. A researcher investigates whether single people who own pets are generally happier than singles without pets. A group of non-pet owners is compared to pet owners using a mood inventory. The pet owners are matched one to one with the non-pet owners for income, number of close friendships, and general health. The data are as follows:

MATCHED PAIR	NON-PET OWNER	PET OWNER
A	10	15
B	9	9
C	11	10
D	11	19
E	5	17
F	9	15

Is there a significant difference in the mood scores for non-pet owners versus pet owners? Test with $\alpha = .05$ for two tails.

25. In the Preview section of this chapter, we discussed a report describing how the accuracy of Olympic marksmen can be affected by heartbeats. Pelton (1983) reported that Olympic-level marksmen shoot much better if they fire between heartbeats rather than squeezing the trigger during a heartbeat. The small vibration caused by a heartbeat seems to be sufficient to affect the marksmen's aim. The following hypothetical data demonstrate this phenomenon. A sample of $n = 6$ Olympic marksmen fire a series of rounds while a researcher records heartbeats. For each marksman, the total score is recorded for shots fired during heartbeats and for shots fired between heartbeats. Do these data indicate a significant difference? Test with $\alpha = .05$.

PARTICIPANT	DURING HEARTBEATS	BETWEEN HEARTBEATS
A	93	98
B	90	94
C	95	96
D	92	91
E	95	97
F	91	97

26. Sensory isolation chambers are used to examine the effects of mild sensory deprivation. The chamber is a dark, silent tank where participants float on heavily salted water and are thereby deprived of nearly all external stimulation. Sensory deprivation produces deep relaxation and has been shown to produce temporary increases in sensitivity for vision, hearing, touch, and even taste. The following data represent hearing threshold scores for a group of participants who were tested before and immediately after 1 hour of deprivation. A lower score indicates more sensitive hearing. Do these data indicate that deprivation has a significant effect on hearing threshold? Test at the .05 level of significance.

PARTICIPANT	BEFORE	AFTER
A	31	30
B	34	31
C	29	29
D	33	29
E	35	32
F	32	34
G	35	28

27. A researcher studies the effect of a drug (MAO inhibitor) on the number of nightmares occurring in veterans with post-traumatic stress disorder (PTSD). A sample of PTSD clients records each incident of a nightmare for 1 month before treatment. Participants are then given the medication for 1 month, and they continue to report each occurrence of a nightmare. For the following hypothetical data, determine whether the MAO inhibitor significantly reduces nightmares. Use the .05 level of significance and a one-tailed test.

NUMBER OF NIGHTMARES	
ONE MONTH BEFORE TREATMENT	ONE MONTH DURING TREATMENT
6	1
10	2
3	0
5	5
7	2

CHAPTER

12

Estimation

Tools You Will Need

The following items are considered essential background material for this chapter. If you doubt your knowledge of any of these items, you should review the appropriate chapter or section before proceeding.

- Single-sample t statistic (Chapter 9)

- Independent-measures t statistic (Chapter 10)

- Related-samples t statistic (Chapter 11)

Preview

Suppose you are asked to describe the "typical" college student in the United States. For example, what is the mean age for this population? On the average, how many hours of sleep do they get each night? What is the mean number of times they have pizza per month? What is the mean amount of time spent studying per week? Notice that we are asking questions about values for population parameters (mean age, mean hours of sleep, and so on). Of course, the population of college students in the United States is large—much too large and unmanageable to study in its entirety.

When you attempt to describe the typical college student, you probably will take a look at students you know on your campus. From this group, you can begin to describe what the population of college students might be like. Notice that you are starting with a sample (the students you know) and then you are making some general statements about the population. For example, the mean age is around 21, the mean number of pizzas ordered per month is 10, the mean amount of time spent studying per week is 16 hours, and so on. The process of using sample data to estimate the values for population parameters is called *estimation*. It is used to make inferences about unknown populations and often serves as a follow-up to hypothesis tests.

As the term *estimation* implies, the sample data provide values that are only approximations (estimates) of the population parameters. Many factors can influence these estimates. One obvious factor is sample size. Suppose that you knew only two other students. How might this affect your estimate? What if you knew 100 other students to use in making your estimates? Another factor is the type of estimate used. Instead of estimating the population mean age of students at precisely 21, why not estimate the mean to be somewhere in the interval between 20 and 23 years? Notice that by using an interval you have made your estimate allowing for a *margin of error*. Accordingly, you can be more certain that your estimate contains the population parameter. In this chapter, we address these and many other questions as we closely examine the process of estimation.

12.1 AN OVERVIEW OF ESTIMATION

In Chapter 8, we introduced hypothesis testing as a statistical procedure that allows researchers to use sample data to draw inferences about populations. Hypothesis testing is probably the most frequently used inferential technique, but it is not the only one. In this chapter, we will examine the process of estimation, which provides researchers with an additional method for using samples as the basis for drawing general conclusions about populations.

The basic principle underlying all of inferential statistics is that samples are representative of the populations from which they come. The most direct application of this principle is the use of sample values as estimators of the corresponding population values—that is, using statistics to estimate parameters. This process is called *estimation*.

DEFINITION   The inferential process of using sample statistics to estimate population parameters is called **estimation**.

The use of samples to estimate populations is quite common. For example, you often hear news reports such as "Sixty percent of the general public approves of the president's new budget plan." Clearly, the percentage that is reported was obtained from a sample (they don't ask everyone's opinion), and this sample statistic is being used as an estimate of the population parameter.

We already have encountered estimation in earlier sections of this book. For example, the formula for sample variance (Chapter 4) was developed so that the sample value would give an accurate and unbiased estimate of the population variance. Now we will examine the process of using sample means as the basis for estimating population means.

Before we begin the actual process of estimation, a few general points should be kept in mind. First, a sample will not give a perfect picture of the whole population. A sample is expected to be representative of the population, but there always will be some differences between the sample and the entire population. These differences are referred to as *sampling error*. Second, there are two distinct ways of making estimates. Suppose, for example, you are asked to estimate the age of the authors of this book. If you look in the frontmatter of the book, just before the Contents, you will find pictures of Gravetter and Wallnau. We are roughly the same age, so pick either one of us and estimate how old we are. Note that you could make your estimate using a single value (for example, Gravetter appears to be 55 years old) or you could use a range of values (for example, Wallnau seems to be between 45 and 55 years old). The first estimate, using a single number, is called a *point estimate*.

DEFINITION

For a **point estimate**, you use a single number as your estimate of an unknown quantity.

Point estimates have the advantage of being very precise; they specify a particular value. On the other hand, you generally do not have much confidence that a point estimate is correct. For example, most of you would not be willing to bet that Gravetter is exactly 55 years old.

The second type of estimate, using a range of values, is called an *interval estimate*. Interval estimates do not have the precision of point estimates, but they do give you more confidence. For example, it would be reasonably safe for you to bet that Wallnau is between 40 and 60 years old. At the extreme, you would be very confident betting that Wallnau is between 20 and 70 years old. Note that there is a tradeoff between precision and confidence. As the interval gets wider and wider, your confidence grows. At the same time, however, the precision of the estimate gets worse. We will be using samples to make both point and interval estimates of a population mean. Because the interval estimates are associated with confidence, they usually are called *confidence intervals*.

DEFINITIONS

For an **interval estimate**, you use a range of values as your estimate of an unknown quantity.

When an interval estimate is accompanied by a specific level of confidence (or probability), it is called a **confidence interval**.

Estimation is used in the same general situations in which we have already used hypothesis testing. In fact, there is an estimation procedure that accompanies each of the hypothesis tests we presented in the preceding chapters. Figure 12.1 shows an example of a research situation in which either hypothesis testing or estimation could be used. The figure shows a population with an unknown mean (the population after treatment). A sample is selected from the unknown population. The goal of estimation is to use the sample data to obtain an estimate of the unknown population mean.

FIGURE 12.1

The basic research situation for either hypothesis testing or estimation. The goal is to use the sample data to answer questions about the unknown population mean after treatment.

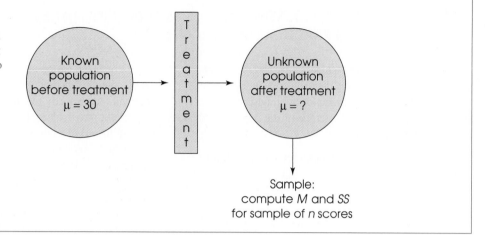

COMPARISON OF HYPOTHESIS TESTS AND ESTIMATION

You should recognize that the situation shown in Figure 12.1 is the same situation in which we have used hypothesis tests in the past. In many ways, hypothesis testing and estimation are similar. They both make use of sample data and either z-scores or t statistics to find out about unknown populations. But these two inferential procedures are designed to answer different questions. Using the situation shown in Figure 12.1 as an example, we could use a hypothesis test to evaluate the effect of the treatment. The test would determine whether or not the treatment has any effect. Notice that this is a yes-no question. The null hypothesis says, "No, there is no treatment effect." The alternative hypothesis says, "Yes, there is a treatment effect."

The goal of estimation, on the other hand, is to determine the value of the population mean after treatment. Essentially, estimation will determine *how much* effect the treatment has (see Box 12.1). If, for example, we obtained a point estimate of $\mu = 38$ for the population after treatment, we could conclude that the effect of the treatment is to increase scores by an average of 8 points (from the original mean of $\mu = 30$ to the post-treatment mean of $\mu = 38$).

WHEN TO USE ESTIMATION

There are three situations in which estimation commonly is used:

1. Estimation is used after a hypothesis test where H_0 is rejected. Remember that when H_0 is rejected, the conclusion is that the treatment does have an effect. The next logical question would be, How much effect? This is exactly the question that estimation is designed to answer.

2. Estimation is used when you already know that there is an effect and simply want to find out how much. For example, the city school board probably knows that a special reading program will help students. However, they want to be sure that the effect is big enough to justify the cost. Estimation is used to determine the size of the treatment effect.

| BOX 12.1 | HYPOTHESIS TESTING VERSUS ESTIMATION: STATISTICAL SIGNIFICANCE VERSUS PRACTICAL SIGNIFICANCE |

As we already noted, hypothesis tests tend to involve a yes-no decision. Either we decide to reject H_0, or we fail to reject H_0. The language of hypothesis testing reflects this process. The outcome of the hypothesis test is one of two conclusions:

There is no evidence for a treatment effect (fail to reject H_0)

or

There is a statistically significant effect (H_0 is rejected)

For example, a researcher studies the effect of a new drug on people with high cholesterol. In hypothesis testing, the question is whether or not the drug has a significant effect on cholesterol levels. Suppose the hypothesis test revealed that the drug did produce a significant decrease in cholesterol. The next question might be, How much of a reduction occurs? This question calls for estimation, in which the size of a treatment effect for the population is estimated.

Estimation can be of great practical importance because the presence of a "statistically significant" effect does not necessarily mean the results are large enough for use in practical applications. Consider the following possibility: Before drug treatment, the sample of patients had a mean cholesterol level of 225. After drug treatment, their cholesterol reading was 210. When analyzed, this 15-point change reached statistical significance (H_0 was rejected). Although the hypothesis test revealed that the drug produced a *statistically significant* change, it may not be *clinically significant*. That is, a cholesterol level of 210 is still quite high. In estimation, we would estimate the population mean cholesterol level for patients who are treated with the drug. This estimated value may reveal that even though the drug does in fact reduce cholesterol levels, it does not produce a large enough change (notice we are looking at a "how much" question) to make it of any practical value. Thus, the hypothesis test might reveal that an effect occurred, but estimation indicates it is small and of little *practical significance* in real-world applications.

3. Estimation is used when you simply want some basic information about an unknown population. Suppose, for example, you want to know about the political attitudes of students at your college. You could use a sample of students as the basis for estimating the population mean.

THE LOGIC OF ESTIMATION   As we have noted, estimation and hypothesis testing are both *inferential* statistical techniques that involve using sample data as the basis for drawing conclusions about an unknown population. More specifically, a researcher begins with a question about an unknown population parameter. To answer the question, a sample is obtained, and a sample statistic is computed. In general, *statistical inference* involves using sample statistics to help answer questions about population parameters. The general logic underlying the processes of estimation and hypothesis testing is based on the fact that each population parameter has a corresponding sample statistic. In addition, you usually can compute a standard error that measures how much discrepancy is expected, on average, between the statistic and the parameter. For example, a sample mean, M, corresponds to the population mean, μ, with a standard error measured by σ_M or s_M.

In the preceding four chapters we presented four different situations for hypothesis testing: the z-score test, the single-sample t, the independent-measures t, and the repeated-measures t. Although it is possible to do estimation in each of these four situations, we will only consider estimation for the three t statistics. Recall that the general structure of a t statistic is as follows:

$$t = \frac{\begin{array}{c}\text{sample mean} \\ \text{(or mean difference)}\end{array} - \begin{array}{c}\text{population mean} \\ \text{(or mean difference)}\end{array}}{\text{estimated standard error}}$$

This general formula is used both for hypothesis testing and for estimation. In each case, the population mean (or mean difference) is unknown. The purpose for a hypothesis test is to evaluate a hypothesis about the unknown population parameter. The purpose for estimation is to determine the value for the unknown population parameter. Because hypothesis testing and estimation have different goals, they will follow different logical paths. These different paths are outlined as follows:

HYPOTHESIS TEST	ESTIMATION
Goal: To test a hypothesis about a population parameter—usually the null hypothesis, which states that the treatment has no effect.	*Goal:* To estimate the value of an unknown population parameter—usually the value for an unknown population mean.
A. For a hypothesis test, you begin by hypothesizing a value for the unknown population parameter. This value is specified in the null hypothesis.	A. For estimation, you do not attempt to calculate t. Instead, you begin by estimating what the t value ought to be. The strategy for making this estimate is to select a "reasonable" value for t. (*Note:* You are not just picking a value for t; rather, you are estimating where the sample is located in the distribution.)
B. The hypothesized value is substituted into the formula, and *the value* for t *is computed.*	
C. If the hypothesized value produces a "reasonable" value for t, we conclude that the hypothesis was "reasonable," and we fail to reject H_0. If the result is an extreme value for t, H_0 is rejected.	B. As with hypothesis testing, a "reasonable" value for t is defined as a high-probability outcome located near the center of the distribution (see Figure 12.2).
D. A "reasonable" value for t is defined by its location in a distribution. In general, "reasonable" values are high-probability outcomes in the center of the distribution. Extreme values with low probability are considered "unreasonable" (see Figure 12.2).	C. The "reasonable" value for t is substituted into the formula, and *the value for the unknown population parameter is computed.*
	D. Because you used a "reasonable" value for t in the formula, it is assumed that the computation will produce a "reasonable" estimate of the population parameter.

FIGURE 12.2

For estimation or hypothesis testing, the distribution of t statistics is divided into two sections: the middle of the distribution, consisting of high-probability outcomes that are considered "reasonable," and the extreme tails of the distribution, consisting of low-probability, "unreasonable" outcomes.

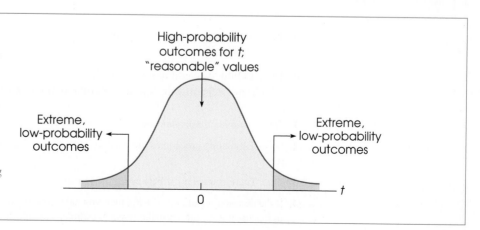

Because the goal of the estimation process is to compute a value for the unknown population mean or mean difference, it usually is easier to regroup the terms in the *t* formula so that the population value is isolated on one side of the equation. In algebraic terms, we are solving the equation for the unknown population parameter. The result takes the following form:

$$\begin{matrix} \text{population mean} \\ \text{(or mean difference)} \end{matrix} = \begin{matrix} \text{sample mean} \\ \text{(or mean difference)} \end{matrix} \pm t(\text{estimated standard error})$$

(12.1)

This is the general equation that we will use for estimation. Consider the following two points about Equation 12.1:

1. On the right-hand side of the equation, the values for the sample mean and the estimated standard error can be computed directly from the sample data. Thus, only the value of *t* is unknown. If we can determine this missing value, then we can use the equation to calculate the unknown population mean.

2. Although the specific value for the *t* statistic cannot be determined, we do know what the entire distribution of *t* statistics looks like. We can use the distribution to *estimate* what the *t* statistic ought to be.

   a. For a point estimate, the best bet is to use $t = 0$, the exact center of the distribution. There is no reason to suspect that the sample data are biased (either above average or below average), so $t = 0$ is a sensible value. Also, $t = 0$ is the most likely value, with probabilities decreasing steadily as you move away from zero toward the tails of the distribution.

   b. For an interval estimate, we will use a range of *t* values around zero. For example, to be 90% confident that our estimation is correct, we will simply use the range of *t* values that forms the middle 90% of the distribution. Note that we are estimating that the sample data correspond to a *t* statistic somewhere in the middle 90% of the *t* distribution.

Once we have estimated a value for *t*, then we have all the numbers on the right-hand side of the equation and we can calculate a value for the unknown population mean. Because one of the numbers on the right-hand side is an estimated value, the population mean that we calculate is also an estimated value.

LEARNING CHECK

1. Estimation procedures basically address a "yes-no" question: whether or not a treatment effect exists. (True or false?)

2. Estimation is primarily used to estimate the value of sample statistics. (True or false?)

3. In general, as precision of an interval estimate increases, confidence decreases. (True or false?)

4. Explain why it would *not* be sensible to use estimation after a hypothesis test in which the decision was "fail to reject the null hypothesis."

ANSWERS

1. False. Estimation asks *how much* difference exists.

2. False. Estimation procedures are used to estimate the value for unknown population parameters.

3. True. There is a trade-off between precision and confidence.

4. If the decision is fail to reject H_0, then you have concluded that there is no treatment effect. In this case, it would not make sense to estimate how much effect there is.

12.2   ESTIMATION WITH THE *t* STATISTIC

In the preceding three chapters, we introduced three different versions of the *t* statistic: the single-sample *t* in Chapter 9, the independent-measures *t* in Chapter 10, and the repeated-measures *t* in Chapter 11. Although the three *t* statistics were introduced in the context of hypothesis testing, they all can be adapted for use in estimation. As we saw in the previous section, the general form of the *t* equation for estimation is as follows:

$$\begin{array}{c}\text{population mean}\\ \text{(or mean difference)}\end{array} = \begin{array}{c}\text{sample mean}\\ \text{(or mean difference)}\end{array} \pm t(\text{estimated standard error})$$

With the single-sample *t*, we will estimate an unknown population mean, μ, using a sample mean, *M*. The estimation formula for the single-sample *t* is

$$\mu = M \pm ts_M \qquad\qquad (12.2)$$

With the independent-measures *t*, we will estimate the size of the difference between two population means, $\mu_1 - \mu_2$, using the difference between two sample means, $M_1 - M_2$. The estimation formula for the independent-measures *t* is

$$\mu_1 - \mu_2 = M_1 - M_2 \pm ts_{(M_1 - M_2)} \qquad\qquad (12.3)$$

Finally, the repeated-measures *t* statistic will be used to estimate the mean difference for the general population, μ_D, using the mean difference for a sample, M_D. The estimation formula for the repeated-measures *t* is

$$\mu_D = M_D \pm ts_{M_D} \qquad\qquad (12.4)$$

To use the *t* statistic formulas for estimation, we must determine all of the values on the right-hand side of the equation (including an estimated value for *t*) and then use these numbers to compute an estimated value for the population mean or mean difference. Specifically, you first compute the sample mean (or mean difference) and the estimated standard error from the sample data. Next, you estimate a value, or a range of values, for *t*. More precisely, you are estimating where the sample data are located in the *t* distribution. These values complete the right-hand side of the equation and allow you to compute an estimated value for the mean (or the mean difference). The following examples demonstrate the estimation procedure with each of the three *t* statistics.

ESTIMATION OF μ FOR SINGLE-SAMPLE STUDIES

In Chapter 9, we introduced single-sample studies and hypothesis testing with the *t* statistic. Now we will use a single-sample study to estimate the value for μ, using point and interval estimates.

EXAMPLE 12.1

A researcher is evaluating the effectiveness of a special summer reading program for grade-school children. Over the past several years, the researcher has determined that students entering the second grade in the city school district have an average reading achievement score of $\mu = 81$. Although it is reasonable to assume that the special summer reading program will improve reading scores, the question is, How much? To answer this question, a sample of $n = 25$ students is obtained. The students are admitted to the special reading program during the summer after they finish the first grade, and they all take the reading achievement test when they enter the second grade. The average score for the sample is $M = 88$ with $SS = 2400$. The researcher would like to use the data to estimate how much effect the special program has on reading achievement scores. Specifically, the researcher would like a point estimate and an 80% confidence interval estimate of the population mean score for students who take the special program. The overall structure for this research study is shown in Figure 12.3.

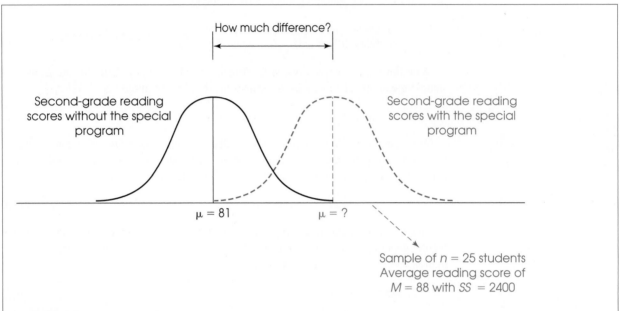

FIGURE 12.3

The structure of the research study described in Example 12.1. The goal is to use the sample to estimate the population mean for students with the special summer program. We can then estimate how much effect the special program has on students' reading scores.

In this example, we are using a single sample to estimate the mean for a single population. In this case, the estimation formula for the single-sample t is

$$\mu = M \pm ts_M$$

To use the equation, we must first compute the estimated standard error and then determine the estimated value(s) to be used for the t statistic.

Compute s^2 and s_M To compute the estimated standard error, it is first necessary to calculate the sample variance. Using the information provided, we obtain

$$s^2 = \frac{SS}{n - 1} = \frac{2400}{24} = 100$$

The estimated standard error is

$$s_M = \sqrt{\frac{s^2}{n}} = \sqrt{\frac{100}{25}} = 2$$

The point estimate   As noted earlier, a point estimate involves selecting a single value for *t*. Because the *t* distribution is always symmetrically distributed with a mean of zero, we will always use $t = 0$ as the best choice for a point estimate. Using the sample data and the estimate of $t = 0$, we obtain

$$\mu = M \pm ts_M$$
$$= 88 \pm 0(2)$$
$$= 88$$

This is our point estimate of the population mean. Note that we simply have used the sample mean, *M*, to estimate the population mean, μ. The sample is the only information that we have about the population, and it provides an unbiased estimate of the population mean (Chapter 7, page 206). That is, on average, the sample mean provides an accurate representation of the population mean. Based on this point estimate, our conclusion is that the special summer reading program will increase reading scores from an average of $\mu = 81$ to an average of $\mu = 88$. We are estimating that the program will have a 7-point effect on reading scores for the general population.

To have 80% in the middle there must be 20% (or .20) in the tails. To find the *t* values, look under two tails, .20 in the *t* table.

The interval estimate   For an interval estimate, select a range of *t* values that is determined by the level of confidence. In this example, we want 80% confidence in our estimate of μ. Therefore, we will estimate that the *t* statistic is located somewhere in the middle 80% of the *t* distribution. With $df = n - 1 = 24$, the middle 80% of the distribution is bounded by *t* values of $+1.318$ and -1.318 (see Figure 12.4). Using the sample data and this estimated range of *t* values, we obtain

$$\mu = M \pm ts_M = 88 \pm 1.318(2) = 88 \pm 2.636$$

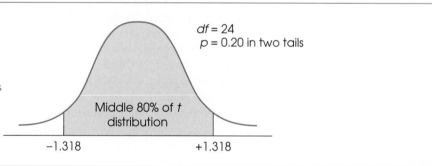

FIGURE 12.4

The 80% confidence interval with $df = 24$ is constructed using *t* values of $t = -1.318$ and $t = +1.318$. The *t* values are obtained from the table using 20% (0.20) as the proportion remaining in the two tails of the distribution.

df = 24
p = 0.20 in two tails

Middle 80% of *t* distribution

−1.318   +1.318

At one end of the interval we obtain 90.636 (88 + 2.636), and at the other end of the interval we obtain 85.364 (88 − 2.636). Thus, we obtain an interval estimate for μ that ranges from 85.364 to 90.636. Our conclusion is that after the special summer program, the population will have a mean reading score between 85.364 and 90.636. In addition, we are 80% confident that the population mean will be within this interval. The confidence comes from the fact that the calculation was based on only one assumption. Specifically, we assumed that the *t* statistic was located between $+1.318$ and -1.318, and we are 80% confident that this assumption is correct because 80% of all the possible *t* values are located in this interval.

Figure 12.5 provides a visual presentation of the results from Example 12.1. The original population mean is shown along with the two estimates (point and interval)

FIGURE 12.5

A representation of the estimates made in Example 12.1. Before a special summer reading program, the population mean reading score was $\mu = 81$. Based on a sample, the special program is estimated to increase the mean to $\mu = 88$ (point estimate), or somewhere between 85.364 and 90.636 (interval estimate).

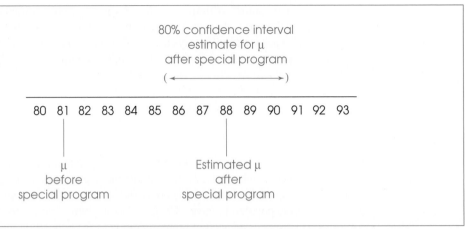

for the new mean after the special reading program is administered. The estimates clearly indicate that the program should increase scores, and they provide a clear indication of how large the increase should be.

Interpretation of the confidence interval   In the preceding example, we computed an 80% confidence interval to estimate an unknown population mean. We obtained an interval ranging from 85.364 to 90.636, and we are 80% confident that the unknown population mean is located within this interval. You should note, however, that the 80% confidence applies to the *process of computing* the interval rather than the specific end points for the interval. For example, if we repeated the process over and over, we could eventually obtain hundreds of different samples (each with $n = 25$ scores) and we could calculate hundreds of different confidence intervals. However, each interval is computed in the same way and each interval is based on the same t values of ± 1.318. As long as the individual sample produces a t statistic between 1.381 and -1.318 (which will happen 80% of the time), then the confidence interval will contain the population mean. On the other hand, if the sample produces a t value that is not between 1.318 and -1.318 (which will happen 20% of the time), then the confidence interval will not contain the population mean. Thus, out of all the different confidence intervals that we could calculate, 80% will actually contain the population mean and 20% will not contain the mean.

ESTIMATION OF $\mu_1 - \mu_2$ FOR INDEPENDENT-MEASURES STUDIES

The independent-measures t statistic uses the data from two separate samples to evaluate the mean difference between two populations. In Chapter 10, we used this statistic to answer a yes-no question: Is there any difference between the two population means? With estimation, we ask, *How much* difference? In this case, the independent-measures t statistic is used to estimate the value of $\mu_1 - \mu_2$. The following example demonstrates the process of estimation with the independent-measures t statistic.

E X A M P L E   1 2 . 2

Recent studies have allowed psychologists to establish definite links between specific foods and specific brain functions. For example, lecithin (found in soybeans, eggs, and liver) has been shown to increase the concentration of certain brain chemicals that help regulate memory and motor coordination. This experiment is designed to demonstrate the importance of this particular food substance.

The experiment involves two separate samples of newborn rats (an independent-measures experiment). The 10 rats in the first sample are given a normal diet containing standard amounts of lecithin. The 5 rats in the other sample are fed a special diet, which contains almost no lecithin. After 6 months, each of the rats is tested on a specially designed learning problem that requires both memory and motor coordination. The purpose of the experiment is to demonstrate the deficit in performance that results from lecithin deprivation. The score for each animal is the number of errors it makes before it solves the learning problem. The data from this experiment are as follows:

REGULAR DIET	NO-LECITHIN DIET
$n = 10$	$n = 5$
$M = 25$	$M = 33$
$SS = 250$	$SS = 140$

Because we fully expect that there will be a significant difference between these two treatments, we will not run the hypothesis test (although you should be able to do it). We want to use these data to obtain an estimate of the size of the difference between the two population means; that is, how much does lecithin affect learning performance? We will use a point estimate and a 95% confidence interval.

The basic equation for estimation with an independent-measures experiment is

$$\mu_1 - \mu_2 = (M_1 - M_2) \pm ts_{(M_1 - M_2)}$$

The first step is to obtain the known values from the sample data. The sample mean difference is easy; one group averaged $M = 25$, and the other averaged $M = 33$, so there is an 8-point difference. Note that it is not important whether we call this a $+8$ or a -8 difference. In either case, the size of the difference is 8 points, and the regular diet group scored lower. Because it is easier to do arithmetic with positive numbers, we will use

$$M_1 - M_2 = 8$$

Compute the standard error   To find the standard error, we first must pool the two variances:

$$s_p^2 = \frac{SS_1 + SS_2}{df_1 + df_2} = \frac{250 + 140}{9 + 4} = \frac{390}{13} = 30$$

Next, the pooled variance is used to compute the standard error:

$$s_{(M_1 - M_2)} = \sqrt{\frac{s_p^2}{n_1} + \frac{s_p^2}{n_2}} = \sqrt{\frac{30}{10} + \frac{30}{5}} = \sqrt{3 + 6} = \sqrt{9} = 3$$

Recall that this standard error combines the error from the first sample and the error from the second sample. Because the first sample is much larger, $n = 10$, it should have less error. This difference shows up in the formula. The larger sample contributes an error of 3 points, and the smaller sample contributes 6 points, which combine for a total error of 9 points under the square root.

Sample 1 has $df = 9$, and sample 2 has $df = 4$. The t statistic has $df = 9 + 4 = 13$.

Estimate the value(s) for t   The final value needed on the right-hand side of the equation is t. The data from this experiment would produce a t statistic with $df = 13$. With

FIGURE 12.6

The distribution of t values with $df = 13$. Note that t values pile up around zero and that 95% of the values are located between -2.160 and $+2.160$.

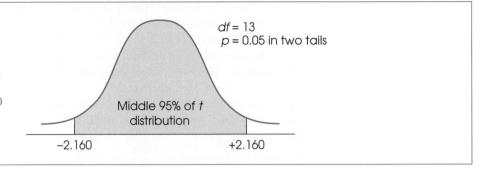

$df = 13$
$p = 0.05$ in two tails

Middle 95% of t distribution

-2.160 $+2.160$

13 degrees of freedom, we can sketch the distribution of all the possible t values. This distribution is shown in Figure 12.6. The t statistic for our data is somewhere in this distribution. The problem is to estimate where. For a point estimate, the best bet is to use $t = 0$. This is the most likely value, located exactly in the middle of the distribution. To gain more confidence in the estimate, you can select a range of t values. For 95% confidence, for example, you would estimate that the t statistic is somewhere in the middle 95% of the distribution. Checking the table, you find that the middle 95% is bounded by values of $t = +2.160$ and $t = -2.160$.

Using these t values and the sample values computed earlier, we now can estimate the magnitude of the performance deficit caused by lecithin deprivation.

Compute the point estimate   For a point estimate, use the single-value (point) estimate of $t = 0$:

$$\mu_1 - \mu_2 = (M_1 - M_2) \pm ts_{(M_1 - M_2)}$$
$$= 8 \pm 0(3)$$
$$= 8$$

Note that the result simply uses the sample mean difference to estimate the population mean difference. The conclusion is that lecithin deprivation produces an average of 8 more errors on the learning task. (Based on the fact that the non-deprived animals averaged around 25 errors, an 8-point increase would mean a performance deficit of approximately 30%.)

Construct the interval estimate   For an interval estimate, or confidence interval, use the range of t values. With 95% confidence, at one extreme,

$$\mu_1 - \mu_2 = (M_1 - M_2) + ts_{(M_1 - M_2)}$$
$$= 8 + 2.160(3)$$
$$= 8 + 6.48$$
$$= 14.48$$

and at the other extreme,

$$\mu_1 - \mu_2 = (M_1 - M_2) - ts_{(M_1 - M_2)}$$
$$= 8 - 2.160(3)$$
$$= 8 - 6.48$$
$$= 1.52$$

This time we conclude that the effect of lecithin deprivation is to increase errors, with an average increase somewhere between 1.52 and 14.48 errors. We are 95% confident of this estimate because our only estimation was the location of the *t* statistic, and we used the middle 95% of all the possible *t* values.

Note that the result of the point estimate is to say that lecithin deprivation will increase errors by *exactly* 8 points. To gain confidence, you must lose precision and say that errors will increase by *around* 8 points (for 95% confidence, we say that the average increase will be 8 ± 6.48).

ESTIMATION OF μ_D FOR REPEATED-MEASURES STUDIES

Finally, we turn our attention to the repeated-measures study. Remember that this type of study has a single sample of subjects, which is measured in two different treatment conditions. By finding the difference between the score for treatment 1 and the score for treatment 2, we can determine a difference score for each subject.

$$D = X_2 - X_1$$

The mean for the sample of D scores, M_D, is used to estimate the population mean μ_D, which is the mean for the entire population of difference scores.

EXAMPLE 12.3

A pharmaceutical company has developed a new medication to lower cholesterol and would like to determine how much effect the new drug has. A random sample of $n = 16$ patients who have been diagnosed with high cholesterol is obtained. Cholesterol levels are recorded for all 16 individuals, who then are given the drug for a 6-week period. After taking the new medication for 6 weeks, each individual's cholesterol is measured again and the difference between the first score and the second score is recorded. For this sample, cholesterol levels declined by an average of $M_D = 21$ points with $SS = 1215$. The company would like to use these data to estimate the drug's effect for the general population, using a point estimate and a 90% confidence interval estimate for μ_D.

You should recognize that this study requires a repeated-measures *t* statistic. For estimation, the repeated-measures *t* equation is as follows:

$$\mu_D = M_D \pm t s_{M_D}$$

The sample mean is $M_D = 21$, so all that remains is to compute the estimated standard error and estimate the appropriate value(s) for *t*.

Compute the standard error   To find the standard error, we first must compute the sample variance:

$$s^2 = \frac{SS}{n-1} = \frac{1215}{15} = 81$$

Now the estimated standard error is

$$s_{M_D} = \sqrt{\frac{s^2}{n}} = \sqrt{\frac{81}{16}} = \frac{9}{4} = 2.25$$

To complete the estimate of μ_D, we must identify the value of *t*. We will consider the point estimate and the interval estimate separately.

Compute the point estimate   To obtain a point estimate, a single value of t is selected to approximate the location of M_D. Remember that the t distribution is symmetrical and bell-shaped with a mean of zero (see Figure 12.7). Because $t = 0$ is the most frequently occurring value in the distribution, this is the t value used for the point estimate. Using this value in the estimation formula gives

$$\mu_D = M_D \pm ts_{M_D} = 21 \pm 0(2.25) = 21$$

For this example, our best estimate is that the new medication will lower cholesterol levels in the general population by an average of $\mu_D = 21$ points. As noted several times before, the sample mean, $M_D = 21$, provides the best point estimate of μ_D.

Construct the interval estimate   The psychologist also wanted to make an interval estimate in order to be 90% confident that the interval contains the value of μ_D. To get the interval, it is necessary to determine what t values form the boundaries of the middle 90% of the t distribution. To use the t distribution table, we first must determine the proportion associated with the tails of this distribution. With 90% in the middle, the remaining area in both tails must be 10%, or $p = .10$. Also note that our sample has $n = 16$ scores, so the t statistic will have $df = n - 1 = 15$. Using $df = 15$ and $p = 0.10$ for two tails, you should find the values $+1.753$ and -1.753 in the t table. These values form the boundaries for the middle 90% of the t distribution. (See Figure 12.7.) We are confident that the t value for our sample is in this range because 90% of all the possible t values are there. Using these values in the estimation formula, we obtain the following: On one end of the interval,

$$\mu_D = M_D - ts_{M_D}$$
$$= 21 - 1.753(2.25)$$
$$= 21 - 3.94$$
$$= 17.06$$

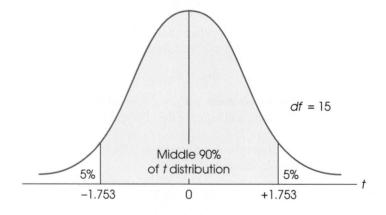

FIGURE 12.7

The t values for the 90% confidence interval are obtained by consulting the t tables for $df = 15, p = 0.10$ for two tails.

$df = 15$

Middle 90%
of t distribution

5%   5%

-1.753 0 $+1.753$ t

and on the other end of the interval,

$$\mu_D = 21 + 1.753(2.25)$$
$$= 21 + 3.94$$
$$= 24.94$$

Therefore, the company can conclude that the new medication will reduce cholesterol levels in the general population by an average of around 21 points. The company can be 90% confident that the average reduction will be between 17.06 and 24.94 points.

1. A professor notices that students who get an A in physics have high grade point averages (GPAs) in their engineering courses. The professor selects a sample of $n = 16$ engineering majors who have earned As in physics. The mean GPA in engineering courses for this sample is $M = 3.30$ with $SS = 10$. Use this sample to find the 99% confidence interval for the population mean.

2. A psychologist studies the change in mood from the follicular phase (prior to ovulation) to the luteal phase (after ovulation) during the menstrual cycle. In a repeated-measures study, a sample of $n = 9$ women take a mood questionnaire during each phase. On average, the participants show an increase in dysphoria (negative moods) of $M_D = 18$ points with $SS = 152$. Determine the 95% confidence interval for population mean change in mood.

3. In families with several children, the first-born tend to be more reserved and serious, whereas the last-born tend to be more outgoing and happy-go-lucky. A psychologist is using a standardized personality inventory to measure the magnitude of this difference. Two samples are used: 8 first-born children and 8 last-born children. Each child is given the personality test. The results are as follows:

FIRST-BORN	LAST-BORN
$M = 11.4$	$M = 13.9$
$SS = 26$	$SS = 30$

  a. Use these sample statistics to make a point estimate of the population mean difference in personality for first-born versus last-born children.

  b. Make an interval estimate of the population mean difference so that you are 80% confident that the true mean difference is in your interval.

ANSWERS   1. $s^2 = 0.667$, $s_M = .20$, $df = 15$, $t = \pm 2.947$; estimate that μ is between 2.71 and 3.89.

2. $s^2 = 19$, $s_{M_D} = 1.45$, $df = 8$, $t = \pm 2.306$; estimate that μ_D is between 14.66 and 21.34.

3. a. For a point estimate, use the sample mean difference: $M_1 - M_2 = 2.5$ points.

  b. Pooled variance $= 4$, estimated standard error $= 1$, $df = 14$, $t = \pm 1.345$. The 80% confidence interval is 1.16 to 3.85.

12.3 A FINAL LOOK AT ESTIMATION

FACTORS AFFECTING THE WIDTH OF A CONFIDENCE INTERVAL

Two characteristics of the confidence interval should be noted. First, notice what happens to the width of the interval when you change the level of confidence (the percent confidence). To gain more confidence in your estimate, you must increase the width of the interval. Conversely, to have a smaller interval, you must give up confidence. This is the basic trade-off between precision and confidence that was discussed earlier. In the estimation formula, the percentage of confidence influences the width of the interval by way of the t value. The larger the level of confidence (the percentage), the larger the t value and the larger the interval. This relationship can be seen in Figure 12.7. In the figure, we have identified the middle 90% of the t distribution in order to find a 90% confidence interval. It should be obvious that if we were to increase the confidence level to 95%, it would be necessary to increase the range of t values and thereby increase the width of the interval.

Second, note what would happen to the interval width if you had a different sample size. This time the basic rule is as follows: The bigger the sample (n), the smaller the interval. This relationship is straightforward if you consider the sample size as a measure of the amount of information. A bigger sample gives you more information about the population and allows you to make a more precise estimate (a narrower interval). The sample size controls the magnitude of the standard error in the estimation formula. As the sample size increases, the standard error decreases, and the interval gets smaller.

With t statistics, the sample size has an additional effect on the width of a confidence interval. Remember that the exact shape of the t distribution depends on degrees of freedom. As the sample size gets larger, df also gets larger, and the t values associated with any specific percentage of confidence get smaller. This fact simply enhances the general relationship that the larger a sample, the smaller a confidence interval.

ESTIMATION, EFFECT SIZE, AND HYPOTHESIS TESTS

The process of estimation, especially the estimation of mean differences, provides a relatively simple and direct method for evaluating effect size. For example, the outcome of the study in Example 12.3 indicates that the new medication will reduce cholesterol levels by an estimated 21 points. In this case, the estimation process produces a very clear and understandable indication of how large the treatment effect actually is.

In addition to describing the size of a treatment effect, estimation can be used to get an indication of the "significance" of the effect. Example 12.2 presented an independent-measures research study examining the effect of lecithin on problem-solving performance for rats. Based on the results of this study, it was estimated that the mean difference in performance produced by lecithin was $\mu_1 - \mu_2 = 8$ points. The 95% confidence interval estimated the mean difference to be between 1.52 points and 14.48 points. The confidence interval estimate is shown in Figure 12.8. In addition to the confidence interval for $\mu_1 - \mu_2$, we have marked the spot where the mean difference is equal to zero. You should recognize that a mean difference of zero is exactly what

FIGURE 12.8

The 95% confidence interval for the population mean difference ($\mu_1 - \mu_2$) from Example 12.2. Note that $\mu_1 - \mu_2 = 0$ is excluded from the confidence interval, indicating that a zero difference is not an acceptable value (H_0 would be rejected in a hypothesis test).

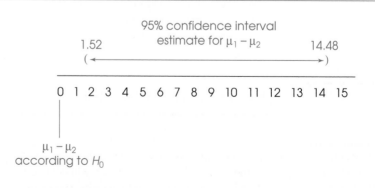

would be predicted by the null hypothesis if we were doing a hypothesis test. You also should realize that a zero difference ($\mu_1 - \mu_2 = 0$) is *outside* the 95% confidence interval. In other words, $\mu_1 - \mu_2 = 0$ is not an acceptable value if we want 95% confidence in our estimate. This conclusion is equivalent to rejecting H_0 with $\alpha = .05$. On the other hand, if a mean difference of zero was included within the confidence interval, then we would have to accept $\mu_1 - \mu_2 = 0$ as a reasonable conclusion, which is the same as failing to reject H_0.

LEARNING CHECK

1. If all other factors are held constant, an 80% confidence interval will be wider than a 90% confidence interval. (True or false?)

2. If all other factors are held constant, a confidence interval computed from a sample of $n = 25$ will be wider than a confidence interval from a sample of $n = 100$. (True or false?)

3. A 99% confidence interval for a population mean difference (μ_D) extends from -1.50 to $+3.50$. If a repeated-measures hypothesis test with two tails and $\alpha = .01$ were conducted using the same data, the decision would be to fail to reject the null hypothesis. (True or false?)

ANSWERS

1. False. Greater confidence requires a wider interval.

2. True. The smaller sample will produce a wider interval.

3. True. The value $\mu_D = 0$ is included within the 99% confidence interval, which means that it is an acceptable value with $\alpha = .01$ and would not be rejected.

SUMMARY

1. Estimation is a procedure that uses sample data to obtain an estimate of a population mean or mean difference. The estimate can be either a point estimate (single value) or an interval estimate (range of values). Point estimates have the advantage of precision, but they do not give much confidence. Interval estimates provide confidence, but you lose precision as the interval grows wider.

2. Estimation and hypothesis testing are similar processes: Both use sample data to answer questions about populations. However, these two procedures are designed to answer different questions. Hypothesis testing will tell you whether or not a treatment effect exists (yes or no). Estimation will tell you how much treatment effect there is.

3. The estimation process begins by solving the t-statistic equation for the unknown population mean (or mean difference).

$$\begin{array}{c} \text{population mean} \\ \text{(or mean difference)} \end{array} = \begin{array}{c} \text{sample mean} \\ \text{(or mean difference)} \\ \pm\ t(\text{estimated standard error}) \end{array}$$

Except for the value of t, the numbers on the right-hand side of the equation are all obtained from the sample data. By using an estimated value for t, you can then compute an estimated value for the population mean (or mean difference). For a point estimate, use $t = 0$. For an interval estimate, first select a level of confidence and then look up the corresponding range of t values from the t-distribution table. For example, for 90% confidence, use the range of t values that determine the middle 90% of the distribution.

4. For a single-sample study, the mean from one sample is used to estimate the mean for the corresponding population.

$$\mu = M \pm t s_M$$

For an independent-measures study, the means from two separate samples are used to estimate the mean difference between two populations.

$$(\mu_1 - \mu_2) = (M_1 - M_2) \pm t s_{(M_1 - M_2)}$$

For a repeated-measures study, the mean from a sample of difference scores (D values) is used to estimate the mean difference for the general population.

$$\mu_D = M_D \pm t s_{M_D}$$

5. The width of a confidence interval is an indication of its precision: A narrow interval is more precise than a wide interval. The interval width is influenced by the sample size and the level of confidence.
   a. As sample size (n) gets larger, the interval width gets smaller (greater precision).
   b. As the percentage confidence increases, the interval width gets larger (less precision).

RESOURCES

WebTUTOR

If you are using WebTutor along with this book, there is a WebTutor section corresponding to this chapter. The WebTutor contains a brief summary of Chapter 12, hints for learning the concepts and the formulas for estimation, cautions about common errors, and sample exam items including solutions.

The statistics computer package SPSS can be used to compute confidence intervals for all three of the t tests (single-sample t, independent-measures t, and repeated-measures t). As you have seen in the previous three chapters, the confidence interval is included automatically in the output from each of the hypothesis tests. SPSS uses a 95% confidence interval as the default value, but you can select any level of confidence by clicking on **Options** before your final OK. Within the Options box, you can change the 95% value to another number of your choice. For more information on each of the three t tests, see the SPSS sections in Chapters 9, 10, and 11.

KEY TERMS

estimation   point estimate   interval estimate   confidence interval

FOCUS ON PROBLEM SOLVING

1. Although hypothesis tests and estimation are similar in some respects, remember that they are separate statistical techniques. A hypothesis test is used to determine whether

or not there is evidence for a treatment effect. Estimation is used to determine how much effect a treatment has.

2. When students perform a hypothesis test and estimation with the same set of data, a common error is to take the t statistic from the hypothesis test and use it in the estimation formula. For estimation, the t value is determined by the level of confidence and must be looked up in the appropriate table.

3. Now that you are familiar with several different formulas for hypothesis tests and estimation, one problem will be determining which formula is appropriate for each set of data. When the data consist of a single sample selected from a single population, the appropriate statistic is the single-sample t. For an independent-measures design, you will always have two separate samples. In a repeated-measures design, there is only one sample, but each individual is measured twice so that difference scores can be computed.

DEMONSTRATION 12.1

ESTIMATION WITH A SINGLE-SAMPLE t STATISTIC

A sample of $n = 16$ is randomly selected from a population with unknown parameters. For the following sample data, estimate the value of μ using a point estimate and a 90% confidence interval:

Sample data: 13, 10, 8, 13, 9, 14, 12, 10, 11, 10,
15, 13, 7, 6, 15, 10

Note that we have a single sample and we do not know the value for σ. Thus, the single-sample t statistic should be used for these data. The formula for estimation is

$$\mu = M \pm ts_M$$

STEP 1   Compute the sample mean.
The sample mean is the basis for our estimate of μ. For these data,

$$\Sigma X = 13 + 10 + 8 + 13 + 9 + 14 + 12 + 10 +$$
$$11 + 10 + 15 + 13 + 7 + 6 + 15 + 10$$

$$= 176$$

$$M = \frac{\Sigma X}{n} = \frac{176}{16} = 11$$

STEP 2   Compute the estimated standard error, s_M.
To compute the estimated standard error, we must first find the value for SS and the sample variance.
Sum of squares. We will use the definitional formula for SS. The following table demonstrates the computations:

X	X − M	(X − M)²
13	13 − 11 = +2	4
10	10 − 11 = −1	1
8	8 − 11 = −3	9
13	13 − 11 = +2	4
9	9 − 11 = −2	4
14	14 − 11 = +3	9
12	12 − 11 = +1	1
10	10 − 11 = −1	1
11	11 − 11 = 0	0
10	10 − 11 = −1	1
15	15 − 11 = +4	16
13	13 − 11 = +2	4
7	7 − 11 = −4	16
6	6 − 11 = −5	25
15	15 − 11 = +4	16
10	10 − 11 = −1	1

To obtain SS, we sum the squared deviation scores in the last column.

$$SS = \Sigma(X - M)^2 = 112$$

Variance. The sample variance is computed for these data.

$$s^2 = \frac{SS}{n - 1} = \frac{112}{16 - 1} = \frac{112}{15} = 7.47$$

Estimated standard error. We can now determine the estimated standard error.

$$s_M = \sqrt{\frac{s^2}{n}} = \sqrt{\frac{7.47}{16}} = \sqrt{0.467} = 0.68$$

STEP 3   Compute the point estimate for μ.

For a point estimate, we use $t = 0$. Using the estimation formula, we obtain

$$\mu = M \pm ts_M$$

$$= 11 \pm 0(0.68)$$

$$= 11 \pm 0 = 11$$

The point estimate for the population mean is $\mu = 11$.

STEP 4   Determine the confidence interval for μ.

For these data, we want the 90% confidence interval. Therefore, we will use a range of t values that form the middle 90% of the distribution. For this demonstration, degrees of freedom are

$$df = n - 1 = 16 - 1 = 15$$

If we are looking for the middle 90% of the distribution, then 10% ($p = 0.10$) would lie in both tails outside of the interval. To find the t values, we look up $p = 0.10$, two tails, for $df = 15$ in the t-distribution table. The t values for the 90% confidence interval are $t = \pm 1.753$.

Using the estimation formula, one end of the confidence interval is

$$\mu = M = ts_M$$
$$= 11 - 1.753(0.68)$$
$$= 11 - 1.19 = 9.81$$

For the other end of the confidence interval, we obtain

$$\mu = M + ts_M$$
$$= 11 + 1.753(0.68)$$
$$= 11 + 1.19 = 12.19$$

Thus, the 90% confidence interval for μ is from 9.81 to 12.19.

DEMONSTRATION 12.2

ESTIMATION WITH THE INDEPENDENT-MEASURES t STATISTIC

Samples are taken from two school districts, and knowledge of American history is tested with a short questionnaire. For the following sample data, estimate the amount of mean difference between the students of these two districts. Specifically, provide a point estimate and a 95% confidence interval for $\mu_1 - \mu_2$.

District A scores: 18,   15,   24,   15

District B scores: 9,   12,   13,   6

STEP 1   Compute the sample means.
The estimate of population mean difference ($\mu_1 - \mu_2$) is based on the sample mean difference ($M_1 - M_2$).
For district A,

$$\Sigma X = 18 + 15 + 24 + 15 = 72$$

$$M_1 = \frac{\Sigma X}{n} = \frac{72}{4} = 18$$

For district B,

$$\Sigma X = 9 + 12 + 13 + 6 = 40$$

$$M_2 = \frac{\Sigma X}{n} = \frac{40}{4} = 10$$

STEP 2   Calculate the estimated standard error for mean difference, $s_{(M_1-M_2)}$.
To compute the estimated standard error, we first need to determine the values of SS for both samples and pooled variance.
Sum of squares. The computations for sum of squares, using the definitional formula, are shown for both samples in the following tables:

DISTRICT A			DISTRICT B		
X	X − M	(X − M)²	X	X − M	(X − M)²
18	18 − 18 = 0	0	9	9 − 10 = −1	1
15	15 − 18 = −3	9	12	12 − 10 = +2	4
24	24 − 18 = +6	36	13	13 − 10 = +3	9
15	15 − 18 = −3	9	6	6 − 10 = −4	16

For district A,

$$SS_1 = \Sigma(X - M)^2 = 0 + 9 + 36 + 9 = 54$$

For district B,

$$SS_2 = \Sigma(X - M)^2 = 1 + 4 + 9 + 16 = 30$$

Pooled variance. For pooled variance, we use the SS and df values from both samples. For district A, $df_1 = n_1 - 1 = 3$. For district B, $df_2 = n_2 - 1 = 3$. Pooled variance is

$$s_p^2 = \frac{SS_1 + SS_2}{df_1 + df_2} = \frac{54 + 30}{3 + 3} = \frac{84}{6} = 14$$

Estimated standard error. The estimated standard error for mean difference can now be calculated.

$$s_{(M_1 - M_2)} = \sqrt{\frac{s_p^2}{n_1} + \frac{s_p^2}{n_2}} = \sqrt{\frac{14}{4} + \frac{14}{4}} = \sqrt{3.5 + 3.5}$$
$$= \sqrt{7} = 2.65$$

STEP 3   Compute the point estimate for $\mu_1 - \mu_2$.

For the point estimate, we use a t value of zero. Using the sample means and estimated standard error from previous steps, we obtain

$$\mu_1 - \mu_2 = (M_1 - M_2) \pm ts_{(M_1 - M_2)}$$
$$= (18 - 10) \pm 0(2.65)$$
$$= 8 \pm 0 = 8$$

STEP 4   Determine the confidence interval for $\mu_1 - \mu_2$.

For the independent-measures t statistic, degrees of freedom are determined by

$$df = n_1 + n_2 - 2$$

For these data, df is

$$df = 4 + 4 - 2 = 6$$

With a 95% level of confidence, 5% of the distribution falls in the tails outside the interval. Therefore, we consult the t-distribution table for $p = 0.05$, two tails, with $df = 6$. The

t values from the table are $t = \pm2.447$. On one end of the confidence interval, the population mean difference is

$$\mu_1 - \mu_2 = (M_1 - M_2) - ts_{(M_1-M_2)}$$
$$= (18 - 10) - 2.447(2.65)$$
$$= 8 - 6.48$$
$$= 1.52$$

On the other end of the confidence interval, the population mean difference is

$$\mu_1 - \mu_2 = (M_1 - M_2) + ts_{(M_1-M_2)}$$
$$= (18 - 10) + 2.447(2.65)$$
$$= 8 + 6.48$$
$$= 14.48$$

Thus, the 95% confidence interval for population mean difference is from 1.52 to 14.48.

PROBLEMS

1. Explain how the purpose of estimation differs from the purpose of a hypothesis test.

2. Explain why it would *not* be reasonable to use estimation after a hypothesis test where the decision was "fail to reject H_0."

3. Explain how each of the following factors affects the width of a confidence interval:
   a. Increasing the sample size
   b. Increasing the sample variability
   c. Increasing the level of confidence (the percentage of confidence)

4. For the following studies, state whether estimation or hypothesis testing is required. Also, is an independent- or a repeated-measures t statistic appropriate?
   a. An educator wants to determine how much mean difference can be expected for the population in SAT scores following an intensive review course. Two samples are selected. The first group takes the review course, and the second receives no treatment. SAT scores are subsequently measured for both groups.
   b. A psychiatrist would like to determine whether a new medication has any effect on psychotic symptoms. A sample of patients is assessed and then placed on the drug therapy for 2 weeks. The severity of their symptoms is assessed again at the end of the treatment.

   c. A researcher would like to determine how much people's moods are affected by seasonal changes. A sample of 100 adults is obtained and each individual is given a mood-analysis questionnaire in the summer and again in the winter.
   d. A researcher would like to determine whether participation in sports has an effect on self-esteem for high school students. The researcher obtains a sample of 50 students who are active in varsity sports and a comparison sample of 50 students who do not participate in any high school sports. Each student is given a self-esteem questionnaire.

5. A college administrator would like to determine how much time students spend on homework assignments during a typical week. A questionnaire is sent to a sample of $n = 25$ students and their responses indicate a mean of $M = 7.4$ hours per week with $SS = 216$.
   a. Use the data to make a point estimate of the mean amount of homework for the entire student population.
   b. Make an interval estimate of the population mean so that you are 90% confident the real mean is in your interval.

6. A sample of $n = 16$ scores is obtained from an unknown population. The sample has a mean of $M = 46$ with $SS = 6000$.
   a. Use the sample data to make an 80% confidence interval estimate of the unknown population mean.

b. Make a 90% confidence interval estimate of μ.

c. Make a 95% confidence interval estimate of μ.

d. In general, how is the width of a confidence interval related to the percentage of confidence?

7. A sample is obtained from an unknown population. The sample mean is $M = 34$ with a sample variance of $s^2 = 36$.

a. Assuming $n = 4$, use the data to make a 90% confidence interval estimate of the unknown population mean.

b. Assuming $n = 16$, use the data to make a 90% confidence interval estimate of μ.

c. Assuming $n = 36$, use the data to make a 90% confidence interval estimate of μ.

d. In general, how is the width of a confidence interval related to the size of the sample?

8. A researcher has constructed a 90% confidence interval of 87 ± 10, based on a sample of $n = 25$ scores. Note that this interval is 20 points wide (from 77 to 97). What would happen to the width of the interval if the sample size was increased (for example, to $n = 100$)? What would happen to the width if the level of confidence was increased (for example, to 95%)? In each case, assume that other factors, such as sample variance, are relatively constant.

9. The manufacturers of a popular medicinal herb claim that their product increases the speed and accuracy of mental processing. To test this claim, a researcher obtains a random sample of $n = 20$ adults. Half of the adults are given the herb for 30 days and the other half take a placebo. Then, all the participants are given a decision-making test and their response times are recorded. Those taking the herb had an average time of $M = 24.5$ seconds with $SS = 43$, and those with the placebo had an average time of $M = 28.1$ seconds with $SS = 47$. Use these data to

a. Make a point estimate of population mean difference in response time.

b. Make an interval estimate of the mean difference so you are 80% confident that the real mean difference is in your interval.

10. To determine how well people are complying with posted speed limits, the State Highway Patrol measured automobile speeds on a section of highway where the posted limit is 55 mph. For a sample of $n = 100$ cars during daylight hours and good weather conditions, they obtained an average speed of $M = 62$ miles per hour with $SS = 2475$. Using these data,

a. Make a point estimate of the average speed for the general population for this section of highway.

b. Make an interval estimate of the population mean speed so you are 90% confident that the real mean is in your interval.

11. In an extension of the highway speed study in Problem 10, the researchers also wanted to determine how much people adjust speed in response to weather conditions. In addition to the sample of speeds obtained in Problem 10, they also recorded speeds for $n = 100$ cars during daylight hours when there was moderate rain. For this second sample, they obtained a mean of $M = 58.5$ mph with $SS = 2970$. Use the two samples to estimate how much difference there is in highway speeds between dry conditions and rainy conditions.

a. Make a point estimate of the population mean difference.

b. Make an interval estimate of the mean difference so you are 95% confident that the true mean difference is in your interval.

12. It is well known that the ability to maintain focused attention declines over time. For this reason, people who must maintain a high level of attention (such as air traffic controllers) require frequent rest breaks. To determine how much decline occurs during the first 30 minutes of focused attention, a researcher tests a sample of $n = 25$ participants on a visual detection task. The participants are required to watch a blank TV screen and respond each time a small dot of light appears. The researcher recorded the number of errors during the first 15 minutes and during the second 15 minutes, and recorded a difference score for each participant. On average, this sample made $M_D = 5.7$ more errors during the second 15-minute period than during the first, with $SS = 864$. Using these data,

a. Make a point estimate of the mean decline in performance for the general population.

b. Make an interval estimate of the mean difference so you are 95% confident that the true mean is in your interval.

c. Based on the interval estimate from part (b), do these data indicate a *significant* change in performance (assume a two-tailed test with $\alpha = .05$)?

13. A developmental psychologist would like to determine how much fine motor skill improves for children from age 3 to age 4. A random sample of $n = 15$ three-year-old children and a second sample of $n = 15$ four-year-olds are obtained. Each child is given a manual dexterity test that measures fine motor skills. The average score for the older children was $M = 40.6$ with $SS = 430$ and the average for the younger children was $M = 35.4$ with $SS = 410$. Using these data,

a. Make a point estimate of the population mean difference in fine motor skills.

b. Make an interval estimate so you are 95% confident that the real mean difference is in your interval.

c. Make an interval estimate so you are 99% confident that the real mean difference is in your interval.

d. Based on your answers from (b) and (c), do these data indicate a significant change using a two-tailed test with $\alpha = .05$? Is the difference significant with $\alpha = .01$?

14. An educational psychologist has observed that children seem to lose interest and enthusiasm for school as they progress through the elementary grades. To measure the extent of this phenomenon, the psychologist selects a sample of $n = 15$ second-grade children and a sample of $n = 15$ fifth-graders. Each child is given a questionnaire measuring his/her attitude toward school. Higher scores indicate a more positive attitude. The second-grade children average $M = 85$ with $SS = 1620$, and the fifth-graders average $M = 71$ with $SS = 1740$. Use these data to estimate how much the enthusiasm for school declines from second to fifth grade. Make a point estimate and a 90% confidence interval estimate of the mean difference.

15. The counseling center at the college offers a short course in study skills for students who are having academic difficulty. To evaluate the effectiveness of this course, a sample of $n = 25$ students is selected, and each student's grade point average is recorded for the semester before the course and for the semester immediately following the course. On average, these students show an increase of $M_D = 0.72$ with $SS = 24$. Use these data to estimate how much effect the course has on grade point average. Make a point estimate and a 95% confidence interval estimate of the mean difference.

16. A psychologist has developed a new personality questionnaire for measuring self-esteem and would like to estimate the population parameters for the test scores. The questionnaire is administered to a sample of $n = 25$ participants. This sample has an average score of $M = 43$ with $SS = 2400$.
a. Provide an estimate for the population variance.
b. Make a point estimate for the population mean.
c. Make an interval estimate of μ so you are 90% confident that the value for μ is in your interval.

17. A common test of short-term memory requires participants to repeat a random string of digits that was presented a few seconds earlier. The number of digits is increased on each trial until the person begins to make mistakes. The longest string that can be reported accurately determines the participant's score. The following data were obtained from a sample of $n = 11$ participants. The scores are 7, 9, 8, 10, 8, 6, 7, 8, 7, 6, 5.
a. Compute the mean and variance for the sample.
b. Use the data to make a point estimate of the population mean.
c. Make an 80% confidence interval estimate of μ.

18. Most adolescents experience a growth spurt when they are between 12 and 15 years old. This period of dramatic growth generally occurs around age 12 for girls and around age 14 for boys. A researcher studying physical development selected a random sample of $n = 9$ girls and a second sample of $n = 16$ boys and recorded the gain in height (in millimeters) between the 14th birthday and 15th birthday for each participant. The girls showed an average gain of $M = 40$ millimeters with $SS = 1152$, and the boys gained an average of $M = 95$ millimeters with $SS = 2160$.
a. Estimate the population mean growth in 1 year for 14-year-old boys. Make a point estimate and an 80% confidence interval estimate.
b. Estimate the population mean growth in 1 year for 14-year-old girls. Make a point estimate and an 80% confidence interval estimate.
c. Estimate the mean difference in growth for boys versus girls during this 1-year-period. Again, make a point estimate and an 80% confidence interval estimate.

19. A therapist has demonstrated that five sessions of relaxation training significantly reduced anxiety levels for a sample of $n = 16$ clients. However, the therapist is concerned about the long-term effects of the training. Six months after therapy is completed, the patients are recalled, and their anxiety levels are measured again. On average, the anxiety scores for these patients are $M_D = 5.5$ points higher after six months than they had been at the end of therapy. The difference scores had $SS = 960$. Use these data to estimate the mean amount of relapse that occurs after therapy ends. Make a point estimate and an 80% confidence interval estimate of the population mean difference.

20. In a classic study of problem solving, Katona (1940) demonstrated that people who are required to figure out a problem on their own will learn more than people who are given the solution. One group of participants is presented with a series of problems that they must figure out. A second group sees the same problems but gets an explanation of the solutions. Later, both groups are given a new set of similar problems and their problem-solving scores are as follows:

ON THEIR OWN	GIVEN SOLUTIONS
$n = 10$	$n = 15$
$M = 78$	$M = 66$
$SS = 1420$	$SS = 2030$

a. Make a point estimate of the mean difference between the two training methods.
b. Make a 95% confidence interval estimate of the mean difference.

21. A researcher would like to estimate how much reaction time is affected by a common over-the-counter cold medication. The researcher measures reaction time for a sample of $n = 36$ participants. Each is then given a dose of the cold medication and reaction time is measured again. For this sample, reaction time increased after the medication by an average of $M_D = 24$ milliseconds with $s = 8$.

   a. Make a point estimate of the mean difference in reaction time caused by the medicine.

   b. Make a 95% confidence interval estimate of the population mean difference.

22. Problem 18 in Chapter 10 presented hypothetical data showing that elderly people who own dogs are significantly less likely to pay visits to their doctors than those who do not own pets. The data are presented again here.

DOCTOR VISITS DURING PAST YEAR	
CONTROL GROUP	DOG OWNERS
12	8
10	5
6	9
9	4
15	6
12	
14	

Use these data to estimate how much difference there is in the number of doctor visits for people with dogs compared to those without.

   a. Make a point estimate of the population mean difference.

   b. Make a 90% confidence interval estimate of the mean difference.

23. Problem 25 in Chapter 11 presented data showing that Olympic marksmen score significantly higher for shots fired between heartbeats than they do for shots fired during heartbeats. The scores for the two conditions are as follows:

PARTICIPANT	DURING HEARTBEATS	BETWEEN HEARTBEATS
A	93	98
B	90	94
C	95	96
D	92	91
E	95	97
F	91	97

Use these data to estimate how much a marksman's score changes for shots fired during heartbeats compared with shots fired between heartbeats.

   a. Make a point estimate of the population mean difference.

   b. Make a 95% confidence interval estimate of the mean difference.

C H A P T E R

13

Introduction to Analysis of Variance

Tools You Will Need

The following items are considered essential background material for this chapter. If you doubt your knowledge of any of these items, you should review the appropriate chapter or section before proceeding.

- Variability (Chapter 4)
  - Sum of squares
  - Sample variance
  - Degrees of freedom
- Introduction to hypothesis testing (Chapter 8)
  - The logic of hypothesis testing
- Independent-measures t statistic (Chapter 10)

Preview

"But I read the chapter four times! How could I possibly have failed the exam!"

Most of you probably have had the experience of reading a textbook and suddenly realizing that you have no idea of what was said on the past few pages. Although you have been reading the words, your mind has wandered off, and the meaning of the words has never reached memory. In an influential paper on human memory, Craik and Lockhart (1972) proposed a *levels of processing* theory of memory that can account for this phenomenon. In general terms, this theory says that all perceptual and mental processing leaves behind a memory trace. However, the quality of the memory trace depends on the level or the depth of the processing. If you superficially skim the words in a book, your memory also will be superficial. On the other hand, when you think about the meaning of the words and try to understand what you are reading, the result will be a good, substantial memory that should serve you well on exams. In general, deeper processing results in better memory.

Rogers, Kuiper, and Kirker (1977) conducted an experiment demonstrating the effect of levels of processing. Participants in this experiment were shown lists of words and asked to answer questions about each word. The questions were designed to require different levels of processing, from superficial to deep. In one experimental condition, participants were simply asked to judge the physical characteristics of each printed word ("Is it printed in capital letters or small letters?") A second condition asked about the sound of each word ("Does it rhyme with 'boat'?"). In a third condition, participants were required to process the meaning of each word ("Does it have the same meaning as 'attractive'?"). The final condition required participants to understand each word and relate its meaning to themselves ("Does this word describe you?"). After going through the complete list, all participants were given a surprise memory test. As you can see in Figure 13.1, deeper processing resulted in better memory. Remember that the participants were not trying to memorize the words; they were simply reading through the list answering questions. However, the more they processed and understood the words, the better they recalled the words on the test.

In terms of human memory, the Rogers, Kuiper, and Kirker experiment is notable because it demonstrates the importance of "self" in memory. You are most likely to remember material that is directly related to you. In terms of statistics, however, this study is notable because it com-

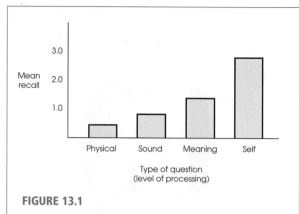

FIGURE 13.1

Mean recall as a function of the level of processing.

Rogers, T. B., Kuiper, N. A., & Kirker, W. S. (1977). Self-reference and the encoding of personal information. *Journal of Personality and Social Psychology, 35*, 677–688. Copyright (1977) by the American Psychological Association. Adapted by permission of the author.

pares four different treatment conditions in a single experiment. Although it may seem like a small step to go from two treatments (as in the *t* tests in Chapters 10 and 11) to four treatments, there is a tremendous gain in experimental sophistication. Suppose, for example, that Rogers, Kuiper, and Kirker had decided to examine only two levels of processing: one based on physical characteristics and one based on sound. In this simplified experiment, they would have made only *one* comparison: physical versus sound. In the real experiment, however, they used four conditions and were able to make *six* different comparisons:

Physical versus sound

Physical versus meaning

Physical versus self

Sound versus meaning

Sound versus self

Meaning versus self

To gain this experimental sophistication, there is some cost. Specifically, you no longer can use the familiar *t* tests we encountered in Chapters 10 and 11. Instead, you now must learn a new statistical technique that is designed for experiments consisting of two or more sets of data. This new, general-purpose procedure is called analysis of variance.

13.1 INTRODUCTION

Analysis of variance (ANOVA) is a hypothesis-testing procedure that is used to evaluate mean differences between two or more treatments (or populations). As with all inferential procedures, ANOVA uses sample data as the basis for drawing general conclusions about populations. It may appear that analysis of variance and *t* tests are simply two different ways of doing exactly the same job: testing for mean differences. In some respects, this is true—both tests use sample data to test hypotheses about population means. However, ANOVA has a tremendous advantage over *t* tests. Specifically, *t* tests are limited to situations in which there are only two treatments to compare. The major advantage of ANOVA is that it can be used to compare *two or more treatments*. Thus, ANOVA provides researchers with much greater flexibility in designing experiments and interpreting results.

Figure 13.2 shows a typical research situation for which analysis of variance would be used. Note that the study involves three samples representing three populations. The goal of the analysis is to determine whether the mean differences observed among the samples provide enough evidence to conclude that there are mean differences among the three populations. Specifically, we must decide between two interpretations:

1. There really are no differences between the populations (or treatments). The observed differences between samples are simply due to chance (sampling error).

2. The populations (or treatments) really do have different means, and these population mean differences are part of the reason that the samples have different means.

You should recognize that these two interpretations correspond to the two hypotheses (null and alternative) that are part of the general hypothesis-testing procedure.

Before we continue, it is necessary to introduce some terminology that is used to describe the research situation shown in Figure 13.2. Recall (from Chapter 1) that when a researcher manipulates a variable to create treatment conditions, the variable is called an *independent variable*. For example, Figure 13.2 could represent a study examining

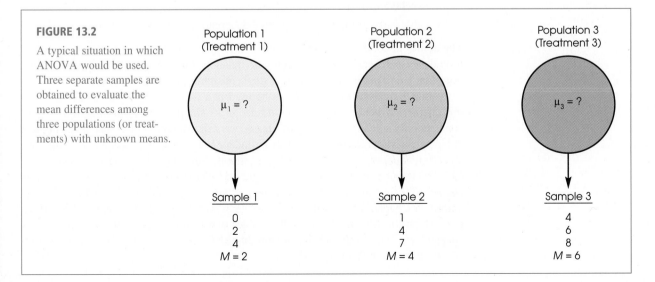

FIGURE 13.2

A typical situation in which ANOVA would be used. Three separate samples are obtained to evaluate the mean differences among three populations (or treatments) with unknown means.

Population 1 (Treatment 1)	Population 2 (Treatment 2)	Population 3 (Treatment 3)
$\mu_1 = ?$	$\mu_2 = ?$	$\mu_3 = ?$

Sample 1	Sample 2	Sample 3
0	1	4
2	4	6
4	7	8
$M = 2$	$M = 4$	$M = 6$

behavior in three different temperature conditions that have been created by the researcher. On the other hand, when a researcher uses a nonmanipulated variable to designate groups, the variable is called a *quasi-independent variable*. For example, the three groups in Figure 13.2 could represent 6-year-old, 8-year-old, and 10-year-old children. In the context of analysis of variance, an independent variable or a quasi-independent variable is called a *factor*. Thus, Figure 13.2 could represent a study in which temperature is the factor being evaluated or it could represent a study in which age is the factor being examined.

DEFINITION

In analysis of variance, the variable (independent or quasi-independent) that designates the groups being compared is called a **factor**.

In addition, the individual groups or treatment conditions that are used to make up a factor are called the *levels* of the factor. For example, a study that examined performance under three different temperature conditions would have three levels of temperature.

DEFINITION

The individual conditions or values that make up a factor are called the **levels** of the factor.

Like the *t* tests presented in Chapters 10 and 11, ANOVA can be used with either an independent-measures or a repeated-measures design. Recall that an independent-measures design means that there is a separate sample for each of the treatments (or populations) being compared. In a repeated-measures design, on the other hand, the same sample is tested in all of the different treatment conditions. In addition, ANOVA can be used to evaluate the results from a research study that involves more than one factor. For example, a researcher may want to compare two different therapy techniques, examining their immediate effectiveness as well as the persistence of their effectiveness over time. In this situation, the research study could involve two different groups of participants, one for each therapy, and measure each group at several different points in time. The structure of this design is shown in Figure 13.3. Notice that the study uses two factors, one independent-measures factor and one repeated-measures factor:

1. Therapy technique: A separate group is used for each technique (independent measures).

2. Time: Each group is tested at three different times (repeated measures).

The ability to combine different factors and to mix different designs within one study provides researchers with the flexibility to develop studies that address scientific questions that could not be answered by a single design using a single factor.

Although analysis of variance can be used in a wide variety of research situations, this chapter introduces analysis of variance in its simplest form. Specifically, we will consider only *single-factor* designs. That is, we will examine studies that have only one independent variable (or only one quasi-independent variable). Second, we will consider only *independent-measures* designs; that is, studies that use a separate sample for each treatment condition. The basic logic and procedures that are presented in this chapter will form the foundation for more complex applications of ANOVA. For example, in Chapter 14 we will extend the analysis to single-factor, repeated-measures designs, and in Chapter 15 we will consider two-factor designs. But for now, in this chapter, we will limit our discussion of ANOVA to *single-factor*, *independent-measures* research studies.

FIGURE 13.3

A research design with two factors. The research study uses two factors: One factor uses two levels of therapy technique (I versus II), and the second factor uses three levels of time (before, after, and 6 months after). Also notice that the therapy factor uses two separate groups (independent measures) and the time factor uses the same group for all three levels (repeated measures).

		TIME	
	Before Therapy	After Therapy	6 Months After Therapy
Therapy I (Group 1)	Scores for group 1 measured before Therapy I	Scores for group 1 measured after Therapy I	Scores for group 1 measured 6 months after Therapy I
Therapy II (Group 2)	Scores for group 2 measured before Therapy II	Scores for group 2 measured after Therapy II	Scores for group 2 measured 6 months after Therapy II

(THERAPY TECHNIQUE)

STATISTICAL HYPOTHESES FOR ANOVA

The following example will be used to introduce the statistical hypotheses for ANOVA. Suppose that a psychologist examined learning performance under three temperature conditions: 50°, 70°, and 90°. Three samples of subjects are selected, one sample for each treatment condition. The purpose of the study is to determine whether room temperature affects learning performance. In statistical terms, we want to decide between two hypotheses: the null hypothesis (H_0), which says that temperature has no effect, and the alternative hypothesis (H_1), which states that temperature does affect learning. In symbols, the null hypothesis states

$$H_0: \quad \mu_1 = \mu_2 = \mu_3$$

In words, the null hypothesis states that temperature has no effect on performance. That is, the population means for the three temperature conditions are all the same. In general, H_0 states that there is no treatment effect. Once again, notice that the hypotheses are always stated in terms of population parameters, even though we use sample data to test them.

For the alternative hypothesis, we may state that

H_1:   At least one population mean is different from another.

In general, H_1 states that the treatment conditions are not all the same; that is, there is a real treatment effect.

Notice that we have not given any specific alternative hypothesis. This is because many different alternatives are possible, and it would be tedious to list them all. One alternative, for example, would be that the first two populations are identical, but that the third is different. Another alternative states that the last two means are the same, but that the first is different. Other alternatives might be

H_1: $\mu_1 \neq \mu_2 \neq \mu_3$ (All three means are different.)

H_1: $\mu_1 = \mu_3$, but μ_2 is different

We should point out that a researcher typically entertains only one (or at most a few) of these alternative hypotheses. Usually a theory or the outcomes of previous studies will dictate a specific prediction concerning the treatment effect. For the sake of simplicity, we will state a general alternative hypothesis rather than try to list all possible specific alternatives.

THE TEST STATISTIC FOR ANOVA

The test statistic for ANOVA is very similar to the t statistics used in earlier chapters. For the t statistic, we computed a ratio with the following structure:

$$t = \frac{\text{obtained difference between sample means}}{\text{difference expected by chance (error)}}$$

For analysis of variance, the test statistic is called an F-ratio and has the following structure:

$$F = \frac{\text{variance (differences) between sample means}}{\text{variance (differences) expected by chance (error)}}$$

Note that the F-ratio is based on *variance* instead of sample mean *difference*. The reason for this change is that ANOVA is used when there are more than two sample means and it is impossible to compute a sample mean difference. For example, if there are only two samples and they have means of $M = 20$ and $M = 30$, then it is easy to show that there is a 10-point difference between the sample means. However, if we add a third sample with a mean of $M = 35$, the concept of a sample mean difference becomes difficult to define and impossible to calculate. The solution to this problem is to use variance to define and measure the size of the differences among the sample means. Consider the following two sets of sample means:

SET 1	SET 2
$M_1 = 20$	$M_1 = 28$
$M_2 = 30$	$M_2 = 30$
$M_3 = 35$	$M_3 = 31$

If you compute the variance for the three numbers in each set, then the variance you obtain for set 1 is $s^2 = 58.33$ and the variance for set 2 is $s^2 = 2.33$. Notice that the two variances provide an accurate representation of the size of the differences. In set 1 there are relatively large differences between sample means and the variance is relatively large. In set 2 the mean differences are small and the variance is small. Thus, the variance in the numerator of the F-ratio provides a single number that describes the differences between all the sample means.

In much the same way, the variance in the denominator of the F-ratio and the standard error in the denominator of the t statistic both measure the differences that would be expected by chance. In the t statistic, the standard error measures standard distance or standard deviation. In the F-ratio, this standard deviation (s) is simply converted to a variance (s^2).

Finally, you should realize that the t statistic and the F-ratio provide the same basic information. In each case, the numerator of the ratio measures the actual difference obtained from the sample data, and the denominator measures the difference that would be expected by chance. With either the F-ratio or the t statistic, a large value provides evidence that the sample mean difference is a result of more than chance (see Box 13.1).

BOX

13.1   TYPE I ERRORS AND MULTIPLE-HYPOTHESIS TESTS

If we already have t tests for comparing mean differences, you might wonder why analysis of variance is necessary. Why create a whole new hypothesis-testing procedure that simply duplicates what the t tests can already do? The answer to this question is based in a concern about Type I errors.

Remember that each time you do a hypothesis test, you select an alpha level that determines the risk of a Type I error. With $\alpha = .05$, for example, there is a 5%, or a 1-in-20, risk of a Type I error. Thus, for every 20 hypothesis tests, you expect to make one Type I error. The more tests you do, the more risk there is of a Type I error. For this reason, researchers often make a distinction between the *testwise* alpha level and the *experimentwise* alpha level. The testwise alpha level is simply the alpha level you select for each individual hypothesis test. The experimentwise alpha level is the total probability of a Type I error accumulated from all of the separate tests in the experiment. As the number of sepa-

rate tests increases, so does the experimentwise alpha level.

For an experiment involving three treatments, you would need three separate t tests to compare all of the mean differences:

Test 1 compares treatment I versus treatment II.

Test 2 compares treatment I versus treatment III.

Test 3 compares treatment II versus treatment III.

The three separate tests accumulate to produce a relatively large experimentwise alpha level. The advantage of analysis of variance is that it performs all three comparisons simultaneously in the same hypothesis test. Thus, no matter how many different means are being compared, ANOVA uses one test with one alpha level to evaluate the mean differences and thereby avoids the problem of an inflated experimentwise alpha level.

13.2 THE LOGIC OF ANALYSIS OF VARIANCE

The formulas and calculations required in ANOVA are somewhat complicated, but the logic that underlies the whole procedure is fairly straightforward. Therefore, this section will give a general picture of analysis of variance before we start looking at the details. We will introduce the logic of ANOVA with the help of the hypothetical data in Table 13.1. These data represent the results of an independent-measures experiment comparing learning performance under three temperature conditions.

TABLE 13.1

Hypothetical data from an experiment examining learning performance under three temperature conditions*

TREATMENT 1 50° (SAMPLE 1)	TREATMENT 2 70° (SAMPLE 2)	TREATMENT 3 90° (SAMPLE 3)
0	4	1
1	3	2
3	6	2
1	3	0
0	4	0
$M = 1$	$M = 4$	$M = 1$

*Note that there are three separate samples, with $n = 5$ in each sample. The dependent variable is the number of problems solved correctly.

One obvious characteristic of the data in Table 13.1 is that the scores are not all the same. In everyday language, the scores are different; in statistical terms, the scores are variable. Our goal is to measure the amount of variability (the size of the differences) and to explain where it comes from.

The first step is to determine the total variability for the entire set of data. To compute the total variability, we will combine all the scores from all the separate samples to obtain one general measure of variability for the complete experiment. Once we have measured the total variability, we can begin to break it apart into separate components. The word *analysis* means dividing into smaller parts. Because we are going to analyze variability, the process is called *analysis of variance*. This analysis process divides the total variability into two basic components.

1. Between-Treatments Variance. Looking at the data in Table 13.1, we clearly see that much of the variability in the scores is due to general differences between treatment conditions. For example, the scores in the 70° condition tend to be much higher ($M = 4$) than the scores in the 50° condition ($M = 1$). We will calculate the variance between treatments to provide a measure of the overall differences between treatment conditions. Notice that the variance between treatments is really measuring the differences between sample means.

2. Within-Treatment Variance. In addition to the general differences between treatment conditions, there is variability within each sample. Looking again at Table 13.1, we see that the scores in the 70° condition are not all the same; they are variable. The within-treatments variance will provide a measure of the variability inside each treatment condition.

Analyzing the total variability into these two components is the heart of analysis of variance. We will now examine each of the components in more detail.

BETWEEN-TREATMENTS VARIANCE

Remember that calculating variance is simply a method for measuring how big the differences are for a set of numbers. When you see the term *variance*, you can automatically translate it into the term *differences*. Thus, the between-treatments variance simply measures how much difference exists between the treatment conditions.

In addition to measuring the differences between treatments, the overall goal of ANOVA is to evaluate the differences between treatments. Specifically, the purpose for the analysis is to distinguish between two alternative explanations:

1. The differences between treatments are simply due to chance.
2. The differences between treatments are significantly greater than can be explained by chance alone; that is, the differences have been caused by the treatment effects.

Thus, there are always two possible explanations for the difference (or variance) that exists between treatments:

1. Treatment Effect. The differences are *caused by the treatments*. For the data in Table 13.1, the scores in sample 1 were obtained in a 50° room and the scores in sample 2 were obtained in a 70° room. It is possible that the difference between samples is caused by the different temperatures.

2. Chance. The differences are *simply due to chance*. If there is no treatment effect at all, you would still expect some differences between samples. The samples consist of different individuals with different scores, and it should not be surprising that differences exist between samples just by chance. In general, "chance differences" can be defined as unplanned and unpredictable differences that are not caused or explained by any action on the part of the researcher. Researchers commonly identify two primary sources for chance differences.

a. *Individual differences:* Individual subjects enter a research study with different characteristics. Although it is reasonable to expect that different subjects will produce different scores, it is impossible to predict exactly what the differences will be.

b. *Experimental error:* Whenever you make a measurement, there is potential for some degree of error. Thus, if you measure the same individual under the same conditions, it is possible that you will obtain two different measurements. Because these differences are unexplained and unpredictable, they are considered to be chance occurrences.

Thus, when we compute the between-treatments variance, we are measuring differences that could be caused by a treatment effect or could simply be due to chance. To demonstrate that there really is a treatment effect, we must establish that the differences between treatments are bigger than would be expected by chance alone. To accomplish this goal, we will determine how big the differences are when there is no treatment effect involved; that is, we will measure how much difference (or variance) occurs by chance alone. To measure chance differences, we compute the variance within treatments.

WITHIN-TREATMENTS VARIANCE

Inside each treatment condition, we have a set of individuals who are treated exactly the same; that is, the researcher does not do anything that would cause these individuals to have different scores. In Table 13.1, for example, the data show that five individuals were tested in a 70° room (treatment 2). Although these five individuals were all treated exactly the same, their scores are different. Why are the scores different? The answer is that the differences within a treatment are simply due to chance.

Thus, the within-treatments variance provides a measure of how much difference is reasonable to expect just by chance. In particular, the within-treatments variance measures the differences that exist when there is no treatment that could cause differences.

Figure 13.4 shows the overall analysis of variance and identifies the sources of variability that are measured by each of the two basic components.

FIGURE 13.4

The independent-measures analysis of variance partitions, or analyzes, the total variability into two components: variance between treatments and variance within treatments.

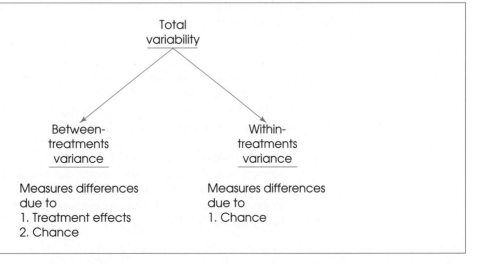

THE *F*-RATIO: THE TEST STATISTIC FOR ANOVA

Once we have analyzed the total variability into two basic components (between treatments and within treatments), we simply compare them. The comparison is made by computing a statistic called an *F-ratio*. For the independent-measures ANOVA, the *F*-ratio has the following structure:

$$F = \frac{\text{variance between treatments}}{\text{variance within treatments}} \tag{13.1}$$

When we express each component of variability in terms of its sources (see Figure 13.4), the structure of the *F*-ratio is

$$F = \frac{\text{treatment effect} + \text{differences due to chance}}{\text{differences due to chance}} \tag{13.2}$$

The value obtained for the *F*-ratio will help determine whether or not any treatment effects exist. Consider the following two possibilities:

1. When the treatment has no effect, then the differences between treatments (numerator) are entirely due to chance. In this case, the numerator and the denominator of the *F*-ratio are both measuring chance differences and should be roughly the same size. With the numerator and denominator roughly equal, the *F*-ratio should have a value around 1.00. In terms of the formula, when the treatment effect is zero, we obtain

$$F = \frac{0 + \text{differences due to chance}}{\text{differences due to chance}}$$

Thus, an *F*-ratio near 1.00 indicates that the differences between treatments (numerator) are about the same as the differences that are expected by chance (the denominator). With an *F*-ratio near 1.00, we will conclude that there is no evidence to suggest that the treatment has any effect.

2. When the treatment does have an effect, causing differences between samples, then the between-treatment differences (numerator) should be larger than those due to chance (denominator). In this case, the numerator of the *F*-ratio should be noticeably larger than the denominator, and we should obtain an *F*-ratio noticeably larger than 1.00. Thus, a large *F*-ratio indicates that the differences between treatments are greater than expected by chance; that is, the treatment does have a significant effect.

In more general terms, the denominator of the *F*-ratio measures only uncontrolled and unexplained (often called *unsystematic*) variability. For this reason, the denominator of the *F*-ratio is called the *error term*. The numerator of the *F*-ratio always includes the same unsystematic variability as in the error term, but it also includes any systematic differences caused by the treatment effect. The goal of ANOVA is to find out whether or not a treatment effect exists.

DEFINITION

For ANOVA, the denominator of the *F*-ratio is called the **error term**. The error term provides a measure of the variance due to chance. When the treatment effect is zero (H_0 is true), the error term measures the same sources of variance as the numerator of the *F*-ratio, so the value of the *F*-ratio is expected to be nearly equal to 1.00.

1. ANOVA is a statistical procedure that compares two or more treatment conditions for differences in variance. (True or false?)

2. In ANOVA, what value is expected, on the average, for the F-ratio when the null hypothesis is true?

3. What happens to the value of the F-ratio if differences between treatments are increased? What happens to the F-ratio if variability inside the treatments is increased?

4. In ANOVA, the total variability is partitioned into two parts. What are these two variability components called, and how are they used in the F-ratio?

1. False. Although ANOVA uses variance in the computations, the purpose of the test is to evaluate differences in *means* between treatments.

2. When H_0 is true, the expected value for the F-ratio is 1.00 because the top and bottom of the ratio are both measuring the same variance.

3. As differences between treatments increase, the F-ratio will increase. As variability within treatments increases, the F-ratio will decrease.

4. The two components are between-treatments variability and within-treatments variability. Between-treatments variance is the numerator of the F-ratio, and within-treatments variance is the denominator.

13.3 ANOVA NOTATION AND FORMULAS

Because ANOVA most often is used to examine data from more than two treatment conditions (and more than two samples), we will need a notational system to help keep track of all the individual scores and totals. To help introduce this notational system, we will use the hypothetical data from Table 13.1 again. The data are reproduced in Table 13.2 along with some of the notation and statistics that will be described.

1. The letter k is used to identify the number of treatment conditions—that is, the number of levels of the factor. For an independent-measures study, k also specifies the number of separate samples. For the data in Table 13.2, there are three treatments, so $k = 3$.

2. The number of scores in each treatment is identified by a lowercase letter n. For the example in Table 13.2, $n = 5$ for all the treatments. If the samples are of different sizes, you can identify a specific sample by using a subscript. For example, n_2 is the number of scores in treatment 2.

3. The total number of scores in the entire study is specified by a capital letter N. When all the samples are the same size (n is constant), $N = kn$. For the data in Table 13.2, there are $n = 5$ scores in each of the $k = 3$ treatments, so $N = 3(5) = 15$.

Because ANOVA formulas require ΣX for each treatment and ΣX for the entire set of scores, we have introduced new notation (T and G) to help identify which ΣX is being used. Remember: T stands for *treatment total*, and G stands for *grand total*.

4. The total (ΣX) for each treatment condition is identified by the capital letter T. The total for a specific treatment can be identified by adding a numerical subscript to the T. For example, the total for the second treatment in Table 13.2 is $T_2 = 20$.

TABLE 13.2

Hypothetical data from an experiment examining learning performance under three temperature conditions*

	TEMPERATURE CONDITIONS		
1 50°	2 70°	3 90°	
0	4	1	$\Sigma X^2 = 106$
1	3	2	$G = 30$
3	6	2	$N = 15$
1	3	0	$k = 3$
0	4	0	
$T_1 = 5$	$T_2 = 20$	$T_3 = 5$	
$SS_1 = 6$	$SS_2 = 6$	$SS_3 = 4$	
$n_1 = 5$	$n_2 = 5$	$n_3 = 5$	
$M_1 = 1$	$M_2 = 4$	$M_3 = 1$	

*Summary values and notation for an analysis of variance are also presented.

5. The sum of all the scores in the research study (the grand total) is identified by G. You can compute G by adding up all N scores or by adding up the treatment totals: $G = \Sigma T$.

6. Although there is no new notation involved, we also have computed SS and M for each sample, and we have calculated ΣX^2 for the entire set of $N = 15$ scores in the study. These values are given in Table 13.2 and will be important in the formulas and calculations for ANOVA.

Finally, we should note that there is no universally accepted notation for analysis of variance. Although we are using Gs and Ts, for example, you may find that other sources use other symbols.

ANOVA FORMULAS

Because analysis of variance requires extensive calculations and many formulas, one common problem for students is simply keeping track of the different formulas and numbers. Therefore, we will examine the general structure of the procedure and look at the organization of the calculations before we introduce the individual formulas.

1. The final calculation for ANOVA is the F-ratio, which is composed of two variances:

$$F = \frac{\text{variance between treatments}}{\text{variance within treatments}}$$

2. Recall that variance for sample data has been defined as

$$\text{sample variance} = s^2 = \frac{SS}{df}$$

Therefore, we will need to compute an SS and a df for the variance between treatments (numerator of F), and we will need another SS and df for the variance within treatments (denominator of F). To obtain these SS and df values, we must go through two separate analyses: First, compute SS for the total study, and analyze it into two components (between and within). Then compute df for the total study, and analyze it into two components (between and within).

Thus, the entire process of analysis of variance will require nine calculations: three values for *SS*, three values for *df*, two variances (between and within), and a final *F*-ratio. However, these nine calculations are all logically related and are all directed toward finding the final *F*-ratio. Figure 13.5 shows the logical structure of ANOVA calculations.

FIGURE 13.5

The structure and sequence of calculations for the analysis of variance.

The final goal for the analysis of variance is an *F*-ratio	$F = \dfrac{\text{Variance between treatments}}{\text{Variance within treatments}}$	
Each variance in the *F*-ratio is computed as *SS/df*	$\begin{array}{l}\text{Variance} \\ \text{between} \\ \text{treatments}\end{array} = \dfrac{SS\,\text{between}}{df\,\text{between}}$	$\begin{array}{l}\text{Variance} \\ \text{within} \\ \text{treatments}\end{array} = \dfrac{SS\,\text{within}}{df\,\text{within}}$
To obtain each of the *SS* and *df* values, the total variability is analyzed into the two components	*SS* total *SS* between *SS* within	*df* total *df* between *df* within

ANALYSIS OF SUM OF SQUARES (*SS*)

The ANOVA requires that we first compute a total sum of squares and then partition this value into two components: between treatments and within treatments. This analysis is outlined in Figure 13.6. We will examine each of the three components separately.

1. Total Sum of Squares, SS_{total}. As the name implies, SS_{total} is the sum of squares for the entire set of *N* scores. We calculate this value by using the computational formula for *SS*:

$$SS = \Sigma X^2 - \frac{(\Sigma X)^2}{N}$$

FIGURE 13.6

Partitioning the sum of squares (*SS*) for the independent-measures analysis of variance.

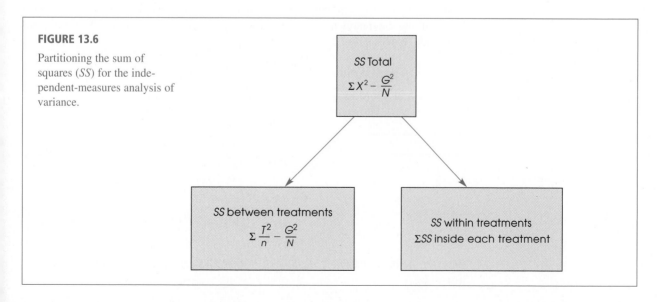

SS Total

$\Sigma X^2 - \dfrac{G^2}{N}$

SS between treatments

$\Sigma \dfrac{T^2}{n} - \dfrac{G^2}{N}$

SS within treatments

ΣSS inside each treatment

To make this formula consistent with the ANOVA notation, we substitute the letter G in place of ΣX and obtain

$$SS_{total} = \Sigma X^2 - \frac{G^2}{N} \tag{13.3}$$

Applying this formula to the set of data in Table 13.2, we obtain

$$SS_{total} = 106 - \frac{30^2}{15}$$

$$= 106 - 60$$

$$= 46$$

2. Within-Treatments Sum of Squares, $SS_{within\ treatments}$. Now we are looking at the variability inside each of the treatment conditions. We already have computed the SS within each of the three treatment conditions (Table 13.2): $SS_1 = 6$, $SS_2 = 6$, and $SS_3 = 4$. To find the overall within-treatment sum of squares, we simply add these values together:

$$SS_{within\ treatments} = \Sigma SS_{inside\ each\ treatment} \tag{13.4}$$

For the data in Table 13.2, this formula gives

$$SS_{within\ treatments} = 6 + 6 + 4$$

$$= 16$$

3. Between-Treatments Sum of Squares, $SS_{between\ treatments}$. Before we introduce the equation for $SS_{between\ treatments}$, consider what we have found so far. The total variability for the data in Table 13.2 is $SS_{total} = 46$. We intend to partition this total into two parts (see Figure 13.6). One part, $SS_{within\ treatments}$, has been found to be equal to 16. This means that $SS_{between\ treatments}$ must be equal to 30 in order for the two parts (16 and 30) to add up to the total (46). The equation for the between-treatments sum of squares should produce a value of $SS_{between\ treatments} = 30$.

Recall that the variability between treatments is measuring the differences between treatment means. Conceptually, the most direct way of measuring the amount of variability among the treatment means is to compute the sum of squares for the set of sample means. Specifically,

$$SS_{between\ treatments,} = n(SS_{means}) \tag{13.5}$$

For the data in Table 13.2, the sample means are 1, 4, and 1, and $n = 5$. The calculation of SS_{means} is shown as follows:

MEAN (X)	X^2
1	1
4	16
1	1
6	18

$$SS = \Sigma X^2 - \frac{(\Sigma X)^2}{n}$$

$$= 18 - \frac{(6)^2}{3}$$

$$= 18 - 12 = 6$$

Thus, for the data in Table 13.2, $SS_{means} = 6$ and $SS_{between\ treatments} = n(SS_{means}) = 5(6) = 30$.

The intent of Equation 13.5 is to emphasize that $SS_{\text{between treatments}}$ measures the mean differences. In this case, we are measuring the differences among three treatment means. However, Equation 13.5 can only be used when all of the samples are exactly the same size (equal ns), and the equation can be very awkward, especially when the treatment means are not whole numbers. Therefore, we will use a computational formula for $SS_{\text{between treatments}}$ that uses the treatment totals (T) instead of the treatment means.

$$SS_{\text{between treatments}} = \Sigma \frac{T^2}{n} - \frac{G^2}{N} \qquad (13.6)$$

Notice the similarity between this formula and the one for SS_{total} (Equation 13.3). When computing SS_{total} we are measuring the variability (differences) for a set of scores or X values, and the formula begins by squaring each of the X values. Now, we are measuring the variability (differences) for a set of treatment totals or T values, and the formula begins by squaring each of the T values. Also, notice that $SS_{\text{between treatments}}$ is intended to measure the differences between treatments, and the treatment totals (the T values) provide an accurate method to measure and describe differences between treatments.

Using this new formula with the data in Table 13.2, we obtain

$$SS_{\text{between treatments}} = \frac{5^2}{5} + \frac{20^2}{5} + \frac{5^2}{5} - \frac{30^2}{15}$$

$$= 5 + 80 + 5 - 60$$

$$= 90 - 60$$

$$= 30$$

At this point in the analysis, we have computed all three of the SS values, and it is appropriate to verify our calculations by checking to see that the two components add up to the total. Using the data from Table 13.2,

$$SS_{\text{total}} = SS_{\text{between treatments}} + SS_{\text{within treatments}}$$

$$46 = 30 + 16$$

The formula for each SS and the relationships among these three values are shown in Figure 13.6.

THE ANALYSIS OF DEGREES OF FREEDOM (df)

The analysis of degrees of freedom (df) follows the same pattern as the analysis of SS. First, we will find df for the total set of N scores, and then we will partition this value into two components: degrees of freedom between treatments and degrees of freedom within treatments. In computing degrees of freedom, there are two important considerations to keep in mind:

1. Each df value is associated with a specific SS value.

2. Normally, the value of df is obtained by counting the number of items that were used to calculate SS and then subtracting 1. For example, if you compute SS for a set of n scores, then $df = n - 1$.

With this in mind, we will examine the degrees of freedom for each part of the analysis.

1. Total Degrees of Freedom, df_{total}. To find the df associated with SS_{total}, you must first recall that this SS value measures variability for the entire set of N scores. Therefore, the df value is

$$df_{\text{total}} = N - 1 \tag{13.7}$$

For the data in Table 13.2, the total number of scores is $N = 15$, so the total degrees of freedom are

$$df_{\text{total}} = 15 - 1$$
$$= 14$$

2. Within-Treatments Degrees of Freedom, $df_{\text{within treatments}}$. To find the df associated with $SS_{\text{within treatments}}$, we must look at how this SS value is computed. Remember, we first find SS inside of each of the treatments and then add these values together. Each of the treatment SS values measures variability for the n scores in the treatment, so each SS will have $df = n - 1$. When all these individual treatment values are added together, we obtain

$$df_{\text{within treatments}} = \Sigma(n - 1) = \Sigma df_{\text{in each treatment}} \tag{13.8}$$

For the experiment we have been considering, each treatment has $n = 5$ scores. This means there are $n - 1 = 4$ degrees of freedom inside each treatment. Because there are three different treatment conditions, this gives a total of 12 for the within-treatments degrees of freedom. Notice that this formula for df simply adds up the number of scores in each treatment (the n values) and subtracts 1 for each treatment. If these two stages are done separately, you obtain

$$df_{\text{within treatments}} = N - k \tag{13.9}$$

(Adding up all the n values gives N. If you subtract 1 for each treatment, then altogether you have subtracted k because there are k treatments.) For the data in Table 13.2, $N = 15$ and $k = 3$, so

$$df_{\text{within treatments}} = 15 - 3$$
$$= 12$$

3. Between-Treatments Degrees of Freedom, $df_{\text{between treatments}}$. The df associated with $SS_{\text{between treatments}}$ can be found by considering the SS formula. This SS formula measures the variability for the set of treatment totals. To find $df_{\text{between treatments}}$, simply count the number of T values and subtract 1. Because the number of treatments is specified by the letter k, the formula for df is

$$df_{\text{between treatments}} = k - 1 \tag{13.10}$$

For the data in Table 13.2, there are three different treatment conditions (three T values), so the between-treatments degrees of freedom are computed as follows:

$$df_{\text{between treatments}} = 3 - 1$$
$$= 2$$

Notice that the two parts we obtained from this analysis of degrees of freedom add up to equal the total degrees of freedom:

$$df_{\text{total}} = df_{\text{within treatments}} + df_{\text{between treatments}}$$

$$14 = 12 + 2$$

The complete analysis of degrees of freedom is shown in Figure 13.7.

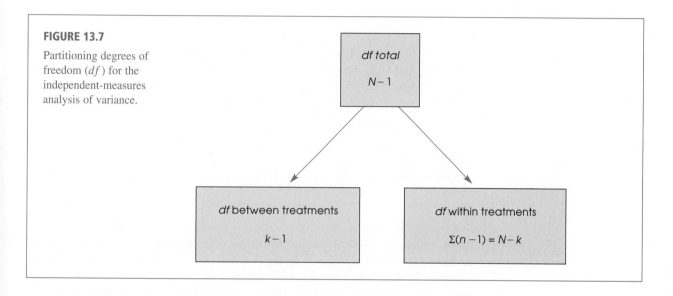

FIGURE 13.7

Partitioning degrees of freedom (df) for the independent-measures analysis of variance.

CALCULATION OF VARIANCES (*MS*) AND THE *F*-RATIO

The next step in the analysis of variance procedure is to compute the variance between treatments and the variance within treatments in order to calculate the F-ratio (see Figure 13.5).

In ANOVA, it is customary to use the term *mean square*, or simply *MS*, in place of the term *variance*. Recall (from Chapter 4) that variance is defined as the mean of the squared deviations. In the same way that we use *SS* to stand for the sum of the squared deviations, we now will use *MS* to stand for the mean of the squared deviations. For the final F-ratio we will need an *MS* (variance) between treatments for the numerator and an *MS* (variance) within treatments for the denominator. In each case

$$MS \ (\text{variance}) = s^2 = \frac{SS}{df} \tag{13.11}$$

For the data we have been considering,

$$MS_{\text{between treatments}} = s^2_{\text{between treatments}} = \frac{SS_{\text{between treatments}}}{df_{\text{between treatments}}} = \frac{30}{2} = 15$$

and

$$MS_{\text{within treatments}} = s^2_{\text{within treatments}} = \frac{SS_{\text{within treatments}}}{df_{\text{within treatments}}} = \frac{16}{12} = 1.33$$

We now have a measure of the variance (or differences) between the treatments and a measure of the variance within the treatments. The *F*-ratio simply compares these two variances:

$$F = \frac{MS_{\text{between treatments}}}{MS_{\text{within treatments}}}$$

(13.12)

For the experiment we have been examining, the data give an *F*-ratio of

$$F = \frac{15}{1.33} = 11.28$$

It is useful to organize the results of the analysis in one table called an *ANOVA summary table*. The table shows the source of variability (between treatments, within treatments, and total variability), *SS*, *df*, *MS*, and *F*. For the previous computations, the ANOVA summary table is constructed as follows:

SOURCE	SS	df	MS	
Between treatments	30	2	15	$F = 11.28$
Within treatments	16	12	1.33	
Total	46	14		

Although these tables are no longer commonly used in published reports, they do provide a concise method for presenting the results of an analysis. (Note that you can conveniently check your work: Adding the first two entries in the *SS* column (30 + 16) yields the total *SS*. The same applies to the *df* column.) When using analysis of variance, you might start with a blank ANOVA summary table and then fill in the values as they are calculated. With this method, you will be less likely to "get lost" in the analysis, wondering what to do next.

For this example, the obtained value of $F = 11.28$ indicates that the numerator of the *F*-ratio is substantially bigger than the denominator. If you recall the conceptual structure of the *F*-ratio as presented in Equations 13.1 and 13.2, the *F* value we obtained indicates that the differences between treatments are more than 11 times bigger than what would be expected by chance. Stated in terms of the experimental variables, it appears that temperature does have an effect on learning performance. However, to properly evaluate the *F*-ratio, we must examine the *F* distribution.

13.4   THE DISTRIBUTION OF *F*-RATIOS

In analysis of variance, the *F*-ratio is constructed so that the numerator and denominator of the ratio are measuring exactly the same variance when the null hypothesis is true (see Equation 13.2). In this situation, we expect the value of *F* to be around 1.00. The problem now is to define precisely what we mean by "around 1.00." What values are considered to be close to 1.00, and what values are far away? To answer this question, we need to look at all the possible *F* values—that is, the *distribution of F-ratios*.

Before we examine this distribution in detail, you should note two obvious characteristics:

1. Because F-ratios are computed from two variances (the numerator and denominator of the ratio), F values always will be positive numbers. Remember that variance is always positive.

2. When H_0 is true, the numerator and denominator of the F-ratio are measuring the same variance. In this case, the two sample variances should be about the same size, so the ratio should be near 1. In other words, the distribution of F-ratios should pile up around 1.00.

With these two factors in mind, we can sketch the distribution of F-ratios. The distribution is cut off at zero (all positive values), piles up around 1.00, and then tapers off to the right (see Figure 13.8). The exact shape of the F distribution depends on the degrees of freedom for the two variances in the F-ratio. You should recall that the precision of a sample variance depends on the number of scores or the degrees of freedom. In general, the variance for a large sample (large df) provides a more accurate estimate of the population variance. Because the precision of the MS values depends on df, the shape of the F distribution also will depend on the df values for the numerator and denominator of the F-ratio. With very large df values, nearly all the F-ratios will be clustered very near to 1.00. With the smaller df values, the F distribution is more spread out.

FIGURE 13.8

The distribution of F-ratios with $df = 2, 12$. Of all the values in the distribution, only 5% are larger than $F = 3.88$, and only 1% are larger than $F = 6.93$.

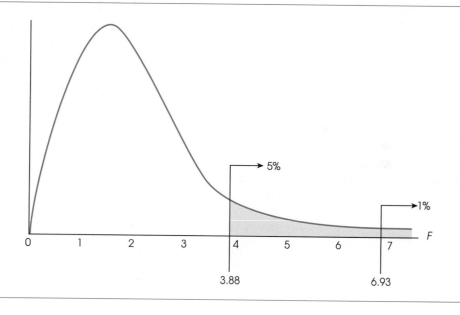

THE F DISTRIBUTION TABLE   For analysis of variance, we expect F near 1.00 if H_0 is true, and we expect a large value for F if H_0 is not true. In the F distribution, we need to separate those values that are reasonably near 1.00 from the values that are significantly greater than 1.00. These critical values are presented in an F distribution table in Appendix B, page 693. A portion of the F distribution table is shown in Table 13.3. To use the table, you must know the df values for the F-ratio (numerator and denominator), and you must know the alpha level for the hypothesis test. It is customary for an F table to have the df values for the numerator of the F-ratio printed across the top of the table. The df values for the denominator of F are printed in a column on the left-hand side. For the temperature experiment we have been considering, the numerator of the F-ratio (between treatments) has $df = 2$, and the denominator of the F-ratio (within treatments) has $df = 12$. This F-ratio is said to have "degrees of freedom equal to 2 and 12." The degrees of freedom would be written as $df = 2, 12$. To use the table, you would first find $df = 2$ across the top of the table and $df = 12$ in the first column. When you line up these two values, they point to a pair of numbers in the middle of the table. These numbers give the critical cutoffs for $\alpha = .05$ and $\alpha = .01$. With $df = 2, 12$, for example, the numbers in the table are 3.88 and 6.93. These values indicate that the most unlikely 5% of the distribution ($\alpha = .05$) begins at a value of 3.88. The most extreme 1% of the distribution begins at a value of 6.93 (see Figure 13.8).

In the temperature experiment, we obtained an F-ratio of 11.28. According to the critical cutoffs in Figure 13.8, this value is extremely unlikely (it is in the most extreme 1%). Therefore, we would reject H_0 with α set at either .05 or .01 and conclude that temperature does have a significant effect on learning performance.

TABLE 13.3

A portion of the F distribution table. Entries in roman type are critical values for the .05 level of significance, and bold type values are for the .01 level of significance. The critical values for $df = 2, 12$ have been highlighted (see text).

DEGREES OF FREEDOM: DENOMINATOR	DEGREES OF FREEDOM: NUMERATOR					
	1	2	3	4	5	6
10	4.96	4.10	3.71	3.48	3.33	3.22
	10.04	**7.56**	**6.55**	**5.99**	**5.64**	**5.39**
11	4.84	3.98	3.59	3.36	3.20	3.09
	9.65	**7.20**	**6.22**	**5.67**	**5.32**	**5.07**
12	4.75	3.88	3.49	3.26	3.11	3.00
	9.33	**6.93**	**5.95**	**5.41**	**5.06**	**4.82**
13	4.67	3.80	3.41	3.18	3.02	2.92
	9.07	**6.70**	**5.74**	**5.20**	**4.86**	**4.62**
14	4.60	3.74	3.34	3.11	2.96	2.85
	8.86	**6.51**	**5.56**	**5.03**	**4.69**	**4.46**

LEARNING CHECK

1. Calculate SS_{total}, $SS_{between\ treatments}$, and $SS_{within\ treatments}$ for the following set of data:

TREATMENT 1	TREATMENT 2	TREATMENT 3	
$n = 10$	$n = 10$	$n = 10$	$N = 30$
$T = 10$	$T = 20$	$T = 30$	$G = 60$
$SS = 27$	$SS = 16$	$SS = 23$	$\Sigma X^2 = 206$

2. A researcher uses an ANOVA to compare three treatment conditions with a sample of $n = 8$ in each treatment. For this analysis, find df_{total}, $df_{between\ treatments}$, and $df_{within\ treatments}$.

3. With $\alpha = .05$, what value forms the boundary for the critical region in the distribution of F-ratios with $df = 2, 24$?

ANSWERS

1. $SS_{total} = 86$; $SS_{between\ treatments} = 20$; $SS_{within\ treatments} = 66$

2. $df_{total} = 23$; $df_{between\ treatments} = 2$; $df_{within\ treatments} = 21$

3. The critical value is 3.40.

13.5 EXAMPLES OF HYPOTHESIS TESTING WITH ANOVA

Although we have now seen all the individual components of ANOVA, the following example demonstrates the complete ANOVA process using the standard four-step procedure for hypothesis testing.

EXAMPLE 13.1   The data depicted in Table 13.4 were obtained from an independent-measures experiment designed to measure the effectiveness of three pain relievers (A, B, and C). A fourth group that received a placebo (sugar pill) also was tested.

TABLE 13.4

The effect of drug treatment on the amount of time (in seconds) a stimulus is endured.

PLACEBO	DRUG A	DRUG B	DRUG C	
3	4	6	7	$N = 20$
0	3	3	6	$G = 60$
2	1	4	5	$\Sigma X^2 = 262$
0	1	3	4	
0	1	4	3	
$T = 5$	$T = 10$	$T = 20$	$T = 25$	
$SS = 8$	$SS = 8$	$SS = 6$	$SS = 10$	

The purpose of the analysis is to determine whether these sample data provide evidence of any significant differences among the four drugs. The dependent variable is the amount of time (in seconds) that subjects can withstand a painfully hot stimulus.

Before we begin the hypothesis test, note that we have already computed several summary statistics for the data in Table 13.4. Specifically, the treatment totals (T) and SS values are shown for each sample, and the grand total (G) as well as N and ΣX^2 are shown for the entire set of data. Having these summary values will simplify the computations in the hypothesis test, and we suggest that you always compute these summary statistics before you begin an analysis of variance.

STEP 1   The first step is to state the hypotheses and select an alpha level:

H_0: $\mu_1 = \mu_2 = \mu_3 = \mu_4$ (There is no treatment effect.)

H_1:   At least one of the treatment means is different.

We will use $\alpha = .05$.

STEP 2   To locate the critical region for the F-ratio, we first must determine degrees of freedom for $MS_{\text{between treatments}}$ and $MS_{\text{within treatments}}$ (the numerator and denominator of F). For these data, the total degrees of freedom are

Often it is easier to postpone finding the critical region until after step 3, where you compute the df values as part of the calculations for the F-ratio.

$$df_{\text{total}} = N - 1$$
$$= 20 - 1$$
$$= 19$$

Analyzing this total into two components, we obtain

$$df_{\text{between treatments}} = k - 1$$

$$= 4 - 1$$

$$= 3$$

$$df_{\text{within treatments}} = \Sigma df_{\text{inside each treatment}} = 4 + 4 + 4 + 4 = 16$$

The F-ratio for these data have $df = 3, 16$. The distribution of all the possible F-ratios with $df = 3, 16$ is presented in Figure 13.9. Note that an F-ratio larger than 3.24 is extremely rare ($p < .05$) if H_0 is true.

FIGURE 13.9

The distribution of F-ratios with $df = 3, 16$. The critical value for $\alpha = .05$ is $F = 3.24$.

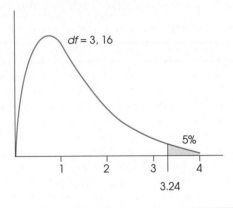

STEP 3   To compute the F-ratio for these data, you must go through the series of calculations outlined in Figure 13.5. The calculations can be summarized as follows:

    a. Analyze the SS to obtain $SS_{\text{between treatments}}$ and $SS_{\text{within treatments}}$.

    b. Use the SS values and the df values (from step 2) to calculate the two variances, $MS_{\text{between treatments}}$ and $MS_{\text{within treatments}}$.

    c. Finally, use the two MS values (variances) to compute the F-ratio.

Analysis of SS. First, we will compute the total SS and then the two components, as indicated in Figure 13.6.

$$SS_{\text{total}} = \Sigma X^2 - \frac{G^2}{N}$$

$$= 262 - \frac{60^2}{20}$$

$$= 262 - 180$$

$$= 82$$

$$SS_{\text{within treatments}} = \Sigma SS_{\text{inside each treatment}} = 8 + 8 + 6 + 10 = 32$$

$$SS_{\text{between treatments}} = \Sigma \frac{T^2}{n} - \frac{G^2}{N}$$

$$= \frac{5^2}{5} + \frac{10^2}{5} + \frac{20^2}{5} + \frac{25^2}{5} - \frac{60^2}{20}$$

$$= 5 + 20 + 80 + 125 - 180$$

$$= 50$$

Calculation of mean squares. Now we must compute the variance or *MS* for each of the two components:

The *df* values ($df_{between}$ = 3 and df_{within} = 16) were computed in step 2 when we located the critical region.

$$MS_{between\ treatments} = \frac{SS_{between\ treatments}}{df_{between\ treatments}} = \frac{50}{3} = 16.67$$

$$MS_{within\ treatments} = \frac{SS_{within\ treatments}}{df_{within\ treatments}} = \frac{32}{16} = 2.00$$

Calculation of F. We compute the *F*-ratio:

$$F = \frac{MS_{between\ treatments}}{MS_{within\ treatments}} = \frac{16.67}{2.00} = 8.33$$

STEP 4 Finally, we make the statistical decision. The *F* value we obtained, *F* = 8.33, is in the critical region (see Figure 13.9). It is very unlikely ($p < .05$) that we will obtain a value this large if H_0 is true. Therefore, we reject H_0 and conclude that there is a significant treatment effect.

Example 13.1 demonstrated the complete, step-by-step application of the ANOVA procedure. There are two additional points that can be made using this example.

First, you should look carefully at the statistical decision. We have rejected H_0 and concluded that not all the treatments are the same. But we have not determined which ones are different. Is drug A different from the placebo? Is drug A different from drug B? Unfortunately, these questions remain unanswered. We do know that at least one difference exists (we rejected H_0), but additional analysis is necessary to find out exactly where this difference is. We address this problem in Section 13.6.

Second, as noted earlier, all of the components of the analysis (the *SS, df, MS,* and *F*) can be presented together in one summary table. The summary table for the analysis in Example 13.1 is as follows:

SOURCE	SS	df	MS	
Between treatments	50	3	16.67	F = 8.33
Within treatments	32	16	2.00	
Total	82	19		

Although these tables are very useful for organizing the components of an analysis of variance, they are not commonly used in published reports. In the following section, we present the current method for reporting the results from an ANOVA.

MEASURING EFFECT SIZE FOR ANALYSIS OF VARIANCE

As we have noted previously, a *significant* mean difference simply indicates that the difference observed in the sample data is very unlikely to have occurred just by chance. Thus, the term significant does not necessarily mean *large*, it simply means larger than due by chance. To provide an indication of how large the effect actually is, it is recommended that researchers report a measure of effect size in addition to the measure of significance.

For analysis of variance, the simplest and most direct way to measure effect size is to compute r^2, the percentage of variance accounted for. In simpler terms, r^2 measures how much of the differences between scores is accounted for by the differences between treatments. For analysis of variance, the calculation and the concept of r^2 is extremely straightforward.

$$r^2 = \frac{SS_{\text{between treatments}}}{SS_{\text{between treatments}} + SS_{\text{within treatments}}} = \frac{SS_{\text{between treatments}}}{SS_{\text{total}}} \qquad (13.13)$$

Remember that $SS_{\text{between treatments}}$ measures the differences between treatments. $SS_{\text{within treatments}}$ measures the rest of the variability in the scores; that is, $SS_{\text{within treatments}}$ measures the variability that is not explained by differences between the treatments. With this in mind, it should be clear that r^2 measures the proportion of the total variability that is accounted for by the differences between treatments. For the data from Example 13.1, we obtain

$$r^2 = \frac{50}{82} = 0.61 \quad \text{(or 61\%)}$$

In published reports of research results, the r^2 value computed for an analysis of variance is usually called η^2 (the Greek letter *eta* squared).

IN THE LITERATURE:
REPORTING THE RESULTS OF ANALYSIS OF VARIANCE

The APA format for reporting the results of ANOVA begins with a presentation of the treatment means and standard deviations in the narrative of the article, a table, or a graph. These descriptive statistics are not needed in the calculations of the actual analysis of variance, but you can easily determine the treatment means from n and $T(M = T/n)$ and the standard deviations from the SS of each treatment $[s = \sqrt{SS/(n-1)}]$. Next, report the results of the ANOVA. For the study described in Example 13.1, the report might state the following:

The means and standard deviations are presented in Table 1. The analysis of variance revealed a significant difference, $F(3, 16) = 8.33, p < .05, \eta^2 = 0.61$.

TABLE 1

Amount of time (seconds) the stimulus was endured

	PLACEBO	DRUG A	DRUG B	DRUG C
M	1.00	2.00	4.00	5.00
SD	1.41	1.41	1.22	1.58

Note how the F-ratio is reported. In this example, degrees of freedom for between and within treatments are $df = 3, 16$, respectively. These values are placed in parentheses immediately following the symbol F. Next, the calculated value for F is reported, followed by the probability of committing a Type I error and the measure of effect size.

When an analysis of variance is done using a computer program, the F-ratio is usually accompanied by an exact value for p. A computer analysis for the data in Example 13.1 produced a p value of $p = 0.001$. Whenever an exact value is available, it should be reported rather than the less precise $p < .05$. For this example, the report would conclude, "The analysis of variance revealed a significant difference, $F(3, 16) = 8.33$, $p = .001$, $\eta^2 = 0.61$." ❏

A CONCEPTUAL VIEW OF ANOVA

Because analysis of variance requires relatively complex calculations, students encountering this statistical technique for the first time often tend to be overwhelmed by the formulas and arithmetic and lose sight of the general purpose for the analysis. The following two examples are intended to minimize the role of the formulas and shift attention back to the conceptual goal of the ANOVA process.

EXAMPLE 13.2

The following data represent the outcome of an experiment using two separate samples to evaluate the mean difference between two treatment conditions. Take a minute to look at the data and, without doing any calculations, try to predict the outcome of an ANOVA for these values. Specifically, predict what values should be obtained for $SS_{\text{between treatments}}$, $MS_{\text{between treatments}}$, and the F-ratio. If you do not "see" the answer after 20 or 30 seconds, try reading the hints that follow the data.

TREATMENT I	TREATMENT II	
4	2	$N = 8$
0	1	$G = 16$
1	0	$\Sigma X^2 = 56$
3	5	
$T = 8$	$T = 8$	
$SS = 10$	$SS = 14$	

If you are having trouble predicting the outcome of the ANOVA, read the following hints, and then go back and look at the data.

Hint 1:   Remember: $SS_{\text{between treatments}}$ and $MS_{\text{between treatments}}$ provide a measure of how much difference there is *between* treatment conditions.

Hint 2:   Find the mean or total (T) for each treatment, and determine how much difference there is between the two treatments.

You should realize by now that the data have been constructed so that there is zero difference between treatments. The two sample means (and totals) are identical, so $SS_{\text{between treatments}} = 0$, $MS_{\text{between treatments}} = 0$, and the F-ratio is zero.

Conceptually, the numerator of the F-ratio always measures how much difference exists between treatments. In Example 13.2, we constructed an extreme set of scores with zero difference. However, you should be able to look at any set of data and quickly compare the means (or totals) to determine whether there are big differences between treatments or small differences between treatments.

Being able to estimate the magnitude of between-treatment differences is a good first step in understanding ANOVA and should help you to predict the outcome of an analysis of variance. However, the *between-treatment* differences are only one part of the analysis. You must also understand the *within-treatment* differences that form the denominator of the F-ratio. The following example is intended to demonstrate the concepts underlying $SS_{\text{within treatments}}$ and $MS_{\text{within treatments}}$. In addition, the example should give you a better understanding of how the between-treatment differences and the within-treatment differences act together within the ANOVA.

EXAMPLE 13.3 The purpose of this example is to present a visual image for the concepts of between-treatments variability and within-treatments variability. In the example, we will compare two hypothetical outcomes for the same experiment. In each case, the experiment uses two separate samples to evaluate the mean difference between two treatments. The following data represent the two outcomes, which we will call experiment A and experiment B.

EXPERIMENT A		EXPERIMENT B	
TREATMENT		TREATMENT	
I	II	I	II
8	12	4	12
8	13	11	9
7	12	2	20
9	11	17	6
8	13	0	16
9	12	8	18
7	11	14	3
$M = 8$	$M = 12$	$M = 8$	$M = 12$
$s = 0.82$	$s = 0.82$	$s = 6.35$	$s = 6.35$

The data from experiment A are displayed in a frequency distribution graph in Figure 13.10(a). In the figure, we have indicated the *between-treatments* difference by showing the distance between the two means. We also have represented the *within-treatment* differences by using the range of scores for each separate sample. Clearly, the between-treatments value is substantially greater than the within-treatments value. This observation is confirmed by computing the F-ratio for experiment A. (You may check the calculations by performing the ANOVA yourself.)

$$F = \frac{\text{between-treatments difference}}{\text{within-treatments differences}} = \frac{MS_{\text{between}}}{MS_{\text{within}}} = \frac{56}{0.667} = 83.96$$

An F-ratio of $F = 83.96$ is sufficient to reject the null hypothesis with $\alpha = .05$, so we conclude that there is a significant difference between the two treatments. Notice that the statistical conclusion agrees with the simple observation that the data [Figure 13.10(a)] show two distinct sets of scores, and it is easy to see that there is a clear difference between the two treatment conditions.

The data from experiment B present a very different picture. These data are shown in Figure 13.10(b). Again, the *between-treatments* difference is represented by the distance between the treatment means, and the *within-treatments* differences are indicated by the range of scores within each treatment condition. Now, the between-treatments value is small in comparison to the within-treatments differences. Calculating the F-ratio confirms this observation.

$$F = \frac{\text{between-treatments difference}}{\text{within-treatments differences}} = \frac{MS_{\text{between}}}{MS_{\text{within}}} = \frac{56}{40.33} = 1.39$$

For experiment B, the F-ratio is not large enough to reject the null hypothesis, so we conclude that there is no significant difference between the two treatments. Once again, the statistical conclusion is consistent with the appearance of the data in Figure 13.10(b). Looking at the figure, we see that the scores from the two samples appear to be intermixed randomly with no clear distinction between treatments.

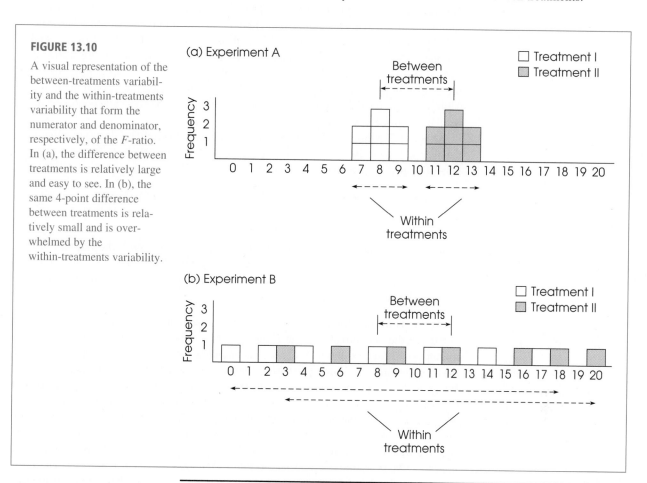

FIGURE 13.10

A visual representation of the between-treatments variability and the within-treatments variability that form the numerator and denominator, respectively, of the F-ratio. In (a), the difference between treatments is relatively large and easy to see. In (b), the same 4-point difference between treatments is relatively small and is overwhelmed by the within-treatments variability.

MS_{within} AND POOLED VARIANCE

You may have recognized that the two research outcomes presented in Example 13.3 are similar to those presented earlier in Example 10.4 in Chapter 10. Both examples are intended to demonstrate the role of variance in a hypothesis test. Both examples show that large values for sample variance can obscure any patterns in the data and reduce the potential for finding significant differences between means.

For the independent-measures t statistic in Chapter 10, the sample variance contributed directly to the standard error in the bottom of the t formula. Now, the sample variance contributes directly to the value of $MS_{within \; treatments}$ in the bottom of the F-ratio. In the t-statistic and in the F-ratio the variances from the separate samples are pooled together to create one average value for sample variance. For the independent-measures t statistic, we pooled two samples together to compute

$$\text{pooled variance} = s_p^2 = \frac{SS_1 + SS_2}{df_1 + df_2}$$

Now, in analysis of variance, we are combining two or more samples to calculate

$$MS_{\text{within treatments}} = \frac{SS_{\text{within treatments}}}{df_{\text{within treatments}}} = \frac{\Sigma SS}{\Sigma df} = \frac{SS_1 + SS_2 + SS_3 + \cdots}{df_1 + df_2 + df_3 + \cdots}$$

Notice that the concept of pooled variance is the same whether you have exactly two samples or more than two samples. In either case, you simply add the SS values and divide by the sum of the df values. The result is an average of all the different sample variances. As always, when variance is large, it means that the scores are scattered far and wide and it is difficult to identify sample means or mean differences (see Figure 13.10). In general, you can think of variance as measuring the amount of "noise" or "confusion" in the data. With large variance there is a lot of noise and confusion and it is difficult to see any clear patterns.

Although Examples 13.2 and 13.3 present somewhat simplified demonstrations with exaggerated data, the general point of the examples is to help you *see* what happens when you perform an analysis of variance. Specifically:

1. The numerator of the F-ratio ($MS_{\text{between treatments}}$) simply measures how much difference exists between the treatment means. The bigger the mean differences, the bigger is the F-ratio.

2. The denominator of the F-ratio ($MS_{\text{within treatments}}$) measures the variance of the scores inside each treatment. As seen in Example 13.3, a large amount of variance can make it difficult to see a mean difference. In general, the greater the variance, the smaller is the F-ratio.

At this point we can expand on a definition that appeared earlier. On page 404 we noted that the denominator of the F-ratio is commonly called the *error term*. This terminology is based on the fact that the denominator measures the amount of variability that is unexplained and unpredictable—that is, the amount of variability due to "error." For an independent-measures design, the error term is always determined by the amount of variability *within treatments*. Within each treatment, all the subjects are treated exactly the same. In other words, the researcher does not do anything to cause variability among the scores within a treatment condition. Therefore, any differences that exist within a treatment can be explained only by chance or error. By contrast, a researcher hopes to obtain systematic and predictable differences between treatments. The error term is used by the researcher as a benchmark or standard for evaluating the differences between treatments. If the between-treatments differences ($MS_{\text{between treatments}}$) are substantially greater than the error term, then the researcher can confidently conclude that the differences between treatments are due to more than chance.

D E F I N I T I O N   In analysis of variance, the MS value in the denominator of the F-ratio is called the **error term**. This MS value is intended to measure the amount of error variability—that is, variability in the data for which there is no systematic or predictable explanation. These unexplained differences from one score to another are assumed to be the result of chance. The error term is used as a standard for determining whether or not the differences between treatments (measured by $MS_{\text{between treatments}}$) are greater than would be expected just by chance.

AN EXAMPLE WITH UNEQUAL SAMPLE SIZES

In the previous examples, all the samples were exactly the same size (equal n's). However, the formulas for ANOVA can be used when the sample size varies within an experiment. With unequal sample sizes, you must take care to be sure that each value of n is matched with the proper T value in the equations. You also should note that the general ANOVA procedure is most accurate when used to examine experimental data with equal sample sizes. Therefore, researchers generally try to plan experiments with equal n's. However, there are circumstances in which it is impossible or impractical to have an equal number of subjects in every treatment condition. In these situations, ANOVA still provides a valid test, especially when the samples are relatively large and when the discrepancy between sample sizes is not extreme.

The following example demonstrates an ANOVA with samples of different sizes.

EXAMPLE 13.4

A psychologist conducts a research study to compare learning performance for three species of monkeys. The animals are tested individually on a delayed-response task. A raisin is hidden in one of three containers while the animal watches from its cage window. A shade is then pulled over the window for 1 minute to block the view. After this delay period, the monkey is allowed to respond by tipping over one container. If its response is correct, the monkey is rewarded with the raisin. The number of trials it takes before the animal makes five consecutive correct responses is recorded. The researcher used all of the available animals from each species, which resulted in unequal sample sizes (n). The data are summarized in Table 13.5.

TABLE 13.5

The performance of different species of monkeys on a delayed-response task

VERVET	RHESUS	BABOON	
$n = 4$	$n = 10$	$n = 6$	$N = 20$
$M = 9$	$M = 14$	$M = 4$	$G = 200$
$T = 36$	$T = 140$	$T = 24$	$\Sigma X^2 = 3400$
$SS = 200$	$SS = 500$	$SS = 320$	

STEP 1 State the hypotheses, and select the alpha level.

$$H_0: \quad \mu_1 = \mu_2 = \mu_3$$

$H_1:$ At least one population is different.

$$\alpha = .05$$

STEP 2 Locate the critical region. To find the critical region, we first must determine the df values for the F-ratio:

$$df_{total} = N - 1 = 20 - 1 = 19$$

$$df_{between\ treatments} = k - 1 = 3 - 1 = 2$$

$$df_{within\ treatments} = N - k = 20 - 3 = 17$$

The F-ratio for these data will have $df = 2, 17$. With $\alpha = .05$, the critical value for the F-ratio is 3.59.

STEP 3   Compute the F-ratio. First, compute SS for all three parts of the analysis:

$$SS_{total} = \Sigma X^2 - \frac{G^2}{N}$$

$$= 3400 - \frac{200^2}{20}$$

$$= 3400 - 2000$$

$$= 1400$$

$$SS_{between\ treatments} = \Sigma \frac{T^2}{n} - \frac{G^2}{N} = \frac{T_1^2}{n_1} + \frac{T_2^2}{n_2} + \frac{T_3^2}{n_3} - \frac{G^2}{N}$$

$$= \frac{36^2}{4} + \frac{140^2}{10} + \frac{24^2}{6} - \frac{200^2}{20}$$

$$= 324 + 1960 + 96 - 2000$$

$$= 380$$

$$SS_{within\ treatments} = \Sigma SS_{inside\ each\ treatment}$$

$$= 200 + 500 + 320$$

$$= 1020$$

Finally, compute the MS values and the F-ratio:

$$MS_{between\ treatments} = \frac{SS}{df} = \frac{380}{2} = 190$$

$$MS_{within\ treatments} = \frac{SS}{df} = \frac{1020}{17} = 60$$

$$F = \frac{MS_{between\ treatments}}{MS_{within\ treatments}} = \frac{190}{60} = 3.17$$

STEP 4   Make a decision. Because the obtained F-ratio is not in the critical region, we fail to reject H_0 and conclude that these data do not provide evidence of significant differences among the three populations of monkeys in terms of average learning performance.

LEARNING CHECK   1. The following data summarize the results of an experiment using three separate samples to compare three treatment conditions:

TREATMENT 1	TREATMENT 2	TREATMENT 3	
$n = 5$	$n = 5$	$n = 5$	$\Sigma X^2 = 325$
$T = 5$	$T = 10$	$T = 30$	
$SS = 45$	$SS = 25$	$SS = 50$	

Do these data provide evidence of any significant mean differences among the treatments? Test with $\alpha = .05$.

2. A researcher reports an F-ratio with $df = 2, 30$ for an independent-measures analysis of variance. How many treatment conditions were compared in the experiment? How many subjects participated in the experiment?

ANSWERS 1. The following summary table presents the results of the analysis:

SOURCE	SS	df	MS	
Between	70	2	35	$F = 3.5$
Within	120	12	10	
Total	190	14		

The critical value for F is 3.88. The obtained value for F is not in the critical region, and we fail to reject H_0.

2. There were 3 treatment conditions ($df_{\text{between treatments}} = k - 1 = 2$). A total of $N = 33$ individuals participated ($df_{\text{within treatments}} = 30 = N - k$).

13.6 POST HOC TESTS

As noted earlier, the primary advantage of ANOVA (compared to t tests) is it allows researchers to test for significant mean differences when there are *more than two* treatment conditions. Analysis of variance accomplishes this feat by comparing all the individual mean differences simultaneously within a single test. Unfortunately, the process of combining several mean differences into a single test statistic creates some difficulty when it is time to interpret the outcome of the test. Specifically, when you obtain a significant F-ratio (reject H_0), it simply indicates that somewhere among the entire set of mean differences there is at least one that is greater than would be expected by chance. In other words, the overall F-ratio only tells you that a significant difference exists; it does not tell exactly which means are significantly different and which are not.

Consider, for example, a research study that uses three samples to compare three treatment conditions. Suppose that the three sample means are $M_1 = 3$, $M_2 = 5$, and $M_3 = 10$. In this hypothetical study there are three mean differences:

1. There is a 2-point difference between M_1 and M_2.
2. There is a 5-point difference between M_2 and M_3.
3. There is a 7-point difference between M_1 and M_3.

If an analysis of variance was used to evaluate these data, a significant F-ratio would indicate that at least one of the sample mean differences is too large to be reasonably explained just by chance and therefore indicates a significant difference between treatments. In this example, the 7-point difference is the biggest of the three and, therefore, it must indicate a significant difference between the first treatment and the third treatment ($\mu_1 \neq \mu_3$). But what about the 5-point difference? Is it also large enough to be significant? And what about the 2-point difference between M_1 and M_2? Is it also significant? The purpose of *post hoc tests* is to answer these questions.

DEFINITION **Post hoc tests** (or **posttests**) are additional hypothesis tests that are done after an analysis of variance to determine exactly which mean differences are significant and which are not.

As the name implies, post hoc tests are done after an analysis of variance. More specifically, these tests are done after ANOVA when

1. You reject H_0 and

2. There are three or more treatments ($k \geq 3$).

Rejecting H_0 indicates that at least one difference exists among the treatments. With $k = 3$ or more, the problem is to find where the differences are. Note that when you have two treatments, rejecting H_0 indicates that the two means are not equal ($\mu_1 \neq \mu_2$). In this case there is no question about *which* means are different, and there is no need to do posttests.

POSTTESTS AND TYPE I ERRORS

In general, a post hoc test enables you to go back through the data and compare the individual treatments two at a time. In statistical terms, this is called making *pairwise comparisons*. For example, with $k = 3$, we would compare μ_1 versus μ_2, then μ_2 versus μ_3, and then μ_1 versus μ_3. In each case, we are looking for a significant mean difference. The process of conducting pairwise comparisons involves performing a series of separate hypothesis tests, and each of these tests includes the risk of a Type I error. As you do more and more separate tests, the risk of a Type I error accumulates and is called the *experimentwise alpha level* (see Box 13.1).

DEFINITION

The **experimentwise alpha level** is the overall probability of a Type I error that accumulates over a series of separate hypothesis tests. Typically, the experiment-wise alpha level is substantially greater than the value of alpha used for any one of the individual tests.

We have seen, for example, that a research study with three treatment conditions produces three separate mean differences, each of which could be evaluated using a post hoc test. If each test uses $\alpha = .05$, then there is a 5% risk of a Type I error for the first posttest, another 5% risk for the second test, and one more 5% risk for the third test. Although the probability of error does not simply sum across the three tests, it should be clear that increasing the number of separate tests will definitely increase the total, experimentwise probability of a Type I error.

Whenever you are conducting posttests, you must be concerned about the experimentwise alpha level. Statisticians have worked with this problem and have developed several methods for trying to control Type I errors in the context of post hoc tests.

PLANNED VERSUS UNPLANNED COMPARISONS

Statisticians often distinguish between what are called *planned* and *unplanned* comparisons. As the name implies, *planned comparisons* refer to specific mean differences that are relevant to specific hypotheses the researcher had in mind before the study was conducted. For example, suppose a researcher suspects that a 70° room is the best environment for human problem solving. Any warmer or colder temperature will interfere with performance. The researcher conducts a study comparing three different temperature conditions: 60°, 70°, and 80°. The researcher is predicting that performance in the 70° condition will be significantly better than in the 60° condition (comparison 1) and that 70° will be significantly better than 80° (comparison 2). Because these specific comparisons were planned before the data were collected, many statisticians would argue that the two specific posttests could be conducted without concern about inflating the risk of a Type I error. When the primary interest is in a few specific comparisons, the overall F-ratio for all the groups may be viewed as relatively unimportant and the overall ANOVA may not even be conducted. Instead, individual tests (*t* tests or two-group ANOVAs) could be done to evaluate the specific, planned comparisons. Although the planned comparisons could all be done with a standard alpha level, it is often recommended that researchers protect against an inflated experimentwise alpha level by dividing α equally among the planned comparisons. For example, with two comparisons, an alpha level of .05 would be divided equally so that α = .025 is used for each comparison and the overall risk of a Type I error is limited to .05 (.025 for the first test and .025 for the second). This technique for controlling the experimentwise alpha level is referred to as the Dunn test.

Unplanned comparisons, on the other hand, typically involve sifting through the data (a "fishing expedition") by conducting a large number of comparisons in the hope that a significant difference will turn up. In this situation it is necessary that a researcher take deliberate steps to limit the risk of a Type I error. Fortunately, many different procedures for conducting post hoc tests have been developed, each using its own technique to control the experimentwise alpha level. We will examine two commonly used procedures: Tukey's HSD test and the Scheffé test. Incidentally, the safest plan (and the authors' recommendation) for posttests is to use one of the special procedures (such as Tukey or Scheffé) for *all* posttests whether they are planned or unplanned.

TUKEY'S HONESTLY SIGNIFICANT DIFFERENCE (HSD) TEST

The first post hoc test we will consider is Tukey's HSD test. We have selected Tukey's HSD test because it is a commonly used test in psychological research. Tukey's test allows you to compute a single value that determines the minimum difference between treatment means that is necessary for significance. This value, called the *honestly significant difference*, or HSD, is then used to compare any two treatment conditions. If the mean difference exceeds Tukey's HSD, you conclude that there is a significant difference between the treatments. Otherwise, you cannot conclude that the treatments are significantly different. The formula for Tukey's HSD is

$$HSD = q\sqrt{\frac{MS_{\text{within treatments}}}{n}} \tag{13.14}$$

The *q* value used in Tukey's HSD test is called a Studentized range statistic.

where the value of *q* is found in Table B.5 (Appendix B, page 696), $MS_{\text{within treatments}}$ is the within-treatments variance from the ANOVA, and *n* is the number of scores in each treatment. Tukey's test requires that the sample size, *n*, be the same for all treatments. To locate the appropriate value of *q*, you must know the number of treatments in the overall experiment (*k*) and the degrees of freedom for $MS_{\text{within treatments}}$ (the error term in the *F*-ratio) and select an alpha level (generally the same α used for the ANOVA).

EXAMPLE 13.5

To demonstrate the procedure for conducting post hoc tests with Tukey's HSD, we will use the hypothetical data shown in Table 13.6. The data represent the results of a study comparing scores in three different treatment conditions. Note that the table displays summary statistics for each sample and the results from the overall ANOVA.

TABLE 13.6

Hypothetical results from a research study comparing three treatment conditions. Summary statistics are presented for each treatment along with the outcome from the analysis of variance.

TREATMENT A	TREATMENT B	TREATMENT C
n = 9	*n* = 9	*n* = 9
T = 27	*T* = 49	*T* = 63
M = 3.00	*M* = 5.44	*M* = 7.00

SOURCE	SS	df	MS
Between	73.19	2	36.60
Within	96.00	24	4.00
Total	169.19	26	
Overall *F*(2, 24) = 9.15			

With $\alpha = .05$ and $k = 3$ treatments, the value of q for the test is $q = 3.53$ (check the table). Therefore, Tukey's HSD is

$$HSD = q\sqrt{\frac{MS_{\text{within treatments}}}{n}} = 3.53\sqrt{\frac{4.00}{9}} = 2.36$$

Thus, the mean difference between any two samples must be at least 2.36 to be significant. Using this value, we can make the following conclusions:

1. Treatment A is significantly different from treatment B ($M_A - M_B = 2.44$).
2. Treatment A is also significantly different from treatment C ($M_A - M_C = 4.00$).
3. Treatment B is not significantly different from treatment C ($M_B - M_C = 1.56$).

THE SCHEFFÉ TEST

Because it uses an extremely cautious method for reducing the risk of a Type I error, the *Scheffé test* has the distinction of being one of the safest of all possible post hoc tests. The Scheffé test uses an *F*-ratio to test for a significant difference between any two treatment conditions. The numerator of the *F*-ratio is an *MS* between treatments that is calculated using *only the two treatments you want to compare*. The denominator is the same *MS* within treatments that was used for the overall ANOVA. The "safety factor" for the Scheffé test comes from the following two considerations:

1. Although you are comparing only two treatments, the Scheffé test uses the value of k from the original experiment to compute *df* between treatments. Thus, *df* for the numerator of the *F*-ratio is $k - 1$.

2. The critical value for the Scheffé *F*-ratio is the same as was used to evaluate the *F*-ratio from the overall ANOVA. Thus, Scheffé requires that every posttest satisfy the same criterion that was used for the complete analysis of variance. The following example uses the data from Table 13.6 to demonstrate the Scheffé posttest procedure.

EXAMPLE 13.6

Remember that the Scheffé procedure requires a separate $SS_{\text{between treatments}}$, $MS_{\text{between treatments}}$, and *F*-ratio for each comparison being made. We begin with a comparison of treatment A ($T = 27$) versus treatment B ($T = 49$). The first step is to compute $SS_{\text{between treatments}}$ for these two groups.

The value of *G* is obtained by adding the two treatment totals being compared. In this case, $G = 27 + 49 = 76$.

$$SS_{\text{between treatments}} = \Sigma\frac{T^2}{n} - \frac{G^2}{N}$$

$$= \frac{27^2}{9} + \frac{49^2}{9} - \frac{76^2}{18}$$

$$= 81 + 266.78 - 320.89$$

$$= 26.89$$

Although we are comparing only two groups, these two were selected from a study consisting of $k = 3$ samples. The Scheffé test uses the overall study to determine the degrees of freedom between treatments. Therefore, $df_{\text{between treatments}} = 3 - 1 = 2$, and the *MS* between treatments is

$$MS_{\text{between treatments}} = \frac{SS_{\text{between treatments}}}{df_{\text{between treatments}}} = \frac{26.89}{2} = 13.45$$

Finally, the Scheffé procedure uses the error term from the overall ANOVA to compute the F-ratio. In this case, $MS_{\text{within treatments}} = 4.00$ with $df_{\text{within treatments}} = 24$. Thus, the Scheffé test produces an F-ratio of

$$F_{\text{A verus B}} = \frac{MS_{\text{between treatments}}}{MS_{\text{within treatments}}} = \frac{13.45}{4.00} = 3.36$$

With $df = 2, 24$ and $\alpha = .05$, the critical value for F is 3.40 (see Table B.4). Therefore, our obtained F-ratio is not in the critical region, and we must conclude that these data show no significant difference between treatment A and treatment B.

The second comparison involves treatment B ($T = 49$) versus treatment C ($T = 63$). This time the data produce $SS_{\text{between treatments}} = 10.89$, $MS_{\text{between treatments}} = 5.45$, and $F(2, 24) = 1.36$ (check the calculations for yourself). Once again the critical value for F is 3.40, so we must conclude that the data show no significant difference between treatment B and treatment C.

The final comparison is treatment A ($T = 27$) versus treatment C ($T = 63$). This time the data produce $SS_{\text{between treatments}} = 72$, $MS_{\text{between treatments}} = 36$, and $F(2, 24) = 9.00$ (check the calculations for yourself). Once again the critical value for F is 3.40, and this time we conclude that the data show a significant difference.

Thus, the Scheffé posttest indicates that the only significant difference is between treatment A and treatment C.

There are two interesting points to be made from the posttest outcomes presented in the preceding two examples. First, the Scheffé test was introduced as being one of the safest of the posttest techniques because it provides the greatest protection from Type I errors. To provide this protection, the Scheffé test simply requires a larger sample mean difference before it is willing to conclude that the difference is significant. In Example 13.5 we found that the difference between treatment A and treatment B was large enough to be significant according to Tukey's test, but this same difference failed to reach significance according to Scheffé (Example 13.6). The discrepancy between the results is an example of Scheffé's extra demands: The Scheffé test simply requires more evidence and, therefore, is less likely to make a Type I error.

The second point concerns the pattern of results from the three Scheffé tests in Example 13.6. You may have noticed that the posttests produce what are apparently contradictory results. Specifically, the tests show no significant difference between A and B and they show no significant difference between B and C. This combination of outcomes might lead you to suspect that there is no significant difference between A and C. However, the test did show a significant difference. The answer to this apparent contradiction lies in the criterion of statistical significance. The differences between A and B and between B and C are too small to satisfy the criterion of significance. However, when these differences are combined, the total difference between A and C is large enough to meet the criterion for significance.

LEARNING CHECK

1. With $k = 2$ treatments, are post hoc tests necessary when the null hypothesis is rejected? Explain why or why not.

2. An analysis of variance comparing three treatments produces an overall F-ratio with $df = 2, 27$. If the Scheffé test was used to compare two of the three treatments, then the Scheffé F-ratio would also have $df = 2, 27$. (True or false?)

3. Using the data and the results from Example 13.1,

    a. Use Tukey's HSD test to determine whether there is a significant mean difference between drug B and the placebo. Use $\alpha = .05$.

    b. Use the Scheffé test to determine whether there is a significant mean difference between drug B and the placebo. Use $\alpha = .05$.

ANSWERS   1. No. Post hoc tests are used to determine which treatments are different. With only two treatment conditions, there is no uncertainty as to which two treatments are different.

2. True

3. a. For this test, $q = 4.05$ and HSD = 2.55. There is a 3-point mean difference between drug B and the placebo, which is large enough to be significant.

    b. The Scheffé $F = 3.75$, which is greater than the critical value of 3.24. Conclude that the mean difference between drug B and the placebo is significant.

13.7   THE RELATIONSHIP BETWEEN ANOVA AND *t* TESTS

When you have data from an independent-measures experiment with only two treatment conditions, you can use either a *t* test (Chapter 10) or an independent-measures ANOVA. In practical terms, it makes no difference which you choose. These two statistical techniques always will result in the same statistical decision. In fact the two methods use many of the same calculations and are very closely related in several other respects. The basic relationship between *t* statistics and *F*-ratios can be stated in an equation:

$$F = t^2$$

This relationship can be explained by first looking at the structure of the formulas for *F* and *t*.

The structure of the *t* statistic compares the actual difference between the samples (numerator) with the standard difference that would be expected by chance (denominator).

$$t = \frac{\text{obtained difference between sample means}}{\text{difference expected by chance (error)}}$$

The structure of the *F*-ratio also compares differences between sample means versus differences due to chance or error.

$$F = \frac{\text{variance (differences) between sample means}}{\text{variance (differences) expected by chance (error)}}$$

However, the numerator and denominator of the *F*-ratio measure variances, or mean squared differences. Therefore, we can express the *F*-ratio as follows:

$$F = \frac{(\text{differences between samples})^2}{(\text{differences expected by chance})^2}$$

The fact that the *t* statistic is based on differences and the *F*-ratio is based on *squared* differences leads to the basic relationship $F = t^2$.

There are several other points to consider in comparing the *t* statistic to the *F*-ratio.

1. It should be obvious that you will be testing the same hypotheses whether you choose a *t* test or an ANOVA. With only two treatments, the hypotheses for either test are

   $H_0: \quad \mu_1 = \mu_2$

   $H_1: \quad \mu_1 \neq \mu_2$

2. The degrees of freedom for the *t* statistic and the *df* for the denominator of the *F*-ratio ($df_{\text{within treatments}}$) are identical. For example, if you have two samples, each with six scores, the independent-measures *t* statistic will have *df* = 10, and the *F*-ratio will have *df* = 1, 10. In each case, you are adding the *df* from the first sample ($n - 1$) and the *df* from the second sample ($n - 1$).

3. The distribution of *t* and the distribution of *F*-ratios match perfectly if you take into consideration the relationship $F = t^2$. Consider the *t* distribution with *df* = 18 and the corresponding *F* distribution with *df* = 1, 18 that are presented in Figure 13.11. Notice the following relationships:

   a. If each of the *t* values is squared, then all of the negative values will become positive. As a result, the whole left-hand side of the *t* distribution (below zero) will be flipped over to the positive side. This creates a nonsymmetrical, positively skewed distribution—that is, the *F* distribution.

   b. For $\alpha = .05$, the critical region for *t* is determined by values greater than $+2.101$ or less than -2.101. When these boundaries are squared, you get

   $\pm2.101^2 = 4.41$

   Notice that 4.41 is the critical value for $\alpha = .05$ in the *F* distribution. Any value that is in the critical region for *t* will end up in the critical region for *F*-ratios after it is squared.

FIGURE 13.11

The distribution of *t* statistics with *df* = 18 and the corresponding distribution of *F*-ratios with *df* = 1, 18. Notice that the critical values for $\alpha = .05$ are $t = \pm2.101$ and that $F = 2.101^2 = 4.41$.

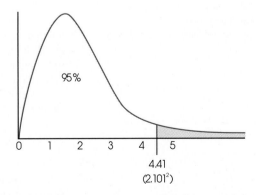

ASSUMPTIONS FOR THE INDEPENDENT-MEASURES ANOVA

The independent-measures ANOVA requires the same three assumptions that were necessary for the independent-measures t hypothesis test:

1. The observations within each sample must be independent (see page 251).
2. The populations from which the samples are selected must be normal.
3. The populations from which the samples are selected must have equal variances (homogeneity of variance).

Ordinarily, researchers are not overly concerned with the assumption of normality, especially when large samples are used, unless there are strong reasons to suspect the assumption has not been satisfied. The assumption of homogeneity of variance is an important one. If a researcher suspects it has been violated, it can be tested by Hartley's F-max test for homogeneity of variance (Chapter 10, page 331).

LEARNING CHECK

1. A researcher uses an independent-measures t test to evaluate the mean difference obtained in a research study, and obtains a t statistic of $t = 3.00$. If the researcher had used an analysis of variance to evaluate the results, the F-ratio would be $F = 9.00$. (True or false?)

2. An ANOVA produces an F-ratio with $df = 1, 34$. Could the data have been analyzed with a t test? What would be the degrees of freedom for the t statistic?

ANSWERS

1. True. $F = t^2$

2. If the F-ratio has $df = 1, 34$, then the experiment compared only two treatments, and you could use a t statistic to evaluate the data. The t statistic would have $df = 34$.

SUMMARY

1. Analysis of variance (ANOVA) is a statistical technique that is used to test for mean differences among two or more treatment conditions. The null hypothesis for this test states that in the general population there are no mean differences among the treatments. The alternative states that at least one mean is different from another.

2. The test statistic for analysis of variance is a ratio of two variances called an F-ratio. The variances in the F-ratio are called mean squares, or MS values. Each MS is computed by

$$MS = \frac{SS}{df}$$

3. For the independent-measures analysis of variance, the F-ratio is

$$F = \frac{MS_{\text{between treatments}}}{MS_{\text{within treatments}}}$$

The $MS_{\text{between treatments}}$ measures differences between the treatments by computing the variability of the treatment means or totals. These differences are assumed to be produced by

a. Treatment effects (if they exist)
b. Differences due to chance

The $MS_{\text{within treatments}}$ measures variability inside each of the treatment conditions. Because individuals inside a treatment condition are all treated exactly the same, any differences within treatments cannot be caused by treatment effects. Thus, the within-treatments MS is produced only by differences due to chance. With these factors in mind, the F-ratio has the following structure:

$$F = \frac{\text{treatment effect} + \text{differences due to chance}}{\text{differences due to chance}}$$

When there is no treatment effect (H_0 is true), the numerator and the denominator of the F-ratio are measuring the same variance, and the obtained ratio should be near 1.00. If there is a significant treatment effect, the numerator of the ratio should be larger than the denominator, and the obtained F value should be much greater than 1.00.

4. The formulas for computing each SS, df, and MS value are presented in Figure 13.12, which also shows the general structure for the analysis of variance.

5. The F-ratio has two values for degrees of freedom, one associated with the MS in the numerator and one associated with the MS in the denominator. These df values are used to find the critical value for the F-ratio in the F distribution table.

6. Effect size for the independent-measures ANOVA is measured by computing eta squared, the percentage of variance accounted for by the treatment effect.

$$\eta^2 = \frac{SS_{\text{between treatments}}}{SS_{\text{between treatments}} + SS_{\text{within treatments}}}$$

$$= \frac{SS_{\text{between treatments}}}{SS_{\text{total}}}$$

7. When the decision from an analysis of variance is to reject the null hypothesis and when the experiment contained more than two treatment conditions, it is necessary to continue the analysis with a post hoc test, such as Tukey's HSD test or the Scheffé test. The purpose of these tests is to determine exactly which treatments are significantly different and which are not.

FIGURE 13.12

Formulas for ANOVA.

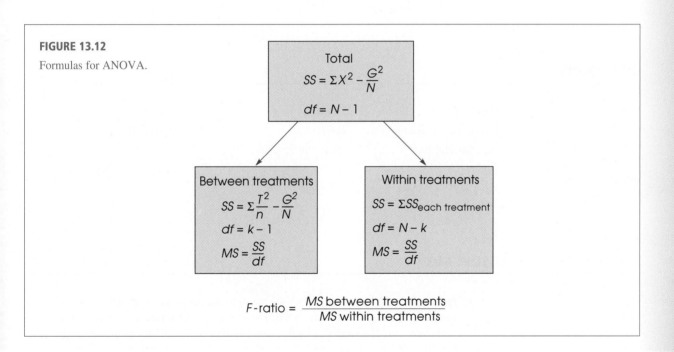

$$F\text{-ratio} = \frac{MS \text{ between treatments}}{MS \text{ within treatments}}$$

RESOURCES

WebTUTOR

If you are using WebTutor along with this book, there is a WebTutor section corresponding to this chapter. The WebTutor contains a brief summary of Chapter 13, hints for learning the concepts and the formulas for analysis of variance, cautions about common errors, and sample exam items including solutions.

The statistics computer package SPSS can be used to conduct the single-factor, independent-measures analysis of variance that is presented in this chapter. The data are entered in a stacked format in the data matrix, which means that all the scores are entered in a single column and a second column is used to designate the treatment condition for each score. For example, enter all the scores in the first column, and in the second column enter a 1 beside each score from the first treatment, enter a 2 beside each score from the second treatment, and so on. Then, click **Analyze** on the tool bar, select **Compare Means**, and click on **One-Way ANOVA**. Enter the column label for the scores in the Dependent List box, and enter the column label for the treatment numbers in the Factor box. Before you conduct the ANOVA, you can click on the **Post Hoc** box and select the Scheffé test and/or the Tukey test (or others). Also, you can click on the **Options** box and select Descriptives (to get descriptive statistics for each sample) or you may want to select the Homogeneity of Variance test or a Plot showing the treatment means. Then click **OK**. If you selected the Descriptives option, SPSS will produce a table showing descriptive statistics for each of the samples along with a summary table showing the results from the analysis of variance.

KEY TERMS

analysis of variance (ANOVA)	treatment effect	factor	distribution of F-ratios
	individual differences	levels	post hoc tests
between-treatments variability	experimental error	mean square (MS)	Scheffé test
within-treatments variability	F-ratio	ANOVA summary table	Tukey's HSD test
	error term	eta squared (η^2)	

FOCUS ON PROBLEM SOLVING

1. The words and labels used to describe the different components of variance can help you remember the ANOVA formulas. For example, *total* refers to the total experiment. Therefore, the SS_{total} and df_{total} values are based on the whole set of N scores. The word *within* refers to the variability inside (within) the treatment groups. Thus, the value for $SS_{within\ treatments}$ is based on an SS value from each group, computed from the scores *within* each group. Finally, *between* refers to the variability (or differences) between treatments. The $SS_{between\ treatments}$ component measures the differences between treatments (T_1 versus T_2, and so on), and $df_{between\ treatments}$ is simply the number of T values (k) minus 1.

2. When you are computing SS and df values, always calculate all three components (total, between, and within) separately; then check your work by making sure that the *between-treatments* and *within-treatments* components add up to the *total*.

3. Remember that an F-ratio has two separate values for df: a value for the numerator and one for the denominator. Property reported, the $df_{between\ treatments}$ value is stated first. You will need both df values when consulting the F distribution table for the critical

F value. You should recognize immediately that an error has been made if you see an F-ratio reported with a single value for df.

4. When you encounter an F-ratio and its df values reported in the literature, you should be able to reconstruct much of the original experiment. For example, if you see "$F(2, 36) = 4.80$," you should realize that the experiment compared $k = 3$ treatment groups (because $df_{\text{between treatments}} = k - 1 = 2$), with a total of $N = 39$ subjects participating in the experiment (because $df_{\text{within treatments}} = N - k = 36$).

DEMONSTRATION 13.1

ANALYSIS OF VARIANCE

A human factors psychologist studied three computer keyboard designs. Three samples of individuals were given material to type on a particular keyboard, and the number of errors committed by each participant was recorded. The data are as follows:

KEYBOARD A	KEYBOARD B	KEYBOARD C	
0	6	6	$N = 15$
4	8	5	$G = 60$
0	5	9	$\Sigma X^2 = 356$
1	4	4	
0	2	6	
$T = 5$	$T = 25$	$T = 30$	
$SS = 12$	$SS = 20$	$SS = 14$	

Are these data sufficient to conclude that there are significant differences in typing performance among the three keyboard designs?

STEP 1 State the hypotheses, and specify the alpha level.

The null hypothesis states that there is no difference among the keyboards in terms of number of errors committed. In symbols, we would state

$$H_0: \quad \mu_1 = \mu_2 = \mu_3 \quad \text{(Type of keyboard used has no effect.)}$$

As noted previously in this chapter, there are a number of possible statements for the alternative hypothesis. Here we state the general alternative hypothesis:

$$H_1: \quad \text{At least one of the treatment means is different.}$$

That is, the type of keyboard has an effect on typing performance. We will set alpha at $\alpha = .05$.

STEP 2 Locate the critical region.

To locate the critical region, we must obtain the values for $df_{\text{between treatments}}$ and $df_{\text{within treatments}}$.

$$df_{\text{between treatments}} = k - 1 = 3 - 1 = 2$$
$$df_{\text{within treatments}} = N - k = 15 - 3 = 12$$

The F-ratio for this problem will have $df = 2, 12$. Consult the F-distribution table for $df = 2$ in the numerator and $df = 12$ in the denominator. The critical F value for $\alpha = .05$ is $F = 3.88$. The obtained F ratio must exceed this value to reject H_0.

STEP 3 Perform the analysis.

The analysis involves the following steps:

1. Perform the analysis of SS.
2. Perform the analysis of df.
3. Calculate mean squares.
4. Calculate the F-ratio.

Perform the analysis of SS. We will compute SS_{total} followed by its two components.

$$SS_{total} = \Sigma X^2 - \frac{G^2}{N} = 356 - \frac{60^2}{15} = 356 - \frac{3600}{15}$$

$$= 356 - 240 = 116$$

$$SS_{within\ treatments} = \Sigma SS_{inside\ each\ treatment}$$

$$= 12 + 20 + 14$$

$$= 46$$

$$SS_{between\ treatments} = \Sigma \frac{T^2}{n} - \frac{G^2}{N}$$

$$= \frac{5^2}{5} + \frac{25^2}{5} + \frac{30^2}{5} - \frac{60^2}{15}$$

$$= \frac{25}{5} + \frac{625}{5} + \frac{900}{5} - \frac{3600}{15}$$

$$= 5 + 125 + 180 - 240$$

$$= 70$$

Analyze degrees of freedom. We will compute df_{total}. Its components, $df_{between\ treatments}$ and $df_{within\ treatments}$, were previously calculated (step 2).

$$df_{total} = N - 1 = 15 - 1 = 14$$

$$df_{between\ treatments} = 2$$

$$df_{within\ treatments} = 12$$

Calculate the MS values. The values for $MS_{between\ treatments}$ and $MS_{within\ treatments}$ are determined.

$$MS_{between\ treatments} = \frac{SS_{between\ treatments}}{df_{between\ treatments}} = \frac{70}{2} = 35$$

$$MS_{within\ treatments} = \frac{SS_{within\ treatments}}{df_{within\ treatments}} = \frac{46}{12} = 3.83$$

Compute the F-ratio. Finally, we can compute F.

$$F = \frac{MS_{between\ treatments}}{MS_{within\ treatments}} = \frac{35}{3.83} = 9.14$$

STEP 4   Make a decision about H_0, and state a conclusion.

The obtained F of 9.14 exceeds the critical value of 3.88. Therefore, we can reject the null hypothesis. The type of keyboard used has a significant effect on the number of errors committed, $F(2, 12) = 9.14$, $p < .05$. The following table summarizes the results of the analysis:

SOURCE	SS	df	MS	
Between treatments	70	2	35	$F = 9.14$
Within treatmens	46	12	3.83	
Total	116	14		

DEMONSTRATION 13.2

COMPUTING EFFECT SIZE FOR ANALYSIS OF VARIANCE

We will compute eta squared (η^2), the percentage of variance explained, for the data that were analyzed in Demonstration 13.1. The data produced a between-treatments SS of 70 and a total SS of 116. Thus,

$$\eta^2 = \frac{SS_{\text{between treatments}}}{SS_{\text{total}}} = \frac{70}{116} = 0.60 \quad \text{(or 60\%)}$$

PROBLEMS

1. Explain why the expected value for an F-ratio is equal to 1.00 when there is no treatment effect.

2. Describe the similarities between an F-ratio and a t statistic.

3. What happens to the value of the F-ratio in analysis of variance if the differences between sample means increase? What happens to the value of F if the sample variances increase?

4. Explain why you should use ANOVA instead of several t tests to evaluate mean differences when an experiment consists of three or more treatment conditions.

5. Describe *when* and *why* post hoc tests are used. Explain why you would not need to do post hoc tests for an experiment with only $k = 2$ treatment conditions.

6. The following data represent the results from an independent-measures experiment comparing three treatment conditions. Use an analysis of variance with $\alpha = .05$ to determine whether these data are sufficient to conclude that there are significant differences between the treatments.

	TREATMENT			
I	II	III		
2	6	6	$N = 12$
4	6	10	$G = 60$
2	2	10	$\Sigma X^2 = 408$
0	6	6	
$T = 8$	$T = 20$	$T = 32$	
$SS = 8$	$SS = 12$	$SS = 16$	

7. In the previous problem the differences between treatments were significantly greater than would be expected by chance, and the statistical decision was to reject the null hypothesis. For the following data, we have taken the original scores from Problem 6 and *reduced the differences between treatments.* Specifically, we increased the scores in treatment I by adding 1 point to each individual and we decreased the scores in treatment III by subtracting 1 point from each individual.

a. Before you begin any calculations, predict how the changes in the data should influence the outcome of the analysis. That is, how will the F-ratio for these data compare with the F-ratio obtained in Problem 6?

b. Use an analysis of variance with $\alpha = .05$ to determine whether there are any significant differences among the three treatments.

TREATMENT			
I	II	III	
3	6	5	$N = 12$
5	6	9	$G = 60$
3	2	9	$\Sigma X^2 = 368$
1	6	5	
$T = 12$	$T = 20$	$T = 28$	
$SS = 8$	$SS = 12$	$SS = 16$	

8. A researcher reports an F-ratio with $df = 3, 24$ for an independent-measures research study.

a. How many treatment conditions were compared in the study?

b. How many subjects participated in the entire study?

9. The following data were obtained in a study using three separate samples to compare three different treatments.

a. Use an analysis of variance with $\alpha = .05$ to determine whether there are any significant differences among the treatments.

b. Compute the value for η^2 for these data.

TREATMENT			
I	II	III	
4	3	8	$N = 12$
3	1	4	$G = 48$
5	3	6	$\Sigma X^2 = 238$
4	1	6	
$T = 16$	$T = 8$	$T = 24$	
$SS = 2$	$SS = 4$	$SS = 8$	

10. In Problem 9 the data show significant differences between the treatments. To construct the following data, we started with the numbers in Probem 9 but we eliminated the differences between treatments. Specifically, we added 2 points to each score in treatment II and we subtracted 2 points from each score in treatment III. The result is that all three treatments now total $T = 16$ and have a mean of $M = 4$. (That is, there are no differences between treatments.) Using the new data, compute the three SS values for the analysis of variance. You should find that all of the

variance is now within the treatments ($SS_{\text{between treatments}} = 0$).

TREATMENT			
I	II	III	
4	5	6	$N = 12$
3	3	2	$G = 48$
5	5	4	$\Sigma X^2 = 206$
4	3	4	
$T = 16$	$T = 16$	$T = 16$	
$SS = 2$	$SS = 4$	$SS = 8$	

11. For this problem, we started with the data from Problem 9 and removed all of the differences within treatments. Using the following data, compute the three SS values for the analysis of variance. You should find that all of the variance is now between the treatments ($SS_{\text{within treatments}} = 0$).

TREATMENT			
I	II	III	
4	2	6	$N = 12$
4	2	6	$G = 48$
4	2	6	$\Sigma X^2 = 224$
4	2	6	
$T = 16$	$T = 8$	$T = 24$	
$SS = 0$	$SS = 0$	$SS = 0$	

12. A developmental psychologist is examining problem-solving ability for grade school children. Random samples of 5-year-old, 6-year-old, and 7-year-old children are obtained with $n = 3$ in each sample. Each child is given a standardized problem-solving task, and the psychologist records the number of errors. The data are as follows:

5-YEAR-OLDS	6-YEAR-OLDS	7-YEAR-OLDS	
5	6	0	
4	4	1	$G = 30$
6	2	2	$\Sigma X^2 = 138$
$T = 15$	$T = 12$	$T = 3$	
$SS = 2$	$SS = 8$	$SS = 2$	

a. Use these data to test whether there are any significant differences among the three age groups. Use $\alpha = .05$.

b. Use the Scheffé test to determine which groups are different.

13. Earlier in the chapter (page 422) we noted that the MS within treatments is an average of the sample variances. When the samples are all the same size (equal n's), $MS_{\text{within treatments}}$ is exactly equal to the average of the sample variance. To demonstrate this fact, compute the variance for each of the three samples in Problem 12. Then compute the average of the three sample variances and compare your result to $MS_{\text{within treatments}}$ that you obtained from the analysis of variance.

14. The following summary table presents the results of an ANOVA from an experiment comparing four treatment conditions with a sample of $n = 10$ in each treatment. Complete all missing values in the table. (*Hint:* Start with the *df* column.)

SOURCE	SS	df	MS	
Between treatments	____	____	10	$F =$ ____
Within treatments	____	____	____	
Total	174	____		

15. A pharmaceutical company has developed a drug that is expected to reduce hunger. To test the drug, three samples of rats are selected with $n = 12$ in each sample. The first sample receives the drug every day. The second sample is given the drug once a week, and the third sample receives no drug at all. The dependent variable is the amount of food eaten by each rat over a 1-month period. These data are analyzed by an analysis of variance, and the results are reported in the following summary table. Fill in all missing values in the table. (*Hint:* Start with the *df* column.)

SOURCE	SS	df	MS	
Between treatments	____	____	____	$F = 6$
Within treatments	66	____	____	
Total	____	____		

16. A common science-fair project involves testing the effects of music on the growth of plants. For one of these projects, a sample of 24 newly sprouted bean plants is obtained. These plants are randomly assigned to four treatments, with $n = 6$ in each group. The four conditions are rock, heavy metal, country, and classical music. The dependent variable is the height of each plant after 2 weeks. The data from this experiment were examined using an ANOVA, and the results are summarized in the following table. Fill in all missing values.

SOURCE	SS	df	MS	
Between treatments	60	____	____	$F =$ ____
Within treatments	____	____	2	
Total	____	____		

17. The following data are from an experiment comparing three treatment conditions with a separate sample of $n = 4$ in each treatment.

	TREATMENT			
I	II	III		
0	1	8	$G = 48$
4	5	5	$\Sigma X^2 = 284$
0	4	6	
4	2	9	
$T = 8$	$T = 12$	$T = 28$	
$SS = 16$	$SS = 10$	$SS = 10$	

a. Use an ANOVA with $\alpha = .05$ to determine whether there are any significant differences among the three treatments.
b. Compute η^2 for these data.

18. You should have obtained a significant F-ratio for the ANOVA in Problem 17. In particular, the scores in treatment III are very different from the scores in the other samples. If we move the odd sample (treatment III) closer to the other two, it should greatly reduce the F-ratio and the value for η^2. To construct the following data, we subtracted 3 points from each score in treatment III. The result is that the samples are now closer together and the differences between treatments have been reduced.

	TREATMENT			
I	II	III		
0	1	5	$G = 36$
4	5	2	$\Sigma X^2 = 152$
0	4	3	
4	2	6	
$T = 8$	$T = 12$	$T = 16$	
$SS = 16$	$SS = 10$	$SS = 10$	

a. Use an ANOVA with $\alpha = .05$ to determine whether there are any significant differences among the three treatments. Compare the outcome with the results in Problem 17.
b. Compute η^2 for these data. Compare the outcome with the result from Problem 17.

19. A psychologist would like to examine how the rate of presentation affects people's ability to memorize a list of words. A list of 20 words is prepared. For one group of participants the list is presented at the rate of one word every $\frac{1}{2}$ second. The next group gets one word every second. The third group has one word every 2 seconds, and the fourth group has one word every 3 seconds. After the list is presented, the psychologist asks each person to recall the entire list. The dependent variable is the number of errors in recall. The data from this experiment are as follows:

$\frac{1}{2}$ SECOND	1 SECOND	2 SECONDS	3 SECONDS	
4	0	3	0	
6	2	1	2	$G = 32$
2	2	2	1	$\Sigma X^2 = 104$
4	0	2	1	
$T = 16$	$T = 4$	$T = 8$	$T = 4$	
$SS = 8$	$SS = 4$	$SS = 2$	$SS = 2$	

  a. Can the psychologist conclude that the rate of presentation has a significant effect on memory? Test at the .05 level.
  b. Use the Tukey HSD test to determine which rates of presentation are statistically different and which are not.

20. A clinical psychologist has noted that autistic children seem to respond to treatment better if they are in a familiar environment. To evaluate the influence of environment, the psychologist selects a group of 15 autistic children who are currently in treatment and randomly divides them into three groups. One group continues to receive treatment in the clinic as usual. For the second group, treatment sessions are conducted entirely in the child's home. The third group gets half of the treatment in the clinic and half at home. After 6 weeks, the psychologist evaluates the progress for each child. The data are as follows:

CLINIC	HOME	BOTH	
1	2	4	$N = 15$
1	7	1	$G = 45$
5	2	2	$\Sigma X^2 = 191$
2	4	2	
1	5	6	
$T = 10$	$T = 20$	$T = 15$	
$SS = 12$	$SS = 18$	$SS = 16$	

Do the data indicate any significant differences between the three settings? Test at the .05 level of significance.

21. A psychologist would like to examine the relative effectiveness of three therapy techniques for treating mild phobias. A sample of $N = 15$ individuals who display a moderate fear of spiders is obtained. These individuals are randomly assigned (with $n = 5$) to each of the three therapies. The dependent variable is a measure of reported fear of spiders after therapy. The data are as follows:

THERAPY A	THERAPY B	THERAPY C	
8	3	1	$G = 45$
5	3	0	$\Sigma X^2 = 221$
5	0	1	
7	2	2	
5	2	1	
$T = 30$	$T = 10$	$T = 5$	
$SS = 8$	$SS = 6$	$SS = 2$	

  a. Do the data indicate that there are any significant differences among the three therapies? Test at the .05 level of significance.
  b. Use Tukey's HSD posttest to determine exactly which treatments are different. Test at the .05 level of significance.

22. The following data represent three separate samples tested in three different treatment conditions.
  a. Using only the first two samples (treatments 1 and 2), use an ANOVA with $\alpha = .05$ to test for a mean difference between the two treatments.
  b. Now use all three samples to test for mean differences among the three treatment conditions, again using $\alpha = .05$.
  c. You should find that the first two treatments are significantly different (part a), but that there are no significant differences when the third treatment is included in the analysis (part b). How can you explain this outcome? (*Hint:* Compute the three sample means. How large is the mean difference when you are comparing only treatment 1 versus treatment 2? How large are the mean differences, on the average, when you are comparing all three treatments?)

TREATMENT		
1	2	3
0	6	4
4	6	0
0	2	4
0	6	4
$T = 4$	$T = 20$	$T = 12$
$SS = 12$	$SS = 12$	$SS = 12$

23. Use an analysis of variance with $\alpha = .05$ to determine whether the following data provide evidence of any significant differences among the three treatments:

TREATMENT 1	TREATMENT 2	TREATMENT 3	
$n = 4$	$n = 5$	$n = 6$	$N = 15$
$T = 2$	$T = 10$	$T = 18$	$G = 30$
$SS = 13$	$SS = 21$	$SS = 26$	$\Sigma X^2 = 135$

24. First-born children tend to develop language skills faster than their younger siblings. One possible explanation for this phenomenon is that first-borns have undivided attention from their parents. If this explanation is correct, then it is also reasonable that twins should show slower language development than single children and that triplets should be even slower. Davis (1937) found exactly this result. The following hypothetical data demonstrate the relationship. The dependent variable is a measure of language skill at age 3 for each child. Do the data indicate any significant differences? Test with $\alpha = .05$.

SINGLE CHILD	TWIN	TRIPLET
8	4	4
7	6	4
10	7	7
6	4	2
9	9	3

25. There have been a number of studies on the adjustment problems ("jet lag") people experience when traveling across time zones (see Moore-Ede, Sulzman, & Fuller, 1982). Jet lag always seems to be worse when traveling east. Consider this hypothetical study. A researcher examines how many days it takes a person to adjust after taking a long flight. One group of people flies east across the time zones (California to New York), a second group of 6 travels west (New York to California), and a third group of 6 takes a long flight within one time zone (San Diego to Seattle). For the following data, create a table that displays the mean and the standard deviation for each sample. Then perform an analysis of variance to determine whether jet lag varies for type of travel. Use the .05 level of significance.

WESTBOUND	EASTBOUND	SAME TIME ZONE
2	6	1
1	4	0
3	6	1
3	8	1
2	5	0
4	7	0

C H A P T E R

14

Repeated-Measures Analysis of Variance (ANOVA)

Tools You Will Need

The following items are considered essential background material for this chapter. If you doubt your knowledge of any of these items, you should review the appropriate chapter or section before proceeding.

- Introduction to analysis of variance (Chapter 13)

- The logic of analysis of variance
  - ANOVA notation and formulas
  - Distribution of F-ratios

- Repeated-measures design (Chapter 11)

Preview

In the early 1970s, it was discovered that the brain manufactures and releases morphine-like substances called endorphins (see Snyder, 1977). Among many possible functions, these substances are thought to act as natural painkillers. For example, acupuncture may relieve pain because the acupuncture needles stimulate a release of endorphins in the brain. It also is thought that endorphins are responsible for reducing pain for long-distance runners and may produce the "runner's high."

Gintzler (1980) studied changes in pain threshold during pregnancy and examined how these changes are related to endorphin activity. Gintzler tested pregnant rats in several sessions spaced at regular intervals throughout their pregnancies (a rat's pregnancy lasts only 3 weeks). In a test session, a rat received a series of foot shocks that increased in intensity. The intensity of shock that elicited a jump response was recorded as the index of pain threshold. In general, the rats showed less sensitivity to shock (increased thresholds) as the pregnancy progressed. The change in threshold from one session to another was gradual, up until just a day or two before birth of the pups. At that point, there was an abrupt increase in pain threshold. Gintzler also found evidence that the change in pain sensitivity was due to enhanced activity of endorphins. Perhaps a natural pain-killing mechanism prepares these animals for the stress of giving birth.

You should recognize Gintzler's experiment as an example of a repeated-measures design: Each rat was observed repeatedly at several different times during the course of pregnancy. You also should recognize that it would be inappropriate to use the repeated-measures t statistic to evaluate the data because each subject is measured in *more than* two conditions. As noted in Chapter 13, whenever the number of levels of a treatment is greater than two, analysis of variance (ANOVA) should be performed rather than multiple t tests. Because each t test has a risk of Type I error, doing several t tests would result in an unacceptably large experimentwise alpha level, making the probability of a Type I error for the entire set of tests undesirably high. The ANOVA, on the other hand, provides a single test statistic (the F-ratio) for the experiment.

In this chapter, we will examine how the repeated-measures design is analyzed with ANOVA. As you will see, many of the notational symbols and computations are the same as those used for the independent-measures ANOVA. In fact, your best preparation for this chapter is a good understanding of the basic ANOVA procedure presented in Chapter 13.

14.1 OVERVIEW

In the preceding chapter, we introduced analysis of variance (ANOVA) as a hypothesis-testing procedure that is used to evaluate mean differences among two or more treatment conditions. The primary advantage of analysis of variance (compared to t tests) is that it allows researchers to test for significant mean differences in situations where there are *more than two* treatment conditions. The F-ratio in the ANOVA compares all the different sample means in a single test statistic using a single alpha level. If the t statistic was used in a situation with more than two sample means, it would require multiple t tests to evaluate all the mean differences. Because each t test involves a risk of a Type I error, doing multiple tests would inflate the overall, experimentwise alpha level and produce an unacceptably high risk of a Type I error.

Chapter 13 introduced the general logic underlying ANOVA and presented the equations used for analyzing data from a single-factor, independent-measures research study. Recall that the term *single factor* indicates that the study involves only one independent variable (or quasi-independent variable). Also, the term *independent-measures* indicates that the study uses a separate sample for each of the treatment conditions being compared. In this chapter we will extend the analysis of variance procedure to research situations using *repeated-measures* designs. A repeated-measures study uses a single sample, with the same set of individuals measured in all of the different treatment conditions.

THE SINGLE-FACTOR, REPEATED-MEASURES DESIGN

In this chapter we will introduce the formulas and procedures for ANOVA as it is used to evaluate mean differences from a single-factor, repeated-measures research design. Table 14.1 shows two sets of data representing common examples of single-factor, repeated-measures designs. Table 14.1(a), for example, shows data from a research study examining visual perception. The participants' task is to detect a very faint visual stimulus under three different conditions: (1) When there is no distraction, (2) when there is visual distraction from flashing lights, and (3) when there is auditory distraction from banging noises. You should notice that this study is an experiment because the researcher is manipulating an independent variable (the type of distraction). You should also realize that this is a repeated-measures study because the same sample is being measured in all three treatment conditions.

The single-factor, repeated-measures design can also be used with nonexperimental research where the different conditions are not created by manipulating an independent variable. Table 14.1(b) shows results from a study examining the effectiveness of a clinical therapy for treating depression. In this example, measurements of depression were obtained for a single sample of patients when they first entered therapy, when therapy is completed, and again 6 months later to evaluate the long-term effectiveness of the therapy. The goal of the study is to evaluate mean differences in the level of depression as a function of the therapy. Notice that the three sets of data are all obtained from the same sample of participants; this is the essence of a repeated-measures design. In this case, a nonmanipulated variable (time) is the single factor in the study, and the three measurement times define the three levels of the factor. Another common example of this type of design is found in developmental psychology where the participants' age is the quasi-independent variable. For example, a psychologist could examine the

TABLE 14.1

Two sets of data representing typical examples of single-factor, repeated-measures research designs

(a) Data from an experimental study evaluating the effects of different types of distraction on the performance of a visual detection task.

	VISUAL DETECTION SCORES		
PARTICIPANT	NO DISTRACTION	VISUAL DISTRACTION	AUDITORY DISTRACTION
A	47	22	41
B	57	31	52
C	38	18	40
D	45	32	43

(b) Data from a nonexperimental design evaluating the effectiveness of a clinical therapy for treating depression.

	DEPRESSION SCORES		
PARTICIPANT	BEFORE THERAPY	AFTER THERAPY	6-MONTH FOLLOW UP
A	71	53	55
B	62	45	44
C	82	56	61
D	77	50	46
E	81	54	55

development of vocabulary skill by measuring vocabulary for a sample of 3-year-old children, then measuring the same children again at age 4 and at age 5. Again, this is a nonexperimental study (age is not manipulated), with the individuals measured at three different times.

| HYPOTHESES FOR THE REPEATED-MEASURES ANOVA | The hypotheses for the repeated-measures ANOVA are exactly the same as those for the independent-measures ANOVA presented in Chapter 13. Specifically, the null hypothesis states that for the general population there are no mean differences among the treatment conditions being compared. In symbols, |

$$H_0: \quad \mu_1 = \mu_2 = \mu_3 = \cdots$$

According to the null hypothesis, on average, all of the treatments have exactly the same effect. As a consequence of H_0, any differences that may exist among the sample means are not caused by the treatments but rather are simply due to chance.

The alternative hypothesis states that there are mean differences among the treatment conditions. Rather than specifying exactly which treatments are different, we will use a generic version of H_1 which simply states that differences exist:

$$H_1: \quad \text{At least one treatment mean } (\mu) \text{ is different from another.}$$

Notice that the alternative says that, on average, the treatments do have different effects. Thus, the treatment conditions may be responsible for causing mean differences among the samples. As always, the goal of the analysis of variance is to use the sample data to determine which of the two hypotheses is more likely to be correct.

| THE *F*-RATIO FOR THE REPEATED-MEASURES ANOVA | The test statistic for the repeated-measures analysis of variance has the same structure that was used for the independent-measures ANOVA in Chapter 13. In each case, the F-ratio compares the actual mean differences between treatments with the amount of difference that would be expected just by chance. As always, the F-ratio uses variance to measure the size of the differences. Thus, the F-ratio for the repeated-measures ANOVA has the general structure |

$$F = \frac{\text{variance (differences) between treatments}}{\text{variance (differences) expected by chance/error}}$$

A large value for the F-ratio indicates that the differences between treatments are greater than would be expected by chance or error alone. If the F-ratio is larger than the critical value in the F distribution table, we can conclude that the differences between treatments are *significantly* larger than would be due to chance.

Although the structure of the F-ratio is the same for independent-measures and repeated-measures designs, there is a fundamental difference between the two designs that produces a corresponding difference in the two F-ratios. Specifically, one of the characteristics of a repeated-measures design is that it eliminates or removes the variance caused by individual differences. This point was first made when we introduced the repeated-measures design in Chapter 11 (page 356), but we will repeat it briefly now.

First, recall that the term *individual differences* refers to subject characteristics such as age, personality, and gender that vary from one person to another and may influence the measurements that you obtain for each person. In some research designs it is possible that the subjects assigned to one treatment will have higher scores simply because they have different characteristics than the subjects assigned to another treatment. For example, the subjects in one treatment may be smarter or older or taller than those in another treatment. In this case, the mean difference between treatments is not necessarily caused by the treatments; instead, it may be caused by individual differences. With a repeated-measures design, however, you never need to worry about this problem. In a repeated-measures study, the subjects in one treatment are exactly the same as the subjects in every other treatment. In terms of the F-ratio for a repeated-measures design, the variance between treatments (the numerator) does not contain any individual differences.

A repeated-measures design also allows you to remove individual differences from the variance in the denominator of the F-ratio. Because the same individuals are measured in every treatment condition, it is possible to measure the size of the individual differences. In Table 14.1(a), for example, subject B has scores that average 10 points higher than subject A. Because there is a consistent difference between subjects across all the treatment conditions, we can be reasonably confident that the 10-point difference is not simply chance or random error, but rather is a systematic and predictable measure of the individual differences between these two subjects. Thus, in a repeated-measures research study, individual differences are not random and unpredictable; rather, they can be measured and separated from other sources of error.

Because individual differences can be eliminated or removed from a repeated-measures study, the structure of the final F-ratio can be modified as follows:

$$F = \frac{\text{variance/differences between treatments (without individual differences)}}{\text{variance/differences expected by chance (with individual differences removed)}}$$

The process of removing individual differences is an important part of the procedure for a repeated-measures analysis of variance.

THE LOGIC OF THE REPEATED-MEASURES ANOVA

The general purpose of a repeated-measures analysis of variance is to determine whether the differences that are found between treatment conditions are significantly greater than would be expected by chance. In the numerator of the F-ratio, the between-treatments variance measures the size of the actual mean differences between the treatment conditions in a research study. The variance in the denominator of the F-ratio is intended to measure how much difference is reasonable to expect just by chance. In this section we will examine the elements that make up each of these two variances.

Variance between treatments: the numerator of the F-ratio   Logically, any differences that are found between treatment conditions can be explained by only two factors:

1. Treatment Effect. It is possible that the different treatment conditions really do have different effects and, therefore, cause the individuals' scores in one condition to be higher (or lower) than in another. Remember that the purpose for the research study is to determine whether or not a treatment effect exists.

2. Error or Chance. Even if there is no treatment effect, it is still possible for the scores in one treatment condition to be different from the scores in another. For example, suppose that I measure your IQ score on a Monday morning. A week later I come back and measure your IQ again under exactly the same conditions. Will you get exactly the same IQ score both times? In fact, minor differences between the two measurement situations will probably cause you to end up with two different scores. For example, for one of the IQ tests you might be more tired, or hungry, or worried, or distracted than you were on the other test. These differences can cause your scores to vary. The same thing can happen in a repeated-measures research study. The same individuals are being measured two different times and, even though there may be no difference between the two treatment conditions, you can still end up with different scores. Because these differences are unsystematic and unpredictable, they are classified as chance or experimental error.

Thus, it is possible that any differences (or variance) found between treatments could be caused by treatment effects, and it is possible that the differences could simply be due to chance. On the other hand, it is *impossible* that the differences between treatments are caused by individual differences. Because the repeated-measures design uses exactly the same individuals in every treatment condition, individual differences are *automatically eliminated* from the variance between treatments in the numerator of the F-ratio.

Variance due to chance/error: the denominator of the F-ratio   The goal of the analysis of variance is to determine whether the mean differences that are observed in the data are greater than would be expected simply by chance (sampling error). To accomplish this goal, the denominator of the F-ratio is intended to measure how much difference (or variance) is reasonable to expect by chance alone. This means that we must measure the variance that is *not caused* by treatment differences or by any other systematic differences.

We will eliminate treatment differences exactly as we did for the independent-measures ANOVA; specifically, we will use the variance that exists within treatments. Recall from Chapter 13 that within each treatment, all the individuals are treated exactly the same, so any differences that exist within treatments cannot be caused by treatment effects.

In a repeated-measures design, it is also possible that individual differences can cause systematic differences between scores; for example, one individual may score consistently higher than another. To eliminate the variance caused by individual differences, we will literally measure the individual differences and then subtract them from the variability within treatments.

In summary, the denominator of the F-ratio is intended to measure the variance (or differences) that exist simply by chance or sampling error. To find this variance, we begin with the variability of the scores and then remove any systematic differences caused by the treatments and any systematic differences caused by individual differences. As a result, the repeated-measures F-ratio has the following structure:

$$F = \frac{\text{variance between treatments}}{\text{variance due to chance/error}}$$

$$= \frac{\text{treatment effect} + \text{chance/error (excluding individual diff's.)}}{\text{chance/error (excluding individual diff's.)}} \tag{14.1}$$

When there is no treatment effect, the F-ratio is balanced because the numerator and denominator are both measuring exactly the same variance. In this case, the F-ratio should have a value near 1.00. When research results produce an F-ratio near 1.00, we conclude that there is no treament effect and we fail to reject the null hypothesis. On the other hand, when a treatment effect does exist, it will contribute only to the numerator and should produce a large value for the F-ratio. Thus, a large value for F indicates that there is a real treatment effect and means that we should reject the null hypothesis.

1. What sources contribute to between-treatments variability for the repeated-measures design?

2. What sources of variability contribute to within-treatments variability?

3. **a.** Describe the structure of the F-ratio for a repeated-measures ANOVA.

   b. Compare it to the F-ratio structure for the independent-measures ANOVA (Chapter 13). How do they differ?

1. Treatment effect, error/chance

2. Individual differences, error/chance

3. **a.** $F = \dfrac{\text{treatment effect} + \text{error/chance}}{\text{error/chance}}$

   b. For the independent-measures ANOVA, individual differences contribute variability to both between-treatments variability and the error term of the F-ratio.

14.2 NOTATION AND FORMULAS FOR REPEATED-MEASURES ANOVA

The overall structure of the repeated-measures analysis of variance is shown in Figure 14.1. Note that the ANOVA can be viewed as a two-stage process. In the first stage, the total variability is partitioned into two components: variance between treatments and variance within treatments. This stage is identical to the analysis that we conducted for an independent-measures design in Chapter 13. In fact, the first stage of the repeated-measures ANOVA will use exactly the same notation and the same formulas that were presented in Chapter 13.

The second stage of the analysis is intended to remove the individual differences from the denominator of the F-ratio. In the second stage, we begin with the variance within treatments and then measure and subtract out the individual differences. The remaining variance, often called the *residual variance* or *error variance*, provides a measure of how much variance is reasonable to expect by chance after the individual differences have been removed. The second stage of the analysis is what differentiates the repeated-measures ANOVA from the independent-measures ANOVA. Specifically, the repeated-measures design requires that the individual differences be removed.

DEFINITION   In a repeated-measures analysis of variance, the denominator of the F-ratio is called the **residual variance** or the **error variance** and measures how much variance is to be expected just by chance after the individual differences have been removed.

NOTATION FOR THE REPEATED-MEASURES ANOVA   We will use the data in Table 14.2 to introduce the notation for the repeated-measures ANOVA. The data represent the results of a study comparing three brands of pain reliever as well as a fourth condition in which the participants receive a placebo (sugar pill). Note that the same group of $n = 5$ individuals is tested in all four conditions.

FIGURE 14.1

The partitioning of variability for a repeated-measures experiment.

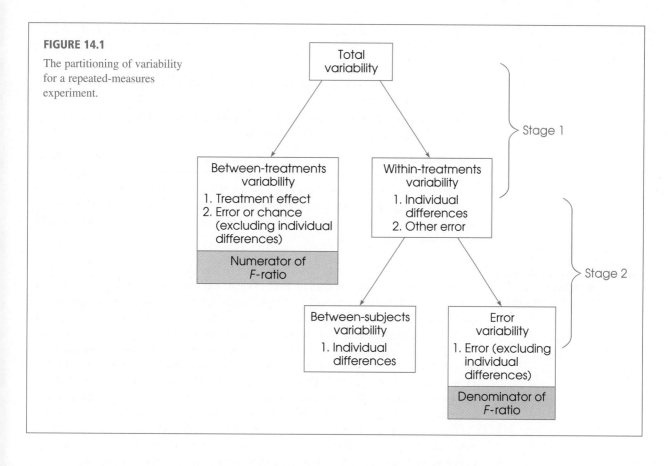

You should recognize that most of the notation in Table 14.2 is identical to the notation used in an independent-measures analysis (Chapter 13). The following list summarizes the familiar notation:

1. The letter k identifies the number of treatment conditions. For Table 14.2, $k = 4$.

2. The number of scores in each treatment is identified by the lowercase letter n. For a repeated-measures study, n is the same for all treatments because the same sample is used in all treatments. In Table 14.2, $n = 5$.

3. The total number of scores in the entire study is identified by an uppercase N. For these data, $N = 20$.

TABLE 14.2

The effect of drug treatment on the amount of time (in seconds) a stimulus is endured.

PERSON	PLACEBO	DRUG A	DRUG B	DRUG C	PERSON TOTALS	
A	3	4	6	7	$P = 20$	$n = 5$
B	0	3	3	6	$P = 12$	$k = 4$
C	2	1	4	5	$P = 12$	$N = 20$
D	0	1	3	4	$P = 8$	$G = 60$
E	0	1	4	3	$P = 8$	$\Sigma X^2 = 262$
	$T = 5$	$T = 10$	$T = 20$	$T = 25$		
	$SS = 8$	$SS = 8$	$SS = 6$	$SS = 10$		

Remember that we are using T and G to help differentiate ΣX for treatments and ΣX for the entire set of scores.

4. The sum of all the scores in the study is G. The value of G corresponds to ΣX for all N scores and specifies the grand total for the experiment. For Table 14.2, $G = 60$.

5. The sum of the scores in each treatment condition is identified by the letter T, for treatment total.

6. We also identify the sum of squares for each treatment (SS) and the sum of the squared scores (ΣX^2) for the entire study.

The repeated-measures ANOVA introduces only one new notational symbol. The letter P is used to represent the total of all the scores for each individual in the study. You can think of the P values as "Person totals" or "Participant totals." In Table 14.2, for example, participant A had scores of 3, 4, 6, and 7 for a total of $P = 20$. The P values will be used to define and measure the magnitude of the individual differences in the second stage of the analysis.

FORMULAS FOR THE REPEATED-MEASURES ANOVA

We will continue to use the data in Table 14.2 to introduce and to demonstrate the formulas for the repeated-measures ANOVA. Again, most of the formulas are the same as those that were introduced in Chapter 13. In fact, the formulas for the first stage of the repeated-measures ANOVA are identical to the formulas for the independent-measures analysis. To help demonstrate the similarity between the independent-measures ANOVA and the repeated-measures ANOVA, the numerical values in Table 14.2 are identical to the values used in Example 13.1 (page 415). As we introduce the formulas and computations for the repeated-measures ANOVA, we encourage you to look back at Example 13.1 and note the similarities and differences between the two analyses.

The final step in the analysis of variance is the calculation of an F-ratio, which is a ratio of two variances. Recall that in the context of ANOVA, the two variances are called *mean squares* or MS values. The numerator of the F-ratio measures the variance (or differences) between treatments and is represented as $MS_{\text{between treatments}}$. The denominator measures the variance due to chance and is usually represented as MS_{error}.

$$F = \frac{\text{variance between treatments}}{\text{variance due to chance/error}} = \frac{MS_{\text{between treatments}}}{MS_{\text{error}}} \tag{14.2}$$

As always, each of the variances or MS values is computed as

$$s^2 = \frac{SS}{df}$$

Therefore, we will need an SS value and a df value for each of the variances in the F-ratio. To obtain these values, we will use the pattern shown in Figure 14.1 to conduct a complete analysis of the sum of squares and a separate analysis of the degrees of freedom. These two analyses are demonstrated in the following sections.

ANALYSIS OF THE SUM OF SQUARES (SS)

Remember: The repeated-measures ANOVA can be viewed as a two-stage process. The first stage separates the total SS into two components: between treatments and within treatments. This stage of the analysis is identical to the process used for an independent-measures design (Chapter 13). The second stage removes the individual differences from the denominator of the F-ratio. The complete analysis of SS is shown in Figure 14.2.

FIGURE 14.2

The partitioning of sum of squares (*SS*) for a repeated-measures analysis of variance.

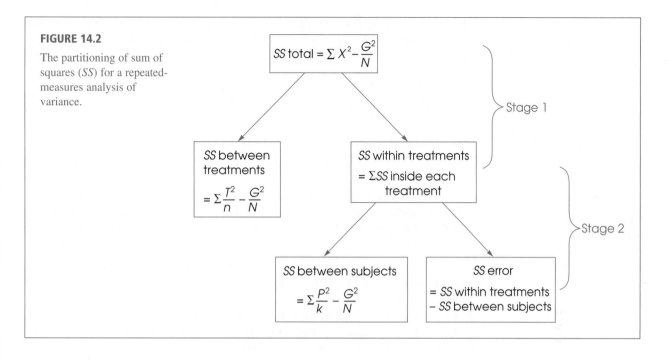

Stage 1   The first step in the analysis of *SS* is to compute SS_{total}, which measures the total variability for the entire set of *N* scores.

$$SS_{\text{total}} = \Sigma X^2 - \frac{G^2}{N} \qquad (14.3)$$

Using this formula for the data in Table 14.2, we obtain

$$SS_{\text{total}} = 262 - \frac{60^2}{20}$$

$$= 262 - 180$$

$$= 82$$

The next step is to partition SS_{total} into two components: between treatments and within treatments (see Figure 14.2). The $SS_{\text{between treatments}}$ measures the mean differences among the treatment conditions and is computed using the treatment totals (*T*'s). The formula for this *SS* value is

$$SS_{\text{between treatments}} = \Sigma \frac{T^2}{n} - \frac{G^2}{N} \qquad (14.4)$$

For the data in Table 14.2, we obtain

$$SS_{\text{between treatments}} = \frac{5^2}{5} + \frac{10^2}{5} + \frac{20^2}{5} + \frac{25^2}{5} - \frac{60^2}{20}$$

$$= 5 + 20 + 80 + 125 - 180$$

$$= 50$$

The $SS_{\text{within treatments}}$ measures the variability inside the treatment conditions. To compute this value, we simply add the SS values from each of the separate treatment conditions.

$$SS_{\text{within treatments}} = \Sigma SS_{\text{inside each treatment}} \tag{14.5}$$

For the data in Table 14.2, we obtain

$$SS_{\text{within treatments}} = 8 + 8 + 6 + 10$$

$$= 32$$

To check your calculations, you should always compute all three SS values and verify that the two components add up to the total.

This completes the first stage of the analysis (see Figure 14.2). Notice that the formulas and computations up to this point are identical to those of the independent-measures ANOVA. You also should note that the two SS components (between and within) add up to the total SS.

Stage 2   The second stage of the analysis involves removing the individual differences from the denominator of the F-ratio. Because the same individuals are used in every treatment, it is possible to measure the size of the individual differences. For the data in Table 14.2, for example, person A tends to have the highest scores and participants D and E tend to have the lowest scores. These individual differences are reflected in the P values, or person totals in the right-hand column. We will use these P values to calculate an SS between subjects in much the same way that we used the treatment totals, the T values, to compute the SS between treatments. Specifically, the formula for SS between subjects is

$$SS_{\text{between subjects}} = \Sigma \frac{P^2}{k} - \frac{G^2}{N} \tag{14.6}$$

Notice that the formula for the between-subjects SS has exactly the same structure as the formula for the between-treatments SS. In this case we use the person totals (P values) instead of the treatment totals (T values). Each P value is squared and divided by the number of scores that were added to obtain the total. In this case, each person has k scores, one for each treatment. Box 14.1 presents another demonstration of the similarity of the formulas for SS between subjects and SS between treatments. For the data in Table 14.2,

$$SS_{\text{between subjects}} = \frac{20^2}{4} + \frac{12^2}{4} + \frac{12^2}{4} + \frac{8^2}{4} + \frac{8^2}{4} - \frac{60^2}{20}$$

$$= 100 + 36 + 36 + 16 + 16 - 180$$

$$= 24$$

The value of $SS_{\text{between subjects}}$ provides a measure of the size of the individual differences—that is, the differences between subjects. In the second stage of the analysis we simply subtract out the individual differences to obtain the measure of error that will form the denominator of the F-ratio. Thus, the final step in the analysis of SS is

$$SS_{\text{error}} = SS_{\text{within treatments}} - SS_{\text{between subjects}} \tag{14.7}$$

For the data in Table 14.2,

$$SS_{\text{error}} = 32 - 24 = 8$$

BOX

14.1 $SS_{\text{between subjects}}$ and $SS_{\text{between treatments}}$

The data for a repeated-measures study are normally presented in a matrix, with the treatment conditions determining the columns and the participants defining the rows. The data in Table 14.2 demonstrate this normal presentation. The calculation of $SS_{\text{between treatments}}$ provides a measure of the differences between treatment conditions—that is, a measure of the mean differences between the *columns* in the data matrix. For the data in Table 14.2, the column totals are 5, 10, 20, and 25. These values are variable, and $SS_{\text{between treatments}}$ measures the amount of variability.

The following table reproduces the data from Table 14.2, but now we have turned the data matrix on its side so that the people define the columns and the treatment conditions define the rows.

In this new format, the differences between the columns represent the between-subjects variability. The

column totals are now P values (instead of T values) and the number of scores in each column is now identified by k (instead of n). With these changes in notation, the formula for $SS_{\text{between subjects}}$ has exactly the same structure as the formula for $SS_{\text{between treatments}}$. If you examine the two equations (14.4 and 14.5), the similarity should be clear.

	A	B	PERSON C	D	E	
Placebo	3	0	2	0	0	$T = 5$
Drug A	4	3	1	1	1	$T = 10$
Drug B	6	3	4	3	4	$T = 20$
Drug C	7	6	5	4	3	$T = 25$
	$P = 20$	$P = 12$	$P = 12$	$P = 8$	$P = 8$	

ANALYSIS OF DEGREES OF FREEDOM (df)

Each of the SS values in analysis of variance has a corresponding degrees of freedom. In general terms, SS measures the variability for a set of items, such as scores, treatment totals, person totals, and so on, and the df value for each SS is determined by the number of items minus one. Thus, for a set of n scores, $df = n - 1$. The df formulas for the first stage of the repeated-measures analysis are identical to those in the independent-measures ANOVA. Figure 14.3 presents the complete analysis of degrees of freedom. You may want to refer to this figure as you read the following section on df formulas.

FIGURE 14.3

The partitioning of degrees of freedom for a repeated-measures experiment.

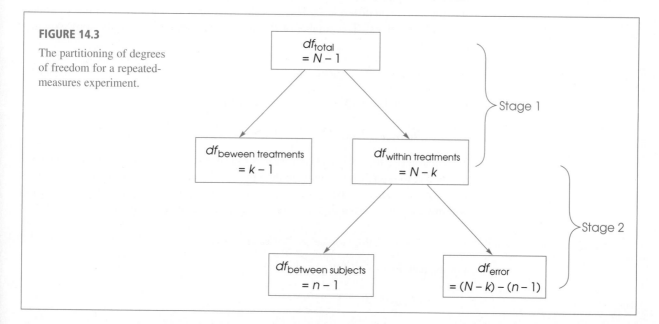

Stage 1   Remember that the total SS measures the variability for the entire set of N scores. The total df is computed by

$$df_{total} = N = 1 \tag{14.8}$$

For the data in Table 14.2,

$$df_{total} = 20 - 1$$
$$= 19$$

The between-treatments SS measures the variability for the set of treatment totals (T values). The number of totals is represented by k, so the corresponding df is

$$df_{between\ treatments} = k = 1 \tag{14.9}$$

For the data in Table 14.2, there are $k = 4$ treatments and

$$df_{between\ treatments} = 4 - 1 = 3$$

The within-treatments SS was obtained by summing the SS values for each of the treatment conditions. Each of these SS values has $df = n - 1$, and $df_{within\ treatments}$ is obtained by summing these df values.

$$df_{within\ treatments} = \Sigma df = \Sigma(n - 1) = N - k \tag{14.10}$$

For the data in Table 14.2, there are $n = 5$ scores in each treatment, so

$$df_{within\ treatments} = 4 + 4 + 4 + 4 = 16$$

This completes the first stage of the analysis of degrees of freedom (see Figure 14.3). As always, you should check that the two components (between and within) add up to the total.

Stage 2   In the second stage of the analysis the between-subjects SS is used to measure the individual differences. This SS value measures the variability for the set of person totals (P values). The number of P values is represented by n, so the corresponding df is

$$df_{between\ subjects} = n - 1 \tag{14.11}$$

For the data in Table 14.2, there are $n = 5$ subjects and

$$df_{between\ subjects} = 5 - 1 = 4$$

The final step in the analysis is to find df for the error variance. Remember that the variability due to error is not computed directly, but rather is defined as the residual that is left when you subtract between-subjects variability (individual differences) from the within-treatments variability. We obtain df_{error} in exactly the same manner.

$$df_{error} = df_{within\ treatments} - df_{between\ subjects} \tag{14.12}$$

For the data in Table 14.2,

$$df_{error} = 16 - 4 = 12$$

CALCULATION OF THE VARIANCES (*MS* VALUES) AND THE *F*-RATIO

The final calculation in the analysis is the *F*-ratio, which is a ratio of two variances. Each variance is called a mean square or *MS*, and is obtained by dividing the appropriate *SS* by its corresponding *df* value. The *MS* in the numerator of the *F*-ratio measures the size of the differences between treatments and is calculated as

$$MS_{\text{between treatments}} = \frac{SS_{\text{between treatments}}}{df_{\text{between treatments}}} \tag{14.13}$$

For the data in Table 14.2,

$$MS_{\text{between treatments}} = \frac{50}{3} = 16.67$$

The denominator of the *F*-ratio measures how much difference is reasonable to expect just by chance, after the individual differences have been removed. This is the error variance or the residual obtained in stage 2 of the analysis.

$$MS_{\text{error}} = \frac{SS_{\text{error}}}{df_{\text{error}}} \tag{14.14}$$

For the data in Table 14.2,

$$MS_{\text{error}} = \frac{8}{12} = 0.67$$

Finally, the *F*-ratio is computed as

$$F = \frac{MS_{\text{between treatments}}}{MS_{\text{error}}} \tag{14.15}$$

For the data in Table 14.2,

$$F = \frac{16.67}{0.67} = 24.88$$

Once again, notice that the repeated-measures ANOVA uses MS_{error} in the denominator of the *F*-ratio. This *MS* value is obtained in the second stage of the analysis, after the individual differences have been removed. As a result, individual differences are completely eliminated from the repeated-measures *F*-ratio, so that the general structure is

$$F = \frac{\text{treatment effect} + \text{chance/error (without individual differences)}}{\text{chance/error (without individual differences)}}$$

For the data we have been examining, the *F* ratio is $F = 24.88$, indicating that the differences between treatments (numerator) are almost 25 times bigger than you would expect just by chance (denominator). A ratio this large would seem to provide clear evidence that there is a real treatment effect. To verify this conclusion you must consult the *F*-distribution table to determine the appropriate critical value for the test. The degrees of freedom for the *F*-ratio are determined by the two variances that form the numerator and the denominator. For a repeated-measures ANOVA, the *df* values for the *F*-ratio are reported as

$$df = df_{\text{between treatments}}, df_{\text{error}}$$

For the example we are considering, the F-ratio has $df = 2, 12$ ("degrees of freedom equal two and twelve"). Using the F-distribution table (page 693) with $\alpha = .05$, the critical value is $F = 3.88$, and with $\alpha = .01$ the critical value is $F = 6.93$. Our obtained F-ratio, $F = 24.88$, is well beyond either of the critical values, so we can conclude that the differences between treatments are *significantly* greater than expected by chance using either $\alpha = .05$ or $\alpha = .01$.

LEARNING CHECK

1. $SS_{\text{within treatments}} - SS_{\text{between subjects}} = $ _____

2. What two df components are associated with the repeated-measures F-ratio? How are they computed?

3. For the following set of data, compute all of the SS components for a repeated-measures ANOVA:

	TREATMENT				
SUBJECT	1	2	3	4	
A	2	2	2	2	$G = 32$
B	4	0	0	4	$\Sigma X^2 = 96$
C	2	0	2	0	
D	4	2	2	4	

4. Which two MS components are used to form the F-ratio of the repeated-measures ANOVA? How are they computed?

ANSWERS

1. SS_{error}

2. Between-treatments df and error df; $df_{\text{between treatments}} = k - 1$; $df_{\text{error}} = (N - k) - (n - 1)$

3. $SS_{\text{total}} = 32$, $SS_{\text{between treatments}} = 10$, $SS_{\text{within treatments}} = 22$, $SS_{\text{between subjects}} = 8$, $SS_{\text{error}} = 14$

4. Between-treatments MS and error MS;

$$MS_{\text{between treatments}} = \frac{SS_{\text{between treatments}}}{df_{\text{between treatments}}}$$

$$MS_{\text{error}} = \frac{SS_{\text{error}}}{df_{\text{error}}}$$

14.3 TESTING HYPOTHESES WITH THE REPEATED-MEASURES ANOVA

In this section we present a complete example of hypothesis testing with a repeated-measures study. We will accomplish this task in the four steps that should be familiar by now: (1) State the hypotheses, and set the alpha level; (2) compute df, and locate the critical region; (3) compute the test statistic; and (4) evaluate the hypotheses.

EXAMPLE 14.1

A school psychologist would like to test the effectiveness of a behavior-modification technique in controlling classroom outbursts of unruly children. A teacher is instructed to use the response-cost technique. Every time a child disrupts the class, he or she is told that the behavior has cost him or her 10 minutes of free time. That is, the free-time period is shortened for each unruly act. For a sample of $n = 4$ children, the number of outbursts is measured for a day before the treatment is initiated and again 1 week, 1 month, and 6 months after the response-cost technique began. Note that the measurements taken after the response-cost technique is administered serve as a long-term follow-up on the effectiveness of the treatment. This underscores the usefulness of the repeated-measures design in evaluating the effectiveness of clinical treatments. The data are summarized in Table 14.3.

TABLE 14.3

The effect of response-cost treatment on the number of outbursts in class after different periods of time

PARTICIPANT	BEFORE TREATMENT	1 WEEK LATER	1 MONTH LATER	6 MONTHS LATER	P
A	8	2	1	1	12
B	4	1	1	0	6
C	6	2	0	2	10
D	8	3	4	1	16
	$T_1 = 26$	$T_2 = 8$	$T_3 = 6$	$T_4 = 4$	
	$SS_1 = 11$	$SS_2 = 2$	$SS_3 = 9$	$SS_4 = 2$	
	$n = 4$ $k = 4$	$N = 16$	$G = 44$	$\Sigma X^2 = 222$	

STEP 1

State the hypotheses, and select an alpha level. According to the null hypothesis, the response-cost technique will be ineffective in producing a change in the number of classroom disruptions. In symbols, the null hypothesis states that

$$H_0: \quad \mu_{\text{before}} = \mu_{1 \text{ week}} = \mu_{1 \text{ month}} = \mu_{6 \text{ months}}$$

Note that the hypothesis refers to the general population, not simply the $n = 4$ children who are participating in the study.

The alternative hypothesis may take on many forms. Somewhere among the four levels of the treatment there will be a difference. Because there are a number of possibilities, the alternative hypothesis states that

$$H_1: \quad \text{At least one mean is different.}$$

For the level of significance, the experimenter selects $\alpha = .05$. In other words, the researcher is willing to take only a 5% chance of committing a Type I error.

STEP 2

Find df, and locate the critical region. For a repeated-measures study, the degrees of freedom associated with the F-ratio are between-treatments df and error df. In this example,

We have computed only those df values that are needed to evaluate the F-ratio. However, it is useful to compute all df components of ANOVA in order to check the computations (see Figure 14.3).

$$df_{\text{between treatments}} = k - 1$$
$$= 4 - 1$$
$$= 3$$
$$df_{\text{error}} = (N - k) - (n - 1)$$
$$= (16 - 4) - (4 - 1)$$
$$= 12 - 3$$
$$= 9$$

FIGURE 14.4

The critical region in the
F distribution for $\alpha = .05$
and $df = 3, 9$.

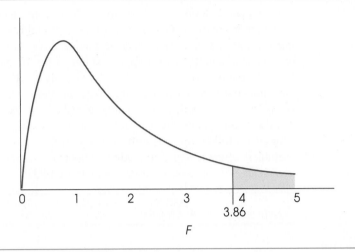

Therefore, for this F-ratio, $df = 3, 9$. Consulting the F-distribution table for $\alpha = .05$, we observe that the critical region begins with $F = 3.86$. Figure 14.4 illustrates the distribution. An obtained F-ratio that exceeds this critical value justifies rejecting H_0.

STEP 3   Compute the test statistic. To compute the F-ratio, we must first calculate the values for SS, df, and MS. This analysis begins by finding the value for total SS.

$$SS_{total} = \Sigma X^2 - \frac{G^2}{N}$$

$$= 222 - \frac{44^2}{16}$$

$$= 222 - 121$$

$$= 101$$

Total variability is partitioned into between-treatments SS and within-treatments SS. For between-treatments variability, we get

$$SS_{between\ treatments} = \Sigma \frac{T^2}{n} - \frac{G^2}{N}$$

$$= \frac{26^2}{4} + \frac{8^2}{4} + \frac{6^2}{4} + \frac{4^2}{4} - \frac{44^2}{16}$$

$$= 169 + 16 + 9 + 4 - 121$$

$$= 77$$

The first stage of the partitioning of variability is completed by computing within-treatments SS:

At this point of the analysis, the work may be checked to see whether total SS equals between-treatments SS plus within-treatments SS.

$$SS_{within\ treatments} = \Sigma SS_{inside\ each\ treatment}$$

$$= 11 + 2 + 9 + 2$$

$$= 24$$

In the second stage of the analysis, the individual differences are measured and removed from the denominator of the F-ratio. This is accomplished by first calculating between-subjects SS and then subtracting it from within-treatments SS. The residual will be error SS:

$$SS_{\text{between subjects}} = \Sigma \frac{P^2}{k} - \frac{G^2}{N}$$

$$= \frac{12^2}{4} + \frac{6^2}{4} + \frac{10^2}{4} + \frac{16^2}{4} - \frac{44^2}{16}$$

$$= 36 + 9 + 25 + 64 - 121$$

$$= 13$$

$$SS_{\text{error}} = SS_{\text{within treatments}} - SS_{\text{between subjects}}$$

$$= 24 - 13$$

$$= 11$$

Finally, we can compute the MS values and then the F-ratio. The repeated-measures F-ratio uses between-treatments MS in the numerator and error MS in the denominator. These are readily obtained by dividing SS by the appropriate number of degrees of freedom:

$$MS_{\text{between treatments}} = \frac{SS_{\text{between treatments}}}{df_{\text{between treatments}}}$$

$$= \frac{77}{3} = 25.67$$

$$MS_{\text{error}} = \frac{SS_{\text{error}}}{df_{\text{error}}}$$

$$= \frac{11}{9} = 1.22$$

$$F = \frac{MS_{\text{between treatments}}}{MS_{\text{error}}}$$

$$= \frac{25.67}{1.22} = 21.04$$

STEP 4   Evaluate the hypotheses. The obtained F-ratio is 21.04. This value falls in the critical region that begins at 3.86. The statistical decision is to reject the null hypothesis. The school psychologist may conclude that the response-cost technique had an effect on the number of disruptions. A report might state:

> There was a significant effect of response-cost training on the number of outbursts exhibited by the children, $F(3, 9) = 21.04, p < .05$.

The analysis of variance summary table is shown in Table 14.4.

TABLE 14.4

Analysis of variance summary for Example 14.1

SOURCE	SS	df	MS	F
Between treatments	77	3	25.67	21.04
Within treatments	24	12		
Between subjects	13	3		
Error	11	9	1.22	
Total	101	15		

MEASURING EFFECT SIZE FOR THE REPEATED-MEASURES ANOVA

The most common method for measuring effect size with analysis of variance is to compute the percentage of variance that is explained by the treatment differences. The percentage of explained variance is usually identified by r^2, but in the context of analysis of variance it is commonly identified as η^2 (eta squared). In Chapter 13, for the independent-measures design, we computed η^2 as

$$\eta^2 = \frac{SS_{\text{between treatments}}}{SS_{\text{between treatments}} + SS_{\text{within treatments}}} = \frac{SS_{\text{between treatments}}}{SS_{\text{total}}}$$

The intent is to measure how much of the total variability is explained by the differences between treatments. With a repeated-measures design however, there is another component that can explain some of the variability in the data. Specifically, part of the variability is caused by differences between individuals. In Table 14.2, for example, person D consistently scored higher than person B. This consistent difference explains some of the variability in the data. When computing the size of the treatment effect, it is customary to remove any variability that can be explained by other factors, and then compute the percentage of the remaining variability that can be explained by the treatment effects. Thus, for a repeated-measures ANOVA, the variability from the individual differences is removed before computing η^2. As a result, η^2 is computed as

$$\eta^2 = \frac{SS_{\text{between treatments}}}{SS_{\text{total}} - SS_{\text{between subjects}}} \tag{14.16}$$

Because Equation 14.16 computes a percentage that is not based on the total variability of the scores (one part, $SS_{\text{between subjects}}$, is removed), the result is often called a *partial* eta squared.

The general goal of Equation 14.16 is to calculate a percentage of the variability that has not already been explained by other factors. Thus, the denominator of Equation 14.16 is limited to variability from the treatment differences and variability that is exclusively due to chance or error. With this in mind, an equivalent version of the η^2 formula is

$$\eta^2 = \frac{SS_{\text{between treatments}}}{SS_{\text{between treatments}} + SS_{\text{error}}} \tag{14.17}$$

In this new version of the eta-squared formula, the denominator consists of the variability that is explained by the treatment differences plus the other *unexplained* variability. Using either formula, the data from Example 14.1 produce

$$\eta^2 = \frac{77}{88} = 0.875 \quad \text{(or 87.5\%)}$$

This result means that 87.5% of the variability in the data (except for the individual differences) is accounted for by the differences between treatments.

IN THE LITERATURE
REPORTING THE RESULTS OF A REPEATED-MEASURES ANOVA

As described in Chapter 13 (page 418), the format for reporting ANOVA results in journal articles consists of

1. A summary of descriptive statistics (at least treatment means and standard deviations, and tables or graphs as needed) and
2. A concise statement of the outcome of the analysis of variance and a measure of effect size.

For the study in Example 14.1, the report could state:

> The means and standard deviations for the sample of children are shown in Table 1. A repeated-measures analysis of variance revealed that the response-cost technique produced a significant decrease in classroom disruptions over the course of the study, $F(3, 9) = 21.04$, $p < .05$, $\eta^2 = 0.875$.
>
> **TABLE 1**
>
> Number of outbursts over course of treatment
>
> | | BEFORE TREATMENT | 1 WEEK LATER | 1 MONTH LATER | 6 MONTHS LATER |
> |---|---|---|---|---|
> | M | 6.5 | 2.0 | 1.5 | 1.0 |
> | SD | 1.91 | 0.82 | 1.73 | 0.82 |

For the descriptive statistics, the APA format uses SD in place of s for the sample standard deviation. The report of the ANOVA results indicates a *significant* change. That is, the null hypothesis was rejected. Then the F-ratio specifies the degrees of freedom values (between-treatments $df = 3$, error $df = 9$), as well as the probability of committing a Type I error ($p < \alpha$ or $p < .05$). Finally, the measure of effect size is reported.

When an analysis of variance is conducted on a computer, the printout usually includes an exact value of p along with the F-ratio. In this case, the exact p value should be reported instead of using $p < \alpha$. Occasionally, the exact p value is so small that the computer rounds the value to zero. For the data in Example 14.1, a computer analysis produced $p = 0.000$. Because the actual probability is really not zero in this case, you should report a significant result with $p < .001$.   ❏

POST HOC TESTS WITH REPEATED MEASURES

Recall that ANOVA provides an overall test of significance for the treatment. When the null hypothesis is rejected, it indicates only that there is a difference between at least two of the treatment means. If $k = 2$, it is obvious where the difference lies in the experiment. However, when k is greater than 2, the situation becomes more complex. To determine exactly where significant differences exist, the researcher must follow the ANOVA with post hoc tests. In Chapter 13, we used Tukey's HSD and the Scheffé test to make these multiple comparisons among treatment means. These two procedures attempt to control the overall alpha level by making adjustments for the number of potential comparisons.

For a repeated-measures ANOVA, Tukey's HSD and the Scheffé test can be used in the exact same manner as was done for the independent-measures ANOVA, *provided* that you substitute MS_{error} in place of $MS_{within\ treatments}$ in the formulas and use df_{error} in place of $df_{within\ treatments}$ when locating the critical value in a statistical table. Note that statisticians are not in complete agreement about the appropriate error term in post hoc tests for repeated-measures designs (for a discussion, see Keppel, 1973, or Keppel & Zedeck, 1989).

14.4   ADVANTAGES OF THE REPEATED-MEASURES DESIGN

When we first encountered the repeated-measures design (Chapter 11), we noted that this type of research study has certain advantages and disadvantages (pages 355–358). On the bright side, a repeated-measures study may be desirable if the supply of participants is limited. A repeated-measures study is economical in that the experimenter can get by using fewer participants. However, the disadvantages may be very great. These take the form of carry-over effects or progressive error, such as fatigue, that can make the interpretation of the data very difficult.

Now that we have introduced the repeated-measures ANOVA, we can examine another advantage—namely, the elimination of the role of variability due to individual differences. Consider the structure of the F-ratio for both the independent- and the repeated-measures designs.

$$F = \frac{\text{treatment effect} + \text{chance/error}}{\text{chance/error}}$$

In each case, the goal of the analysis is to determine whether or not the data provide evidence for a treatment effect. If there is no treatment effect, then the F-ratio should produce a value near 1.00. On the other hand, the existence of a real treatment effect should make the numerator substantially larger than the denominator and result in a large value for the F-ratio.

For the independent-measures design, the magnitude of the chance/error component of the variances includes individual differences as well as other random, unexplained sources of error. Thus, for the independent-measures ANOVA, the F-ratio has the following structure:

$$F = \frac{\text{treatment effect} + (\text{individual differences and other error})}{\text{individual differences and other error}}$$

For the repeated-measures ANOVA, the individual differences are eliminated or subtracted out, and the resulting F-ratio is structured as follows:

$$F = \frac{\text{treatment effect} + \text{chance/error (excluding individual differences)}}{\text{chance/error (excluding individual differences)}}$$

The removal of individual differences from the analysis becomes an advantage in situations where very large individual differences exist among the participants being studied.

When individual differences are extraordinarily large, the presence of a treatment effect may be masked if an independent-measures study is performed. In this case, a repeated-measures design would be more sensitive in detecting a treatment effect because individual differences do not influence the value of the F-ratio.

This point will become evident in the following example. Suppose that an experiment is performed in two ways, with an independent-measures design and a repeated-measures experiment. Also, let's suppose that we know how much variability is accounted for by the different sources of variance. For example,

$$\text{treatment effect} = 10 \text{ units of variance}$$

$$\text{individual differences} = 1000 \text{ units of variance}$$

$$\text{other error} = 1 \text{ unit of variance}$$

Notice that a very large amount of the variability in the experiment is due to individual differences. By comparing the F-ratios of both types of experiments, we will be able to see a fundamental difference between the two types of experimental designs. For the independent-measures experiment, we obtain

$$F = \frac{\text{treatment effect} + \text{individual differences} + \text{error}}{\text{individual differences} + \text{error}}$$

$$= \frac{10 + 1000 + 1}{1000 + 1}$$

$$= \frac{1011}{1001}$$

$$= 1.01$$

However, the repeated-measures F-ratio provides a different outcome:

$$F = \frac{\text{treatment effect} + \text{error}}{\text{error}}$$

$$= \frac{10 + 1}{1}$$

$$= \frac{11}{1}$$

$$= 11$$

All things (sources of variability) being equal, the repeated-measures F-ratio is larger. In this example, the F-ratio is much larger for the repeated-measures study because the individual differences, which are extremely large, have been removed (see Box 14.2). In the independent-measures ANOVA, the presence of a treatment effect is obscured by the influence of individual differences. This problem is remedied by the repeated-measures design in which variability due to individual differences has been partitioned out of the analysis. When the individual differences are large, a repeated-measures experiment may provide a more sensitive test for a treatment effect. In statistical terms, a repeated-measures test has more *power* than an independent-measures test; that is, it is more likely to detect a real treatment effect.

BOX

14.2   **DATA THAT PRODUCE A LARGE MS_{error}**

To get some idea of a situation that results in a large MS_{error}, it is useful to examine hypothetical data in the form of a graph. Suppose that an experimenter examines the effect of amount of reward on maze performance in rats. A sample of $n = 4$ rats that have learned to solve a maze is subsequently tested in all four reward conditions: 2, 4, 6, and 8 grams of food reward. The experimenter measures the speed with which they solve the maze after experiencing the new amount of reward. The broken line in Figure 14.5 represents a graph of the means of each treatment. The speed of maze running seems to increase with amount of reward.

FIGURE 14.5

The effect of amount of reward on running speed. Treatment means are depicted by the broken line. Individual scores for each subject at each level of reward are shown by solid lines.

Figure 14.5 also shows a set of hypothetical data for each individual (solid lines) that would produce the treatment means (broken line). Notice that there are individual differences in these data. The subject totals (P values) differ, indicating that some subjects generally run faster than others. Also note that the individual differences are consistent from one treatment to the next. For example, in all four treatment conditions, subject 1 is fastest, and subject 2 is slowest. This means that much of the variability within treatments is due to consistent, predictable individual differences. When the individual differences are subtracted out of the analysis, the result is a very small value for MS_{error}. Remember

that a small value for MS_{error} will tend to produce a large value for the F-ratio, indicating a significant difference between treatments. The consistency in these data can be described in another way that may help you to understand the F-ratio. For the data in Figure 14.5, the effect of the treatment is consistent for all subjects: Every rat shows an increase in speed when the amount of reward is increased. Because the treatment effect is very consistent, you should expect to find a significant difference between treatments.

Figure 14.6 depicts another possibility for individual subjects. Although these data will produce the same treatment means as the previous example (Figure 14.5), the treatment effect is no longer consistent across subjects. Now when the amount of reward is increased, some subjects run faster, and some run slower. Because the treatment effect is not consistent, you should not expect to find significant differences between treatments. Also note that the data in Figure 14.6 do not show consistent individual differences from one treatment to another. For example, subject 2 is the slowest rat in the first treatment, but is the fastest in the second treatment. Because there are no consistent individual differences, most of the variability within treatments is due to experimental error. As a result, MS_{error} will be large and will tend to produce a relatively small F-ratio.

FIGURE 14.6

The effect of amount of reward on running speed. The treatment means are depicted by the broken line. Individual scores for each subject at each level of reward are shown by solid lines.

14.5 ASSUMPTIONS OF THE REPEATED-MEASURES ANOVA

The basic assumptions for the repeated-measures ANOVA are identical to those required for the independent-measures ANOVA.

1. The observations within each treatment condition must be independent (see page 251).
2. The population distribution within each treatment must be normal. (As before, the assumption of normality is important only with small samples.)
3. The variances of the population distributions for each treatment should be equivalent.

For the repeated-measures ANOVA, there is an additional assumption, called homogeneity of covariance. Basically, it refers to the requirement that the relative standing of each subject be maintained in each treatment condition. This assumption will be violated if the effect of the treatment is not consistent for all of the subjects or if carryover effects exist for some, but not other, subjects. This issue is a very complex one and beyond the scope of this book. However, methods do exist for dealing with violations of this assumption (for a discussion, see Keppel, 1973).

LEARNING CHECK

1. It has been suggested that pupil size increases during emotional arousal. A researcher would, therefore, like to see whether the increase in pupil size is a function of the type of arousal (pleasant versus aversive). A random sample of five participants is selected for the study. Each participant views *all* three stimuli: neutral, pleasant, and aversive photographs. The neutral photograph portrays a plain brick building. The pleasant photograph consists of a young man and woman sharing a large ice cream cone. Finally, the aversive stimulus is a graphic photograph of an automobile accident. Upon viewing each stimulus, the pupil size is measured (in millimeters) with sophisticated equipment. The data are as follows. Perform an ANOVA, and state a conclusion about the findings.

	STIMULUS		
PARTICIPANT	NEUTRAL	PLEASANT	AVERSIVE
A	4	8	3
B	3	6	3
C	2	5	2
D	3	3	6
E	3	8	1

ANSWERS

SOURCE	SS	df	MS	F
Between treatments	30	2	15	4.29
Within treatments	34	12		
Between subjects	6	4		
Error	28	8	3.5	
Total	64	14		

$F(2, 8) = 4.29$. For an alpha level of .05, the obtained F-ratio fails to reach statistical significance. The null hypothesis cannot be rejected. There is not sufficient evidence for an effect of stimulus type on pupil size.

SUMMARY

1. The repeated-measures ANOVA is used to evaluate the mean differences obtained in a research study comparing two or more treatment conditions using the same sample of individuals in each condition. The test statistic is an F-ratio, where the numerator measures the variance (differences) between treatments and the denominator measures the variance (differences) due to chance or error.

$$F = \frac{MS_{\text{between treatments}}}{MS_{\text{error}}}$$

2. The first stage of the repeated-measures ANOVA is identical to the independent-measures analysis and separates the total variability into two components: between-treatments and within-treatments. Because a repeated measures design uses the same subjects in every treatment condition, the differences between treatments cannot be caused by individual differences. Thus, individual differences are automatically eliminated from the between-treatments variance in the numerator of the F-ratio.

3. In the second stage of the repeated-measures analysis, individual differences are computed and removed from the denominator of the F-ratio. To remove the individual differences, you first compute the variability between subjects (SS and df) and then subtract these values from the corresponding within-treatments values. The residual provides a measure of error excluding individual differences, which is the appropriate denominator for the repeated-measures F-ratio.

4. Effect size for the repeated-measures ANOVA is measured by computing eta squared, the percentage of variance accounted for by the treatment effect. For the repeated-measures ANOVA

$$\eta^2 = \frac{SS_{\text{between treatments}}}{SS_{\text{total}} - SS_{\text{between subjects}}}$$

$$= \frac{SS_{\text{between treatments}}}{SS_{\text{between treatments}} + SS_{\text{error}}}$$

Because part of the variability (the SS due to individual differences) is removed before computing η^2, this measure of effect size is often called a partial eta squared.

5. When the obtained F-ratio is significant (that is, H_0 is rejected), it indicates that a significant difference lies between at least two of the treatment conditions. To determine exactly where the difference lies, post hoc comparisons may be made. Post hoc tests, such as Tukey's HSD, use MS_{error} rather than $MS_{\text{within treatments}}$ and df_{error} instead of $df_{\text{within treatments}}$.

6. A repeated-measures ANOVA eliminates the influence of individual differences from the analysis. If individual differences are extremely large, a treatment effect might be masked in an independent-measures experiment. In this case, a repeated-measures design might be a more sensitive test for a treatment effect.

RESOURCES

WebTUTOR

If you are using WebTutor along with this book, there is a WebTutor section corresponding to this chapter. The WebTutor contains a brief summary of Chapter 14, hints for learning the concepts and formulas for the repeated-measures analysis of variance, cautions about common errors, and sample exam items including solutions.

The statistics computer package SPSS can be used to conduct the single-factor, repeated-measures analysis of variance that is presented in this chapter. The data are entered in the data matrix with the scores for each treatment condition in a separate column (with the scores for each individual in the same row). Then, click **Analyze** on the tool bar, select **General Linear Model**, and click on **Repeated Measures**. SPSS will present a box titled Repeated-Measures Define Factors. Within the box, the Within-Subjects Factor Name should already contain FACTOR1. If not, type in Factor 1. Then, enter the number of levels (number of treatment conditions) in the next box, click on **Add**, then click **Define**. One by one, move the column labels for your treatment conditions into the Within Subjects Variables box. (Highlight the column label on the left and click the arrow to move it into the box.) Before you conduct the ANOVA, you can click on the Options box and select Descriptives (to get descriptive statistics for each sample). Then click **OK**. If you selected the Descriptives Option, SPSS will produce a table showing the mean and standard deviation for each treatment condition. The rest of the SPSS output is relatively complex and includes a lot of statistical information that goes well beyond the scope of this book. However, if you focus on the table showing *Test of Within-Subjects Effects*, the top line of the FACTOR1 box and the top line of the Error(FACTOR1) box will show the sum of squares, the degrees of freedom, and the mean square for the numerator and denominator of the F-ratio, as well as the value of the F-ratio and the level of significance (p value).

KEY TERMS

between-treatments variability	error variability	individual differences	mean squares
within-treatments variability	treatment effect	error	F-ratio
between-subjects variability			

FOCUS ON PROBLEM SOLVING

1. Before you begin a repeated-measures ANOVA, complete all the preliminary calculations needed for the ANOVA formulas. This requires that you find the total for each treatment (Ts), the total for each person (Ps), the grand total (G), the SS for each treatment condition, and ΣX^2 for the entire set of N scores. As a partial check on these calculations, be sure that the T values add up to G and that the P values have a sum of G.

2. To help remember the structure of repeated-measures ANOVA, keep in mind that a repeated-measures experiment eliminates the contribution of individual differences. There are no individual differences contributing to the numerator of the F-ratio ($MS_{\text{between treatments}}$) because the same individuals are used for all treatments. Therefore, you must also eliminate individual differences in the denominator. This is accomplished by partitioning within-treatments variability into two components:

between-subjects variability and error variability. It is the *MS* value for error variability that is used in the denominator of the *F*-ratio.

3. As with any ANOVA, it helps to organize your work so you do not get lost. Compute the *SS* values first, followed by the *df* values. You can then calculate the *MS* values and the *F*-ratio. Filling in the columns on an ANOVA summary table (for example, page 460) as you do the computations will help guide your way through the problem.

4. Be careful when using *df* values to find the critical *F* value. Remember that df_{error} (*not* $df_{within\ treatments}$) is used for the denominator.

DEMONSTRATION 14.1

REPEATED-MEASURES ANOVA

The following data were obtained from a research study examining the effect of sleep deprivation on motor skills performance. A sample of five participants was tested on a motor-skills task after 24 hours of sleep deprivation, tested again after 36 hours, and tested once more after 48 hours. The dependent variable is the number of errors made on the motor-skills task.

PARTICIPANT	24 HOURS	36 HOURS	48 HOURS
A	0	0	6
B	1	3	5
C	0	1	5
D	4	5	9
E	0	1	5

Do these data indicate that the number of hours of sleep deprivation has a significant effect on motor skills performance?

STEP 1 *State the hypotheses, and specify alpha.* The null hypothesis states that there are no differences among the three deprivation conditions. In symbols,

$$H_0: \quad \mu_1 = \mu_2 = \mu_3$$

The general form of the alternative hypothesis states that there are differences among the conditions.

$$H_1: \quad \text{At least one of the treatment means is different.}$$

We will set alpha at $\alpha = .05$.

STEP 2 *Locate the critical region.* To locate the critical region, we must obtain the *df* values for the *F*-ratio—specifically, $df_{between\ treatments}$ for the numerator and df_{error} for the denominator. (Often it is easier to postpone this step until the analysis of the *df* values in step 3.)

$$df_{between\ treatments} = k - 1 = 3 - 1 = 2$$

$$df_{error} = (N - k) - (n - 1)$$

$$= (15 - 3) - (5 - 1)$$

$$= 12 - 4$$

$$= 8$$

Thus, the final F-ratio will have $df = 2, 8$. With $\alpha = .05$, the critical value for the F-ratio is $F = 4.46$. The obtained F-ratio must exceed this critical value to reject H_0.

STEP 3   *Perform the analysis.* The complete repeated-measures ANOVA can be divided into a series of stages:

1. Compute the summary statistics for the data. This involves calculating T and SS for each treatment condition, obtaining G and ΣX^2 for the entire set of scores, and finding the P totals for each person.

2. Perform the first stage of the analysis: Separate the total variability (SS and df) into the between- and within-treatment components.

3. Perform the second stage of the analysis: Separate the within-treatment variability (SS and df) into the between-subjects and error components.

4. Calculate the mean squares for the F-ratio.

5. Calculate the F-ratio.

Compute summary statistics. We will use the computational formula to obtain SS for each treatment condition. These calculations will also provide numerical values for the treatment totals (T), G, and ΣX^2.

24 HOURS		36 HOURS		48 HOURS	
X	X²	X	X²	X	X²
0	0	0	0	6	36
1	1	3	9	5	25
0	0	1	1	5	25
4	16	5	25	9	81
0	0	1	1	5	25
$\Sigma X = 5$	$\Sigma X^2 = 17$	$\Sigma X = 10$	$\Sigma X^2 = 36$	$\Sigma X = 30$	$\Sigma X^2 = 192$

For the 24-hour condition,

$$T = \Sigma X = 5$$

$$SS = \Sigma X^2 - \frac{(\Sigma X)^2}{n} = 17 - \frac{5^2}{5} = 17 - 5 = 12$$

For the 36-hour condition,

$$T = \Sigma X = 10$$

$$SS = \Sigma X^2 - \frac{(\Sigma X)^2}{n} = 36 - \frac{10^2}{5} = 36 - 20 = 16$$

For the 48-hour condition,

$$T = \Sigma X = 30$$

$$SS = \Sigma X^2 - \frac{(\Sigma X)^2}{n} = 192 - \frac{30^2}{5} = 192 - 180 = 12$$

The grand total is obtained by summing the three treatment totals.

$$G = \Sigma X = 5 + 10 + 30 = 45$$

The value of ΣX^2 for the entire set of scores is obtained by summing the ΣX^2 values from each of the treatment conditions.

$$\Sigma X^2 = 17 + 36 + 192 = 245$$

The person totals, P values, are obtained by summing the three scores for each individual.

$$P_1 = 0 + 0 + 6 = 6$$

$$P_2 = 1 + 3 + 5 = 9$$

$$P_3 = 0 + 1 + 5 = 6$$

$$P_4 = 4 + 5 + 9 = 18$$

$$P_5 = 0 + 1 + 5 = 6$$

Stage 1 of the analysis. We begin by analyzing SS:

$$SS_{total} = \Sigma X^2 - \frac{G^2}{N} = 245 - \frac{45^2}{15} = 245 - 135 = 110$$

$$SS_{between\ treatments} = \Sigma \frac{T^2}{n} - \frac{G^2}{N} = \frac{5^2}{5} + \frac{10^2}{5} + \frac{30^2}{5} - \frac{45^2}{15}$$

$$= 5 + 20 + 180 - 135$$

$$= 205 - 135$$

$$= 70$$

$$SS_{within\ treatments} = \Sigma SS_{each\ treatment} = 12 + 16 + 12 = 40$$

For this stage, the *df* values are

$$df_{total} = N - 1 = 15 - 1 = 14$$

$$df_{between\ treatments} = k - 1 = 3 - 1 = 2$$

$$df_{within\ treatments} = N - k = 15 - 3 = 12$$

Stage 2 of the analysis. We begin by analyzing SS_{within}:

$$SS_{between\ subjects} = \Sigma \frac{P^2}{k} - \frac{G^2}{N}$$

$$= \frac{6^2}{3} + \frac{9^2}{3} + \frac{6^2}{3} + \frac{18^2}{3} + \frac{6^2}{3} - \frac{45^2}{15}$$

$$= 171 - 135$$

$$= 36$$

$$SS_{error} = SS_{within\ treatments} - SS_{between\ subjects}$$

$$= 40 - 36$$

$$= 4$$

For stage 2, the *df* values are

$$df_{between\ subjects} = n - 1 = 5 - 1 = 4$$

$$df_{error} = (N - k) - (n - 1)$$

$$= 12 - 4 = 8$$

Calculate the two MS values.

$$MS_{\text{between treatments}} = \frac{SS_{\text{between treatments}}}{df_{\text{between treatments}}} = \frac{70}{2} = 35$$

$$MS_{\text{error}} = \frac{SS_{\text{error}}}{df_{\text{error}}} = \frac{4}{8} = 0.50$$

Calculate the F-ratio.

$$F = \frac{MS_{\text{between treatments}}}{MS_{\text{error}}} = \frac{35}{0.50} = 70.00$$

STEP 4 *Make a decision about H_0, and state a conclusion.* The obtained F-ratio, $F = 70.00$, exceeds the critical value of 4.46. Therefore, we reject the null hypothesis. We conclude that the number of hours of sleep deprivation has a significant effect on the number of errors committed, $F(2, 8) = 70.00$, $p < .05$. The following table summarizes the results of the analysis:

SOURCES	SS	df	MS	
Between treatments	70	2	35	$F = 70.00$
Within treatments	40	12		
Between subjects	36	4		
Error	4	8	0.50	
Total	110	14		

DEMONSTRATION 14.2

EFFECT SIZE FOR THE REPEATED-MEASURES ANOVA

We will compute η^2, the percentage of variance explained by the treatment differences, for the data in Demonstration 14.1. Using Equation 14.16, we obtain

$$\eta^2 = \frac{SS_{\text{between treatments}}}{SS_{\text{total}} - SS_{\text{between subjects}}} = \frac{70}{110 - 36} = \frac{70}{74} = 0.95 \quad (\text{or } 95\%)$$

Equivalently, Equation 14.17 produces

$$\eta^2 = \frac{SS_{\text{between treatments}}}{SS_{\text{between treatments}} + SS_{\text{error}}} = \frac{70}{70 + 4} = \frac{70}{74} = 0.95 \quad (\text{or } 95\%)$$

PROBLEMS

1. What advantages does a repeated-measures design have over an independent-measures design?

2. How does the denominator of the F-ratio (the error term) differ for a repeated-measures analysis of variance compared to an independent-measures ANOVA?

3. What process takes place during the second stage of the repeated-measures analysis of variance? Why is this process necessary?

4. A researcher conducts an experiment comparing three treatment conditions. The data consist of $n = 20$ scores for each treatment condition.
   a. If the researcher used an independent-measures design, then how many individuals were needed for the study and what are the df values for the F-ratio?
   b. If the researcher used a repeated-measures design, then how many individuals were needed for the study and what are the df values for the F-ratio?

5. A researcher conducts a repeated-measures study using a sample of $n = 15$ participants to evaluate the differences among four treatment conditions. If the results are examined with an ANOVA, what are the df values for the F-ratio?

6. A researcher reports an F-ratio with $df = 2, 24$ for a repeated-measures ANOVA.
   a. How many treatment conditions were evaluated in this experiment?
   b. How many subjects participated in this experiment?

7. A researcher reports an F-ratio with $df = 1, 38$ from a repeated-measures experiment.
   a. How many treatment conditions were compared in this experiment?
   b. How many subjects participated in the experiment?

8. A researcher used an analysis of variance to evaluate the results from a single-factor repeated-measures experiment. The reported F-ratio was $F(3, 48) = 5.80$.
   a. How many different treatments were compared in this experiment?
   b. How many subjects participated in the experiment?

9. The following data were obtained in a repeated-measures study using one sample of $n = 4$ individuals to compare three different treatments.
   a. Use an analysis of variance with $\alpha = .05$ to determine whether there are any significant differences among the treatments.
   b. Compute the value for η^2 for these data.

	TREATMENT			
PERSON	I	II	III	
A	4	3	8	$N = 12$
B	4	1	4	$G = 48$
C	3	3	6	$\Sigma X^2 = 238$
D	5	1	6	
	$T = 16$	$T = 8$	$T = 24$	
	$SS = 2$	$SS = 4$	$SS = 8$	

10. The following data were obtained from a repeated-measures study comparing three treatment conditions.

	TREATMENT			
SUBJECT	I	II	III	P
A	6	8	10	24	$G = 48$
B	5	5	5	15	$\Sigma X^2 = 294$
C	1	2	3	6	
D	0	1	2	3	
	$T = 12$	$T = 16$	$T = 20$	
	$SS = 26$	$SS = 30$	$SS = 38$	

Use a repeated-measures analysis of variance with $\alpha = .05$ to determine whether these data are sufficient to demonstrate significant differences between the treatments.

11. In Problem 10 the data show large and consistent differences between subjects. For example, subject A has the largest score in every treatment and subject D always has the smallest score. In the second stage of the ANOVA, the large individual differences get subtracted out of the denominator of the F-ratio, which results in a larger value for F.

    The following data were created by using the same numbers that appeared in Problem 10. However, we have now eliminated the consistent individual differences by scrambling the scores within each treatment.

	TREATMENT			
SUBJECT	I	II	III	P
A	6	2	3	11	$G = 48$
B	5	1	5	11	$\Sigma X^2 = 294$
C	0	5	10	15	
D	1	8	2	11	
	$T = 12$	$T = 16$	$T = 20$	
	$SS = 26$	$SS = 30$	$SS = 38$	

   a. Use a repeated-measures analysis of variance with $\alpha = .05$ to determine whether these data are sufficient to demonstrate significant differences between the treatments.
   b. Explain how the results of this analysis compare with the results from Problem 10.

12. It has been demonstrated that when participants must memorize a list of words serially (in the order of presentation), words at the beginning and end of the list are remembered better than words in the middle. This observation has been called the *serial-position effect*. The following data represent the number of errors

made in recall of the first eight, second eight, and last eight words in the list:

	SERIAL POSITION				
PERSON	FIRST	MIDDLE	LAST	P	
A	1	5	0	6	$\Sigma X^2 = 164$
B	3	7	2	12	
C	5	6	1	12	
D	3	2	1	6	
	$T = 12$	$T = 20$	$T = 4$		
	$SS = 8$	$SS = 14$	$SS = 2$		

a. Compute the mean number of errors for each position, and draw a graph of the data.
b. Is there evidence for a significant effect of serial position? Test at the .05 level of significance. Based on the ANOVA, explain the results of the study.

13. In Problem 12 the data show moderate differences from one person to another. In the following data, we have exaggerated the individual differences by adding 4 points to each score for participant B. Except for this change, the data are identical to the scores in Problem 12.
a. Keeping in mind that the repeated-measures ANOVA removes individual differences, predict how the change in the data will affect the outcome of the analysis.
b. Evaluate the mean differences for the new data using a repeated-measures ANOVA with $\alpha = .05$ and compare the outcome with the results from Problem 12.

	SERIAL POSITION				
PERSON	FIRST	MIDDLE	LAST	P	
A	1	5	0	6	$\Sigma X^2 = 308$
B	7	11	6	24	
C	5	6	1	12	
D	3	2	1	6	
	$T = 16$	$T = 24$	$T = 8$		
	$SS = 20$	$SS = 42$	$SS = 22$		

14. A researcher uses a repeated-measures study to examine the effectiveness of a therapy procedure for treating depression. A sample of $n = 10$ depressed patients is obtained and each individual is given a depression test immediately before therapy, immediately after therapy, and once again 3 months after therapy. The data are examined with an ANOVA, and portions of

the results are presented in the following table. Fill in the missing values. (*Hint:* Start with the *df* values.)

SOURCE	SS	df	MS	
Between treatments	___	___	___	$F = 5.00$
Within treatments	500	___		
Between subjects	___	___		
Error	___	___	10	
Total	___	___		

15. A teacher studies the effectiveness of a reading skills course on comprehension. A sample of $n = 15$ students is studied. The instructor assesses their comprehension with a standardized reading test. The test is administered at the beginning of the course and at the end of the course. The instructor uses analysis of variance to determine whether or not a significant change has occurred in the students' reading performance. The following summary table presents a portion of the ANOVA results. Provide the missing values in the table. (Start with the *df* values.)

SOURCE	SS	df	MS	
Between treatments	___	___	24	$F = 4$
Within treatments	120	___		
Between subjects	___	___		
Error	___	___	___	
Total	___	___		

16. The following summary table presents the results of an ANOVA from a repeated-measures experiment comparing four treatment conditions with a sample of $n = 10$ subjects. Fill in all missing values in the table.

SOURCE	SS	df	MS	
Between treatments	___	___	10	$F = $ ___
Within treatments	___	___		
Between subjects	12	___		
Error	___	___	___	
Total	150	___		

17. A computer manufacturer would like to compare four different designs for a keyboard. A sample of eight secretaries is selected, and each spends 10 minutes testing each of the four keyboards and then rates its performance. The manufacturer would like to know whether there are any significant differences among the four designs. The data from this study are examined using an analysis of variance. The results are

shown in the following summary table. Fill in all missing values.

SOURCE	SS	df	MS	
Between treatments	270	___	___	$F = 9$
Within treatments	___	___		
Between subjects	___	___		
Error	___	___	___	
Total	680	___		

18. A psychologist is asked by a dog food manufacturer to determine whether animals will show a preference among three new food mixes recently developed. The psychologist takes a sample of $n = 6$ dogs. They are deprived of food overnight and presented simultaneously with three bowls of the mixes on the next morning. After 10 minutes, the bowls are removed, and the amount of food (in ounces) consumed is determined for each type of mix. The data are as follows:

SUBJECT	I	II	III	P	
A	3	2	1	6	$G = 48$
B	0	5	1	6	$\Sigma X^2 = 196$
C	2	4	3	9	
D	0	7	5	12	
F	0	3	3	6	
G	1	3	5	9	

	MIX		
	$T = 6$	$T = 24$	$T = 18$
	$SS = 8$	$SS = 16$	$SS = 16$

a. Use an analysis of variance with $\alpha = .05$ to determine whether there are any significant differences among the treatments.

b. Compute η^2 to measure the effect size for these data.

19. The following data represent idealized results from an experiment comparing two treatments. Notice that the mean for treatment II is 4 points higher than the mean for treatment I. Also notice that this 4-point treatment effect is perfectly consistent for all participants.

	TREATMENT		
PERSON	I	II	P
A	1	5	6
B	4	8	12
C	7	11	18
D	4	8	12
E	6	10	16
F	2	6	8

$T = 24$	$T = 48$	
$M = 4$	$M = 8$	
$SS = 26$	$SS = 26$	

a. Calculate the within-treatments SS for these data.

b. Calculate the between-subjects SS for these data. You should find that all of the within-treatments variability is accounted for by variability between subjects; that is, $SS_{\text{within treatments}} = SS_{\text{between subjects}}$. For these data, the treatment effect is perfectly consistent across subjects, and there is no error variability ($SS_{\text{error}} = 0$).

20. The following data are from an experiment comparing three different treatment conditions:

A	B	C	
0	1	2	$N = 15$
2	5	5	$\Sigma X^2 = 354$
1	2	6	
5	4	9	
2	8	8	
$T = 10$	$T = 20$	$T = 30$	
$SS = 14$	$SS = 30$	$SS = 30$	

a. If the experiment uses an *independent-measures design*, can the researcher conclude that the treatments are significantly different? Test at the .05 level of significance.

b. If the experiment is done with a *repeated-measures design*, should the researcher conclude that the treatments are significantly different? Set alpha at .05 again.

c. Explain why the results are different in the analyses of parts (a) and (b).

21. A psychologist studies the effect of practice on maze learning for rats. Rats are tested in the maze in one daily session for 4 days. The psychologist records the number of errors made in each daily session. The data are as follows:

RAT	1	2	3	4	P	
			SESSION			
1	3	1	0	0	4	$\Sigma X^2 = 78$
2	3	2	2	1	8	
3	6	3	1	2	12	

$T = 12 \quad T = 6 \quad T = 3 \quad T = 3$
$SS = 6 \quad SS = 2 \quad SS = 2 \quad SS = 2$

a. Use an analysis of variance with $\alpha = .05$ to determine whether there is a significant practice effect.
b. Compute η^2 to measure the effect size for these data.

22. A repeated-measures experiment comparing only two treatments can be evaluated with either a t statistic or an ANOVA. As we found with the independent-measures design, the t test and the ANOVA will produce equivalent conclusions, and the two test statistics are related by the equation $F = t^2$.

The following data are from a repeated-measures study:

SUBJECT	TREATMENT 1	TREATMENT 2	DIFFERENCE
1	2	4	+2
2	1	3	+2
3	0	10	+10
4	1	3	+2

a. Use a repeated-measures t statistic with $\alpha = .05$ to determine whether or not the data provide evidence of a significant difference between the two treatments.
(*Caution:* ANOVA calculations are done with the X values, but for t you use the difference scores.)
b. Use a repeated-measures ANOVA with $\alpha = .05$ to evaluate the data. (You should find $F = t^2$.)

23. An educational psychologist is studying student motivation in elementary school. A sample of $n = 5$ students is followed over 3 years from fourth grade to sixth grade. Each year the students complete a questionnaire measuring their motivation and enthusiasm for school. The psychologist would like to know whether there are significant changes in motivation

across the three grade levels. The data from this study are as follows:

STUDENT	FOURTH GRADE	FIFTH GRADE	SIXTH GRADE
A	4	3	1
B	8	6	4
C	5	3	3
D	7	4	2
E	6	4	0

a. Compute the mean motivation score for each grade level.
b. Use an ANOVA to determine whether there are any significant differences in motivation among the three grade levels. Use the .05 level of significance.

24. The following data represent a second sample of $n = 5$ students from the motivation study described in Problem 23.

STUDENT	FOURTH GRADE	FIFTH GRADE	SIXTH GRADE
A	10	0	4
B	0	10	0
C	4	6	0
D	14	0	6
E	2	4	0

a. Compute the mean motivation score for each grade level.
b. Use an ANOVA to determine whether there are any significant differences in motivation among the three grade levels. Use the .05 level of significance.
c. You should find that the means for these data are identical to the means obtained in Problem 23. How do you explain the fact that the two ANOVAs produce different results?

25. A habituation task is often used to test memory for infants. In the habituation procedure, a stimulus is shown to the infant for a brief period, and the researcher records how much time the infant spends looking at the stimulus. This same process is repeated again and again. If the infant begins to lose interest in the stimulus (decreases the time looking at it), the researcher can conclude that the infant "remembers" the earlier presentations and is demonstrating habituation to an old familiar stimulus. Hypothetical data from a habituation experiment are as follows:

	AMOUNT OF TIME (IN SEC.) ATTENDING TO THE STIMULUS		
INFANT	FIRST PRESENTATION	SECOND PRESENTATION	THIRD PRESENTATION
A	112	81	20
B	97	35	42
C	82	58	27
D	104	70	39
E	78	51	46

a. Use an analysis of variance with $\alpha = .01$ to determine whether there is a significant change in behavior for successive presentations.
b. Compute η^2 to measure the effect size for these data.

26. When a stimulus is presented continuously and it does not vary in intensity, the individual will eventually perceive the stimulus as less intense or not perceive it at all. This phenomenon is known as sensory adaptation. Years ago, Zigler (1932) studied adaptation for skin (cutaneous) sensation by placing a small weight on part of the body and measuring how much time lapsed until participants reported they felt nothing at

all. Suppose that a researcher does a similar study, comparing adaptation for four regions of the body for a sample of $n = 7$ participants. A 500-milligram weight is gently placed on the region, and the latency (in seconds) for a report that it is no longer felt is recorded for each participant. The data are as follows:

	AREA OF STIMULATION			
PARTICIPANT	BACK OF HAND	LOWER BACK	MIDDLE OF PALM	CHIN BELOW LOWER LIP
1	6.5	4.6	10.2	12.1
2	5.8	3.5	9.7	11.8
3	6.0	4.2	9.9	11.5
4	6.7	4.7	8.1	10.7
5	5.2	3.6	7.9	9.9
6	4.3	3.5	9.0	11.3
7	7.4	4.8	10.8	12.6

a. Use an analysis of variance with $\alpha = .01$ to determine whether there are any significant differences among the four regions.
b. Compute η^2 to measure the effect size for these data.

C H A P T E R

15

Two-Factor Analysis of Variance (Independent Measures)

Tools You Will Need

The following items are considered essential background material for this chapter. If you doubt your knowledge of any of these items, you should review the appropriate chapter or section before proceeding.

- Introduction to analysis of variance (Chapter 13)
  - The logic of analysis of variance
  - ANOVA notation and formulas
  - Distribution of *F*-ratios

Preview

Imagine that you are seated at your desk, ready to take the final exam in statistics. Just before the exams are handed out, a television crew appears and sets up a camera and lights aimed directly at you. They explain they are filming students during exams for a television special. You are told to ignore the camera and go ahead with your exam.

Would the presence of a TV camera affect your performance on an exam? For some of you, the answer to this question is "definitely yes" and for others, "probably not." In fact, both answers are right; whether or not the TV camera affects performance depends on your personality. Some of you would become terribly distressed and self-conscious, while others really could ignore the camera and go on as if everything were normal.

In an experiment that duplicates the situation we have described, Shrauger (1972) tested participants on a concept formation task. Half the participants worked alone (no audience), and half worked with an audience of people who claimed to be interested in observing the experiment. Shrauger also divided the participants into two groups on the basis of personality: those high in self-esteem and those low in self-esteem. The dependent variable for this experiment was the number of errors on the concept formation task. Data similar to those obtained by Shrauger are shown in Figure 15.1. Notice that the audience had no effect on the high-self-esteem participants. However, the low-self-esteem participants made nearly twice as many errors with an audience as when working alone.

We have presented Shrauger's study as an introduction to research studies that have two independent variables. In this study, the independent variables are

1. Audience (present or absent)
2. Self-esteem (high or low)

The results of this study indicate that the effect of one variable (audience) *depends on* another variable (self-esteem).

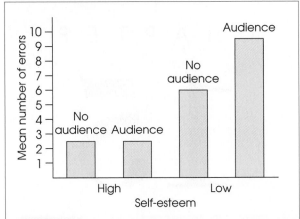

FIGURE 15.1

Results of an experiment examining the effect of an audience on the number of errors made on a concept formation task for participants who are rated either high or low in self-esteem. Notice that the effect of the audience depends on the self-esteem of the participants.

Shrauger, J. S. (1972). Self-esteem and reactions to being observed by others. *Journal of Personality and Social Psychology*, *23*, 192–200. Copyright 1972 by the American Psychological Association. Adapted by permission of the author.

You should realize that it is quite common to have experimental variables that interact in this way. For example, a particular drug may have a profound effect on some patients and have no effect whatsoever on others. Some children develop normally in a single-parent home, while others show serious difficulties. In general, the effects of a particular treatment often depend on other factors. To determine whether two variables are interdependent, it is necessary to examine both variables together in a single study. In this chapter, we will introduce the experimental techniques that are used for studies with two independent variables.

15.1 OVERVIEW

In most research situations, the goal is to examine the relationship between two variables. Typically, the research study will attempt to isolate the two variables in order to eliminate or reduce the influence of any outside variables that may distort the relationship being studied. A typical experiment, for example, will focus on one independent variable (which is expected to influence behavior) and one dependent variable (which is a measure of the behavior). In real life, however, variables rarely exist in isolation.

That is, behavior usually is influenced by a variety of different variables acting and interacting simultaneously. To examine these more complex, real-life situations, researchers often design research studies that include more than one independent variable. Thus, researchers will systematically change two (or more) variables and then observe how the changes influence another (dependent) variable.

In the preceding two chapters, we examined analysis of variance for *single-factor* research designs—that is, designs that included only one independent variable or only one quasi-independent variable. When a research study involves more than one factor, it is called a *factorial design*. In this chapter, we consider the simplest version of a factorial design. Specifically, we will examine analysis of variance as it applies to research studies with exactly two factors. In addition, we will limit our discussion to studies that use a separate sample for each treatment condition—that is, independent-measures designs. Finally, we will consider research designs only where the sample size (n) is the same for all treatment conditions. In the terminology of ANOVA, this chapter will examine *two-factor*, *independent-measures*, *equal n designs*. The following example demonstrates the general structure of this kind of research study.

EXAMPLE 15.1 Most of us find it difficult to think clearly or to work efficiently on hot summer days. If you listen to people discussing this problem, you will occasionally hear comments like "It's not the heat; it's the humidity." To evaluate this claim scientifically, you would need to design a study in which both heat and humidity are manipulated within the same experiment and then observe behavior under a variety of different heat and humidity combinations. Table 15.1 shows the structure of a hypothetical study intended to examine the heat-versus-humidity question. Note that the study involves two independent variables: The temperature (heat) is varied from 70° to 80° to 90°, and the humidity is varied from low to high. The two independent variables are used to create a *matrix* with the different values of temperature defining the columns and the different levels of humidity defining the rows. The resulting two-by-three matrix shows six different combinations of the variables, producing six treatment conditions. Thus, the research study would require six separate samples, one for each of the *cells* or boxes in the matrix. The dependent variable for our study would be a measure of thinking/working proficiency (such as performance on a problem-solving task) for people observed in each of the six conditions.

TABLE 15.1

The structure of a two-factor experiment presented as a matrix. The factors are humidity and temperature. There are two levels for the humidity factor (low and high), and there are three levels for the temperature factor (70°, 80°, and 90°).

		FACTOR *B*: TEMPERATURE		
		70° ROOM	80° ROOM	90° ROOM
FACTOR *A*: HUMIDITY	LOW HUMIDITY	Scores for $n = 15$ participants tested in a 70° room with low humidity	Scores for $n = 15$ participants tested in an 80° room with low humidity	Scores for $n = 15$ participants tested in a 90° room with low humidity
	HIGH HUMIDITY	Scores for $n = 15$ participants tested in a 70° room with high humidity	Scores for $n = 15$ participants tested in an 80° room with high humidity	Scores for $n = 15$ participants tested in a 90° room with high humidity

The two-factor analysis of variance will test for mean differences in research studies that are structured like the heat-and-humidity example in Table 15.1. For this example, the two-factor ANOVA tests for

1. Mean difference between the two humidity levels.

2. Mean differences between the three temperature levels.

3. Any other mean differences that may result from unique combinations of a specific temperature and a specific humidity level. (For example, high humidity may be especially disruptive when the temperature is also high.)

Thus, the two-factor ANOVA combines three separate hypothesis tests in one analysis. Each of these three tests will be based on its own F-ratio computed from the data. The three F-ratios will all have the same basic structure:

$$F = \frac{\text{variance (differences) between sample means}}{\text{variance (differences) expected by chance}}$$

As always in ANOVA, a large value for the F-ratio indicates that the sample mean differences are greater than expected by chance. To determine whether the obtained F-ratios are *significantly* greater than expected by chance, we will need to compare each F-ratio with the critical values found in the F-distribution table in Appendix B.

15.2 MAIN EFFECTS AND INTERACTIONS

As noted in the previous section, a two-factor ANOVA actually involves three distinct hypothesis tests. In this section, we examine these three tests in more detail.

Traditionally, the two independent variables in a two-factor experiment are identified as factor A and factor B. For the experiment presented in Table 15.1, humidity is factor A, and temperature is factor B. The goal of the experiment is to evaluate the mean differences that may be produced by either of these factors independently or by the two factors acting together.

MAIN EFFECTS   One purpose of the experiment is to determine whether differences in humidity (factor A) result in differences in performance. To answer this question, we will compare the mean score for all participants tested with low humidity versus the mean score for all participants tested with high humidity. Note that this process evaluates the mean difference between the rows in Table 15.1.

To make this process more concrete, we have presented a set of hypothetical data in Table 15.2. This table shows the mean score for each of the treatment conditions (cells) as well as the mean for each column (each temperature) and for each row (humidity level). These data indicate that participants in the low-humidity condition (the top row) obtained an average score of $M = 80$. This overall mean was obtained by computing the average of the three means in the top row. In contrast, high humidity resulted in a mean score of $M = 70$ (the overall mean for the bottom row). The difference between these means constitutes what is called the *main effect* for humidity, or the *main effect for factor A*.

Similarly, the main effect for factor B (temperature) is defined by the mean differences among columns of the matrix. For the data in Table 15.2, the two groups of participants tested with a temperature of 70° obtained an overall mean score of $M = 80$.

TABLE 15.2

Hypothetical data from an experiment examining two different levels of humidity (factor A) and three different levels of temperature (factor B)

	70°	80°	90°	
Low humidity	$M = 85$	$M = 80$	$M = 75$	$M = 80$
High humidity	$M = 75$	$M = 70$	$M = 65$	$M = 70$
	$M = 80$	$M = 75$	$M = 70$	

Participants tested with the temperature at 80° averaged only $M = 75$, and those tested at 90° achieved a mean score of $M = 70$. The differences among these means constitute the *main effect* for temperature, or the *main effect for factor B*.

DEFINITION

The mean differences among the levels of one factor are referred to as the **main effect** of that factor. When the design of the research study is represented as a matrix with one factor determining the rows and the second factor determining the columns, then the mean differences among the rows describe the main effect of one factor, and the mean differences among the columns describe the main effect for the second factor.

You should realize that the mean differences among columns or rows simply *describe* the main effects for a two-factor study. As we have observed in earlier chapters, the existence of sample mean differences does not necessarily imply that the differences are *statistically significant*. In the case of a two-factor study, any main effects that are observed in the data must be evaluated with a hypothesis test to determine whether or not they are statistically significant effects. Unless the hypothesis test demonstrates that the main effects are significant, you must conclude that the observed mean differences are simply the result of sampling error.

The evaluation of main effects will make up two of the three hypothesis tests contained in a two-factor ANOVA. We will state hypotheses concerning the main effect of factor A and the main effect of factor B and then calculate two separate F-ratios to evaluate the hypotheses.

For the example we are considering, factor A involves the comparison of two different levels of humidity. The null hypothesis would state that there is no difference between the two levels; that is, humidity has no effect on performance. In symbols,

$$H_0: \quad \mu_{A_1} = \mu_{A_2}$$

The alternative hypothesis is that the two different levels of humidity do produce different scores:

$$H_1: \quad \mu_{A_1} \neq \mu_{A_2}$$

To evaluate these hypotheses, we will compute an F-ratio that compares the actual mean differences between the two humidity levels versus the amount of difference that would be expected by chance or sampling error.

$$F = \frac{\text{variance (differences) between the means for factor } A}{\text{variance (differences) expected by chance/error}}$$

$$F = \frac{\text{variance (differences) between row means}}{\text{variance (differences) expected by chance/error}}$$

Similarly, factor B involves the comparison of the three different temperature conditions. The null hypothesis states that overall there are no differences in mean performance among the three temperatures. In symbols,

$$H_0: \quad \mu_{B_1} = \mu_{B_2} = \mu_{B_3}$$

As always, the alternative hypothesis states that there are differences:

$$H_1: \quad \text{At least one mean is different from another.}$$

Again, the F-ratio will compare the obtained mean differences among the three temperature conditions versus the amount of difference that would be expected by chance.

$$F = \frac{\text{variance (differences) between the means for factor } B}{\text{variance (differences) expected by chance/error}}$$

$$F = \frac{\text{variance (differences) between column means}}{\text{variance (differences) expected by chance/error}}$$

INTERACTIONS   In addition to evaluating the main effect of each factor individually, the two-factor ANOVA allows you to evaluate other mean differences that may result from unique combinations of the two factors. For example, specific combinations of heat and humidity may have effects that are different from the overall effects of heat or humidity acting alone. Any "extra" mean differences that are not explained by the main effects are called an *interaction between factors*. The real advantage of combining two factors within the same study is the ability to examine the unique effects caused by an interaction.

DEFINITION   An **interaction between two factors** occurs whenever the mean differences between individual treatment conditions, or cells, are different from what would be predicted from the overall main effects of the factors.

To make the concept of an interaction more concrete, we will reexamine the data shown in Table 15.2. For these data, there is no interaction; that is, there are no extra mean differences that are not explained by the main effects. For example, within each temperature condition (each column of the matrix) the participants scored 10 points higher in the low-humidity condition than in the high-humidity condition. This 10-point mean difference is exactly what is predicted by the overall main effect for humidity.

Now consider the data shown in Table 15.3. These new data show exactly the same main effects that existed in Table 15.2 (the column means and the row means have not

TABLE 15.3

Hypothetical data from an experiment examining two different levels of humidity (factor A) and three different temperature conditions (factor B). (These data show the same main effects as the data in Table 15.2, but the individual treatment means have been modified to produce an interaction.)

	70°	80°	90°	
Low humidity	$M = 80$	$M = 80$	$M = 80$	$M = 80$
High humidity	$M = 80$	$M = 70$	$M = 60$	$M = 70$
	$M = 80$	$M = 75$	$M = 70$	

been changed). But now there is an interaction between the two factors. For example, when the temperature is 90° (third column), there is a 20-point difference between the low-humidity and high-humidity conditions. This 20-point difference cannot be explained by the 10-point main effect for humidity. Also, when the temperature is 70° (first column), the data show no difference between the two humidity conditions. Again, this zero difference is not what would be expected based on the 10-point main effect for humidity. The extra, unexplained mean differences are an indication that there is an interaction between the two factors.

To evaluate the interaction, the two-factor ANOVA first identifies mean differences that cannot be explained by the main effects. After the extra mean differences are identified, they are evaluated by an F-ratio with the following structure:

$$F = \frac{\text{variance (mean differences) not explained by main effects}}{\text{variance (differences) expected by chance/error}}$$

The null hypothesis for this F-ratio simply states that there is no interaction:

H_0:   There is no interaction between factors A and B. All the mean differences between treatment conditions are explained by the main effects of the two factors.

The alternative hypothesis is that there is an interaction between the two factors:

H_1:   There is an interaction between factors. The mean differences between treatment conditions are not what would be predicted from the overall main effects of the two factors.

MORE ABOUT INTERACTIONS

In the previous section, we introduced the concept of an interaction as the unique effect produced by two factors working together. This section presents two alternative definitions of an interaction. These alternatives are intended to help you understand the concept of an interaction and to help you identify an interaction when you encounter one in a set of data. You should realize that the new definitions are equivalent to the original and simply present slightly different perspectives on the same concept.

The first new perspective on the concept of an interaction focuses on the notion of independence for the two factors. More specifically, if the two factors are independent, so that one factor does not influence the effect of the other, then there will be no interaction. On the other hand, when the two factors are not independent, so that the effect of one factor is influenced by the other, then there is an interaction. You should realize that the notion of independence is consistent with our earlier discussion of interactions. If one factor influences the effect of the other, then unique combinations of the factors will produce unique effects.

DEFINITION   When the effect of one factor depends on the different levels of a second factor, then there is an **interaction between the factors**.

Returning to the data in Table 15.2, you will notice that the size of the humidity effect (top row versus bottom row) *does not depend on* the temperature. For these data, the change in humidity shows the same 10-point effect for all three levels of temperature. Thus, the humidity effect does not depend on temperature, and there is no interaction. Now consider the data in Table 15.3. This time, the effect of changing humidity *depends on* the temperature, and there is an interaction. At 70°, for example, there is no difference between the high and low humidity conditions. However, there is a 10-point difference between high and low humidity when the temperature is 80°, and there is a 20-point effect at 90°. Thus, the effect of humidity depends on the temperature, which means that there is an interaction between the two factors.

The second alternative definition of an interaction is obtained when the results of a two-factor study are presented in a graph. In this case, the concept of an interaction can be defined in terms of the pattern displayed in the graph. Figure 15.2 shows the two sets of data we have been considering. The original data from Table 15.2, where there is no interaction, are presented in Figure 15.2(a). To construct this figure, we selected one of the factors to be displayed on the horizontal axis; in this case, the different levels of temperature are displayed. The dependent variable, mean level of performance, is shown on the vertical axis. Note that the figure actually contains two separate graphs: The top line shows the relationship between temperature and mean performance when the humidity is low, and the bottom line shows the relationship when the humidity is high. In general, the picture in the graph matches the structure of the data matrix; the columns of the matrix appear as values along the X-axis, and the rows of the matrix appear as separate lines in the graph.

FIGURE 15.2

(a) Graph showing the data from Table 15.2, where there is no interaction. (b) Graph showing the data from Table 15.3, where there is an interaction.

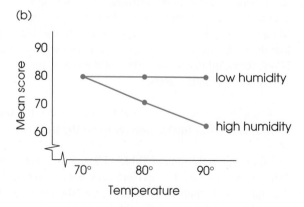

For this particular set of data, Figure 15.2(a), note that the two lines are parallel; that is, the distance between lines is constant. In this case, the distance between lines reflects the 10-point difference in mean performance between high and low humidity, and this 10-point difference is the same for all three temperature conditions.

Now look at a graph that is obtained when there is an interaction in the data. Figure 15.2(b) shows the data from Table 15.3. This time, note that the lines in the graph are not parallel. The distance between the lines changes as you scan from left to right. For these data, the distance between the lines corresponds to the humidity effect—that is, the mean difference in performance for low humidity versus high humidity. The fact that this difference depends on temperature indicates an interaction between factors.

DEFINITION

When the results of a two-factor study are presented in a graph, the existence of nonparallel lines (lines that cross or converge) indicates an **interaction between the two factors**.

The $A \times B$ interaction typically is called "A by B" interaction. If there is an interaction of temperature and humidity, it may be called the "temperature by humidity" interaction.

For many students, the concept of an interaction is easiest to understand using the perspective of interdependency; that is, an interaction exists when the effects of one variable *depend* on another factor. However, the easiest way to identify an interaction within a set of data is to draw a graph showing the treatment means. The presence of nonparallel lines is an easy way to spot an interaction.

INDEPENDENCE OF MAIN EFFECTS AND INTERACTIONS

The two-factor ANOVA consists of three hypothesis tests, each evaluating specific mean differences: the A effect, the B effect, and the $A \times B$ interaction. As we have noted, these are three *separate* tests, but you should also realize that the three tests are *independent*. That is, the outcome for any one of the three tests is totally unrelated to the outcome for either of the other two. Thus, it is possible for data from a two-factor study to display any possible combination of significant and/or nonsignificant main effects and interactions. The data sets in Table 15.4 show several possibilities.

Table 15.4(a) shows data with mean differences between levels of factor A (an A effect) but no mean differences for factor B or for an interaction. To identify the A effect, notice that the overall mean for A_1 (the top row) is 10 points higher than the overall mean for A_2 (the bottom row). This 10-point difference is the main effect for factor A. To evaluate the B effect, notice that both columns have exactly the same overall mean, indicating no difference between levels of factor B; hence, there is no B effect. Finally, the absence of an interaction is indicated by the fact that the overall A effect (the 10-point difference) is constant within each column; that is, the A effect *does not depend* on the levels of factor B. (Alternatively, the data indicate that the overall B effect is constant within each row.)

Table 15.4(b) shows data with an A effect and a B effect but no interaction. For these data, the A effect is indicated by the 10-point mean difference between rows, and the B effect is indicated by the 20-point mean difference between columns. The fact that the 10-point A effect is constant within each column indicates no interaction.

Finally, Table 15.4(c) shows data that display an interaction but no main effect for factor A or for factor B. For these data, note that there is no mean difference between rows (no A effect) and no mean difference between columns (no B effect). However, within each row (or within each column), there are mean differences. The "extra" mean differences within the rows and columns cannot be explained by the overall main effects and therefore indicate an interaction.

TABLE 15.4

Three sets of data showing different combinations of main effects and interaction for a two-factor study. (The numerical value in each cell of the matrices represents the mean value obtained for the sample in that treatment condition.)

(a) Data showing a main effect for factor A but no B effect and no interaction

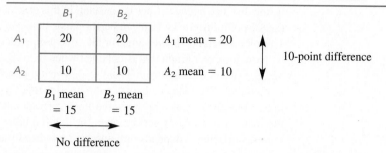

(b) Data showing main effects for both factor A and factor B but no interaction

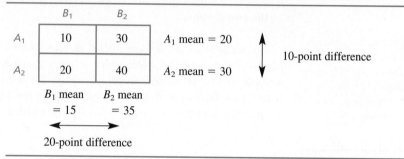

(c) Data showing no main effect for either factor but an interaction

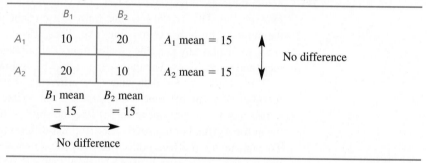

B O X

15.1   **GRAPHING RESULTS FROM A TWO-FACTOR DESIGN**

One of the best ways to get a quick overview of the results from a two-factor study is to present the data in a graph. Because the graph must display the means obtained for *two* independent variables (two factors), constructing the graph can be a bit more complicated than constructing the single-factor graphs we presented in Chapter 3 (pages 93–95). The following step-by-step procedure should help you construct graphs and interpret graphs showing results from a two-factor study.

1. The easiest way to begin is with a matrix showing the set of means for all the individual treatment combinations. For this demonstration, we will use the following data, representing the results from a two-factor study with two levels of factor A and three levels of

factor B. The subscripts indicate the levels for both factors. The numerical values inside the matrix are the means for the different treatment combinations. Notice that with a 2×3 design we have a total of $2 \times 3 = 6$ separate means.

	FACTOR B		
	B_1	B_2	B_3
FACTOR A A_1	10	40	20
A_2	30	50	30

2. The values for the dependent variable (the treatment means) are always shown on the vertical axis. For these data, the means range from 10 to 50, and this range of values is displayed on the vertical axis in Figure 15.3.

3. Next, you select one of the factors and display the different values (levels) for that factor on the horizontal axis. Although you may choose either one of the two factors, the best advice is to select a factor for which the different levels are measured on an interval or ratio scale. The reason for this suggestion is that an interval or ratio scale will permit you to construct a *line graph*, which usually provides a better picture of the results. If neither of the factors is measured on an interval or a ratio scale, then you should use a *bar graph* to display the results. (Recall from Chapters 2 and 3 that line graphs (polygons) are used for interval or ratio scales and bar graphs are used for nominal or ordinal scales.)

4. *Line graphs:* In Figure 15.3(a), we have assumed that factor B is an interval or a ratio variable, and the three levels for this factor are listed on the horizontal axis. Directly above the B_1 value on the horizontal axis, we have placed two dots corresponding to the two means in the B_1 column of the data matrix. Similarly, we have placed two dots above B_2 and another two dots above B_3. Finally, we have drawn a line connecting the three dots corresponding to level 1 of factor A (the three means in the top row of the data matrix). We have also drawn a second line that con-

nects the three dots corresponding to level 2 of factor A. These lines are labeled A_1 and A_2 in the figure.

Bar graphs: Figure 15.3(b) also shows the three levels of factor B displayed on the horizontal axis. This time, however, we assume that factor B is measured on a nominal or ordinal scale, and the result is a bar graph. Directly above the B_1 value, we have drawn two bars so that the height of the bars correspond to the two means in the B_1 column of the data matrix. Similarly, we have drawn two bars above B_2 and two more bars above B_3. Finally, the three bars corresponding to level 1 of factor A (the top row of the data matrix) are all colored (or shaded) to differentiate them from the three bars for level 2 of factor A.

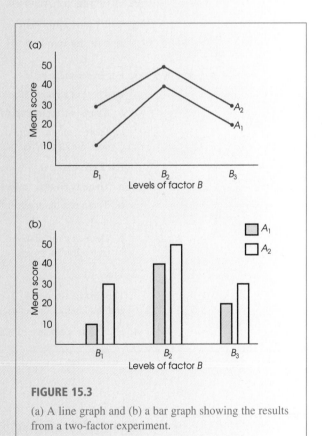

FIGURE 15.3

(a) A line graph and (b) a bar graph showing the results from a two-factor experiment.

LEARNING CHECK

1. Each of the following matrices represents a possible outcome of a two-factor experiment. For each experiment:

   a. Describe the main effect for factor A.

   b. Describe the main effect for factor B.

   c. Does there appear to be an interaction between the two factors?

EXPERIMENT I

	B_1	B_2
A_1	$M = 10$	$M = 20$
A_2	$M = 30$	$M = 40$

EXPERIMENT II

	B_1	B_2
A_1	$M = 10$	$M = 30$
A_2	$M = 20$	$M = 20$

2. In a graph showing the means from a two-factor experiment, parallel lines indicate that there is no interaction. (True or false?)

3. A two-factor ANOVA consists of three hypothesis tests. What are they?

4. It is impossible to have an interaction unless you also have main effects for at least one of the two factors. (True or false?)

ANSWERS

1. For Experiment I:

   a. There is a main effect for factor A; the scores in A_2 average 20 points higher than in A_1.

   b. There is a main effect for factor B; the scores in B_2 average 10 points higher than in B_1.

   c. There is no interaction; there is a constant 20-point difference between A_1 and A_2 that does not depend on the levels of factor B.

   For Experiment II:

   a. There is no main effect for factor A; the scores in A_1 and in A_2 both average 20.

   b. There is a main effect for factor B; on average, the scores in B_2 are 10 points higher than in B_1.

   c. There is an interaction. The difference between A_1 and A_2 depends on the level of factor B. (There is a $+10$ difference in B_1 and a -10 difference in B_2.)

2. True

3. The two-factor ANOVA evaluates the main effect for factor A, the main effect for factor B, and the interaction between the two factors.

4. False. The existence of main effects and interactions is completely independent.

15.3 NOTATION AND FORMULAS

STRUCTURE OF THE TWO-FACTOR ANALYSIS

The two-factor ANOVA is composed of three distinct hypothesis tests:

1. The main effect of factor A (often called the A-effect). Assuming that factor A is used to define the rows of the matrix, the main effect of factor A evaluates the mean differences between rows.

2. The main effect of factor B (called the B-effect). Assuming that factor B is used to define the columns of the matrix, the main effect of factor B evaluates the mean differences between columns.

3. The interaction (called the $A \times B$ interaction). The interaction evaluates mean differences between treatment conditions that are not predicted from the overall main effects from factor A or factor B.

For each of these three tests, we are looking for mean differences between treatments that are larger than would be expected by chance. In each case, the magnitude of the treatment effect will be evaluated by an F-ratio. All three F-ratios have the same basic structure:

$$F = \frac{\text{variance (mean differences) between treatments}}{\text{variance (differences) expected by chance/error}}$$

The general structure of the two-factor ANOVA is shown in Figure 15.4. Note that the overall analysis is divided into two stages. In the first stage, the total variability is separated into two components: between-treatments variability and within-treatments variability. This first stage is identical to the single-factor analysis of variance introduced in Chapter 13 with each cell in the two-factor matrix viewed as a separate treatment condition. The second stage of the analysis partitions the between-treatments variability into three components: differences attributed to factor A, differences attributed to factor B, and any remaining mean differences that define the interaction. These three components will form the basis for the three F-ratios in the analysis.

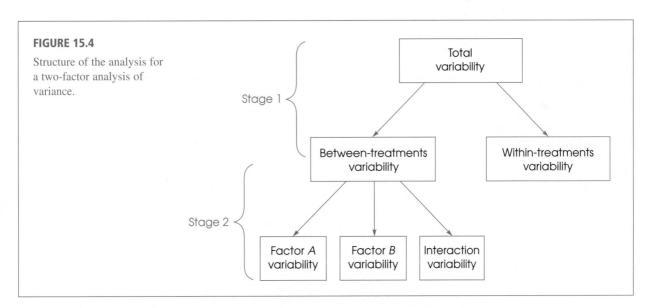

FIGURE 15.4

Structure of the analysis for a two-factor analysis of variance.

The goal of this analysis is to compute the variance values needed for the three F-ratios. We will need three between-treatments variances (one for factor A, one for factor B, and one for the interaction), and we will need a within-treatments variance. Each of these variances (or mean squares) will be determined by a sum of squares value (SS) and a degrees of freedom value (df):

Remember that in ANOVA a variance is called a mean square, or MS.

$$\text{mean square} = MS = \frac{SS}{df}$$

FORMULAS   The hypothetical data shown in Table 15.5 will be used to demonstrate the two-factor ANOVA. The data are representative of many studies examining the relationship between arousal and performance. The general result of these studies is that increasing the level of arousal (or motivation) tends to improve the level of performance. For very difficult tasks, however, increasing arousal beyond a certain point tends to lower the level of performance. (This relationship is generally known as the Yerkes–Dodson law). The data are displayed in a matrix where the levels of task difficulty make up the rows and the levels of arousal make up the columns. There are two levels of task difficulty (easy and difficult), which we have designated factor A, and three levels of arousal (low, medium, high), which we have designated as factor B. For the easy task, note that performance scores increase consistently as arousal increases. For the difficult task, on the other hand, performance peaks at a medium level of arousal and drops when arousal is increased to a high level. Note that the data matrix has a total of six *cells* or treatment conditions with a separate sample of $n = 5$ participants in each condition. Most of the notation should be familiar from the single-factor ANOVA presented in Chapter 13. Specifically, the treatment totals are identified by T values, the total number of scores in the entire study is $N = 30$, and the grand total (sum) of all 30 scores is $G = 120$. In addition to these familiar values, we have included the totals for each row and for each column in the matrix. The goal of the ANOVA is to determine whether the mean differences observed in the data are significantly greater than would be expected just by chance.

TABLE 15.5

Hypothetical data for a two-factor research study comparing two levels of task difficulty (easy and hard) and three levels of arousal (low, medium, and high). The study involves a total of six different treatment conditions with $n = 5$ participants in each condition.

		FACTOR B AROUSAL LEVEL			
		LOW	MEDIUM	HIGH	
		3	2	9	
		1	5	9	
		1	9	13	
		6	7	6	$T_{\text{ROW1}} = 90$
	EASY	4	7	8	
		$M = 3$	$M = 6$	$M = 9$	
		$T = 15$	$T = 30$	$T = 45$	
		$SS = 18$	$SS = 28$	$SS = 26$	$N = 30$
FACTOR A					$G = 120$
TASK DIFFICULTY		0	3	0	$\Sigma X^2 = 840$
		2	8	0	
		0	3	0	
		0	3	5	$T_{\text{ROW2}} = 30$
	DIFFICULT	3	3	0	
		$M = 1$	$M = 4$	$M = 1$	
		$T = 5$	$T = 20$	$T = 5$	
		$SS = 8$	$SS = 20$	$SS = 20$	

$T_{\text{COL1}} = 20 \quad T_{\text{COL2}} = 50 \quad T_{\text{COL3}} = 50$

STAGE 1 OF THE ANALYSIS

The first stage of the two-factor analysis separates the total variability into two basic components: between-treatments and within-treatments. The formulas for this stage are identical to the formulas used in the single-factor ANOVA in Chapter 13. The formulas and the calculations for the data in Table 15.5 are as follows:

Total variability

$$SS_{total} = \Sigma X^2 - \frac{G^2}{N} \tag{15.1}$$

For these data,

$$SS_{total} = 840 - \frac{120^2}{30}$$

$$= 840 - 480$$

$$= 360$$

This SS value measures the variability for all $N = 30$ scores and has degrees of freedom given by

$$df_{total} = N - 1 \tag{15.2}$$

For the data in Table 15.5, $df_{total} = 29$.

Between-treatments variability   Remember that each treatment condition corresponds to a cell in the matrix. With this in mind, the between-treatments SS is computed as

$$SS_{between\ treatments} = \Sigma \frac{T^2}{n} - \frac{G^2}{N} \tag{15.3}$$

For the data in Table 15.5, there are six treatments (six T values), and the between-treatments SS is

$$SS_{between\ treatments} = \frac{15^2}{5} + \frac{30^2}{5} + \frac{45^2}{5} + \frac{5^2}{5} + \frac{20^2}{5} + \frac{5^2}{5} - \frac{120^2}{30}$$

$$= 45 + 180 + 405 + 5 + 80 + 5 - 480$$

$$= 240$$

The between-treatments df value is determined by the number of treatments (or the number of T values) minus one. For a two-factor study, the number of treatments is equal to the number of cells in the matrix. Thus,

$$df_{between\ treatments} = \text{number of cells} - 1 \tag{15.4}$$

For these data, $df_{between\ treatments} = 5$.

Within-treatments variability   To compute the variance within treatments, we will use exactly the same formulas that were introduced in Chapter 13. Specifically, you first compute SS and $df = n - 1$ for each of the individual treatments (each separate sample). Then the within-treatments SS is defined as

$$SS_{\text{within treatments}} = \Sigma SS_{\text{each treatment}}$$

And the within-treatments df is defined as

$$df_{\text{within treatments}} = \Sigma df_{\text{each treatment}}$$

For the data in Table 15.5,

$$SS_{\text{within treatments}} = 18 + 28 + 26 + 8 + 20 + 20$$
$$= 120$$

$$df_{\text{within treatments}} = 4 + 4 + 4 + 4 + 4 + 4$$
$$= 24$$

This completes the first stage of the analysis. You should note that the two components add to equal the total for both SS values and df values.

$$SS_{\text{between treatments}} + SS_{\text{within treatments}} = SS_{\text{total}}$$
$$240 + 120 = 360$$

$$df_{\text{between treatments}} + df_{\text{within treatments}} = df_{\text{total}}$$
$$5 + 24 = 29$$

STAGE 2 OF THE ANALYSIS   The second stage of the analysis will determine the numerators for the three F-ratios. Specifically, this stage will determine the between-treatments variance for factor A, factor B, and the interaction.

   1. Factor A.   The main effect for factor A evaluates the mean differences between the levels of factor A. For this example, factor A defines the rows of the matrix, so we are evaluating the mean differences between rows. To compute the SS for factor A, we will calculate a between-treatment SS using the row totals exactly the same as we computed $SS_{\text{between treatments}}$ using the treatment totals (T values) earlier. For factor A, the row totals are 90 and 30, and each total was obtained by adding up 15 scores. Therefore,

$$SS_{\text{factor } A} = \Sigma \frac{T^2_{\text{ROW}}}{n_{\text{ROW}}} - \frac{G^2}{N} \tag{15.5}$$

For our data,

$$SS_{\text{factor } A} = \frac{90^2}{15} + \frac{30^2}{15} - \frac{120^2}{30}$$
$$= 540 + 60 - 480$$
$$= 120$$

Factor A involves two treatments (or two rows), easy and difficult, so the df value is

$$df_{\text{factor } A} = \text{number of rows} - 1 \tag{15.6}$$
$$= 2 - 1$$
$$= 1$$

2. Factor B. The calculations for factor B follow exactly the same pattern that was used for factor A, except for substituting columns in place of rows. The main effect for factor B evaluates the mean differences between the levels of factor B, which define the columns of the matrix.

$$SS_{\text{factor } B} = \Sigma \frac{T^2_{\text{COL}}}{n_{\text{COL}}} - \frac{G^2}{N} \tag{15.7}$$

For our data, the column totals are 20, 50, and 50, and each total was obtained by summing 10 scores. Thus,

$$SS_{\text{factor } B} = \frac{20^2}{10} + \frac{50^2}{10} + \frac{50^2}{10} - \frac{120^2}{30}$$

$$= 40 + 250 + 250 - 480$$

$$= 60$$

$$df_{\text{factor } B} = \text{number of columns} - 1 \tag{15.8}$$

$$= 3 - 1$$

$$= 2$$

3. The $A \times B$ Interaction. The $A \times B$ interaction is defined as the "extra" mean differences not accounted for by the main effects of the two factors. We will use this definition to find the SS and df values for the interaction by simple subtraction. Specifically, the between-treatments variability is partitioned into three parts: the A effect, the B effect, and the interaction (see Figure 15.4). We have already computed the SS and df values for A and B, so we can find the interaction values by subtracting to find out how much is left. Thus,

$$SS_{A \times B} = SS_{\text{between treatments}} - SS_{\text{factor } A} - SS_{\text{factor } B} \tag{15.9}$$

For our hypothetical data,

$$SS_{A \times B} = 240 - 120 - 60$$

$$= 60$$

Similarly,

$$df_{A \times B} = df_{\text{between treatments}} - df_{\text{factor } A} - df_{\text{factor } B} \tag{15.10}$$

$$= 5 - 1 - 2$$

$$= 2$$

MEAN SQUARES AND F-RATIOS FOR THE TWO-FACTOR ANALYSIS

The two-factor ANOVA consists of three separate hypothesis tests with three separate F-ratios. The denominator for each F-ratio is intended to measure the variance (differences) that would be expected just by chance. As we saw in Chapter 13, the within-treatments variance forms the denominator for an independent-measures design. Remember that inside each treatment all the individuals are treated exactly the same, which means any differences that occur are simply due to chance (see Chapter 13, page 403). The within-treatments variance is called a *mean square*, or *MS*, and is computed as follows:

$$MS_{\text{within treatments}} = \frac{SS_{\text{within treatments}}}{df_{\text{within treatments}}}$$

For the data in Table 15.5,

$$MS_{\text{within treatments}} = \frac{120}{24} = 5.00$$

This value will form the denominator for all three F-ratios.

The numerators of the three F-ratios all measured variance or differences between treatments: differences between levels of factor A, or differences between levels of factor B, or extra differences that are attributed to the $A \times B$ interaction. These three variances are computed as follows:

$$MS_A = \frac{SS_A}{df_A} \qquad MS_B = \frac{SS_B}{df_B} \qquad MS_{A \times B} = \frac{SS_{A \times B}}{df_{A \times B}}$$

For the data in Table 15.5, the three MS values are

$$MS_A = \frac{SS_A}{df_A} = \frac{120}{1} = 120 \qquad MS_B = \frac{SS_B}{df_B} = \frac{60}{2} = 30$$

$$MS_{A \times B} = \frac{SS_{A \times B}}{df_{A \times B}} = \frac{60}{2} = 30$$

Finally, the three F-ratios are

$$F_A = \frac{MS_A}{MS_{\text{within treatments}}} = \frac{120}{5} = 24.00$$

$$F_B = \frac{MS_B}{MS_{\text{within treatments}}} = \frac{30}{5} = 6.00$$

$$F_{A \times B} = \frac{MS_{A \times B}}{MS_{\text{within treatments}}} = \frac{30}{5} = 6.00$$

A decision concerning the significance of each test would require consulting the F-distribution table for the appropriate alpha level and degrees of freedom for each of the three F-ratios.

LEARNING CHECK

1. The following data summarize the results from a two-factor independent-measures experiment:

		FACTOR B		
		B_1	B_2	B_3
FACTOR A	A_1	$n = 10$ $T = 0$ $SS = 30$	$n = 10$ $T = 10$ $SS = 40$	$n = 10$ $T = 20$ $SS = 50$
	A_2	$n = 10$ $T = 40$ $SS = 60$	$n = 10$ $T = 30$ $SS = 50$	$n = 10$ $T = 20$ $SS = 40$

   a. Calculate the totals for each level of factor A, and compute SS for factor A.

b. Calculate the totals for factor B, and compute SS for this factor. (*Note:* You should find that the totals for B are all the same, so there is no variability for this factor.)

c. Given that the between-treatments (or between-cells) SS is equal to 100, what is the SS for the interaction?

d. Calculate the within-treatments SS, df, and MS for these data. (*Note:* $MS_{\text{within treatments}}$ would be the denominator for each of the three F-ratios.)

A N S W E R S   **1. a.** The totals for factor A are 30 and 90, and each total is obtained by summing 30 scores. $SS_A = 60$.

b. All three totals for factor B are equal to 40. Because they are all the same, there is no variability, and $SS_B = 0$.

c. The interaction is determined by differences that remain after the main effects have been accounted for. For these data,

$$SS_{A \times B} = SS_{\text{between treatments}} - SS_A - SS_B$$

$$= 100 - 60 - 0$$

$$= 40$$

d. $SS_{\text{within treatments}} = \Sigma SS_{\text{each cell}} = 30 + 40 + 50 + 60 + 50 + 40$

$$= 270$$

$df_{\text{within treatments}} = \Sigma df_{\text{each cell}} = 9 + 9 + 9 + 9 + 9 + 9$

$$= 54$$

$$MS_{\text{within treatments}} = \frac{SS_{\text{within treatments}}}{df_{\text{within treatments}}} = \frac{270}{54} = 5.00$$

15.4 AN EXAMPLE OF THE TWO-FACTOR ANOVA

Example 15.2 presents the complete hypothesis-testing procedure for a two-factor independent-measures study.

E X A M P L E   1 5 . 2   In 1968, Schachter published an article in *Science* reporting a series of experiments on obesity and eating behavior. One of these studies examined the hypothesis that obese individuals do not respond to internal, biological signals of hunger. In simple terms, this hypothesis says that obese individuals tend to eat whether or not their bodies are actually hungry.

In Schachter's study, participants were led to believe that they were taking part in a "taste test." All participants were told to come to the experiment without eating for several hours beforehand. The study used two independent variables or factors:

1. Weight (obese versus non-obese participants)

2. Full stomach versus empty stomach (half the participants were given a full meal, as much as they wanted, after arriving at the experiment, and half were left hungry)

All participants were then invited to taste and rate five different types of crackers. The dependent variable was the number of crackers eaten by each participant.

The prediction for this study was that the obese participants would eat the same amount of crackers whether or not they were full. The non-obese participants were expected to eat more with empty stomachs and less with full stomachs. Notice that the primary prediction of this study is that there will be an interaction between weight and fullness. That is, it is predicted that the effect of fullness *will depend on* the weight of the participants (non-obese or obese).

Hypothetical data similar to those obtained by Schachter are presented in Table 15.6. For this analysis, we will identify weight as factor *A* and fullness as factor *B*.

TABLE 15.6

Results from an experiment examining the eating behavior of non-obese and obese individuals who have either a full or an empty stomach*

		FACTOR B: FULLNESS	
		EMPTY STOMACH	FULL STOMACH
FACTOR A: WEIGHT	NON-OBESE	$n = 20$ $M = 22$ $T = 440$ $SS = 1540$	$n = 20$ $M = 15$ $T = 300$ $SS = 1270$
	OBESE	$n = 20$ $M = 17$ $T = 340$ $SS = 1320$	$n = 20$ $M = 18$ $T = 360$ $SS = 1260$

$T_{non-obese} = 740$

$T_{obese} = 700$

$G = 1440$
$N = 80$
$\Sigma X^2 = 31{,}836$

$T_{empty} = 780 \quad T_{full} = 660$

*The dependent variable is the number of crackers eaten in a taste test (hypothetical data).

STEP 1   State the hypotheses, and select α. For factor *A*, the null hypothesis states that there is no difference in the amount eaten for non-obese versus obese participants. In symbols,

$$H_0: \quad \mu_{A_1} = \mu_{A_2}$$
$$H_1: \quad \mu_{A_1} \neq \mu_{A_2}$$

For factor *B*, the null hypothesis states that the amount eaten will be the same for full-stomach participants as for empty-stomach participants. In symbols,

$$H_0: \quad \mu_{B_1} = \mu_{B_2}$$
$$H_1: \quad \mu_{B_1} \neq \mu_{B_2}$$

For the $A \times B$ interaction, the null hypothesis can be stated two different ways. First, the difference in eating between the full-stomach and empty-stomach conditions will be the same for non-obese and obese participants. Second, the difference in eating between the non-obese and obese participants will be the same for the full-stomach and empty-stomach conditions. In more general terms,

H_0:   The effect of factor *A* does not depend on the levels of factor *B* (and *B* does not depend on *A*).

H_1:   The effect of one factor does depend on the levels of the other factor.

We will use $\alpha = .05$ for all tests.

STEP 2   Locate the critical region. To locate the critical values for each of the three F-ratios, we first must determine the df values. For these data (Table 15.6),

$$df_{total} = N - 1 = 79$$

$$df_{between\ treatments} = (\text{number of cells}) - 1 = 3$$

$$df_{within\ treatments} = \Sigma(n - 1) = 76$$

$$df_A = (\text{number of levels of } A) - 1 = 1$$

$$df_B = (\text{number of levels of } B) - 1 = 1$$

$$df_{A \times B} = df_{between\ treatments} - df_A - df_B = 1$$

Notice that the F distribution table has no entry for $df = 1, 76$. A close and conservative estimate of this critical value may be obtained by using $df = 1, 70$ (critical $F = 3.98$). Whenever there is no entry for the df value of the error term, use the nearest smaller df value in the table.

Thus, all three F-ratios will have $df = 1, 76$. With $\alpha = .05$, the critical F value is 3.98 for all three tests.

STEP 3   Use the data to compute the F-ratios. First, we will analyze the SS values:

$$SS_{total} = \Sigma X^2 - \frac{G^2}{N} = 31{,}836 - \frac{1440^2}{80}$$

$$= 31{,}836 - 25{,}920$$

$$= 5916$$

$$SS_{between\ treatments} = \Sigma\frac{T^2}{n} - \frac{G^2}{N}$$

$$= \frac{440^2}{20} + \frac{300^2}{20} + \frac{340^2}{20} + \frac{360^2}{20} - \frac{1440^2}{80}$$

$$= 26{,}440 - 25{,}920$$

$$= 520$$

$$SS_{within\ treatments} = \Sigma SS_{inside\ each\ cell}$$

$$= 1540 + 1270 + 1320 + 1266$$

$$= 5396$$

For factor A (weight), the two row totals are $T_{non\text{-}obese} = 740$ and $T_{obese} = 700$. Each row total is obtained by summing 40 scores. Using these values, the SS for factor A is

$$SS_{factor\ A} = \Sigma\frac{T_{ROW}^2}{n_{ROW}} - \frac{G^2}{N}$$

$$= \frac{740^2}{40} + \frac{700^2}{40} - \frac{1440^2}{80}$$

$$= 25{,}940 - 25{,}920$$

$$= 20$$

For factor B (fullness), the two column totals are $T_{empty} = 780$ and $T_{full} = 660$. Each of these column totals is the sum of 40 scores. Using these values, the SS for factor B is

$$SS_{factor\ B} = \Sigma \frac{T_{COLUMN}^2}{n_{COLUMN}} - \frac{G^2}{N}$$

$$= \frac{780^2}{40} + \frac{660^2}{40} - \frac{1440^2}{80}$$

$$= 26{,}100 - 25{,}920$$

$$= 180$$

$$SS_{A \times B} = SS_{between\ treatments} - SS_A - SS_B$$

$$= 520 - 20 - 180$$

$$= 320$$

The MS values needed for the F-ratios are

$$MS_A = \frac{SS_A}{df_A} = \frac{20}{1} = 20$$

$$MS_B = \frac{SS_B}{df_B} = \frac{180}{1} = 180$$

$$MS_{A \times B} = \frac{SS_{A \times B}}{df_{A \times B}} = \frac{320}{1} = 320$$

$$MS_{within\ treatments} = \frac{SS_{within\ treatments}}{df_{within\ treatments}} = \frac{5396}{76} = 71$$

Finally, the F-ratios are

$$F_A = \frac{MS_A}{MS_{within\ treatments}} = \frac{20}{71} = 0.28$$

$$F_B = \frac{MS_B}{MS_{within\ treatments}} = \frac{180}{71} = 2.54$$

$$F_{A \times B} = \frac{MS_{A \times B}}{MS_{within\ treatments}} = \frac{320}{71} = 4.51$$

STEP 4   Make decisions. For these data, factor A (weight) has no significant effect; $F(1, 76) = 0.28$. Statistically, there is no difference in the number of crackers eaten by non-obese versus obese participants.

Similarly, factor B (fullness) has no significant effect; $F(1, 76) = 2.54$. Statistically, the number of crackers eaten by full participants is no different from the number eaten by hungry participants. (*Note:* This conclusion concerns the combined group of non-obese and obese participants. The interaction concerns these two groups separately. See page 500 for information concerning the interpretation of results when there is a significant interaction.)

These data produce a significant interaction; $F(1, 76) = 4.51, p < .05$. This means that the effect of fullness does depend on weight. Specifically, the degree of fullness did affect the non-obese participants, but it had no effect on the obese participants.

As we saw in Chapter 13, the results from an ANOVA can be organized in a summary table, which shows all the components of the analysis (SS, df, etc.) as well as the final F-ratios. For this example, the summary table would be as follows:

SOURCE	SS	df	MS	F
Between treatments	520	3		
Factor A (weight)	20	1	20	0.28
Factor B (fullness)	180	1	180	2.54
$A \times B$ interaction	320	1	320	4.51
Within treatments	5396	76	71	
Total	5916	79		

MEASURING EFFECT SIZE FOR THE TWO-FACTOR ANOVA

The general technique for measuring effect size with an analysis of variance is to compute a value for η^2, the percentage of variance that is explained by the treatment effects. For a two-factor ANOVA, we will compute three separate values for eta squared: one measuring how much of the variance is explained by the main effect for factor A, one for factor B, and a third for the interaction. As we did with the repeated-measures ANOVA (page 460) we will remove any variability that can be explained by other sources before we calculate the percentage for each of the three specific effects. Thus, for example, before we compute the η^2 for factor A, we will remove the variability that is explained by factor B and the variability explained by the interaction. The resulting equation is,

$$\text{for factor } A, \eta^2 = \frac{SS_A}{SS_{\text{total}} - SS_B - SS_{A \times B}} \tag{15.11}$$

Note that the denominator of Equation 15.11 consists of the variability that is explained by factor A and the other *unexplained* variability. Thus, an equivalent version of the equation is,

$$\text{for factor } A, \eta^2 = \frac{SS_A}{SS_A + SS_{\text{within treatments}}} \tag{15.12}$$

Similarly, the η^2 formulas for factor B and for the interaction are as follows:

$$\text{for factor } B, \eta^2 = \frac{SS_B}{SS_{\text{total}} - SS_A - SS_{A \times B}} = \frac{SS_B}{SS_B + SS_{\text{within treatments}}} \tag{15.13}$$

$$\text{for } A \times B, \eta^2 = \frac{SS_{A \times B}}{SS_{\text{total}} - SS_A - SS_B} = \frac{SS_{A \times B}}{SS_{A \times B} + SS_{\text{within treatments}}} \tag{15.14}$$

Because each of the η^2 equations computes a percentage that is not based on the total variability of the scores, the results are often called *partial* eta squares. For the data in Example 15.2, the equations produce the following values:

$$\eta^2 \text{ for factor } A \text{ (weight)} = \frac{20}{5916 - 180 - 320} = \frac{20}{5416} = 0.004 \quad \text{(or 0.4\%)}$$

$$\eta^2 \text{ for factor } B \text{ (fullness)} = \frac{180}{5916 - 20 - 320} = \frac{180}{5576} = 0.032 \quad \text{(or 3.2\%)}$$

$$\eta^2 \text{ for } A \times B = \frac{320}{5916 - 20 - 180} = \frac{320}{5716} = 0.056 \quad \text{(or 5.6\%)}$$

IN THE LITERATURE
REPORTING THE RESULTS OF A TWO-FACTOR ANOVA

The APA format for reporting the results of a two-factor analysis of variance follows the same basic guidelines as the single-factor report. First, the means and standard deviations are reported. Because a two-factor design typically involves several treatment conditions, these descriptive statistics usually are presented in a table or a graph. Next, the results of all three hypothesis tests (*F*-ratios) are reported. For the research study in Example 15.2, the report would have the following form:

> The means and standard deviations are presented in Table 1. The two-factor analysis of variance showed no significant main effect for the weight factor, $F(1, 76) = 0.28$, $p > .05$, $\eta^2 = 0.004$; and no significant main effect for the fullness factor, $F(2, 76) = 2.54$, $p > .05$, $\eta^2 = 0.032$; but the interaction between weight and fullness was significant, $F(1, 76) = 4.51$, $p < .05$, $\eta^2 = 0.056$.
>
> **TABLE 1**
>
> Mean number of crackers eaten in each treatment condition
>
> | | | FULLNESS | |
> |---|---|---|---|
> | | | EMPTY STOMACH | FULL STOMACH |
> | WEIGHT | NON-OBESE | $M = 22.0$ $SD = 9.00$ | $M = 15.0$ $SD = 8.18$ |
> | | OBESE | $M = 17.0$ $SD = 8.34$ | $M = 18.0$ $SD = 8.16$ |

You should recognize the elements of this report as being identical to other examples reporting the results from an analysis of variance. Each of the obtained *F*-ratios is reported with its *df* values in parentheses. The term *significant* is used to indicate that the null hypothesis was rejected (the mean differences are greater than what would be expected by chance), and the term *not significant* indicates that the test failed to reject the null hypothesis. The *p* value reported with each *F*-ratio reflects the alpha level used for the test. For example, $p < .05$ indicates that the probability is less than .05 that the obtained mean difference is simply due to chance or sampling error. When an exact value of *p* is available, usually from a computer program, it should be reported instead of using $p < \alpha$. Finally, the measure of effect size is reported for each test. ❏

INTERPRETING INTERACTIONS AND MAIN EFFECTS: SIMPLE MAIN EFFECTS

When a two-factor ANOVA produces a significant interaction, you should be very cautious about accepting the main effects (whether significant or not significant) at face value. In particular, a significant interaction can distort, conceal, or exaggerate the main effects. Therefore, the best advice is to start with the interaction, not the main effects, as the basis for interpreting the results of the study.

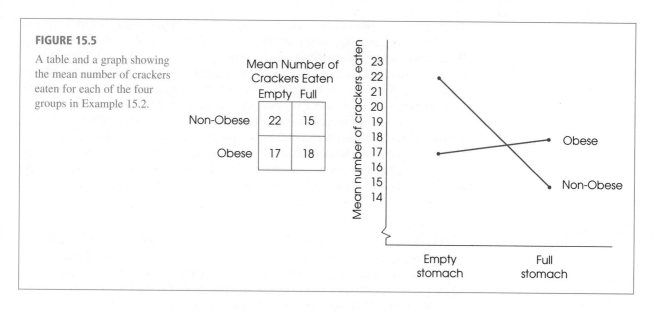

FIGURE 15.5

A table and a graph showing the mean number of crackers eaten for each of the four groups in Example 15.2.

The data from Example 15.2 provide a good demonstration of how a significant interaction can influence the interpretation of main effects. The data have been summarized in a table and a graph in Figure 15.5. For these data, the ANOVA resulted in a significant interaction, but no significant main effect for factor *A* (non-obese/obese) or for factor *B* (full/empty stomach). If you ignore the interaction and focus on the main effects, you might be tempted to conclude that there is no significant difference in eating behavior between non-obese and obese participants and that being full versus being hungry has no effect on eating behavior. Looking more closely at the data, however, it is immediately apparent that the main effects do not tell the whole story. Although the obese participants seem to eat the same amount whether they are full or hungry, the non-obese participants show that hunger has a large effect on eating behavior. You should realize that this discrepancy between non-obese and obese participants is exactly what Schachter predicted, and it is the source of the significant interaction.

Again, whenever a two-factor ANOVA produces a significant interaction, you should use the interaction, not the main effects, as the basis for interpreting the results. Often (as in the Schachter experiment), an interaction is the primary research prediction. In this case, the existence of a significant interaction and the *pattern* of the interaction may be sufficient for interpreting the results. In Schachter's study, for example, the data show the predicted pattern in the form of a significant interaction. In this case, there may not be any need for additional statistical analysis.

On the other hand, a researcher may want more than a simple description of an interaction. Specifically, a researcher may want to use a statistical analysis to evaluate the separate components of the interaction. For example, the data in Figure 15.5 appear to show that non-obese participants eat more when they are hungry than when they are full; but do they eat *significantly* more? Also, the data indicate that with full stomachs the obese participants tend to eat more than the non-obese participants, but is this difference *significant*? In each case, we are asking questions about specific mean differences within the interaction. Because the questions are phrased in terms of statistical significance, you should realize that additional hypothesis tests will be needed to obtain answers. These additional statistical tests are discussed in the following section.

Testing simple main effects   The existence of a significant interaction indicates that the effect (mean differences) for one factor depends on the levels of the second factor. When the data are presented in a matrix showing treatment means, a significant interaction indicates that the mean differences within one column (or row) show a different pattern than the mean differences within another column (or row). In this case, a researcher may want to perform a separate analysis for each of the individual columns (or rows). In effect, the researcher is separating the two-factor experiment into a series of separate single-factor experiments. The process of testing the significance of mean differences within one column (or one row) of a two-factor design is called testing *simple main effects*. To demonstrate this process, we will once again use the data from the Schachter study (Example 15.2), which are summarized in Figure 15.5.

EXAMPLE 15.3   For this demonstration, we will test for significant mean differences within each row of the two-factor data matrix. That is, we will test for significant mean differences between full stomach versus empty stomach for the non-obese participants and then repeat the test for the obese participants. In terms of the two-factor notational system, we will test the simple main effects of factor B for each level of factor A.

For the non-obese participants   We will begin by considering only the non-obese participants. Because we are restricting the data to the top row of the data matrix, the data effectively have been reduced to a single-factor study comparing only two treatment conditions. Therefore, the analysis is essentially a single-factor ANOVA duplicating the procedure presented in Chapter 13. To facilitate the change from a two-factor to a single-factor analysis, the data for the non-obese participants (top row of the matrix) are reproduced as follows, using the notation for a single-factor study:

EMPTY STOMACH	FULL STOMACH	
$n = 20$	$n = 20$	$N = 40$
$T = 440$	$T = 300$	$G = 740$

STEP 1   State the hypothesis. For this restricted set of the data, the null hypothesis would state that there is no difference between the mean for the empty stomach condition and the mean for the full stomach condition. In symbols,

$$H_0: \quad \mu_{empty} = \mu_{full} \text{ for the non-obese participants}$$

STEP 2   To evaluate this hypothesis, we will use an F-ratio where the numerator, $MS_{between\ treatments}$, is determined by the mean differences between these two groups, and the denominator consists of $MS_{within\ treatments}$ from the original analysis of variance. Thus, the F-ratio will have the structure

$$F = \frac{\text{variance (differences) for the means in row 1}}{\text{variance (differences) expected by chance}}$$

$$= \frac{MS_{between\ treatments} \text{ for the two treatments in row 1}}{MS_{within\ treatments} \text{ from the original ANOVA}}$$

To compute $MS_{\text{between treatments}}$, we begin with the two treatment totals $T = 440$ and $T = 300$. Each of these totals is based on $n = 20$ scores, and the two totals sum to a grand total of $G = 740$. $SS_{\text{between treatments}}$ for the two treatments is

$$SS_{\text{between treatments}} = \Sigma \frac{T^2}{n} - \frac{G^2}{N}$$

$$= \frac{440^2}{20} + \frac{300^2}{20} - \frac{740^2}{40}$$

$$= 9680 + 4500 - 13690$$

$$= 490$$

Because this SS value is based on only two treatments, it has $df = 1$. Therefore, $MS_{\text{between treatments}}$ is

$$MS_{\text{between treatments}} = \frac{SS_{\text{between treatments}}}{df_{\text{between treatments}}} = \frac{490}{1} = 490$$

Remember that the *F*-ratio uses $MS_{\text{within treatments}}$ from the original ANOVA. This $MS = 71$ with $df = 76$.

Using $MS_{\text{within treatments}} = 71$ from the original two-factor analysis, the final F-ratio is

$$F = \frac{MS_{\text{between treatments}}}{MS_{\text{within treatments}}} = \frac{490}{71} = 6.90$$

Note that this F-ratio has the same df values (1, 76) as the test for factor B main effects (empty versus full) in the original ANOVA. Therefore, the critical value for the F-ratio will be the same as the one we used in the original ANOVA. With $df = 1, 76$, the critical value is 3.98. In this case, our F-ratio far exceeds the critical value, so we conclude that the two treatments, empty versus full, are significantly different for the non-obese participants.

For the obese participants   The test for the obese participants follows exactly the same pattern. First, the data for the obese participants are as follows:

EMPTY STOMACH	FULL STOMACH	
$n = 20$	$n = 20$	$N = 40$
$T = 340$	$T = 360$	$G = 700$

For these data,

$$SS_{\text{between treatments}} = \Sigma \frac{T^2}{n} - \frac{G^2}{N} = \frac{340^2}{20} + \frac{360^2}{20} - \frac{700^2}{40}$$

$$= 5780 + 6480 - 12250$$

$$= 10$$

Again, we are comparing only two treatment conditions, so $df = 1$ and

$$MS_{\text{between treatments}} = \frac{SS_{\text{between treatments}}}{df_{\text{between treatments}}} = \frac{10}{1} = 10$$

Thus, for the obese participants, the final F-ratio is

$$F = \frac{MS_{\text{between treatments}}}{MS_{\text{within treatments}}} = \frac{10}{71} = 0.14$$

As before, this F-ratio has $df = 1, 76$ and is compared with the critical value, $F = 3.98$. This time the data indicate that there is no significant difference between empty stomach versus full stomach for the obese participants.

As a final note, we should point out that the evaluation of simple main effects is used to account for the interaction as well as the overall main effect for one factor. In Example 15.2, the significant interaction indicates that the effect of fullness (factor B) depends on the participants' weights (factor A). The evaluation of the simple main effects demonstrates this dependency. Specifically, fullness has a significant effect on eating behavior for the non-obese participants, but has no significant effect for the obese participants. Thus, the analysis of simple main effects provides a detailed evaluation of the effects of one factor *including its interaction with a second factor*.

The fact that the simple main effects for one factor encompass both the interaction and the overall main effect of the factor can be seen if you consider the SS values. For this demonstration,

SIMPLE MAIN EFFECTS	OVERALL ANOVA
$SS_{\text{fullness for non-obese}} = 490$	$SS_{A \times B} = 320$
$SS_{\text{fullness for obese}} = 10$	$SS_B = 180$
Total $SS = 500$	Total $SS = 500$

Notice that the total variability from the simple main effects of fullness (factor B) completely accounts for the total variability of factor B and the $A \times B$ interaction.

15.5   ASSUMPTIONS FOR THE TWO-FACTOR ANOVA

The validity of the analysis of variance presented in this chapter depends on the same three assumptions we have encountered with other hypothesis tests for independent-measures designs (the t test in Chapter 10 and the single-factor ANOVA in Chapter 13):

1. The observations within each sample must be independent (see page 251).
2. The populations from which the samples are selected must be normal.
3. The populations from which the samples are selected must have equal variances (homogeneity of variance).

As before, the assumption of normality generally is not a cause for concern, especially when the sample size is relatively large. The homogeneity of variance assumption is more important, and if it appears that your data fail to satisfy this requirement, you should conduct a test for homogeneity before you attempt the ANOVA. Hartley's F-max test (see page 331) allows you to use the sample variances from your data to determine whether there is evidence for any differences among the population variances.

HIGHER-ORDER
FACTORIAL DESIGNS

The basic concepts of the two-factor ANOVA can be extended to more complex designs involving three or more factors. A three-factor design, for example, might look at academic performance scores for two different teaching methods (factor A), for boys versus girls (factor B), and for first-grade versus second-grade classes (factor C). The logic of the analysis and many of the formulas from the two-factor ANOVA are simply extended to the situation with three (or more) factors. In a three-factor experiment, for example, you would evaluate the main effects for each of the three factors, and you would evaluate a set of two-way interactions: $A \times B$, $B \times C$, and $A \times C$. In addition, however, the extra factor introduces the potential for a three-way interaction: $A \times B \times C$.

The general logic for defining and interpreting higher-order interactions follows the pattern set by two-way interactions. For example, a two-way interaction, $A \times B$, means that the effect of factor A depends on the levels of factor B. Extending this definition, a three-way interaction, $A \times B \times C$, indicates that the two-way interaction between A and B depends on the levels of factor C. Although you may have a good understanding of two-way interactions and you may grasp the general idea of a three-way interaction, most people have great difficulty comprehending or interpreting a four-way (or more) interaction. For this reason, factorial experiments involving three or more factors can produce very complex results that are difficult to understand and thus often have limited practical value.

SUMMARY

1. A research study with two independent variables is called a two-factor design. Such a design can be diagrammed as a matrix by listing the levels of one factor across the top and the levels of the other factor down the side. Each *cell* in the matrix corresponds to a specific combination of the two factors.

2. Traditionally, the two factors are identified as factor A and factor B. The purpose of the analysis of variance is to determine whether there are any significant mean differences among the treatment conditions or cells in the experimental matrix. These treatment effects are classified as follows:

   a. The A effect: Differential effects produced by the different levels of factor A.

   b. The B effect: Differential effects produced by the different levels of factor B.

   c. The $A \times B$ interaction: Differences that are produced by unique combinations of A and B. An interaction exists when the effect of one factor depends on the levels of the other factor.

3. The two-factor analysis of variance produces three F-ratios: one for factor A, one for factor B, and one for the $A \times B$ interaction. Each F-ratio has the same basic structure:

$$F = \frac{MS_{\text{treatment effect}}(\text{either } A \text{ or } B \text{ or } A \times B)}{MS_{\text{within treatments}}}$$

The formulas for the SS, df, and MS values for the two-factor ANOVA are presented in Figure 15.6.

4. Effect size for each main effect and for the interaction is computed as a percentage of variance explained by the specific main effect or interaction. In each case, the variability that is explained by other sources is removed before the percentage is computed.

$$\text{For factor } A, \ \eta^2 = \frac{SS_A}{SS_{\text{total}} - SS_B - SS_{A \times B}}$$

$$= \frac{SS_A}{SS_A + SS_{\text{within treatments}}}$$

$$\text{For factor } B, \ \eta^2 = \frac{SS_B}{SS_{\text{total}} - SS_A - SS_{A \times B}}$$

$$= \frac{SS_B}{SS_B + SS_{\text{within treatments}}}$$

$$\text{For } A \times B, \ \eta^2 = \frac{SS_{A \times B}}{SS_{\text{total}} - SS_A - SS_B}$$

$$= \frac{SS_{A \times B}}{SS_{A \times B} + SS_{\text{within treatments}}}$$

FIGURE 15.6

The analysis of variance for an independent-measures two-factor design.

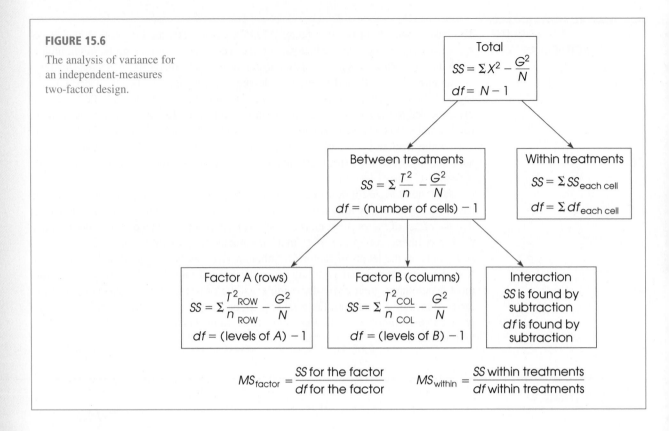

RESOURCES

WebTUTOR

If you are using WebTutor along with this book, there is a WebTutor section corresponding to this chapter. The WebTutor contains a brief summary of Chapter 15, hints for learning the concepts and formulas for the two-factor analysis of variance, cautions about common errors, and sample exam items including solutions.

SPSS

The statistics computer package SPSS can be used to conduct the two-factor, independent-measures analysis of variance that is presented in this chapter. The scores are entered into the SPSS data matrix in a stacked format, which means that all the scores from all the different treatment conditions are entered in a single column. A second column is used to designate the level of factor *A* for each score, and a third column is used to designate the level of factor *B*. For example, enter all the scores in the first column. In the second column, enter a 1 beside each score from A1, enter a 2 beside each score from A2, and so on. In the third column, enter a 1 beside each score from B1, a 2 beside each score from B2, and so on. Then click **Analyze** on the tool bar, select **General Linear Model**, and click on **Univariant**. Enter the column level for the scores in the Dependent Variable box, and enter the two column labels for the two factors in the Fixed Factors box. Before you conduct the ANOVA, you can click on the Options box and select Descriptives (to get descriptive statistics for each sample). Then click **OK**. If you selected the Descriptives Option, SPSS will produce a table showing the means and standard deviations for each treatment condition. The results of the ANOVA are shown in a summary table in which each factor is identified by its column label. (Note that the summary table includes some extra values, such as *Corrected Model* and *Intercept*, that are not covered in this text.)

KEY TERMS

| two-factor experiment | matrix | cells | main effect | interaction | simple main effect |

FOCUS ON PROBLEM SOLVING

1. Before you begin a two-factor ANOVA, take time to organize and summarize the data. It is best if you summarize the data in a matrix with rows corresponding to the levels of one factor and columns corresponding to the levels of the other factor. In each cell of the matrix, show the number of scores (n), the total and mean for the cell, and the SS within the cell. Also compute the row totals and column totals that will be needed to calculate main effects.

2. To draw a graph of the result from a two-factor experiment, first prepare a matrix with each cell containing the mean for that treatment condition.

Next, list the levels of factor B on the X-axis (B_1, B_2, etc.), and put the scores (dependent variable) on the Y-axis. Starting with the first row of the matrix, place a dot above each B level so that the height of the dot corresponds to the cell mean. Connect the dots with a line, and label the line A_1. In the same way, construct a separate line for each level of factor A (that is, a separate line for each row of the matrix). In general, your graph will be easier to draw and easier to understand if the factor with the larger number of levels is placed on the X-axis (factor B in this example).

3. The concept of an interaction is easier to grasp if you sketch a graph showing the means for each treatment. Remember that parallel lines indicate no interaction. Crossing or converging lines indicate that the effect of one treatment depends on the levels of the other treatment you are examining. This indicates that an interaction exists between the two treatments.

4. For a two-factor ANOVA, there are three separate F-ratios. These three F-ratios use the same error term in the denominator (MS_{within}). On the other hand, these F-ratios will have different numerators and may have different df values associated with each of these numerators. Therefore, be careful when you look up the critical F values in the table. The two factors and the interaction may have different critical F values.

5. As we have mentioned in previous ANOVA chapters, it helps tremendously to organize your computations; start with SS and df values, and then compute MS values and F-ratios. Once again, using an ANOVA summary table (see page 499) can be of great assistance.

DEMONSTRATION 15.1

TWO-FACTOR ANOVA

The following data are from an experiment examining the phenomenon of encoding specificity. According to this psychological principle, recall of information will be best if the testing conditions are the same as the conditions that existed at the time of learning.

The experiment involves presenting a group of students with a lecture on an unfamiliar topic. One week later the students are given a test on the lecture material. To manipulate the conditions at the time of learning, some students receive the lecture in a large classroom, and some hear the lecture in a small classroom. For those students who were lectured in the large room, one-half are tested in the same large room, and the others are changed to the small room for testing. Similarly, one-half of the students who were lectured in the small room are tested in the same small room, and the other half are tested in the large room. Thus, the experiment involves four groups of participants in a two-factor design, as shown in the following table. The score for each participant is the number of correct answers on the test.

		TESTING CONDITION	
		LARGE TESTING ROOM	SMALL TESTING ROOM
LECTURE CONDITION	LARGE LECTURE ROOM	15 20 11 18 16	5 8 1 1 5
	SMALL LECTURE ROOM	1 4 2 5 8	22 15 20 17 16

Do these data indicate that the size of the lecture room and/or testing room has a significant effect on test performance?

STEP 1   *State the hypotheses, and specify* α. The two-factor ANOVA evaluates three separate sets of hypotheses:

1. The main effect of lecture-room size: Is there a significant difference in test performance for students who received the lecture in a large room versus students who received the lecture in a small room? With lecture-room size identified as factor A, the null hypothesis states that there is no difference between the two room sizes. In symbols,

$$H_0: \quad \mu_{A_1} = \mu_{A_2}$$

The alternative hypothesis states that there is a difference between the two lecture-room sizes.

$$H_1: \quad \mu_{A_1} \neq \mu_{A_2}$$

2. The main effect of testing-room size: Is there a significant difference in test performance for students who were tested in a large room versus students who were tested in a small room? With testing-room size identified as factor B, the null hypothesis states that there is no difference between the two room sizes. In symbols,

$$H_0: \quad \mu_{B_1} = \mu_{B_2}$$

The alternative hypothesis states that there is a difference between the two testing-room sizes.

$$H_1: \quad \mu_{B_1} \neq \mu_{B_2}$$

3. The interaction between lecture-room size and testing-room size: The null hypothesis states that there is no interaction:

H_0:   The effect of testing-room size does not depend on the size of the lecture room.

The alternative hypothesis states that there is an interaction:

H_1:   The effect of testing-room size does depend on the size of the lecture room.

Notice that the researcher is not predicting any main effects for this study: There is no prediction that one room size is better than another for either learning or testing. However, the researcher is predicting that there will be an interaction. Specifically, the small testing room should be better for students who learned in the small room, and the large testing room should be better for those who learned in the large room. Remember: The principle of encoding specificity states that the better the match between testing and learning conditions, the better the recall. We will set alpha at $\alpha = .05$.

STEP 2   *Locate the critical region.* To locate the critical region, we must obtain the df values for each of the three F-ratios. Specifically, we will need df_A, df_B, and $df_{A \times B}$ for the numerators and $df_{\text{within treatments}}$ for the denominator. (Often it is easier to postpone this step until the analysis of the df values in step 3.)

$$df_{\text{total}} = N - 1 = 19$$

$$df_{\text{between treatments}} = (\text{number of cells}) - 1 = 3$$

$$df_{\text{within treatments}} = \Sigma(n - 1) = 16$$

$$df_A = (\text{number of levels of } A) - 1 = 1$$

$$df_B = (\text{number of levels of } B) - 1 = 1$$

$$df_{A \times B} = df_{\text{between treatments}} - df_A - df_B = 1$$

Thus, all three F-ratios will have $df = 1, 16$. With $\alpha = .05$, the critical value for each F-ratio is $F = 4.49$. For each test, the obtained F-ratio must exceed this critical value to reject H_0.

STEP 3   *Perform the analysis.* The complete two-factor ANOVA can be divided into a series of stages:

1. Compute the summary statistics for the data. This involves calculating the total and SS for each treatment condition, finding the A totals and B totals for the rows and columns, respectively, and obtaining G and ΣX^2 for the entire set of scores.
2. Perform the first stage of the analysis: Separate the total variability (SS and df) into the between- and within-treatments components.
3. Perform the second stage of the analysis: Separate the between-treatments variability (SS and df) into the A effect, B effect, and interaction components.
4. Calculate the mean squares for the F-ratios.
5. Calculate the F-ratios.

Compute the summary statistics. We will use the computational formula to obtain SS for each treatment condition. These calculations will also provide numerical values for the row and column totals as well as G and ΣX^2.

LARGE TESTING LARGE LECTURE		SMALL TESTING LARGE LECTURE	
X	X^2	X	X^2
15	225	5	25
20	400	8	64
11	121	1	1
18	324	1	1
16	256	5	25

$$\Sigma X = 80 \qquad \Sigma X^2 = 1326$$

$$SS = \Sigma X^2 - \frac{(\Sigma X)^2}{n}$$

$$= 1326 - \frac{80^2}{5}$$

$$= 1326 - 1280$$

$$= 46$$

$$T = \Sigma X = 80$$

$$\Sigma X = 20 \qquad \Sigma X^2 = 116$$

$$SS = \Sigma X^2 - \frac{(\Sigma X)^2}{n}$$

$$= 116 - \frac{20^2}{5}$$

$$= 116 - 80$$

$$= 36$$

$$T = \Sigma X = 20$$

LARGE TESTING SMALL LECTURE		SMALL TESTING SMALL LECTURE	
X	X^2	X	X^2
1	1	22	484
4	16	15	225
2	4	20	400
5	25	17	289
8	64	16	256

$$\Sigma X = 20 \qquad \Sigma X^2 = 110$$

$$SS = \Sigma X^2 - \frac{(\Sigma X)^2}{n}$$

$$= 110 - \frac{20^2}{5}$$

$$= 110 - 80$$

$$= 30$$

$$T = \Sigma X = 20$$

$$\Sigma X = 90 \qquad \Sigma X^2 = 1654$$

$$SS = \Sigma X^2 - \frac{(\Sigma X)^2}{n}$$

$$= 1654 - \frac{90^2}{5}$$

$$= 1654 - 1620$$

$$= 34$$

$$T = \Sigma X = 90$$

The column totals (factor B) are 100 and 110. The row totals (factor A) are 100 and 110. The grand total for these data is $G = 210$, and ΣX^2 for the entire set can be obtained by summing the ΣX^2 values for the four treatment conditions.

$$\Sigma X^2 = 1326 + 116 + 110 + 1654 = 3206$$

For this study, there are two levels for factor A and for factor B and there are $n = 5$ scores in each of the four conditions, so $N = 20$.

Perform stage 1 of the analysis. We begin by analyzing SS into two basic components:

$$SS_{\text{total}} = \Sigma X^2 - \frac{G^2}{N} = 3206 - \frac{210^2}{20} = 3206 - 2205 = 1001$$

$$SS_{\text{between treatments}} = \Sigma \frac{T^2}{n} - \frac{G^2}{N} = \frac{80^2}{5} + \frac{20^2}{5} + \frac{20^2}{5} + \frac{90^2}{5} - \frac{210^2}{20}$$

$$= 1280 + 80 + 80 + 1620 - 2205$$

$$= 3060 - 2205$$

$$= 885$$

$$SS_{\text{within treatments}} = \Sigma SS_{\text{each treatment}} = 46 + 36 + 30 + 34 = 146$$

For this stage, the *df* values are

$$df_{\text{total}} = 19$$

$$df_{\text{between treatments}} = 3$$

$$df_{\text{within treatments}} = 16$$

Perform stage 2 of the analysis. We begin by analyzing $SS_{\text{between treatments}}$.

The row totals (factor A) are 100 and 110, and each total is the sum of 10 scores. Thus, SS for factor A is

$$SS_{\text{factor } A} = \Sigma \frac{T^2_{\text{ROW}}}{n_{\text{ROW}}} - \frac{G^2}{N}$$

$$= \frac{100^2}{10} + \frac{110^2}{10} - \frac{210^2}{20}$$

$$= 1000 + 1210 - 2205$$

$$= 5$$

The column totals (factor B) are 100 and 110, and each total is the sum of 10 scores. Thus, SS for factor B is

$$SS_{\text{factor } B} = \Sigma \frac{T^2_{\text{COLUMN}}}{n_{\text{COLUMN}}} - \frac{G^2}{N}$$

$$= \frac{100^2}{10} + \frac{110^2}{10} - \frac{210^2}{20}$$

$$= 1000 + 1210 - 2205$$

$$= 5$$

$$SS_{A \times B} = SS_{\text{between treatments}} - SS_A - SS_B$$

$$= 855 - 5 - 5$$

$$= 845$$

For stage 2, the *df* values are

$$df_A = 1$$

$$df_B = 1$$

$$df_{A \times B} = 1$$

Calculate the MS values.

$$MS_A = \frac{SS_A}{df_A} = \frac{5}{1} = 5$$

$$MS_B = \frac{SS_B}{df_B} = \frac{5}{1} = 5$$

$$MS_{A \times B} = \frac{SS_{A \times B}}{df_{A \times B}} = \frac{845}{1} = 845$$

$$MS_{\text{within treatments}} = \frac{SS_{\text{within treatments}}}{df_{\text{within treatments}}} = \frac{146}{16} = 9.125$$

Calculate the F-ratios. For factor A (lecture-room size),

$$F = \frac{MS_A}{MS_{\text{within treatments}}} = \frac{5}{9.125} = 0.55$$

For factor B (testing-room size),

$$F = \frac{MS_B}{MS_{\text{within treatments}}} = \frac{5}{9.125} = 0.55$$

For the $A \times B$ interaction,

$$F = \frac{MS_{A \times B}}{MS_{\text{within treatments}}} = \frac{845}{9.125} = 92.60$$

STEP 4 *Make a decision about each H_0, and state conclusions.* For factor A, lecture-room size, the obtained F-ratio, $F = 0.55$, is not in the critical region. Therefore, we fail to reject the null hypothesis. We conclude that the size of the lecture room does not have a significant effect on test performance, $F(1, 16) = 0.55$, $p > .05$.

For factor B, testing-room size, the obtained F-ratio, $F = 0.55$, is not in the critical region. Therefore, we fail to reject the null hypothesis. We conclude that the size of the testing room does not have a significant effect on test performance, $F(1, 16) = 0.55$, $p > .05$.

For the $A \times B$ interaction, the obtained F-ratio, $F = 92.60$, exceeds the critical value of $F = 4.46$. Therefore, we reject the null hypothesis. We conclude that there is a significant interaction between lecture-room size and testing-room size, $F(1, 16) = 92.60$, $p < .05$.

Notice that the significant interaction means that you must be cautious interpreting the main effects (see page 500). In this experiment, for example, the size of the testing room (factor B) does have an effect on performance, depending on which room was used for the lecture. Specifically, performance is higher when the testing room and lecture room match, and performance is lower when the lecture and testing occur in different rooms.

The following table summarizes the results of the analysis:

SOURCE	SS	df	MS	
Between treatments	855	3		
A (lecture room)	5	1	5	F = 0.55
B (testing room)	5	1	5	F = 0.55
A × B interaction	845	1	845	F = 92.60
Within treatments	146	16	9.125	
Total	1001	19		

DEMONSTRATION 15.2

MEASURING EFFECT SIZE FOR THE TWO-FACTOR ANOVA

Effect size for each main effect and for the interaction is measured by eta squared (η^2), the percentage of variance explained by the specific main effect or interaction. In each case, the variability that is explained by other sources is removed before the percentage is computed. For the two-factor ANOVA in Demonstration 15.1,

$$\text{For factor } A, \eta^2 = \frac{SS_A}{SS_{\text{total}} - SS_B - SS_{A \times B}} = \frac{5}{1001 - 5 - 845} = 0.033 \quad \text{(or 3.3\%)}$$

$$\text{For factor } B, \eta^2 = \frac{SS_B}{SS_{\text{total}} - SS_A - SS_{A \times B}} = \frac{5}{1001 - 5 - 845} = 0.033 \quad \text{(or 3.3\%)}$$

$$\text{For } A \times B, \eta^2 = \frac{SS_{A \times B}}{SS_{\text{total}} - SS_A - SS_B} = \frac{845}{1001 - 5 - 5} = 0.853 \quad \text{(or 85.3\%)}$$

PROBLEMS

1. Define each of the following terms:
   a. Factor
   b. Level
   c. Two-factor study

2. The structure of a two-factor study can be presented as a matrix with one factor determining the rows and the second factor determining the columns. With this structure in mind, identify the three separate hypothesis tests that make up a two-factor ANOVA, and explain the purpose of each test.

3. Briefly explain what happens during the second stage of the two-factor ANOVA.

4. Each of the following sets of data represent the outcome from a two-factor study evaluating males and females (factor A) and two different ages (factor B). For each set of data, determine whether the data indicate a main effect for factor A (overall mean difference between males and females); a main effect for factor B (overall mean difference between the two ages); and an interaction (mean differences not explained by the main effects).

a.

	FACTOR B	
	AGE 5	AGE 6
MALE	$M = 10$	$M = 10$
FEMALE	$M = 20$	$M = 20$

FACTOR A

b.

	FACTOR B	
	AGE 5	AGE 6
MALE	$M = 10$	$M = 20$
FEMALE	$M = 20$	$M = 30$

FACTOR A

c.

	FACTOR B	
	AGE 5	AGE 6
MALE	$M = 10$	$M = 20$
FEMALE	$M = 10$	$M = 40$

FACTOR A

5. The results of a two-factor experiment are examined using an ANOVA, and the researcher reports an

F-ratio for factor A with $df = 1, 54$ and an F-ratio for factor B with $df = 2, 108$. Explain why this report cannot be correct.

6. A psychologist conducts a two-factor study comparing the effectiveness of two different therapy techniques (factor A) for treating mild and severe phobias (factor B). The dependent variable is a measure of fear for each participant. If the study uses $n = 10$ participants in each of the four conditions, then identify the df values for each of the three F-ratios.
   a. What are the df values for the F-ratio for factor A?
   b. What are the df values for the F-ratio for factor B?
   c. What are the df values for the F-ratio for the $A \times B$ interaction?

7. The following matrix presents the results of a two-factor experiment with two levels of factor A, two levels of factor B, and $n = 10$ subjects in each treatment condition. Each value in the matrix is the mean score for the subjects in that treatment condition. Notice that one of the mean values is missing.

		FACTOR B	
		B_1	B_2
FACTOR A	A_1	10	40
	A_2	30	?

a. What value should be assigned to the missing mean so that the resulting data would show no main effect for factor A?
   b. What value should be assigned to the missing mean so that the data would show no main effect for factor B?
   c. What value should be assigned to the missing mean so that the data would show no interaction?

8. The following data are from a two-factor experiment with $n = 10$ subjects in each treatment condition (each cell):

		FACTOR B	
		B_1	B_2
FACTOR A	A_1	$T = 40$ $SS = 70$	$T = 10$ $SS = 80$
	A_2	$T = 30$ $SS = 73$	$T = 20$ $SS = 65$

$$\Sigma X^2 = 588$$

a. Use a two-factor ANOVA with $\alpha = .05$ to evaluate the main effects and the interaction.

b. Calculate the effect size (η^2) for each main effect and for the interaction.

9. The following data were obtained from an independent-measures experiment using $n = 5$ subjects in each treatment condition:

	FACTOR B	
	B_1	B_2
A_1	$T = 15$ $SS = 80$	$T = 25$ $SS = 90$
A_2	$T = 5$ $SS = 70$	$T = 55$ $SS = 80$

(FACTOR A on the left side, rows A_1 and A_2)

$$\Sigma X^2 = 1100$$

a. Compute the means for each cell, and draw a graph showing the results of this experiment. Your graph should be similar to those shown in Figure 15.2.

b. Just from looking at your graph, do you think there is a main effect for factor A? What about factor B? Does there appear to be an interaction?

c. Use an analysis of variance with $\alpha = .05$ to evaluate these data.

10. Most sports injuries are immediate and obvious, like a broken leg. However, some can be more subtle, like the neurological damage that may occur when soccer players repeatedly head a soccer ball. To examine long-term effects of repeated heading, Downs and Abwender (2002) examined two different age groups of soccer players and swimmers. The dependent variable was performance on a conceptual thinking task. Following are hypothetical data, similar to the research results.

a. Use a two-factor ANOVA with $\alpha = .05$ to evaluate the main effects and interaction.

b. Calculate the effects size (η^2) for the main effects and the interaction.

c. Briefly describe the outcome of the study.

		FACTOR B: AGE	
		COLLEGE	OLDER
FACTOR A: SPORT	SOCCER	$n = 20$ $M = 9$ $T = 180$ $SS = 380$	$n = 20$ $M = 4$ $T = 80$ $SS = 390$
	SWIMMING	$n = 20$ $M = 9$ $T = 180$ $SS = 350$	$n = 20$ $M = 8$ $T = 160$ $SS = 400$

11. A researcher conducted a two-factor research study using two levels of factor A and four levels of factor B with a separate sample of $n = 6$ subjects in each of the eight treatment conditions (cells). The following table summarizes the results of the analysis, but it is not complete. Fill in the missing values. (*Hint:* Start with the *df* values.)

SOURCE	SS	df	MS	
Between treatments	280	——		
Main effect for A	——	——	——	$F =$ ——
Main effect for B	——	——	48	$F =$ ——
$A \times B$ interaction	120	——	——	$F =$ ——
Within treatments	——	——	——	
Total	600	——		

12. A researcher conducted a two-factor research study using two levels of factor A and three levels of factor B with a separate sample of $n = 10$ subjects in each of the six treatment conditions (cells). The following table summarizes the results of the analysis, but it is not complete. Fill in the missing values. (*Hint:* Start with the *df* values.)

SOURCE	SS	df	MS	
Between treatments	60	——		
Main effect for A	——	——	0	$F =$ ——
Main effect for B	——	——	——	$F =$ ——
$A \times B$ interaction	40	——	——	$F =$ ——
Within treatments	——	——	2	
Total	——	——		

13. A researcher conducted a two-factor research study using two levels of factor A and two levels of factor B with a separate sample of $n = 12$ subjects in each of the four treatment conditions (cells). The following table summarizes the results of the analysis, but it is not complete. Fill in the missing values. (*Hint:* Start with the *df* values.)

SOURCE	SS	df	MS	
Between treatments	——	——		
Main effect for A	——	——	——	$F = 4.00$
Main effect for B	——	——	3	$F =$ ——
$A \times B$ interaction	15	——	——	$F =$ ——
Within treatments	132	——	——	
Total	——	——		

14. The following data are from a study examining the extent to which different personality types are affected by distraction. Individuals were selected to represent two different personality types: introverts and extroverts. Half the individuals in each group were tested on a monotonous task in a relatively quiet, calm room. The individuals in the other half of each group were tested in a noisy room filled with distractions. The dependent variable was the number of errors committed by each individual. The results of this study are as follows:

		FACTOR B (PERSONALITY)	
		INTROVERT	EXTROVERT
FACTOR A (DISTRACTION)	QUIET	$n = 5$ $T = 10$ $SS = 15$	$n = 5$ $T = 10$ $SS = 25$
	NOISY	$n = 5$ $T = 20$ $SS = 10$	$n = 5$ $T = 40$ $SS = 30$

$$\Sigma X^2 = 520$$

a. Use a two-factor ANOVA with $\alpha = .05$ to evaluate these results.
b. Does distraction (quiet versus noisy) have a significant effect on the performance of the introverted participants? Test the simple main effect using $\alpha = .05$.
c. Does distraction (quiet versus noisy) have a significant effect on the performance of the extroverted participants? Test the simple main effect using $\alpha = .05$.

15. Many species of animals communicate using odors. A researcher suspects that specific chemicals contained in the urine of male rats can influence the behavior of other males in the colony. The researcher predicts that male rats will become anxious and more active if they think they are in territory that has been marked by another male. Also, it is predicted that these chemicals will have no effect on female rats. To test this theory, the researcher obtains samples of 15 male and 15 female rats. One-third of each group is tested in a sterile cage. Another one-third of each group is tested in a cage that has been painted with a small amount of the chemicals. The rest of the rats are tested in a cage that has been painted with a large amount of the chemicals. The dependent variable is the activity level of each rat. The data from this experiment are as follows:

		FACTOR B (AMOUNT OF CHEMICAL)		
		NONE	SMALL	LARGE
FACTOR A (SEX)	MALE	$n = 5$ $T = 10$ $SS = 15$	$n = 5$ $T = 20$ $SS = 19$	$n = 5$ $T = 30$ $SS = 31$
	FEMALE	$n = 5$ $T = 10$ $SS = 19$	$n = 5$ $T = 10$ $SS = 21$	$n = 5$ $T = 10$ $SS = 15$

$$\Sigma X^2 = 460$$

Use an ANOVA with $\alpha = .05$ to test the researcher's predictions. Explain the results.

16. It has been demonstrated in a variety of experiments that memory is best when the conditions at the time of testing are identical to the conditions at the time of learning. This phenomenon is called *encoding specificity* because the specific cues that you use to learn (or encode) new information are the best possible cues to help you recall the information at a later time. In an experimental demonstration of encoding specificity, Tulving and Osler (1968) prepared a list of words to be memorized. For each word on the list, they selected an associated word to serve as a cue. For example, if the word *queen* were on the list, the word *lady* could be a cue. Four groups of subjects participated in the experiment. One group was given the cues during learning and during the recall test. Another group received the cues only during recall. A third group received the cues only during learning, and the final group was not given any cues at all. The dependent variable was the number of words correctly recalled. Data similar to Tulving and Osler's results are as follows:

		CUES AT LEARNING	
		YES	NO
CUES AT RECALL	YES	$n = 10$ $M = 3$ $SS = 22$	$n = 10$ $M = 1$ $SS = 15$
	NO	$n = 10$ $M = 1$ $SS = 16$	$n = 10$ $M = 1$ $SS = 19$

$$\Sigma X^2 = 192$$

a. Use an ANOVA with $\alpha = .05$ to evaluate the data. Describe the results.
b. Calculate the effect size (η^2) for each main effect and for the interaction.

17. The following data show the results of a two-factor experiment with $n = 10$ in each treatment condition (cell):

		FACTOR B	
		B_1	B_2
FACTOR A	A_1	$M = 4$ $SS = 40$	$M = 2$ $SS = 50$
	A_2	$M = 3$ $SS = 50$	$M = 1$ $SS = 40$

$$\Sigma X^2 = 480$$

a. Sketch a graph showing the results of this experiment. (See Box 15.1 for examples.)

b. Look at your graph: does there appear to be an $A \times B$ interaction? Does factor A appear to have any effect? Does factor B appear to have any effect?

c. Evaluate these data using an ANOVA with $\alpha = .05$.

18. When participants are presented with a list of items to be remembered, there are two common methods for testing memory. A *recall* test simply asks each participant to report as many items as he/she can remember. In a *recognition* test, participants are shown a second list of items (including some items from the original list as well as some new items) and asked to identify which items they remember having seen on the first list. Typically, recall performance is very poor for young children, but improves gradually with age. Recognition performance, on the other hand, is relatively good for all ages. The following data are intended to demonstrate this phenomenon. There are $n = 10$ children in each of the six samples, and the dependent variable is a measure of the number of items correctly remembered for each child. Use an ANOVA with $\alpha = .05$ to evaluate these data.

	AGE 2	AGE 6	AGE 10
RECALL TEST	$M = 3$ $SS = 16$	$M = 7$ $SS = 19$	$M = 12$ $SS = 14$
RECOGNITION TEST	$M = 15$ $SS = 21$	$M = 16$ $SS = 20$	$M = 17$ $SS = 18$

19. Hyperactivity in children usually is treated by counseling, by drugs, or by both. The following data are from an experiment designed to evaluate the effectiveness of these different treatments. The dependent variable is a measure of attention span (how long each child was able to concentrate on a specific task).

	DRUG	NO DRUG
COUNSELING	$n = 10$ $T = 140$ $SS = 40$	$n = 10$ $T = 80$ $SS = 36$
NO COUNSELING	$n = 10$ $T = 120$ $SS = 45$	$n = 10$ $T = 100$ $SS = 59$

$$\Sigma X^2 = 5220$$

a. Use an ANOVA with $\alpha = .05$ to evaluate these data.

b. Do the data indicate that the drug has a significant effect? Does the counseling have an effect? Describe these results in terms of the effectiveness of the drug and counseling and their interaction.

20. In the Preview section of this chapter, we presented an experiment that examined the effect of an audience on the performance of two different personality types. Data from this experiment are as follows. The dependent variable is the number of errors made by each participant.

		ALONE	AUDIENCE
SELF-ESTEEM	HIGH	3 6 2 2 4 7	9 4 5 8 4 6
	LOW	7 7 2 6 8 6	10 14 11 15 11 11

a. Use an ANOVA with $\alpha = .05$ to evaluate the data. Describe the effect of the audience and the effect of self-esteem on performance.

b. Calculate the effect size (η^2) for each main effect and for the interaction.

21. For the following set of data:

a. Compute the cell means and place them in an $A \times B$ matrix (see Box 15.1, page 486).

b. Using the cell means, row means, and column means, state whether or not you think there will be an interaction of factors A and B, a main effect for factor A, and a main effect for factor B. (See Problem 4.)

c. Perform a two-factor analysis of variance for the data. Use the .01 level of significance for all hypothesis tests in this analysis.

		FACTOR B		
		B_1	B_2	B_3
FACTOR A	A_1	1 3 1 4	4 3 3 6	8 6 8 10
	A_2	8 6 6 8	1 6 8 1	1 4 1 4

22. The following data are from a two-factor independent-measures experiment.

		FACTOR B	
		B_1	B_2
FACTOR A	A_1	7 12 10 9 12	1 0 4 3 2
	A_2	1 4 2 3 5	8 11 11 9 16

a. For these data, compute the cell means and place them in an $A \times B$ matrix.
b. Looking at the cell means, row means, and column means, describe whether or not there might be an interaction, a main effect for factor A, and a main effect for factor B.
c. Perform a two-factor analysis of variance with $\alpha = .01$.
d. Describe the results of the analysis.

23. The following data are from a study examining the influence of a specific hormone on eating behavior. Three different drug doses were used, including a control condition (no drug), and the study measured eating behavior for males and females. The dependent variable was the amount of food consumed over a 48-hour period.

	NO DRUG	SMALL DOSE	LARGE DOSE
MALES	1 6 1 1 1	7 7 11 4 6	3 1 1 6 4
FEMALES	0 3 7 5 5	0 0 0 5 0	0 2 0 0 3

Use an ANOVA with $\alpha = .05$ to evaluate these data, and describe the results (i.e., the drug effect, the sex difference, and the interaction). Test the simple main effects at the .05 level of significance.

24. Individuals who are identified as having an antisocial personality disorder also tend to have reduced physiological responses to painful or anxiety-producing stimuli. In everyday terms, these individuals show a limited physical response to fear, guilt, or anxiety. One way of measuring this response is with the galvanic skin response (GSR). Normally, when a person is aroused, there is an increase in perspiration, which causes a measurable reduction in the electrical resistance on the skin. The following data represent the results of a study measuring GSR for non-antisocial and antisocial individuals in regular and stress-provoking situations:

	AROUSAL LEVEL			
	BASELINE		STRESS	
NON-ANTISOCIAL PERSONALITY	27 26 19 27 21	25 23 24 26 19	15 20 14 19 13	21 18 21 24 17
ANTISOCIAL PERSONALITY	24 28 23 25 21	26 29 27 22 25	29 27 22 20 28	22 25 20 29 21

a. Compute the cell means, and sketch a graph of the results.
b. Use an ANOVA with $\alpha = .05$ to evaluate these data.
c. Explain the findings of this study.

25. The process of interference is assumed to be responsible for much of forgetfulness in human memory. New information going into memory interferes with the information that already is there. One demonstration of interference examines the process of forgetting while people are asleep versus while they are awake. Because there should be less interference during sleep, there also should be less forgetting. The following data are the results from an experiment examining four groups of participants. All participants were given a list of words to remember. Then half of the participants went to sleep, and the others stayed awake. Within both the asleep and the awake groups, half of the participants were tested after 2 hours, and the rest were tested after 8 hours. The dependent variable is the number of words correctly recalled.

	DELAY OF MEMORY TEST			
	2 HOURS		8 HOURS	
ASLEEP	5	7	6	4
	6	10	8	4
	4	5	5	7
	3	8	10	8
	6	5	7	6
AWAKE	3	2	1	2
	4	3	0	1
	5	4	0	2
	3	2	1	0
	2	4	1	1

a. Use a two-factor ANOVA to examine these data.
b. Describe and explain the interaction.

Correlations and Nonparametric Tests

Back in Chapter 1 we stated that the primary goal of science is to establish relationships between variables. Up to now, the statistics we have presented all attempt to accomplish this goal by comparing *groups of scores* using *means and variances* as the basic statistical measures. Typically, one variable is used to define the groups, and a second variable is measured to obtain a set of scores within each group. Means and variances are then computed for the scores, and the sample means are used to test hypotheses about population means. If the hypothesis test indicates a significant mean difference, then we conclude that there is a relationship between the variables.

However, many research situations do not involve comparing groups and do not produce data that allow you to calculate means and variances. For example, a researcher can investigate the relationship between two variables (for example, IQ and creativity) by measuring both variables within a single group of individuals. Also, the measurement procedure may not produce numerical scores. For example, participants can indicate their color preferences simply by picking a favorite color or by ranking several choices. Without numerical scores, it is impossible to calculate means and variances. Instead, the data consist of proportions or frequencies. For example, a research study may investigate what proportion of people select red as their favorite color, and whether this proportion is different for introverted people and extroverted people.

Notice that these new research situations are still asking questions about the relationships between variables, and they are still using sample data to make inferences about populations. However, they are no longer comparing groups and they are no longer based on means and variances. In this part, we introduce the statistical methods that have been developed for these other kinds of research.

C H A P T E R

16

Correlation and Regression

Tools You Will Need

The following items are considered essential background material for this chapter. If you doubt your knowledge of any of these items, you should review the appropriate chapter or section before proceeding.

- Sum of squares (*SS*) (Chapter 4)
  - Computational formula
  - Definitional formula

- *z*-scores (Chapter 5)

- Hypothesis testing (Chapter 8)

Preview

Having been a student and taken exams for much of your life, you probably have noticed a curious phenomenon. In every class, there are some students who zip through exams and turn in their papers while everyone else is still on page 1. Other students cling to their exams and are still working frantically when the instructor announces that time is up and demands that all papers be turned in. Have you wondered what grades these students receive? Are the students who finish first the best in the class, or are they completely unprepared and simply accepting their failure? Are the *A* students the last to finish because they are compulsively checking and rechecking their answers? To help answer these questions, we carefully observed a recent exam and recorded the amount of time each student spent and the grade each student received. These data are shown in Figure 16.1. Note that we have listed time along the *X*-axis and grade on the *Y*-axis. Each student is identified by a point on the graph that is located directly above the student's time and directly across from the student's grade. Also note that we have drawn a line through the middle of the data points in Figure 16.1. The line helps make the relationship between time and grade more obvious. The graph shows that the highest grades tend to go to the students who finished their exams early. Students who held their papers to the bitter end tended to have low grades.

In statistical terms, these data show a correlation between time and grade. In this chapter, we will see how correlations are used to measure and describe relationships. Just as a sample mean provides a concise description of an entire sample, a correlation provides a description of a relationship. We also will look at how correlations are used and interpreted. For example, now that you have seen the relationship between time and grades, don't you think it might be a good idea to start turning in your exam papers a little sooner? Wait and see.

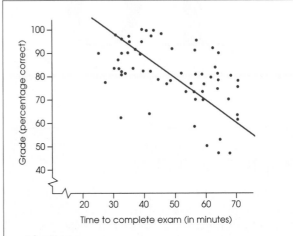

FIGURE 16.1

The relationship between exam grade and time needed to complete the exam. Notice the general trend in these data: Students who finish the exam early tend to have better grades.

16.1 OVERVIEW

Correlation is a statistical technique that is used to measure and describe a relationship between two variables. Usually the two variables are simply observed as they exist naturally in the environment—there is no attempt to control or manipulate the variables. For example, a researcher interested in the relationship between sugar consumption and activity level could measure (and record) the types of food eaten by a group of preschool children, and then observe their activity levels on the playground. Notice that the researcher is not trying to manipulate the children's diet or activity level, but is simply observing what occurs naturally. You also should notice that a correlation requires two scores for each individual (one score from each of the two variables). These scores normally are identified as *X* and *Y*. The pairs of scores can be listed in a table, or they can be presented graphically in a scatterplot (see Figure 16.2). In the scatterplot, the

X values are placed on the horizontal axis of a graph, and the *Y* values are placed on the vertical axis. Each individual is then identified by a single point on the graph so that the coordinates of the point (the *X* and *Y* values) match the individual's *X* score and *Y* score. The value of the scatterplot is that it allows you to see any patterns or trends that exist in the data (see Figure 16.2).

FIGURE 16.2

The same set of *n* = 6 pairs of scores (*X* and *Y* values) is shown in a table and in a scatterplot. Notice that the scatterplot allows you to see the relationship between *X* and *Y*.

Person	X	Y
A	1	1
B	1	3
C	3	2
D	4	5
E	6	4
F	7	5

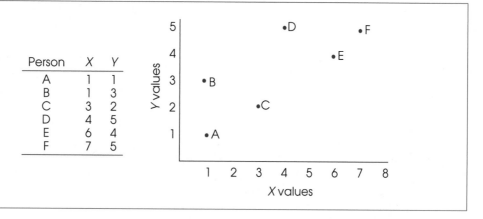

THE CHARACTERISTICS OF A RELATIONSHIP

A correlation measures three characteristics of the relationship between *X* and *Y*. These three characteristics are as follows:

1. The Direction of the Relationship. Correlations can be classified into two basic categories: positive and negative.

DEFINITIONS

In a **positive correlation**, the two variables tend to move in the same direction: As the value of the *X* variable increases from one individual to another, the *Y* variable also tends to increase; when the *X* variable decreases, the *Y* variable also decreases.

In a **negative correlation**, the two variables tend to go in opposite directions. As the *X* variable increases, the *Y* variable decreases. That is, it is an inverse relationship.

The direction of a relationship is identified by the sign of the correlation. A positive value (+) indicates a positive relationship; a negative value (−) indicates a negative relationship. The following example provides a description of positive and negative relationships.

EXAMPLE 16.1

Remember that the actual data appear as *points* in the figures. The dashed lines have been added as visual aids to help make the direction of the relationship easier to see.

Suppose that you run the drink concession at the football stadium. After several seasons, you begin to notice a relationship between the temperature at game time and the beverages you sell. Specifically, you have noted that when the temperature is high, you tend to sell a lot of beer. When the temperature is low, you sell relatively little beer [see Figure 16.3(a)]. This is an example of a positive correlation. At the same time, you have noted a relationship between temperature and coffee sales: On cold days, you sell much more coffee than on hot days [see Figure 16.3(b)]. This is an example of a negative relationship.

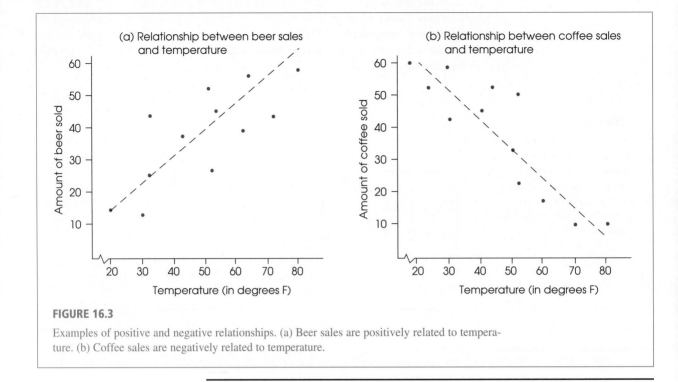

FIGURE 16.3

Examples of positive and negative relationships. (a) Beer sales are positively related to temperature. (b) Coffee sales are negatively related to temperature.

2. The Form of the Relationship. In the preceding coffee and beer examples, the relationships tend to have a linear form; that is, the points in the scatterplot tend to form a straight line. Notice that we have drawn a line through the middle of the data points in each figure to help show the relationship. The most common use of correlation is to measure straight-line relationships. However, you should note that other forms of relationship do exist and that there are special correlations used to measure them. (We will examine alternatives in Sections 16.5 and 16.6.) Figure 16.4(a) shows the relationship between reaction time and age. In this scatterplot, there is a curved relationship. Reaction time improves with age until the late teens, when it reaches a peak; after that, reaction time starts to get worse. Figure 16.4(b) shows the typical dose-response relationship. Again, this is not a straight-line relationship. In this graph, the elevation in mood increases rather rapidly with dose increases. However, beyond a certain dose, the amount of improvement in mood levels off. Many types of measurements exist for correlation. In general, each type is designed to evaluate a specific form of relationship. In this text, we will concentrate on the correlation that measures linear relationships.

3. The Degree of the Relationship. Finally, a correlation measures how well the data fit the specific form being considered. For example, a linear correlation measures how well the data points fit on a straight line. The degree of relationship is measured by the numerical value of the correlation. A *perfect correlation* always is identified by a correlation of 1.00 and indicates a perfect fit. At the other extreme, a correlation of 0 indicates no fit at all. Intermediate values represent the degree to which the data points approximate the perfect fit. The numerical value of the correlation also reflects

A correlation of -1.00 also indicates a perfect fit. The direction of the relationship (positive or negative) should be considered separately from the degree of the relationship.

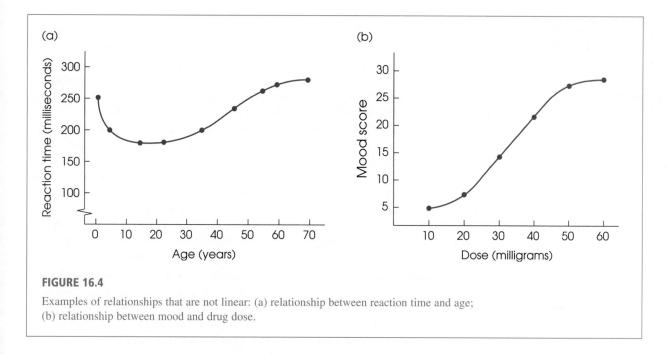

FIGURE 16.4

Examples of relationships that are not linear: (a) relationship between reaction time and age;
(b) relationship between mood and drug dose.

the degree to which there is a consistent, predictable relationship between the two variables. Again, a correlation of 1.00 (or −1.00) indicates a perfectly consistent relationship.

Examples of different values for linear correlations are shown in Figure 16.5. Notice that in each example we have sketched a line around the data points. This line, called an *envelope* because it encloses the data, often helps you to see the overall trend in the data.

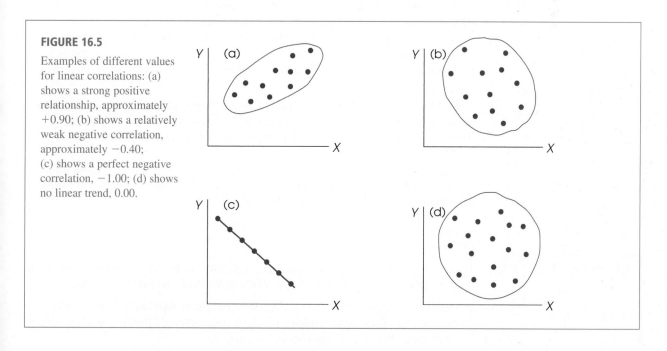

FIGURE 16.5

Examples of different values for linear correlations: (a) shows a strong positive relationship, approximately +0.90; (b) shows a relatively weak negative correlation, approximately −0.40; (c) shows a perfect negative correlation, −1.00; (d) shows no linear trend, 0.00.

Although correlations have a number of different applications, a few specific examples are presented next to give an indication of the value of this statistical measure.

1. Prediction. If two variables are known to be related in some systematic way, it is possible to use one of the variables to make accurate predictions about the other. For example, when you applied for admission to college, you were required to submit a great deal of personal information, including your scores on the Scholastic Achievement Test (SAT). College officials want this information so they can predict your chances of success in college. It has been demonstrated over several years that SAT scores and college grade point averages are correlated. Students who do well on the SAT tend to do well in college; students who have difficulty with the SAT tend to have difficulty in college. Based on this relationship, the college admissions office can make a prediction about the potential success of each applicant. You should note that this prediction is not perfectly accurate. Not everyone who does poorly on the SAT will have trouble in college. That is why you also submit letters of recommendation, high school grades, and other information with your application.

2. Validity. Suppose that a psychologist develops a new test for measuring intelligence. How could you show that this test truly is measuring what it claims; that is, how could you demonstrate the validity of the test? One common technique for demonstrating validity is to use a correlation. If the test actually is measuring intelligence, then the scores on the test should be related to other measures of intelligence—for example, standardized IQ tests, performance on learning tasks, problem-solving ability, and so on. The psychologist could measure the correlation between the new test and each of these other measures of intelligence in order to demonstrate that the new test is valid.

3. Reliability. In addition to evaluating the validity of a measurement procedure, correlations are used to determine reliability. A measurement procedure is considered reliable to the extent that it produces stable, consistent measurements. That is, a reliable measurement procedure will produce the same (or nearly the same) scores when the same individuals are measured under the same conditions. For example, if your IQ were measured as 113 last week, you would expect to obtain nearly the same score if your IQ were measured again this week. One way to evaluate reliability is to use correlations to determine the relationship between two sets of measurements. When reliability is high, the correlation between two measurements should be strong and positive.

4. Theory Verification. Many psychological theories make specific predictions about the relationship between two variables. For example, a theory may predict a relationship between brain size and learning ability; a developmental theory may predict a relationship between the parents' IQs and the child's IQ; a social psychologist may have a theory predicting a relationship between personality type and behavior in a social situation. In each case, the prediction of the theory could be tested by determining the correlation between the two variables.

LEARNING CHECK

1. If the world were fair, would you expect a positive or negative relationship between grade point average (X) and weekly studying hours (Y) for college students?

2. Data suggest that, on the average, children from large families have lower IQs than children from small families. Do these data indicate a positive or a negative relationship between family size and average IQ?

3. If you are measuring linear relationship, correlations of $+0.50$ and -0.50 are equally good in terms of how well the data fit on a straight line. (True or false?)

4. It is impossible to have a correlation greater than $+1.00$ or less than -1.00. (True or false?)

ANSWERS

1. Positive. More hours studying should be associated with higher grade point averages.

2. Negative

3. True. The degree of fit is measured by the magnitude of the correlation independent of sign.

4. True. Correlations are always from $+1.00$ to -1.00.

16.2 THE PEARSON CORRELATION

By far the most common correlation is the Pearson correlation (or the Pearson product-moment correlation).

DEFINITION

The **Pearson correlation** measures the degree and direction of linear relationship between two variables.

The Pearson correlation is identified by the letter r. Conceptually, this correlation is computed by

$$r = \frac{\text{degree to which } X \text{ and } Y \text{ vary together}}{\text{degree to which } X \text{ and } Y \text{ vary separately}}$$

$$= \frac{\text{covariability of } X \text{ and } Y}{\text{variability of } X \text{ and } Y \text{ separately}}$$

When there is a perfect linear relationship, every change in the X variable is accompanied by a corresponding change in the Y variable. In Figure 16.5(c), for example, every time the value of X increases, there is a perfectly predictable decrease in Y. The result is a perfect linear relationship, with X and Y always varying together. In this case, the covariability (X and Y together) is identical to the variability of X and Y separately, and the formula produces a correlation of -1.00. At the other extreme, when there is no linear relationship, a change in the X variable does not correspond to any predictable change in Y. In this case, there is no covariability, and the resulting correlation is zero.

THE SUM OF PRODUCTS OF DEVIATIONS

To calculate the Pearson correlation, it is necessary to introduce one new concept: the *sum of products* of deviations, or *SP*. In the past, we used a similar concept *SS* (the sum of squared deviations) to measure variability for a single variable. Now, we will use *SP* to measure the amount of covariability between two variables. The value for *SP* can be calculated with either a definitional formula or a computational formula.

The *definitional formula* for the sum of products is

$$SP = \Sigma(X - M_X)(Y - M_Y) \tag{16.1}$$

where M_X is the mean for the X scores and M_Y is the mean for the Ys.

This formula instructs you to find the X deviation and the Y deviation for each individual, then multiply the deviations to obtain the product for each individual, then add up the products. Notice that the terms in the formula define the value being calculated: the sum of the products of the deviations.

The *computational formula* for the sum of products is

Caution: The n in this formula refers to the number of pairs of scores.

$$SP = \Sigma XY - \frac{\Sigma X \Sigma Y}{n} \tag{16.2}$$

Because the computational formula uses the original scores (X and Y values), it usually results in easier calculations than those required with the definitional formula. However, both formulas will always produce the same value for SP.

You may have noted that the formulas for SP are similar to the formulas you have learned for SS (sum of squares). The relationship between the two sets of formulas is described in Box 16.1. The following example demonstrates the calculation of SP with both formulas.

EXAMPLE 16.2   The same set of $n = 4$ pairs of scores will be used to calculate SP, first using the definitional formula and then using the computational formula.

For the definitional formula, you need deviation scores for each of the X values and each of the Y values. Note that the mean for the Xs is $M_X = 3$ and that the mean for the Ys is $M_Y = 5$. The deviations and the products of deviations are shown in the following table:

Caution: The signs ($+$ and $-$) are critical in determining the sum of products, SP.

SCORES		DEVIATIONS		PRODUCTS
X	Y	$X - M_X$	$Y - M_Y$	$(X - M_X)(Y - M_Y)$
1	3	-2	-2	$+4$
2	6	-1	$+1$	-1
4	4	$+1$	-1	-1
5	7	$+2$	$+2$	$+4$
				$+6 = SP$

For these scores, the sum of the products of the deviations is $SP = +6$.

For the computational formula, you need the X value, the Y value, and the XY product for each individual. Then you find the sum of the Xs, the sum of the Ys, and the sum of the XY products. These values are as follows:

X	Y	XY	
1	3	3	
2	6	12	
4	4	16	
5	7	35	
12	20	66	Totals

BOX

16.1   COMPARING THE *SP* AND *SS* FORMULAS

It will help you to learn the formulas for *SP* if you note the similarity between the two *SP* formulas and the corresponding formulas for *SS* that were presented in Chapter 4. The definitional formula for *SS* is

$$SS = \Sigma(X - M)^2$$

In this formula, you must square each deviation, which is equivalent to multiplying it by itself. With this in mind, the formula can be rewritten as

$$SS = \Sigma(X - M)(X - M)$$

The similarity between the *SS* formula and the *SP* formula should be obvious—the *SS* formula uses squares and the *SP* formula uses products. This same relation

ship exists for the computational formulas. For *SS*, the computational formula is

$$SS = \Sigma X^2 - \frac{(\Sigma X)^2}{n}$$

As before, each squared value can be rewritten so that the formula becomes

$$SS = \Sigma XX - \frac{\Sigma X \Sigma X}{n}$$

Again, note the similarity in structure between the *SS* formula and the *SP* formula. If you remember that *SS* uses squares and *SP* uses products, the two new formulas for the sum of products should be easy to learn.

Substituting the sums in the formula gives

$$SP = \Sigma XY - \frac{\Sigma X \Sigma Y}{n}$$

$$= 66 - \frac{12(20)}{4}$$

$$= 66 - 60$$

$$= 6$$

Note that both formulas produce the same result, *SP* = 6.

CALCULATION OF THE PEARSON CORRELATION

As noted earlier, the Pearson correlation consists of a ratio comparing the covariability of *X* and *Y* (the numerator) with the variability of *X* and *Y* separately (the denominator). In the formula for the Pearson *r*, we will use *SP* to measure the covariability of *X* and *Y*. The variability of *X* and *Y* will be measured by computing *SS* for the *X* scores and *SS* for the *Y* scores separately. With these definitions, the formula for the Pearson correlation becomes

Note that you *multiply SS* for *X* by *SS* for *Y* in the denominator of the Pearson formula.

$$r = \frac{SP}{\sqrt{SS_X SS_Y}} \tag{16.3}$$

The following example demonstrates the use of this formula with a simple set of scores.

EXAMPLE 16.3

X	Y
0	1
10	3
4	1
8	2
8	3

The Pearson correlation is computed for the set of $n = 5$ pairs of scores shown in the margin.

Before starting any calculations, it is useful to put the data in a scatterplot and make a preliminary estimate of the correlation. These data have been graphed in Figure 16.6. Looking at the scatterplot, it appears that there is a very good (but not perfect) positive correlation. You should expect an approximate value of $r = +.8$ or $+.9$. To find the Pearson correlation, we will need SP, SS for X, and SS for Y. The following table presents the calculations for each of these values using the definitional formulas. (Note that the mean for the X values is $M_X = 6$ and the mean for the Y scores is $M_Y = 2$.)

SCORES		DEVIATIONS		SQUARED DEVIATIONS		PRODUCTS
X	Y	$X - M_X$	$Y - M_Y$	$(X - M_X)^2$	$(Y - M_Y)^2$	$(X - M_X)(Y - M_Y)$
0	1	−6	−1	36	1	+6
10	3	+4	+1	16	1	+4
4	1	−2	−1	4	1	+2
8	2	+2	0	4	0	0
8	3	+2	+1	4	1	+2
				64 = SS for X	4 = SS for Y	+14 = SP

By using these values, the Pearson correlation is

$$r = \frac{SP}{\sqrt{SS_X SS_Y}} = \frac{14}{\sqrt{64(4)}}$$

$$= \frac{14}{16} = +0.875$$

Note that the value we obtained for the correlation is perfectly consistent with the pattern shown in Figure 16.6. First, the positive value of the correlation indicates that the points are clustered around a line that slopes up to the right. Second, the high value for the correlation (near 1.00) indicates that the points are very tightly clustered close to the line. Thus, the value of the correlation describes the relationship that exists in the data.

FIGURE 16.6

Scatterplot of the data from Example 16.3.

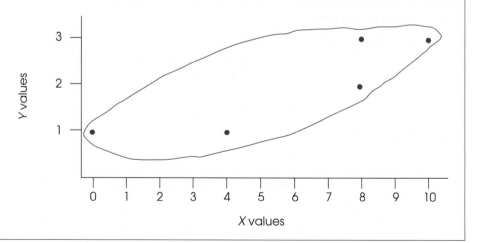

THE PEARSON CORRELATION AND z-SCORES

The Pearson correlation measures the relationship between an individual's location in the X distribution and his or her location in the Y distribution. For example, a positive correlation means that individuals who score high on X also tend to score high on Y. Similarly, a negative correlation indicates that individuals with high X scores tend to have low Y scores.

Recall from Chapter 5 that z-scores provide a precise way to identify the location of an individual score within a distribution. Because the Pearson correlation measures the relationship between locations and because z-scores are used to specify locations, the formula for the Pearson correlation can be expressed entirely in terms of z-scores:

$$r = \frac{\Sigma z_X z_Y}{n} \tag{16.4}$$

In this formula, z_X identifies each individual's position within the X distribution, and z_Y identifies the position within the Y distribution. The product of the z-scores (like the product of the deviation scores) determines the strength and direction of the correlation.

Because z-scores are considered to be the best way to describe a location within a distribution, Equation 16.4 often is considered to be the best way to define the Pearson correlation. However, this formula requires a lot of tedious calculations (changing each score to a z-score), so it rarely is used to calculate a correlation.

Note: Computing a correlation with z-scores (Equation 16.4) requires that you use the *population* standard deviation for both the X values and the Y values. If you compute z-scores with the *sample* standard deviations (using $df = n - 1$), then Equation 16.4 will not produce the correct correlation. However, the correlation produced by the equation can be corrected simply by multiplying the value by $n/(n - 1)$.

LEARNING CHECK

1. Describe what is measured by a Pearson correlation.

2. Can *SP* ever have a value less than zero?

3. Calculate the sum of products of deviations (*SP*) for the following set of scores. Use the definitional formula and then the computational formula. Verify that you get the same answer with both formulas.

X	Y
1	0
3	1
7	6
5	2
4	1

Remember: It is useful to sketch a scatterplot and make an estimate of the correlation before you begin calculations.

4. Compute the Pearson correlation for the following data:

X	Y
2	9
1	10
3	6
0	8
4	2

ANSWERS   1. The Pearson correlation measures the degree and direction of linear relationship between two variables.

2. Yes. *SP* can be positive, negative, or zero depending on the relationship between *X* and *Y*.

3. $SP = 19$

4. $r = -\dfrac{16}{20} = -0.80$

16.3   UNDERSTANDING AND INTERPRETING THE PEARSON CORRELATION

When you encounter correlations, there are four additional considerations that you should bear in mind:

1. Correlation simply describes a relationship between two variables. It does not explain why the two variables are related. Specifically, a correlation should not and cannot be interpreted as proof of a cause-and-effect relationship between the two variables.

2. The value of a correlation can be affected greatly by the range of scores represented in the data.

3. One or two extreme data points, often called *outliers*, can have a dramatic effect on the value of a correlation.

4. When judging how "good" a relationship is, it is tempting to focus on the numerical value of the correlation. For example, a correlation of $+.5$ is halfway between 0 and 1.00 and therefore appears to represent a moderate degree of relationship. However, a correlation should not be interpreted as a proportion. Although a correlation of 1.00 does mean that there is a 100% perfectly predictable relationship between *X* and *Y*, a correlation of .5 does not mean that you can make predictions with 50% accuracy. To describe how accurately one variable predicts the other, you must square the correlation. Thus, a correlation of $r = .5$ means that one variable *partially* predicts the other, but the predictable portion is only $r^2 = .5^2 = 0.25$ (or 25%) of the total variability.

We will now discuss each of these four points in detail.

CORRELATION AND CAUSATION

One of the most common errors in interpreting correlations is to assume that a correlation necessarily implies a cause-and-effect relationship between the two variables. (Even Pearson blundered by asserting causation from correlational data [Blum, 1978]). We constantly are bombarded with reports of relationships: Cigarette smoking is related to heart disease; alcohol consumption is related to birth defects; carrot consumption is related to good eyesight. Do these relationships mean that cigarettes cause heart disease or carrots cause good eyesight? The answer is *no*. Although there may be a causal relationship, the simple existence of a correlation does not prove it. This point should become clear in the following hypothetical example.

E X A M P L E   1 6 . 4

Suppose that we select a variety of different cities and towns throughout the United States and measure the number of serious crimes (*Y* variable) and the number of churches (*X* variable) for each. A scatterplot showing hypothetical data for this study is presented in Figure 16.7. Notice that this scatterplot shows a strong, positive correlation between churches and crime. You also should note that these are realistic data. It is reasonable that the smaller towns would have less crime and fewer churches and that the large cities would have large values for both variables. Does this relationship mean that churches cause crime? Does it mean that crime causes churches? It should be clear that the answer to both is no. Although a strong correlation exists between churches and crime, the real cause of the relationship is the size of the population.

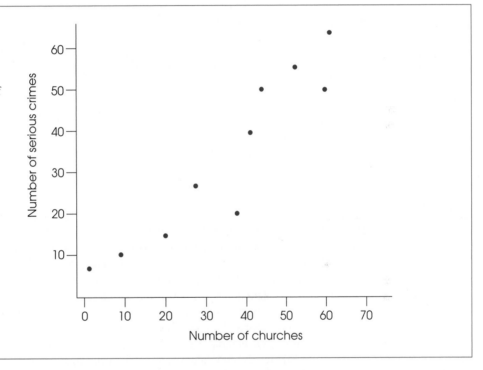

FIGURE 16.7

Hypothetical data showing the logical relationship between the number of churches and the number of serious crimes for a sample of U.S. cities.

CORRELATION AND RESTRICTED RANGE

Whenever a correlation is computed from scores that do not represent the full range of possible values, you should be cautious in interpreting the correlation. Suppose, for example, that you are interested in the relationship between IQ and creativity. If you select a sample of your fellow college students, your data probably will represent only a limited range of IQ scores (most likely from 110 to 130). The correlation within this restricted range could be completely different from the correlation that would be obtained from a full range of IQ scores. Two extreme examples are shown in Figure 16.8.

Figure 16.8(a) shows an example of a strong positive relationship between *X* and *Y* when the entire range of scores is considered. However, this relationship is obscured when the data are limited to a *restricted range*. In Figure16.8(b), there is no consistent relationship between *X* and *Y* for the full range of scores. However, when the range of *X* values is restricted, the data show a strong positive relationship.

To be safe, you should not generalize any correlation beyond the range of data represented in the sample. For a correlation to provide an accurate description for the general population, there should be a wide range of X and Y values in the data.

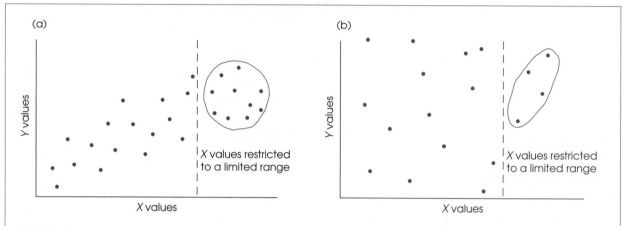

FIGURE 16.8

(a) In this example, the full range of X and Y values shows a strong, positive correlation, but the restricted range of scores produces a correlation near zero. (b) An example in which the full range of X and Y values shows a correlation near zero, but the scores in the restricted range produce a strong, positive correlation.

OUTLIERS   An outlier is an individual with X and/or Y values that are substantially different (larger or smaller) from the values obtained for the other individuals in the data set. The data point of a single outlier can have a dramatic influence on the value obtained for the correlation. This effect is illustrated in Figure 16.9. Figure 16.9(a) shows a set of $n = 5$ data points for which the correlation between X and Y variables is nearly zero (actually $r = -0.08$). In Figure 16.9(b), one extreme data point (14, 12) has been added to the original data set. When this outlier is included in the analysis, a strong, positive correlation emerges (now $r = +0.85$). Note that the single outlier drastically alters the value for the correlation and thereby can affect one's interpretation of the relationship between variables X and Y. Without the outlier, one would conclude there is no relationship between the two variables. With the extreme data point, $r = +0.85$ implies that as X increases, Y will increase—and do so consistently. The problem of outliers is a good reason why you should always look at a scatterplot, instead of simply basing your interpretation on the numerical value of the correlation. If you only "go by the numbers," you might overlook the fact that one extreme data point inflated the size of the correlation.

CORRELATION AND THE STRENGTH OF THE RELATIONSHIP   A correlation measures the degree of relationship between two variables on a scale from 0 to 1.00. Although this number provides a measure of the degree of relationship, many researchers prefer to square the correlation and use the resulting value to measure the strength of the relationship.

One of the common uses of correlation is for prediction. If two variables are correlated, you can use the value of one variable to predict the other. For example, college admissions officers do not just guess which applicants are likely to do well; they use

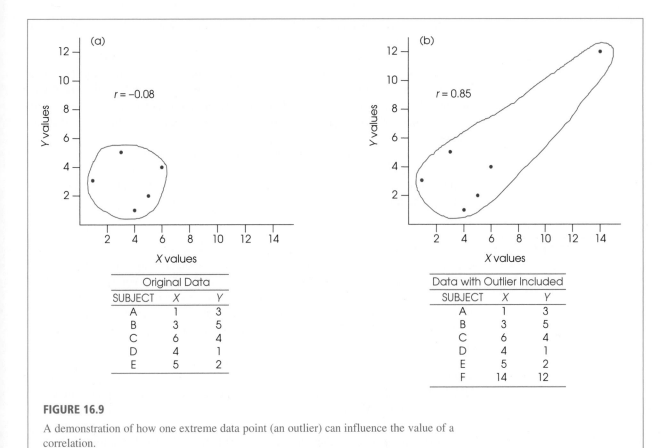

FIGURE 16.9

A demonstration of how one extreme data point (an outlier) can influence the value of a correlation.

other variables (SAT scores, high school grades, etc.) to predict which students are most likely to be successful. These predictions are based on correlations. By using correlations, the admissions officers expect to make more-accurate predictions than would be obtained by chance. In general, the squared correlation (r^2) measures the gain in accuracy that is obtained from using the correlation for prediction. The squared correlation measures the proportion of variability in the data that is explained by the relationship between X and Y. It is sometimes called the *coefficient of determination*.

DEFINITION   The value r^2 is called the **coefficient of determination** because it measures the proportion of variability in one variable that can be determined from the relationship with the other variable. A correlation of $r = 0.80$ (or -0.80), for example, means that $r^2 = 0.64$ (or 64%) of the variability in the Y scores can be predicted from the relationship with X.

In earlier chapters (see pages 292, 321, and 350) we introduced r^2 as a method for measuring effect size for research studies where mean differences were used to compare treatments. Specifically, we measured how much of the variance in the scores was accounted for by the differences between treatments. In experimental terminology, r^2 measures how much of the variance in the dependent variable is accounted for by the independent variable. Now we are doing the same thing, except that there is no independent or dependent variable. Instead, we simply have two variables, X and Y, and we use r^2 to measure how much of the variance in one variable can be determined from its relationship with the other variable.

Just as r^2 was used to evaluate effect size for mean differences in Chapters 9, 10, and 11, r^2 can now be used to evaluate the size or strength of the correlation. The same standards that were introduced in Table 9.3 (page 295), apply to both uses of the r^2 measure. Specifically, an r^2 value of 0.01 indicates a small effect or a small correlation, r^2 of 0.09 indicates a medium correlation, and r^2 of 0.25 or larger indicates a large correlation.

A more detailed discussion of the coefficient of determination is presented in Section 16.7. For now, you simply should realize that whenever two variables are consistently related, it is possible to use one variable to predict the values of the second variable. The following example demonstrates this concept.

EXAMPLE 16.5   Figure 16.10 shows three sets of data representing different degrees of linear relationship. The first set of data [Figure 16.10(a)] shows the relationship between IQ and shoe size. In this case, the correlation is $r = 0$ (and $r^2 = 0$), and you have no ability to predict a person's IQ based on his or her shoe size. Knowing a person's shoe size provides no information (0%) about the person's IQ. In this case, the two measurements are completely independent and there is no overlap between the information they provide.

Now consider the data in Figure 16.10(b). These data show a reasonably good, positive correlation, $r = +0.80$, between SAT scores and college grade point averages (GPA). Students who score high on the SAT tend to have higher grades than students with low SAT scores. From this relationship, it is possible to predict a student's GPA based on his or her SAT score. However, you should realize that the prediction is not perfect. Although students with high SAT scores *tend* to have high GPAs, this is not always true. Thus, knowing a student's SAT score provides some information about the student's grades, or knowing a student's grades provides some information about the student's SAT score. Although we are measuring two separate variables (SAT and GPA), the two scores for each student actually provide much the

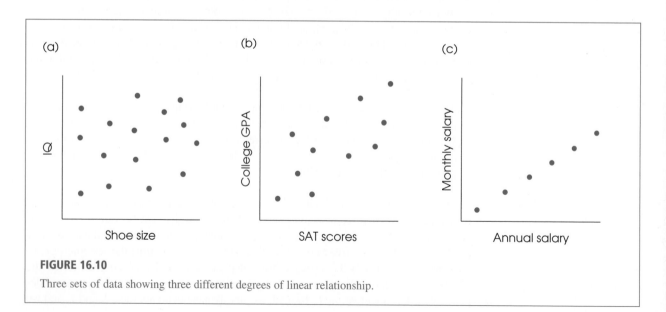

FIGURE 16.10

Three sets of data showing three different degrees of linear relationship.

same information. In other words, there is some overlap or redundancy between the two measures. With a correlation of $r = +0.80$, we obtain $r^2 = 0.64$, which means that 64% of the variability in the students' GPAs is predicted or determined by the students' SAT scores.

Finally, consider the data in Figure 16.10(c). This time we show a perfect linear relationship ($r = +1.00$) between monthly salary and yearly salary for a group of college employees. With $r = 1.00$ and $r^2 = 1.00$, there is 100% predictability. If you know a person's monthly salary, you can predict perfectly the person's annual salary. In this case, there is complete overlap between the two measures because a person's monthly salary provides exactly the same information as the person's annual salary.

One final comment concerning the interpretation of a correlation is presented in Box 16.2.

16.4 HYPOTHESIS TESTS WITH THE PEARSON CORRELATION

The Pearson correlation is generally computed for sample data. As with most sample statistics, however, a sample correlation often is used to answer questions about the general population. That is, the sample correlation is used as the basis for drawing inferences about the corresponding population correlation. For example, a psychologist would like to know whether there is a relationship between IQ and creativity. This is a general question concerning a population. To answer the question, a sample would be selected, and the sample data would be used to compute the correlation value. You should recognize this process as an example of inferential statistics: using samples to draw inferences about populations. In the past, we have been concerned primarily with using sample means as the basis for answering questions about population means. In this section, we examine the procedures for using a sample correlation as the basis for testing hypotheses about the corresponding population correlation.

THE HYPOTHESES

The basic question for this hypothesis test is whether or not a correlation exists in the population. The null hypothesis is "No, there is no correlation in the population" or "The population correlation is zero." The alternative hypothesis is "Yes, there is a real, nonzero correlation in the population." Because the population correlation is traditionally represented by ρ (the Greek letter rho), these hypotheses would be stated in symbols as

Directional hypotheses for a "one-tailed" test would specify either a positive correlation ($\rho > 0$) or a negative correlation ($\rho < 0$).

H_0: $\rho = 0$ (There is no population correlation.)

H_1: $\rho \neq 0$ (There is a real correlation.)

The correlation from the sample data (r) will be used to evaluate these hypotheses. As always, samples are not expected to be identical to the populations from which they come; there will be some discrepancy (sampling error) between a sample statistic and the corresponding population parameter. Specifically, you should always expect some error between a sample correlation and the population correlation it represents. One implication of this fact is that even when there is no correlation in the population ($\rho = 0$), you are still likely to obtain a nonzero value for the sample correlation. This is particularly

BOX
16.2   REGRESSION TOWARD THE MEAN

Consider the following problem.

Explain why the rookie of the year in major-league baseball usually does not perform as well in his second season.

Notice that this question does not appear to be statistical or mathematical in nature. However, the answer to the question is directly related to the statistical concepts of correlation and regression. Specifically, there is a simple observation about correlations known as *regression toward the mean*.

D E F I N I T I O N   When there is a less-than-perfect correlation between two variables, extreme scores (high or low) on one variable tend to be paired with the less extreme scores (more toward the mean) on the second variable. This fact is called **regression toward the mean.**

Figure 16.11 shows a scatterplot with a less-than-perfect correlation between two variables. The data points in this figure might represent batting averages for baseball rookies in 2000 (variable 1) and batting averages for the same players in 2001 (variable 2). Because the correlation is less than perfect, the highest scores on variable 1 are generally *not* the highest scores on variable 2. In baseball terms, the rookies with the highest averages in 2000 do not have the highest averages in 2001.

Remember that a correlation does not explain *why* one variable is related to the other; it simply says that there is a relationship. The correlation cannot explain why the best rookie does not perform as well in his second year. But, because the correlation is not perfect, it is a statistical fact that extremely high scores in one year generally will *not* be paired with extremely high scores in the next year.

Regression toward the mean often poses a problem for interpreting experimental results. Suppose, for example, that you want to evaluate the effects of a special preschool program for disadvantaged children. You select a sample of children who score extremely low on an academic performance test. After participating in your preschool program, these children score significantly higher on the test. Why did their scores improve? One answer is that the special program helped. But an alternative answer is regression toward the mean. If there is a less-than-perfect correlation between scores on the first test and scores on the second test (which is usually the case), individuals with extremely low scores on test 1 will tend to have higher scores on test 2. It is a statistical fact of life, not necessarily the result of any special program.

Now try using the concept of regression toward the mean to explain the following phenomena:

1. You have a truly outstanding meal at a restaurant. However, when you go back with a group of friends, you find that the food is disappointing.

2. You have the highest score on exam I in your statistics class, but score only a little above average on exam II.

FIGURE 16.11

A demonstration of regression toward the mean. The figure shows a scatterplot for a set of data with a less-than-perfect correlation. Notice that the highest scores on variable 1 (extreme right-hand points) are not the highest scores on variable 2, but are displaced downward toward the mean. Also, the lowest scores on variable 1 (extreme left-hand points) are not the lowest scores on variable 2, but are displaced upward toward the mean.

true for small samples. Figure 16.12 illustrates how a small sample from a population with a near-zero correlation could result in a correlation that deviates from zero. The colored dots in the figure represent the entire population and the three circled dots represent a random sample. Note that the three sample points show a relatively good, positive correlation even though there is no linear trend ($\rho = 0$) for the population.

FIGURE 16.12

Scatterplot of a population of X and Y values with a near-zero correlation. However, a small sample of $n = 3$ data points from this population shows a relatively strong, positive correlation. Data points in the sample are circled.

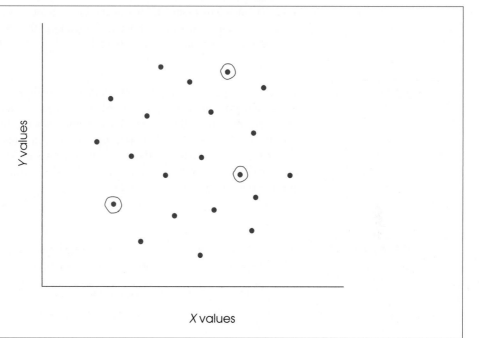

The purpose of the hypothesis test is to decide between the following two alternatives:

1. The nonzero sample correlation is simply due to chance. That is, there is no correlation in the population, and the sample value is simply the result of sampling error. This is the situation specified by H_0.

2. The nonzero sample correlation accurately represents a real, nonzero correlation in the population. This is the alternative stated in H_1.

DEGREES OF FREEDOM FOR THE CORRELATION TEST

The hypothesis test for the Pearson correlation has degrees of freedom defined by $df = n - 2$. An intuitive explanation for this value is that a sample with only $n = 2$ data points has no degrees of freedom. Specifically, with only two points, the sample must produce a perfect correlation of $r = +1.00$ or $r = -1.00$. For example, if you pick any two points in Figure 16.12, the two points will fit perfectly on a straight line. As a result, the sample correlation is free to vary only when the data set contains more than two points. Thus, $df = n - 2$.

THE HYPOTHESIS TEST

The table lists critical values in terms of degrees of freedom: $df = n - 2$. Remember to subtract 2 when using this table.

Although it is possible to conduct the hypothesis test by computing either a t statistic or an F-ratio, the computations have been completed and are summarized in Table B.6 in Appendix B. The table is based on the concept that if the population correlation is zero, then a sample should produce a correlation around zero. Specifically, the table shows the range of sample correlations that are likely to be obtained from a population when the correlation is zero. To use the table, you need to know the sample size (n) and the alpha level. With a sample size of $n = 20$ and an alpha level of .05, for example, you locate $df = n - 2 = 18$ in the left-hand column and the value .05 for either one tail or two tails across the top of the table. For $n = 20$ and $\alpha = .05$ for a two-tailed test, the table shows a value of 0.444. This means that the sample correlation ought to have a value between $+0.444$ and -0.444. It is very unlikely ($\alpha = .05$) to obtain a sample correlation outside this range if the population correlation is zero. Therefore, a sample correlation beyond ±0.444 will lead to rejecting the null hypothesis. The following examples demonstrate the use of the table.

EXAMPLE 16.6   A researcher is using a regular, two-tailed test with $\alpha = .05$ to determine whether or not a nonzero correlation exists in the population. A sample of $n = 30$ individuals is obtained. With $\alpha = .05$ and $n = 30$, the table lists a value of 0.361. Thus, the sample correlation (independent of sign) must have a value greater than or equal to 0.361 to reject H_0 and conclude that there is a significant correlation in the population. Any sample correlation between 0.361 and -0.361 is considered within the realm of sampling error and, therefore, not significant.

EXAMPLE 16.7   This time the researcher is using a directional, one-tailed test to determine whether or not there is a positive correlation in the population.

$$H_0: \quad \rho \leq 0 \quad \text{(not positive)}$$

$$H_1: \quad \rho > 0 \quad \text{(positive)}$$

With $\alpha = .05$ and a sample of $n = 30$, the table lists a value of 0.306 for a one-tailed test. Thus, the researcher must obtain a sample correlation that is positive (as predicted) and has a value greater than or equal to 0.306 to reject H_0 and conclude that there is a significant positive correlation in the population.

IN THE LITERATURE
REPORTING CORRELATIONS

When correlations are computed, the results are reported using APA format. The statement should include the sample size, the calculated value for the correlation, whether or not it is a statistically significant relationship, the probability level, and the type of test used (one- or two-tailed). For example, a correlation might be reported as follows:

> A correlation for the data revealed that amount of education and annual income were significantly related, $r = +.65$, $n = 30$, $p < .01$, two tails.

Sometimes a study might look at several variables, and correlations between all possible variable pairings are computed. Suppose, for example, that a study measured people's annual income, amount of education, age, and intelligence. With four variables, there are six possible pairings. Correlations were computed for every pair of variables. These results are most easily reported in a table called a *correlation matrix*, using footnotes to indicate which correlations are significant. For example, the report might state:

> The analysis examined the relationships between income, amount of education, age, and intelligence for $n = 30$ participants. The correlations between pairs of variables are reported in Table 1. Significant correlations are noted in the table.
>
> **TABLE 1**
>
> Correlation matrix for income, amount of education, age, and intelligence
>
> | | EDUCATION | AGE | IQ |
> |---|---|---|---|
> | Income | +.65* | +.41** | +.27 |
> | Education | | +.11 | +38** |
> | Age | | | −.02 |
>
> $n = 30$
> *$p < .01$, two tails
> **$p < .05$, two tails

LEARNING CHECK

1. A researcher obtains a correlation of $r = -.41$ for a sample of $n = 25$ individuals. Does this sample provide sufficient evidence to conclude that there is a significant, nonzero correlation in the population? Assume a nondirectional test with $\alpha = .05$.

2. For a sample of $n = 20$, how large a correlation is needed to conclude at the .05 level that there is a nonzero correlation in the population? Assume a nondirectional test.

3. As sample size gets smaller, what happens to the magnitude of the correlation necessary for significance? Explain why this occurs.

ANSWERS

1. Yes. For $n = 25$, the critical value is $r = .396$. The sample value is in the critical region.

2. For $n = 20$, the critical value is $r = .444$.

3. As the sample size gets smaller, the magnitude of the correlation needed for significance gets larger. With a small sample, it is easy to get a relatively good correlation just by chance (see Figure 16.12). Therefore, a small sample requires a very large correlation before you can be confident that there is a real (nonzero) relationship in the population.

16.5 THE SPEARMAN CORRELATION

The Pearson correlation specifically measures the degree of linear relationship between two variables. It is the most commonly used measure of relationship and is used with data from an interval or a ratio scale of measurement. However, other correlation measures have been developed for nonlinear relationships and for other types of data (scales of measurement). One of these useful measures is called the *Spearman correlation*. The Spearman correlation is used in two situations.

First, the Spearman correlation is designed to measure the relationship between variables measured on an ordinal scale of measurement. Recall that in Chapter 1 we noted that an ordinal scale of measurement involves placing observations into rank order. Rank-order data are fairly common because they are often easier to obtain than interval or ratio scale data. For example, a teacher may feel confident about rank-ordering students' leadership abilities but would find it difficult to measure leadership on some other scale.

In addition to measuring relationships for ordinal data, the Spearman correlation can be used as a valuable alternative to the Pearson correlation, even when the original raw scores are on an interval or a ratio scale. As we have noted, the Pearson correlation measures the degree of *linear* relationship between two variables—that is, how well the data points fit on a straight line. However, a researcher often expects the data to show a *consistent* relationship but not necessarily a linear relationship (for example, see Figure 16.4 and the discussion on page 524). Consider the situation in which a researcher is investigating the relationship between amount of practice (X) and level of performance (Y). For these data, the researcher expects a strong, positive relationship: More practice leads to better performance. However, the expected relationship probably does not fit a linear form, so a Pearson correlation would not be appropriate (see Figure 16.13). In this situation, the Spearman correlation can be used to obtain a measure of the consistency of relationship, independent of its specific form.

FIGURE 16.13

Hypothetical data showing the relationship between practice and performance. Although this relationship is not linear, there is a consistent positive relationship. An increase in performance tends to accompany an increase in practice.

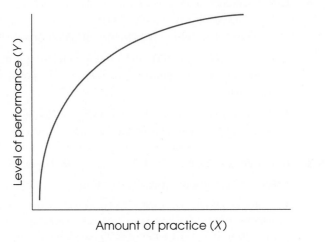

The reason that the Spearman correlation measures consistency, rather than form, comes from a simple observation: When two variables are consistently related, their ranks will be linearly related. For example, a perfectly consistent positive relationship means that every time the X variable increases, the Y variable also increases. Thus, the smallest value of X is paired with the smallest value of Y, the second-smallest value of X is paired with the second-smallest value of Y, and so on. This phenomenon is demonstrated in the following example.

EXAMPLE 16.8

TABLE 16.1

Scores for Example 16.8.

PERSON	X	Y
A	4	9
B	2	2
C	10	10
D	3	8

TABLE 16.2

Ranks for Example 16.8.

PERSON	X RANK	Y RANK
A	3	3
B	1	1
C	4	4
D	2	2

The data in Table 16.1 represent X and Y scores for a sample of $n = 4$ people. Note that person B has the lowest X score and the lowest Y score. Similarly, person D has the second-lowest score for both X and Y, person A has the third-lowest scores, and person C has the highest scores. These data show a perfectly consistent relationship: Each increase in X is accompanied by an increase in Y. However, the relationship is not linear, as can be seen in the graph of the data in Figure 16.14(a).

Now, we convert the raw scores to ranks: The lowest X is assigned a rank of 1, the next lowest X a rank of 2, and so on. This procedure is repeated for the Y scores. What happens when the X and Y scores are converted to ranks? Again, person B has the lowest X and Y scores, so this individual is ranked first on both variables. Similarly, person D is ranked second on both variables, person A is ranked third, and person C is ranked fourth (Table 16.2). When the ranks are plotted on a graph [see Figure 16.14(b)], the result is a perfect linear relationship.

The preceding example has demonstrated that a consistent relationship among scores produces a linear relationship when the scores are converted to ranks. Thus, if you want to measure the consistency of a relationship for a set of scores, you can simply convert the scores to ranks and then use the Spearman correlation to measure the correlation for the ranked data. The degree of relationship for the ranks (the Spearman correlation) provides a measure of the degree of consistency for the original scores.

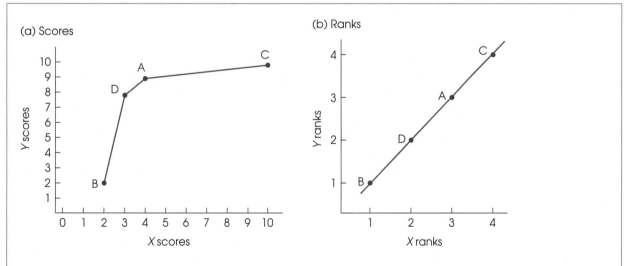

FIGURE 16.14

Scatterplots showing (a) the scores and (b) the ranks for the data in Example 16.8. Notice that there is a consistent, positive relationship between the X and Y scores, although it is not a linear relationship. Also notice that the scatterplot of the ranks shows a perfect linear relationship.

To summarize, the Spearman correlation measures the relationship between two variables when both are measured on ordinal scales (ranks). There are two general situations in which the Spearman correlation is used:

1. Spearman is used when the original data are ordinal—that is, when the X and Y values are ranks.

2. Spearman is used when a researcher wants to measure the consistency of a relationship between X and Y, independent of the specific form of the relationship. In this case, the original scores are first converted to ranks, then the Spearman correlation is used to measure the relationship for the ranks. Incidentally, when there is a consistently one-directional relationship between two variables, the relationship is said to be *monotonic*. Thus, the Spearman correlation can be used to measure the degree of monotonic relationship between two variables.

The word *monotonic* describes a sequence that is consistently increasing (or decreasing). Like the word *monotonous*, it means constant and unchanging.

CALCULATION OF THE SPEARMAN CORRELATION

The calculation of the Spearman correlation is remarkably simple, provided you know how to compute a Pearson correlation. First, be sure that you have ordinal data (ranks) for the X and Y scores. If necessary, you may have to place X scores and/or Y scores in rank order if they are not already. Ranking is accomplished as follows: The smallest X is assigned a rank of 1, the next smallest a rank of 2, and so on. Then the same is done for the Y scores. (Note that the X and Y scores are ranked separately.) Finally, using the same formula from the Pearson correlation, the Spearman correlation is computed *for the ranks* of X and Y.

That's all there is to it. When you use the Pearson correlation formula for ordinal data, the result is called a Spearman correlation. The Spearman correlation is identified by the symbol r_S to differentiate it from the Pearson correlation. The complete process of computing the Spearman correlation, including ranking scores, is demonstrated in Example 16.9.

EXAMPLE 16.9

The following data show a nearly perfect monotonic relationship between X and Y. When X increases, Y tends to decrease, and there is only one reversal in this general trend. To compute the Spearman correlation, we first rank the X and Y values, and we then compute the Pearson correlation for the ranks.

We have listed the X values in order so that the trend is easier to recognize.

ORIGINAL DATA		RANKS		
X	Y	X	Y	XY
3	12	1	5	5
4	5	2	3	6
5	6	3	4	12
10	4	4	2	8
13	3	5	1	5
				$36 = \Sigma XY$

FIGURE 16.15

Scatterplots showing (a) the scores and (b) the ranks for the data in Example 16.9.

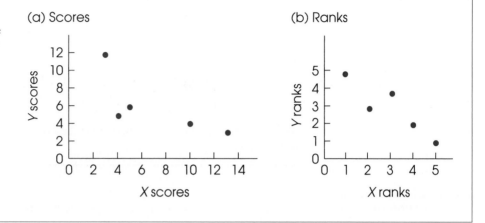

The scatterplots for the original data and the ranks are shown in Figure 16.15. To compute the correlation, we will need *SS* for *X*, *SS* for *Y*, and *SP*. Remember that all these values are computed with the ranks, not the original scores.

The *X* ranks are simply the integers 1, 2, 3, 4, and 5. These values have $\Sigma X = 15$ and $\Sigma X^2 = 55$. The *SS* for the *X* ranks is

$$SS_X = \Sigma X^2 - \frac{(\Sigma X)^2}{n}$$

$$= 55 - \frac{(15)^2}{5}$$

$$= 55 - 45$$

$$= 10$$

Note that the ranks for *Y* are identical to the ranks for *X*; that is, they are the integers 1, 2, 3, 4, and 5. Therefore, the *SS* for *Y* will be identical to the *SS* for *X*:

$$SS_Y = 10$$

To compute the *SP* value, we need ΣX, ΣY, and ΣXY for the ranks. The *XY* values are listed in the table with the ranks, and we already have found that both the *X*s and the *Y*s have a sum of 15. Using these values, we obtain

$$SP = \Sigma XY - \frac{(\Sigma X)(\Sigma Y)}{n}$$

$$= 36 - \frac{(15)(15)}{5}$$

$$= 36 - 45$$

$$= -9$$

The final Spearman correlation is

$$r_S = \frac{SP}{\sqrt{(SS_X)(SS_Y)}}$$

$$= \frac{-9}{\sqrt{10(10)}}$$

$$= -0.9$$

The Spearman correlation indicates that the data show a strong (nearly perfect) negative trend.

RANKING TIED SCORES

When you are converting scores into ranks for the Spearman correlation, you may encounter two (or more) identical scores. Whenever two scores have exactly the same value, their ranks should also be the same. This is accomplished by the following procedure:

1. List the scores in order from smallest to largest. Include tied values in the list.
2. Assign a rank (first, second, etc.) to each position in the ordered list.
3. When two (or more) scores are tied, compute the mean of their ranked positions, and assign this mean value as the final rank for each score.

The process of finding ranks for tied scores is demonstrated here. These scores have been listed in order from smallest to largest.

SCORES	RANK POSITION	FINAL RANK	
3	1	1.5	Mean of 1 and 2
3	2	1.5	
5	3	3	
6	4	5	
6	5	5	Mean of 4, 5, and 6
6	6	5	
12	7	7	

Note that this example has seven scores and uses all seven ranks. For $X = 12$, the largest score, the appropriate rank is 7. It cannot be given a rank of 6 because that rank has been used for the tied scores.

SPECIAL FORMULA FOR THE SPEARMAN CORRELATION

After the original X values and Y values have been ranked, the calculations necessary for SS and SP can be greatly simplified. First, you should note that the X ranks and the Y ranks are really just a set of integers: 1, 2, 3, 4, . . . , n. To compute the mean for these integers, you can locate the midpoint of the series by $M = (n + 1)/2$. Similarly, the SS for this series of integers can be computed by

$$SS = \frac{n(n^2 - 1)}{12} \qquad \text{(Try it out.)}$$

Also, because the X ranks and the Y ranks are the same values, the SS for X will be identical to the SS for Y.

Because calculations with ranks can be simplified and because the Spearman correlation uses ranked data, these simplifications can be incorporated into the final calculations for the Spearman correlation. Instead of using the Pearson formula after ranking the data, you can put the ranks directly into a simplified formula:

Caution: In this formula, you compute the value of the fraction and then subtract from 1. The first 1 is not part of the fraction.

$$r_S = 1 - \frac{6\Sigma D^2}{n(n^2 - 1)} \tag{16.5}$$

where D is the difference between the X rank and the Y rank for each individual. This special formula will produce the same result that would be obtained from the Pearson formula. However, you should note that this special formula can be used only after the scores have been converted to ranks and only when there are no ties among the ranks. If there are relatively few tied ranks, the formula still may be used, but it loses accuracy as the number of ties increases. The application of this formula is demonstrated in the following example.

EXAMPLE 16.10   To demonstrate the special formula for the Spearman correlation, we will use the same data that were presented in Example 16.9. The ranks for these data are shown again here:

RANKS		DIFFERENCE	
X	Y	D	D^2
1	5	4	16
2	3	1	1
3	4	1	1
4	2	−2	4
5	1	−4	16
			$38 = \Sigma D^2$

Using the special formula for the Spearman correlation, we obtain

$$r_S = 1 - \frac{6\Sigma D^2}{n(n^2 - 1)}$$

$$= 1 - \frac{6(38)}{5(25 - 1)}$$

$$= 1 - \frac{228}{120}$$

$$= 1 - 1.90$$

$$= -0.90$$

Notice that this is exactly the same answer that we obtained in Example 16.9, using the Pearson formula on the ranks.

TESTING THE SIGNIFICANCE OF THE SPEARMAN CORRELATION   Testing a hypothesis for the Spearman correlation is similar to the procedure used for the Pearson r. The basic question is whether or not a correlation exists in the population. The sample correlation could be due to chance, or perhaps it reflects an actual relationship between the variables in the population. For the Pearson correlation, the

Greek letter rho (ρ) was used for the population correlation. For the Spearman, ρ_S is used for the population parameter. Note that this symbol is consistent with the sample statistic, r_S. The null hypothesis states that there is no correlation (no monotonic relationship) between the variables for the population, or in symbols:

$$H_0: \quad \rho_S = 0 \quad \text{(The population correlation is zero.)}$$

The alternative hypothesis predicts that a nonzero correlation exists in the population, which can be stated in symbols as

$$H_1: \quad \rho_S \neq 0 \quad \text{(There is a real correlation.)}$$

To determine whether the Spearman correlation is statistically significant (that is, H_0 should be rejected), consult Table B.7. This table is similar to the one used to determine the significance of Pearson's r (Table B.6); however, the first column is sample size (n) rather than degrees of freedom.

EXAMPLE 16.11   An industrial psychologist selects a sample of $n = 15$ employees. These employees are ranked in order of work productivity by their manager. They also are ranked by a peer. The Spearman correlation computed for these data revealed a correlation of $r_S = .60$. Using Table B.7 with $n = 15$ and $\alpha = .05$, a correlation of $\pm .45$ is needed to reject H_0. The observed correlation for the sample easily surpasses this critical value. The correlation between manager and peer ratings is statistically significant.

LEARNING CHECK   1. Describe what is measured by a Spearman correlation, and explain how this correlation is different from the Pearson correlation.

2. Identify the two procedures that can be used to compute the Spearman correlation.

3. Compute the Spearman correlation for the following set of scores:

X	Y
2	7
12	38
9	6
10	19

ANSWERS   1. The Spearman correlation measures the consistency of the direction of the relationship between two variables. The Spearman correlation does not depend on the form of the relationship, whereas the Pearson correlation measures how well the data fit a linear form.

2. After the X and Y values have been ranked, you can compute the Spearman correlation by using either the special formula or the Pearson formula.

3. $r_S = 0.80$

16.6 OTHER MEASURES OF RELATIONSHIP

Although the Pearson correlation is the most commonly used method for evaluating the relationship between two variables, there are many other measures of relationship that exist. We already examined one such alternative, the Spearman correlation. Like the Spearman correlation, many of the alternative measures are simply special applications of the Pearson formula. In this section, we examine two more correlational methods. Notice that each alternative simply applies the Pearson formula to data with special characteristics.

THE POINT-BISERIAL CORRELATION

The point-biserial correlation is used to measure the relationship between two variables in situations where one variable is measured on an interval or ratio scale ("regular" scores), but the *second variable has only two different values*. Variables with only two values are called *dichotomous variables*. Some examples of dichotomous variables are

1. Male versus female
2. College graduate versus not a college graduate
3. First-born child versus later-born child
4. Success versus failure on a particular task
5. Over 30 years old versus under 30 years old

It is customary to use the numerical values 0 and 1, but any two different numbers would work equally well and would not affect the value of the correlation.

To compute the point-biserial correlation, the dichotomous variable is first converted to numerical values by assigning a value of zero (0) to one category and a value of one (1) to the other category. Then the regular Pearson correlation formula is used with the converted data. This process is demonstrated in the following example.

EXAMPLE 16.12

The following data represent measurements of attitude (X) and gender (Y) for each individual in a sample of $n = 8$. In the original data, one variable consists of two categories: male and female. In the converted data, the categories are assigned numerical values: male = 0 and female = 1. The point-biserial correlation is obtained by using the Pearson formula for the converted data.

Because the assignment of numerical values 0 and 1 to male and female is arbitrary, the sign (+ or −) of the point-biserial correlation is unimportant.

ORIGINAL DATA		CONVERTED DATA	
ATTITUDE SCORE X	GENDER Y	ATTITUDE SCORE X	GENDER Y
5	Male	5	0
0	Female	0	1
4	Male	4	0
1	Male	1	0
1	Female	1	1
2	Male	2	0
3	Female	3	1
0	Female	0	1

THE RELATIONSHIP BETWEEN THE POINT-BISERIAL CORRELATION AND THE INDEPENDENT-MEASURES *t* HYPOTHESIS TEST

You may have noticed that the data for a point-biserial correlation are identical to the data used for an independent-measures *t* hypothesis test (Chapter 10). The dichotomous variable can be used to separate the sample into two independent groups. For example, the data from Example 16.12 could be reorganized and presented as follows:

MALES	FEMALES
5	0
4	1
1	3
2	0

With the data in this form, the independent-measures *t* (or ANOVA) would be used to evaluate the mean difference between the two groups.

In many respects, the point-biserial correlation and the hypothesis tests (*t* or ANOVA) are evaluating the same thing. Both statistical techniques are examining the relationship between two variables—for this example, the relationship between sex and attitude. Specifically,

1. The correlation measures the *strength* of the relationship between the two variables. A large correlation (near 1.00) would indicate that there is a consistent, predictable relationship between gender and attitude. In particular, the value of r^2 measures how much of the variability in the attitude scores can be predicted on the basis of gender (see Box 16.3).

2. The *t* test evaluates the *significance* of the relationship. The hypothesis test determines whether the mean difference in attitude between males and females is greater than can reasonably be explained by chance alone.

As we noted in Chapter 10 (page 321), the outcome of the hypothesis test and the value of r^2 are often reported together. The *t* value measures statistical significance and r^2 measures the effect size. Also, as we noted in Chapter 10, the values for *t* and *r* are directly related. In fact, either one can be calculated from the other by the equations

$$r^2 = \frac{t^2}{t^2 + df} \quad \text{and} \quad t^2 = \frac{r^2}{(1 - r^2)/df}$$

where *df* is the degrees of freedom for the *t* statistic.

Note that r^2 is determined entirely by the size of the correlation, whereas *t* is influenced by the size of the correlation and the size of the sample. For example, a correlation of $r = 0.30$ produces $r^2 = 0.09$ (9%) no matter how large the sample may be. On the other hand, if the point-biserial correlation is computed for a sample of 10 people (for example, 5 males and 5 females), a correlation of $r = 0.30$ produces a nonsignificant value of $t = 0.791$. If the sample size is increased to 100 people (50 males and 50 females), the same correlation produces a significant *t* value of $t = 9.68$. Although *t* and *r* are related, they are measuring different things.

THE PHI-COEFFICIENT

When both variables (*X* and *Y*) measured for each individual are dichotomous, the correlation between the two variables is called the phi-coefficient. To compute phi (ϕ), you follow a two-step procedure:

1. Convert each of the dichotomous variables to numerical values by assigning a 0 to one category and a 1 to the other category for each of the variables.

BOX

16.3   **THE POINT-BISERIAL CORRELATION AND r^2**

In Chapters 10 and 11 we introduced r^2 as a measure of effect size. Specifically, r^2 measures the percentage of variance that is explained by the difference between two groups. The data in Example 16.12 provide an excellent opportunity to demonstrate this concept.

First, when the data are organized as follows, you can use the Pearson correlation formula to calculate the point-biserial correlation. For these data, the point-biserial correlation is $r = 0.5773$. (Check it for yourself.)

MALES	FEMALES
5	0
4	1
1	3
2	0
$M = 3$	$M = 1$
$SS = 10$	$SS = 6$

Notice that when the scores are separated by gender, we obtain $SS = 10$ for the males and $SS = 6$ for the females. Thus, when the groups are separated, the total variability is $SS = 16$ ($10 + 6$). Also notice that before the groups were separated, the SS for the attitude scores was $SS = 24$ (see the calculations for the point-biserial correlation). Finally, notice that grouping the scores by gender has reduced the variability by 8 points: from $SS = 24$ before the groups were separated to $SS = 16$ after separation. The 8 points is exactly one-third (33.33%) of the original variability, $SS = 24$, for the attitude scores.

ATTITUDE X	GENDER Y		
5	0	$SS_X = 24$	
0	1	$SS_Y = 2$	$r = \dfrac{4}{\sqrt{(24)(2)}} = 0.5773$
4	0	$SP = 4$	
1	0		
1	1		
2	0		
3	1		
0	1		

Once again, the value of r^2 (in this case, $r^2 = 33.33\%$) measures how much of the variability in the scores is predicted or explained by the difference between the two groups.

A correlation of $r = 0.5773$ produces $r^2 = 0.3333$, indicating that 33.33% (or one-third) of the variability in the attitude scores is predicted by differences in gender. To demonstrate this percentage, we will separate the scores into two groups, males and females.

---

2. Use the regular Pearson formula with the converted scores.

This process is demonstrated in the following example.

EXAMPLE 16.13   A researcher is interested in examining the relationship between birth-order position and personality. A random sample of $n = 8$ individuals is obtained, and each individual is classified in terms of birth-order position as first-born or only child versus later-born. Then each individual's personality is classified as either introvert or extrovert.

The original measurements are then converted to numerical values by the following assignments:

BIRTH ORDER	PERSONALITY
1st or only child $= 0$	Introvert $= 0$
Later-born child $= 1$	Extrovert $= 1$

The original data and the converted scores are as follows:

ORIGINAL DATA		CONVERTED SCORES	
BIRTH ORDER X	PERSONALITY Y	BIRTH ORDER X	PERSONALITY Y
1st	Introvert	0	0
3rd	Extrovert	1	1
Only	Extrovert	0	1
2nd	Extrovert	1	1
4th	Extrovert	1	1
2nd	Introvert	1	0
Only	Introvert	0	0
3rd	Extrovert	1	1

The Pearson correlation formula would then be used with the converted data to compute the phi-coefficient.

Because the assignment of numerical values is arbitrary (either category could be designated 0 or 1), the sign of the resulting correlation is meaningless. As with most correlations, the *strength* of the relationship is best described by the value of r^2, the coefficient of determination, which measures how much of the variability in one variable is predicted or determined by the association with the second variable.

We also should note that although the phi-coefficient can be used to assess the relationship between two dichotomous variables, the more common statistical procedure is a chi-square statistic, which is examined in Chapter 17.

LEARNING CHECK

1. Define a *dichotomous* variable.

2. The following data represent job-related stress scores for a sample of $n = 8$ individuals. These people also are classified by salary level.

   a. Convert the data into a form suitable for the point-biserial correlation.

   b. Compute the point-biserial correlation for these data.

SALARY OVER $40,000	SALARY UNDER $40,000
8	4
7	2
5	1
3	3

ANSWERS

1. A dichotomous variable has only two possible values.

2. Salary level is a dichotomous variable and can be coded as 0 = over $40,000, 1 = under $40,000 for each participant. The resulting point-biserial correlation is −0.719.

16.7 INTRODUCTION TO REGRESSION

Earlier in this chapter, we introduced the Pearson correlation as a technique for describing and measuring the linear relationship between two variables. Figure 16.16 presents hypothetical data showing the relationship between SAT scores and college grade point average (GPA). Note that the figure shows a good, but not perfect, positive relationship. Also, note that we have drawn a line through the middle of the data points. This line serves several purposes:

a. The line makes the relationship between SAT and GPA easier to see.

b. The line identifies the center, or "central tendency," of the relationship, just as the mean describes central tendency for a set of scores. Thus, the line provides a simplified description of the relationship. For example, if the data points were removed, the straight line would still give a general picture of the relationship between SAT and GPA.

c. Finally, the line can be used for prediction. The line establishes a precise relationship between each X value (SAT score) and a corresponding Y value (GPA). For example, an SAT score of 620 corresponds to a GPA of 3.25 (see Figure 16.16). Thus, the college admissions office could use the straight-line relationship to predict that a student entering college with an SAT score of 620 should achieve a college GPA of approximately 3.25.

Our goal in this section is to develop a procedure that identifies and defines the straight line that provides the best fit for any specific set of data. You should realize that this straight line does not have to be drawn on a graph; it can be presented in a simple equation. Thus, our goal is to find the equation for the line that best describes the relationship for a set of X and Y data.

FIGURE 16.16

Hypothetical data showing the relationship between SAT scores and GPA with a regression line drawn through the data points. The regression line defines a precise, one-to-one relationship between each X value (SAT score) and its corresponding Y value (GPA).

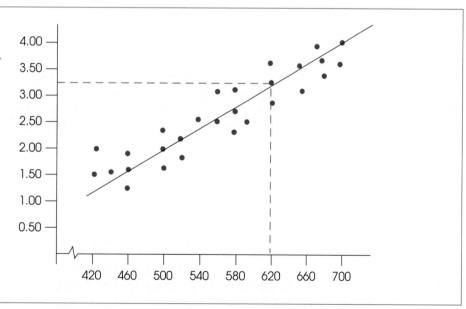

LINEAR EQUATIONS   In general, a *linear relationship* between two variables X and Y can be expressed by the equation

$$Y = bX + a \qquad\qquad (16.6)$$

where a and b are fixed constants.

For example, a local tennis club charges a fee of $5 per hour plus an annual membership fee of $25. With this information, the total cost of playing tennis can be computed using a *linear equation* that describes the relationship between the total cost (Y) and the number of hours (X).

$$Y = 5X + 25$$

Note that a positive slope means that Y increases when X increases, and a negative slope indicates that Y decreases when X increases.

In the general linear equation, the value of b is called the *slope*. The slope determines how much the Y variable will change when X is increased by one point. For the tennis club example, the slope is $b = \$5$ and indicates that your total cost will increase by $5 for each hour you play. The value of a in the general equation is called the Y-intercept because it determines the value of Y when $X = 0$. (On a graph, the a value identifies the point where the line intercepts the Y-axis.) For the tennis club example, $a = \$25$; there is a $25 charge even if you never play tennis.

Figure 16.17 shows the general relationship between cost and number of hours for the tennis club example. Notice that the relationship results in a straight line. To obtain this graph, we picked any two values of X and then used the equation to compute the corresponding values for Y. For example,

when $X = 10$:	when $X = 30$:
$Y = bX + a$	$Y = bX + a$
$= \$5(10) + \$25$	$= \$5(30) + \$25$
$= \$50 + \$25$	$= \$150 + \$25$
$= \$75$	$= \$175$

When drawing a graph of a linear equation, it is wise to compute and plot at least three points to be certain you have not made a mistake.

Next, these two points are plotted on the graph: one point at $X = 10$ and $Y = 75$, the other point at $X = 30$ and $Y = 175$. Because two points completely determine a straight line, we simply drew the line so that it passed through these two points.

Because a straight line can be extremely useful for describing a relationship between two variables, a statistical technique has been developed that provides a standardized method for determining the best-fitting straight line for any set of data. The statistical procedure is regression, and the resulting straight line is called the regression line.

D E F I N I T I O N   The statistical technique for finding the best-fitting straight line for a set of data is called **regression**, and the resulting straight line is called the **regression line**.

The goal for regression is to find the best-fitting straight line for a set of data. To accomplish this goal, however, it is first necessary to define precisely what is meant by "best fit." For any particular set of data, it is possible to draw lots of different straight lines that all appear to pass through the center of the data points. Each of these lines can be defined by a linear equation of the form $Y = bX + a$, where b and a are constants that determine the slope and Y-intercept of the line, respectively. Each individual line has its own unique values for b and a. The problem is to find the specific line that provides the best fit to the actual data points.

FIGURE 16.17

Relationship between total cost and number of hours playing tennis. The tennis club charges a $25 membership fee plus $5 per hour. The relationship is described by a linear equation:

total cost =
  $5 (number of hours) + $25

$Y = bX + a$

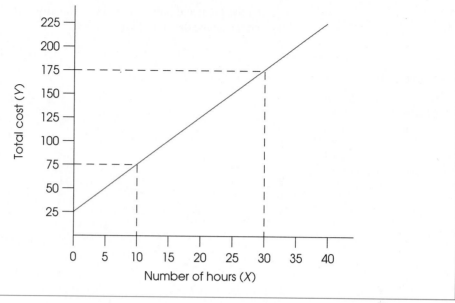

---

1. Identify the slope and Y-intercept for the following linear equation:

   $$Y = -3X + 7$$

2. Use the linear equation $Y = 2X - 7$ to determine the value of Y for each of the following values of X: 1, 3, 5, 10.

3. If the slope constant (*b*) in a linear equation is positive, then a graph of the equation will be a line tilted from lower left to upper right. (True or false?)

ANSWERS   1. Slope = −3 and Y-intercept = +7.

2.

X	Y
1	−5
3	−1
5	3
10	13

3. True. A positive slope indicates that Y increases (goes up in the graph) when X increases (goes to the right in the graph).

THE LEAST-SQUARES
SOLUTION

To determine how well a line fits the data points, the first step is to define mathematically the distance between the line and each data point. For every X value in the data, the linear equation will determine a Y value on the line. This value is the predicted Y and is called $\hat{Y}$ ("Y hat"). The distance between this predicted value and the actual Y value in the data is determined by

$$\text{distance} = Y - \hat{Y}$$

Notice that we simply are measuring the vertical distance between the actual data point (Y) and the predicted point on the line. This distance measures the error between the line and the actual data (see Figure 16.18).

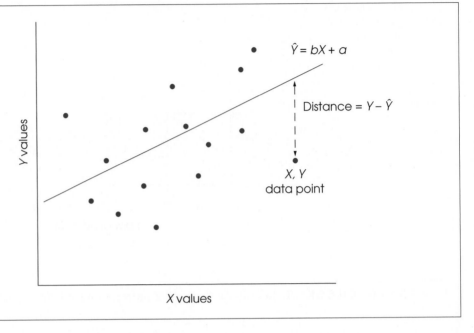

FIGURE 16.18

The distance between the actual data point (Y) and the predicted point on the line ($\hat{Y}$) is defined as $Y - \hat{Y}$. The goal of regression is to find the equation for the line that minimizes these distances.

Because some of these distances will be positive and some will be negative, the next step is to square each distance in order to obtain a uniformly positive measure of error. Finally, to determine the total error between the line and the data, we sum the squared errors for all of the data points. The result is a measure of overall squared error between the line and the data:

$$\text{total squared error} = \Sigma(Y - \hat{Y})^2$$

Now we can define the *best-fitting* line as the one that has the smallest total squared error. For obvious reasons, the resulting line is commonly called the *least-squared-error* solution.

In symbols, we are looking for a linear equation of the form

$$\hat{Y} = bX + a$$

For each value of X in the data, this equation will determine the point on the line ($\hat{Y}$) *that gives the best prediction of Y.* The problem is to find the specific values for a and b that will make this the best-fitting line.

The calculations that are needed to find this equation require calculus and some sophisticated algebra, so we will not present the details of the solution. The results, however, are relatively straightforward, and the solutions for b and a are as follows:

A commonly used alternative formula for the slope is

$$b = r\frac{s_Y}{s_X}$$

where s_X and s_Y are the standard deviations for X and Y, respectively.

$$b = \frac{SP}{SS_X} \tag{16.7}$$

where SP is the sum of products and SS_X is the sum of squares for the X scores.

$$a = M_Y - bM_X \tag{16.8}$$

Note that these two formulas determine the linear equation that provides the best prediction of Y values. This equation is called the regression equation for Y.

DEFINITION

The *regression equation for Y* is the linear equation

$$\hat{Y} = bX + a$$

where the constants b and a are determined by Equations 16.7 and 16.8, respectively. This equation results in the least-squared error between the data points and the line.

You should notice that the values of SS and SP are needed in the formulas for b and a just as they are needed to compute the Pearson correlation. An example demonstrating the calculation and use of this best-fitting line is presented now.

EXAMPLE 16.14

The following table presents X and Y scores for a sample of $n = 5$ individuals. These data will be used to demonstrate the procedure for determining the linear regression equation for predicting Y values.

X	Y	$X - M_X$	$Y - M_Y$	$(X - M_X)(Y - M_Y)$	$(X - M_X)^2$
7	11	2	5	10	4
4	3	-1	-3	3	1
6	5	1	-1	-1	1
3	4	-2	-2	4	4
5	7	0	1	0	0
				$16 = SP$	$10 = SS_X$

For these data, $\Sigma X = 25$, so $M_X = 5$. Also, $\Sigma Y = 30$, so $M_Y = 6$. These means have been used to compute the deviation scores for each X and Y value. The final two columns show the products of the deviation scores and the squared deviations for X. Based on these values,

$$SP = \Sigma(X - M_X)(Y - M_Y) = 16$$
$$SS_X = \Sigma(X - M_X)^2 = 10$$

Our goal is to find the values for b and a in the linear equation so that we obtain the best-fitting straight line for these data.

By using Equations 16.7 and 16.8, the solutions for b and a are

$$b = \frac{SP}{SS_X} = \frac{16}{10} = 1.6$$

$$a = M_Y - bM_X$$
$$= 6 - 1.6(5)$$
$$= 6 - 8$$
$$= -2$$

The resulting regression equation is

$$\hat{Y} = 1.6X - 2$$

The original data and the regression line are shown in Figure 16.19.

FIGURE 16.19

The scatterplot for the data in Example 16.14 is shown with the best-fitting straight line. The predicted Y values ($\hat{Y}$) are on the regression line. Unless the correlation is perfect ($+1.00$ or -1.00), there will be some error between the actual Y values and the predicted Y values. The larger the correlation is, the less the error will be.

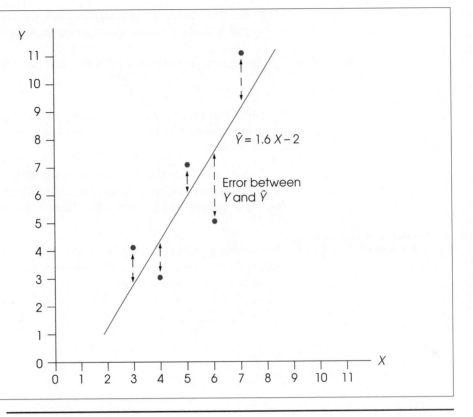

As we noted at the beginning of this section, one common use of regression equations is for prediction. For any given value of X, we can use the equation to compute a predicted value for Y. For the equation from Example 16.13, an individual with an X score of $X = 5$ would be predicted to have a Y score of

$$\hat{Y} = 1.6X - 2$$
$$= 1.6(5) - 2$$
$$= 8 - 2$$
$$= 6$$

Although regression equations can be used for prediction, there are two cautions that should be considered whenever you are interpreting the predicted values:

1. The predicted value is not perfect (unless $r = +1.00$ or -1.00). If you examine Figure 16.19, it should be clear that the data points do not fit perfectly on the line. In general, there will be some error between the predicted Y values (on the line) and the actual data. Although the amount of error will vary from point to point, on the average the errors will be directly related to the magnitude of the correlation. With a correlation near 1.00 (or -1.00), the data points will generally be close to the line (small error), but as the correlation gets nearer to zero, the magnitude of the error will increase. The statistical procedure for measuring this error is described in the following section.

2. The regression equation should not be used to make predictions for X values that fall outside the range of values covered by the original data. For Example 16.14, the X values ranged from $X = 3$ to $X = 7$, and the regression equation was calculated as the best-fitting line within this range. Because you have no information about the X-Y relationship outside this range, the equation should not be used to predict Y for any X value lower than 3 or greater than 7.

LEARNING CHECK

1. Sketch a scatterplot for the following data—that is, a graph showing the X, Y data points:

X	Y
1	4
3	9
5	8

   a. Find the regression equation for predicting Y and X. Draw this line on your graph. Does it look like the best-fitting line?

   b. Use the regression equation to find the predicted Y value corresponding to each X in the data.

ANSWERS

1. a. $SS_X = 8$, $SP = 8$, $b = 1$, $a = 4$.
      The equation is

   $$\hat{Y} = X + 4$$

   b. The predicted Y values are 5, 7, and 9.

THE STANDARD ERROR OF ESTIMATE

It is possible to determine a best-fitting regression equation for any set of data by simply using the formulas already presented. The linear equation you obtain is then used to generate predicted Y values for any known value of X. However, it should be clear that the accuracy of this prediction depends on how well the points on the line correspond to the actual data points—that is, the amount of error between the predicted values, $\hat{Y}$, and the actual scores, Y values. Figure 16.20 shows two different sets of data that have exactly the same regression equation. In one case, there is a perfect correlation ($r = +1$) between X and Y, so the linear equation fits the data perfectly. For the second set of data, the predicted Y values on the line only approximate the real data points.

A regression equation, by itself, allows you to make predictions, but it does not provide any information about the accuracy of the predictions. To measure the precision of the regression, it is customary to compute a standard error of estimate.

DEFINITION

The **standard error of estimate** gives a measure of the standard distance between a regression line and the actual data points.

Conceptually, the standard error of estimate is very much like a standard deviation: Both provide a measure of standard distance. Also note that the calculation of the standard error of estimate is very similar to the calculation of standard deviation.

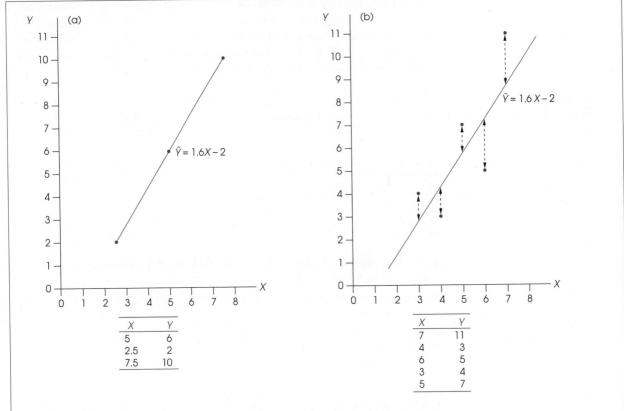

FIGURE 16.20

(a) Scatterplot showing data points that perfectly fit the regression equation $\hat{Y} = 1.6X - 2$. Note that the correlation is $r = 1.00$. (b) Scatterplot for the data from Example 16.14. Notice that there is error between the actual data points and the predicted Y values of the regression line.

To calculate the standard error of estimate, we first will find a sum of squared deviations (SS). Each deviation will measure the distance between the actual Y value (from the data) and the predicted Y value (from the regression line). This sum of squares is commonly called SS_{error} because it measures the sum of squared distances, or errors, between the actual data and the predicted values:

$$SS_{\text{error}} = \Sigma(Y - \hat{Y})^2 \qquad\qquad (16.9)$$

The obtained SS value is then divided by its degrees of freedom to obtain a measure of variance. This procedure should be very familiar:

$$\text{Variance} = \frac{SS}{df}$$

The degrees of freedom for the standard error of estimate are $df = n - 2$. The reason for having $n - 2$ degrees of freedom, rather than the customary $n - 1$, is that we now are measuring deviations from a line rather than deviations from a mean. Recall that it is necessary to know SP to find the slope of the regression line (the value of b in the equation). To calculate SP, you must know the means for both the X and the Y scores. Specifying these two means places two restrictions on the variability of the data, with the result that the scores have only $n - 2$ degrees of freedom. (A more intuitive explanation for the fact that the SS_{error} has $df = n - 2$ comes from the simple observation that it takes exactly two points to determine a straight line. If there are only two data points, they always fit perfectly on a straight line, so there will be no error. It is only when you have more than two points that there is some freedom in determining the best-fitting line.)

The final step in the calculation of the standard error of estimate is to take the square root of the variance in order to obtain a measure of standard distance. The final equation is

$$\text{standard error of estimate} = \sqrt{\frac{SS_{error}}{df}} = \sqrt{\frac{\Sigma(Y - \hat{Y})^2}{n - 2}} \tag{16.10}$$

The following example demonstrates the calculation of this standard error.

EXAMPLE 16.15   The data in Example 16.14 will be used to demonstrate the calculation of the standard error of estimate. These data have the regression equation

$$\hat{Y} = 1.6X - 2$$

Using this regression equation, we have computed the predicted Y value, the error, and the squared error for each individual in the data.

DATA		PREDICTED Y VALUES $\hat{Y} = 1.6X - 2$	ERROR $Y - \hat{Y}$	SQUARED ERROR $(Y - \hat{Y})^2$
X	Y			
7	11	9.2	1.8	3.24
4	3	4.4	−1.4	1.96
6	5	7.6	−2.6	6.76
3	4	2.8	1.2	1.44
5	7	6.0	1.0	1.00
				$14.40 = SS_{error}$

For these data, the sum of the squared errors is $SS_{error} = 14.40$. With $n = 5$, the data have $df = n - 2 = 3$, so the standard error of estimate is

$$\text{standard error of estimate} = \sqrt{\frac{SS_{error}}{df}} = \sqrt{\frac{14.40}{3}} = 2.19$$

Remember: The standard error of estimate provides a measure of how accurately the regression equation predicts the Y values. In this case, the standard distance between the actual data points and the regression line is measured by standard error of estimate = 2.19.

RELATIONSHIP BETWEEN THE STANDARD ERROR AND THE CORRELATION

It should be clear from Example 16.15 that the standard error of estimate is directly related to the magnitude of the correlation between X and Y. If the correlation is near 1.00 (or -1.00), the data points will be clustered close to the line, and the standard error of estimate will be small. As the correlation gets nearer to zero, the line will provide less accurate predictions, and the standard error of estimate will grow larger. Earlier (page 535) we observed that squaring the correlation provides a measure of the accuracy of prediction: r^2 is called the coefficient of determination because it determines what proportion of the variability in Y is predicted by the relationship with X. Figures 16.21 and 16.22 should help make this concept more concrete.

FIGURE 16.21

A set of 9 data points (X and Y values) with a correlation of 0.80. The colored lines in part (a) show deviations from the mean for Y. For these data, $SS_Y = 240$. In part (b) the colored lines show deviations from the regression line. For these data, $SS_{error} = 86.4$. The regression line reduces SS value by $r^2 = 0.64$ or 64%.

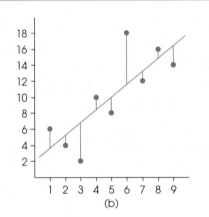

(a)   (b)

In Figure 16.21 we show a scatterplot for a set of data with a correlation of $r = 0.80$. In part (a) we have drawn a horizontal line showing the mean for the Y scores and a colored vertical line showing the deviation from the mean for each value of Y. For these data, the sum of the squared deviations is $SS_Y = 240$. Part (b) shows the same data, but now we have drawn in the regression line and a set of vertical lines showing the deviation from the regression line for each value of Y. The vertical lines in part (b) represent the error that still remains after the correlation has been taken into account. For these data $SS_{error} = 86.4$. In this example, the deviations (colored lines) in part (b) are noticeably smaller than the deviations from the mean in part (a). The size of the correlation determines exactly how much the deviations are reduced. For these data $r = 0.80$, so $r^2 = 0.64$ or 64% of the variability in the Y scores is explained by the relationship with X. The remaining 36% is not explained by the correlation and remains as error. In this example, we started with $SS_Y = 240$ in part (a) and by using the correlations we reduced the variability by 64% to obtain $SS_{error} = 86.4$ in part (b). Note that $SS_{error} = 86.4$ is exactly 36% of the original variability, $SS_Y = 240$.

Now consider the data shown in Figure 16.22. This time we show a scatterplot for a set of data with a correlation of only $r = 0.30$. Again, part (a) shows deviations from the mean for each value of Y. For these data, $SS_Y = 240$. Part (b) shows deviations from the regression line. For these data, $SS_{error} = 218.4$. In this case, there is *not* a noticeable reduction in variability. With a correlation of $r = 0.30$, the variability is reduced by only $r^2 = 0.09$ or 9%. As a result, the variability that remains after accounting for the correlation is still 91% of the original variability. In this case, $SS_{error} = 218.4$ is exactly 91% of the original $SS_Y = 240$.

FIGURE 16.22

A set of 9 data points (X and Y values) with a correlation of 0.30. The colored lines in part (a) show deviations from the mean for Y. For these data, $SS_Y = 240$. In part (b) the colored lines show deviations from the regression line. For these data, $SS_{error} = 218.4$. The regression line reduces the SS value by $r^2 = 0.09$ or only 9%.

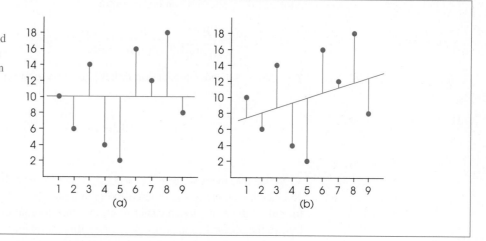

(a)   (b)

Because r^2 measures the predicted portion of the variability in Y, we can use the expression $1 - r^2$ to measure the unpredicted, or error, portion. With a correlation of $r = 0.80$, for example, the predicted portion of the Y variability is

$$r^2 = 0.64 \quad \text{(or 64\%)}$$

The remaining, unpredicted portion is

$$1 - r^2 = 1 - 0.64 = 0.36 \quad \text{(or 36\%)}$$

The unpredicted variability is exactly what we have defined as error variability in the standard error of estimate. Because the total variability in Y is measured by SS_Y, the error portion can be calculated as

$$SS_{error} = (1 - r^2)SS_Y \tag{16.11}$$

Notice that when $r = 1.00$, there is no error (perfect prediction), and that as the correlation approaches zero, the error will grow larger. By using this formula for SS_{error}, the standard error of estimate can be computed as

$$\text{standard error of estimate} = \sqrt{\frac{SS_{error}}{df}} = \sqrt{\frac{(1 - r^2)SS_Y}{n - 2}}$$

The following example demonstrates this new formula.

EXAMPLE 16.16   The same data used in Examples 16.14 and 16.15 are reproduced in the following table:

X	Y	$X - M_X$	$Y - M_Y$	$(X - M_X)^2$	$(Y - M_Y)^2$	$(X - M_X)(Y - M_Y)$
7	11	2	5	4	25	10
4	3	−1	−3	1	9	3
6	5	1	−1	1	1	−1
3	4	−2	−2	4	4	4
5	7	0	1	0	1	0
				$SS_X = 10$	$SS_Y = 40$	$SP = 16$

For these data, the Pearson correlation is

$$r = \frac{SP}{\sqrt{SS_X SS_Y}} = \frac{16}{\sqrt{10(40)}} = \frac{16}{20} = 0.80$$

With $SS_Y = 40$ and a correlation of 0.80, the error portion of the Y variability is

$$SS_{error} = (1 - r^2)SS_Y = (1 - 0.64)(40)$$
$$= 0.36(40)$$
$$= 14.40$$

Notice that the new formula for SS_{error} produces exactly the same value that we obtained by summing the squared errors in Example 16.15. Also note that this new formula is generally much easier to use because it requires only the correlation value (r) and the SS for Y. The primary point of this example, however, is that the standard error of estimate is closely related to the value of the correlation. With a large correlation (near $+1.00$ or -1.00), the data points will be close to the regression line, and the standard error of estimate will be small. As a correlation gets smaller (near zero), the data points move away from the regression line, and the standard error of estimate gets larger.

Because it is possible to have the same regression line for sets of data that have different correlations, it is important to examine r^2 and the standard error of estimate. The regression equation simply describes the best-fitting line and is used for making predictions. However, r^2 and the standard error of estimate indicate how accurate these predictions will be.

LEARNING CHECK

1. Use the following set of data:

X	Y	
1	4	$SS_X = 10$
2	1	$SS_Y = 90$
3	7	$SP = 24$
4	13	
5	10	

   a. Find the regression equation for predicting Y from X.
   b. Use the regression equation to find the predicted Y value for each X in the data.
   c. Find the error $(Y - \hat{Y})$ for each data point. Square each error value, and sum the results to find SS_{error}.
   d. Calculate the Pearson correlation for these data.
   e. Use the correlation and SS_Y to compute SS_{error}.

2. Assuming that all other factors are held constant, what happens to the standard error of estimate as the correlation between X and Y moves toward zero?

ANSWERS   **1. a.** $\hat{Y} = 2.4X - 0.2$

b. c.

Y	$\hat{Y}$	ERROR	ERROR^2
4	2.2	1.8	3.24
1	4.6	−3.6	12.96
7	7.0	0	0
13	9.4	3.6	12.96
10	11.8	−1.8	3.24
			32.40 = SS_{error}

d. $r = 0.80$

e. $(1 - r^2)SS_Y = 0.36(90) = 32.40 = SS_{error}$

2. The standard error of estimate would get larger.

SUMMARY

1. A correlation measures the relationship between two variables, X and Y. The relationship is described by three characteristics:

a. *Direction*. A relationship can be either positive or negative. A positive relationship means that X and Y vary in the same direction. A negative relationship means that X and Y vary in opposite directions. The sign of the correlation ($+$ or $-$) specifies the direction.

b. *Form*. The most common form for a relationship is a straight line. However, special correlations exist for measuring other forms. The form is specified by the type of correlation used. For example, the Pearson correlation measures linear form.

c. *Degree*. The magnitude of the correlation measures the degree to which the data points fit the specified form. A correlation of 1.00 indicates a perfect fit, and a correlation of 0 indicates no degree of fit.

2. The most commonly used correlation is the Pearson correlation, which measures the degree of linear relationship. The Pearson correlation is identified by the letter r and is computed by

$$r = \frac{SP}{\sqrt{SS_X SS_Y}}$$

In this formula, SP is the sum of products of deviations and can be calculated with either a definitional formula or a computational formula:

definitional formula: $SP = \Sigma(X - M_X)(Y - M_Y)$

computational formula: $SP = \Sigma XY - \dfrac{\Sigma X \Sigma Y}{n}$

3. The Pearson correlation and z-scores are closely related because both are concerned with the location of individuals within a distribution. When X and Y scores are transformed into z-scores, the Pearson correlation can be computed by

$$r = \frac{\Sigma z_X z_Y}{n}$$

4. A correlation between two variables should not be interpreted as implying a causal relationship. Simply because X and Y are related does not mean that X causes Y or that Y causes X.

5. When the X or Y values used to compute a correlation are limited to a relatively small portion of the potential range, you should exercise caution in generalizing the value of the correlation. Specifically, a limited range of values can either obscure a strong relationship or exaggerate a poor relationship.

6. To evaluate the strength of a relationship, you should square the value of the correlation. The resulting value, r^2, is called the *coefficient of determination* because it measures the portion of the variability in one variable that can be predicted using the relationship with the second variable.

7. The Spearman correlation (r_S) measures the consistency of direction in the relationship between X and Y—that is, the degree to which the relationship is one-directional, or monotonic. The Spearman correlation is computed by a two-stage process:

a. Rank the X scores and the Y scores separately.

b. Compute the Pearson correlation using the ranks.

Note: After the X and Y values are ranked, you may use a special formula to determine the Spearman correlation:

$$r_S = 1 - \frac{6\Sigma D^2}{n(n^2 - 1)}$$

where D is the difference between the X rank and the Y rank for each individual. The formula is accurate only when there are no tied scores in the data.

8. Special correlations exist to measure relationships between variables with specific characteristics. The Spearman correlation is used when both variables are measured on ordinal scales. The point-biserial correlation is used when one of the two variables is dichotomous, and the phi-coefficient is used when both variables are dichotomous.

9. When there is a general linear relationship between two variables, X and Y, it is possible to construct a linear equation that allows you to predict the Y value corresponding to any known value of X:

$$\text{predicted } Y \text{ value} = \hat{Y} = bX + a$$

The technique for determining this equation is called regression. By using a *least-squares* method to mini-mize the error between the predicted Y values and the actual Y values, the best-fitting line is achieved when the linear equation has

$$b = \frac{SP}{SS_X} \quad \text{and} \quad a = M_Y - bM_X$$

10. The linear equation generated by regression (called the *regression equation*) can be used to compute a predicted Y value for any value of X. The accuracy of the prediction is measured by the standard error of estimate, which provides a measure of the standard distance (or error) between the predicted Y values on the line and the actual data points. The standard error of estimate is computed by

$$\text{standard error of estimate} = \sqrt{\frac{SS_{\text{error}}}{n - 2}}$$

where SS_{error} may be computed directly from the error scores,

$$SS_{\text{error}} = \Sigma(Y - \hat{Y})^2$$

or as a proportion of the original Y variability,

$$SS_{\text{error}} = (1 - r^2)SS_Y$$

RESOURCES

If you are using WebTutor along with this book, there is a WebTutor section corresponding to this chapter. The WebTutor contains a brief summary of Chapter 16, hints for learning the concepts and formulas for correlation and regression, cautions about common errors, and sample exam items including solutions.

SPSS

The statistics computer package SPSS can be used to compute correlations and find regression equations. First, the data are entered into two columns in the data matrix with the X values in one column and the Y values in a second column. For a correlation, click **Analyze** on the tool bar, select **Correlate**, and click on **Bivariate**. Next you must enter the labels for the two data columns into the Variables box. The Pearson box should be checked, but you can click the Spearman box if you want to compute a Spearman correlation (SPSS will convert the scores to ranks). Then click **OK**. SPSS will produce a correlation matrix showing all the possible correlations. You want the correlation of X and Y, which is contained in the upper right corner (or the lower left). The output includes the significance level (p value) for the correlation hypothesis test.

   To find a regression equation, enter the X and Y values into two columns, click **Analyze** on the tool bar, select **Regression**, and click on **Linear**. Enter the column label for Y into the Dependent box and the column label for X in the Independent(s) box. Then click **OK**. In the output, the *Model Summary* box includes the value for the correlation and r^2, as well as the standard error of estimate for the regression equation. The *Coefficients* box includes the Y-intercept and the slope for the equation in column B.

KEY TERMS

correlation	sum of products (*SP*)	phi-coefficient	least-squares solution
positive correlation	restricted range	linear relationship	regression equation for Y
negative correlation	coefficient of determination	linear equation	regression
perfect correlation	Spearman correlation	slope	regression line
Pearson correlation	point-biserial correlation	Y-intercept	standard error of estimate

FOCUS ON PROBLEM SOLVING

1 A correlation always has a value from $+1.00$ to -1.00. If you obtain a correlation outside this range, then you have made a computational error.

2. When interpreting a correlation, do not confuse the sign ($+$ or $-$) with its numerical value. The sign and numerical value must be considered separately. Remember that the sign indicates the direction of the relationship between X and Y. On the other hand, the numerical value reflects the strength of the relationship or how well the points approximate a linear (straight-line) relationship. Therefore, a correlation of -0.90 is just as strong as a correlation of $+0.90$. The signs tell us that the first correlation is an inverse relationship.

3. Before you begin to calculate a correlation, you should sketch a scatterplot of the data and make an estimate of the correlation. (Is it positive or negative? Is it near 1 or near 0?) After computing the correlation, compare your final answer with your original estimate.

4. The definitional formula for the sum of products (*SP*) should be used only when you have a small set (n) of scores and the means for X and Y are both whole numbers. Otherwise, the computational formula will produce quicker, easier, and more accurate results.

5. For computing a correlation, n is the number of individuals (and, therefore, the number of *pairs* of X and Y values).

6. When using the special formula for the Spearman correlation, remember that the fraction is computed separately and then subtracted from 1. Students often include the 1 as a part of the numerator, or they get so absorbed in computing the fractional part of the equation that they forget to subtract it from 1. Be careful using this formula.

7. When computing a Spearman correlation, be sure that both X and Y values have been ranked. Sometimes the data will consist of one variable already ranked with the other variable on an interval or a ratio scale. If one variable is ranked, do not forget to rank the other. When interpreting a Spearman correlation, remember that it measures how monotonic (consistent) the relationship is between X and Y.

8. To draw a graph from a linear equation, choose any three values for X, put each value in the equation, and calculate the corresponding values for Y. Then plot the three X, Y points on the graph. It is a good idea to use $X = 0$ for one of the three values because this will give you the Y-intercept. You can get a quick idea of what the graph should look like if you know the Y-intercept and the slope. Remember: The Y-intercept is the point where the line crosses the Y-axis, and the slope identifies the tilt of the line. For example, suppose that the Y-intercept is 5 and the slope is -3. The line would pass through the point 0, 5, and its slope indicates that the Y value goes down 3 points each time X increases by 1.

9. Rather than memorizing the formula for the Y-intercept in the regression equation, simply remember that the graphed line of the regression equation always goes through the point M_X, M_Y. Therefore, if you plug the mean value for X (M_X) into the regression equation, the result equals the mean value for Y (M_Y).

$$M_Y = bM_X + a$$

If you simply solve this equation for a, you get the formula for the Y-intercept.

$$a = M_Y - bM_X$$

DEMONSTRATION 16.1

CORRELATION AND REGRESSION

For the following data, calculate the Pearson correlation, and find the regression equation:

PERSON	X	Y
A	0	4
B	2	1
C	8	10
D	6	9
E	4	6

STEP 1   *Sketch a scatterplot.* We have constructed a scatterplot for the data (Figure 16.23) and placed an envelope around the data points to make a preliminary estimate of the correlation. Note that the envelope is narrow and elongated. This indicates that the correlation is large—perhaps 0.80 to 0.90. Also, the correlation is positive because increases in X are generally accompanied by increases in Y.

We can sketch a straight line through the middle of the envelope and data points. Now we can roughly approximate the slope and Y-intercept of the best-fitting line. This is only an educated guess, but it will tell us what values are reasonable when we actually compute the regression line. The line has a positive slope (as X increases, Y increases), and it intersects the Y-axis in the vicinity of $+2$.

FIGURE 16.23

The scatterplot for the data of Demonstration 16.1. An envelope is drawn around the points to estimate the magnitude of the correlation. A line is drawn through the middle of the envelope to roughly estimate the Y-intercept for the regression equation.

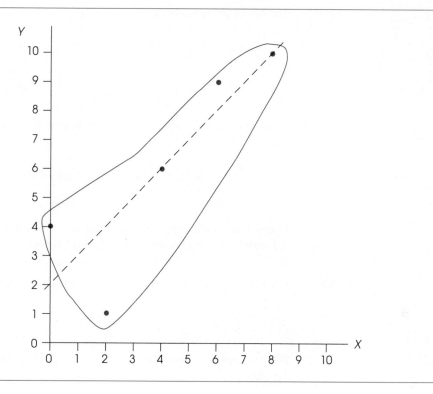

STEP 2   *Obtain the values for SS and SP.* To compute the Pearson correlation, we must find the values for SS_X, SS_Y, and SP. These values are needed for the regression equation as well. The following table illustrates these calculations with the computational formulas for SS and SP:

X	Y	X^2	Y^2	XY
0	4	0	16	0
2	1	4	1	2
8	10	64	100	80
6	9	36	81	54
4	6	16	36	24
$\Sigma X = 20$	$\Sigma Y = 30$	$\Sigma X^2 = 120$	$\Sigma Y^2 = 234$	$\Sigma XY = 160$

For SS_X, we obtain

$$SS_X = \Sigma X^2 - \frac{(\Sigma X)^2}{n} = 120 - \frac{20^2}{5} = 120 - \frac{400}{5} = 120 - 80$$

$$= 40$$

For Y, the sum of squares is

$$SS_Y = \Sigma Y^2 - \frac{(\Sigma Y)^2}{n} = 234 - \frac{30^2}{5} = 234 - \frac{900}{5} = 234 - 180$$

$$= 54$$

The sum of products equals

$$SP = \Sigma XY - \frac{\Sigma X \Sigma Y}{n} = 160 - \frac{20(30)}{5} = 160 - \frac{600}{5} = 160 - 120$$

$$= 40$$

STEP 3   *Compute the Pearson correlation.* For these data, the Pearson correlation is

$$r = \frac{SP}{\sqrt{SS_X SS_Y}} = \frac{40}{\sqrt{40(54)}} = \frac{40}{\sqrt{2160}} = \frac{40}{46.48}$$

$$= 0.861$$

In step 1, our preliminary estimate for the correlation was between $+0.80$ and $+0.90$. The calculated correlation is consistent with this estimate.

STEP 4   *Compute the values for the regression equation.* The general form of the regression equation is

$$\hat{Y} = bX + a$$

We will need to compute the values for the slope (b) of the line and the Y-intercept (a). For slope, we obtain

$$b = \frac{SP}{SS_X} = \frac{40}{40} = +1$$

The formula for the Y-intercept is

$$a = M_Y - bM_X$$

Thus, we will need the values for the sample means. For these data, the sample means are

$$M_X = \frac{\Sigma X}{n} = \frac{20}{5} = 4$$

$$M_Y = \frac{\Sigma Y}{n} = \frac{30}{5} = 6$$

Now we can compute the Y-intercept:

$$a = 6 - 1(4) = 6 - 4 = 2$$

Finally, the regression equation is

$$\hat{Y} = bX + a$$

$$= 1X + 2$$

or

$$\hat{Y} = X + 2$$

DEMONSTRATION 16.2

THE SPEARMAN CORRELATION

The following data will be used to demonstrate the calculation of the Spearman correlation. Both X and Y values are measurements on interval scales.

X	Y
5	12
7	18
2	9
15	14
10	13

STEP 1   Rank the X and Y values.
Remember that the X values and Y values are ranked separately.

X SCORE	Y SCORE	X RANK	Y RANK
5	12	2	2
7	18	3	5
2	9	1	1
15	14	5	4
10	13	4	3

STEP 2   Use the special Spearman formula to compute the correlation.
The special Spearman formula requires that you first find the difference (D) between the X rank and the Y rank for each individual and then square the differences and find the sum of the squared differences.

X RANK	Y RANK	D	D^2
2	2	0	0
3	5	2	4
1	1	0	0
5	4	−1	1
4	3	−1	1
			$6 = \Sigma D^2$

Using this value in the Spearman formula, we obtain

$$r_S = 1 - \frac{6\Sigma D^2}{n(n^2 - 1)}$$

$$= 1 - \frac{6(6)}{5(24)}$$

$$= 1 - \frac{36}{120}$$

$$= 1 - 0.30$$

$$= 0.70$$

There is a positive relationship between X and Y for these data. The Spearman correlation is fairly high, which indicates a very consistent positive relationship.

PROBLEMS

1. What information is provided by the sign ($+$ or $-$) of the Pearson correlation?

2. What information is provided by the numerical value of the Pearson correlation?

3. What is the major distinction between the Pearson and Spearman correlations?

4. How is a point-biserial correlation like an independent-measures t test?

5. For each of the following sets of scores, calculate SP using the definitional formula and then using the computational formula:

SET 1		SET 2	
X	Y	X	Y
1	2	3	5
3	2	6	1
5	3	4	2
7	5	7	0

6. For the following set of data,

X	Y
1	0
2	2
4	6
5	8
3	4

a. Sketch a graph showing the location of the five X, Y points.
b. Just looking at your graph, estimate the value of the Pearson correlation.
c. Compute the Pearson correlation for this data set.

7. For this problem, we have used the same X and Y values that appeared in Problem 6, but we have changed the X, Y pairings:

X	Y
1	6
2	8
4	2
5	0
3	4

a. Sketch a graph showing these reorganized data.
b. Estimate the Pearson correlation just by looking at your graph.
c. Compute the Pearson correlation. (*Note:* Much of the calculation for this problem was done already in Problem 6.)

If you compare the results of Problem 6 and Problem 7, you will see that the correlation measures the relationship between X and Y. These two problems use the same X and Y values, but they differ in the way X and Y are related.

8. With a very small sample, a single point can have a large influence on the magnitude of a correlation. For the following data set,

X	Y
0	1
10	3
4	1
8	2
8	3

a. Sketch a graph showing the X, Y points.
b. Estimate the value of the Pearson correlation.
c. Compute the Pearson correlation.
d. Now we will change the value of one of the points. For the first individual in the sample ($X = 0$ and $Y = 1$), change the Y value to $Y = 6$. What happens to the graph of the X, Y points? What happens to the Pearson correlation? Compute the new correlation.

9. In the following data, there are three scores (X, Y, and Z) for each of the $n = 5$ individuals:

X	Y	Z
3	5	5
4	3	2
2	4	6
1	1	3
0	2	4

a. Sketch a graph showing the relationship between X and Y. Compute the Pearson correlation between X and Y.
b. Sketch a graph showing the relationship between Y and Z. Compute the Pearson correlation between Y and Z.
c. Given the results of parts (a) and (b), what would you predict for the correlation between X and Z?
d. Sketch a graph showing the relationship between X and Z. Compute the Pearson correlation for these data.
e. What general conclusion can you make concerning relationships among correlations? If X is related to Y and Y is related to Z, does this necessarily mean that X is related to Z?

10. a. Compute the Pearson correlation for the following set of data:

X	Y
0	3
1	1
2	5
3	9
4	7

b. Add 5 points to each X value, and compute the Pearson correlation again.
c. When you add a constant to each score, what happens to SS for X and Y? What happens to SP? What happens to the correlation between X and Y?
d. Now multiply each X in the original data by 3, and calculate the Pearson correlation once again.
e. When you multiply by a constant, what happens to SS for X and Y? What happens to SP? What happens to the correlation between X and Y?

11. A researcher has developed a new test of self-esteem. To evaluate the reliability of the test, the researcher obtains a sample of $n = 8$ participants. Each individual takes the test on a Monday morning, then returns 2 weeks later to take the test again. The two scores for each individual are reported in the following table. Use a Pearson correlation to measure the degree of consistency between the first and second measure for this sample.

FIRST TEST	SECOND TEST
13	15
5	4
12	13
11	11
9	10
14	13
8	8
8	6

12. It is well known that similarity in attitudes, beliefs, and interests plays an important role in interpersonal attraction (see Byrne, 1971, for example). Thus, correlations for attitudes between married couples should be strong. Suppose a researcher developed a questionnaire that measures how liberal or conservative one's attitudes are. Low scores indicate that the person has liberal attitudes, whereas high scores indicate conservatism. The following hypothetical data are scores for married couples.

COUPLE	WIFE	HUSBAND
A	11	14
B	6	7
C	18	15
D	4	7
E	1	3
F	10	9
G	5	9
H	3	3

Compute the Pearson correlation for these data, and determine whether or not there is a significant correlation between attitudes for husbands and wives. Set alpha at .05, two tails.

13. To measure the relationship between anxiety level and test performance, a psychologist obtains a sample of $n = 6$ college students from an introductory statistics course. The students are asked to come to the laboratory 15 minutes before the final exam. In the lab, the psychologist records physiological measures of anxiety (heart rate, skin resistance, blood pressure, etc.) for each participant. In addition, the psychologist obtains the exam score for each participant. Compute the Pearson correlation for the following data:

STUDENT	ANXIETY RATING	EXAM SCORE
A	5	80
B	2	88
C	7	80
D	7	79
E	4	86
F	5	85

14. Researchers who measure reaction time for human participants often observe a relationship between the reaction time scores and the number of errors that the participants commit. This relationship is known as the *speed-accuracy trade-off*. The following data are from a reaction time study where the researcher recorded the average reaction time (milliseconds) and the total number of errors for each individual in a sample of $n = 8$ participants.

PARTICIPANT	REACTION TIME	ERRORS
A	184	10
B	213	6
C	234	2
D	197	7
E	189	13
F	221	10
G	237	4
H	192	9

a. Compute the Pearson correlation for the data.
b. In words, describe the speed-accuracy trade-off.

15. A high school counselor would like to know whether there is a relationship between mathematical skill and verbal skill. A sample of $n = 25$ students is selected, and the counselor records achievement test scores in mathematics and English for each student. The Pearson correlation for this sample is $r = +0.50$. Do these data provide sufficient evidence for a real rela-

tionship in the population? Test at the .05 level, two tails.

16. A psychologist would like to determine whether there is any consistent relationship between intelligence and creativity. A random sample of $n = 18$ people is obtained, and the psychologist administers a standardized IQ test and a creativity test to each individual. Using these data, the psychologist obtains a Pearson correlation of $r = +0.20$ between IQ and creativity.
a. Do the sample data provide sufficient evidence to conclude that a real (nonzero) correlation exists in the population? Test at the .05 level of significance.
b. If the same correlation, $r = 0.20$, was obtained for a sample of $n = 102$ people, what decision would be made about the population correlation?

17. Rank the following scores, and compute the Spearman correlation between X and Y:

X	Y
7	19
2	4
11	34
15	28
32	104

18. While grading essay exams, a professor noticed huge differences in the writing skills of the students. To investigate a possible cause for the differences, the professor asked the students how many English courses they had completed. The number of courses completed and the professor's ranking of the essay are reported for each student in the following table.

QUALITY OF ESSAY (PROFESSOR'S RANKING)	NUMBER OF ENGLISH COURSES
1 (best)	7
2	4
3	1
4	3
5	1
6	1
7	0
8	2

Compute the Spearman correlation between writing ability and English background. (*Note:* You must convert the data before computing the correlation.) Is this correlation statistically significant? Test with alpha set at .05.

19. A common concern for students (and teachers) is the assignment of grades for essays or term papers. Because there are no absolute right or wrong answers, these grades must be based on a judgment of quality. To demonstrate that these judgments actually are reliable, an English instructor asked a colleague to rank-order a set of term papers. The ranks and the instructor's grades for these papers are as follows:

RANK	GRADE
1	A
2	B
3	A
4	B
5	B
6	C
7	D
8	C
9	C
10	D
11	E

a. Calculate the Spearman correlation for these data. *(Note:* You must convert the letter grades to ranks.)
b. Based on this correlation, does it appear that there is reasonable agreement between these two instructors in their judgment of the papers? Test with $\alpha = .05$.

20. In the following data, X and Y are related by the equation $Y = X^2$:

X	Y
0	0
1	1
2	4
3	9
4	16
5	25

a. Sketch a graph showing the relationship between X and Y. Describe the relationship shown in your graph.
b. Compute the Spearman correlation for these data.

21. To test the effectiveness of a new studying strategy, a psychologist randomly divides a sample of 8 students into two groups, with $n = 4$ in each group. The students in one group receive training in the new studying strategy. Then all students are given 30 minutes to study a chapter from a history textbook before they take a quiz on the chapter. The quiz scores for the two groups are as follows:

TRAINING	NO TRAINING
9	4
7	7
6	3
10	6

a. Convert these data into a form suitable for the point-biserial correlation. (Use $X = 1$ for training, $X = 0$ for no training, and the quiz score for Y.)
b. Calculate the point-biserial correlation for these data.

22. The data in Problem 21 showed a mean difference of 3 points between the training group and the no-training group ($M = 8$ versus $M = 5$). Because there is a 3-point difference, you would expect to find that quiz scores are related to training (there is a nonzero correlation). Now consider the following data, where there is no mean difference between the two groups:

TRAINING	NO TRAINING
2	4
5	7
6	3
7	6

a. Without doing any calculations, estimate the point-biserial correlation for these data.
b. Convert the data to a form suitable for the point-biserial, and compute the correlation.

23. A researcher would like to evaluate the relationship between a person's age and his or her preference between two leading brands of cola. In a sample of 12 people, the researcher found that 5 out of 8 people 30 years or older preferred brand A and only 1 out of 4 people under 30 years old preferred brand A.

a. Convert the data to a form suitable for computing the phi-coefficient. (Code the two age categories as 0 and 1 for the X variable, and code the preferred brand of soft drink as 0 and 1 for the Y variable.)
b. Compute the phi-coefficient for the data.

24. For the following set of data, find the linear regression equation for predicting Y from X:

X	Y
0	9
2	9
4	7
6	3

25. For the following set of data,

X	Y
1	2
4	7
3	5
2	1
5	14
3	7

  a. Compute the Pearson correlation.
  b. Find the linear regression equation for predicting Y from X.
  c. Use the regression equation to compute the predicted Y for each value of X.
  d. Compute the difference between the actual Y value and the predicted Y value and calculate SS_{error} for these data.
  e. Use the correlation (from part a) and SS_Y to calculate SS_{error} for these data.

26. Find the regression equation for the following data:

X	Y
3	12
0	8
4	18
2	12
1	8

27. A set of $n = 6$ pairs of X and Y values has a Pearson correlation of $r = +0.60$ and $SS_Y = 100$. If you are using these data as the basis for a regression equation,

  a. On the average, how much error would you expect if the regression equation were used to predict the Y score for a specific individual? That is, find the standard error of estimate.
  b. How much error would you expect if the sample size were $n = 102$ instead of $n = 6$?

28. A college professor claims that the scores on the first exam provide an excellent indication of how students will perform throughout the term. To test this claim, first-exam scores and final grades were recorded for a sample of $n = 12$ students in an introductory psychology class. The data are as follows:

FIRST EXAM	FINAL GRADE
62	74
73	93
88	68
82	79
85	91
77	72
94	96
65	61
91	92
74	82
85	93
98	95

  a. Is the professor right? Is there a significant correlation between scores on the first exam and final grades? Test with $\alpha = .01$.
  b. How accurately do the exam scores predict final grades? Compute the standard error of estimate.

CHAPTER

17

The Chi-Square Statistic: Tests for Goodness of Fit and Independence

Tools You Will Need

The following items are considered essential background material for this chapter. If you doubt your knowledge of any of these items, you should review the appropriate chapter or section before proceeding.

- Proportions (math review, Appendix A)
- Frequency distributions (Chapter 2)

Preview

Imagine that you are walking into an amusement park and someone gives you a handbill advertising an entertainment program scheduled for that evening: DON'T MISS TONIGHT'S SHOW. Assuming that you are not an avid collector of handbills, you are now stuck with a piece of paper that you don't want. As you continue to walk down the path, you realize that there are no trash cans in sight. What do you do with the handbill?

In a study examining littering in public places, Cialdini, Reno, and Kallgren (1990) looked at people's behavior in the situation we have just described. One specific factor the researchers considered was the amount of litter already on the ground. After receiving the handbill, the participants entered a path that had been prepared with either 0, 1, 2, 4, 8, or 16 handbills scattered around, apparently discarded by other visitors. Did the amount of existing litter influence littering by the participants?

The data from this study are shown in Table 17.1 and seem to indicate that littering behavior is directly related to the amount of preexisting litter. When the path had 0 or 1 pieces of litter, only 14% of the participants (17 out of 119) dropped their handbills. When the amount of litter was increased to 2 or 4 pieces, the rate of littering increased to 24% (28 out of 119 participants). And, when the path was heavily littered with 8 or 16 handbills, the data show that 41% of the participants (49 out of 120 people) dropped their own handbills and contributed to the mess.

Although the Cialdini study involves a dependent variable (littering behavior) and an independent variable (the amount of preexisting litter), you should realize that this experiment is different from any we have considered in the past. Specifically, the Cialdini study does not produce a numerical score for each subject. Instead, each individual is simply classified into one of two categories (littering or not littering) and the data consist of *frequencies* showing how many individuals are in each category. Because there are no numerical scores, it is impossible to calculate a mean or a variance or a standard deviation for the sample data. Therefore, it is impossible to use any of the familiar hypothesis tests (such as a *t* statistic or ANOVA) to determine whether or not there are *significant* differences between the treatment conditions.

Fortunately, statistical techniques have been developed specifically to analyze and interpret frequency data. In this chapter we will introduce a statistic called *chi-square* and two hypothesis tests based on the chi-square statistic. Unlike earlier tests that require numerical scores (*X* values), the chi-square tests use frequencies and proportions from a sample to test hypotheses about the corresponding population.

TABLE 17.1

The number of people littering or not littering depending on the amount of litter already on the ground.

	AMOUNT OF EXISTING LITTER		
	0 OR 1 PIECES	2 OR 4 PIECES	8 OR 16 PIECES
LITTERING	17	28	49
NOT LITTERING	102	91	71

Cialdini, R. B., Reno, R. R., and Kallgren, C. A. (1990). A focus theory of normative conduct: Recycling the concept of norms to reduce littering in public places. *Journal of Personality and Social Psychology, 58,* 1015–1020. Copyright © 1990 by the American Psychological Association.

17.1 PARAMETRIC AND NONPARAMETRIC STATISTICAL TESTS

All the statistical tests we have examined thus far are designed to test hypotheses about specific population parameters. For example, we used *t* tests to assess hypotheses about μ and later about $\mu_1 - \mu_2$. In addition, these tests typically make assumptions about the shape of the population distribution and about other population parameters. Recall that for analysis of variance the population distributions are assumed to be normal and homogeneity of variance is required. Because these tests all concern parameters and require assumptions about parameters, they are called *parametric tests*.

Another general characteristic of parametric tests is that they require a numerical score for each individual in the sample. The scores then are added, squared, averaged, and otherwise manipulated using basic arithmetic. In terms of measurement scales, parametric tests require data from an interval or a ratio scale (see Chapter 1).

Often, researchers are confronted with experimental situations that do not conform to the requirements of parametric tests. In these situations, it may not be appropriate to use a parametric test. Remember: When the assumptions of a test are violated, the test may lead to an erroneous interpretation of the data. Fortunately, there are several hypothesis-testing techniques that provide alternatives to parametric tests. These alternatives are called *nonparametric tests*.

In this chapter, we will introduce two commonly used examples of nonparametric tests. Both tests are based on a statistic known as chi-square and both tests use sample data to evaluate hypotheses about the proportions or relationships that exist within populations. Note that the two chi-square tests, like most nonparametric tests, do not state hypotheses in terms of a specific parameter, and they make few (if any) assumptions about the population distribution. For the latter reason, nonparametric tests sometimes are called *distribution-free tests*.

One of the most obvious differences between parametric and nonparametric tests is the type of data they use. All of the parametric tests that we have examined so far require numerical scores. For nonparametric tests, on the other hand, the subjects are usually just classified into categories such as Democrat and Republican, or High, Medium, and Low IQ. Notice that these classifications involve measurement on nominal or ordinal scales, and they do not produce numerical values that can be used to calculate means and variances. Instead, the data for many nonparametric tests are simply frequencies; for example, the number of Democrats and the number of Republicans in a sample of $n = 100$ registered voters.

Finally, you should be warned that nonparametric tests generally are not as sensitive as parametric tests; nonparametric tests are more likely to fail in detecting a real difference between two treatments. Therefore, whenever the experimental data give you a choice between a parametric and a nonparametric test, you always should choose the parametric alternative.

17.2 THE CHI-SQUARE TEST FOR GOODNESS OF FIT

Parameters such as the mean and the standard deviation are the most common way to describe a population, but there are situations in which a researcher has questions about the proportions or relative frequencies for a distribution. For example,

How does the number of women lawyers compare with the number of men in the profession?

Of the three leading brands of soft drinks, which is preferred by most Americans? Which brands are second and third, and how big are the differences in popularity among the three?

To what extent are different ethnic groups represented in the population of your college or university?

The name of the test comes from the Greek letter χ (chi, pronounced "kye"), which is used to identify the test statistic.

Notice that each of the preceding examples asks a question about proportions in the population. The chi-square test for goodness of fit is specifically designed to answer this type of question. In general terms, this chi-square test is a hypothesis-testing procedure that uses the proportions obtained for a sample distribution to test hypotheses about the corresponding proportions in the population distribution.

DEFINITION   The **chi-square test for goodness of fit** uses sample data to test hypotheses about the shape or proportions of a population distribution. The test determines how well the obtained sample proportions fit the population proportions specified by the null hypothesis.

Recall from Chapter 2 that a frequency distribution is defined as a tabulation of the number of individuals located in each category of the scale of measurement. In a frequency distribution graph, the categories that make up the scale of measurement are listed on the X-axis. In a frequency distribution table, the categories are listed in the first column. With chi-square tests, however, it is customary to present the scale of measurement as a series of boxes, with each box corresponding to a separate category on the scale. The frequency corresponding to each category is simply presented as a number written inside the box. Figure 17.1 shows how a distribution of eye colors for a set of $n = 40$ students can be presented as a graph, a table, or a series of boxes. Notice that the scale of measurement for this example consists of four categories of eye color (blue, brown, green, other).

FIGURE 17.1

Distribution of eye colors for a sample of $n = 40$ individuals. The same frequency distribution is shown as a bar graph, as a table, and with the frequencies written in a series of boxes.

THE NULL HYPOTHESIS FOR THE GOODNESS-OF-FIT TEST   For the chi-square test of goodness of fit, the null hypothesis specifies the proportion (or percentage) of the population in each category. For example, a hypothesis might state that 90% of all lawyers are men and only 10% are women. The simplest way of presenting this hypothesis is to put the hypothesized proportions in the series of boxes representing the scale of measurement:

	MEN	WOMEN
H_0:	90%	10%

Although it is conceivable that a researcher could choose any proportions for the null hypothesis, there usually is some well-defined rationale for stating a null hypothesis. Generally H_0 will fall into one of the following categories:

1. No Preference. The null hypothesis often states that there is no preference among the different categories. In this case, H_0 states that the population is divided equally among the categories. For example, a hypothesis stating that there is no preference among the three leading brands of soft drinks would specify a population distribution as follows:

	BRAND X	BRAND Y	BRAND Z
H_0:	$\frac{1}{3}$	$\frac{1}{3}$	$\frac{1}{3}$

(Preferences in the population are equally divided among the three soft drinks.)

The no-preference hypothesis is used in situations where a researcher wants to determine whether there are any preferences among categories or any variations from one category to another.

2. No Difference From a Comparison Population. The null hypothesis can state that the frequency distribution for one population is not different from the distribution that is known to exist for another population. For example, suppose it is known that 60% of Americans favor the president's foreign policy and 40% are in opposition. A researcher might wonder whether this same pattern of attitudes exists among Europeans. The null hypothesis would state that there is no difference between the two populations and would specify that the Europeans are distributed as follows:

	FAVOR	OPPOSE
H_0:	60%	40%

(Proportions for the European population are not different from the American proportions.)

The no-difference hypothesis is used when a specific population distribution is already known. You may have a known distribution from an earlier time, and the question is whether there has been any change in the distribution. Or you may have a known distribution for one population, and the question is whether a second population has the same distribution.

Because the null hypothesis for the goodness-of-fit test specifies an exact distribution for the population, the alternative hypothesis (H_1) simply states that the population distribution has a different shape from that specified in H_0. If the null hypothesis states that the population is equally divided among three categories, the alternative hypothesis will say that the population is not divided equally.

THE DATA FOR THE GOODNESS-OF-FIT TEST

The data for a chi-square test are remarkably simple. There is no need to calculate a sample mean or SS; you just select a sample of n individuals and count how many are in each category. The resulting values are called *observed frequencies*. The symbol for observed frequency is f_o. For example, the following data represent observed frequencies for a sample of $n = 40$ participants. Each person was given a personality questionnaire and classified into one of three personality categories: A, B, or C.

CATEGORY A	CATEGORY B	CATEGORY C	
15	19	6	$n = 40$

Notice that each individual in the sample is classified into one and only one of the categories. Thus, the frequencies in this example represent three completely separate groups of individuals: 15 who were classified as category A, 19 classified as B, and 6 classified as C. Also note that the observed frequencies add up to the total sample size: $\Sigma f_o = n$.

DEFINITION

The **observed frequency** is the number of individuals from the sample who are classified in a particular category. Each individual is counted in one and only one category.

EXPECTED FREQUENCIES

The general goal of the chi-square test for goodness of fit is to compare the data (the observed frequencies) with the null hypothesis. The problem is to determine how well the data fit the distribution specified in H_0—hence the name *goodness of fit*.

The first step in the chi-square test is to construct a hypothetical sample that represents how the sample distribution would look if it was in perfect agreement with the proportions stated in the null hypothesis. Suppose, for example, that the null hypothesis states that the population is distributed in three categories with the following proportions:

CATEGORY A	CATEGORY B	CATEGORY C
25%	50%	25%

If this hypothesis is correct, how would you expect a random sample of $n = 40$ individuals to be distributed among the three categories? It should be clear that your best strategy is to predict that 25% of the sample would be in category A, 50% would be in category B, and 25% would be in category C. To find the exact frequency expected for each category, multiply the sample size (n) by the proportion (or percentage) from the null hypothesis. For this example, you would expect

25% of 40 = 0.25(40) = 10 individuals in category A

50% of 40 = 0.50(40) = 20 individuals in category B

25% of 40 = 0.25(40) = 10 individuals in category C

The frequency values predicted from the null hypothesis are called *expected frequencies*. The symbol for expected frequency is f_e, and the expected frequency for each category is computed by

$$\text{expected frequency} = f_e = pn \qquad (17.1)$$

where p is the proportion stated in the null hypothesis and n is the sample size.

DEFINITION

The **expected frequency** for each category is the frequency value that is predicted from the null hypothesis and the sample size (n). The expected frequencies define an ideal, *hypothetical* sample distribution that would be obtained if the sample proportions were in perfect agreement with the proportions specified in the null hypothesis.

Note that the no-preference null hypothesis will always produce equal f_e values for all categories because the proportions (p) are the same for all categories. On the other hand, the no-difference null hypothesis typically will not produce equal values for the expected frequencies because the hypothesized proportions typically vary from one category to another. You also should note that the expected frequencies are calculated, hypothetical values and the numbers that you obtain may be decimals or fractions. The observed frequencies, on the other hand, always represent real individuals and always will be whole numbers.

THE CHI-SQUARE STATISTIC

The general purpose of any hypothesis test is to determine whether the sample data support or refute a hypothesis about the population. In the chi-square test for goodness of fit, the sample is expressed as a set of observed frequencies (f_o values), and the null hypothesis is used to generate a set of expected frequencies (f_e values). The *chi-square statistic* simply measures how well the data (f_o) fit the hypothesis (f_e). The symbol for the chi-square statistic is χ^2. The formula for the chi-square statistic is

$$\text{chi-square} = \chi^2 = \Sigma \frac{(f_o - f_e)^2}{f_e} \qquad (17.2)$$

As the formula indicates, the value of chi-square is computed by the following steps:

1. Find the difference between f_o (the data) and f_e (the hypothesis) for each category.
2. Square the difference. This ensures that all values are positive.
3. Next, divide the squared difference by f_e. A justification for this step is given in Box 17.1.
4. Finally, sum the values from all the categories.

BOX

17.1   THE CHI-SQUARE FORMULA

The numerator of the chi-square formula is fairly easy to understand. Specifically, the numerator simply measures how much difference there is between the data (the f_o values) and the hypothesis (represented by the f_e values). A large value indicates that the data do not fit the hypothesis, and will lead us to reject the hypothesis.

The denominator of the formula, on the other hand, is not so easy to understand. Why must we divide by f_e before we sum the category values? The answer to this question is that the obtained discrepancy between f_o and f_e is viewed as *relatively* large or *relatively* small depending on the size of the expected frequency. This point is demonstrated in the following analogy.

Suppose you were going to throw a party and you *expected* 1000 people to show up. However, at the party

you counted the number of guests and *observed* that 1040 actually showed up. Forty more guests than expected are no major problem when all along you were planning for 1000. There will still probably be enough beer and potato chips for everyone. On the other hand, suppose you had a party and you expected 10 people to attend but instead 50 actually showed up. Forty more guests in this case spell big trouble. How "significant" the discrepancy is depends in part on what you were originally expecting. With very large expected frequencies, allowances are made for more error between f_o and f_e. This is accomplished in the chi-square formula by dividing the squared discrepancy for each category, $(f_o - f_e)^2$, by its expected frequency.

It should be clear from the chi-square formula that the numerical value of chi-square is a measure of the discrepancy between the observed frequencies (data) and the expected frequencies (H_0). When there are large differences between f_o and f_e, the value of chi-square will be large, and we will conclude that the data do not fit the hypothesis. Thus, a large value for chi-square will lead us to reject H_0. On the other hand, when the observed frequencies are very close to the expected frequencies, chi-square will be small, and we will conclude that there is a very good fit between the data and the hypothesis. Thus, a small chi-square value indicates that we should fail to reject H_0. To decide whether a particular chi-square value is "large" or "small," we must refer to a *chi-square distribution*. This distribution is the set of chi-square values for all the possible random samples when H_0 is true. Much like other distributions we have examined (*t* distribution, *F* distribution), the chi-square distribution is a theoretical distribution with well-defined characteristics. Some of these characteristics are easy to infer from the chi-square formula.

1. The formula for chi-square involves adding squared values, so you can never obtain a negative value. Thus, all chi-square values are zero or larger.

2. When H_0 is true, you expect the data (f_o values) to be close to the hypothesis (f_e values). Thus, we expect chi-square values to be small when H_0 is true.

These two factors suggest that the typical chi-square distribution will be positively skewed (see Figure 17.2). Note that small values, near zero, are expected when H_0 is true and large values (in the right-hand tail) are very unlikely. Thus, unusually large values of chi-square will form the critical region for the hypothesis test.

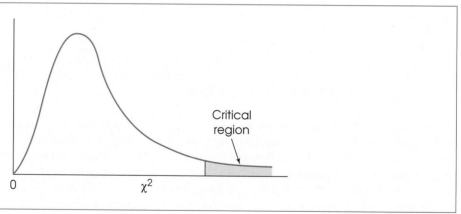

FIGURE 17.2

Chi-square distributions are positively skewed. The critical region is placed in the extreme tail, which reflects large chi-square values.

Although the typical chi-square distribution is positively skewed, there is one other factor that plays a role in the exact shape of the chi-square distribution—the number of categories. Recall that the chi-square formula requires that you sum values from every category. The more categories you have, the more likely it is that you will obtain a large sum for the chi-square value. On average, chi-square will be larger when you are summing over 10 categories than when you are summing over only 3 categories. As a result, there is a whole family of chi-square distributions, with the exact shape of each distribution determined by the number of categories used in the study. Technically, each specific chi-square distribution is identified by degrees of freedom (*df*), rather than the number of categories. For the goodness-of-fit test, the degrees of freedom are determined by

$$df = C - 1 \tag{17.3}$$

Caution: The *df* for a chi-square test is *not* related to sample size (*n*), as it is in most other tests.

where *C* is the number of categories. A brief discussion of this *df* formula is presented in Box 17.2. Figure 17.3 shows the general relationship between *df* and the shape of the chi-square distribution. Note that the peak in the chi-square distribution (the mode) gets larger and larger as the value for *df* increases.

BOX

17.2   A CLOSER LOOK AT DEGREES OF FREEDOM

Degrees of freedom for the chi-square test literally measure the number of free choices that exist when you are determining the null hypothesis or the expected frequencies. For example, when you are classifying individuals into three categories, you have exactly two free choices in stating the null hypothesis. You may select any two proportions for the first two categories, but then the third proportion is determined. If you hypothesize 25% in the first category and 50% in the second category, then the third category must be 25% in order to account for 100% of the population. In general, you are free to select proportions for all but one of the categories, but then the final proportion is determined by the fact that the entire set must total 100%. Thus, you have $C - 1$ free choices, where *C* is the number of categories: Degrees of freedom, *df*, equal $C - 1$.

The same restriction holds when you are determining the expected frequencies. Again, suppose you have three categories and a sample of $n = 40$ individuals. If you specify expected frequencies of $f_e = 10$ for the first category and $f_e = 20$ for the second category, then you must use $f_e = 10$ for the final category.

CATEGORY A	CATEGORY B	CATEGORY C
10	20	???

$n = 40$

As before, you may distribute the sample freely among the first $C - 1$ categories, but then the final category is determined by the total number of individuals in the sample.

FIGURE 17.3

The shape of the chi-square distribution for different values of *df*. As the number of categories increases, the peak (mode) of the distribution has a larger chi-square value.

$df = 1$

$df = 5$

$df = 9$

0 χ^2

LOCATING THE CRITICAL REGION FOR A CHI-SQUARE TEST

Recall that a large value for the chi-square statistic indicates a big discrepancy between the data and the hypothesis, and suggests that we reject H_0. To determine whether a particular chi-square value is significantly large, you must consult the table entitled The Chi-Square Distribution (Appendix B). A portion of the chi-square table is shown in Table 17.2. The first column lists df values for the chi-square test, and the top row of the table lists proportions (alpha levels) in the extreme right-hand tail of the distribution. The numbers in the body of the table are the critical values of chi-square. The table shows, for example, that in a chi-square distribution with $df = 3$, only 5% (.05) of the values are larger than 7.81, and only 1% (.01) are larger than 11.34.

TABLE 17.2

A portion of the table of critical values for the chi-square distribution.

df	PROPORTION IN CRITICAL REGION				
	0.10	0.05	0.025	0.01	0.005
1	2.71	3.84	5.02	6.63	7.88
2	4.61	5.99	7.38	9.21	10.60
3	6.25	7.81	9.35	11.34	12.84
4	7.78	9.49	11.14	13.28	14.86
5	9.24	11.07	12.83	15.09	16.75
6	10.64	12.59	14.45	16.81	18.55
7	12.02	14.07	16.01	18.48	20.28
8	13.36	15.51	17.53	20.09	21.96
9	14.68	16.92	19.02	21.67	23.59

EXAMPLE OF THE CHI-SQUARE TEST FOR GOODNESS OF FIT

We will use the same step-by-step process for testing hypotheses with chi-square as we used for other hypothesis tests. In general, the steps consist of stating the hypotheses, locating the critical region, computing the test statistic, and making a decision about H_0. The following example demonstrates the complete process of hypothesis testing with the goodness-of-fit test.

EXAMPLE 17.1

A psychologist examining art appreciation selected an abstract painting that had no obvious top or bottom. Hangers were placed on the painting so that it could be hung with any one of the four sides at the top. The painting was shown to a sample of $n = 50$ participants, and each was asked to hang the painting in whatever orientation looked best. The following data indicate how many times each of the four sides was placed at the top:

TOP UP (CORRECT)	BOTTOM UP	LEFT SIDE UP	RIGHT SIDE UP
18	17	7	8

The question for the hypothesis test is whether there are any preferences among the four possible orientations. Are any of the orientations selected more (or less) often than would be expected simply by chance?

STEP 1   State the hypotheses and select an alpha level. The hypotheses can be stated as follows:

H_0:   In the general population, there is no preference for any specific orientation. Thus, the four possible orientations are selected equally often, and the population distribution has the following proportions:

TOP UP (CORRECT)	BOTTOM UP	LEFT SIDE UP	RIGHT SIDE UP
25%	25%	25%	25%

H_1:   In the general population, one or more of the orientations is preferred over the others.

We will use $\alpha = .05$.

STEP 2   Locate the critical region. For this example, the value for degrees of freedom is

$$df = C - 1 = 4 - 1 = 3$$

For $df = 3$ and $\alpha = .05$, the table of critical values for chi-square indicates that the critical χ^2 has a value of 7.81. The critical region is sketched in Figure 17.4.

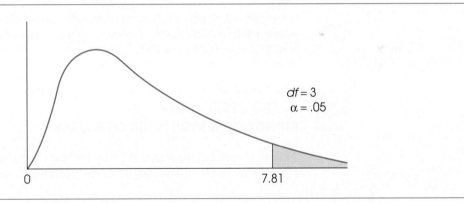

FIGURE 17.4

For Example 17.1, the critical region begins at a chi-square value of 7.81.

$df = 3$
$\alpha = .05$

0   7.81

STEP 3   Calculate the chi-square statistic. The calculation of chi-square is actually a two-stage process. First, you must compute the expected frequencies from H_0 and then calculate the value of the chi-square statistic. For this example, the null hypothesis specifies a proportion of $p = 25\%$ $\left(\text{or } p = \frac{1}{4}\right)$ for each of the four categories, and the sample size is $n = 50$. Thus, the expected frequency for each category is

 Expected frequencies are computed and may be decimal values. Observed frequencies are always whole numbers.

$$f_e = pn = \tfrac{1}{4}(50) = 12.5$$

The observed frequencies and the expected frequencies are presented in Table 17.3.

TABLE 17.3

The observed frequencies and the expected frequencies for the chi-square test in Example 17.1.

OBSERVED FREQUENCIES	TOP UP (CORRECT)	BOTTOM UP	LEFT SIDE UP	RIGHT SIDE UP
	18	17	7	8

EXPECTED FREQUENCIES	TOP UP (CORRECT)	BOTTOM UP	LEFT SIDE UP	RIGHT SIDE UP
	12.5	12.5	12.5	12.5

Using these values, the chi-square statistic may now be calculated.

$$\chi^2 = \Sigma \frac{(f_o - f_e)^2}{f_e}$$

$$= \frac{(18 - 12.5)^2}{12.5} + \frac{(17 - 12.5)^2}{12.5} + \frac{(7 - 12.5)^2}{12.5} + \frac{(8 - 12.5)^2}{12.5}$$

$$= \frac{30.25}{12.5} + \frac{20.25}{12.5} + \frac{30.25}{12.5} + \frac{20.25}{12.5}$$

$$= 2.42 + 1.62 + 2.42 + 1.62$$

$$= 8.08$$

STEP 4   State a decision and a conclusion. The obtained chi-square value is in the critical region. Therefore, H_0 is rejected and the researcher may conclude that the four orientations are not equally likely to be preferred. Instead, there are significant differences among the four orientations, with some selected more often and others less often than would be expected by chance.

IN THE LITERATURE
REPORTING THE RESULTS FOR CHI-SQUARE

APA style specifies the format for reporting the chi-square statistic in scientific journals. For the results of Example 17.1, the report might state:

> The participants showed significant preferences among the four orientations for hanging the painting, $\chi^2(3, n = 50) = 8.08, p < .05$.

Note that the form of the report is similar to that of other statistical tests we have examined. Degrees of freedom are indicated in parentheses following the chi-square symbol. Also contained in the parentheses is the sample size (n). This additional information is important because the degrees of freedom value is based on the number of categories (C), not sample size. Next, the calculated value of chi-square is presented, followed by the probability that a Type I error has been committed. Because the null hypothesis was rejected, the probability is *less than* the alpha level.

Additionally, the report should provide the observed frequencies (f_o) for each category. This information may be presented in a simple sentence or in a table (see, for example, Table 17.3). ❑

GOODNESS OF FIT AND THE
SINGLE-SAMPLE _t_ TEST

We began this chapter with a general discussion of the difference between parametric tests and nonparametric tests. In this context, the chi-square test for goodness of fit is an example of a nonparametric test; that is, it makes no assumptions about the parameters of the population distribution, and it does not require data from an interval or ratio scale. In contrast, the single-sample _t_ test introduced in Chapter 9 is an example of a parametric test: It assumes a normal population, it tests hypotheses about the population mean (a parameter), and it requires numerical scores that can be summed, squared, divided, and so on.

Although the chi-square test and the single-sample _t_ are clearly distinct, they are also very similar. In particular, both tests are intended to use the data from a single sample to test hypotheses about a single population.

When the sample data consist of numerical scores, it is usually most appropriate to compute the sample mean and then use a _t_ test to evaluate a hypothesis about the population mean. For example, a researcher could measure IQ scores for a sample of registered voters and then use a _t_ test to test a hypothesis about the mean IQ for the general population of voters. On the other hand, when the individuals in the sample are simply classified into nonnumerical categories, you probably would use a chi-square test to evaluate a hypothesis about the population proportions. For example, a researcher could measure gender for a sample of registered voters, recording the proportion of males and females in the sample. A chi-square test would then be the appropriate method for using the sample proportions to evaluate a hypothesis about the corresponding population proportions.

LEARNING CHECK

1. A researcher uses a chi-square test for goodness of fit with a sample of $n = 120$ people to determine whether there are any preferences among four different brands of chocolate-chip cookies. What is the _df_ value for the chi-square statistic?

2. Is it possible for expected frequencies to be fractions or decimal values?

3. A researcher has developed three different designs for a computer keyboard. A sample of $n = 60$ participants is obtained, and each individual tests all three keyboards and identifies his or her favorite. The frequency distribution of preferences is as follows:

DESIGN A	DESIGN B	DESIGN C	
23	12	25	$n = 60$

Use a chi-square test for goodness of fit with $\alpha = .05$ to determine whether there are any significant preferences among the three designs.

ANSWERS

1. With four categories, $df = 3$.

2. Yes. Expected frequencies are computed and may be fractions or decimal values.

3. The null hypothesis states that there are no preferences; one-third of the population would prefer each design. With $df = 2$, the critical value is 5.99. The expected frequencies are all 20, and the chi-square statistic is 4.90. Fail to reject the null hypothesis. There are no significant preferences.

17.3   THE CHI-SQUARE TEST FOR INDEPENDENCE

The chi-square statistic may also be used to test whether or not there is a relationship between two variables. In this situation, each individual in the sample is measured or classified on two separate variables. For example, a group of students could be classified in terms of personality (introvert, extrovert) and in terms of color preference (red, yellow, green, or blue). Usually, the data from this classification are presented in the form of a matrix, where the rows correspond to the categories of one variable and the columns correspond to the categories of the second variable. Table 17.4 presents some hypothetical data for a sample of $n = 200$ students who have been classified by personality and color preference. The number in each box, or cell, of the matrix depicts the frequency of that particular group. In Table 17.4, for example, there are 10 students who were classified as introverted and who selected red as their preferred color. To obtain these data, the researcher first selects a random sample of $n = 200$ students. Each student is then given a personality test, and each student is asked to select a preferred color from among the four choices. Notice that the classification is based on the measurements for each student; the researcher does not assign students to categories. Also, notice that the data consist of frequencies, not scores, from a sample. These sample data will be used to test a hypothesis about the corresponding population frequency distribution. Once again, we will be using the chi-square statistic for the test, but in this case, the test is called the chi-square *test for independence*.

In the context of a chi-square test, the frequency matrix is often called a contingency table.

TABLE 17.4

Color preferences according to personality types.

	RED	YELLOW	GREEN	BLUE	
INTROVERT	10	3	15	22	50
EXTROVERT	90	17	25	18	150
	100	20	40	40	$n = 200$

THE NULL HYPOTHESIS FOR THE TEST FOR INDEPENDENCE

The null hypothesis for the chi-square test for independence states that the two variables being measured are independent; that is, for each individual, the value obtained for one variable is not related to (or influenced by) the value for the second variable. This general hypothesis can be expressed in two different conceptual forms, each viewing the data and the test from slightly different perspectives. The data in Table 17.4 describing color preference and personality will be used to present both versions of the null hypothesis.

H_0 **version 1**   For this version of H_0, the data are viewed as a single sample with each individual measured on two variables. The goal of the chi-square test is to evaluate the relationship between the two variables. For the example we are considering, the goal is to determine whether or not there is a consistent, predictable relationship between personality and color preference. That is, if I know your personality, will it help me to predict your color preference? The null hypothesis states that there is no relationship. The alternative hypothesis, H_1, states that there is a relationship between the two variables.

H_0:   For the general population of students, there is no relationship between color preference and personality.

This version of H_0 demonstrates the similarity between the chi-square test for independence and a correlation. In each case, the data consist of two measurements (X and Y) for each individual, and the goal is to evaluate the relationship between the two variables. The correlation, however, requires numerical scores for X and Y. The chi-square test, on the other hand, simply uses frequencies for individuals classified into categories.

H_0 version 2   For this version of H_0, the data are viewed as two (or more) separate samples representing two (or more) separate populations. The goal of the chi-square test is to determine whether or not there are significant differences between the populations. For the example we are considering, the data in Table 17.4 would be viewed as a sample of $n = 50$ introverts (top row) and a separate sample of $n = 150$ extroverts (bottom row). The chi-square test will determine whether the distribution of color preferences for introverts is significantly different from the distribution of color preferences for extroverts. From this perspective, the null hypothesis is stated as follows:

> H_0:   In the general population of students, the distribution of color preferences has the same shape (same proportions) for both categories of personality.

This version of H_0 demonstrates the similarity between the chi-square test and an independent-measures t test (or ANOVA). In each case, the data consist of two (or more) separate samples that are being used to test for differences between two (or more) populations. With the t statistic (or ANOVA) we used sample means to test for differences between population means. The chi-square test, on the other hand, uses sample proportions to test for differences between population proportions. The null hypothesis states that there are no differences between population proportions. The alternative hypothesis, H_1, states that the populations do not have the same shape (same proportions). For the example we are considering, H_1 states that the distribution of color preferences for introverts is different from the distribution of color preferences for extroverts.

Equivalence of H_0 version 1 and H_0 version 2   Although we have presented two different statements of the null hypothesis, you should realize that these two versions are equivalent. The first version of H_0 states that color preference is not related to personality. If this hypothesis is correct, then the distribution of color preferences should not depend on personality. In other words, the distribution of color preferences should be the same for introverts and extroverts, which is the second version of H_0.

For example, if we found that 60% of the introverts preferred red, then H_0 would predict that we also should find that 60% of the extroverts prefer red. In this case, knowing that an individual prefers red does not help you predict his/her personality. Notice that finding the *same proportions* indicates *no relationship*.

On the other hand, if the proportions were different, it would suggest that there is a relationship. For example, if red were preferred by 60% of the extroverts but only 10% of the introverts, then there is a clear, predictable relationship between personality and color preference. (If I know your personality, I can predict your color preference.) Thus, finding *different proportions* means that there is *a relationship* between the two variables.

D E F I N I T I O N

Two variables are **independent** when there is no consistent, predictable relationship between them. In this case, the frequency distribution for one variable is not related to (or dependent on) the categories of the second variable. As a result, when two variables are independent, the frequency distribution for one variable will have the same shape for all categories of the second variable.

Thus, stating that there is no relationship between two variables (version 1 of H_0) is equivalent to stating that the distributions have equal proportions (version 2 of H_0).

OBSERVED AND EXPECTED FREQUENCIES

The chi-square test for independence uses the same basic logic that was used for the goodness-of-fit test. First, a sample is selected, and each individual is classified or categorized. Because the test for independence considers two variables, every individual is classified on both variables, and the resulting frequency distribution is presented as a two-dimensional matrix (see Table 17.4). As before, the frequencies in the sample distribution are called observed frequencies and are identified by the symbol f_o.

The next step is to find the expected frequencies, or f_e values, for this chi-square test. As before, the expected frequencies define an ideal hypothetical distribution that is in perfect agreement with the null hypothesis. Once the expected frequencies are obtained, we will compute a chi-square statistic to determine how well the data (observed frequencies) fit the null hypothesis (expected frequencies).

Although you can use either version of the null hypothesis to find the expected frequencies, the logic of the process is much easier when you use H_0 stated in terms of equal proportions. For the example we are considering, the null hypothesis states

H_0:   The frequency distribution of color preference has the same shape (same proportions) for both categories of personality.

To find the expected frequencies, we first determine the overall distribution of color preferences and then apply this distribution to both categories of personality. Table 17.5 shows an empty matrix corresponding to the data from Table 17.4. Notice that the empty matrix includes all of the row totals and the column totals from the original sample data. The row totals and the column totals are essential for computing the expected frequencies.

TABLE 17.5

An empty frequency distribution matrix showing only the row totals and column totals. (These numbers describe the basic characteristics of the sample from Table 17.4.)

	RED	YELLOW	GREEN	BLUE	
INTROVERT					50
EXTROVERT					150
	100	20	40	40	

The column totals for the matrix describe the overall distribution of color preferences. For these data, 100 people selected red as their preferred color. Because the total sample consists of 200 people, it is easy to determine that the proportion selecting red is 100 out of 200 or 50%. The complete set of color preference proportions is as follows:

100 out of 200 = 50% prefer red

20 out of 200 = 10% prefer yellow

40 out of 200 = 20% prefer green

40 out of 200 = 20% prefer blue

The row totals in the matrix define the two samples of personality types. For example, the matrix shows a total of 50 introverts (the top row) and a sample of 150 extroverts (the bottom row). According to the null hypothesis, both personality groups should have the same proportions for color preferences. To find the expected frequencies, we simply apply the overall distribution of color preferences to each sample. Beginning with the sample of 50 introverts in the top row, we obtain expected frequencies of

$$50\% = 0.50 \text{ choose red:} \quad f_e = 0.50(50) = 25$$

$$10\% = 0.10 \text{ choose yellow:} \quad f_e = 0.10(50) = 5$$

$$20\% = 0.20 \text{ choose green:} \quad f_e = 0.20(50) = 10$$

$$20\% = 0.20 \text{ choose blue:} \quad f_e = 0.20(50) = 10$$

Using exactly the same proportions for the sample of 150 extroverts in the bottom row, we obtain expected frequencies of

$$50\% = 0.50 \text{ choose red:} \quad f_e = 0.50(150) = 75$$

$$10\% = 0.10 \text{ choose yellow:} \quad f_e = 0.10(150) = 15$$

$$20\% = 0.20 \text{ choose green:} \quad f_e = 0.20(150) = 30$$

$$20\% = 0.20 \text{ choose blue:} \quad f_e = 0.20(150) = 30$$

The complete set of expected frequencies are shown in Table 17.6. Notice that the row totals and the column totals for the expected frequencies are the same as those for the original data (the observed frequencies) in Table 17.4.

TABLE 17.6

Expected frequencies corresponding to the data in Table 17.4. (This is the distribution predicted by the null hypothesis.)

	RED	YELLOW	GREEN	BLUE	
INTROVERT	25	5	10	10	50
EXTROVERT	75	15	30	30	150
	100	20	40	40	

A SIMPLE FORMULA FOR DETERMINING EXPECTED FREQUENCIES

Although you should understand that expected frequencies are derived directly from the null hypothesis and the sample characteristics, it is not necessary to go through extensive calculations in order to find f_e values. In fact, there is a simple formula that determines f_e for any cell in the frequency distribution table:

$$f_e = \frac{f_c f_r}{n} \tag{17.4}$$

where f_c is the frequency total for the column (column total), f_r is the frequency total for the row (row total), and n is the number of individuals in the entire sample. To demonstrate this formula, we will compute the expected frequency for introverts selecting yellow in Table 17.6. First, note that this cell is located in the top row and second column in the table. The column total is $f_c = 20$, the row total is $f_r = 50$, and the sample size is $n = 200$. Using these values in Equation 17.4, we obtain

$$f_e = \frac{f_c f_r}{n} = \frac{20(50)}{200} = 5$$

Notice that this is identical to the expected frequency we obtained using percentages from the overall distribution.

THE CHI-SQUARE STATISTIC AND DEGREES OF FREEDOM

The chi-square test of independence uses exactly the same chi-square formula as the test for goodness of fit:

$$\chi^2 = \Sigma \frac{(f_o - f_e)^2}{f_e}$$

As before, the formula measures the discrepancy between the data (f_o values) and the hypothesis (f_e values). A large discrepancy will produce a large value for chi-square and will indicate that H_0 should be rejected. To determine whether a particular chi-square statistic is significantly large, you must first determine degrees of freedom (df) for the statistic and then consult the chi-square distribution in the appendix. For the chi-square test of independence, degrees of freedom are based on the number of cells for which you can freely choose expected frequencies. Recall that the f_e values are partially determined by the sample size (n) and by the row totals and column totals from the original data. These various totals restrict your freedom in selecting expected frequencies. This point is illustrated in Table 17.7. Once three of the f_e values have been selected, all the other f_e values in the table are also determined. In general, the row totals and the column totals restrict the final choices in each row and column. Thus, we may freely choose all but one f_e in each row and all but one f_e in each column. The total number of f_e values that you can freely choose is $(R - 1)(C - 1)$, where R is the number of rows and C is the number of columns. The degrees of freedom for the chi-square test of independence are given by the formula

$$df = (R - 1)(C - 1) \tag{17.5}$$

TABLE 17.7

Degrees of freedom and expected frequencies. (Once three values have been selected, all the remaining expected frequencies are determined by the row totals and the column totals. This example has only three free choices, so $df = 3$.)

RED	YELLOW	GREEN	BLUE	
25	5	10	?	50
?	?	?	?	150
100	20	40	40	

Further discussion of the relationships among chi-square and other statistical procedures is presented in Section 17.5.

Before we begin a step-by-step example of the chi-square test for independence, we ask you to consider a general research situation that demonstrates the similarities among the chi-square test for independence, the Pearson correlation, and the independent-measures t hypothesis test. For this demonstration, we will assume that a researcher is investigating the relationship between academic performance and self-esteem for 10-year-old children. Notice that the researcher has identified two variables, and the general research question concerns the relationship between them. Depending on how the researcher decides to measure the two variables, a correlation, an independent-measures t statistic, or a chi-square test for independence will be the appropriate statistical procedure for evaluating the relationship.

For example, if the researcher obtained a numerical score for both variables, the resulting data would permit the calculation of a Pearson correlation (see Chapter 16). In this case, the data would appear as follows:

PARTICIPANT	ACADEMIC PERFORMANCE SCORE	SELF-ESTEEM SCORE
A	94	31
B	78	26
C	81	27
D	65	23
.	.	.
.	.	.
.	.	.

The researcher would compute means, SS values, and SP for the data, and the Pearson correlation would describe the degree and direction of the relationship between academic performance and self-esteem.

On the other hand, if the researcher measured academic performance by simply classifying individuals into two categories, high and low, and then obtained a numerical score for each individual's self-esteem, the resulting data would be appropriate for an independent-measures t test (see Chapter 10). In this case, the data would appear as follows:

ACADEMIC PERFORMANCE		
HIGH (SAMPLE 1)	LOW (SAMPLE 2)	
31	26	
29	23	Self-esteem
33	25	scores
.	.	
.	.	

Now the researcher has two separate samples and would proceed by computing the mean and SS for each sample. The independent-measures t statistic would be used to determine whether or not there is a significant difference in self-esteem between high academic achievers and low academic achievers. A significant difference would indicate that self-esteem does depend on academic performance; that is, there is a relationship between the two variables.

Finally, the researcher may choose simply to classify individuals into categories for both variables. For example, each student could be classified as either high or low for academic performance and classified as high, medium, or low for self-esteem. The resulting data would produce a frequency distribution that could be displayed in a matrix such as the data shown in Table 17.8. Notice that these data do not involve any numerical scores, but rather consist of a set of frequencies appropriate for a chi-square test.

The following example will use the data in Table 17.8 to demonstrate the chi-square test for independence. Before we begin the chi-square test, however, please note once again that the Pearson correlation, the independent-measures t, and the chi-square test for independence all serve the same general purpose; that is, they all are intended to evaluate the relationship between two variables. Thus, the chi-square test for independence can be viewed as an alternative to the other statistical procedures, specifically, an alternative that is available when research data consist of frequencies instead of numerical scores.

TABLE 17.8

A frequency distribution showing the level of self-esteem according to the level of academic performance for a sample of $n = 150$ ten-year-old children.

		LEVEL OF SELF-ESTEEM			
		HIGH	MEDIUM	LOW	
ACADEMIC PERFORMANCE	HIGH	17	32	11	60
	LOW	13	43	34	90
		30	75	45	$n = 150$

EXAMPLE 17.2   Once again, our researcher is investigating the relationship between academic performance and self-esteem. A sample of $n = 150$ ten-year-old children is obtained, and each child is classified by level of academic performance and level of self-esteem. The frequency distribution for this sample, the set of observed frequencies, is shown in Table 17.8.

STEP 1   State the hypotheses, and select a level of significance. According to the null hypothesis, the two variables are independent. This general hypothesis can be stated in two different ways:

Version 1

H_0:   In the general population, there is no relationship between academic performance and self-esteem.

This version of H_0 emphasizes the similarity between the chi-square test and a correlation. The corresponding alternative hypothesis would state:

H_1:   There is a consistent, predictable relationship between academic performance and self-esteem.

Version 2

H_0:   In the general population, the distribution of self-esteem is the same for high and low academic performers.

The corresponding alternative hypothesis would state:

H_1:   The distribution for self-esteem for high academic performers is different from the distribution for low academic performers.

The second version of H_0 emphasizes the similarity between the chi-square test and the independent-measures t test.

Remember that the two versions for the hypotheses are equivalent. The choice between them is largely determined by how the researcher wants to describe the outcome. For example, a researcher may want to emphasize the *relationship* between variables or may want to emphasis the *difference* between groups.

For this test, we will use $\alpha = .05$.

STEP 2   Determine the degrees of freedom, and locate the critical region. For the chi-square test for independence,

$$df = (R - 1)(C - 1)$$

Therefore, for this study,

$$df = (2 - 1)(3 - 1) = 2$$

With $df = 2$ and $\alpha = .05$, the critical value for chi-square is 5.99 (see Table B.8, page 699).

STEP 3   Determine the expected frequencies, and compute the chi-square statistic. The following table shows an empty matrix with the same row totals and column totals as the original data. The calculation of expected frequencies requires that this table be filled in so that the resulting values provide an ideal frequency distribution that perfectly represents the null hypothesis.

		LEVEL OF SELF-ESTEEM			
		HIGH	MEDIUM	LOW	
ACADEMIC PERFORMANCE	HIGH				60
	LOW				90
		30	75	45	$n = 150$

The column totals indicate that 30 out of 150 participants are classified as high self-esteem. This value represents a proportion of $\frac{30}{150}$ or 20% of the participants. Similarly, $\frac{75}{150} = 50\%$ are medium self-esteem, and $\frac{45}{150} = 30\%$ are low self-esteem. The null hypothesis (version 2) states that this distribution of self-esteem is the same for high and low performers. Therefore, we simply apply the same proportions to each group to obtain the expected frequencies. For the 60 students classified as high academic performers, it is expected that

20% of 60 = 12 students would have high self-esteem

50% of 60 = 30 students would have medium self-esteem

30% of 60 = 18 students would have low self-esteem

For the low academic performers, it is expected that

20% of 90 = 18 students would have high self-esteem

50% of 90 = 45 students would have medium self-esteem

30% of 90 = 27 students would have low self-esteem

These expected frequencies are summarized in Table 17.9.

TABLE 17.9

The expected frequencies (f_e values) that would be predicted if academic performance and self-esteem were completely independent.

		LEVEL OF SELF-ESTEEM			
		HIGH	MEDIUM	LOW	
ACADEMIC PERFORMANCE	HIGH	12	30	18	60
	LOW	18	45	27	90
		30	75	45	$n = 150$

The chi-square statistic is now used to measure the discrepancy between the data (the observed frequencies in Table 17.8) and the null hypothesis that was used to generate the expected frequencies in Table 17.9.

$$\chi^2 = \frac{(17 - 12)^2}{12} + \frac{(32 - 30)^2}{30} + \frac{(11 - 18)^2}{18}$$

$$+ \frac{(13 - 18)^2}{18} + \frac{(43 - 45)^2}{45} + \frac{(34 - 27)^2}{27}$$

$$= 2.08 + 0.13 + 2.72 + 1.39 + 0.09 + 1.81$$

$$= 8.22$$

STEP 4   Make a decision regarding the null hypothesis and the outcome of the study. The obtained chi-square value exceeds the critical value (5.99). Therefore, the decision is to reject the null hypothesis. In the literature, this would be reported as a significant result with $\chi^2(2, n = 150) = 8.22, p < .05$. According to version 1 of H_0, this means that we have decided that there is a significant relationship between academic performance and self-esteem. Expressed in terms of version 2 of H_0, the data show a significant difference between the distribution of self-esteem for high academic performers versus low academic performers. To describe the details of the significant result, you must compare the original data (Table 17.8) with the expected frequencies from the null hypothesis (Table 17.9). Looking at the two tables, it should be clear that the high performers had higher self-esteem than would be expected if the two variables were independent and the low performers had lower self-esteem than would be expected.

LEARNING CHECK   1. A researcher suspects that color blindness is inherited by a sex-linked gene. This possibility is examined by looking for a relationship between gender and color vision. A sample of 1000 people is tested for color blindness, and then they are classified according to their sex and color vision status (normal, red-green blind, other color blindness). Is color blindness related to gender? The data are as follows:

OBSERVED FREQUENCIES OF COLOR VISION STATUS ACCORDING TO SEX

	NORMAL COLOR VISION	RED-GREEN COLOR BLINDNESS	OTHER COLOR BLINDNESS	TOTALS
MALE	320	70	10	400
FEMALE	580	10	10	600
TOTALS	900	80	20	

a. State the hypotheses.

b. Determine the value for *df*, and locate the critical region.

c. Compute the f_e values and then chi-square.

d. Make a decision regarding H_0.

ANSWERS   **1. a.** H_0: In the population, there is no relationship between gender and color vision.
H_1: In the population, gender and color vision are related.

b. $df = 2$; critical $\chi^2 = 5.99$ for $\alpha = .05$.

c. f_e values are as follows:

EXPECTED FREQUENCIES

	NORMAL	RED-GREEN	OTHER
MALE	360	32	8
FEMALE	540	48	12

Obtained $\chi^2 = 83.44$

d. Reject H_0.

17.4 ASSUMPTIONS AND RESTRICTIONS FOR CHI-SQUARE TESTS

To use a chi-square test for goodness of fit or a test of independence, several conditions must be satisfied. For any statistical test, violation of assumptions and restrictions will cast doubt on the results. For example, the probability of committing a Type I error may be distorted when assumptions of statistical tests are not satisfied. Some important assumptions and restrictions for using chi-square tests are the following:

1. Independence of Observations. This is *not* to be confused with the concept of independence between *variables* as seen in the test of independence (Section 17.3). One consequence of independent observations is that each observed frequency is generated by a different subject. A chi-square test would be inappropriate if a person could produce responses that can be classified in more than one category or contribute more than one frequency count to a single category. (See page 251 for more information on independence.)

2. Size of Expected Frequencies. A chi-square test should not be performed when the expected frequency of any cell is less than 5. The chi-square statistic can be distorted when f_e is very small. Consider the chi-square computations for a single cell. Suppose that the cell has values of $f_e = 1$ and $f_o = 5$. The contribution of this cell to the total chi-square value is

$$\text{cell} = \frac{(f_o - f_e)^2}{f_e} = \frac{(5 - 1)^2}{1} = \frac{4^2}{1} = 16$$

Now consider another instance, in which $f_e = 10$ and $f_o = 14$. The difference between the observed and the expected frequencies is still 4, but the contribution of this cell to the total chi-square value differs from that of the first case:

$$\text{cell} = \frac{(f_o - f_e)^2}{f_e} = \frac{(14 - 10)^2}{10} = \frac{4^2}{10} = 1.6$$

It should be clear that a small f_e value can have a great influence on the chi-square value. This problem becomes serious when f_e values are less than 5. When f_e is very small, what would otherwise be a minor discrepancy between f_o and f_e will now result in large chi-square values. The test is too sensitive when f_e values are extremely small. One way to avoid small expected frequencies is to use large samples.

17.5   SPECIAL APPLICATIONS OF THE CHI-SQUARE TESTS

At the beginning of this chapter, we introduced the chi-square tests as examples of nonparametric tests. Although nonparametric tests serve a function that is uniquely their own, they also can be viewed as alternatives to the common parametric techniques that were examined in earlier chapters. In general, nonparametric tests are used as substitutes for parametric techniques in situations where one of the following occurs:

1. The data do not meet the assumptions needed for a standard parametric test.
2. The data consist of nominal or ordinal measurements, so that it is impossible to compute standard descriptive statistics such as the mean and standard deviation.

In this section, we will examine some of the relationships between chi-square tests and the parametric procedures for which they may substitute.

CHI-SQUARE AND THE PEARSON CORRELATION   The chi-square test for independence and the Pearson correlation are both statistical techniques intended to evaluate the relationship between two variables. The type of data obtained in a research study determines which of these two statistical procedures is appropriate.

The Pearson correlation is used to evaluate a relationship in situations where both variables consist of numerical values; that is, X and Y are numbers obtained from measurement on interval or ratio scales. The chi-square test for independence, on the other hand, is used when the data are obtained by classifying individuals into categories, usually determined by a nominal or ordinal scale of measurement. Although both statistical techniques are used to measure the relationship between two variables, the distinction between them is determined by the type of data that a researcher has obtained. For the chi-square test, the data are frequencies, not scores. For the Pearson correlation, the scores are numerical values.

THE PHI-COEFFICIENT   In Chapter 16 (page 550), we introduced the phi-coefficient as a correlational statistic for measuring the degree of association between two dichotomous variables. A dichotomous variable is one for which there are exactly two categories of measurement;

for example, gender can be classified as male or female; people's opinions concerning a new law can be classified as "for" or "against." The phi-coefficient allows you to compute a correlation measuring the degree of relationship between two such variables.

The same data that are used to compute the phi-coefficient can be reorganized into a matrix of frequencies that is suitable for the chi-square test for independence. This process is shown in Figure 17.5. In the figure, the original data show that each individual is classified according to gender (male, female) and according to opinion concerning a proposal to ban smoking in all college buildings (for, against). To compute the phi-coefficient, the original data are transformed into numerical values by substituting values of 0 and 1 for the two categories of each variable, and the Pearson formula is used. The transformed data are shown in the lower left-hand side of Figure 17.5. On the other hand, the original two variables can be used to create a 2×2 matrix, with the categories of one variable determining the two columns and the categories of the second variable determining the two rows of the matrix. This process is demonstrated in the lower right-hand side of Figure 17.5. The number in each cell of the matrix is the frequency, or number of individuals classified in that cell. The chi-square statistic is computed from the frequency data using the procedures described in Section 17.3. The value for the phi-coefficient (ϕ) may be computed directly from chi-square by the following formula:

$$\phi = \sqrt{\frac{\chi^2}{n}}$$

(17.6)

FIGURE 17.5

Hypothetical data showing the first four individuals in a sample where each person is classified on two dichotomous variables. The original data (top of figure) can be reorganized into a form suitable for computing the phi-coefficient (lower left) or into a form suitable for computing a chi-square test for independence (lower right).

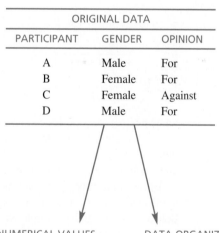

ORIGINAL DATA

PARTICIPANT	GENDER	OPINION
A	Male	For
B	Female	For
C	Female	Against
D	Male	For

DATA CODED AS NUMERICAL VALUES (0 AND 1) FOR COMPUTING THE PHI-COEFFICIENT

PARTICIPANT	GENDER	OPINION
A	0	0
B	1	0
C	1	1
D	0	0

DATA ORGANIZED IN A FREQUENCY DISTRIBUTION FOR COMPUTING CHI-SQUARE

	OPINION	
	FOR	AGAINST
MALE	2	0
FEMALE	1	1

Although the phi-coefficient and the chi-square value are mathematically related, they measure different characteristics of a relationship and are in some respects independent. The phi-coefficient is a correlation and measures the *strength* of the relationship. On the other hand, chi-square is a hypothesis test and evaluates the *significance* of the relationship (Is it more than can be reasonably explained by chance?). In particular, the phi-coefficient is influenced only by the proportions that exist in the data, but chi-square is influenced by the proportions and by the sample size. This distinction is demonstrated in the following example.

EXAMPLE 17.3 The following data show a frequency distribution evaluating the relationship between gender and preference between two candidates for student president.

CANDIDATE

	A	B
Male	5	10
Female	10	5

Note that the data show that males prefer candidate B by a 2-to-1 margin and females prefer candidate A by 2 to 1. Also note that the sample includes a total of 15 males and 15 females. We will not perform all the arithmetic here, but these data produce chi-square equal to 3.33 (which is not significant) and a phi-coefficient of 0.3332.

Next we will keep exactly the same proportions in the data, but double all of the frequencies. The resulting data are as follows:

CANDIDATE

	A	B
Male	10	20
Female	20	10

Once again, males prefer candidate B by 2 to 1 and females prefer candiate A by 2 to 1. However, the sample now contains 30 males and 30 females. For these new data, the value of chi-square is 6.66, twice as big as it was before (and now significant with $\alpha = .05$), but the value of the phi-coefficient is still 0.3332.

Because the proportions are the same for the two samples, the value of the phi-coefficient is unchanged. However, the larger sample provides more convincing evidence than the smaller sample, so the larger sample is more likely to produce a significant result.

INDEPENDENT-MEASURES *t* AND ANOVA

The independent-measures *t* test and ANOVA are statistical procedures used to examine the relationship between an independent variable and a dependent variable. Both tests require that the scores for the dependent variable consist of numerical values measured on an interval or a ratio scale. The chi-square test for independence often can be used as a substitute for *t* or ANOVA, particularly in the following situations:

1. The independent variable is actually a quasi-independent variable (Chapter 1) consisting of distinct subject groups (men versus women; 8-year-olds versus 10-year-olds).

2. The dependent variable involves classifying individuals into nominal or ordinal categories.

For example, suppose that a researcher is interested in examining the difference in vocabulary skills between 4-year-old boys and girls. The data for this study would require two independent samples (boys versus girls). If each child's vocabulary skill were measured by a numerical score, the mean difference between boys and girls could be evaluated using either an independent-measures t test or an ANOVA. On the other hand, if each child were simply classified as being either high, medium, or low with respect to vocabulary skill, the data would be suitable for a chi-square test. Examples of both types of data are shown in Table 17.10.

TABLE 17.10

Two possible sets of data from a study comparing vocabulary skills for 4-year-old boys versus girls. [In data set A, vocabulary skill is measured by numerical scores suitable for an independent-measures t (or ANOVA) hypothesis test. In data set B, each child's vocabulary skill is classified into one of three categories (high, medium, low), and the numbers represent the frequency, or number of children in each category.]

DATA SET A VOCABULARY SCORES	
BOYS	GIRLS
18	20
4	9
21	24
17	18
10	5
3	11
.	.
.	.
.	.

DATA SET B FREQUENCY DISTRIBUTION OF VOCABULARY SKILL		
	BOYS	GIRLS
HIGH	6	11
MEDIUM	12	14
LOW	8	7

THE MEDIAN TEST FOR INDEPENDENT SAMPLES

The median test provides a nonparametric alternative to the independent-measures t test (or ANOVA) to determine whether or not there are significant differences among two or more independent samples. The null hypothesis for the median test states that the different samples come from populations that share a common median (no differences). The alternative hypothesis states that the samples come from populations that are different and do not share a common median.

The logic behind the median test is that whenever several different samples are selected from the same population distribution, roughly half of the scores in each sample should be above the population median and roughly half should be below. That is, all the separate samples should be distributed around the same median. On the other hand, if the samples come from populations with different medians, then the scores in some samples will be consistently higher and the scores in other samples will be consistently lower.

The first step in conducting the median test is to combine all the scores from the separate samples and then find the median for the combined group (see Chapter 3, page 82, for instructions for finding the median). Next, a matrix is constructed with a column for each of the separate samples and two rows: one for individuals above the median and one for individuals below the median. Finally, for each sample, you count how many individuals scored above the combined median and how many scored below. These values are the observed frequencies that are entered in the matrix.

The frequency distribution matrix is evaluated using a chi-square test for independence. The expected frequencies and a value for chi-square are computed exactly as described in Section 17.3. A significant value for chi-square indicates that the discrepancy between the individual sample distributions is greater than would be expected by chance.

The median test is demonstrated in the following example.

EXAMPLE 17.4   The following data represent self-esteem scores obtained from a sample of $n = 40$ children. The children are then separated into three groups based on their level of academic performance (high, medium, low). The median test will evaluate whether there is a significant relationship between self-esteem and level of academic performance.

SELF-ESTEEM SCORES FOR CHILDREN
AT THREE LEVELS OF ACADEMIC PERFORMANCE

HIGH		MEDIUM				LOW	
22	14	22	13	24	20	11	19
19	18	18	22	10	16	13	15
12	21	19	15	14	19	20	16
20	18	11	18	11	10	10	18
23	20	12	19	15	12	15	11

The median for the combined group of $n = 40$ scores is $X = 17$ (exactly 20 scores are above this value and 20 are below). For the first sample, 8 out of 10 scores are above the combined median. For the second sample, 9 out of 20 are above the median, and for the third sample, only 3 out of 10 are above the median. These observed frequencies are shown in the following matrix:

ACADEMIC PERFORMANCE

	HIGH	MEDIUM	LOW
ABOVE MEDIAN	8	9	3
BELOW MEDIAN	2	11	7

The expected frequencies for this test are as follows:

ACADEMIC PERFORMANCE

	HIGH	MEDIUM	LOW
ABOVE MEDIAN	5	10	5
BELOW MEDIAN	5	10	5

The chi-square statistic is

$$\chi^2 = \frac{9}{5} + \frac{9}{5} + \frac{1}{10} + \frac{1}{10} + \frac{4}{5} + \frac{4}{5} = 5.40$$

With $df = 2$ and $\alpha = .05$, the critical value for chi-square is 5.99. The obtained chi-square of 5.40 does not fall in the critical region, so we would fail to reject the null hypothesis. These data do not provide sufficient evidence to conclude that there are significant differences among the self-esteem distributions for these three groups of students.

A few words of caution are in order concerning the interpretation of the median test. First, the median test is *not* a test for mean differences. Remember: The mean for a distribution can be strongly affected by a few extreme scores. Therefore, the mean and the median for a distribution are not necessarily the same, and they may not even be related. The results from a median test *cannot* be interpreted as indicating that there is (or is not) a difference between means.

Second, you may have noted that the median test does not directly compare the median from one sample with the median from another. Thus, the median test is not a test for significant differences between medians. Instead, this test compares the distribution of scores for one sample versus the distribution for another sample. If the samples are distributed evenly around a common point (the group median), the test will conclude that there is no significant difference. On the other hand, finding a significant difference simply indicates that the samples are not distributed evenly around the common median. Thus, the best interpretation of a significant result is that there is a *difference in the distributions* of the samples.

17.6 MEASURING EFFECT SIZE FOR THE CHI-SQUARE TEST FOR INDEPENDENCE

A hypothesis test, like the chi-square test for independence, evaluates the statistical significance of the results from a research study. Specifically, the intent of the test is to determine whether the patterns or relationships observed in the sample data are likely to have occurred simply by chance; that is, without any corresponding patterns or relationships in the population. Tests of significance are influenced not only by the size or strength of the treatment effects but also by the size of the samples. As a result, even a small effect can be statistically significant if it is observed in a very large sample. Because a significant effect does not necessarily mean a large effect, it is generally recommended that the outcome of a hypothesis test be accompanied by a measure of the effect size. This general recommendation also applies to the chi-square test for independence.

THE PHI-COEFFICIENT AND CRAMÉR'S *V*   As noted in the previous section, when a chi-square test for independence involves two dichotomous variables (the data form a 2×2 matrix), it is possible to compute a phi-coefficient as a correlational alternative to the chi-square value. Because phi (ϕ) is a correlation, it measures the strength of the relationship (rather than the significance) and therefore can be viewed as a measure of effect size. The interpretation of ϕ follows the same standards used to evaluate a correlation (see Table 9.3, page 295): 0.10 is a small effect, 0.30 is a medium effect, and 0.50 is a large effect. Occasionally, the value of ϕ^2 is reported as a percentage of variance accounted for, exactly the same as r^2.

When the chi-square test involves a matrix larger than 2×2, there is a modification of the phi-coefficient, known as Cramér's V, that can be used to measure effect size.

$$V = \sqrt{\frac{\chi^2}{n(df^*)}} \qquad\qquad (17.7)$$

Note that the formula for Cramér's V (17.7) is identical to the formula for the phi-coefficient (17.6) except for the addition of df^* in the denominator. The df^* value is *not* the same as the degrees of freedom for the chi-square test, but it is related. You should recall that the chi-square test for independence has $df = (R - 1)(C - 1)$, where R is the number of rows in the table and C is the number of columns. For Cramér's V, the value of df^* is the smaller of either $(R - 1)$ or $(C - 1)$.

Cohen (1988) has also suggested standards for interpreting Cramér's V that are shown in Table 17.11. Note that when $df^* = 1$, as in a 2×2 matrix, the criteria for interpreting V are exactly the same as the criteria for interpreting a regular correlation or a phi-coefficient.

We will use the results from Example 17.2 (page 596) to demonstrate the calculation of Cramér's V. The example evaluated the relationship between academic performance and self-esteem. There were two levels of academic performance and three levels of self-esteem, producing a 2×3 table with a total of $n = 150$ participants. The data produced $\chi^2 = 8.22$. Using these values we obtain

$$V = \sqrt{\frac{\chi^2}{n(df^*)}} = \sqrt{\frac{8.22}{150(1)}} = \sqrt{0.055} = 0.23$$

According to Cohen's guidelines (Table 17.11), this value indicates a small relationship.

TABLE 17.11

Standards for interpreting Cramér's V as proposed by Cohen (1988).

For $df^* = 1$	$0.10 < V < 0.30$	Small effect
	$0.30 < V < 0.50$	Medium effect
	$V > 0.50$	Large effect
For $df^* = 2$	$0.07 < V < 0.21$	Small effect
	$0.21 < V < 0.35$	Medium effect
	$V > 0.35$	Large effect
For $df^* = 3$	$0.06 < V < 0.17$	Small effect
	$0.17 < V < 0.29$	Medium effect
	$V > 0.29$	Large effect

SUMMARY

1. Chi-square tests are a type of nonparametric technique that tests hypotheses about the form of the entire frequency distribution. Two types of chi-square tests are the test for goodness of fit and the test for independence. The data for these tests consist of the frequency of observations that fall into various categories of a variable.

2. The test for goodness of fit compares the frequency distribution for a sample to the frequency distribution that is predicted by H_0. The test determines how well the observed frequencies (sample data) fit the expected frequencies (data predicted by H_0).

3. The expected frequencies for the goodness-of-fit test are determined by

$$\text{expected frequency} = f_e = pn$$

where p is the hypothesized proportion (according to H_0) of observations falling into a category and n is the size of the sample.

4. The chi-square statistic is computed by

$$\text{chi-square} = \chi^2 = \Sigma \frac{(f_o - f_e)^2}{f_e}$$

where f_o is the observed frequency for a particular category and f_e is the expected frequency for that category. Large values for χ^2 indicate that there is a large discrepancy between the observed (f_o) and the expected (f_e) frequencies and may warrant rejection of the null hypothesis.

5. Degrees of freedom for the test for goodness of fit are

$$df = C - 1$$

where C is the number of categories in the variable. Degrees of freedom measure the number of categories for which f_e values can be freely chosen. As can be seen from the formula, all but the last f_e value to be determined are free to vary.

6. The chi-square distribution is positively skewed and begins at the value of zero. Its exact shape is determined by degrees of freedom.

7. The test for independence is used to assess the relationship between two variables. The null hypothesis states that the two variables in question are independent of each other. That is, the frequency distribution for one variable does not depend on the categories of the second variable. On the other hand, if a relationship does exist, then the form of the distribution for one variable will depend on the categories of the other variable.

8. For the test for independence, the expected frequencies for H_0 can be directly calculated from the marginal frequency totals,

$$f_e = \frac{f_c f_r}{n}$$

where f_c is the total column frequency and f_r is the total row frequency for the cell in question.

9. Degrees of freedom for the test for independence are computed by

$$df = (R - 1)(C - 1)$$

where R is the number of row categories and C is the number of column categories.

10. For the test of independence, a large chi-square value means there is a large discrepancy between the f_o and f_e values. Rejecting H_0 in this test provides support for a relationship between the two variables.

11. Both chi-square tests (for goodness of fit and independence) are based on the assumption that each observation is independent of the others. That is, each observed frequency reflects a different individual, and no individual can produce a response that would be classified in more than one category or more than one frequency in a single category.

12. The chi-square statistic is distorted when f_e values are small. Chi-square tests, therefore, are restricted to situations in which f_e values are 5 or greater. The test should not be performed when the expected frequency of any cell is less than 5.

13. The test for independence also can be used to test whether two sample distributions share a common median. Observed frequencies are obtained by classifying each individual as above or below the median value for the combined samples. A significant value for chi-square indicates a difference between the two sample distributions.

14. The effect size for a chi-square test for independence can be measured computing a phi-coefficient for data that form a 2×2 matrix or computing Cramér's V for a matrix that is larger than 2×2.

$$\text{phi} = \sqrt{\frac{\chi^2}{n}} \qquad \text{Cramér's } V = \sqrt{\frac{\chi^2}{n(df^*)}}$$

where df^* is the smaller of $(R - 1)$ and $(C - 1)$. Both phi and Cramér's V are evaluated using the criteria in Table 17.11.

RESOURCES

If you are using WebTutor along with this book, there is a WebTutor section correspon-ding to this chapter. The WebTutor contains a brief summary of Chapter 17, hints for learning the concepts and formulas for the chi-square tests, cautions about common errors, and sample exam items including solutions.

SPSS

The statistics computer package SPSS can be used to perform both of the chi-square tests presented in this chapter. However, the program requires that the score(s) for each partici-pant be entered individually. This system works well for preexisting data sheets (for ex-ample, data from a large survey that have already been entered into the computer). However, it can be very tedious if you simply want to use SPSS for a single problem.

The test for goodness of fit was demonstrated using Example 17.1 in which partici-pants were asked to determine the orientation for an abstract painting (deciding which side should go up). The example used a total of $n = 50$ participants. To perform the chi-square test with SPSS, you would first need to enter the 50 scores into one column of the data matrix. The scores are numerically coded—for example, enter a 1 for each of the 18 peo-ple who placed the top side up, enter a 2 for each of the 17 people who placed the bottom side up, and so on. Then click **Analyze** on the tool bar, select **Nonparametric Tests**, and click on **Chi-square**. Next you must enter the label for the data column into the Test Variables List box. Notice that you may either use the **All Categories Equal** option for the Expected Frequencies, or you can select **Values** and enter specific numbers. Then click **OK**. SPSS will produce a table listing the observed and expected frequencies, and a second table containing the calculated value of chi-square, the degrees of freedom, and the level of significance (the p value).

The test for independence was demonstrated using Example 17.2, which examined the relationship between academic performance and self-esteem for a sample of $n = 150$ students. To perform the chi-square test with SPSS, you would need to enter the academic performance score for each of the 150 participants in one column (coded as high = 1 and low = 2), and enter the same participant's self-esteem score (high = 1, medium =2, low = 3) in a second column. Note that you are entering a total of 300 scores. Then click **Analyze** on the tool bar, select **Descriptive Statistics**, and click on **Crosstabs**. Enter the column label for the academic scores into the Rows box, and enter the column label for the self-esteem scores into the Columns box. Then click on **Statistics**. Select **Chi-Square**, click **Continue**, and click **OK**. The SPSS output will include a cross-tabulation table showing the matrix of observed frequencies, and a table of chi-square tests in which you should focus on the Pearson chi-square. The table includes the calculated chi-square value, the degrees of freedom, and the level of significance (p value).

KEY TERMS

goodness-of-fit test	chi-square statistic	median test	parametric test
observed frequencies	distribution of chi-square	test for independence	nonparametric test
expected frequencies	Cramér's V		

FOCUS ON PROBLEM SOLVING

1. The expected frequencies that you calculate must satisfy the constraints of the sample. For the goodness-of-fit test, $\Sigma f_e = \Sigma f_o = n$. For the test of independence, the row totals and column totals for the expected frequencies should be identical to the corresponding totals for the observed frequencies.

2. It is entirely possible to have fractional (decimal) values for expected frequencies. Observed frequencies, however, are always whole numbers.

3. Whenever $df = 1$, the difference between observed and expected frequencies ($f_o - f_e$) will be identical (the same value) for all cells. This makes the calculation of chi-square easier.

4. Although you are advised to compute expected frequencies for all categories (or cells), you should realize that it is not essential to calculate all f_e values separately. Remember that df for chi-square identifies the number of f_e values that are free to vary. Once you have calculated that number of f_e values, the remaining f_e values are determined. You can get these remaining values by subtracting the calculated f_e values from their corresponding row or column totals.

5. Remember that, unlike previous statistical tests, the degrees of freedom (df) for a chi-square test are *not* determined by the sample size (n). Be careful!

DEMONSTRATION 17.1

TEST FOR INDEPENDENCE

A manufacturer of watches would like to examine preferences for digital versus analog watches. A sample of $n = 200$ people is selected, and these individuals are classified by age and preference. The manufacturer would like to know whether there is a relationship between age and watch preference. The observed frequencies (f_o) are as follows:

		PREFERENCE		
		DIGITAL	ANALOG	UNDECIDED
AGE	UNDER 30	90	40	10
	OVER 30	10	40	10

STEP 1   *State the hypotheses, and select an alpha level.* The null hypothesis states that there is no relationship between the two variables.

H_0:   Preference is independent of age. That is, the frequency distribution of preference has the same form for people under 30 as for people over 30.

The alternative hypothesis states that there is a relationship between the two variables.

H_1:   Preference is related to age. That is, the type of watch preferred depends on a person's age.

We will set alpha to $\alpha = .05$.

STEP 2   *Locate the critical region.* Degrees of freedom for the chi-square test for independence are determined by

$$df = (C - 1)(R - 1)$$

For these data,

$$df = (3 - 1)(2 - 1) = 2(1) = 2$$

For $df = 2$ with $\alpha = .05$, the critical chi-square value is 5.99. Thus, our obtained chi-square must exceed 5.99 to be in the critical region and to reject H_0.

STEP 3   *Compute the test statistic.* Computing the chi-square statistic requires the following preliminary calculations:

1. Obtain the row and column totals.
2. Calculate expected frequencies.

Row and column totals. We start by determining the row and column totals from the original observed frequencies, f_o.

	DIGITAL	ANALOG	UNDECIDED	TOTALS
UNDER 30	90	40	10	140
OVER 30	10	40	10	60
COLUMN TOTALS	100	80	20	$n = 200$

Expected frequencies, f_e. For the test for independence, the following formula is used to obtain expected frequencies:

$$f_e = \frac{f_c f_r}{n}$$

For people under 30, we obtain the following expected frequencies:

$$f_e = \frac{100(140)}{200} = \frac{14,000}{200} = 70 \text{ for digital}$$

$$f_e = \frac{80(140)}{200} = \frac{11,200}{200} = 56 \text{ for analog}$$

$$f_e = \frac{20(140)}{200} = \frac{2800}{200} = 14 \text{ for undecided}$$

For individuals over 30, the expected frequencies are as follows:

$$f_e = \frac{100(60)}{200} = \frac{6000}{200} = 30 \text{ for digital}$$

$$f_e = \frac{80(60)}{200} = \frac{4800}{200} = 24 \text{ for analog}$$

$$f_e = \frac{20(60)}{200} = \frac{1200}{200} = 6 \text{ for undecided}$$

The following table summarizes the expected frequencies:

	DIGITAL	ANALOG	UNDECIDED
UNDER 30	70	56	14
OVER 30	30	24	6

The chi-square statistic. The chi-square statistic is computed from the formula

$$\chi^2 = \Sigma \frac{(f_o - f_e)^2}{f_e}$$

That is, we must

1. Find the $f_o - f_e$ difference for each cell.
2. Square these differences.
3. Divide the squared differences by f_e.
4. Sum the results of 3.

The following table summarizes these calculations:

CELL	f_o	f_e	$(f_o - f_e)$	$(f_o - f_e)^2$	$(f_o - f_e)^2/f_e$
Under 30—digital	90	70	20	400	5.71
Under 30—analog	40	56	−16	256	4.57
Under 30—undecided	10	14	−4	16	1.14
Over 30—digital	10	30	−20	400	13.33
Over 30—analog	40	24	16	256	10.67
Over 30—undecided	10	6	4	16	2.67

Finally, we can sum the last column to get the chi-square value.

$$\chi^2 = 5.71 + 4.57 + 1.14 + 13.33 + 10.67 + 2.67$$

$$= 38.09$$

STEP 4 *Make a decision about H_0, and state the conclusion.* The chi-square value is in the critical region. Therefore, we can reject the null hypothesis. There is a relationship between watch preference and age, $\chi^2(2, n = 200) = 38.09$, $p < .05$.

PROBLEMS

1. Parametric tests (such as t or ANOVA) differ from nonparametric tests (such as chi-square) primarily in terms of the assumptions they require and the data they use. Explain these differences.

2. A major snack-food company has started an advertising campaign claiming that it makes "America's favorite potato chip." To test this claim, a researcher assembles a sample of 90 volunteers and asks each individual to taste the company's chips along with two other brands. Each individual is asked to select a favorite from the three brands. The following data show the number of individuals selecting each brand. Do the data provide significant support for any preferences among the three brands? Test at the .05 level of significance.

BRAND A	BRAND B	BRAND C
38	28	24

3. Automobile insurance is much more expensive for teenage drivers than for older drivers. To justify this cost difference, insurance companies claim that the younger drivers are much more likely to be involved in costly accidents. To test this claim, a researcher obtains information about registered drivers from the department of motor vehicles and selects a sample of $n = 300$ accident reports from the police department. The motor vehicle department reports the percentage of registered drivers in each age category as follows: 16% are under age 20; 28% are 20 to 29 years old, and 56% are age 30 or older. The number of accident reports for each age group is as follows:

UNDER AGE 20	AGE 20–29	AGE 30 OR OLDER
68	92	140

Do the data indicate that the distribution of accidents for the three age groups is significantly different from the distribution of drivers? Test with $\alpha = .05$.

4. A social psychologist suspects that people who serve on juries tend to be much older than citizens in the general population. The psychologist obtains voter registration records and finds that 20% of registered voters are between 18 and 29 years old, 45% are between 30 and 49 years old, and 35% are age 50 or older. The psychologist also monitors jury composition over several weeks and observes the following distribution of ages for actual juries:

AGE CATEGORIES FOR JURORS

18–29	30–49	50 AND OVER
12	36	32

Are these data sufficient to conclude that the age distribution for jurors is significantly different from the distribution for the population of registered voters? Test with $\alpha = .05$. (*Note:* Jurors are selected from the population of registered voters.)

5. A developmental psychologist would like to determine whether infants display any color preferences. A stimulus consisting of four color patches (red, green, blue, and yellow) is projected onto the ceiling above a crib. Infants are placed in the crib, one at a time, and the psychologist records how much time an infant spends looking at each of the four colors. The color that receives the most attention during a 100-second test period is identified as the preferred color for that infant. The preferred colors for a sample of 60 infants are shown in the following table:

RED	GREEN	BLUE	YELLOW
21	11	18	10

Do the data indicate any significant preferences among the four colors? Test at the .05 level of significance.

6. A researcher would like to examine the factors that influence a person's decision when buying a new car. The researcher interviews 60 people who are shopping for a new car and asks each person to select the single most important factor that determines which car they will eventually buy. The frequency distribution for the sample is as follows:

FACTORS INFLUENCING DECISION

STYLE	SAFETY	RELIABILITY
12	16	32

Do the data indicate any significant preferences among the three factors, or are the three factors equally important in the general population? Test at the .05 level of significance.

7. A researcher is investigating the physical characteristics that influence whether or not a person's face is judged as beautiful. The researcher selects a photograph of a woman and then creates two modifications of the photo by (1) moving the eyes slightly farther apart and (2) moving the eyes slightly closer together. The original photograph and the two modifications are then shown to a sample of $n = 150$ college students, and each student is asked to select the "most beautiful" of the three faces. The distribution of responses was as follows:

ORIGINAL PHOTO	EYES MOVED APART	EYES MOVED TOGETHER
51	72	27

Do the data indicate any significant preferences among the three versions of the photograph? Test at the .05 level of significance.

8. A researcher would like to determine whether any particular age group has a greater risk of influenza-related death. A sample of 50 such cases is categorized according to the victim's age. The observed frequencies are as follows:

NUMBER OF FLU-RELATED DEATHS

UNDER 30	30–60	OVER 60
5	5	40

Note that in the city from which the sample was selected, 30% of the population is in the "under 30" bracket, 40% is in "30–60," and 30% is in "over 60." (This information should help in determining f_e values.) Can the investigator conclude that risk differs with age? Test with the .05 level of significance.

9. A professor in the psychology department would like to determine whether there has been a significant change in grading practices over the years. It is known that the overall grade distribution for the department in 1985 had 14% As, 26% Bs, 31% Cs, 19% Ds, and 10% Fs. A sample of $n = 200$ psychology students from last semester produced the following grade distribution:

A	B	C	D	F
32	61	64	31	12

Do the data indicate a significant change in the grade distribution? Test at the .05 level of significance.

10. A researcher obtained a random sample of $n = 60$ students to determine whether there were any significant preferences among three leading brands of colas. Each student tasted all three brands and then selected his or her favorite. The resulting frequency distribution is as follows:

BRAND A	BRAND B	BRAND C
28	14	18

Are the data sufficient to indicate any preferences among the three brands? Test with $\alpha = .05$.

11. Suppose the researcher from the previous problem repeated the cola-preference study using twice as many students and obtaining observed frequencies that exactly doubled the original values. The resulting data are as follows:

BRAND A	BRAND B	BRAND C
56	28	36

a. Use a chi-square test to determine whether the data indicate any significant preferences among the three brands.

b. You should find that the conclusion from the hypothesis test is that there are significant preferences (reject H_0). However, in Problem 10 the decision was to fail to reject H_0. How do you explain the difference between the two tests?

12. A psychology professor is trying to decide which textbook to use for next year's introductory class. To help make the decision, the professor asks the current students to review three texts and identify which one they prefer. The distribution of preferences for the current class is as follows:

BOOK 1	BOOK 2	BOOK 3
52	41	27

Do the data indicate any significant preferences among the three books? Test with $\alpha = .05$.

13. In Problem 12, a professor asked students to rate three textbooks to determine whether there were any preferences among them. Although the data appear to indicate an overall preference for book 1, the professor would like to know whether this opinion is shared by students with different levels of academic ability. To answer this question, the median grade was used to separate the students into two groups: the upper half and the lower half of the class. The distribution of preferences for these two groups is as follows:

	BOOK 1	BOOK 2	BOOK 3	
UPPER HALF	17	31	12	60
LOWER HALF	35	10	15	60
	52	41	27	

Do the data indicate that the distribution of preferences for students in the upper half of the class is significantly different from the distribution for students in the lower half? Test at the .05 level of significance.

14. Gender differences in dream content are well documented (see Winget & Kramer, 1979). Suppose a researcher studies aggression content in the dreams of men and women. Each participant reports his or her most recent dream. Then each dream is judged by a panel of experts to have low, medium, or high aggression content. The observed frequencies are shown in the following matrix:

		AGGRESSION CONTENT		
		LOW	MEDIUM	HIGH
GENDER	FEMALE	18	4	2
	MALE	4	17	15

Is there a relationship between gender and the aggression content of dreams? Test with $\alpha = .01$.

15. Recent reports suggest that children who grow up with pets in the home tend to develop resistence to allergies. To test this phenomenon, a researcher interviews a sample of $n = 120$ college students. Each student is asked about pets during childhood and about his or her current allergies. Do the following data indicate a significant relationship between childhood pets and allergies? Test at the .05 level of significance.

NUMBER OF DOGS OR CATS
IN CHILDHOOD HOME

	0	1	2 OR MORE	
NO ALLERGIES	22	50	18	90
ALLERGIES	18	10	2	30
	40	60	20	

16. Research indicates that people who volunteer to participate in research studies tend to have higher intelligence than nonvolunteers. To test this phenomenon, a researcher obtains a sample of 200 high school students. The students are given a description of a psychological research study and asked whether or not they would volunteer to participate. The researcher also obtains an IQ score for each student and classifies the students into high, medium, and low IQ groups. Do the following data indicate a significant relationship between IQ and volunteering? Test at the .05 level of significance.

IQ

	HIGH	MEDIUM	LOW	
VOLUNTEER	43	73	34	150
NOT VOLUNTEER	7	27	16	50
	50	100	50	

17. In the Preview section of this chapter, we presented a study by Cialdini, Reno, and Kallgren (1990) examining how people conform to norms concerning littering. The researchers wanted to determine whether a person's tendency to litter depended on the amount of litter already in the area. People were handed a handbill as they entered an amusement park that already had either 0, 1, 2, 4, 8, or 16 handbills lying on the ground. Then the people were observed to determine whether or not they dropped the handbill on the ground. The frequency data are as follows:

AMOUNT OF EXISTING LITTER

	0 OR 1 PIECE	2 OR 4 PIECES	8 OR 16 PIECES
LITTERING	17	28	49
NOT LITTERING	102	91	71

Do these data indicate a significant tendency for people to litter when there already is litter in the area? That is, does the proportion of litterers depend on the amount of litter? Test at the .05 level of significance.

18. Last fall the college installed a new e-mail system and conducted a series of training sessions to teach students and staff how to use the system. In the spring semester, the college used a survey to determine the level of satisfaction with the new system. In addition to measuring satisfaction, the survey asked whether or not each individual had attended a training session. The results of the survey are as follows:

LEVEL OF SATISFACTION

	VERY SATISFIED	SOME-WHAT SATISFIED	SOME-WHAT DISSAT-ISFIED	VERY DISSAT-ISFIED
ATTENDED TRAINING	15	35	5	5
NO TRAINING	5	45	35	15

Do the data indicate a significant difference in the distribution of satisfaction for those who attended training compared with those who did not? Test with $\alpha = .05$.

19. Darley and Latané (1968) conducted a study examining the factors that influence whether a person will try to help someone in need. Each participant was placed in a preexisting group of either 2 people, 3 people, or 6 people. The participants were led to believe that they were participating in discussion groups. Suddenly, one of the others in the group pretended to suffer a seizure, and the researcher recorded whether or not the participant tried to help. Darley and Latané predicted that the more people available to help, the less an individual participant would feel the responsibility to help. Hypothetical data similar to the study results are as follows:

	NUMBER IN THE GROUP		
	2	3	6
HELPING	11	16	4
NOT HELPING	2	10	9

Do the data indicate a significant relationship between helping behavior and the number of potential helpers available in the group? Test at the .01 level of significance.

20. Friedman and Rosenman (1974) have suggested that personality type is related to heart disease. Specifically, type A people, who are competitive, driven, pressured, and impatient, are more prone to heart disease. On the other hand, type B individuals, who are less competitive and more relaxed, are less likely to have heart disease. Suppose that an investigator would like to examine the relationship between personality type and disease. For a random sample of individuals, personality type is assessed with a standardized test. These individuals are then examined and categorized as to whether they have a heart disorder. The observed frequencies are as follows:

	NO HEART DISEASE	HEART DISEASE
TYPE A	60	110
TYPE B	126	54

Is there a relationship between personality and disorder? Test at the .05 level of significance.

21. A scientist would like to see whether there is a relationship between handedness and eye preference, and selects a random sample of $n = 150$ participants. For each participant, the researcher determines two things: (1) whether the person is left-handed or right-handed and (2) which eye the person prefers to use when looking through a camera viewfinder. The observed frequencies are as follows:

		HAND PREFERENCE	
		LEFT	RIGHT
EYE PREFERENCE	LEFT	20	40
	RIGHT	10	80

Is there a relationship between the two variables? Test at the .01 level of significance.

22. As part of a campaign to demonstrate sex discrimination in salary within the county government, a sample of $n = 200$ employees was selected, and each individual's salary was recorded. The median salary was computed for the sample and then each individual was classified by gender and relationship to the median. The obtained distribution was as follows:

	FEMALE	MALE
ABOVE MEDIAN	28	72
BELOW MEDIAN	52	48

Do these data indicate that the salary distribution for females is significantly different from the distribution for males? Test at the .05 level of significance.

23. A local county is considering a budget proposal that would allocate extra funding toward the renovation of city parks. A survey is conducted to measure public opinion concerning the proposal. A total of 300 individuals respond to the survey: 100 who live within the city limits and 200 from the surrounding suburbs. The frequency distribution is as follows:

		OPINION ABOUT PROPOSAL		
		FAVOR	OPPOSE	
RESIDENCE	CITY	68	32	100
	SUBURB	86	114	200

a. Is there a significant difference in the distribution of opinions for city residents compared to those in the suburbs? Test at the .05 level of significance.
b. The relationship between home location and opinion can also be evaluated using the phi-coefficient. If the phi-coefficient were computed for these data, what value would be obtained for phi?

24. A researcher would like to evaluate the relationship between a person's age and his/her preference between two leading brands of cola. In a sample of 30 people, the researcher found that 15 out of 18 people who were over 30 years old preferred brand A and only 3 out of 12 people under 30 preferred brand A.
a. Are these data sufficient to indicate a significant relationship between age and cola preference? Test at the .05 level of significance.
b. The relationship between age and preference could also be evaluated using the phi-coefficient (Chapter 16). If the phi-coefficient were computed for these data, what value would be obtained for phi?

CHAPTER 18

The Binomial Test

Tools You Will Need

The following items are considered essential background material for this chapter. If you doubt your knowledge of any of these items, you should review the appropriate chapter or section before proceeding.

- Binomial distribution (Chapter 6)
- *z*-score hypothesis tests (Chapter 8)
- Chi-square test for goodness of fit (Chapter 17)

Preview

In 1960, Gibson and Walk designed a classic piece of apparatus to test depth perception. Their device, called a *visual cliff*, consisted of a wide board with a deep drop (the cliff) to one side and a shallow drop on the other side. An infant was placed on the board and then observed to see whether he or she crawled off the shallow side or crawled off the cliff. (*Note:* Infants who moved to the deep side actually crawled onto a sheet of heavy glass, which prevented them from falling. Thus, the deep side only appeared to be a cliff—hence the name *visual cliff*.)

Gibson and Walk reasoned that if infants are born with the ability to perceive depth, they would recognize the deep side and not crawl off the cliff. On the other hand, if depth perception is a skill that develops over time through learning and experience, then infants should not be able to perceive any difference between the shallow and the deep sides.

Out of 27 infants who moved off the board, only 3 ventured onto the deep side at any time during the experiment. Gibson and Walk interpreted these data as convincing evidence that depth perception is innate. The infants showed a systematic preference for the shallow side.

You should notice immediately that the data from this experiment are different from the data that we usually encounter. There are no scores. Gibson and Walk simply counted the number of infants who went off the deep side and the number who went to the shallow side. Still, we would like to use these data to make statistical decisions. Do these sample data provide sufficient evidence to make a confident conclusion about depth perception in the population? Suppose that 8 of the 27 infants had crawled to the deep side. Would you still be convinced that there is a significant preference for the shallow side? What about 12 out of 27?

Notice that we are asking questions about probability and statistical significance. In this chapter, we will examine the statistical techniques designed for use with data similar to those obtained in the visual cliff experiments. Each individual in the sample is classified into one of two possible categories (for example, deep or shallow), and we simply count the number in each category. These frequency data are then used to draw inferences about the general population.

18.1 OVERVIEW

In Chapter 6, we introduced the concept of binomial data. You should recall that binomial data exist whenever a measurement procedure classifies individuals into exactly two distinct categories. For example, the outcomes from tossing a coin can be classified as heads and tails; people can be classified as male or female, plastic products can be classified as recyclable or nonrecyclable. In general, binomial data exist when

1. The measurement scale consists of exactly two categories.
2. Each individual observation in a sample is classified in only one of the two categories.
3. The sample data consist of the frequency or number of individuals in each category.

The traditional notation system for binomial data identifies the two categories as A and B and identifies the probability (or proportion) associated with each category as p and q, respectively. For example, a coin toss results in either heads (A) or tails (B), with probabilities $p = \frac{1}{2}$ and $q = \frac{1}{2}$.

In this chapter, we examine the statistical process of using binomial data for testing hypotheses about the values of p and q for the population. This type of hypothesis test is called a *binomial test*.

DEFINITION   A **binomial test** uses sample data to evaluate hypotheses about the values of p and q for a population consisting of binomial data.

Consider the following two situations:

1. In a sample of $n = 34$ color-blind students, 30 are male, and only 4 are female. Does this sample indicate that color blindness is significantly more common for males in the general population?

2. In 1990, only 10% of American families had incomes below the poverty level. This year, in a sample of 100 families, 19 were below the poverty level. Does this sample indicate that there has been a significant change in the population proportions?

Notice that both of these examples have binomial data (exactly two categories). Although the data are relatively simple, we are asking the same statistical question about significance that is appropriate for a hypothesis test: Do the sample data provide sufficient evidence to make a conclusion about the population?

HYPOTHESES FOR THE BINOMIAL TEST

In the binomial test, the null hypothesis specifies exact values for the population proportions p and q. Theoretically, you could choose any proportions for H_0, but usually there is a clear reason for the values that are selected. The null hypothesis typically falls into one of the following two categories:

1. **Just Chance.** Often the null hypothesis states that the two outcomes A and B occur in the population with the proportions that would be predicted simply by chance. If you were tossing a coin, for example, the null hypothesis might specify $p(\text{heads}) = \frac{1}{2}$ and $p(\text{tails}) = \frac{1}{2}$. Notice that this hypothesis states the usual, chance proportions for a balanced coin. Also notice that it is not necessary to specify both proportions. Once the value of p is identified, the value of q is determined by $1 - p$. For the coin toss example, the null hypothesis would simply state

$$H_0: \quad p = p(\text{heads}) = \frac{1}{2} \qquad \text{(The coin is balanced.)}$$

Similarly, if you were selecting cards from a deck and trying to predict the suit on each draw, the probability of predicting correctly would be $p = \frac{1}{4}$ for any given trial. (With four suits, you have a 1-out-of-4 chance of guessing correctly.) In this case, the null hypothesis would state

$$H_0: \quad p = p(\text{guessing correctly}) = \frac{1}{4} \quad \text{(The outcome is simply due to chance.)}$$

In each case, the null hypothesis simply states that there is nothing unusual about the population; that is, the outcomes are occurring by chance.

2. **No Change or No Difference.** Often you may know the proportions for one population and want to determine whether or not the same proportions apply to a different population. In this case, the null hypothesis would simply specify that there is no difference between the two populations. Suppose that national statistics indicate that 1 out of 12 drivers will be involved in a traffic accident during the next year. Does this same proportion apply to 16-year-olds who are driving for the first time? According to the null hypothesis,

$$H_0: \quad \text{For 16-year-olds, } p = p(\text{accident}) = \frac{1}{12} \qquad \text{(No different from the general population)}$$

Similarly, suppose that last year, 30% of the freshman class failed the college writing test. This year, the college is requiring all freshmen to take a writing course. Will the course have any effect on the number who fail the test? According to the null hypothesis,

H_0:   For this year, $p = p(\text{fail}) = 30\%$ (No different from last year's class)

THE DATA FOR THE
BINOMIAL TEST

For the binomial test, a sample of n individuals is obtained and you simply count how many are classified in category A and how many are classified in category B. We will focus attention on category A and use the symbol X to stand for the number of individuals classified in category A. Recall from Chapter 6 that X can have any value from 0 to n and that each value of X has a specific probability. The distribution of probabilities for each value of X is called the binomial distribution. Figure 18.1 shows an example of a binomial distribution where X is the number of heads obtained in four tosses of a balanced coin.

FIGURE 18.1

A binomial distribution for the number of heads obtained in four tosses of a balanced coin.

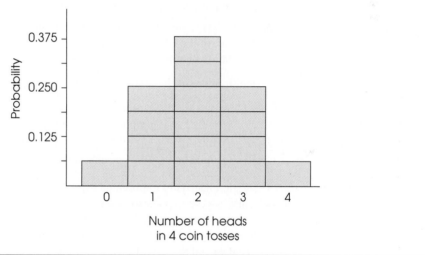

THE TEST STATISTIC FOR
THE BINOMIAL TEST

As we noted in Chapter 6, when the values pn and qn are both equal to or greater than 10, the binomial distribution approximates a normal distribution. This fact is important because it allows us to compute z-scores and use the unit normal table to answer probability questions about binomial events. In particular, when pn and qn are both at least 10, the binomial distribution will have the following properties:

1. The shape of the distribution is approximately normal.
2. The mean of the distribution is $\mu = pn$.
3. The standard deviation of the distribution is

$$\sigma = \sqrt{npq}$$

With these parameters in mind, it is possible to compute a z-score corresponding to each value of X in the binomial distribution.

$$z = \frac{X - \mu}{\sigma} = \frac{X - pn}{\sqrt{npq}} \qquad \text{(See Equation 6.4.)} \qquad (18.1)$$

This is the basic z-score formula that will be used for the binomial test. However, we will modify the formula slightly to make it more compatible with the logic of the binomial hypothesis test. The modification consists of dividing both the numerator and the denominator of the z-score by n. (You should realize that dividing both the numerator and the denominator by the same value does not change the value of the z-score.) The resulting equation is

$$z = \frac{X/n - p}{\sqrt{pq/n}} \qquad (18.2)$$

For the binomial test, the values in this formula are defined as follows:

1. X/n is the proportion of individuals in the sample who are classified in category A.
2. p is the hypothesized value (from H_0) for the proportion of individuals in the population who are classified in category A.
3. $\sqrt{pq/n}$ is the standard error for the sampling distribution of X/n and provides a measure of the standard distance between the sample statistic (X/n) and the population parameter (p).

Thus, the structure of the binomial z-score (Equation 18.2) can be expressed as

$$z = \frac{X/n - p}{\sqrt{pq/n}} = \frac{\begin{array}{c}\text{sample} \\ \text{proportion} \\ \text{(data)}\end{array} - \begin{array}{c}\text{hypothesized} \\ \text{population} \\ \text{proportion}\end{array}}{\text{standard error}}$$

The logic underlying the binomial test is exactly the same as we encountered with the original z-score hypothesis test in Chapter 8. The hypothesis test involves comparing the sample data with the hypothesis. If the data are consistent with the hypothesis, we will conclude that the hypothesis is reasonable. But if there is a big discrepancy between the data and the hypothesis, we will reject the hypothesis. The value of the standard error will provide a benchmark for determining whether the discrepancy between the data and the hypothesis is more than would be expected by chance. The alpha level for the test will provide a criterion for deciding whether or not the discrepancy is significant. The hypothesis-testing procedure is demonstrated in the following section.

LEARNING CHECK

1. In the Preview we described a research study using a visual cliff. State the null hypothesis for this study in words and as a probability value (p) that an infant will crawl off the deep side.

2. If the visual cliff study had used a sample of $n = 15$ infants, would it be appropriate to use the normal approximation to the binomial distribution? Explain why or why not.

3. If the results from the visual cliff study showed that only 3 out of 36 infants crawled off the deep side, what z-score value would be obtained using Equation 18.1?

ANSWERS 1. The null hypothesis states that the probability of choosing between the deep side and the shallow side is just chance: $p(\text{deep side}) = \frac{1}{2}$.

2. The normal approximation to the binomial distribution requires that both pn and qn are at least 10. With $n = 15$, $pn = qn = 7.5$. The normal approximation should not be used.

3. With $X = 3$, $\mu = \frac{1}{2}(36) = 18$, and $\sigma = \sqrt{\left(\frac{1}{2}\right)\left(\frac{1}{2}\right)(36)} = \sqrt{9} = 3$, the z-score is

$$z = -15/3 = -5.00.$$

18.2 THE BINOMIAL TEST

The binomial test follows the same four-step procedure presented earlier with other examples for hypothesis testing. The four steps are summarized as follows.

STEP 1   *State the hypotheses.* In the binomial test, the null hypothesis specifies values for the population proportions p and q. Typically, H_0 will specify a value only for p, the proportion associated with category A. The value of q is directly determined from p by the relationship $q = 1 - p$. Finally, you should realize that the hypothesis, as always, addresses the probabilities or proportions for the *population*. Although we will use a sample to test the hypothesis, the hypothesis itself always concerns a population.

STEP 2   *Locate the critical region.* When both values for pn and qn are greater than or equal to 10, the z-scores defined by Equation 18.1 or 18.2 will form an approximately normal distribution. Thus, the unit normal table can be used to find the boundaries for the critical region. With $\alpha = .05$, for example, you may recall that the critical region is defined as z-score values greater than $+1.96$ or less than -1.96.

STEP 3   *Compute the test statistic (z-score).* At this time, you obtain a sample of n individuals (or events) and count the number of times category A occurs in the sample. The number of occurrences of A in the sample is the X value for Equation 18.1 or 18.2. Because the two z-score equations are equivalent, you may use either one for the hypothesis test. Usually Equation 18.1 is easier to use because it involves larger numbers (fewer decimals) and it is less likely to be affected by rounding error.

STEP 4   *Make a decision.* If the z-score for the sample data is in the critical region, you reject H_0 and conclude that the discrepancy between the sample proportions and the hypothesized population proportions is significantly greater than chance. That is, the data are not consistent with the null hypothesis, so H_0 must be wrong. On the other hand, if the z-score is not in the critical region, you fail to reject H_0.

The following example demonstrates a complete binomial test.

EXAMPLE 18.1   In the Preview section, we described the *visual cliff* experiment designed to examine depth perception in infants. To summarize briefly, an infant is placed on a wide board that appears to have a deep drop on one side and a relatively shallow drop on the other. An infant who is able to perceive depth should avoid the deep side and move toward the shallow side. Without depth perception, the infant should show no preference between the two sides. In the experiment, only 3 out of 27 infants moved onto the deep side. The purpose of the hypothesis test is to determine whether these data demonstrate that infants have a significant preference for the shallow side.

This is a binomial hypothesis-testing situation. The two categories are

$$A = \text{move onto the deep side}$$

$$B = \text{move onto the shallow side}$$

STEP 1   The null hypothesis states that for the general population of infants, there is no preference between the deep and the shallow sides; the direction of movement is determined by chance. In symbols,

$$H_0: \quad p = p(\text{deep side}) = \tfrac{1}{2} \quad \left(\text{and } q = \tfrac{1}{2}\right)$$

$$H_1: \quad p \neq \tfrac{1}{2} \quad \text{(There is a preference.)}$$

We will use $\alpha = .05$.

STEP 2   With a sample of $n = 27$, $pn = 13.5$ and $qn = 13.5$. Both values are greater than 10, so the distribution of z-scores are approximately normal. With $\alpha = .05$, the critical region is determined by boundaries of $z = \pm 1.96$.

STEP 3   For this experiment, the data consist of $X = 3$ out of $n = 27$. Using Equation 18.1, these data produce a z-score value of

$$z = \frac{X - pn}{\sqrt{npq}} = \frac{3 - 13.5}{\sqrt{27\left(\tfrac{1}{2}\right)\left(\tfrac{1}{2}\right)}} = \frac{-10.5}{2.60} = -4.04$$

To use Equation 18.2, you first compute the sample proportion, $X/n = 3/27 = 0.111$. The z-score is then

$$z = \frac{X/n - p}{\sqrt{pq/n}} = \frac{0.111 - 0.5}{\sqrt{\tfrac{1}{2}\left(\tfrac{1}{2}\right)/27}} = \frac{-0.389}{0.096} = -4.05$$

Within rounding error, the two equations produce the same result.

STEP 4   Because the data are in the critical region, our decision is to reject H_0. These data do provide sufficient evidence to conclude that there is a significant preference for the shallow side. Gibson and Walk (1960) interpreted these data as convincing evidence that depth perception is innate.

IN THE LITERATURE
REPORTING THE RESULTS OF A BINOMIAL TEST

Reporting the results of the binomial test typically consists of describing the data and reporting the z-score value and the probability that the results are due to chance. It is also helpful to note that a binomial test was used because z-scores are used in other hypothesis-testing situations (see, for example, Chapter 8). For Example 18.1, the report might state:

> Three out of 27 infants moved to the deep side of the visual cliff. A binomial test revealed that there is a significant preference for the shallow side of the cliff, $z = -4.04$, $p < .05$.

Once more, p is *less than* alpha (.05), indicating the null hypothesis has been rejected. The probability that these results are simply due to chance is less than 5%. ❏

ASSUMPTIONS FOR THE BINOMIAL TEST

The binomial test requires two very simple assumptions:

1. The sample must consist of *independent* observations (see Chapter 8, page 251).

2. The values for pn and qn must both be greater than or equal to 10 to justify using the unit normal table for determining the critical region.

LEARNING CHECK

1. For a binomial test, the null hypothesis always states that $p = \frac{1}{2}$. (True or false?)

2. The makers of brand X beer claim that people like their beer more than the leading brand. The basis for this claim is an experiment in which 64 beer drinkers compared the two brands in a side-by-side taste test. In this sample, 38 preferred brand X, and 26 preferred the leading brand. Do these data support the claim that there is a significant preference? Test at the .05 level.

ANSWERS

1. False

2. H_0: $p = \frac{1}{2} = q$, $X = 38$, $\mu = 32$, $\sigma = 4$, $z = +1.50$, fail to reject H_0. Conclude that there is no evidence for a significant preference.

18.3 THE RELATIONSHIP BETWEEN CHI-SQUARE AND THE BINOMIAL TEST

You may have noticed that the binomial test evaluates the same basic hypotheses as the chi-square test for goodness of fit; that is, both tests evaluate how well the sample proportions fit a hypothesis about the population proportions. When an experiment produces binomial data, these two tests are equivalent, and either one may be used. The relationship between the two tests can be expressed by the equation

$$\chi^2 = z^2$$

where χ^2 is the statistic from the chi-square test for goodness of fit and z is the z-score from the binomial test.

To demonstrate the relationship between the goodness-of-fit test and the binomial test, we will reexamine the data from Example 18.1.

STEP 1 *Hypotheses.* In the visual cliff experiment from Example 18.1, the null hypothesis states that there is no preference between the shallow side and the deep side. For the binomial test, the null hypothesis states

$$H_0: \quad p = p(\text{deep side}) = q = p(\text{shallow side}) = \frac{1}{2}$$

The chi-square test for goodness of fit would state the same hypothesis, specifying the population proportions as

	SHALLOW SIDE	DEEP SIDE
H_0:	$\frac{1}{2}$	$\frac{1}{2}$

STEP 2   *Critical region.* For the binomial test, the critical region is located by using the unit normal table. With $\alpha = .05$, the critical region consists of any z-score value beyond ± 1.96. The chi-square test would have $df = 1$, and with $\alpha = .05$, the critical region consists of chi-square values greater than 3.84. Notice that the basic relationship, $\chi^2 = z^2$, holds:

$$3.84 = (1.96)^2$$

STEP 3   *Test statistic.* For the binomial test (Example 18.1), we obtained a z-score of $z = -4.04$. For the chi-square test, the expected frequencies are

	SHALLOW SIDE	DEEP SIDE
f_e	13.5	13.5

With observed frequencies of 24 and 3, respectively, the chi-square statistic is

$$\chi^2 = \frac{(24 - 13.5)^2}{13.5} + \frac{(3 - 13.5)^2}{13.5}$$

$$= \frac{(10.5)^2}{13.5} + \frac{(-10.5)^2}{13.5}$$

$$= 8.167 + 8.167$$

$$= 16.33$$

With a little rounding error, the values obtained for the z-score and chi-square are related by the equation

$$\chi^2 = z^2$$
$$16.33 = (-4.04)^2$$

STEP 4   *Decision.* Because the critical values for both tests are related by the equation $\chi^2 = z^2$ and the test statistics are related in the same way, these two tests *always* will result in the same statistical conclusion.

18.4   THE SIGN TEST

Although the binomial test can be used in many different situations, there is one specific application that merits special attention. For a repeated-measures study that compares two conditions, it is often possible to use a binomial test to evaluate the results. You should recall that a repeated-measures study involves measuring each individual in two different treatment conditions or at two different points in time. When the measurements produce numerical scores, the researcher can simply subtract to determine the difference between the two scores and then evaluate the data using a repeated-measures t test (see Chapter 11). Occasionally, however, a researcher may record only the *direction* of the difference between the two observations. For example, a clinician may observe patients before therapy and after therapy and simply note whether each patient got better or got worse. Note that there is no measurement of how much change occurred; the clinician is simply recording the direction of change. Also note that the direction of change is a binomial variable; that is, there are only two values. In this situation it is possible to use a binomial test to evaluate the data. Traditionally, the two possible directions of change are coded by signs, with a positive sign indicating an increase and a negative sign indicating a decrease. When the binomial test is applied to signed data, it is called a *sign test*.

An example of signed data is shown in Table 18.1. Notice that the data can be summarized by saying that seven out of eight patients showed a decrease in symptoms after therapy.

TABLE 18.1

Hypothetical data from a research study evaluating the effectiveness of a clinical therapy. For each patient, symptoms are assessed before and after treatment and the data record whether there is an increase or a decrease in symptoms following therapy.

PATIENT	DIRECTION OF CHANGE AFTER TREATMENT
A	− (decrease)
B	− (decrease)
C	− (decrease)
D	+ (increase)
E	− (decrease)
F	− (decrease)
G	− (decrease)
H	− (decrease)

The null hypothesis for the sign test states that there is no difference between the two treatments. Therefore, any change in a participant's score is due to chance. In terms of probabilities, this means that increases and decreases are equally likely, so

$$p = p(\text{increase}) = \tfrac{1}{2}$$

$$q = p(\text{decrease}) = \tfrac{1}{2}$$

A complete example of a sign test follows.

EXAMPLE 18.2   A researcher tests the effect of a diet plan on body weight. A sample of 36 people is selected. Weights are measured at the start of the diet and 1 month later. The observations revealed that 23 people lost weight and 13 people gained weight.

STEP 1   State the hypothesis. The null hypothesis states that the diet has no effect. Any change in weight is due to chance, so increases and decreases are equally likely. Expressed as probabilities, the hypotheses are

$$H_0: \quad p = p(\text{lose weight}) = \tfrac{1}{2}$$

$$q = p(\text{gain weight}) = \tfrac{1}{2}$$

$$H_1: \quad p \neq q \quad \text{(Changes are consistently in one direction or the other.)}$$

Set $\alpha = .05$.

STEP 2   Locate the critical region. With $p = q = \tfrac{1}{2}$ and $n = 36$, the normal approximation to the binomial distribution is appropriate:

$$pn = qn = 18 \quad \text{(Both greater than 10)}$$

With $\alpha = .05$, the critical region is defined as the most extreme 5% of the distribution. This portion is bounded by z-scores of $z = +1.96$ at one extreme and $z = -1.96$ at the other.

STEP 3   Compute the test statistic. For this sample, we have $X = 23$ people who lost weight. This score corresponds to a z-score of

$$z = \frac{X - pn}{\sqrt{npq}} = \frac{23 - 18}{\sqrt{36\left(\frac{1}{2}\right)\left(\frac{1}{2}\right)}} = \frac{5}{3} = 1.67$$

STEP 4   Make a decision. Because the data are not in the critical region, we fail to reject H_0. These data do not provide sufficient evidence to conclude that the diet plan has any consistent effect on weight, $z = 1.67$, $p > .05$.

ZERO DIFFERENCES IN THE SIGN TEST

You should notice that the null hypothesis in the sign test refers only to those individuals who show some difference between treatment 1 versus treatment 2. The null hypothesis states that if there is any change in an individual's score, then the probability of an increase is equal to the probability of a decrease. Stated in this form, the null hypothesis does not consider individuals who show zero difference between the two treatments. As a result, the usual recommendation is that these individuals be discarded from the data and the value of n be reduced accordingly. However, if the null hypothesis is interpreted more generally, it states that there is no difference between the two treatments. Phrased this way, it should be clear that individuals who show no difference actually are supporting the null hypothesis and should not be discarded. Therefore, an alternative approach to the sign test is to divide individuals who show zero differences equally between the positive and negative categories. (With an odd number of zero differences, discard one, and divide the rest evenly.) This alternative results in a more conservative test; that is, the test is more likely to fail to reject the null hypothesis.

EXAMPLE 18.3   It has been demonstrated that stress or exercise causes an increase in the concentration of certain chemicals in the brain called endorphins. Endorphins are similar to morphine and produce a generally relaxed feeling and a sense of well-being. The endorphins may explain the "high" experienced by long-distance runners. To demonstrate this phenomenon, a researcher tested pain tolerance for 40 athletes before and after they completed a mile run. Immediately after running, the ability to tolerate pain increased for 21 of the athletes, decreased for 12, and showed no change for the remaining 7.

Following the standard recommendation for handling zero differences, you would use $n = 33$ for the sign test because only 33 participants showed any difference between the two treatments. The other 7 athletes are eliminated from the sample. With the more conservative approach, only 1 of the 7 athletes who showed no difference would be discarded, and the other 6 would be divided equally between the two categories. This would result in a total of $n = 39$ in the sample data with 24 (21 + 3) in one category and 15 (12 + 3) in the other.

WHEN TO USE THE SIGN TEST

In many cases, data from a repeated-measures experiment can be evaluated using either a sign test or a repeated-measures t test. In general, you should use the t test whenever possible. Because the t test uses the actual difference scores (not just the signs), it makes maximum use of the available information and results in a more powerful test. However, there are some cases in which a t test cannot or should not be used, and in these situations, the sign test can be valuable. Three specific cases where a t test is inappropriate or inconvenient will now be described.

BEFORE	AFTER	DIFFERENCE
20	23	+3
14	39	+25
27	Failed	+??
.	.	.
.	.	.
.	.	.

1. When you have infinite or undetermined scores, a t test is impossible, and the sign test is appropriate. Suppose, for example, that you are evaluating the effects of a sedative drug on problem-solving ability. A sample of rats is obtained, and each animal's performance is measured before and after receiving the drug. Hypothetical data are shown in the margin. Note that the third rat in this sample failed to solve the problem after receiving the drug. Because there is no score for this animal, it is impossible to compute a sample mean, an SS, or a t statistic. However, you could do a sign test because you know that the animal made more errors (an increase) after receiving the drug.

2. Often it is possible to describe the difference between two treatment conditions without precisely measuring a score in either condition. In a clinical setting, for example, a doctor can say whether a patient is improving, growing worse, or showing no change even though the patient's condition is not precisely measured by a score. In this situation, the data are sufficient for a sign test, but you could not compute a t statistic without individual scores.

3. Often a sign test is done as a preliminary check on an experiment before serious statistical analysis begins. For example, a researcher may predict that scores in treatment 2 should be consistently greater than scores in treatment 1. However, examination of the data after 1 week indicates that only 8 of 15 subjects showed the predicted increase. On the basis of these preliminary results, the researcher may choose to reevaluate the experiment before investing additional time.

LEARNING CHECK

1. A researcher used a chi-square test for goodness of fit to determine whether people had any preferences among three leading brands of potato chips. Could the researcher have used a binomial test instead of the chi-square test? Explain why or why not.

2. A researcher used a chi-square test to evaluate preferences between two logo designs for a minor league hockey team. With a sample of $n = 100$ people, the researcher obtained a chi-square of 9.00. If a binomial test had been used instead of chi-square, what value would have been obtained for the z-score?

3. A developmental psychologist is using a behavior-modification program to help control the disruptive behavior of 40 children in a local school. After 1 month, 26 of the children have improved, 10 are worse, and 4 show no change in behavior. On the basis of these data, can the psychologist conclude that the program is working? Test at the .05 level.

ANSWERS

1. The binomial test cannot be used when there are three categories.

2. The z-score would be $\sqrt{9} = 3.00$.

3. Discarding the four participants who showed zero difference, $X = 26$ increases out of $n = 36$; $z = 2.67$; reject H_0; the program is working. If the four participants showing no change are divided between the two groups, then $X = 28$ out of $n = 40$; $z = 2.53$ and H_0 is still rejected.

SUMMARY

1. The binomial test is used with dichotomous data—that is, when each individual in the sample is classified in one of two categories. The two categories are identified as A and B, with probabilities of

$$p(A) = p \quad \text{and} \quad p(B) = q$$

2. The binomial distribution gives the probability for each value of X, where X equals the number of occurrences of category A in a sample of n events. For example, X equals the number of heads in $n = 10$ tosses of a coin.

3. When pn and qn are both at least 10, the binomial distribution is closely approximated by a normal distribution with

$$\mu = pn \qquad \sigma = \sqrt{npq}$$

By using this normal approximation, each value of X has a corresponding z-score:

$$z = \frac{X - \mu}{\sigma} = \frac{X - pn}{\sqrt{npq}} \quad \text{or} \quad z = \frac{X/n - p}{\sqrt{pq/n}}$$

4. The binomial test uses sample data to test hypotheses about the binomial proportions, p and q, for a population. The null hypothesis specifies p and q, and the binomial distribution (or the normal approximation) is used to determine the critical region.

5. One common use of the binomial distribution is for the sign test. This test evaluates the difference between two treatments using the data from a repeated measures design. The difference scores are coded as being either increases ($+$) or decreases ($-$). Without a consistent treatment effect, the increases and decreases should be mixed randomly, so the null hypothesis states that

$$p(\text{increase}) = \tfrac{1}{2} = p(\text{decrease})$$

With dichotomous data and hypothesized values for p and q, this is a binomial test.

RESOURCES

If you are using WebTutor along with this book, there is a WebTutor section corresponding to this chapter. The WebTutor contains a brief summary of Chapter 18, hints for learning the concepts and formulas for the binomial test, cautions about common errors, and sample exam items including solutions.

SPSS

The statistics computer package SPSS can be used to conduct the binomial test presented in this chapter. As with chi-square, however, the program requires that the score for each participant be entered individually. This system works well for preexisting data sheets (for example, data from a large survey that have already been entered into the computer). However, it can be very tedious if you simply want to use SPSS for a single problem.

The scores for the entire sample are entered into one column of the data matrix. The two binomial categories are numerically coded—for example, with 0 and 1. In the visual cliff example from the chapter Preview, only 3 out of 27 infants moved over the deep side. These data would be entered as three 0s and twenty-four 1s. After the data have been entered, click **Analyze** on the tool bar, select **Nonparametric Tests**, and click on **Binomial**. Next you must enter the label for the data column into the Test Variables List box, and set the value for p in the Test Proportion box. Then click **OK**. The output includes the number of individuals in each of the two categories, the proportion in each category, the test proportion that you selected, and a level of significance for either a one- or two-tailed test. Note that the test is based on a z-score but SPSS does not report the z-score value.

KEY TERMS

dichotomous data   binomial distribution   binomial test   sign test

FOCUS ON PROBLEM SOLVING

1. For all binomial tests, the values of p and q must sum to 1.00 (or 100%).

2. Remember that both pn and qn must be at least 10 before you can use the normal distribution to determine critical values for a binomial test.

3. Although the binomial test usually specifies the critical region in terms of z-scores, it is possible to identify the X values that determine the critical region. With $\alpha = .05$, the critical region is determined by z-scores greater than 1.96 or less than -1.96. That is, to be significantly different from what is expected by chance, the individual score must be above (or below) the mean by at least 1.96 standard deviations. For example, in an ESP experiment in which an individual is trying to predict the suit of a playing card for a sequence of $n = 64$ trials, chance probabilities are

$$p = p(\text{right}) = \tfrac{1}{4} \qquad q = p(\text{wrong}) = \tfrac{3}{4}$$

For this example, the binomial distribution has a mean of $pn = \left(\tfrac{1}{4}\right)(64) = 16$ right and a standard deviation of $\sqrt{npq} = \sqrt{64\left(\tfrac{1}{4}\right)\left(\tfrac{3}{4}\right)} = \sqrt{12} = 3.46$. To be significantly different from chance, a score must be above (or below) the mean by at least $1.96(3.46) = 6.78$. Thus, with a mean of 16, an individual would need to score above 22.78 $(16 + 6.78)$ or below 9.22 $(16 - 6.78)$ to be significantly different from chance.

DEMONSTRATION 18.1

THE BINOMIAL TEST

The population of students in the psychology department at State College consists of 60% females and 40% males. Last semester, the Psychology of Gender course had a total of 36 students, of whom 26 were female and only 10 were male. Are the proportions of females and males in this class significantly different from what would be expected by chance from a population with 60% females and 40% males? Test at the .05 level of significance.

STEP 1   *State the hypotheses, and specify alpha.* The null hypothesis states that the male/female proportions for the class are not different from what is expected for a population with these proportions. In symbols,

$$H_0: \quad p = p(\text{female}) = 0.60 \quad \text{and} \quad q = p(\text{male}) = 0.40$$

The alternative hypothesis is that the proportions for this class are different from what is expected for these population proportions.

$$H_1: \quad p \neq 0.60 \quad (\text{and } q \neq 0.40)$$

We will set alpha at $\alpha = .05$.

STEP 2   *Locate the critical region.* Because pn and qn are both greater than 10, we can use the normal approximation to the binomial distribution. With $\alpha = .05$, the critical region is defined as any z-score value greater than $+1.96$ or less than -1.96.

STEP 3   *Calculate the test statistic.* The sample has 26 females out of 36 students, so the sample proportion is

$$\frac{X}{n} = \frac{26}{36} = 0.72$$

The corresponding z-score (using Equation 18.2) is

$$z = \frac{X/n - p}{\sqrt{pq/n}} = \frac{0.72 - 0.60}{\sqrt{\dfrac{0.60(0.40)}{36}}} = \frac{0.12}{0.0816} = 1.47$$

STEP 4   *Make a decision about H_0, and state a conclusion.* The obtained z-score is not in the critical region. Therefore, we fail to reject the null hypothesis. On the basis of these data, you conclude that the male/female proportions in the gender class are not significantly different from the proportions in the psychology department as a whole.

PROBLEMS

1. In 1985 only 8% of the students in the city school district were classified as being learning disabled. A school psychologist suspects that the proportion of learning disabled children has changed dramatically over the years. To demonstrate this point, a random sample of $n = 300$ students is selected. In this sample there are 42 students who have been identified as learning disabled. Is this sample sufficient to indicate that there has been a significant change in the proportion of learning disabled students since 1985? Use the .05 level of significance.

2. A recent report states that 90% of Americans over age 65 say that they would stop and pick up a penny lying on the sidewalk. A researcher suspects that the value of money is very different for teenagers compared with senior citizens. To test this theory, a random sample of $n = 100$ high school students is asked whether or not they would stop and pick up a penny. Only 60 students said that they would pick up the penny. Do these data indicate a significant difference in the proportion of students, compared with seniors, who would pick up a penny? Test with $\alpha = .05$.

3. A recent survey of practicing psychotherapists revealed that 25% of the individuals responding agreed with the statement, "Hypnosis can be used to recover accurate memories of past lives" (Yapko, 1994). A researcher would like to determine whether this same level of belief exists in the general population. A sample of 192 adults is surveyed and 65 believe that hypnosis can be used to recover accurate memories of past lives. Based on these data, can you conclude that beliefs held by the general population are significantly different from beliefs held by psychotherapists? Test with $\alpha = .05$.

4. The probability that an individual in the general population will develop schizophrenia is 0.01 (1 out of 100). If schizophrenia has a genetic component, however, the probability should be much higher for individuals with a parent who has been diagnosed with the disorder. To test this prediction, a researcher obtains a sample of 1000 individuals who have at least one parent with schizophrenia. In the sample, 97 individuals had been diagnosed with the disorder. Do the data indicate that the probability of schizophrenia is significantly different from $p = 0.01$ for individuals with a parent who has been diagnosed? Test with $\alpha = .01$.

5. One of the original methods for testing ESP (extrasensory perception) was Zener cards. Each card shows one of five symbols (square, circle, star, wavy lines, cross). The person being tested must predict the symbol before the card is turned over. Chance performance on this task would produce 1 out of 5 (20%) correct predictions. Is a person who has 15 correct predictions out of 50 attempts performing significantly different from chance? Test with $\alpha = .05$.

6. In a recent study examining color preferences in infants, 30 babies were offered a choice between a red rattle and a green rattle. Twenty-five of the 30 selected the red rattle. Do these data provide evidence for a significant color preference? Test at the .01 level of significance.

7. A college dormitory recently sponsored a taste comparison between two major soft drinks. Of the 64 students who participated, 39 selected brand A, and only 25 selected brand B. Do these data indicate a significant preference? Test at the .05 level of significance.

8. Nationwide, only 6% of the population develops ulcers each year. In a sample of 200 executive vice presidents, 40 had developed ulcers during the previous 12 months. Do these data indicate that vice presidents have an incidence of ulcers different from the general population? Test at the .05 level of significance.

9. Use $\alpha = .05$ to answer each of the following questions. Use the normal approximation to the binomial distribution. In each case, assume that a two-tailed test is being used.
   a. For a true-false test with 20 questions, how many would you have to get right to do significantly better than expected by chance?
   b. How many would you need to get right on a 40-question test?
   c. How many would you need to get right on a 100-question test?

10. A multiple-choice exam has 48 questions, each with four possible answers. What score (X = number correct) is needed on this exam to do significantly better than expected by chance? Assume a one-tailed test with $\alpha = .05$.

11. Graduates from the social work program must pass a state certification exam before they are officially designated as certified social workers. In the past, 20% of the individuals failed the exam. This year, the social work department began a special training course for students preparing to take the exam. In a sample of 60 students who received special training, only 2 failed the exam. Does this sample indicate a significant change in the proportion who fail? Test with $\alpha = .05$.

12. The general college population is composed of 55% females and 45% males. However, there are 23 males and 14 females serving as representatives in the student government. Is the distribution of genders in the student government significantly different from what would be expected from the general population? Test at the .05 level of significance.

13. In 1980, students in the social work department were 80% female and only 20% male. This fall, a survey of students expressing an interest in social work showed 43 females and 32 males. Do these data indicate a significant change in the gender distribution within the social work department? Test at the .05 level of significance.

14. A psychologist is examining the relationship between professional interests and birth-order position for students who grew up in families with exactly 2 children.
   a. A sample of 34 mathematics majors showed 24 first-born children and 10 second-born children. Is this significantly different from a distribution that would be expected by chance? Test with $\alpha = .05$.
   b. A sample of 22 theater majors showed 3 first-born children and 19 second-born children. Is this significantly different from a distribution that would be expected by chance? Test with $\alpha = .05$.

15. A professor who is trying to decide which of two textbooks is better asks a group of students to review the two books and state their preferences. Of the 25 students, 19 preferred book A and 6 preferred book B. Do these data indicate a significant preference? Test with $\alpha = .01$.

16. A researcher is evaluating a new therapy for treating depression. A sample of $n = 28$ depressed patients is obtained and each patient begins the new therapy. After 2 weeks, each patient is asked to evaluate his/her own progress. For this sample, 18 patients reported improvement, 8 claimed that their depression had become worse, and 2 reported no change.

a. Discard the two patients who showed no change and use a sign test to evaluate the effectiveness of the therapy. Use $\alpha = .05$.

b. Divide the no-change patients between the two groups and then use a sign test to evaluate the therapy. Again, test at the .05 level of significance.

17. The habituation technique is one method that is used to examine memory for infants. The procedure involves presenting a stimulus to an infant (usually projected on the ceiling above the crib) for a fixed time period and recording how long the infant spends looking at the stimulus. After a brief delay, the stimulus is presented again. If the infant spends less time looking at the stimulus during the second presentation, it is interpreted as indicating that the stimulus is remembered and, therefore, is less novel and less interesting than it was on the first presentation. This procedure is used with a sample of $n = 30$ two-week-old infants. For this sample, 22 infants spent less time looking at the stimulus during the second presentation than during the first. Do these data indicate a significant difference? Test at the .01 level of significance.

18. Thirty percent of the students in the local elementary school are classified as only children (no siblings). However, in the special program for talented and gifted children, 43 out of 90 students are only children. Is the proportion of only children in the special program significantly different from the proportion for the school? Test at the .05 level of significance.

19. Last year the college counseling center offered a workshop for students who claimed to suffer from extreme exam anxiety. Of the 45 students who attended the workshop, 31 had higher grade point averages this semester than they did last year. Do these data indicate a significant difference from what would be expected by chance? Test at the .01 level of significance.

20. In the general population, 8 out of 10 people can be hypnotized. A researcher suspects that the ability to be hypnotized is partially determined by an individual's personality. A sample of $n = 80$ participants is obtained. All these participants are known to be *field independent*, which means that they tend to rely on internal cues rather than external cues for making judgments. The researcher finds that 51 of the 80 participants can be hypnotized. Do these data indicate that field-independent people are different from the general population in terms of their ability to be hypnotized? Use an alpha of .05.

21. In a study of human memory, Sachs (1967) demonstrated that people recall the meaning of verbal material, but tend to forget the exact word-for-word details. In this study, people read a passage of text. Then the people were shown a test sentence and asked whether or not the identical sentence had appeared in the text. In one condition, the test sentence was phrased differently, but had the same meaning as a sentence that was in the text. For example, the sentence in the text might be "The boy hit the ball," and the test sentence might be "The ball was hit by the boy." If only 27 out of 45 people correctly notice that change in the sentence, can you conclude that their performance is significantly better than that expected by chance? Use a two-tailed test at the .05 level of significance.

22. Trying to fight a drug-resistant bacteria, a researcher tries an experimental drug on infected subjects. Out of 70 monkeys, 42 showed improvement, 22 got worse, and 6 showed no change. Is this researcher working in the right direction? Is there a significant effect of the drug on the infection? Use a two-tailed test at the .05 level of significance.

23. An English professor conducts a special 1-week course in writing skills for freshmen who are poorly prepared for college-level writing. To test the effectiveness of this course, each student in a recent class of $n = 36$ was required to write a short paragraph describing a painting before the course started. At the end of the course, the students were once again required to write a paragraph describing the same painting. The students' paragraphs were then given to another instructor to be evaluated. For 25 students, the writing was judged to be better after the course. For the rest of the class, the first paragraph was judged to be better. Do these results indicate a significant change in performance? Test at the .05 level of significance.

24. Biofeedback training is often used to help people who suffer migraine headaches. A recent study found that 29 out of 50 participants reported a decrease in the frequency and severity of their headaches after receiving biofeedback training. Of the remaining participants in this study, 10 reported that their headaches were worse, and 11 reported no change.

a. Discard the zero-difference participants, and use a sign test with $\alpha = .05$ to determine whether or not the biofeedback produced a significant difference.

b. Divide the zero-difference participants between the two groups, and use a sign test to evaluate the effect of biofeedback training.

25. The sense of smell plays an important role in the taste of food. A favorite demonstration of this fact involves having people hold their noses while tasting slices of apple and onion. Without smell, these two taste much the same. In a sample of 25 people who tried this demonstration, only 15 were able to identify the onion correctly. Is this significantly better than what would be expected by chance? Use a two-tailed test at the .05 level of significance.

CHAPTER

19

Statistical Techniques for Ordinal Data: Mann-Whitney, Wilcoxon, and Kruskal-Wallis Tests

Tools You Will Need

The following items are considered essential background material for this chapter. If you doubt your knowledge of any of these items, you should review the appropriate chapter or section before proceeding.

- Ordinal scales (Chapter 1)
- Probability (Chapter 6)
  - The unit normal table
- Introduction to hypothesis testing (Chapter 8)
- Correlation (Chapter 16)

Preview

People have a passion for ranking things. Everyone wants to know who is number one, and we all tend to describe things in terms of their relationship to others. For example, the Nile is the longest river in the world, the Labrador retriever is the number one registered dog in the United States, the state with the highest average SAT math score is Iowa, *Reader's Digest* is the most popular magazine in the United States, English is the fourth most common native language in the world (after Chinese, Hindi, and Spanish), and Lake Superior is the largest lake in North America and the second largest in the world.

Part of the fascination with ranks is that they are easy to obtain and they are easy to understand. For example, what is your favorite ice-cream flavor? Notice that you do not need to conduct any sophisticated measurement to arrive at an answer. There is no need to rate 31 different flavors on a 100-point scale, and you do not need to worry about how much difference there is between your first and second choices.

You should recall from Chapter 1 that ranking is an example of measurement on an ordinal scale. In general, ordinal scales are less demanding and less sophisticated than the interval or ratio scales that are most commonly used to measure variables. Because ordinal scales are less demanding, they are often easier to use. On the other hand, because they are less sophisticated, they can cause some problems for statistical analysis. Consider the data presented in Table 19.1. The table shows the top 25 cities for singles as ranked by Forbes.com in 2002. The ranking considered the 40 most populated metropolitan areas as defined by the U.S Census, and combined different factors such as culture, nightlife, and the number of other singles to produce an overall rank for each city.

Because the values in Table 19.1 are numbers, it is tempting to try some routine calculations such as means and standard deviations. For example, we could calculate the average score for cities in the East and compare it with the average for cities in the West. However, you should recall that means and standard deviations are measures based on *distance*. The mean for example is a balance point, so the distances above the mean are equal to the distances below. Ranks, on the other hand, do not provide any information about distance. In the table, for example, Boston is ranked #1 and Austin is ranked #2, giving the appearance that these two cities are very close together. However, a detailed examination of how the ranks were obtained shows that Boston and Austin have very different profiles. For example, Boston is ranked very high in terms of culture and nightlife, but Austin is near the bottom in both categories. Austin, on the other hand, is highly ranked in terms of job growth and cost of living for singles, categories in which Boston has poor ranks. Although Boston and Austin appear to be close in the overall ranking, in many ways the two cities are very different. In general, ranks do not tell you how much difference there is between two individuals.

Because ordinal data (ranks) provide limited information they must be used and interpreted carefully. In particular, standard statistical methods such as means, *t* tests, or analysis of variance should not be used when data are measured on an ordinal scale. Fortunately, special statistical techniques have been developed for ordinal data. In this chapter, we introduce some of the statistics that have been designed specifically for ordinal data.

TABLE 19.1

The best cities for singles as ranked by Forbes.com (2002)

RANK	CITY	RANK	CITY
1	Boston	13 (tie)	Phoenix
2	Austin	15	Chicago
3	Washington–Baltimore	15 (tie)	Los Angeles
4	Raleigh–Durham	17	Miami
5	Denver–Boulder	18	Seattle
6	San Francisco–Oakland	19	Orlando
7	San Diego	20	New Orleans
8	Houston	21	Tampa
9	Minneapolis–St. Paul	22	St. Louis
10	Atlanta	23	Milwaukee
11	New York	24	Salt Lake City
12	Dallas–Fort Worth	25	Columbus
13	Philadelphia		

19.1 DATA FROM AN ORDINAL SCALE

Occasionally, a research study will generate data that consist of measurements on an ordinal scale. Recall from Chapter 1 that an ordinal scale simply produces a *rank ordering* for the individuals being measured. For example, a kindergarten teacher may rank children in terms of their maturity, or a business manager may rank employee suggestions in terms of creativity. Whenever a set of data consists of ranks, you should be very cautious about choosing a statistical procedure to describe or to interpret the data. Specifically, most of the commonly used statistical methods such as the mean, the standard deviation, hypothesis tests with the *t* statistic, and the Pearson correlation are generally considered to be inappropriate for ordinal data.

The concern about using traditional statistical methods with ranked data stems from the fact that most statistical procedures begin by calculating a sample mean and computing deviations from the mean. When scores are measured on an interval or a ratio scale, the concepts of the *mean* and *distance* are very well defined. On a ruler, for example, a measurement of 2 inches is exactly halfway between a measurement of 1 inch and a measurement of 3 inches. In this case, the mean $M = 2$ provides a precise central location that defines a balance point that is equidistant from the two scores, $X = 1$ and $X = 3$. However, ordinal measurements do not provide this same degree of precision. Specifically, ordinal values (ranks) tell you only the direction from one score to another, but provide no information about the distance between scores. Thus, you know that first place is better than second or third, but you do not know how much better. In a horse race, for example, you know that the second-place horse is somewhere between the first- and third-place horses, but you do not know exactly where. In particular, a rank of second is not necessarily halfway between first and third.

Because the concepts of *mean* and *distance* are not well defined with ordinal data and because most of the traditional statistical methods are based on these concepts, it generally is considered unwise to use traditional statistics with ordinal data. Therefore, statisticians have developed special techniques that are designed specifically for use with ordinal data. In the remainder of this chapter, we will introduce several of these special statistical methods.

OBTAINING ORDINAL MEASUREMENTS

The process of ranking can be based directly on observations of individuals. For example, you could arrange a group of people in order of height simply by comparing them side by side to see who is tallest, who is next tallest, and so on. This kind of direct ordinal measurement is fairly common because it is easy to do. Notice that you do not need any absolute measurement to complete the ranking; that is, you never need to know anyone's exact height. It is necessary only to make relative judgments—given any two individuals, you must decide who is taller. Because ordinal scales do not require any absolute measurements, they can be used with variables that are not routinely measured. Variables such as beauty or talent can be observed, judged, and ranked, although they may be difficult to define and measure with more-sophisticated scales.

In addition to obtaining ranks by direct observation, a researcher may begin with a set of numerical measurements and convert these scores into ranks. For example, if you had a listing of the actual heights for a group of individuals, you could arrange the numbers in order from greatest to least. This process converts data from an interval or a ratio scale into ordinal measurements. There are a number of reasons for converting scores into ranks, but the following list should give you some indication of why this might be done.

1. Ranks are simpler. If someone asks you how tall your sister is, you could reply with a specific numerical value, such as 5 feet $7\frac{3}{4}$ inches tall. Or you could answer, "She is a little taller than I am." For many situations, the relative answer would be better.

2. The original scores may violate some of the basic assumptions that underlie certain statistical procedures. For example, the *t* tests and analysis of variance assume that the data come from normal distributions. Also, the independent-

measures tests assume that the different populations all have the same variance (the homogeneity-of-variance assumption). If a researcher suspects that the data do not satisfy these assumptions, it may be safer to convert the scores to ranks and use a statistical technique designed for ranks.

3. Occasionally, an experiment will produce an undetermined, or infinite, score. For example, a rat may show no sign of solving a particular maze after hundreds of trials. This animal has an infinite, or undetermined, score. Although there is no absolute score that can be assigned, you can say that this rat has the highest score for the sample and then rank the rest of the scores by their numerical values.

RANKING TIED SCORES

Whenever you are transforming numerical scores into ranks, you may find two or more scores that have exactly the same value. Because the scores are tied, the transformation process should produce ranks that are also tied. The procedure for converting tied scores into tied ranks was presented in Chapter 16 (page 546) when we introduced the Spearman correlation and is repeated briefly here. First, you list the scores in order including tied values. Second, you assign each position in the list an ordinal value (1st, 2nd, and so on). Finally, for any scores that are tied, you compute the mean of the tied ordinal positions, and use the mean value as the final rank. The following set of scores demonstrates this process for a set of $n = 8$ scores.

Original scores:	3	4	4	7	9	9	9	12
Ordinal positions:	1	2	3	4	5	6	7	8
Final ranks:	1	2.5	2.5	4	6	6	6	8

HYPOTHESIS TESTS WITH ORDINAL DATA

The remainder of this chapter examines three hypothesis-testing procedures that are used with ordinal data. Each of the new tests can be viewed as an alternative for one of the standard, parametric tests that was introduced earlier. The three tests and the situations in which they are used are as follows:

1. The Mann-Whitney test uses data from two separate samples to evaluate the difference between two treatment conditions or two populations. The Mann-Whitney test can be viewed as an alternative to the independent-measures t hypothesis test introduced in Chapter 10.

2. The Wilcoxon test uses data from a repeated-measures design to evaluate the difference between two treatment conditions. This test is an alternative to the repeated-measures t test from Chapter 11.

3. The Kruskal-Wallis test uses data from three or more separate samples to evaluate the differences between three or more treatment conditions (or populations). The Kruskal-Wallis test is an alternative to the single-factor, independent-measures analysis of variance introduced in Chapter 13.

In each case, you should realize that the new, ordinal-data tests are back-up procedures that are available in situations where the standard, parametric tests cannot be used. In general, if the data are appropriate for an analysis of variance or one of the t tests, then the standard test is preferred to its ordinal-data alternative.

19.2 THE MANN-WHITNEY *U*-TEST

Recall that a study using two separate samples is called an *independent-measures* study or a *between-subjects* study.

The Mann-Whitney test is designed to use the data from two separate samples to evaluate the difference between two treatments (or two populations). The calculations for this test require that the individual scores in the two samples be rank-ordered. The mathematics of the Mann-Whitney test are based on the following simple observation:

A real difference between the two treatments should cause the scores in one sample to be generally larger than the scores in the other sample. If the two samples are combined and all the scores placed in rank order on a line, then the scores from one sample should be concentrated at one end of the line, and the scores from the other sample should be concentrated at the other end.

On the other hand, if there is no treatment difference, then large and small scores will be mixed evenly in the two samples because there is no reason for one set of scores to be systematically larger or smaller than the other.

This observation is demonstrated in Figure 19.1.

FIGURE 19.1

In part (a), the scores from the two samples are clustered at opposite ends of the rank ordering. In this case, the data suggest a systematic difference between the two treatments. Part (b) shows the two samples intermixed evenly along the scale, indicating no consistent difference between treatments.

(a)

(b)

THE NULL HYPOTHESIS FOR THE MANN-WHITNEY TEST

Because the Mann-Whitney test compares two distributions (rather than two means), the hypotheses tend to be somewhat vague. We will state the hypotheses in terms of a consistent, systematic difference between the two treatments being compared.

H_0: There is no difference between the two treatments. Therefore, there is no tendency for the ranks in one treatment condition to be systematically higher (or lower) than the ranks in the other treatment condition.

H_1: There is a difference between the two treatments. Therefore, the ranks in one treatment condition are systematically higher (or lower) than the ranks in the other treatment condition.

CALCULATION OF THE MANN-WHITNEY U

The first steps in the calculations for the Mann-Whitney test have already been discussed. To summarize,

1. A separate sample is obtained from each of the two treatments. We will use n_A to refer to the number of subjects in sample A and n_B to refer to the number in sample B.

2. These two samples are combined, and the total group of $n_A + n_B$ subjects is rank ordered.

The remaining problem is to decide whether the scores from the two samples are mixed randomly in the rank ordering or whether they are systematically clustered at opposite ends of the scale. This is the familiar question of statistical significance: Are the data simply the result of chance, or has some systematic effect produced these results? We will answer this question exactly as we always have answered it. First, look at all the possible results that could have been obtained. Next, separate these outcomes into two groups:

1. Those results that are reasonably likely to occur by chance

2. Those results that are very unlikely to occur by chance (this is the critical region)

For the Mann-Whitney test, the first step is to identify each of the possible outcomes. This is done by assigning a numerical value to every possible set of sample data. This number is called the Mann-Whitney U. The value of U is computed as if the two samples were two teams of athletes competing in a sports event. Each individual in sample A (the A team) gets one point whenever he or she is ranked ahead of an individual from sample B. The total number of points accumulated for sample A is called U_A. In the same way, a U value, or team total, is computed for sample B. The final Mann-Whitney U is the smaller of these two values. This process is demonstrated in the following example.

EXAMPLE 19.1

We begin with two separate samples with $n = 6$ scores in each.

Sample A (treatment 1):   27   2   9   48   6   15

Sample B (treatment 2):   71   63   18   68   94   8

Next, the two samples are combined, and all 12 scores are placed in rank order.*

---

*For this example, we are assuming that the researcher has obtained a score (X value) for each individual. You should realize, however, that it is not necessary to have a set of previously obtained scores. The researcher could have started with the group of 12 individuals (two samples of $n = 6$) and then have rank-ordered the subjects from 1st to 12th with respect to the variable being measured. The end result, in either case, is that the two samples are combined into one group and all the individuals are then ranked.

Each individual in sample A is assigned 1 point for every score in sample B that has a higher rank.

| | ORDERED SCORES | | POINTS FOR |
RANK	SCORE	SAMPLE	SAMPLE A
1	2	(A)	6 points
2	6	(A)	6 points
3	8	(B)	
4	9	(A)	5 points
5	15	(A)	5 points
6	18	(B)	
7	27	(A)	4 points
8	48	(A)	4 points
9	63	(B)	
10	68	(B)	
11	71	(B)	
12	94	(B)	

Finally, the points from all the individuals in sample A are combined, and the total number of points is computed for the sample. In this example, the total points or the U value for sample A is $U_A = 30$. In the same way, you can compute the U value for sample B. You should obtain $U_B = 6$. As a simple check on your arithmetic, note that

$$U_A + U_B = n_A n_B \tag{19.1}$$

For these data,

$$30 + 6 = 6(6)$$

To avoid errors, it is wise to compute both U values and verify that the sum is equal to $n_A n_B$.

FORMULAS FOR THE MANN-WHITNEY *U*

Because the process of counting points to determine the Mann-Whitney U can be tedious, especially with large samples, there is a formula that will generate the U value for each sample. To use this formula, you combine the samples and rank-order all the scores as before. Then you must find ΣR_A, which is the sum of the ranks for individuals in sample A, and the corresponding ΣR_B for sample B. The U value for each sample is then computed as follows: For sample A,

$$U_A = n_A n_B + \frac{n_A(n_A + 1)}{2} - \Sigma R_A \tag{19.2}$$

and for sample B,

$$U_B = n_A n_B + \frac{n_B(n_B + 1)}{2} - \Sigma R_B \tag{19.3}$$

These formulas are demonstrated using the data from Example 19.1. For sample A, the sum of the ranks is

$$\Sigma R_A = 1 + 2 + 4 + 5 + 7 + 8$$
$$= 27$$

For sample B, the sum of the ranks is

$$\Sigma R_B = 3 + 6 + 9 + 10 + 11 + 12$$
$$= 51$$

By using the special formula, for sample A,

$$U_A = n_A n_B + \frac{n_A(n_A + 1)}{2} - \Sigma R_A$$

$$= 6(6) + \frac{6(7)}{2} - 27$$

$$= 36 + 21 - 27$$

$$= 30$$

For sample B,

$$U_B = n_A n_B + \frac{n_B(n_B + 1)}{2} - \Sigma R_B$$

$$= 6(6) + \frac{6(7)}{2} - 51$$

$$= 36 + 21 - 51$$

$$= 6$$

Notice that these are the same U values we obtained in Example 19.1 using the counting method. The Mann-Whitney U value is the smaller of these two,

$$U = 6$$

HYPOTHESIS TESTS WITH THE MANN-WHITNEY U

Now that we have developed a method for identifying each rank order with a numerical value, the remaining problem is to decide whether the U value provides evidence for a real difference between the two treatment conditions. We will look at each possibility separately.

A large difference between the two treatments will cause all the ranks from sample A to cluster at one end of the scale and all the ranks from sample B to cluster at the other (see Figure 19.1). At the extreme, there will be no overlap between the two samples. In this case, the Mann-Whitney U will be zero because one of the samples will get no points at all. In general, a Mann-Whitney U of zero indicates the greatest possible difference between the two samples. As the two samples become more alike, their ranks begin to intermix, and the U becomes larger. If there is no consistent tendency for one treatment to produce larger scores than the other, then the ranks from the two samples should be intermixed evenly. In terms of a competition between the two samples (the A team versus the B team), a final score of 49 to 51 indicates that the two teams were nearly equal; a final score of 100 to 0 indicates a real difference between the two teams.

Notice that the null hypothesis does not specify any population parameter.

The null hypothesis for the Mann-Whitney test states that there is no systematic difference between the two treatments being compared; that is, there is no real difference between the two populations from which the samples are selected. In this case, the most likely outcome is that the two samples are similar and that the U value is relatively large. On the other hand, a very small value of U, near zero, is evidence that the two samples are very different. Therefore, a U value near zero would tend to refute the null hypothesis. The distribution of all the possible U values has been constructed, and the critical values for $\alpha = .05$ and $\alpha = .01$ are presented in Table B.9 of Appendix B. When sample data produce a U that is *less than or equal to* the table value, we reject H_0.

A complete example of a hypothesis test using the Mann-Whitney U follows.

EXAMPLE 19.2

A psychologist interested in the development of manual dexterity prepared a block-manipulation task for 3-year-old children. A sample of 13 children was obtained, 5 boys and 8 girls. The psychologist recorded the amount of time (in seconds) required by each child to arrange the blocks in a specified pattern. The data are as follows:

Boys: 23 18 29 42 21

Girls: 37 56 39 34 26 104 48 25

STEP 1
The null hypothesis states that there is no consistent difference between the two populations. For this example,

H_0: There is no systematic difference between the solution times for boys versus the times for girls.

H_1: There is a systematic difference.

STEP 2
For a nondirectional test with $\alpha = .05$ and with $n_A = 5$ and $n_B = 8$, the Mann-Whitney table gives a critical value of $U = 6$. If the data produce a U less than or equal to 6, we will reject the null hypothesis.

STEP 3
We will designate the boys as sample A and the girls as sample B. Combining the two samples and arranging the scores in order produce the following result:

Rank	1	2	3	4	5	6	7	8	9	10	11	12	13
Score	18	21	23	25	26	29	34	37	39	42	48	56	104
Sample	A	A	A	B	B	A	B	B	B	A	B	B	B
Points for the girls (B)				2	2		1	1	1		0	0	0

Because the girls (sample B) tend to cluster at the bottom of the rankings, they should have the smaller U. Therefore, we have identified the points for this sample. The girls' point total is $U_B = 7$.

Notice that the larger sample does not always have the larger point total.

It is not necessary to compute U for the boys, but we will continue with this calculation to demonstrate the formula for U. The boys (sample A) have ranks of 1, 2, 3, 6, and 10. The sum is $\Sigma R_A = 22$, so the U_A value is

$$U_A = n_A n_B + \frac{n_A(n_A + 1)}{2} - \Sigma R_A$$

$$= 5(8) + \frac{5(6)}{2} - 22$$

$$= 40 + 15 - 22$$

$$= 33$$

To check our calculations,

$$U_A + U_B = n_A n_B$$

$$33 + 7 = 5(8)$$

$$40 = 40$$

The final U is the smaller of the two values, so the Mann-Whitney U statistic is $U = 7$.

STEP 4   Because $U = 7$ is greater than the critical value of $U = 6$, we fail to reject the null hypothesis. At the .05 level of significance, these data do not provide sufficient evidence to conclude that there is a significant difference in manual dexterity between boys and girls at 3 years of age.

 IN THE LITERATURE
REPORTING THE RESULTS OF A MANN-WHITNEY U-TEST

Unlike many other statistical results, there are no strict rules for reporting the outcome of a Mann-Whitney U-test. APA guidelines, however, do suggest that the report include a summary of the data (including such information as the sample size and the sum of the ranks) and the obtained statistic and p value. For the study presented in Example 19.2, the results could be reported as follows:

> The original scores, measured in seconds, were rank-ordered and a Mann-Whitney U-test was used to compare the ranks for the $n = 5$ boys versus the $n = 8$ girls. The results indicate no significant difference between genders, $U = 7, p > .05$, with the sum of the ranks equal to 22 for the boys and 69 for the girls. ❏

NORMAL APPROXIMATION FOR THE MANN-WHITNEY U

When the two samples are both large (about $n = 20$), the distribution of the Mann-Whitney U statistic tends to approximate a normal shape. In this case, the Mann-Whitney hypotheses can be evaluated using a z-score statistic and the unit normal distribution. You may have noticed that the table of critical values for the Mann-Whitney test does not list values for samples larger than $n = 20$. This is because the normal approximation typically is used with larger samples. The procedure for this normal approximation is as follows:

1. Find the U values for sample A and sample B as before. The Mann-Whitney U is the smaller of these two values.

2. When both samples are relatively large (around $n = 20$ or more), the distribution of the Mann-Whitney U statistic tends to form a normal distribution with

$$\mu = \frac{n_A n_B}{2} \quad \text{and} \quad \sigma = \sqrt{\frac{n_A n_B (n_A + n_B + 1)}{12}}$$

The Mann-Whitney U obtained from the sample data can be located in this distribution using a z-score:

This approximation is intended for data without tied scores or with very few ties. A special formula has been developed for data with many ties and can be found in most advanced statistics texts such as Hays (1981).

$$z = \frac{X - \mu}{\sigma} = \frac{U - \dfrac{n_A n_B}{2}}{\sqrt{\dfrac{n_A n_B (n_A + n_B + 1)}{12}}} \tag{19.4}$$

3. Use the unit normal table to establish the critical region for this z-score. For example, with $\alpha = .05$ the critical values would be ± 1.96.

An example of this normal approximation for the Mann-Whitney U follows.

EXAMPLE 19.3   To demonstrate the normal approximation, we will use the same data that were used in Example 19.2. This experiment tested manual dexterity for 3-year-olds, using a sample of $n = 5$ boys and a sample of $n = 8$ girls. The data produced a value of $U = 7$. (You should realize that when samples are this small, you normally use the Mann-Whitney table rather than the normal approximation. However, we will use the normal approximation so that we can compare the outcome to the result from the regular Mann-Whitney test.)

STEP 1   The normal approximation does not affect the statement of the hypotheses.

H_0: There is no systematic difference between the manual dexterity scores for boys versus the scores for girls.

H_1: There is a systematic difference.

We will use $\alpha = .05$.

STEP 2   The critical region for this test is defined in terms of z-scores and the normal distribution. The unit normal table states that the extreme 5% of this distribution is located beyond z-scores of ± 1.96 (see Figure 19.2).

STEP 3   The sample value of $U = 7$ is used to compute a z-score from the normal approximation formula.

$$z = \frac{U - \dfrac{n_A n_B}{2}}{\sqrt{\dfrac{n_A n_B (n_A + n_B + 1)}{12}}}$$

$$= \frac{7 - \dfrac{5(8)}{2}}{\sqrt{\dfrac{5(8)(5 + 8 + 1)}{12}}}$$

$$= \frac{-13}{\sqrt{\dfrac{560}{12}}}$$

$$= -1.90$$

STEP 4   Because this z-score is not in the critical region, our decision is to fail to reject the null hypothesis. These data do not provide sufficient evidence to conclude that there is a significant difference in manual dexterity between boys and girls at age three.

Notice that we reached the same conclusion with the original Mann-Whitney test in Example 19.2.

ASSUMPTIONS AND CAUTIONS FOR THE MANN-WHITNEY *U*   The Mann-Whitney U-test is a very useful alternative to the independent-measures t test. Because the Mann-Whitney test does not require homogeneity of variance or normal distributions, it can be used in situations where the t test would be inappropriate. However, the U-test does require independent observations, and it assumes that the

FIGURE 19.2

The normal distribution of z-scores used with the normal approximation to the Mann-Whitney U-test. The critical region for $\alpha = .05$ has been shaded.

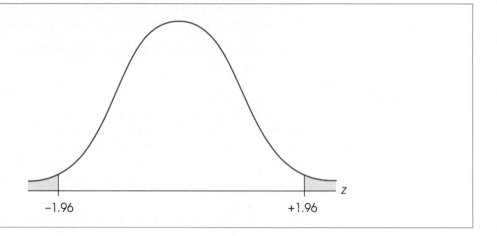

-1.96 $+1.96$ z

dependent variable is continuous. You should recall from Chapter 1 that a continuous scale has an infinite number of distinct points. One consequence of this fact is that it is very unlikely for two individuals to have exactly the same score. This means that there should be few, if any, tied scores in the data. When sample data do have several tied scores, you should suspect that a basic assumption underlying the Mann-Whitney U-test has been violated. In this situation, you should be cautious about using the Mann-Whitney U.

When there are relatively few tied scores in the data, the Mann-Whitney test may be used, but you must follow the standard procedure for ranking tied scores.

LEARNING CHECK

1. Rank the following scores, using tied ranks where necessary: 14, 3, 4, 0, 3, 5, 14, 3.

2. An experiment using $n = 25$ in one sample and $n = 10$ in the other produced a Mann-Whitney U of $U = 50$. Assuming that this is the smaller of the two U values, what was the value of U for the other sample?

3. A developmental psychologist is examining social assertiveness for preschool children. Three- and 4-year-old children are observed for 10 hours in a day-care center. The psychologist records the number of times each child initiates a social interaction with another child. The scores for the sample of four boys and nine girls are as follows:

    Boys' scores: 8   17   14   21

    Girls' scores: 18   25   23   21   34   28   32   30   13

    Use a Mann-Whitney test to determine whether these data provide evidence for a significant difference in social assertiveness between preschool boys and girls. Test at the .05 level.

4. According to the Mann-Whitney table, a value of $U = 30$ is significant (in the critical region) with $\alpha = .05$ when both samples have $n = 11$. If this value is used in the normal approximation, does it produce a z-score in the critical region?

ANSWERS   **1.** Scores: 0   3   3   3   4   5   14   14
            Ranks: 1   3   3   3   5   6   7.5   7.5

2. With $n_A = 25$ and $n_B = 10$, the two U values must total $n_A n_B = 25(10) = 250$. If the smaller value is 50, then the larger value must be 200.

3. For the boys, $U = 31.5$. For the girls, $U = 4.5$. The critical value in the table is 4, so fail to reject H_0.

4. $U = 30$ produces a z-score of $z = -2.00$. This is in the critical region.

19.3 THE WILCOXON SIGNED-RANKS TEST

The Wilcoxon test is designed to evaluate the difference between two treatments, using the data from a repeated-measures experiment. Recall that a repeated-measures study involves only one sample, with each individual in the sample being measured twice: once in the first treatment and once in the second treatment. The difference between the two scores for each individual is computed, resulting in a sample of *difference scores*. The data for the Wilcoxon test consist of the difference scores from the repeated-measures design. The test requires that these differences be ranked from smallest to largest in terms of their *absolute values* (without regard to the sign). This process is demonstrated in the following example.

EXAMPLE 19.4   The following data are from a repeated-measures experiment using a sample of $n = 6$ subjects to compare two treatment conditions:

	TREATMENTS			
SUBJECT	1	2	DIFFERENCE	RANK
1	18	43	+25	6 (Largest)
2	9	14	+5	2
3	21	20	−1	1 (Smallest)
4	30	48	+18	5
5	14	21	+7	3
6	12	4	−8	4

The null hypothesis for the Wilcoxon test simply states that there is no consistent, systematic difference between the two treatments.

H_0:   There is no difference between the two treatments. Therefore, in the general population there is no tendency for the difference scores to be either systematically positive or systematically negative.

H_1:   There is a difference between the two treatments. Therefore, in the general population the difference scores are systematically positive or systematically negative.

If the null hypothesis is true, any differences that exist in the sample data must be due to chance. Therefore, we would expect positive and negative differences to be inter-mixed evenly. On the other hand, a consistent difference between the two treatments should cause the scores in one treatment to be consistently larger than the scores in the other. This should produce difference scores that tend to be consistently positive or consistently negative. The Wilcoxon test uses the signs and the ranks of the difference scores to decide whether or not there is a significant difference between the two treatments.

CALCULATION AND INTERPRETATION OF THE WILCOXON T

As with most nonparametric tests, the calculations for the Wilcoxon are quite simple. After ranking the absolute values of the difference scores as in Example 19.4, you separate the ranks into two groups: those associated with positive differences and those associated with negative differences. Next, find the sum of the ranks for each group. The smaller of these two sums is the test statistic for the Wilcoxon test and is identified by the letter T. For the data in Example 19.4, the ranks for positive differences are 6, 2, 3, and 5. These ranks sum to $\Sigma R = 16$. The ranks for negative differences are 1 and 4, which sum to $\Sigma R = 5$. The smaller of these two sums is 5, so the Wilcoxon T for these data is $T = 5$.

We noted earlier that a strong treatment effect should cause the difference scores to be consistently positive or consistently negative. In the extreme case, all of the differences will be in the same direction. This will produce a Wilcoxon T of zero. For example, when all of the differences are positive, the sum of the negative ranks will be zero. On the other hand, when there is no treatment effect, the signs of the difference scores should be intermixed evenly. In this case, the Wilcoxon T will be relatively large. In general, a small T value (near zero) provides evidence for a real difference between the two treatment conditions. The distribution of all the possible T values has been constructed, and the critical values for $\alpha = .05$ and $\alpha = .01$ are given in Table B.10 of Appendix B. Whenever sample data produce a T that is *less than or equal to* this critical value, we will reject H_0.

TIED SCORES AND ZERO SCORES

Although the Wilcoxon test does not require normal distributions, it does assume that the dependent variable is continuous. As noted with the Mann-Whitney test, this assumption implies that tied scores should be very unlikely. When ties do appear in sample data, you should be concerned that the basic assumption of continuity has been violated, and the Wilcoxon test may not be appropriate. If there are relatively few ties in the data, most researchers assume that the data actually are continuous, but have been crudely measured. In this case, the Wilcoxon test may be used, but the tied values must receive special attention in the calculations.

With the Wilcoxon test, there are two different types of tied scores:

1. A subject may have the same score in treatment 1 and in treatment 2, resulting in a difference score of zero.

2. Two (or more) subjects may have identical difference scores (ignoring the sign of the difference).

Remember: The null hypothesis says there is no difference between the two treatments. Subjects with difference scores of zero tend to support this hypothesis.

When the data include individuals with difference scores of zero, one strategy is to discard these individuals from the analysis and reduce the sample size (n). However, this procedure ignores the fact that a difference score of zero is evidence for retaining the null hypothesis. A better procedure is to divide the zero differences evenly between the positives and negatives. (If you have an odd number of zero differences, discard one and divide the rest evenly.) This second procedure will tend to increase ΣR for both the positive and the negative ranks, which increases the final value of T and makes it more likely that H_0 will be retained.

When you have ties among the difference scores, each of the tied scores should be assigned the average of the tied ranks. This procedure was presented in detail in an earlier section of this chapter (see p. 636).

HYPOTHESIS TESTS WITH THE WILCOXON *T*

Following is a complete example of the Wilcoxon test using data containing both types of tied scores.

EXAMPLE 19.5

The local Red Cross has conducted an intensive campaign to increase blood donations. This campaign has been concentrated in 10 local businesses. In each company, the goal was to increase the percentage of employees who participate in the blood donation program. Figures showing the percentages of participation from last year (before the campaign) and from this year are as follows. We will use the Wilcoxon test to decide whether these data provide evidence that the campaign had a significant impact on blood donations. Note that the 10 companies are listed in rank order according to the absolute value of the difference scores.

COMPANY	PERCENTAGE OF PARTICIPATION			RANK DISCARDING ZEROS	RANK INCLUDING ZEROS
	BEFORE	AFTER	DIFFERENCE		
A	18	18	0	—	1.5
B	24	24	0	—	1.5
C	31	30	−1	1	3
D	28	24	−4	2	4
E	17	24	+7	3	5
F	16	24	+8	4	6
G	15	26	+11	5.5	7.5
H	18	29	+11	5.5	7.5
I	20	36	+16	7	9
J	9	28	+19	8	10

Note: We will conduct the Wilcoxon test using the recommendation that difference scores of zero be discarded. Following this test, we will examine what would happen if the difference scores of zero were included.

STEP 1   The null hypothesis states that the campaign had no effect. Therefore, any differences are due to chance, and there should be no consistent pattern.

STEP 2   The two companies with difference scores of zero are discarded, and n is reduced to 8. With $n = 8$ and $\alpha = .05$, the critical value for the Wilcoxon test is $T = 3$. A sample value that is less than or equal to 3 will lead us to reject H_0.

STEP 3   For these data, the positive differences have ranks of 3, 4, 5.5, 5.5, 7, and 8:

$$\Sigma R_+ = 33$$

The negative differences have ranks of 1 and 2:

$$\Sigma R_- = 3$$

The Wilcoxon T is the smaller of these sums, so $T = 3$.

STEP 4   The T value from the data is in the critical region. This value is very unlikely to occur by chance ($p < .05$); therefore, we reject H_0 and conclude that there is a significant change in participation after the Red Cross campaign.

Note: If we include the difference scores of zero in this test, then $n = 10$, and with $\alpha = .05$, the critical value for the Wilcoxon T is 8. Because the difference scores of zero are tied for first and second in the ordering, each is given a rank of 1.5. One of these ranks is assigned to the positive group and one to the negative group. As a result, the sums are

$$\Sigma R_+ = 1.5 + 5 + 6 + 7.5 + 7.5 + 9 + 10$$

$$= 46.5$$

and

$$\Sigma R_- = 1.5 + 3 + 4$$

$$= 8.5$$

The Wilcoxon T is the smaller of these two sums, $T = 8.5$. Because this T value is larger than the critical value, we fail to reject H_0 and conclude that these data do not indicate a significant change in blood donor participation.

By including the difference scores of zero in the test, we have changed the statistical conclusion. Remember: The difference scores of zero are an indication that the null hypothesis is correct. When these scores are considered, the test is more likely to retain H_0.

IN THE LITERATURE
REPORTING THE RESULTS OF A WILCOXON *T*-TEST

As with the Mann-Whitney U-test, there is no specified format for reporting the results of a Wilcoxon T-test. It is suggested, however, that the report include a summary of the data and the value obtained for the test statistic as well as the p value. It also is suggested that the report describe the treatment of zero-difference scores in the analysis. For the study presented in Example 19.5, the report could be as follows:

> Two companies showing no change in participation were discarded prior to analysis. The remaining eight companies were rank-ordered by the magnitude of the change in level of participation, and a Wilcoxon T was used to evaluate the data. The results show a significant increase in participation following the campaign, $T = 3$, $p < .05$, with the ranks for increases totaling 33 and the ranks for decreases totaling 3.   ❑

NORMAL APPROXIMATIONS FOR THE WILCOXON *T*-TEST

When a sample is relatively large, the values for the Wilcoxon T statistic tend to form a normal distribution. In this situation, it is possible to perform the test using a z-score statistic and the normal distribution rather than looking up a T value in the Wilcoxon table. When the sample size is greater than 20, the normal approximation is very accurate and can be used. For samples larger than $n = 50$, the Wilcoxon table typically does not provide any critical values, so the normal approximation must be used. The procedure for the normal approximation is as follows:

1. Find the total for the positive ranks and the total for the negative ranks as before. The Wilcoxon T is the smaller of the two values.

2. With n greater than 20, the Wilcoxon T values form a normal distribution with a mean of

$$\mu = \frac{n(n + 1)}{4}$$

and a standard deviation of

$$\sigma = \sqrt{\frac{n(n + 1)(2n + 1)}{24}}$$

The Wilcoxon T from the sample data corresponds to a z-score in this distribution defined by

$$z = \frac{X - \mu}{\sigma} = \frac{T - \dfrac{n(n + 1)}{4}}{\sqrt{\dfrac{n(n + 1)(2n + 1)}{24}}} \tag{19.5}$$

3. The unit normal table is used to determine the critical region for the z-score. For example, the critical values are ± 1.96 for $\alpha = .05$.

EXAMPLE 19.6 Although the normal approximation is intended for samples with at least $n = 20$ individuals, we will demonstrate the calculations with the same data that were used to introduce the Wilcoxon T in Example 19.5.

To evaluate the data, we first discarded the two companies with zero differences, which reduced the number to $n = 8$ and produced a value of $T = 3$. This value was exactly equal to the criterion for significance, and we concluded that there was a significant difference at the .05 level. When the normal approximation is applied to the same data, we obtain

$$\mu = \frac{n(n + 1)}{4} = \frac{8(9)}{4} = 18$$

$$\sigma = \sqrt{\frac{n(n + 1)(2n + 1)}{24}} = \sqrt{\frac{8(9)(17)}{24}} = \sqrt{51} = 7.14$$

With these values, the obtained $T = 3$ corresponds to a z-score of

$$z = \frac{T - \mu}{\sigma} = \frac{3 - 18}{7.14} = \frac{-15}{7.14} = -2.10$$

With critical boundaries of ± 1.96, the obtained z-score is close to the boundary, but it is enough to be significant at the .05 level. Note that this is exactly the same conclusion we reached using the Wilcoxon T table in Example 19.5.

The second analysis in Example 19.5 included the two companies with zero differences and split them equally between the positive ranks and the negative ranks. In this case we had $n = 10$ companies and the analysis produced $T = 8.5$. Although this value was close to the critical boundary, it was not sufficient to conclude that there was a significant difference at the .05 level. When the normal approximation is applied to the same data, we obtain

$$\mu = \frac{n(n+1)}{4} = \frac{10(11)}{4} = 27.5$$

$$\sigma = \sqrt{\frac{n(n+1)(2n+1)}{24}} = \sqrt{\frac{10(11)(21)}{24}} = \sqrt{96.25} = 9.81$$

With these values, the obtained $T = 8.5$ corresponds to a z-score of

$$z = \frac{T - \mu}{\sigma} = \frac{8.5 - 27.5}{9.81} = \frac{-19}{9.81} = -1.94$$

With critical boundaries of ± 1.96, the obtained z-score is close to the boundary, but it is not quite enough to be significant at the .05 level. Again, this is the same conclusion we reached using the Wilcoxon T table in Example 19.5.

LEARNING CHECK

1. A physician is testing the effectiveness of a new arthritis drug by measuring patients' grip strength before and after they receive the drug. The difference scores for 10 patients are as follows: $+3$, $+46$, $+16$, -2, $+38$, $+14$, 0 (no change), -8, $+25$, and $+41$. Each score is the difference in strength, with a positive value indicating a stronger grip after receiving the drug. Use a Wilcoxon test to determine whether these data provide sufficient evidence to conclude that the drug has a significant effect. Test at the .05 level.

ANSWER

1. The Wilcoxon $T = 4$. Because one patient showed no change, n is reduced to 9, and the critical value is $T = 5$. Therefore, we reject H_0 and conclude that there is a significant effect.

19.4   THE KRUSKAL-WALLIS TEST

The Kruskal-Wallis test is used to evaluate differences between three or more treatment conditions (or populations) using data from an independent-measures design. You should recognize that this test is an alternative to the single-factor analysis of variance introduced in Chapter 13. However, the analysis of variance requires numerical scores that can be used to calculate means and variances. The Kruskal-Wallis test, on the other hand, simply requires that you are able to rank-order the individuals for the variable being measured. You also should recognize that the Kruskal-Wallis test is similar to the Mann-Whitney test introduced earlier in this chapter. However, the Mann-Whitney test is limited to comparing only two treatments, whereas the Kruskal-Wallis test is used to compare three or more treatments.

THE DATA FOR A KRUSKAL-WALLIS TEST

The Kruskal-Wallis test requires three or more separate samples. The samples can represent different treatment conditions or they can represent different pre-existing populations. For example, a researcher may want to examine how social input affects creativity. Children are asked to draw pictures under three different conditions:

(1) working alone without supervision, (2) working in groups where the children are encouraged to examine and criticize each other's work, and (3) working alone but with frequent supervision and comments from a teacher. Three separate samples are used to represent the three treatment conditions with $n = 6$ children in each condition. At the end of the study, the researcher collects the drawings from all 18 children and rank-orders the complete set of drawings in terms of creativity. The purpose for the study is to determine whether one treatment condition produces drawings that are ranked consistently higher (or lower) than another condition. Notice that the researcher does not need to determine an absolute creativity score for each painting. Instead, the data consist of relative measures; that is, the researcher must decide which painting shows the most creativity, which shows the second most creativity, and so on.

The creativity study that was just described is an example of a research study comparing different treatment conditions. It also is possible that the three groups could be defined by a subject variable so that the three samples represent different populations. For example, a researcher could obtain drawings from a sample of 5-year-old children, a sample of 6-year-old children, and a third sample of 7-year-old children. Again, the Kruskal-Wallis test would begin by rank-ordering all of the drawings to determine whether one age group showed significantly more (or less) creativity than another.

Finally, the Kruskal-Wallis test can be used if the original, numerical data are converted into ordinal values. The following example demonstrates how a set of numerical scores is transformed into ranks to be used in a Kruskal-Wallis analysis.

EXAMPLE 19.7

Table 19.2(a) shows the original data from an independent-measures study comparing three treatment conditions. To prepare the data for a Kruskal-Wallis test, the complete set of original scores is rank-ordered using the standard procedure for ranking tied scores. Table 19.2(b) shows the transformed data where each of the original scores is replaced by its ordinal rank. The Kruskal-Wallis test uses the set of ranks as shown in Table 19.2(b).

TABLE 19.2

Preparing a set of data for analysis using the Kruskal-Wallis test. The original data consisting of numerical scores are shown in table (a). The original scores are listed in order and assigned ranks using the standard procedure for ranking tied scores. The ranks are then substituted for the original scores to create the set of ordinal data shown in table (b).

(a) ORIGINAL DATA NUMERICAL SCORES				FIND THE RANK FOR EACH SCORE.			(b) ORDINAL DATA RANKS		
I	II	III		ORIGINAL NUMERICAL SCORES	ORDINAL RANKS		I	II	III
14	2	26	$\Rightarrow$	2	1	$\Rightarrow$	9	1	15
3	14	8		3	2		2	9	5
21	9	14		5	3.5		14	6	9
5	12	19		5	3.5		3.5	7	12
16	5	20		8	5		11	3.5	13
				9	6				
				12	7				
				14	9				
				14	9				
				14	9				
				16	11				
				19	12				
				20	13				
				21	14				
				26	15				

THE NULL HYPOTHESIS FOR THE KRUSKAL-WALLIS TEST

As with the other tests for ordinal data, the null hypothesis for the Kruskal-Wallis test tends to be somewhat vague. In general, the null hypothesis states that there are no differences among the treatments being compared. Somewhat more specifically, H_0 states that there is no tendency for the ranks in one treatment condition to be systematically higher (or lower) than the ranks in any other condition. Generally, we will use the concept of "systematic differences" to phrase the statement of H_0 and H_1. Thus, the hypotheses for the Kruskal-Wallis test will be phrased as follows:

H_0: There is no tendency for the ranks in any treatment condition to be systematically higher or lower than the ranks in any other treatment condition. There are no differences between treatments.

H_1: The ranks in at least one treatment condition are systematically higher (or lower) than the ranks in another treatment condition. There are differences between treatments.

FORMULAS AND NOTATION FOR THE KRUSKAL-WALLIS TEST

The first stage of the Kruskal-Wallis test involves obtaining an ordinal measurement (a rank) for each individual. This process was demonstrated in the preceding section (see Table 19.2) and is summarized as follows:

1. The data consist of three (or more) separate samples representing different treatment conditions or different populations.
2. The separate samples are combined and the entire group is rank-ordered.
3. The individuals are regrouped into the original separate samples and the score for each subject is the ordinal rank that was obtained from step 2.

Table 19.3 presents a set of ordinal data along with the notation that will be used in the Kruskal-Wallis formula. These values are the same ranks that were obtained in Example 19.6. The notation is relatively simple and involves the following values:

1. The ranks in each treatment are summed to obtain a total or T value for that treatment condition. The T values will be used in the Kruskal-Wallis formula.
2. The number of subjects in each treatment condition is identified by a lower-case n.
3. The total number of subjects in the entire study is identified by an uppercase N.

TABLE 19.3

An example of ordinal data including the notation used for a Kruskal-Wallis test. These are the same data that were presented in Table 19.2, and they include some tied measurements that appear as tied ranks.

ORDINAL DATA RANKS			
I	II	III	
9	1	15	$N = 15$
2	9	5	
14	6	9	
3.5	7	12	
11	3.5	13	
$T_1 = 39.5$	$T_2 = 26.5$	$T_3 = 54$	
$n_1 = 5$	$n_2 = 5$	$n_3 = 5$	

The Kruskal-Wallis formula produces a statistic that is usually identified with the letter H and has approximately the same distribution as chi-square with degrees of freedom defined by the number of treatment conditions minus one. For the data in Table 19.3, there are 3 treatment conditions so the formula produces a chi-square value with $df = 2$. The formula for the Kruskal-Wallis statistic is

$$H = \frac{12}{N(N + 1)} \left(\Sigma \frac{T^2}{n} \right) - 3(N + 1) \tag{19.6}$$

Using the data in Table 19.3, the Kruskal-Wallis formula produces a chi-square value of

$$H = \frac{12}{15(16)} \left(\frac{39.5^2}{5} + \frac{26.5^2}{5} + \frac{54^2}{5} \right) - 3(16)$$

$$= 0.05(312.05 + 140.45 + 583.2) - 48$$

$$= 0.05(1035.7) - 48$$

$$= 51.785 - 48$$

$$= 3.785$$

With $df = 2$, the chi-square table lists a critical value of 5.99 for $\alpha = .05$. Because the obtained chi-square value (3.785) is not greater than the critical value, our statistical decision is to fail to reject H_0. The data do not provide sufficient evidence to conclude that there are significant differences among the three treatments.

IN THE LITERATURE
REPORTING THE RESULTS OF A KRUSKAL-WALLIS TEST

As with the Mann-Whitney and the Wilcoxon tests, there is no standard format for reporting the outcome of a Kruskal-Wallis test. However, the report should provide a summary of the data, the value obtained for the chi-square statistic, as well as the value of df, N, and the p value. For the Kruskal-Wallis test that we just completed, the results could be reported as follows:

> After ranking the individual scores, a Kruskal-Wallis test was used to evaluate differences among the three treatments. The outcome of the test indicated no significant differences among the treatment conditions, $H = 3.785$ (2, $N = 15$), $p > .05$. ❏

There is one assumption for the Kruskal-Wallis test that is necessary to justify using the chi-square distribution to identify critical values for H. Specifically, each of the treatment conditions must contain at least five scores. If any sample has fewer than five scores, you can still compute the Kruskal-Wallis H statistic but you must consult a special table to evaluate the significance (Siegel, 1956).

SUMMARY

1. The Mann-Whitney test, the Wilcoxon test, and the Kruskal-Wallis test are nonparametric alternatives to the independent-measures t, the repeated-measures t, and the single-factor analysis of variance, respectively. These tests do not require normal distributions or homogeneity of variance. However, the tests do require that the data be rank-ordered, and they assume that the dependent variable is continuously distributed. For all three tests, the null hypothesis states that there are no consistent, systematic differences between the treatments being compared.

2. The Mann-Whitney U can be computed either by a counting process or by a formula. A small value of U (near zero) is evidence of a difference between the two treatments. With the counting procedure, U is determined by the following:
   a. The scores from the two samples are combined and ranked from smallest to largest.
   b. Each individual in sample A is awarded one point for every member of sample B with a larger rank.
   c. U_A equals the total points for sample A. U_B is the total for sample B. The Mann-Whitney U is the smaller of these two values.

   In formula form,

   $$U_A = n_A n_B + \frac{n_A(n_A + 1)}{2} = \Sigma R_A$$

   $$U_B = n_A n_B + \frac{n_B(n_B + 1)}{2} - \Sigma R_B$$

3. For large samples, larger than those normally presented in the Mann-Whitney table, the normal distribution can be used to evaluate the difference between the two treatments. This normal approximation to the Mann-Whitney is used as follows:
   a. Find the value of U as before.
   b. Convert the Mann-Whitney U to a z-score by the formula

   $$z = \frac{U - \dfrac{n_A n_B}{2}}{\sqrt{\dfrac{n_A n_B(n_A + n_B + 1)}{12}}}$$

   c. If this z-score is in the critical region of the unit normal distribution, reject the null hypothesis.

4. The test statistic for the Wilcoxon test is called a T score. A small value of T (near zero) provides evidence of a difference between the two treatments. T is computed as follows:
   a. Compute a difference score (treatment 1 versus treatment 2) for each individual in the sample.
   b. Rank these difference scores from smallest to largest without regard to the signs.
   c. Sum the ranks for the positive differences, and sum the ranks for the negative differences. T is the smaller of these two sums.

5. For large samples, $n \geq 20$, The Wilcoxon T is converted into a z-score by the formula

   $$z = \frac{X - \mu}{\sigma} = \frac{T - \dfrac{n(n + 1)}{4}}{\sqrt{\dfrac{n(n + 1)(2n + 1)}{24}}}$$

   The unit normal table can then be used to find critical values for different alpha levels.

6. The Kruskal-Wallis test is used to evaluate differences among three or more treatment conditions using a separate sample for each treatment. The test statistic, H, is equivalent to a chi-square statistic with degrees of freedom equal to the number of treatment conditions minus one. The calculation of the test statistic involves the following:
   a. The individuals from all the separate samples are combined and ranked.
   b. The sum of the ranks, T, is computed for each treatment.
   c. The test statistic is computed as

   $$H = \frac{12}{N(N + 1)}\left(\Sigma \frac{T^2}{n}\right) - 3(N + 1)$$

   where n is the number of scores in each individual sample and N is the number of scores for all samples combined.

RESOURCES

If you are using WebTutor along with this book, there is a WebTutor section correspon-ding to this chapter. The WebTutor contains a brief summary of Chapter 19, hints for learning the Mann-Whitney, Wilcoxon, and Kruskal-Wallis tests, cautions about common errors, and sample exam items including solutions.

SPSS

The statistics computer package SPSS can be used to conduct the Mann-Whitney, Wilcoxon, and Kruskal-Wallis tests presented in this chapter. The data are simply entered into the computer and SPSS calculates the ranks and the appropriate test statistic.

For the Mann-Whitney test, the data are entered in a stacked format with all the scores from both groups in one column. A second column is used to identify the group for each score. (Type a 1 beside each score from group one and a 2 beside each score from group two.) Then, click **Analyze** on the tool bar, select **Nonparametric Tests**, and click on **2 Independent Samples**. Next you must enter the label for the data column into the Test Variable List box, and enter the column label for the group identification numbers in the Grouping Variable box. Then click **Define Groups** and enter the numbers that you used to identify the two groups. Click **Continue**, make sure that **Mann-Whitney U** is selected, and click **OK**. The output includes the number of individuals in each group, as well as the mean rank and the sum of the ranks for each group. In addition, the output includes the value for U, the z-score from the normal approximation, and a level of significance for the result.

For the Wilcoxon test, the data are entered into two columns, with the first score for each participant in one column and the second score in the second column. Then, click **Analyze** on the tool bar, select **Nonparametric Tests**, and click on **2 Related Samples**. Highlight both column labels for the scores and move them into the Test Pair(s) List box. Be sure that **Wilcoxon** is selected, and click **OK**. The output includes the number, the mean rank, and the sum of the ranks for both the positive and negative values (the smaller of the two sums is the Wilcoxon T). Also, the z-score for the normal approximation is reported along with the level of significance for the test.

The Kruskal-Wallis test also requires data in a stacked format, with all the scores from all groups in one column. A second column is used to identify the group for each score. (Type a 1 beside each score from group one, a 2 beside each score from group two, and so on.) Then click **Analyze** on the tool bar, select **Nonparametric Tests**, and click on **K Independent Samples**. Next you must enter the label for the data column into the Test Variable List box and enter the column label for the group identification numbers in the Grouping Variable box. Then click **Define Range** and enter the minimum and maximum numbers that you used to identify the groups. Click **Continue**, make sure that **Kruskal-Wallis H** is selected, and click **OK**. The output includes the number of scores and the mean rank for each group, as well as the chi-square statistic with degrees of freedom and the level of significance.

KEY TERMS

Mann-Whitney U

Wilcoxon T

normal approximation for the Mann-Whitney U

normal approximation for the Wilcoxon T

Kruskal-Wallis test

FOCUS ON PROBLEM SOLVING

1. In computing the Mann-Whitney U, it will sometimes be obvious from the data which sample has the smaller U value. Even if this is the case, you should still compute both U values so that your computations can be checked using the formula

$$U_A + U_B = n_A n_B$$

2. Contrary to the previous statistical tests, the Mann-Whitney U and the Wilcoxon T are significant when the obtained value is *equal to or less than* the critical value in the table.

3. For the Wilcoxon T, remember to ignore the signs of the difference scores when assigning ranks to them.

DEMONSTRATION 19.1

THE MANN-WHITNEY *U*-TEST

A local police expert claims to be able to judge an individual's personality on the basis of his or her handwriting. To test this claim, 10 samples of handwriting are obtained: 5 from prisoners convicted of violent crimes and 5 from psychology majors at the college. The expert ranks the handwriting samples from 1st to 10th, with 1 representing the most anti-social personality. The rankings are as follows:

RANKING	SOURCE
1	Prisoner
2	Prisoner
3	Student . . . 3 points
4	Prisoner
5	Prisoner
6	Student . . . 1 point
7	Prisoner
8	Student . . . 0 points
9	Student . . . 0 points
10	Student . . . 0 points

STEP 1   The null hypothesis states that there is no difference between the two populations. For this example, H_0 states that the police expert cannot differentiate the handwriting for prisoners from the handwriting for students.

The alternative hypothesis says there is a discernible difference.

STEP 2   For $\alpha = .05$ and with $n_A = n_B = 5$, the Mann-Whitney table gives a critical value of $U = 2$. If our data produce a U less than or equal to 2, we will reject the null hypothesis.

STEP 3   We designate the students as sample A and the prisoners as sample B. Because the students tended to cluster at the bottom of the rankings, they should have the smaller U. Therefore, we have identified the points for this sample. The students' point total is $U_A = 4$.

Because we have found the smaller of the two U values, it is not necessary to compute U for the sample of prisoners. However, we will continue with this calculation to demonstrate the formula for U. The sample of prisoners has ranks 1, 2, 4, 5, and 7. The sum is $\Sigma R_B = 19$, so the U_B value is

$$U_B = n_A n_B + \frac{n_B(n_B + 1)}{2} - \Sigma R_B$$

$$= 5(5) + \frac{5(6)}{2} - 19$$

$$= 25 + 15 - 19$$

$$= 21$$

To check our calculations,

$$U_A + U_B = n_A n_B$$

$$4 + 21 = 5(5)$$

$$25 = 25$$

The final U is the smaller of the two values, so the Mann-Whitney U statistic is $U = 4$.

STEP 4   Because $U = 4$ is not in the critical region, we fail to reject the null hypothesis. With $\alpha = .05$, these data do not provide sufficient evidence to conclude that there is a discernible difference in handwriting between the two populations.

DEMONSTRATION 19.2

THE WILCOXON SIGNED-RANKS TEST

A researcher obtains a random sample of $n = 7$ individuals and tests each person in two different treatment conditions. The data for this sample are as follows:

PARTICIPANT	TREATMENT 1	TREATMENT 2	DIFFERENCE
1	8	24	+16
2	12	10	−2
3	15	19	+4
4	31	52	+21
5	26	20	−6
6	32	40	+8
7	19	29	+10

STEP 1   *State the hypotheses, and select alpha.* The hypotheses for the Wilcoxon test do not refer to any specific population parameter.

H_0:   There is no systematic difference between the two treatments.

H_1:   There is a consistent difference between the treatments that causes the scores in one treatment to be generally higher than the scores in the other treatment.

We will use $\alpha = .05$.

STEP 2   *Locate the critical region.* A small value for the Wilcoxon T indicates that the difference scores were consistently positive or consistently negative, which indicates a systematic treatment difference. Thus, small values will tend to refute H_0. With $n = 7$ and $\alpha = .05$, the Wilcoxon table shows that a T value of 2 or smaller is needed to reject H_0.

STEP 3   *Compute the test statistic.* The calculation of the Wilcoxon T is very simple, but requires several stages:

a. Ignoring the signs (+ or −), rank the difference scores from smallest to largest.

b. Compute the sum of the ranks for the positive differences and the sum for the negative differences.

c. The Wilcoxon T is the smaller of the two sums.

For these data, we have the following:

DIFFERENCE	RANK	
(+) 16	6	$\Sigma R_+ = 6 + 2 + 7 + 4 + 5 = 24$
(−) 2	1	$\Sigma R_- = 1 + 3 = 4$
(+) 4	2	
(+) 21	7	
(−) 6	3	
(+) 8	4	
(+) 10	5	

The Wilcoxon T is $T = 4$.

STEP 4   *Make a decision.* The obtained T value is not in the critical region. These data are not significantly different from chance. Therefore, we fail to reject H_0 and conclude that there is not sufficient evidence to suggest a systematic difference between the two treatment conditions.

DEMONSTRATION 19.3

THE KRUSKAL-WALLIS TEST

The following data represent the outcome from a research study using three separate samples to compare three treatment conditions. After treatment, the individuals from all three samples were ranked and the sum of the ranks, T, was computed for each treatment condition.

	TREATMENTS		
I	II	III	
8	1	11	$N = 15$
2	7	5	
4	6	9	
10	3	12	
14	13	15	
$T_1 = 38$	$T_2 = 30$	$T_3 = 52$	
$n_1 = 5$	$n_2 = 5$	$n_3 = 5$	

STEP 1   *State the hypotheses and select alpha.* The null hypothesis states that there are no differences among the three treatment conditions.

H_0:   There is no tendency for the ranks in any treatment condition to be systematically higher or lower than the ranks in any other treatment condition. There are no differences among the three treatments.

H_1:   The ranks in at least one treatment condition are systematically higher (or lower) than the ranks in another treatment condition. There are differences among the treatments.

STEP 2   *Locate the critical region.* The Kruskal-Wallis test statistic is equivalent to chi-square with degrees of freedom equal to the number of treatment conditions minus one. In this case, $df = 3 - 1 = 2$. With $\alpha = .05$, the critical value is 5.99. If we obtain an H value greater than 5.99, our decision will be to reject H_0.

STEP 3   *Compute the test statistic.* We have already calculated the sum of the ranks, T, for each treatment condition. For these data, $n = 4$ for each treatment and the three samples combine for a total of $N = 12$ scores. Therefore, the Kruskal-Wallis statistic is

$$H = \frac{12}{N(N + 1)}\left(\Sigma\frac{T^2}{n}\right) - 3(N + 1)$$

$$H = \frac{12}{15(16)}\left(\frac{38^2}{5} + \frac{30^2}{5} + \frac{52^2}{5}\right) - 3(16)$$

$$= \frac{1}{20}(288.8 + 180 + 540.8) - 48$$

$$= \frac{1}{20}(1009.6) - 48$$

$$= 50.48 - 48$$

$$= 2.48$$

STEP 4   *Make a decision.* The H value for these data is not in the critical region. Therefore, we fail to reject H_0 and conclude that the data are not sufficient to show any significant differences among the three treatments.

PROBLEMS

1. Following is a set of numerical scores for a sample of $n = 10$ subjects. Transform the scores into ordinal values and find the rank for each subject.

Subject:	A	B	C	D	E	F	G	H	I	J
Score:	11	14	14	14	18	22	27	35	35	43

2. Following is a set of numerical scores for a sample of $n = 8$ subjects. Transform the scores into ordinal values and find the rank for each subject.

Subject:	A	B	C	D	E	F	G	H
Score:	7	7	7	11	18	21	25	25

3. Under what circumstances is it necessary or advisable to convert interval or ratio scale measurements into ranked data before attempting a hypothesis test?

4. A psychologist suspects that some children learn new skills by watching and playing with their older siblings. To test this theory, the psychologist obtains a sample of 20 four-year-old children, of which 10 have older siblings and 10 do not. The children are all given a motor-skills task, and the psychologist rank-orders the children in terms of their motor skills. The ranks for the children with older siblings sum to $\Sigma R = 85$, and the ranks for the children without older siblings sum to $\Sigma R = 125$. Are these data sufficient to conclude that there is a significant difference between these two groups? Test with $\alpha = .05$.

5. The local state college has started a scholarship program to encourage highly qualified applicants to choose the state college over other schools. To test whether the scholarship program is actually targeting the best students, an introductory psychology teacher compares the course grades for a group of scholarship students and a group of non-scholarship students. Do the following results indicate that the students receiving scholarships have significantly better grades than those who are not awarded scholarships? Test at the .05 level of significance.

Scholarship students: A, A, A−, B+, C+, C

Non-scholarship students   A, B, B, B−, C, C−, C−, D+, D, F

6. A researcher is using an independent-measures design to compare two treatment conditions with a sample of $n = 6$ subjects in each treatment. After treatment, the complete set of 12 subjects is rank-ordered. The distribution of ranks for each treatment is as follows:

Treatment I: 4th, 7th, 8th, 10th, 11th, and 12th

Treatment II: 1st, 2nd, 3rd, 5th, 6th, and 9th

Do these data indicate a significant difference between treatments? Test at the .01 level of significance.

7. One of the advantages of using ranks (ordinal data) instead of interval or ratio scores is that ranks can reduce the impact of one extreme individual score. The following data represent results from an independent-measures study comparing two treatments. Notice that most subjects in treatment 1 have scores around 40, and most subjects in treatment 2 have scores around 15. However, there is one odd individual in treatment 2 with a score of $X = 104$. This extreme score distorts the data so that an independent-measures t test indicates no significant difference between treatments, $t(10) = 0.868$, $p > .05$. When the scores are converted to ranks, however, the effect of the extreme score is much less dramatic. Transform the score into ranks and use a Mann-Whitney test to evaluate the difference between treatments. Test with $\alpha = .05$.

TREATMENT 1	TREATMENT 2
41	10
39	14
37	9
44	17
40	12
45	104
$M = 41.00$	$M = 27.67$
$s = 3.03$	$s = 37.51$

8. Occasionally, researchers will measure a behavior that has extremely large variability. In these situations, the variability of the scores can overwhelm any mean differences in the data so that it is essentially impossible to find significant differences. If the scores are converted to ranks, however, the problems of large variability can be eliminated. Consider the following example. A researcher is examining how success or failure on a task can influence an individual's future performance on the same task. A group of participants is given 10 minutes to work on a problem-solving task. Based on their performance, they are classified as having either succeeded or failed. The participants are then given a second chance to work on a similar task with no time limit. The researcher simply records how long each person persists on the task before stopping. The scores (in minutes) for the two groups are as follows:

Initial success: 13, 20, 14, 7, 19, 23, 38, 10

Initial failure: 2, 5, 7, 3, 6, 5, 4, 32

With the large variability in the scores, a t test would conclude that there is no significant difference between the two groups. However, you can rank the scores and use a Mann-Whitney test to evaluate the difference between the two groups. Perform the Mann-Whitney test using the .05 level of significance.

9. A researcher is trying to determine which of two species of laboratory rats should be housed in the psychology department. A sample of $n = 10$ rats is obtained for each species, and the researcher records the amount of food each rat consumes during a 1-week period. The data are as follows:

Species A: 7, 9, 14, 20, 16, 18, 10, 22, 25, 13

Species B: 24, 19, 21, 26, 21, 29, 13, 28, 32, 17

Do these data indicate that one species eats significantly more than the other? Use a Mann-Whitney test with $\alpha = .05$.

10. A doctor has been collecting data on the birth weight of newborn children for smoking and nonsmoking mothers:

Smoking mothers: 92 oz   111 oz   108 oz
                    120 oz   101 oz

Nonsmoking mothers: 127 oz   118 oz   134 oz
                       136 oz   109 oz   122 oz
                       115 oz   129 oz   113 oz

Do these data indicate a significant difference in birth-weight between these two groups? Use a Mann-Whitney U-test at the .05 level of significance.

11. An instructor teaches two sections of the same statistics course. All students take a common final exam, and the instructor receives a printout of the grades in rank order (lowest to highest). For the morning section with $n = 14$ students, the sum of the ranks is $\Sigma R = 192$. The afternoon section with $n = 10$ students has $\Sigma R = 108$. Do these data indicate a significant difference between the two sections? Test at the .05 level.

12. Psychosis such as schizophrenia often is expressed in the artistic work produced by patients. To test the reliability of this phenomenon, a psychologist collected 10 paintings done by schizophrenic patients and another 10 paintings done by non-schizophrenic college students. A professor in the art department was asked to rank-order all 20 paintings in terms of bizarreness. These ranks are as follows:

Ranks for
schizophrenic patients: 1   3   4   5   6   8   9   11
                           12   14

Ranks for
non-schizophrenic students: 2   7   10   13   15   16
                               17   18   19   20

On the basis of these data, can the psychologist conclude that there is a significant difference between the paintings for these two populations? Test at the .05 level of significance.

13. As part of a product-testing program, a paint manufacturer painted 12 houses in a suburban community. Six of the houses were painted with the company's own product, and the other six were painted with a competitor's paint. After 5 years, a panel of homeowners inspected the 12 houses and ranked them according to how well the paint was holding up. The rankings are as follows:

Ranks for company's paint: 1, 3, 4, 5, 7, 8

Ranks for competitor's paint: 2, 6, 9, 10, 11, 12

Do these data indicate a significant difference between the two brands of paint? Test with $\alpha = .05$.

14. One assumption for parametric tests with independent-measures data is that the different treatment conditions have the same variance (homogeneity-of-variance assumption). However, a treatment effect that increases the mean often will also increase the variability. In this situation, the parametric t test or ANOVA is not justified, and a Mann-Whitney test should be used. The following data represent an example of this situation:

TREATMENT 1 (SAMPLE A)	TREATMENT 2 (SAMPLE B)
1	8
5	20
0	14
2	27
4	6
2	10
3	19

a. Compute the mean and variance for each sample. Note the difference between the two sample variances.
b. Use a Mann-Whitney test, with $\alpha = .05$, to test for a significant difference between the two treatments.

15. A researcher using a repeated-measures study to compare two treatments found that 11 out of 12 subjects showed higher scores in the second treatment than in the first. However, the one subject who did score lower in the second treatment had such a large decrease that the repeated-measures t hypothesis test showed no significant difference between treatments. If the researcher converted the difference scores to ranks and used a Wilcoxon test, would the conclusion be that there is a significant difference between treatments at the .05 level of significance? Explain your answer.

16. In a repeated-measures research study comparing two treatments, if all of the subjects show a decrease in scores for treatment 2 compared with treatment 1, then what is the minimum number of subjects needed to demonstrate a significant difference between treatments at the .05 level of significance?

17. The following data are from a repeated-measures study comparing two treatments.

SUBJECT	TREATMENTS I	TREATMENTS II
A	12	23
B	16	18
C	10	28
D	24	31
E	28	24
F	17	19
G	12	26
H	15	24
I	17	22
J	21	20

a. Find the difference score for each subject and prepare the data for a Wilcoxon test by rank-ordering the difference scores in terms of their absolute values.
b. Use a Wilcoxon test to determine whether these data indicate a significant difference between treatments at the .05 level of significance.

18. Hyperactive children often are treated with a stimulant such as Ritalin to improve their attention spans. In one test of this drug treatment, hyperactive children were given a boring task to work on. A psychologist recorded the amount of time (in seconds) each child spent on the task before becoming distracted. Each child's performance was measured before he or she received the drug and again after the drug was administered. Because the scores on this task are extremely variable, the psychologist decided to convert the data

to ranks. Use a Wilcoxon test with $\alpha = .05$ to determine whether the following data provide evidence that the drug has a significant effect:

CHILD	TREATMENT 1 (WITHOUT THE DRUG)	TREATMENT 2 (WITH THE DRUG)
1	28	135
2	15	309
3	183	150
4	48	224
5	30	25
6	233	345
7	21	43
8	110	188
9	12	15

19. One situation in which ranking can be very useful is when research results produce undetermined scores. Undetermined scores occur when a subject fails to complete a task. The following data represent the time (in seconds) required for participants to unscramble the letters in an anagram. The researcher is comparing the difficulty of 4-letter anagrams and 8-letter anagrams, using a repeated-measures design where each participant is tested on both types of anagrams. Notice that participant F failed to solve the eight-letter anagram during the course of the experiment. Without a score for this participant, it is impossible to compute a mean or use a repeated-measures t test to evaluate the data. However, you can rank the scores and use a Wilcoxon T to compare the two treatment conditions. Perform the test with $\alpha = .05$.

PARTICIPANT	4-LETTER ANAGRAM	8-LETTER ANAGRAM
A	17	48
B	15	31
C	9	52
D	21	20
E	12	18
F	23	(Failed)

20. The school psychologist at an elementary school is conducting a counseling program for disruptive students. To evaluate the program, the students' teacher is asked to rate the change in behavior for each student using a scale from $+10$ (maximum positive change) to -10 (maximum negative change). A rating of 0 indicates no change in behavior. Because the teacher's ratings are subjective evaluations, the psychologist feels more comfortable treating the scores as ordinal values instead of interval or ratio scores. Convert the following data to ranks, and use a Wilcoxon T to evaluate the effectiveness of the counseling:

STUDENT	CHANGE IN BEHAVIOR AS RATED BY TEACHER
A	+1
B	+6
C	0
D	+8
E	−3
F	+10
G	+4
H	+3

21. A new cold remedy contains a chemical that causes drowsiness and disorientation. Part of the testing for this drug involved measuring maze-learning performance for rats. Using a repeated-measures design, individual rats were tested on one maze with the drug and on a similar maze without the drug. The dependent variable is the number of errors before the rat solves the maze. In the drug condition, two rats failed to solve the maze after 200 errors and were simply marked as "failed."

SUBJECT	NO DRUG	DRUG
1	28	125
2	43	90
3	37	(Failed)
4	16	108
5	47	40
6	51	75
7	23	91
8	31	23
9	26	115
10	53	55
11	26	(Failed)
12	32	87

Do these data indicate that the drug has a significant effect on maze-learning performance? Test at the .05 level of significance.

22. For the vast majority of right-handed people, language is controlled in the left hemisphere of the brain. The left hemisphere also manages motor control for the right side of the body. Because language and fine motor control with the right hand are dependent on the same general area of the brain, these two activities can interfere with each other if an individual tries both at the same time. To demonstrate this fact, a psychologist

asked participants to balance a ruler on the index finger of their right hand. The psychologist recorded the amount of time the ruler was balanced under two conditions. In one condition, the participant was allowed to concentrate on balancing the ruler. In the second condition, the participant was required to recite a nursery rhyme while balancing the ruler. The data for this experiment are as follows:

PARTICIPANT	JUST BALANCING	BALANCING AND RECITING
1	43 seconds	15 seconds
2	127 seconds	21 seconds
3	18 seconds	25 seconds
4	28 seconds	6 seconds
5	21 seconds	10 seconds
6	47 seconds	9 seconds
7	12 seconds	14 seconds
8	25 seconds	6 seconds
9	53 seconds	24 seconds
10	17 seconds	11 seconds

On the basis of these data, can the psychologist conclude that the language task (nursery rhyme) significantly affects right-hand motor skill? Use a Wilcoxon test with $\alpha = .05$.

23. Several studies have shown that people who are physically attractive are also judged to have other positive characteristics such as intelligence. To test this observation, a researcher obtains a set of photographs of people and asks a group of participants to classify the individuals in the photos as either very attractive, average, or very unattractive. The researcher then selects 8 photographs from each category and has another group of participants rank-order the people pictured in terms of their apparent intelligence, with a rank of 1 being the most intelligent. The attractive photos received ranks of 1, 2, 5, 7, 8, 11, 15, and 16; the average photos received ranks of 3, 6, 9, 10, 17, 20, 21, and 23; and the unattractive photos received ranks of 4, 12, 13, 14, 18, 19, 22, and 24. Do these data indicate significant differences in perceived intelligence based on appearance? Test at the .05 level of significance.

24. A researcher would like to compare the relative effectiveness of three different techniques for treating phobia. A sample of 15 patients, all of whom suffer from severe phobia, is obtained. The patients are randomly assigned to the three therapies with $n = 5$ in each group. After 2 weeks of therapy, all 15 patients were interviewed by a clinical psychologist who rank-ordered the 15 individuals in order of the severity of their symptoms. The patients in the first therapy had ranks of 1, 3, 4, 5, and 9. The individuals in the second therapy group had ranks of 7, 10, 12, 14, and 15. The patients in the third group had ranks of 2, 6, 8, 11, and 13. Do these data indicate any significant differences among the three therapies? Test at the .05 level of significance.

25. Each week the newspaper publishes the television ratings for the top TV shows for the previous week. Considering only the three major networks, the top 20 shows from last week's ratings were distributed as follows:

ABC: 1st, 4th, 5th, 7th, 13th, and 18th

CBS: 2nd, 6th, 8th, 10th, 11th, 15th, and 20th

NBC: 3rd, 9th, 12th, 14th, 16th, 17th, and 19th

Do these data indicate any significant differences among the networks for the specific week being examined? Test with $\alpha = .05$. (Note that the sample sizes are not all the same.)

26. Davis (1937) found that single children developed language skills faster than twins and that twins developed language faster than triplets. A possible explanation for this phenomenon is that the single children have undivided attention from their parents, whereas twins and triplets must share attention. The following study represents hypothetical results similar to those obtained by Davis. The study involves a sample of 18 children, all of whom are 24 months old. In the sample, 6 are single children, 6 are twins, and 6 are triplets. (The twins and triplets are all from different families.) Each child is observed in a familiar day-care setting, and all 18 children are rank-ordered in terms of verbal skills. The distribution of rankings is as follows:

Single: 1st, 2nd, 5th, 8th, 9th, and 12th

Twin: 4th, 6th, 7th, 11th, 13th, and 16th

Triplet: 3rd, 10th, 14th, 15th, 17th, and 18th

Do these data indicate significant differences among the three categories of children? Test with $\alpha = .05$.

27. A researcher who is interested in the relationship between health and personality obtains a sample of $n = 25$ men between the ages of 30 and 35. The men are all given a physical exam by a doctor who then rank-orders the men in terms of overall health. The men are also given personality tests that classify them as either alpha, beta, or gamma personality types. Alphas are cautious and steady, betas are carefree and outgoing, and gammas tend toward extremes of behavior. The health rankings for the three groups of participants are as follows:

Alphas: 2nd, 6th, 7th, 10th, 13th, 15th, 19th, 21st, 23rd

Betas: 1st, 3rd, 5th, 8th, 11th, 12th, 16th, 17th

Gammas: 4th, 9th, 14th, 18th, 20th, 22nd, 24th, 25th

Do these data indicate significant differences in health among the three personality types? Test at the .01 level of significance.

28. The following data are from a research study using three separate samples to evaluate the differences among three treatment conditions.

TREATMENTS		
I	II	III
22	44	20
17	39	60
31	50	48
18	14	36
34	57	53
12	24	72
28	40	42

a. Transform the scores into ranks and find the sum of the ranks (the T values) for the individuals within each of the three treatments.

b. Use the Kruskal-Wallis test to determine whether there are any significant differences among the three treatments. Test at the .05 level of significance.

APPENDIX A # Basic Mathematics Review

Preview

This appendix reviews some of the basic math skills that are necessary for the statistical calculations presented in this book. Many students already will know some or all of this material. Others will need to do extensive work and review. To help you assess your own skills, we include a skills assessment exam here. You should allow approximately 30 minutes to complete the test. When you finish, grade your test using the answer key on page 685.

Notice that the test is divided into five sections. If you miss more than three questions in any section of the test, you probably need help in that area. Turn to the section of this appendix that corresponds to your problem area. In each section, you will find a general review, examples, and additional practice problems. After reviewing the appropriate section and doing the practice problems, turn to the end of the appendix. You will find another version of the skills assessment exam. If you still miss more than three questions in any section of the exam, continue studying. Get assistance from an instructor or a tutor if necessary. At the end of this appendix is a list of recommended books for individuals who need a more extensive review than can be provided here. We stress that mastering this material now will make the rest of the course much easier.

SKILLS ASSESSMENT PREVIEW EXAM

SECTION 1

(corresponding to Section A.1 of this appendix)

1. $3 + 2 \times 7 = ?$
2. $(3 + 2) \times 7 = ?$
3. $3 + 2^2 - 1 = ?$
4. $(3 + 2)^2 - 1 = ?$
5. $12/4 + 2 = ?$
6. $12/(4 + 2) = ?$
7. $12/(4 + 2)^2 = ?$
8. $2 \times (8 - 2^2) = ?$
9. $2 \times (8 - 2)^2 = ?$
10. $3 \times 2 + 8 - 1 \times 6 = ?$
11. $3 \times (2 + 8) - 1 \times 6 = ?$
12. $3 \times 2 + (8 - 1) \times 6 = ?$

SECTION 2

(corresponding to Section A.2 of this appendix)

1. The fraction $\frac{3}{4}$ corresponds to a percentage of _____ .
2. Express 30% as a fraction.
3. Convert $\frac{12}{40}$ to a decimal.
4. $\frac{2}{13} + \frac{8}{13} = ?$
5. $1.375 + 0.25 = ?$
6. $\frac{2}{5} \times \frac{1}{4} = ?$
7. $\frac{1}{8} + \frac{2}{3} = ?$
8. $3.5 \times 0.4 = ?$
9. $\frac{1}{5} \div \frac{3}{4} = ?$
10. $3.75/0.5 = ?$
11. In a group of 80 students, 20% are psychology majors. How many psychology majors are in this group?
12. A company reports that two-fifths of its employees are women. If there are 90 employees, how many are women?

SECTION 3

(corresponding to Section A.3 of this appendix)

1. $3 + (-2) + (-1) + 4 = ?$
2. $6 - (-2) = ?$
3. $-2 - (-4) = ?$
4. $6 + (-1) - 3 - (-2) - (-5) = ?$
5. $4 \times (-3) = ?$

6. $-2 \times (-6) = ?$
7. $-3 \times 5 = ?$
8. $-2 \times (-4) \times (-3) = ?$
9. $12 \div (-3) = ?$
10. $-18 \div (-6) = ?$
11. $-16 \div 8 = ?$
12. $-100 \div (-4) = ?$

SECTION 4

(corresponding to Section A.4 of this appendix)
For each equation, find the value of X.

1. $X + 6 = 13$
2. $X - 14 = 15$
3. $5 = X - 4$
4. $3X = 12$
5. $72 = 3X$
6. $X/5 = 3$
7. $10 = X/8$
8. $3X + 5 = -4$
9. $24 = 2X + 2$
10. $(X + 3)/2 = 14$
11. $(X - 5)/3 = 2$
12. $17 = 4X - 11$

SECTION 5

(corresponding to Section A.5 of this appendix)

1. $4^3 = ?$
2. $\sqrt{25 - 9} = ?$
3. If $X = 2$ and $Y = 3$, then $XY^3 = ?$
4. If $x = 2$ and $Y = 3$, then $(X + Y)^2 = ?$
5. If $a = 3$ and $b = 2$, then $a^2 + b^2 = ?$
6. $(-3)^3 = ?$
7. $(-4)^4 = ?$
8. $\sqrt{4} \times 4 = ?$
9. $36/\sqrt{9} = ?$
10. $(9 + 2)^2 = ?$
11. $5^2 + 2^3 = ?$
12. If $a = 3$ and $b = -1$, then $a^2 b^3 = ?$

The answers to the skills assessment exam are at the end of the appendix (pages 685–686).

A.1   SYMBOLS AND NOTATION

Table A.1 presents the basic mathematical symbols that you should know, along with examples of their use. Statistical symbols and notation are introduced and explained throughout this book as they are needed. Notation for exponents and square roots is covered separately at the end of this appendix.

Parentheses are a useful notation because they specify and control the order of computations. Everything inside the parentheses is calculated first. For example,

$$(5 + 3) \times 2 = 8 \times 2 = 16$$

Changing the placement of the parentheses also changes the order of calculations. For example,

$$5 + (3 \times 2) = 5 + 6 = 11$$

ORDER OF OPERATIONS   Often a formula or a mathematical expression will involve several different arithmetic operations, such as adding, multiplying, squaring, and so on. When you encounter these situations, you must perform the different operations in the correct sequence. Following is a list of mathematical operations, showing the order in which they are to be performed.

1. Any calculation contained within parentheses is done first.
2. Squaring (or raising to other exponents) is done second.
3. Multiplying and/or dividing is done third. A series of multiplication and/or division operations should be done in order from left to right.
4. Adding and/or subtracting is done fourth.

The following examples demonstrate how this sequence of operations is applied in different situations.

To evaluate the expression

$$(3 + 1)^2 - 4 \times 7/2$$

first, perform the calculation within parentheses:

$$(4)^2 - 4 \times 7/2$$

TABLE A.1

SYMBOL	MEANING	EXAMPLE
$+$	Addition	$5 + 7 = 12$
$-$	Subtraction	$8 - 3 = 5$
$\times, (\ )$	Multiplication	$3 \times 9 = 27, 3(9) = 27$
$\div, /$	Division	$15 \div 3 = 5, 15/3 = 5, \frac{15}{3} = 5$
$>$	Greater than	$20 > 10$
$<$	Less than	$7 < 11$
$\neq$	Not equal to	$5 \neq 6$

Next, square the value as indicated:

$$16 - 4 \times 7/2$$

Then perform the multiplication and division:

$$16 - 14$$

Finally, do the subtraction:

$$16 - 14 = 2$$

A sequence of operations involving multiplication and division should be performed in order from left to right. For example, to compute $12/2 \times 3$, you divide 12 by 2 and then multiply the result by 3:

$$12/2 \times 3 = 6 \times 3 = 18$$

Notice that violating the left-to-right sequence can change the result. For this example, if you multiply before dividing, you will obtain

$$12/2 \times 3 = 12/6 = 2 \qquad \text{(This is wrong.)}$$

A sequence of operations involving only addition and subtraction can be performed in any order. For example, to compute $3 + 8 - 5$, you can add 3 and 8 and then subtract 5:

$$(3 + 8) - 5 = 11 - 5 = 6$$

or you can subtract 5 from 8 and then add the result to 3:

$$3 + (8 - 5) = 3 + 3 = 6$$

A mathematical expression or formula is simply a concise way to write a set of instructions. When you evaluate an expression by performing the calculation, you simply follow the instructions. For example, assume you are given these instructions:

1. First, add 3 and 8.
2. Next, square the result.
3. Next, multiply the resulting value by 6.
4. Finally, subtract 50 from the value you have obtained.

You can write these instructions as a mathematical expression.

1. The first step involves addition. Because addition is normally done last, use parentheses to give this operation priority in the sequence of calculations:

   $$(3 + 8)$$

2. The instruction to square a value is noted by using the exponent 2 beside the value to be squared:

   $$(3 + 8)^2$$

3. Because squaring has priority over multiplication, you can simply introduce the multiplication into the expression:

   $$6 \times (3 + 8)^2$$

4. Addition and subtraction are done last, so simply write in the requested subtraction:

$$6 \times (3 + 8)^2 - 50$$

To calculate the value of the expression, you work through the sequence of operations in the proper order:

$$6 \times (3 + 8)^2 - 50 = 6 \times (11)^2 - 50$$
$$= 6 \times (121) - 50$$
$$= 726 - 50$$
$$= 676$$

As a final note, you should realize that the operation of squaring (or raising to any exponent) applies only to the value that immediately precedes the exponent. For example,

$$2 \times 3^2 = 2 \times 9 = 18 \qquad \text{(Only the 3 is squared.)}$$

If the instructions require multiplying values and then squaring the product, you must use parentheses to give the multiplication priority over squaring. For example, to multiply 2 times 3 and then square the product, you would write

$$(2 \times 3)^2 = (6)^2 = 36$$

LEARNING CHECK

1. Evaluate each of the following expressions:
   a. $4 \times 8/2^2$
   b. $4 \times (8/2)^2$
   c. $100 - 3 \times 12/(6 - 4)^2$
   d. $(4 + 6) \times (3 - 1)^2$
   e. $(8 - 2)/(9 - 8)^2$
   f. $6 + (4 - 1)^2 - 3 \times 4^2$
   g. $4 \times (8 - 3) + 8 - 3$

ANSWERS
1. a. 8   **b.** 64   **c.** 91   **d.** 40   **e.** 6   **f.** -33 **g.** 25

A.2 PROPORTIONS: FRACTIONS, DECIMALS, AND PERCENTAGES

A proportion is a part of a whole and can be expressed as a fraction, a decimal, or a percentage. For example, in a class of 40 students, only 3 failed the final exam.

The proportion of the class that failed can be expressed as a fraction

$$\text{fraction} = \frac{3}{40}$$

or as a decimal value

$$\text{decimal} = 0.075$$

or as a percentage

$$percentage = 7.5\%$$

In a fraction, such as $\frac{3}{4}$, the bottom value (the denominator) indicates the number of equal pieces into which the whole is split. Here the "pie" is split into 4 equal pieces:

If the denominator has a larger value—say, 8—then each piece of the whole pie is smaller:

A larger denominator indicates a smaller fraction of the whole.

The value on top of the fraction (the numerator) indicates how many pieces of the whole are being considered. Thus, the fraction $\frac{3}{4}$ indicates that the whole is split evenly into 4 pieces and that 3 of them are being used:

A fraction is simply a concise way of stating a proportion: "Three out of four" is equivalent to $\frac{3}{4}$. To convert the fraction to a decimal, you divide the numerator by the denominator:

$$\frac{3}{4} = 3 \div 4 = 0.75$$

To convert the decimal to a percentage, simply multiply by 100, and place a percent sign (%) after the answer:

$$0.75 \times 100 = 75\%$$

The U.S. money system is a convenient way of illustrating the relationship between fractions and decimals. "One quarter," for example, is one-fourth $\left(\frac{1}{4}\right)$ of a dollar, and its decimal equivalent is 0.25. Other familiar equivalencies are as follows:

	DIME	QUARTER	50 CENTS	75 CENTS
Fraction	$\frac{1}{10}$	$\frac{1}{4}$	$\frac{1}{2}$	$\frac{3}{4}$
Decimal	0.10	0.25	0.50	0.75
Percentage	10%	25%	50%	75%

FRACTIONS

1. Finding Equivalent Fractions. The same proportional value can be expressed by many equivalent fractions. For example,

$$\frac{1}{2} = \frac{2}{4} = \frac{10}{20} = \frac{50}{100}$$

To create equivalent fractions, you can multiply the numerator and denominator by the same value. As long as both the numerator and the denominator of the fraction are multiplied by the same value, the new fraction will be equivalent to the original. For example,

$$\frac{3}{10} = \frac{9}{30}$$

because both the numerator and the denominator of the original fraction have been multiplied by 3. Dividing the numerator and denominator of a fraction by the same value will also result in an equivalent fraction. By using division, you can reduce a fraction to a simpler form. For example,

$$\frac{40}{100} = \frac{2}{5}$$

because both the numerator and the denominator of the original fraction have been divided by 20.

You can use these rules to find specific equivalent fractions. For example, find the fraction that has a denominator of 100 and is equivalent to $\frac{3}{4}$. That is,

$$\frac{3}{4} = \frac{?}{100}$$

Notice that the denominator of the original fraction must be multiplied by 25 to produce the denominator of the desired fraction. For the two fractions to be equal, both the numerator and the denominator must be multiplied by the same number. Therefore, we also multiply the top of the original fraction by 25 and obtain

$$\frac{3 \times 25}{4 \times 25} = \frac{75}{100}$$

2. Multiplying Fractions. To multiply two fractions, you first multiply the numerators and then multiply the denominators. For example,

$$\frac{3}{4} \times \frac{5}{7} = \frac{3 \times 5}{4 \times 7} = \frac{15}{28}$$

3. Dividing Fractions. To divide one fraction by another, you invert the second fraction and then multiply. For example,

$$\frac{1}{2} \div \frac{1}{4} = \frac{1}{2} \times \frac{4}{1} = \frac{1 \times 4}{2 \times 1} = \frac{4}{2} = \frac{2}{1} = 2$$

4. Adding and Subtracting Fractions. Fractions must have the same denominator before you can add or subtract them. If the two fractions already have a common denominator, you simply add (or subtract as the case may be) *only* the values in the numerators. For example,

$$\frac{2}{5} + \frac{1}{5} = \frac{3}{5}$$

Suppose you divided a pie into five equal pieces (fifths). If you first ate two-fifths of the pie and then another one-fifth, the total amount eaten would be three-fifths of the pie:

If the two fractions do not have the same denominator, you must first find equivalent fractions with a common denominator before you can add or subtract. The product of the two denominators will always work as a common denominator for equivalent fractions (although it may not be the lowest common denominator). For example,

$$\frac{2}{3} + \frac{1}{10} = ?$$

Because these two fractions have different denominators, it is necessary to convert each into an equivalent fraction and find a common denominator. We will use $3 \times 10 = 30$ as the common denominator. Thus, the equivalent fraction of each is

$$\frac{2}{3} = \frac{20}{30} \quad \text{and} \quad \frac{1}{10} = \frac{3}{30}$$

Now the two fractions can be added:

$$\frac{20}{30} + \frac{3}{30} = \frac{23}{30}$$

5. Comparing the Size of Fractions. When comparing the size of two fractions with the same denominator, the larger fraction will have the larger numerator. For example,

$$\frac{5}{8} > \frac{3}{8}$$

The denominators are the same, so the whole is partitioned into pieces of the same size. Five of these pieces are more than three of them:

When two fractions have different denominators, you must first convert them to fractions with a common denominator to determine which is larger. Consider the following fractions:

$$\frac{3}{8} \quad \text{and} \quad \frac{7}{16}$$

If the numerator and denominator of $\frac{3}{8}$ are multiplied by 2, the resulting equivalent fraction will have a denominator of 16:

$$\frac{3}{8} = \frac{3 \times 2}{8 \times 2} = \frac{6}{16}$$

Now a comparison can be made between the two fractions:

$$\frac{6}{16} < \frac{7}{16}$$

Therefore,

$$\frac{3}{8} < \frac{7}{16}$$

DECIMALS

1. Converting Decimals to Fractions. Like a fraction, a decimal represents part of the whole. The first decimal place to the right of the decimal point indicates how many tenths are used. For example,

$$0.1 = \frac{1}{10} \qquad 0.7 = \frac{7}{10}$$

The next decimal place represents $\frac{1}{100}$, the next $\frac{1}{1000}$, the next $\frac{1}{10,000}$, and so on. To change a decimal to a fraction, just use the number without the decimal point for the numerator. Use the denominator that the last (on the right) decimal place represents. For example,

$$0.32 = \frac{32}{100} \qquad 0.5333 = \frac{5333}{10,000} \qquad 0.05 = \frac{5}{100} \qquad 0.001 = \frac{1}{1000}$$

2. Adding and Subtracting Decimals. To add and subtract decimals, the only rule is that you must keep the decimal points in a straight vertical line. For example,

$$
\begin{array}{r}
0.27 \\
+1.326 \\
\hline
1.596
\end{array}
\qquad
\begin{array}{r}
3.595 \\
-0.67 \\
\hline
2.925
\end{array}
$$

3. Multiplying Decimals. To multiply two decimal values, you first multiply the two numbers, ignoring the decimal points. Then you position the decimal point in the answer so that the number of digits to the right of the decimal point is equal to the total number of decimal places in the two numbers being multiplied. For example,

$$
\begin{array}{rl}
1.73 & \text{(two decimal places)} \\
\times 0.251 & \text{(three decimal places)} \\
\hline
173 & \\
865 & \\
346 & \\
\hline
0.43423 & \text{(five decimal places)}
\end{array}
\qquad
\begin{array}{rl}
0.25 & \text{(two decimal places)} \\
\times 0.005 & \text{(three decimal places)} \\
\hline
125 & \\
00 & \\
00 & \\
\hline
0.00125 & \text{(five decimal places)}
\end{array}
$$

4. Dividing Decimals. The simplest procedure for dividing decimals is based on the fact that dividing two numbers is identical to expressing them as a fraction:

$$0.25 \div 1.6 \text{ is identical to } \frac{0.25}{1.6}$$

You now can multiply both the numerator and the denominator of the fraction by 10, 100, 1000, or whatever number is necessary to remove the decimal places. Remember that multiplying both the numerator and the denominator of a fraction by the *same* value will create an equivalent fraction. Therefore,

$$\frac{0.25}{1.6} = \frac{0.25 \times 100}{1.6 \times 100} = \frac{25}{160} = \frac{5}{32}$$

The result is a division problem without any decimal places in the two numbers.

PERCENTAGES

1. Converting a Percentage to a Fraction or a Decimal. To convert a percentage to a fraction, remove the percent sign, place the number in the numerator, and use 100 for the denominator. For example,

$$52\% = \frac{52}{100} \qquad 5\% = \frac{5}{100}$$

To convert a percentage to a decimal, remove the percent sign and divide by 100, or simply move the decimal point two places to the left. For example,

$$83\% = 83. \ = 0.83$$

$$14.5\% = 14.5 = 0.145$$

$$5\% = \ 5. \ = 0.05$$

2. Performing Arithmetic Operations with Percentages. There are situations in which it is best to express percent values as decimals in order to perform certain arithmetic operations. For example, what is 45% of 60? This question may be stated as

$$45\% \times 60 = ?$$

The 45% should be converted to decimal form to find the solution to this question. Therefore,

$$0.45 \times 60 = 27$$

LEARNING CHECK

1. Convert $\frac{3}{25}$ to a decimal.

2. Convert $\frac{3}{8}$ to a percentage.

3. Next to each set of fractions, write "True" if they are equivalent and "False" if they are not:

a. $\frac{3}{8} = \frac{9}{24}$ _____   b. $\frac{7}{9} = \frac{17}{19}$ _____

c. $\frac{2}{7} = \frac{4}{14} =$ _____

4. Compute the following:

   a. $\frac{1}{6} \times \frac{7}{10}$ **b.** $\frac{7}{8} - \frac{1}{2}$ **c.** $\frac{9}{10} \div \frac{2}{3}$ **d.** $\frac{7}{22} + \frac{2}{3}$

5. Identify the larger fraction of each pair:

   a. $\frac{7}{10}, \frac{21}{100}$ **b.** $\frac{3}{4}, \frac{7}{12}$ **c.** $\frac{22}{3}, \frac{19}{3}$

6. Convert the following decimals into fractions:

   a. 0.012   **b.** 0.77   **c.** 0.005

7. $2.59 \times 0.015 = ?$

8. $1.8 \div 0.02 = ?$

9. What is 28% of 45?

ANSWERS   **1.** 0.12   **2.** 37.5%   **3. a.** True   **b.** False   **c.** True

4. a. $\frac{7}{60}$ **b.** $\frac{3}{8}$ **c.** $\frac{27}{20}$ **d.** $\frac{65}{66}$ **5. a.** $\frac{7}{10}$ **b.** $\frac{3}{4}$ **c.** $\frac{22}{3}$

6. a. $\frac{12}{1000} = \frac{3}{250}$ **b.** $\frac{77}{100}$ **c.** $\frac{5}{1000} = \frac{1}{200}$ **7.** 0.03885   **8.** 90   **9.** 12.6

A.3 NEGATIVE NUMBERS

Negative numbers are used to represent values less than zero. Negative numbers may occur when you are measuring the difference between two scores. For example, a researcher may want to evaluate the effectiveness of a propaganda film by measuring people's attitudes with a test both before and after viewing the film:

	BEFORE	AFTER	AMOUNT OF CHANGE
Person A	23	27	+4
Person B	18	15	−3
Person C	21	16	−5

Notice that the negative sign provides information about the direction of the difference: A plus sign indicates an increase in value, and a minus sign indicates a decrease.

Because negative numbers are frequently encountered, you should be comfortable working with these values. This section reviews basic arithmetic operations using negative numbers. You should also note that any number without a sign (+ or −) is assumed to be positive.

1. Adding Negative Numbers. When adding numbers that include negative values, simply interpret the negative sign as subtraction. For example,

$$3 + (-2) + 5 = 3 - 2 + 5 = 6$$

When adding a long string of numbers, it often is easier to add all the positive values to obtain the positive sum and then to add all of the negative values to obtain the negative sum. Finally, you subtract the negative sum from the positive sum. For example,

$$-1 + 3 + (-4) + 3 + (-6) + (-2)$$

positive sum = 6   negative sum = 13

Answer: $6 - 13 = -7$

2. Subtracting Negative Numbers. To subtract a negative number, change it to a positive number, and add. For example,

$$4 - (-3) = 4 + 3 = 7$$

This rule is easier to understand if you think of positive numbers as financial gains and negative numbers as financial losses. In this context, taking away a debt is equivalent to a financial gain. In mathematical terms, taking away a negative number is equivalent to adding a positive number. For example, suppose you are meeting a friend for lunch. You have \$7, but you owe your friend \$3. Thus, you really have only \$4 to spend for lunch. But your friend forgives (takes away) the \$3 debt. The result is that you now have \$7 to spend. Expressed as an equation,

$$\$4 \text{ minus a } \$3 \text{ debt} = \$7$$

$$4 - (-3) = 4 + 3 = 7$$

3. Multiplying and Dividing Negative Numbers. When the two numbers being multiplied (or divided) have the same sign, the result is a positive number. When the two numbers have different signs, the result is negative. For example,

$$3 \times (-2) = -6$$

$$-4 \times (-2) = +8$$

The first example is easy to explain by thinking of multiplication as repeated addition. In this case,

$$3 \times (-2) = (-2) + (-2) + (-2) = -6$$

You add three negative 2s, which results in a total of negative 6. In the second example, we are multiplying by a negative number. This amounts to repeated subtraction. That is,

$$-4 \times (-2) = -(-2) - (-2) - (-2) - (-2)$$

$$= 2 + 2 + 2 + 2 = 8$$

By using the same rule for both multiplication and division, we ensure that these two operations are compatible. For example,

$$-6 \div 3 = -2$$

which is compatible with

$$3 \times (-2) = -6$$

Also,

$$8 \div (-4) = -2$$

which is compatible with

$$-4 \times (-2) = +8$$

LEARNING CHECK

1. Complete the following calculations:
   a. $3 + (-8) + 5 + 7 + (-1) + (-3)$
   b. $5 - (-9) + 2 - (-3) - (-1)$
   c. $3 - 7 - (-21) + (-5) - (-9)$
   d. $4 - (-6) - 3 + 11 - 14$
   e. $9 + 8 - 2 - 1 - (-6)$
   f. $9 \times (-3)$
   g. $-7 \times (-4)$
   h. $-6 \times (-2) \times (-3)$
   i. $-12 \div (-3)$
   j. $18 \div (-6)$

ANSWERS

1. a. 3   b. 20   c. 21   d. 4   e. 20

   f. −27   g. 28   h. −36   i. 4   j. −3

A.4   BASIC ALGEBRA: SOLVING EQUATIONS

An equation is a mathematical statement that indicates two quantities are identical. For example,

$$12 = 8 + 4$$

Often an equation will contain an unknown (or variable) quantity that is identified with a letter or symbol, rather than a number. For example,

$$12 = 8 + X$$

In this event, your task is to find the value of X that makes the equation "true," or balanced. For this example, an X value of 4 will make a true equation. Finding the value of X is usually called *solving the equation*.

To solve an equation, there are two points to keep in mind:

1. Your goal is to have the unknown value (X) isolated on one side of the equation. This means that you need to remove all of the other numbers and symbols that appear on the same side of the equation as the X.

2. The equation will remain balanced, provided you treat both sides exactly the same. For example, you could add 10 points to *both* sides, and the solution (the X value) for the equation would be unchanged.

FINDING THE SOLUTION FOR AN EQUATION

We will consider four basic types of equations and the operations needed to solve them.

1. When X Has a Value Added to It. An example of this type of equation is

$$X + 3 = 7$$

Your goal is to isolate X on one side of the equation. Thus, you must remove the $+3$ on the left-hand side. The solution is obtained by subtracting 3 from *both* sides of the equation:

$$X + 3 - 3 = 7 - 3$$
$$X = 4$$

The solution is $X = 4$. You should always check your solution by returning to the original equation and replacing X with the value you obtained for the solution. For this example,

$$X + 3 = 7$$
$$4 + 3 = 7$$
$$7 = 7$$

2. When X Has a Value Subtracted From It. An example of this type of equation is

$$X - 8 = 12$$

In this example, you must remove the -8 from the left-hand side. Thus, the solution is obtained by adding 8 to *both* sides of the equation:

$$X - 8 + 8 = 12 + 8$$
$$X = 20$$

Check the solution:

$$X - 8 = 12$$
$$20 - 8 = 12$$
$$12 = 12$$

3. When X Is Multiplied by a Value. An example of this type of equation is

$$4X = 24$$

In this instance, it is necessary to remove the 4 that is multiplied by X. This may be accomplished by dividing both sides of the equation by 4:

$$\frac{4X}{4} = \frac{24}{4}$$
$$X = 6$$

Check the solution:

$$4X = 24$$
$$4(6) = 24$$
$$24 = 24$$

4. When X Is Divided by a Value. An example of this type of equation is

$$\frac{X}{3} = 9$$

Now the X is divided by 3, so the solution is obtained by multiplying by 3. Multiplying both sides yields

$$3\left(\frac{X}{3}\right) = 9(3)$$

$$X = 27$$

For the check,

$$\frac{X}{3} = 9$$

$$\frac{27}{3} = 9$$

$$9 = 9$$

SOLUTIONS FOR MORE-COMPLEX EQUATIONS More-complex equations can be solved by using a combination of the preceding simple operations. Remember that at each stage you are trying to isolate X on one side of the equation. For example,

$$3X + 7 = 22$$

$$3X + 7 - 7 = 22 - 7 \qquad \text{(Remove + 7 by subtracting 7 from both sides.)}$$

$$3X = 15$$

$$\frac{3X}{3} = \frac{15}{3} \qquad \text{(Remove 3 by dividing both sides by 3.)}$$

$$X = 5$$

To check this solution, return to the original equation, and substitute 5 in place of X:

$$3X + 7 = 22$$

$$3(5) + 7 = 22$$

$$15 + 7 = 22$$

$$22 = 22$$

Following is another type of complex equation frequently encountered in statistics:

$$\frac{X + 3}{4} = 2$$

First, remove the 4 by multiplying both sides by 4:

$$4\left(\frac{X + 3}{4}\right) = 2(4)$$

$$X + 3 = 8$$

Now remove the $+\,3$ by subtracting 3 from both sides:

$$X + 3 - 3 = 8 - 3$$

$$X = 5$$

To check this solution, return to the original equation, and substitute 5 in place of X:

$$\frac{X + 3}{4} = 2$$

$$\frac{5 + 3}{4} = 2$$

$$\frac{8}{4} = 2$$

$$2 = 2$$

LEARNING CHECK   **1.** Solve for X, and check the solutions:

a. $3X = 18$ **b.** $X + 7 = 9$ **c.** $X - 4 = 18$ **d.** $5X - 8 = 12$

e. $\dfrac{X}{9} = 5$ **f.** $\dfrac{X + 1}{6} = 4$ **g.** $X + 2 = -5$ **h.** $\dfrac{X}{5} = -5$

i. $\dfrac{2X}{3} = 12$ **j.** $\dfrac{X}{3} + 1 = 3$

ANSWERS   **1. a.** $X = 6$ **b.** $X = 2$ **c.** $X = 22$ **d.** $X = 4$ **e.** $X = 45$

f. $X = 23$ **g.** $X = -7$ **h.** $X = -25$ **i.** $X = 18$ **j.** $X = 6$

A.5 EXPONENTS AND SQUARE ROOTS

EXPONENTIAL NOTATION   A simplified notation is used whenever a number is being multiplied by itself. The notation consists of placing a value, called an *exponent*, on the right-hand side of and raised above another number, called a *base*. For example,

$$7^3 \leftarrow \text{exponent}$$
$$\uparrow$$
$$\text{base}$$

The exponent indicates how many times the base is used as a factor in multiplication. Following are some examples:

$$7^3 = 7(7)(7)$$ (Read "7 cubed" or "7 raised to the third power")

$$5^2 = 5(5)$$ (Read "5 squared")

$$2^5 = 2(2)(2)(2)(2)$$ (Read "2 raised to the fifth power")

There are a few basic rules about exponents that you will need to know for this course. They are outlined here.

1. Numbers Raised to One or Zero. Any number raised to the first power equals itself. For example,

$$6^1 = 6$$

Any number (except zero) raised to the zero power equals 1. For example,

$$9^0 = 1$$

2. Exponents for Multiple Terms. The exponent applies only to the base that is just in front of it. For example,

$$XY^2 = XYY$$

$$a^2b^3 = aabbb$$

3. Negative Bases Raised to an Exponent. If a negative number is raised to a power, then the result will be positive for exponents that are even and negative for exponents that are odd. For example,

$$(-4)^3 = -4(-4)(-4)$$
$$= 16(-4)$$
$$= -64$$

and

$$(-3)^4 = -3(-3)(-3)(-3)$$
$$= 9(-3)(-3)$$
$$= 9(9)$$
$$= 81$$

Note: The parentheses are used to ensure that the exponent applies to the entire negative number, including the sign. Without the parentheses there is some ambiguity as to how the exponent should be applied. For example, the expression -3^2 could have two interpretations:

$$-3^2 = (-3)(-3) = 9 \quad \text{or} \quad -3^2 = -(3)(3) = -9$$

4. Exponents and Parentheses. If an exponent is present outside of parentheses, then the computations within the parentheses are done first, and the exponential computation is done last:

$$(3 + 5)^2 = 8^2 = 64$$

Notice that the meaning of the expression is changed when each term in the parentheses is raised to the exponent individually:

$$3^2 + 5^2 = 9 + 25 = 34$$

Therefore,

$$X^2 + Y^2 \neq (X + Y)^2$$

5. Fractions Raised to a Power. If the numerator and denominator of a fraction are each raised to the same exponent, then the entire fraction can be raised to that exponent. That is,

$$\frac{a^2}{b^2} = \left(\frac{a}{b}\right)^2$$

For example,

$$\frac{3^2}{4^2} = \left(\frac{3}{4}\right)^2$$

$$\frac{9}{16} = \frac{3}{4}\left(\frac{3}{4}\right)$$

$$\frac{9}{16} = \frac{9}{16}$$

SQUARE ROOTS The square root of a value equals a number that when multiplied by itself yields the original value. For example, the square root of 16 equals 4 because 4 times 4 equals 16. The symbol for the square root is called a *radical,* $\sqrt{\ }$. The square root is taken for the number under the radical. For example,

$$\sqrt{16} = 4$$

Finding the square root is the inverse of raising a number to the second power (squaring). Thus,

$$\sqrt{a^2} = a$$

For example,

$$\sqrt{3^2} = \sqrt{9} = 3$$

Also,

$$(\sqrt{b})^2 = b$$

For example,

$$(\sqrt{64})^2 = 8^2 = 64$$

Computations under the same radical are performed *before* the square root is taken. For example,

$$\sqrt{9 + 16} = \sqrt{25} = 5$$

Note that with addition (or subtraction) separate radicals yield a different result:

$$\sqrt{9} + \sqrt{16} = 3 + 4 = 7$$

Therefore,

$$\sqrt{X} + \sqrt{Y} \neq \sqrt{X + Y}$$

$$\sqrt{X} - \sqrt{Y} \neq \sqrt{X - Y}$$

If the numerator and denominator of a fraction each have a radical, then the entire fraction can be placed under a single radical:

$$\frac{\sqrt{16}}{\sqrt{4}} = \sqrt{\frac{16}{4}}$$

$$\frac{4}{2} = \sqrt{4}$$

$$2 = 2$$

Therefore,

$$\frac{\sqrt{X}}{\sqrt{Y}} = \sqrt{\frac{X}{Y}}$$

Also, if the square root of one number is multiplied by the square root of another number, then the same result would be obtained by taking the square root of the product of both numbers. For example,

$$\sqrt{9} \times \sqrt{16} = \sqrt{9 \times 16}$$

$$3 \times 4 = \sqrt{144}$$

$$12 = 12$$

Therefore,

$$\sqrt{a} \times \sqrt{b} = \sqrt{ab}$$

LEARNING CHECK

1. Perform the following computations:
   a. $(-6)^3$
   b. $(3 + 7)^2$
   c. $a^3 b^2$ when $a = 2$ and $b = -5$
   d. $a^4 b^3$ when $a = 2$ and $b = 3$
   e. $(XY)^2$ when $X = 3$ and $Y = 5$
   f. $X^2 + Y^2$ when $X = 3$ and $Y = 5$
   g. $(X + Y)^2$ when $X = 3$ and $Y = 5$
   h. $\sqrt{5 + 4}$
   i. $(\sqrt{9})^2$
   j. $\dfrac{\sqrt{16}}{\sqrt{4}}$

ANSWERS

1. a. -216 b. 100   c. 200   d. 432   e. 225

   f. 34   g. 64   h. 3   i. 9   j. 2

PROBLEMS FOR APPENDIX A   Basic Mathematics Review

1. $50/(10 - 8) = ?$

2. $(2 + 3)^2 = ?$

3. $20/10 \times 3 = ?$

4. $12 - 4 \times 2 + 6/3 = ?$

5. $24/(12 - 4) + 2 \times (6 + 3) = ?$

6. Convert $\frac{7}{20}$ to a decimal.

7. Express $\frac{9}{25}$ as a percentage.

8. Convert 0.91 to a fraction.

9. Express 0.0031 as a fraction.

10. Next to each set of fractions, write "True" if they are equivalent and "False" if they are not:

    a. $\dfrac{4}{1000} = \dfrac{2}{100}$ _____

    b. $\dfrac{5}{6} = \dfrac{52}{62}$ _____

    c. $\dfrac{1}{8} = \dfrac{7}{56}$ _____

11. Perform the following calculations:

    a. $\dfrac{4}{5} \times \dfrac{2}{3} = ?$ **b.** $\dfrac{7}{9} \div \dfrac{2}{3} = ?$

    c. $\dfrac{3}{8} + \dfrac{1}{5} = ?$ **d.** $\dfrac{5}{18} - \dfrac{1}{6} = ?$

12. $2.51 \times 0.017 = ?$

13. $3.88 \times 0.0002 = ?$

14. $3.17 + 17.0132 = ?$

15. $5.55 + 10.7 + 0.711 + 3.33 + 0.031 = ?$

16. $2.04 \div 0.2 = ?$

17. $0.36 \div 0.4 = ?$

18. $5 + 3 - 6 - 4 + 3 = ?$

19. $9 - (-1) - 17 + 3 - (-4) + 5 = ?$

20. $5 + 3 - (-8) - (-1) + (-3) - 4 + 10 = ?$

21. $8 \times (-3) = ?$

22. $-22 \div (-2) = ?$

23. $-2(-4) \times (-3) = ?$

24. $84 \div (-4) = ?$

Solve the equations in Problems 25–32 for X.

25. $X - 7 = -2$ **26.** $9 = X + 3$

27. $\dfrac{X}{4} = 11$ **28.** $-3 = \dfrac{X}{3}$

29. $\dfrac{X + 3}{5} = 2$ **30.** $\dfrac{X + 1}{3} = -8$

31. $6X - 1 = 11$ **32.** $2X + 3 = -11$

33. $(-5)^2 = ?$ **34.** $(-5)^3 = ?$

35. If $a = 4$ and $b = 3$, then $a^2 + b^4 = ?$

36. If $a = -1$ and $b = 4$, then $(a + b)^2 = ?$

37. If $a = -1$ and $b = 5$, then $ab^2 = ?$

38. $\dfrac{18}{\sqrt{4}} = ?$ **39.** $\sqrt{\dfrac{20}{5}} = ?$

SKILLS ASSESSMENT FINAL EXAM

SECTION 1

1. $4 + 8/4 = ?$ **2.** $(4 + 8)/4 = ?$

3. $4 \times 3^2 = ?$ **4.** $(4 \times 3)^2 = ?$

5. $10/5 \times 2 = ?$ **6.** $10/(5 \times 2) = ?$

7. $40 - 10 \times 4/2 = ?$ **8.** $(5 - 1)^2/2 = ?$

9. $3 \times 6 - 3^2 = ?$ **10.** $2 \times (6 - 3)^2 = ?$

11. $4 \times 3 - 1 + 8 \times 2 = ?$

12. $4 \times (3 - 1 + 8) \times 2 = ?$

SECTION 2

1. Express $\frac{14}{80}$ as a decimal.

2. Convert $\frac{6}{25}$ to a percentage.

3. Convert 18% to a fraction.

4. $\frac{3}{5} \times \frac{2}{3} = ?$ **5.** $\frac{5}{24} + \frac{5}{6} = ?$

6. $\frac{7}{12} \div \frac{5}{6} = ?$ **7.** $\frac{5}{9} - \frac{1}{3} = ?$

8. $6.11 \times 0.22 = ?$

9. $0.18 \div 0.9 = ?$

10. $8.742 + 0.76 = ?$

11. In a statistics class of 72 students, three-eighths of the students received a B on the first test. How many Bs were earned?

12. What is 15% of 64?

SECTION 3

1. $3 - 1 - 3 + 5 - 2 + 6 = ?$
2. $-8 - (-6) = ?$
3. $2 - (-7) - 3 + (-11) - 20 = ?$
4. $-8 - 3 - (-1) - 2 - 1 = ?$
5. $8(-2) = ?$
6. $-7(-7) = ?$
7. $-3(-2)(-5) = ?$
8. $-3(5)(-3) = ?$
9. $-24 \div (-4) = ?$
10. $36 \div (-6) = ?$
11. $-56/7 = ?$
12. $-7/(-1) = ?$

SECTION 4

Solve for X.

1. $X + 5 = 12$
2. $X - 11 = 3$
3. $10 = X + 4$
4. $4X = 20$
5. $\dfrac{X}{2} = 15$
6. $18 = 9X$
7. $\dfrac{X}{5} = 35$
8. $2X + 8 = 4$

9. $\dfrac{X + 1}{3} = 6$
10. $4X + 3 = -13$
11. $\dfrac{X + 3}{3} = -7$
12. $23 = 2X - 5$

SECTION 5

1. $5^3 = ?$
2. $(-4)^3 = ?$
3. $(-2)^5 = ?$
4. $(-2)^6 = ?$
5. If $a = 4$ and $b = 2$, then $ab^2 = ?$
6. If $a = 4$ and $b = 2$, then $(a + b)^3 = ?$
7. If $a = 4$ and $b = 2$, then $a^2 + b^2 = ?$
8. $(11 + 4)^2 = ?$
9. $\sqrt{7^2} = ?$
10. If $a = 36$ and $b = 64$, then $\sqrt{a + b} = ?$
11. $\dfrac{25}{\sqrt{25}} = ?$
12. If $a = -1$ and $b = 2$, then $a^3 b^4 = ?$

ANSWER KEY   Skills Assessment Exams

PREVIEW EXAM

SECTION 1

1. 17	**2.** 35	**3.** 6
4. 24	**5.** 5	**6.** 2
7. $\dfrac{1}{3}$	**8.** 8	**9.** 72
10. 8	**11.** 24	**12.** 48

SECTION 2

1. 75%	**2.** $\dfrac{30}{100}$, or $\dfrac{3}{10}$	**3.** 0.3
4. $\dfrac{10}{13}$	**5.** 1.625	**6.** $\dfrac{2}{20}$, or $\dfrac{1}{10}$
7. $\dfrac{19}{24}$	**8.** 1.4	**9.** $\dfrac{4}{15}$
10. 7.5	**11.** 16	**12.** 36

SECTION 3

1. 4	**2.** 8	**3.** 2
4. 9	**5.** -12	**6.** 12
7. -15	**8.** -24	**9.** -4
10. 3	**11.** -2	**12.** 25

FINAL EXAM

SECTION 1

1. 6	**2.** 3	**3.** 36
4. 144	**5.** 4	**6.** 1
7. 20	**8.** 8	**9.** 9
10. 18	**11.** 27	**12.** 80

SECTION 2

1. 0.175	**2.** 24%	**3.** $\dfrac{18}{100}$, or $\dfrac{9}{50}$
4. $\dfrac{6}{15}$, or $\dfrac{2}{5}$	**5.** $\dfrac{25}{24}$	**6.** $\dfrac{42}{60}$, or $\dfrac{7}{10}$
7. $\dfrac{2}{9}$	**8** 1.3442	**9.** 0.2
10. 9.502	**11.** 27	**12.** 9.6

SECTION 3

1. 8	**2.** -2	**3.** -25
4. -13	**5.** -16	**6.** 49
7. -30	**8.** 45	**9.** 6
10. -6	**11.** -8	**12.** 7

PREVIEW EXAM

SECTION 4

1. $X = 7$	**2.** $X = 29$	**3.** $X = 9$
4. $X = 4$	**5.** $X = 24$	**6.** $X = 15$
7. $X = 80$	**8.** $X = -3$	**9.** $X = 11$
10. $X = 25$	**11.** $X = 11$	**12.** $X = 7$

SECTION 5

1. 64	**2.** 4	**3.** 54
4. 25	**5.** 13	**6.** -27
7. 256	**8.** 8	**9.** 12
10. 121	**11.** 33	**12.** -9

FINAL EXAM

SECTION 4

1. $X = 7$	**2.** $X = 14$	**3.** $X = 6$
4. $X = 5$	**5.** $X = 30$	**6.** $X = 2$
7. $X = 175$	**8.** $X = -2$	**9.** $X = 17$
10. $X = -4$	**11.** $X = -24$	**12.** $X = 14$

SECTION 5

1. 125	**2.** -64	**3.** -32
4. 64	**5.** 16	**6.** 216
7. 20	**8.** 225	**9.** 7
10. 10	**11.** 5	**12.** -16

SOLUTIONS TO SELECTED PROBLEMS FOR APPENDIX A   Basic Mathematics Review

1. 25	**3.** 6	**17.** 0.9	**19.** 5
5. 21	**6.** 0.35	**21.** -24	**22.** 11
7. 36%	**9.** $\dfrac{31}{10,000}$	**25.** $X = 5$	**28.** $X = -9$
		30. $X = -25$	**31.** $X = 2$
10. b. False		**34.** -125	**36.** 9
11. a. $\dfrac{8}{15}$ **b.** $\dfrac{21}{18}$ **c.** $\dfrac{23}{40}$		**37.** -25	**39.** 2
12. 0.04267	**14.** 20.1832		

SUGGESTED REVIEW BOOKS

There are many basic mathematics review books available if you need a more extensive review than this appendix can provide. The following books are but a few of the many that you may find helpful:

Angel, A. R., and Porter, S. R. (2001). *A Survey of Mathematics With Applications*, 6th ed. Boston: Addison Wesley.

Aufmann, R. N., Barker, V. C. Lockwood, J. S., Bello, I., and Britton, J. R. (2002). *College Mathematics*. Boston: Houghton Mifflin.

Bloomfield, D. I. (1994). *Introductory algebra*. St. Paul, MN: West.

Goodman, A., & Hirsch, L. (1994). *Understanding elementary algebra*. St. Paul, MN: West.

APPENDIX B Statistical Tables

TABLE B.1   THE UNIT NORMAL TABLE*

*Column A lists z-score values. A vertical line drawn through a normal distribution at a z-score location divides the distribution into two sections.

Column B identifies the proportion in the larger section, called the *body*.

Column C identifies the proportion in the smaller section, called the *tail*.

Column D identifies the proportion between the mean and the z-score.

Note: Because the normal distribution is symmetrical, the proportions for negative z-scores are the same as those for positive z-scores.

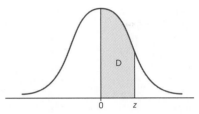

(A)	(B) PROPORTION IN BODY	(C) PROPORTION IN TAIL	(D) PROPORTION BETWEEN MEAN AND z	(A)	(B) PROPORTION IN BODY	(C) PROPORTION IN TAIL	(D) PROPORTION BETWEEN MEAN AND z
z				z			
0.00	.5000	.5000	.0000	0.25	.5987	.4013	.0987
0.01	.5040	.4960	.0040	0.26	.6026	.3974	.1026
0.02	.5080	.4920	.0080	0.27	.6064	.3936	.1064
0.03	.5120	.4880	.0120	0.28	.6103	.3897	.1103
0.04	.5160	.4840	.0160	0.29	.6141	.3859	.1141
0.05	.5199	.4801	.0199	0.30	.6179	.3821	.1179
0.06	.5239	.4761	.0239	0.31	.6217	.3783	.1217
0.07	.5279	.4721	.0279	0.32	.6255	.3745	.1255
0.08	.5319	.4681	.0319	0.33	.6293	.3707	.1293
0.09	.5359	.4641	.0359	0.34	.6331	.3669	.1331
0.10	.5398	.4602	.0398	0.35	.6368	.3632	.1368
0.11	.5438	.4562	.0438	0.36	.6406	.3594	.1406
0.12	.5478	.4522	.0478	0.37	.6443	.3557	.1443
0.13	.5517	.4483	.0517	0.38	.6480	.3520	.1480
0.14	.5557	.4443	.0557	0.39	.6517	.3483	.1517
0.15	.5596	.4404	.0596	0.40	.6554	.3446	.1554
0.16	.5636	.4364	.0636	0.41	.6591	.3409	.1591
0.17	.5675	.4325	.0675	0.42	.6628	.3372	.1628
0.18	.5714	.4286	.0714	0.43	.6664	.3336	.1664
0.19	.5753	.4247	.0753	0.44	.6700	.3300	.1700
0.20	.5793	.4207	.0793	0.45	.6736	.3264	.1736
0.21	.5832	.4168	.0832	0.46	.6772	.3228	.1772
0.22	.5871	.4129	.0871	0.47	.6808	.3192	.1808
0.23	.5910	.4090	.0910	0.48	.6844	.3156	.1844
0.24	.5948	.4052	.0948	0.49	.6879	.3121	.1879

(A)	(B) PROPORTION IN BODY	(C) PROPORTION IN TAIL	(D) PROPORTION BETWEEN MEAN AND z	(A)	(B) PROPORTION IN BODY	(C) PROPORTION IN TAIL	(D) PROPORTION BETWEEN MEAN AND z
z				z			
0.50	.6915	.3085	.1915	1.00	.8413	.1587	.3413
0.51	.6950	.3050	.1950	1.01	.8438	.1562	.3438
0.52	.6985	.3015	.1985	1.02	.8461	.1539	.3461
0.53	.7019	.2981	.2019	1.03	.8485	.1515	.3485
0.54	.7054	.2946	.2054	1.04	.8508	.1492	.3508
0.55	.7088	.2912	.2088	1.05	.8531	.1469	.3531
0.56	.7123	.2877	.2123	1.06	.8554	.1446	.3554
0.57	.7157	.2843	.2157	1.07	.8577	.1423	.3577
0.58	.7190	.2810	.2190	1.08	.8599	.1401	.3599
0.59	.7224	.2776	.2224	1.09	.8621	.1379	.3621
0.60	.7257	.2743	.2257	1.10	.8643	.1357	.3643
0.61	.7291	.2709	.2291	1.11	.8665	.1335	.3665
0.62	.7324	.2676	.2324	1.12	.8686	.1314	.3686
0.63	.7357	.2643	.2357	1.13	.8708	.1292	.3708
0.64	.7389	.2611	.2389	1.14	.8729	.1271	.3729
0.65	.7422	.2578	.2422	1.15	.8749	.1251	.3749
0.66	.7454	.2546	.2454	1.16	.8770	.1230	.3770
0.67	.7486	.2514	.2486	1.17	.8790	.1210	.3790
0.68	.7517	.2483	.2517	1.18	.8810	.1190	.3810
0.69	.7549	.2451	.2549	1.19	.8830	.1170	.3830
0.70	.7580	.2420	.2580	1.20	.8849	.1151	.3849
0.71	.7611	.2389	.2611	1.21	.8869	.1131	.3869
0.72	.7642	.2358	.2642	1.22	.8888	.1112	.3888
0.73	.7673	.2327	.2673	1.23	.8907	.1093	.3907
0.74	.7704	.2296	.2704	1.24	.8925	.1075	.3925
0.75	.7734	.2266	.2734	1.25	.8944	.1056	.3944
0.76	.7764	.2236	.2764	1.26	.8962	.1038	.3962
0.77	.7794	.2206	.2794	1.27	.8980	.1020	.3980
0.78	.7823	.2177	.2823	1.28	.8997	.1003	.3997
0.79	.7852	.2148	.2852	1.29	.9015	.0985	.4015
0.80	.7881	.2119	.2881	1.30	.9032	.0968	.4032
0.81	.7910	.2090	.2910	1.31	.9049	.0951	.4049
0.82	.7939	.2061	.2939	1.32	.9066	.0934	.4066
0.83	.7967	.2033	.2967	1.33	.9082	.0918	.4082
0.84	.7995	.2005	.2995	1.34	.9099	.0901	.4099
0.85	.8023	.1977	.3023	1.35	.9115	.0885	.4115
0.86	.8051	.1949	.3051	1.36	.9131	.0869	.4131
0.87	.8078	.1922	.3078	1.37	.9147	.0853	.4147
0.88	.8106	.1894	.3106	1.38	.9162	.0838	.4162
0.89	.8133	.1867	.3133	1.39	.9177	.0823	.4177
0.90	.8159	.1841	.3159	1.40	.9192	.0808	.4192
0.91	.8186	.1814	.3186	1.41	.9207	.0793	.4207
0.92	.8212	.1788	.3212	1.42	.9222	.0778	.4222
0.93	.8238	.1762	.3238	1.43	.9236	.0764	.4236
0.94	.8264	.1736	.3264	1.44	.9251	.0749	.4251
0.95	.8289	.1711	.3289	1.45	.9265	.0735	.4265
0.96	.8315	.1685	.3315	1.46	.9279	.0721	.4279
0.97	.8340	.1660	.3340	1.47	.9292	.0708	.4292
0.98	.8365	.1635	.3365	1.48	.9306	.0694	.4306
0.99	.8389	.1611	.3389	1.49	.9319	.0681	.4319

| (A) | (B) PROPORTION IN BODY | (C) PROPORTION IN TAIL | (D) PROPORTION BETWEEN MEAN AND z | (A) | (B) PROPORTION IN BODY | (C) PROPORTION IN TAIL | (D) PROPORTION BETWEEN MEAN AND z |
z				z			
1.50	.9332	.0668	.4332	2.00	.9772	.0228	.4772
1.51	.9345	.0655	.4345	2.01	.9778	.0222	.4778
1.52	.9357	.0643	.4357	2.02	.9783	.0217	.4783
1.53	.9370	.0630	.4370	2.03	.9788	.0212	.4788
1.54	.9382	.0618	.4382	2.04	.9793	.0207	.4793
1.55	.9394	.0606	.4394	2.05	.9798	.0202	.4798
1.56	.9406	.0594	.4406	2.06	.9803	.0197	.4803
1.57	.9418	.0582	.4418	2.07	.9808	.0192	.4808
1.58	.9429	.0571	.4429	2.08	.9812	.0188	.4812
1.59	.9441	.0559	.4441	2.09	.9817	.0183	.4817
1.60	.9452	.0548	.4452	2.10	.9821	.0179	.4821
1.61	.9463	.0537	.4463	2.11	.9826	.0174	.4826
1.62	.9474	.0526	.4474	2.12	.9830	.0170	.4830
1.63	.9484	.0516	.4484	2.13	.9834	.0166	.4834
1.64	.9495	.0505	.4495	2.14	.9838	.0162	.4838
1.65	.9505	.0495	.4505	2.15	.9842	.0158	.4842
1.66	.9515	.0485	.4515	2.16	.9846	.0154	.4846
1.67	.9525	.0475	.4525	2.17	.9850	.0150	.4850
1.68	.9535	.0465	.4535	2.18	.9854	.0146	.4854
1.69	.9545	.0455	.4545	2.19	.9857	.0143	.4857
1.70	.9554	.0446	.4554	2.20	.9861	.0139	.4861
1.71	.9564	.0436	.4564	2.21	.9864	.0136	.4864
1.72	.9573	.0427	.4573	2.22	.9868	.0132	.4868
1.73	.9582	.0418	.4582	2.23	.9871	.0129	.4871
1.74	.9591	.0409	.4591	2.24	.9875	.0125	.4875
1.75	.9599	.0401	.4599	2.25	.9878	.0122	.4878
1.76	.9608	.0392	.4608	2.26	.9881	.0119	.4881
1.77	.9616	.0384	.4616	2.27	.9884	.0116	.4884
1.78	.9625	.0375	.4625	2.28	.9887	.0113	.4887
1.79	.9633	.0367	.4633	2.29	.9890	.0110	.4890
1.80	.9641	.0359	.4641	2.30	.9893	.0107	.4893
1.81	.9649	.0351	.4649	2.31	.9896	.0104	.4896
1.82	.9656	.0344	.4656	2.32	.9898	.0102	.4898
1.83	.9664	.0336	.4664	2.33	.9901	.0099	.4901
1.84	.9671	.0329	.4671	2.34	.9904	.0096	.4904
1.85	.9678	.0322	.4678	2.35	.9906	.0094	.4906
1.86	.9686	.0314	.4686	2.36	.9909	.0091	.4909
1.87	.9693	.0307	.4693	2.37	.9911	.0089	.4911
1.88	.9699	.0301	.4699	2.38	.9913	.0087	.4913
1.89	.9706	.0294	.4706	2.39	.9916	.0084	.4916
1.90	.9713	.0287	.4713	2.40	.9918	.0082	.4918
1.91	.9719	.0281	.4719	2.41	.9920	.0080	.4920
1.92	.9726	.0274	.4726	2.42	.9922	.0078	.4922
1.93	.9732	.0268	.4732	2.43	.9925	.0075	.4925
1.94	.9738	.0262	.4738	2.44	.9927	.0073	.4927
1.95	.9744	.0256	.4744	2.45	.9929	.0071	.4929
1.96	.9750	.0250	.4750	2.46	.9931	.0069	.4931
1.97	.9756	.0244	.4756	2.47	.9932	.0068	.4932
1.98	.9761	.0239	.4761	2.48	.9934	.0066	.4934
1.99	.9767	.0233	.4767	2.49	.9936	.0064	.4936

(A)	(B) PROPORTION IN BODY	(C) PROPORTION IN TAIL	(D) PROPORTION BETWEEN MEAN AND z	(A)	(B) PROPORTION IN BODY	(C) PROPORTION IN TAIL	(D) PROPORTION BETWEEN MEAN AND z
z				z			
2.50	.9938	.0062	.4938	2.95	.9984	.0016	.4984
2.51	.9940	.0060	.4940	2.96	.9985	.0015	.4985
2.52	.9941	.0059	.4941	2.97	.9985	.0015	.4985
2.53	.9943	.0057	.4943	2.98	.9986	.0014	.4986
2.54	.9945	.0055	.4945	2.99	.9986	.0014	.4986
2.55	.9946	.0054	.4946	3.00	.9987	.0013	.4987
2.56	.9948	.0052	.4948	3.01	.9987	.0013	.4987
2.57	.9949	.0051	.4949	3.02	.9987	.0013	.4987
2.58	.9951	.0049	.4951	3.03	.9988	.0012	.4988
2.59	.9952	.0048	.4952	3.04	.9988	.0012	.4988
2.60	.9953	.0047	.4953	3.05	.9989	.0011	.4989
2.61	.9955	.0045	.4955	3.06	.9989	.0011	.4989
2.62	.9956	.0044	.4956	3.07	.9989	.0011	.4989
2.63	.9957	.0043	.4957	3.08	.9990	.0010	.4990
2.64	.9959	.0041	.4959	3.09	.9990	.0010	.4990
2.65	.9960	.0040	.4960	3.10	.9990	.0010	.4990
2.66	.9961	.0039	.4961	3.11	.9991	.0009	.4991
2.67	.9962	.0038	.4962	3.12	.9991	.0009	.4991
2.68	.9963	.0037	.4963	3.13	.9991	.0009	.4991
2.69	.9964	.0036	.4964	3.14	.9992	.0008	.4992
2.70	.9965	.0035	.4965	3.15	.9992	.0008	.4992
2.71	.9966	.0034	.4966	3.16	.9992	.0008	.4992
2.72	.9967	.0033	.4967	3.17	.9992	.0008	.4992
2.73	.9968	.0032	.4968	3.18	.9993	.0007	.4993
2.74	.9969	.0031	.4969	3.19	.9993	.0007	.4993
2.75	.9970	.0030	.4970	3.20	.9993	.0007	.4993
2.76	.9971	.0029	.4971	3.21	.9993	.0007	.4993
2.77	.9972	.0028	.4972	3.22	.9994	.0006	.4994
2.78	.9973	.0027	.4973	3.23	.9994	.0006	.4994
2.79	.9974	.0026	.4974	3.24	.9994	.0006	.4994
2.80	.9974	.0026	.4974	3.30	.9995	.0005	.4995
2.81	.9975	.0025	.4975	3.40	.9997	.0003	.4997
2.82	.9976	.0024	.4976	3.50	.9998	.0002	.4998
2.83	.9977	.0023	.4977	3.60	.9998	.0002	.4998
2.84	.9977	.0023	.4977	3.70	.9999	.0001	.4999
2.85	.9978	.0022	.4978	3.80	.99993	.00007	.49993
2.86	.9979	.0021	.4979	3.90	.99995	.00005	.49995
2.87	.9979	.0021	.4979	4.00	.99997	.00003	.49997
2.88	.9980	.0020	.4980				
2.89	.9981	.0019	.4981				
2.90	.9981	.0019	.4981				
2.91	.9982	.0018	.4982				
2.92	.9982	.0018	.4982				
2.93	.9983	.0017	.4983				
2.94	.9984	.0016	.4984				

TABLE B.2   THE *t* DISTRIBUTION

Table entries are values of *t* corresponding to proportions in one tail or in two tails combined.

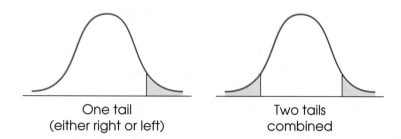

One tail
(either right or left)

Two tails
combined

df	PROPORTION IN ONE TAIL					
	0.25	0.10	0.05	0.025	0.01	0.005
	PROPORTION IN TWO TAILS COMBINED					
	0.50	0.20	0.10	0.05	0.02	0.01
1	1.000	3.078	6.314	12.706	31.821	63.657
2	0.816	1.886	2.920	4.303	6.965	9.925
3	0.765	1.638	2.353	3.182	4.541	5.841
4	0.741	1.533	2.132	2.776	3.747	4.604
5	0.727	1.476	2.015	2.571	3.365	4.032
6	0.718	1.440	1.943	2.447	3.143	3.707
7	0.711	1.415	1.895	2.365	2.998	3.499
8	0.706	1.397	1.860	2.306	2.896	3.355
9	0.703	1.383	1.833	2.262	2.821	3.250
10	0.700	1.372	1.812	2.228	2.764	3.169
11	0.697	1.363	1.796	2.201	2.718	3.106
12	0.695	1.356	1.782	2.179	2.681	3.055
13	0.694	1.350	1.771	2.160	2.650	3.012
14	0.692	1.345	1.761	2.145	2.624	2.977
15	0.691	1.341	1.753	2.131	2.602	2.947
16	0.690	1.337	1.746	2.120	2.583	2.921
17	0.689	1.333	1.740	2.110	2.567	2.898
18	0.688	1.330	1.734	2.101	2.552	2.878
19	0.688	1.328	1.729	2.093	2.539	2.861
20	0.687	1.325	1.725	2.086	2.528	2.845
21	0.686	1.323	1.721	2.080	2.518	2.831
22	0.686	1.321	1.717	2.074	2.508	2.819
23	0.685	1.319	1.714	2.069	2.500	2.807
24	0.685	1.318	1.711	2.064	2.492	2.797
25	0.684	1.316	1.708	2.060	2.485	2.787
26	0.684	1.315	1.706	2.056	2.479	2.779
27	0.684	1.314	1.703	2.052	2.473	2.771
28	0.683	1.313	1.701	2.048	2.467	2.763
29	0.683	1.311	1.699	2.045	2.462	2.756
30	0.683	1.310	1.697	2.042	2.457	2.750
40	0.681	1.303	1.684	2.021	2.423	2.704
60	0.679	1.296	1.671	2.000	2.390	2.660
120	0.677	1.289	1.658	1.980	2.358	2.617
∞	0.674	1.282	1.645	1.960	2.326	2.576

Table III of R. A. Fisher and F. Yates, *Statistical Tables for Biological, Agricultural and Medical Research*, 6th ed. London: Longman Group Ltd., 1974 (previously published by Oliver and Boyd Ltd., Edinburgh). Adapted and reprinted with permission of the Addison Wesley Longman Publishing Co.

TABLE B.3   CRITICAL VALUES FOR THE *F*-MAX STATISTIC*

*The critical values for α = .05 are in lightface type, and for α = .01, they are in boldface type.

					k = NUMBER OF SAMPLES						
n − 1	2	3	4	5	6	7	8	9	10	11	12
4	9.60	15.5	20.6	25.2	29.5	33.6	37.5	41.4	44.6	48.0	51.4
	23.2	**37.**	**49.**	**59.**	**69.**	**79.**	**89.**	**97.**	**106.**	**113.**	**120.**
5	7.15	10.8	13.7	16.3	18.7	20.8	22.9	24.7	26.5	28.2	29.9
	14.9	**22.**	**28.**	**33.**	**38.**	**42.**	**46.**	**50.**	**54.**	**57.**	**60.**
6	5.82	8.38	10.4	12.1	13.7	15.0	16.3	17.5	18.6	19.7	20.7
	11.1	**15.5**	**19.1**	**22.**	**25.**	**27.**	**30.**	**32.**	**34.**	**36.**	**37.**
7	4.99	6.94	8.44	9.70	10.8	11.8	12.7	13.5	14.3	15.1	15.8
	8.89	**12.1**	**14.5**	**16.5**	**18.4**	**20.**	**22.**	**23.**	**24.**	**26.**	**27.**
8	4.43	6.00	7.18	8.12	9.03	9.78	10.5	11.1	11.7	12.2	12.7
	7.50	**9.9**	**11.7**	**13.2**	**14.5**	**15.8**	**16.9**	**17.9**	**18.9**	**19.8**	**21.**
9	4.03	5.34	6.31	7.11	7.80	8.41	8.95	9.45	9.91	10.3	10.7
	6.54	**8.5**	**9.9**	**11.1**	**12.1**	**13.1**	**13.9**	**14.7**	**15.3**	**16.0**	**16.6**
10	3.72	4.85	5.67	6.34	6.92	7.42	7.87	8.28	8.66	9.01	9.34
	5.85	**7.4**	**8.6**	**9.6**	**10.4**	**11.1**	**11.8**	**12.4**	**12.9**	**13.4**	**13.9**
12	3.28	4.16	4.79	5.30	5.72	6.09	6.42	6.72	7.00	7.25	7.48
	4.91	**6.1**	**6.9**	**7.6**	**8.2**	**8.7**	**9.1**	**9.5**	**9.9**	**10.2**	**10.6**
15	2.86	3.54	4.01	4.37	4.68	4.95	5.19	5.40	5.59	5.77	5.93
	4.07	**4.9**	**5.5**	**6.0**	**6.4**	**6.7**	**7.1**	**7.3**	**7.5**	**7.8**	**8.0**
20	2.46	2.95	3.29	3.54	3.76	3.94	4.10	4.24	4.37	4.49	4.59
	3.32	**3.8**	**4.3**	**4.6**	**4.9**	**5.1**	**5.3**	**5.5**	**5.6**	**5.8**	**5.9**
30	2.07	2.40	2.61	2.78	2.91	3.02	3.12	3.21	3.29	3.36	3.39
	2.63	**3.0**	**3.3**	**3.5**	**3.6**	**3.7**	**3.8**	**3.9**	**4.0**	**4.1**	**4.2**
60	1.67	1.85	1.96	2.04	2.11	2.17	2.22	2.26	2.30	2.33	2.36
	1.96	**2.2**	**2.3**	**2.4**	**2.4**	**2.5**	**2.5**	**2.6**	**2.6**	**2.7**	**2.7**

Table 31 of E. Pearson and H. O. Hartley, *Biometrika Tables for Statisticians*, 2nd ed. New York: Cambridge University Press, 1958. Adapted and reprinted with permission of the Biometrika trustees.

TABLE B.4   THE *F* DISTRIBUTION*

*Table entries in lightface type are critical values for the .05 level of significance.
Boldface type values are for the .01 level of significance.

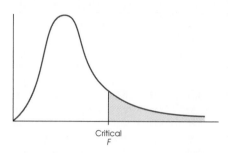

Critical
F

DEGREES OF FREEDOM: DENOMINATOR	DEGREES OF FREEDOM: NUMERATOR														
	1	2	3	4	5	6	7	8	9	10	11	12	14	16	20
1	161	200	216	225	230	234	237	239	241	242	243	244	245	246	248
	4052	**4999**	**5403**	**5625**	**5764**	**5859**	**5928**	**5981**	**6022**	**6056**	**6082**	**6106**	**6142**	**6169**	**6208**
2	18.51	19.00	19.16	19.25	19.30	19.33	19.36	19.37	19.38	19.39	19.40	19.41	19.42	19.43	19.44
	98.49	**99.00**	**99.17**	**99.25**	**99.30**	**99.33**	**99.34**	**99.36**	**99.38**	**99.40**	**99.41**	**99.42**	**99.43**	**99.44**	**99.45**
3	10.13	9.55	9.28	9.12	9.01	8.94	8.88	8.84	8.81	8.78	8.76	8.74	8.71	8.69	8.66
	34.12	**30.92**	**29.46**	**28.71**	**28.24**	**27.91**	**27.67**	**27.49**	**27.34**	**27.23**	**27.13**	**27.05**	**26.92**	**26.83**	**26.69**
4	7.71	6.94	6.59	6.39	6.26	6.16	6.09	6.04	6.00	5.96	5.93	5.91	5.87	5.84	5.80
	21.20	**18.00**	**16.69**	**15.98**	**15.52**	**15.21**	**14.98**	**14.80**	**14.66**	**14.54**	**14.45**	**14.37**	**14.24**	**14.15**	**14.02**
5	6.61	5.79	5.41	5.19	5.05	4.95	4.88	4.82	4.78	4.74	4.70	4.68	4.64	4.60	4.56
	16.26	**13.27**	**12.06**	**11.39**	**10.97**	**10.67**	**10.45**	**10.27**	**10.15**	**10.05**	**9.96**	**9.89**	**9.77**	**9.68**	**9.55**
6	5.99	5.14	4.76	4.53	4.39	4.28	4.21	4.15	4.10	4.06	4.03	4.00	3.96	3.92	3.87
	13.74	**10.92**	**9.78**	**9.15**	**8.75**	**8.47**	**8.26**	**8.10**	**7.98**	**7.87**	**7.79**	**7.72**	**7.60**	**7.52**	**7.39**
7	5.59	4.74	4.35	4.12	3.97	3.87	3.79	3.73	3.68	3.63	3.60	3.57	3.52	3.49	3.44
	12.25	**9.55**	**8.45**	**7.85**	**7.46**	**7.19**	**7.00**	**6.84**	**6.71**	**6.62**	**6.54**	**6.47**	**6.35**	**6.27**	**6.15**
8	5.32	4.46	4.07	3.84	3.69	3.58	3.50	3.44	3.39	3.34	3.31	3.28	3.23	3.20	3.15
	11.26	**8.65**	**7.59**	**7.01**	**6.63**	**6.37**	**6.19**	**6.03**	**5.91**	**5.82**	**5.74**	**5.67**	**5.56**	**5.48**	**5.36**
9	5.12	4.26	3.86	3.63	3.48	3.37	3.29	3.23	3.18	3.13	3.10	3.07	3.02	2.98	2.93
	10.56	**8.02**	**6.99**	**6.42**	**6.06**	**5.80**	**5.62**	**5.47**	**5.35**	**5.26**	**5.18**	**5.11**	**5.00**	**4.92**	**4.80**
10	4.96	4.10	3.71	3.48	3.33	3.22	3.14	3.07	3.02	2.97	2.94	2.91	2.86	2.82	2.77
	10.04	**7.56**	**6.55**	**5.99**	**5.64**	**5.39**	**5.21**	**5.06**	**4.95**	**4.85**	**4.78**	**4.71**	**4.60**	**4.52**	**4.41**
11	4.84	3.98	3.59	3.36	3.20	3.09	3.01	2.95	2.90	2.86	2.82	2.79	2.74	2.70	2.65
	9.65	**7.20**	**6.22**	**5.67**	**5.32**	**5.07**	**4.88**	**4.74**	**4.63**	**4.54**	**4.46**	**4.40**	**4.29**	**4.21**	**4.10**
12	4.75	3.88	3.49	3.26	3.11	3.00	2.92	2.85	2.80	2.76	2.72	2.69	2.64	2.60	2.54
	9.33	**6.93**	**5.95**	**5.41**	**5.06**	**4.82**	**4.65**	**4.50**	**4.39**	**4.30**	**4.22**	**4.16**	**4.05**	**3.98**	**3.86**
13	4.67	3.80	3.41	3.18	3.02	2.92	2.84	2.77	2.72	2.67	2.63	2.60	2.55	2.51	2.46
	9.07	**6.70**	**5.74**	**5.20**	**4.86**	**4.62**	**4.44**	**4.30**	**4.19**	**4.10**	**4.02**	**3.96**	**3.85**	**3.78**	**3.67**
14	4.60	3.74	3.34	3.11	2.96	2.85	2.77	2.70	2.65	2.60	2.56	2.53	2.48	2.44	2.39
	8.86	**6.51**	**5.56**	**5.03**	**4.69**	**4.46**	**4.28**	**4.14**	**4.03**	**3.94**	**3.86**	**3.80**	**3.70**	**3.62**	**3.51**
15	4.54	3.68	3.29	3.06	2.90	2.79	2.70	2.64	2.59	2.55	2.51	2.48	2.43	2.39	2.33
	8.68	**6.36**	**5.42**	**4.89**	**4.56**	**4.32**	**4.14**	**4.00**	**3.89**	**3.80**	**3.73**	**3.67**	**3.56**	**3.48**	**3.36**
16	4.49	3.63	3.24	3.01	2.85	2.74	2.66	2.59	2.54	2.49	2.45	2.42	2.37	2.33	2.28
	8.53	**6.23**	**5.29**	**4.77**	**4.44**	**4.20**	**4.03**	**3.89**	**3.78**	**3.69**	**3.61**	**3.55**	**3.45**	**3.37**	**3.25**
17	4.45	3.59	3.20	2.96	2.81	2.70	2.62	2.55	2.50	2.45	2.41	2.38	2.33	2.29	2.23
	8.40	**6.11**	**5.18**	**4.67**	**4.34**	**4.10**	**3.93**	**3.79**	**3.68**	**3.59**	**3.52**	**3.45**	**3.35**	**3.27**	**3.16**
18	4.41	3.55	3.16	2.93	2.77	2.66	2.58	2.51	2.46	2.41	2.37	2.34	2.29	2.25	2.19
	8.28	**6.01**	**5.09**	**4.58**	**4.25**	**4.01**	**3.85**	**3.71**	**3.60**	**3.51**	**3.44**	**3.37**	**3.27**	**3.19**	**3.07**

TABLE B.4　(continued)

DEGREES OF FREEDOM: DENOMINATOR	DEGREES OF FREEDOM: NUMERATOR														
	1	2	3	4	5	6	7	8	9	10	11	12	14	16	20
19	4.38	3.52	3.13	2.90	2.74	2.63	2.55	2.48	2.43	2.38	2.34	2.31	2.26	2.21	2.15
	8.18	**5.93**	**5.01**	**4.50**	**4.17**	**3.94**	**3.77**	**3.63**	**3.52**	**3.43**	**3.36**	**3.30**	**3.19**	**3.12**	**3.00**
20	4.35	3.49	3.10	2.87	2.71	2.60	2.52	2.45	2.40	2.35	2.31	2.28	2.23	2.18	2.12
	8.10	**5.85**	**4.94**	**4.43**	**4.10**	**3.87**	**3.71**	**3.56**	**3.45**	**3.37**	**3.30**	**3.23**	**3.13**	**3.05**	**2.94**
21	4.32	3.47	3.07	2.84	2.68	2.57	2.49	2.42	2.37	2.32	2.28	2.25	2.20	2.15	2.09
	8.02	**5.78**	**4.87**	**4.37**	**4.04**	**3.81**	**3.65**	**3.51**	**3.40**	**3.31**	**3.24**	**3.17**	**3.07**	**2.99**	**2.88**
22	4.30	3.44	3.05	2.82	2.66	2.55	2.47	2.40	2.35	2.30	2.26	2.23	2.18	2.13	2.07
	7.94	**5.72**	**4.82**	**4.31**	**3.99**	**3.76**	**3.59**	**3.45**	**3.35**	**3.26**	**3.18**	**3.12**	**3.02**	**2.94**	**2.83**
23	4.28	3.42	3.03	2.80	2.64	2.53	2.45	2.38	2.32	2.28	2.24	2.20	2.14	2.10	2.04
	7.88	**5.66**	**4.76**	**4.26**	**3.94**	**3.71**	**3.54**	**3.41**	**3.30**	**3.21**	**3.14**	**3.07**	**2.97**	**2.89**	**2.78**
24	4.26	3.40	3.01	2.78	2.62	2.51	2.43	2.36	2.30	2.26	2.22	2.18	2.13	2.09	2.02
	7.82	**5.61**	**4.72**	**4.22**	**3.90**	**3.67**	**3.50**	**3.36**	**3.25**	**3.17**	**3.09**	**3.03**	**2.93**	**2.85**	**2.74**
25	4.24	3.38	2.99	2.76	2.60	2.49	2.41	2.34	2.28	2.24	2.20	2.16	2.11	2.06	2.00
	7.77	**5.57**	**4.68**	**4.18**	**3.86**	**3.63**	**3.46**	**3.32**	**3.21**	**3.13**	**3.05**	**2.99**	**2.89**	**2.81**	**2.70**
26	4.22	3.37	2.98	2.74	2.59	2.47	2.39	2.32	2.27	2.22	2.18	2.15	2.10	2.05	1.99
	7.72	**5.53**	**4.64**	**4.14**	**3.82**	**3.59**	**3.42**	**3.29**	**3.17**	**3.09**	**3.02**	**2.96**	**2.86**	**2.77**	**2.66**
27	4.21	3.35	2.96	2.73	2.57	2.46	2.37	2.30	2.25	2.20	2.16	2.13	2.08	2.03	1.97
	7.68	**5.49**	**4.60**	**4.11**	**3.79**	**3.56**	**3.39**	**3.26**	**3.14**	**3.06**	**2.98**	**2.93**	**2.83**	**2.74**	**2.63**
28	4.20	3.34	2.95	2.71	2.56	2.44	2.36	2.29	2.24	2.19	2.15	2.12	2.06	2.02	1.96
	7.64	**5.45**	**4.57**	**4.07**	**3.76**	**3.53**	**3.36**	**3.23**	**3.11**	**3.03**	**2.95**	**2.90**	**2.80**	**2.71**	**2.60**
29	4.18	3.33	2.93	2.70	2.54	2.43	2.35	2.28	2.22	2.18	2.14	2.10	2.05	2.00	1.94
	7.60	**5.42**	**4.54**	**4.04**	**3.73**	**3.50**	**3.33**	**3.20**	**3.08**	**3.00**	**2.92**	**2.87**	**2.77**	**2.68**	**2.57**
30	4.17	3.32	2.92	2.69	2.53	2.42	2.34	2.27	2.21	2.16	2.12	2.09	2.04	1.99	1.93
	7.56	**5.39**	**4.51**	**4.02**	**3.70**	**3.47**	**3.30**	**3.17**	**3.06**	**2.98**	**2.90**	**2.84**	**2.74**	**2.66**	**2.55**
32	4.15	3.30	2.90	2.67	2.51	2.40	2.32	2.25	2.19	2.14	2.10	2.07	2.02	1.97	1.91
	7.50	**5.34**	**4.46**	**3.97**	**3.66**	**3.42**	**3.25**	**3.12**	**3.01**	**2.94**	**2.86**	**2.80**	**2.70**	**2.62**	**2.51**
34	4.13	3.28	2.88	2.65	2.49	2.38	2.30	2.23	2.17	2.12	2.08	2.05	2.00	1.95	1.89
	7.44	**5.29**	**4.42**	**3.93**	**3.61**	**3.38**	**3.21**	**3.08**	**2.97**	**2.89**	**2.82**	**2.76**	**2.66**	**2.58**	**2.47**
36	4.11	3.26	2.86	2.63	2.48	2.36	2.28	2.21	2.15	2.10	2.06	2.03	1.98	1.93	1.87
	7.39	**5.25**	**4.38**	**3.89**	**3.58**	**3.35**	**3.18**	**3.04**	**2.94**	**2.86**	**2.78**	**2.72**	**2.62**	**2.54**	**2.43**
38	4.10	3.25	2.85	2.62	2.46	2.35	2.26	2.19	2.14	2.09	2.05	2.02	1.96	1.92	1.85
	7.35	**5.21**	**4.34**	**3.86**	**3.54**	**3.32**	**3.15**	**3.02**	**2.91**	**2.82**	**2.75**	**2.69**	**2.59**	**2.51**	**2.40**
40	4.08	3.23	2.84	2.61	2.45	2.34	2.25	2.18	2.12	2.07	2.04	2.00	1.95	1.90	1.84
	7.31	**5.18**	**4.31**	**3.83**	**3.51**	**3.29**	**3.12**	**2.99**	**2.88**	**2.80**	**2.73**	**2.66**	**2.56**	**2.49**	**2.37**
42	4.07	3.22	2.83	2.59	2.44	2.32	2.24	2.17	2.11	2.06	2.02	1.99	1.94	1.89	1.82
	7.27	**5.15**	**4.29**	**3.80**	**3.49**	**3.26**	**3.10**	**2.96**	**2.86**	**2.77**	**2.70**	**2.64**	**2.54**	**2.46**	**2.35**

TABLE B.4   (continued)

DEGREES OF FREEDOM: DENOMINATOR	DEGREES OF FREEDOM: NUMERATOR														
	1	2	3	4	5	6	7	8	9	10	11	12	14	16	20
44	4.06	3.21	2.82	2.58	2.43	2.31	2.23	2.16	2.10	2.05	2.01	1.98	1.92	1.88	1.81
	7.24	**5.12**	**4.26**	**3.78**	**3.46**	**3.24**	**3.07**	**2.94**	**2.84**	**2.75**	**2.68**	**2.62**	**2.52**	**2.44**	**2.32**
46	4.05	3.20	2.81	2.57	2.42	2.30	2.22	2.14	2.09	2.04	2.00	1.97	1.91	1.87	1.80
	7.21	**5.10**	**4.24**	**3.76**	**3.44**	**3.22**	**3.05**	**2.92**	**2.82**	**2.73**	**2.66**	**2.60**	**2.50**	**2.42**	**2.30**
48	4.04	3.19	2.80	2.56	2.41	2.30	2.21	2.14	2.08	2.03	1.99	1.96	1.90	1.86	1.79
	7.19	**5.08**	**4.22**	**3.74**	**3.42**	**3.20**	**3.04**	**2.90**	**2.80**	**2.71**	**2.64**	**2.58**	**2.48**	**2.40**	**2.28**
50	4.03	3.18	2.79	2.56	2.40	2.29	2.20	2.13	2.07	2.02	1.98	1.95	1.90	1.85	1.78
	7.17	**5.06**	**4.20**	**3.72**	**3.41**	**3.18**	**3.02**	**2.88**	**2.78**	**2.70**	**2.62**	**2.56**	**2.46**	**2.39**	**2.26**
55	4.02	3.17	2.78	2.54	2.38	2.27	2.18	2.11	2.05	2.00	1.97	1.93	1.88	1.83	1.76
	7.12	**5.01**	**4.16**	**3.68**	**3.37**	**3.15**	**2.98**	**2.85**	**2.75**	**2.66**	**2.59**	**2.53**	**2.43**	**2.35**	**2.23**
60	4.00	3.15	2.76	2.52	2.37	2.25	2.17	2.10	2.04	1.99	1.95	1.92	1.86	1.81	1.75
	7.08	**4.98**	**4.13**	**3.65**	**3.34**	**3.12**	**2.95**	**2.82**	**2.72**	**2.63**	**2.56**	**2.50**	**2.40**	**2.32**	**2.20**
65	3.99	3.14	2.75	2.51	2.36	2.24	2.15	2.08	2.02	1.98	1.94	1.90	1.85	1.80	1.73
	7.04	**4.95**	**4.10**	**3.62**	**3.31**	**3.09**	**2.93**	**2.79**	**2.70**	**2.61**	**2.54**	**2.47**	**2.37**	**2.30**	**2.18**
70	3.98	3.13	2.74	2.50	2.35	2.23	2.14	2.07	2.01	1.97	1.93	1.89	1.84	1.79	1.72
	7.01	**4.92**	**4.08**	**3.60**	**3.29**	**3.07**	**2.91**	**2.77**	**2.67**	**2.59**	**2.51**	**2.45**	**2.35**	**2.28**	**2.15**
80	3.96	3.11	2.72	2.48	2.33	2.21	2.12	2.05	1.99	1.95	1.91	1.88	1.82	1.77	1.70
	6.96	**4.88**	**4.04**	**3.56**	**3.25**	**3.04**	**2.87**	**2.74**	**2.64**	**2.55**	**2.48**	**2.41**	**2.32**	**2.24**	**2.11**
100	3.94	3.09	2.70	2.46	2.30	2.19	2.10	2.03	1.97	1.92	1.88	1.85	1.79	1.75	1.68
	6.90	**4.82**	**3.98**	**3.51**	**3.20**	**2.99**	**2.82**	**2.69**	**2.59**	**2.51**	**2.43**	**2.36**	**2.26**	**2.19**	**2.06**
125	3.92	3.07	2.68	2.44	2.29	2.17	2.08	2.01	1.95	1.90	1.86	1.83	1.77	1.72	1.65
	6.84	**4.78**	**3.94**	**3.47**	**3.17**	**2.95**	**2.79**	**2.65**	**2.56**	**2.47**	**2.40**	**2.33**	**2.23**	**2.15**	**2.03**
150	3.91	3.06	2.67	2.43	2.27	2.16	2.07	2.00	1.94	1.89	1.85	1.82	1.76	1.71	1.64
	6.81	**4.75**	**3.91**	**3.44**	**3.14**	**2.92**	**2.76**	**2.62**	**2.53**	**2.44**	**2.37**	**2.30**	**2.20**	**2.12**	**2.00**
200	3.89	3.04	2.65	2.41	2.26	2.14	2.05	1.98	1.92	1.87	1.83	1.80	1.74	1.69	1.62
	6.76	**4.71**	**3.88**	**3.41**	**3.11**	**2.90**	**2.73**	**2.60**	**2.50**	**2.41**	**2.34**	**2.28**	**2.17**	**2.09**	**1.97**
400	3.86	3.02	2.62	2.39	2.23	2.12	2.03	1.96	1.90	1.85	1.81	1.78	1.72	1.67	1.60
	6.70	**4.66**	**3.83**	**3.36**	**3.06**	**2.85**	**2.69**	**2.55**	**2.46**	**2.37**	**2.29**	**2.23**	**2.12**	**2.04**	**1.92**
1000	3.85	3.00	2.61	2.38	2.22	2.10	2.02	1.95	1.89	1.84	1.80	1.76	1.70	1.65	1.58
	6.66	**4.62**	**3.80**	**3.34**	**3.04**	**2.82**	**2.66**	**2.53**	**2.43**	**2.34**	**2.26**	**2.20**	**2.09**	**2.01**	**1.89**
∞	3.84	2.99	2.60	2.37	2.21	2.09	2.01	1.94	1.88	1.83	1.79	1.75	1.69	1.64	1.57
	6.64	**4.60**	**3.78**	**3.32**	**3.02**	**2.80**	**2.64**	**2.51**	**2.41**	**2.32**	**2.24**	**2.18**	**2.07**	**1.99**	**1.87**

Table A14 of *Statistical Methods*, 7th ed. by George W. Snedecor and William G. Cochran. Copyright © 1980 by the Iowa State University Press, 2121 South State Avenue, Ames, Iowa 50010. Reprinted with permission of the Iowa State University Press.

TABLE B.5   THE STUDENTIZED RANGE STATISTIC (q)*

*The critical values for q corresponding to $\alpha = .05$ (lightface type) and $\alpha = .01$ (boldface type).

df FOR ERROR TERM	\multicolumn{11}{c}{k = NUMBER OF TREATMENTS}
	2	3	4	5	6	7	8	9	10	11	12
5	3.64	4.60	5.22	5.67	6.03	6.33	6.58	6.80	6.99	7.17	7.32
	5.70	**6.98**	**7.80**	**8.42**	**8.91**	**9.32**	**9.67**	**9.97**	**10.24**	**10.48**	**10.70**
6	3.46	4.34	4.90	5.30	5.63	5.90	6.12	6.32	6.49	6.65	6.79
	5.24	**6.33**	**7.03**	**7.56**	**7.97**	**8.32**	**8.61**	**8.87**	**9.10**	**9.30**	**9.48**
7	3.34	4.16	4.68	5.06	5.36	5.61	5.82	6.00	6.16	6.30	6.43
	4.95	**5.92**	**6.54**	**7.01**	**7.37**	**7.68**	**7.94**	**8.17**	**8.37**	**8.55**	**8.71**
8	3.26	4.04	4.53	4.89	5.17	5.40	5.60	5.77	5.92	6.05	6.18
	4.75	**5.64**	**6.20**	**6.62**	**6.96**	**7.24**	**7.47**	**7.68**	**7.86**	**8.03**	**8.18**
9	3.20	3.95	4.41	4.76	5.02	5.24	5.43	5.59	5.74	5.87	5.98
	4.60	**5.43**	**5.96**	**6.35**	**6.66**	**6.91**	**7.13**	**7.33**	**7.49**	**7.65**	**7.78**
10	3.15	3.88	4.33	4.65	4.91	5.12	5.30	5.46	5.60	5.72	5.83
	4.48	**5.27**	**5.77**	**6.14**	**6.43**	**6.67**	**6.87**	**7.05**	**7.21**	**7.36**	**7.49**
11	3.11	3.82	4.26	4.57	4.82	5.03	5.20	5.35	5.49	5.61	5.71
	4.39	**5.15**	**5.62**	**5.97**	**6.25**	**6.48**	**6.67**	**6.84**	**6.99**	**7.13**	**7.25**
12	3.08	3.77	4.20	4.51	4.75	4.95	5.12	5.27	5.39	5.51	5.61
	4.32	**5.05**	**5.50**	**5.84**	**6.10**	**6.32**	**6.51**	**6.67**	**6.81**	**6.94**	**7.06**
13	3.06	3.73	4.15	4.45	4.69	4.88	5.05	5.19	5.32	5.43	5.53
	4.26	**4.96**	**5.40**	**5.73**	**5.98**	**6.19**	**6.37**	**6.53**	**6.67**	**6.79**	**6.90**
14	3.03	3.70	4.11	4.41	4.64	4.83	4.99	5.13	5.25	5.36	5.46
	4.21	**4.89**	**5.32**	**5.63**	**5.88**	**6.08**	**6.26**	**6.41**	**6.54**	**6.66**	**6.77**
15	3.01	3.67	4.08	4.37	4.59	4.78	4.94	5.08	5.20	5.31	5.40
	4.17	**4.84**	**5.25**	**5.56**	**5.80**	**5.99**	**6.16**	**6.31**	**6.44**	**6.55**	**6.66**
16	3.00	3.65	4.05	4.33	4.56	4.74	4.90	5.03	5.15	5.26	5.35
	4.13	**4.79**	**5.19**	**5.49**	**5.72**	**5.92**	**6.08**	**6.22**	**6.35**	**6.46**	**6.56**
17	2.98	3.63	4.02	4.30	4.52	4.70	4.86	4.99	5.11	5.21	5.31
	4.10	**4.74**	**5.14**	**5.43**	**5.66**	**5.85**	**6.01**	**6.15**	**6.27**	**6.38**	**6.48**
18	2.97	3.61	4.00	4.28	4.49	4.67	4.82	4.96	5.07	5.17	5.27
	4.07	**4.70**	**5.09**	**5.38**	**5.60**	**5.79**	**5.94**	**6.08**	**6.20**	**6.31**	**6.41**
19	2.96	3.59	3.98	4.25	4.47	4.65	4.79	4.92	5.04	5.14	5.23
	4.05	**4.67**	**5.05**	**5.33**	**5.55**	**5.73**	**5.89**	**6.02**	**6.14**	**6.25**	**6.34**
20	2.95	3.58	3.96	4.23	4.45	4.62	4.77	4.90	5.01	5.11	5.20
	4.02	**4.64**	**5.02**	**5.29**	**5.51**	**5.69**	**5.84**	**5.97**	**6.09**	**6.19**	**6.28**
24	2.92	3.53	3.90	4.17	4.37	4.54	4.68	4.81	4.92	5.01	5.10
	3.96	**4.55**	**4.91**	**5.17**	**5.37**	**5.54**	**5.69**	**5.81**	**5.92**	**6.02**	**6.11**
30	2.89	3.49	3.85	4.10	4.30	4.46	4.60	4.72	4.82	4.92	5.00
	3.89	**4.45**	**4.80**	**5.05**	**5.24**	**5.40**	**5.54**	**5.65**	**5.76**	**5.85**	**5.93**
40	2.86	3.44	3.79	4.04	4.23	4.39	4.52	4.63	4.73	4.82	4.90
	3.82	**4.37**	**4.70**	**4.93**	**5.11**	**5.26**	**5.39**	**5.50**	**5.60**	**5.69**	**5.76**
60	2.83	3.40	3.74	3.98	4.16	4.31	4.44	4.55	4.65	4.73	4.81
	3.76	**4.28**	**4.59**	**4.82**	**4.99**	**5.13**	**5.25**	**5.36**	**5.45**	**5.53**	**5.60**
120	2.80	3.36	3.68	3.92	4.10	4.24	4.36	4.47	4.56	4.64	4.71
	3.70	**4.20**	**4.50**	**4.71**	**4.87**	**5.01**	**5.12**	**5.21**	**5.30**	**5.37**	**5.44**
∞	2.77	3.31	3.63	3.86	4.03	4.17	4.28	4.39	4.47	4.55	4.62
	3.64	**4.12**	**4.40**	**4.60**	**4.76**	**4.88**	**4.99**	**5.08**	**5.16**	**5.23**	**5.29**

Table 29 of E. Pearson and H. Hartley, *Biometrika Tables for Statisticians*, 3rd ed. New York: Cambridge University Press, 1966. Adapted and reprinted with permission of the Biometrika trustees.

TABLE B.6   CRITICAL VALUES FOR THE PEARSON CORRELATION*

*To be significant, the sample correlation, r, must be greater than or equal to the critical value in the table.

	LEVEL OF SIGNIFICANCE FOR ONE-TAILED TEST			
	.05	.025	.01	.005
	LEVEL OF SIGNIFICANCE FOR TWO-TAILED TEST			
$df = n - 2$	.10	.05	.02	.01
1	.988	.997	.9995	.9999
2	.900	.950	.980	.990
3	.805	.878	.934	.959
4	.729	.811	.882	.917
5	.669	.754	.833	.874
6	.622	.707	.789	.834
7	.582	.666	.750	.798
8	.549	.632	.716	.765
9	.521	.602	.685	.735
10	.497	.576	.658	.708
11	.476	.553	.634	.684
12	.458	.532	.612	.661
13	.441	.514	.592	.641
14	.426	.497	.574	.623
15	.412	.482	.558	.606
16	.400	.468	.542	.590
17	.389	.456	.528	.575
18	.378	.444	.516	.561
19	.369	.433	.503	.549
20	.360	.423	.492	.537
21	.352	.413	.482	.526
22	.344	.404	.472	.515
23	.337	.396	.462	.505
24	.330	.388	.453	.496
25	.323	.381	.445	.487
26	.317	.374	.437	.479
27	.311	.367	.430	.471
28	.306	.361	.423	.463
29	.301	.355	.416	.456
30	.296	.349	.409	.449
35	.275	.325	.381	.418
40	.257	.304	.358	.393
45	.243	.288	.338	.372
50	.231	.273	.322	.354
60	.211	.250	.295	.325
70	.195	.232	.274	.302
80	.183	.217	.256	.283
90	.173	.205	.242	.267
100	.164	.195	.230	.254

Table VI of R.A. Fisher and F. Yates, *Statistical Tables for Biological, Agricultural and Medical Research*, 6th ed. London: Longman Group Ltd., 1974 (previously published by Oliver and Boyd Ltd., Edinburgh). Adapted and reprinted with permission of the Addison Wesley Longman Publishing Co.

TABLE B.7   CRITICAL VALUES FOR THE SPEARMAN CORRELATION*

*To be significant, the sample correlation, r_s, must be greater than or equal to the critical value in the table.

	LEVEL OF SIGNIFICANCE FOR A ONE-TAILED TEST			
	.05	.025	.01	.005
	LEVEL OF SIGNIFICANCE FOR A TWO-TAILED TEST			
n	.10	.05	.02	.01
4	1.000			
5	0.900	1.000	1.000	
6	0.829	0.886	0.943	1.000
7	0.714	0.786	0.893	0.929
8	0.643	0.738	0.833	0.881
9	0.600	0.700	0.783	0.833
10	0.564	0.648	0.745	0.794
11	0.536	0.618	0.709	0.755
12	0.503	0.587	0.671	0.727
13	0.484	0.560	0.648	0.703
14	0.464	0.538	0.622	0.675
15	0.443	0.521	0.604	0.654
16	0.429	0.503	0.582	0.635
17	0.414	0.485	0.566	0.615
18	0.401	0.472	0.550	0.600
19	0.391	0.460	0.535	0.584
20	0.380	0.447	0.520	0.570
21	0.370	0.435	0.508	0.556
22	0.361	0.425	0.496	0.544
23	0.353	0.415	0.486	0.532
24	0.344	0.406	0.476	0.521
25	0.337	0.398	0.466	0.511
26	0.331	0.390	0.457	0.501
27	0.324	0.382	0.448	0.491
28	0.317	0.375	0.440	0.483
29	0.312	0.368	0.433	0.475
30	0.306	0.362	0.425	0.467
35	0.283	0.335	0.394	0.433
40	0.264	0.313	0.368	0.405
45	0.248	0.294	0.347	0.382
50	0.235	0.279	0.329	0.363
60	0.214	0.255	0.300	0.331
70	0.190	0.235	0.278	0.307
80	0.185	0.220	0.260	0.287
90	0.174	0.207	0.245	0.271
100	0.165	0.197	0.233	0.257

TABLE B.8   THE CHI-SQUARE DISTRIBUTION*

*The table entries are critical values of χ^2.

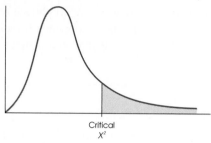

Critical
X^2

df	PROPORTION IN CRITICAL REGION				
	0.10	0.05	0.025	0.01	0.005
1	2.71	3.84	5.02	6.63	7.88
2	4.61	5.99	7.38	9.21	10.60
3	6.25	7.81	9.35	11.34	12.84
4	7.78	9.49	11.14	13.28	14.86
5	9.24	11.07	12.83	15.09	16.75
6	10.64	12.59	14.45	16.81	18.55
7	12.02	14.07	16.01	18.48	20.28
8	13.36	15.51	17.53	20.09	21.96
9	14.68	16.92	19.02	21.67	23.59
10	15.99	18.31	20.48	23.21	25.19
11	17.28	19.68	21.92	24.72	26.76
12	18.55	21.03	23.34	26.22	28.30
13	19.81	22.36	24.74	27.69	29.82
14	21.06	23.68	26.12	29.14	31.32
15	22.31	25.00	27.49	30.58	32.80
16	23.54	26.30	28.85	32.00	34.27
17	24.77	27.59	30.19	33.41	35.72
18	25.99	28.87	31.53	34.81	37.16
19	27.20	30.14	32.85	36.19	38.58
20	28.41	31.41	34.17	37.57	40.00
21	29.62	32.67	35.48	38.93	41.40
22	30.81	33.92	36.78	40.29	42.80
23	32.01	35.17	38.08	41.64	44.18
24	33.20	36.42	39.36	42.98	45.56
25	34.38	37.65	40.65	44.31	46.93
26	35.56	38.89	41.92	45.64	48.29
27	36.74	40.11	43.19	46.96	49.64
28	37.92	41.34	44.46	48.28	50.99
29	39.09	42.56	45.72	49.59	52.34
30	40.26	43.77	46.98	50.89	53.67
40	51.81	55.76	59.34	63.69	66.77
50	63.17	67.50	71.42	76.15	79.49
60	74.40	79.08	83.30	88.38	91.95
70	85.53	90.53	95.02	100.42	104.22
80	96.58	101.88	106.63	112.33	116.32
90	107.56	113.14	118.14	124.12	128.30
100	118.50	124.34	129.56	135.81	140.17

Table 8 of E. Pearson and H. Hartley, *Biometrika Tables for Statisticians*, 3d ed. New York: Cambridge University Press, 1966. Adapted and reprinted with permission of the Biometrika trustees.

TABLE B.9A   CRITICAL VALUES OF THE MANN-WHITNEY U FOR $\alpha = .05$*

*Critical values are provided for a *one-tailed* test at $\alpha = .05$ (lightface type) and for a *two-tailed* test at $\alpha = .05$ (bold-face type). To be significant for any given n_A and n_B, the obtained U must be *equal to* or *less than* the critical value in the table. Dashes (—) in the body of the table indicate that no decision is possible at the stated level of significance and values of n_A and n_B.

n_B \ n_A	1	2	3	4	5	6	7	8	9	10	11	12	13	14	15	16	17	18	19	20
1	—	—	—	—	—	—	—	—	—	—	—	—	—	—	—	—	—	—	0	0
	—	—	—	—	—	—	—	—	—	—	—	—	—	—	—	—	—	—	—	—
2	—	—	—	—	0	0	0	1	1	1	1	2	2	2	3	3	3	4	4	4
	—	—	—	—	—	—	—	**0**	**0**	**0**	**0**	**1**	**1**	**1**	**1**	**1**	**2**	**2**	**2**	**2**
3	—	—	0	0	1	2	2	3	3	4	5	5	6	7	7	8	9	9	10	11
	—	—	—	**0**	**1**	**1**	**2**	**2**	**3**	**3**	**4**	**4**	**5**	**5**	**6**	**6**	**7**	**7**	**8**
4	—	—	0	1	2	3	4	5	6	7	8	9	10	11	12	14	15	16	17	18
	—	—	—	**0**	**1**	**2**	**3**	**4**	**4**	**5**	**6**	**7**	**8**	**9**	**10**	**11**	**11**	**12**	**13**	**13**
5	—	0	1	2	4	5	6	8	9	11	12	13	15	16	18	19	20	22	23	25
	—	—	**0**	**1**	**2**	**3**	**5**	**6**	**7**	**8**	**9**	**11**	**12**	**13**	**14**	**15**	**17**	**18**	**19**	**20**
6	—	0	2	3	5	7	8	10	12	14	16	17	19	21	23	25	26	28	30	32
	—	—	**1**	**2**	**3**	**5**	**6**	**8**	**10**	**11**	**13**	**14**	**16**	**17**	**19**	**21**	**22**	**24**	**25**	**27**
7	—	0	2	4	6	8	11	13	15	17	19	21	24	26	28	30	33	35	37	39
	—	—	**1**	**3**	**5**	**6**	**8**	**10**	**12**	**14**	**16**	**18**	**20**	**22**	**24**	**26**	**28**	**30**	**32**	**34**
8	—	1	3	5	8	10	13	15	18	20	23	26	28	31	33	36	39	41	44	47
	—	**0**	**2**	**4**	**6**	**8**	**10**	**13**	**15**	**17**	**19**	**22**	**24**	**26**	**29**	**31**	**34**	**36**	**38**	**41**
9	—	1	3	6	9	12	15	18	21	24	27	30	33	36	39	42	45	48	51	54
	—	**0**	**2**	**4**	**7**	**10**	**12**	**15**	**17**	**20**	**23**	**26**	**28**	**31**	**34**	**37**	**39**	**42**	**45**	**48**
10	—	1	4	7	11	14	17	20	24	27	31	34	37	41	44	48	51	55	58	62
	—	**0**	**3**	**5**	**8**	**11**	**14**	**17**	**20**	**23**	**26**	**29**	**33**	**36**	**39**	**42**	**45**	**48**	**52**	**55**
11	—	1	5	8	12	16	19	23	27	31	34	38	42	46	50	54	57	61	65	69
	—	**0**	**3**	**6**	**9**	**13**	**16**	**19**	**23**	**26**	**30**	**33**	**37**	**40**	**44**	**47**	**51**	**55**	**58**	**62**
12	—	2	5	9	13	17	21	26	30	34	38	42	47	51	55	60	64	68	72	77
	—	**1**	**4**	**7**	**11**	**14**	**18**	**22**	**26**	**29**	**33**	**37**	**41**	**45**	**49**	**53**	**57**	**61**	**65**	**69**
13	—	2	6	10	15	19	24	28	33	37	42	47	51	56	61	65	79	75	80	84
	—	**1**	**4**	**8**	**12**	**16**	**20**	**24**	**28**	**33**	**37**	**41**	**45**	**50**	**54**	**59**	**63**	**67**	**72**	**76**
14	—	2	7	11	16	21	26	31	36	41	46	51	56	61	66	71	77	82	87	92
	—	**1**	**5**	**9**	**13**	**17**	**22**	**26**	**31**	**36**	**40**	**45**	**50**	**55**	**59**	**64**	**67**	**74**	**78**	**83**
15	—	3	7	12	18	23	28	33	39	44	50	55	61	66	72	77	83	88	94	100
	—	**1**	**5**	**10**	**14**	**19**	**24**	**29**	**34**	**39**	**44**	**49**	**54**	**59**	**64**	**70**	**75**	**80**	**85**	**90**
16	—	3	8	14	19	25	30	36	42	48	54	60	65	71	77	83	89	95	101	107
	—	**1**	**6**	**11**	**15**	**21**	**26**	**31**	**37**	**42**	**47**	**53**	**59**	**64**	**70**	**75**	**81**	**86**	**92**	**98**
17	—	3	9	15	20	26	33	39	45	51	57	64	70	77	83	89	96	102	109	115
	—	**2**	**6**	**11**	**17**	**22**	**28**	**34**	**39**	**45**	**51**	**57**	**63**	**67**	**75**	**81**	**87**	**93**	**99**	**105**
18	—	4	9	16	22	28	35	41	48	55	61	68	75	82	88	95	102	109	116	123
	—	**2**	**7**	**12**	**18**	**24**	**30**	**36**	**42**	**48**	**55**	**61**	**67**	**74**	**80**	**86**	**93**	**99**	**106**	**112**
19	0	4	10	17	23	30	37	44	51	58	65	72	80	87	94	101	109	116	123	130
	—	**2**	**7**	**13**	**19**	**25**	**32**	**38**	**45**	**52**	**58**	**65**	**72**	**78**	**85**	**92**	**99**	**106**	**113**	**119**
20	0	4	11	18	25	32	39	47	54	62	69	77	84	92	100	107	115	123	130	138
	—	**2**	**8**	**13**	**20**	**27**	**34**	**41**	**48**	**55**	**62**	**69**	**76**	**83**	**90**	**98**	**105**	**112**	**119**	**127**

TABLE B.9B   CRITICAL VALUES OF THE MANN-WHITNEY U FOR $\alpha = .01$*

*Critical values are provided for a *one-tailed* test at $\alpha = .01$ (lightface type) and for a *two-tailed* test at $\alpha = .01$ (bold-face type). To be significant for any given n_A and n_B, the obtained U must be *equal to* or *less than* the critical value in the table. Dashes (—) in the body of the table indicate that no decision is possible at the stated level of significance and values of n_A and n_B.

n_B \ n_A	1	2	3	4	5	6	7	8	9	10	11	12	13	14	15	16	17	18	19	20
1	—	—	—	—	—	—	—	—	—	—	—	—	—	—	—	—	—	—	—	—
2	—	—	—	—	—	—	—	—	—	—	—	—	0	0	0	0	0	0	1	1
	—	—	—	—	—	—	—	—	—	—	—	—	—	—	—	—	—	—	**0**	**0**
3	—	—	—	—	—	—	0	0	1	1	1	2	2	2	3	3	4	4	4	5
	—	—	—	—	—	—	—	**0**	**0**	**0**	**1**	**1**	**1**	**2**	**2**	**2**	**2**	**3**	**3**	
4	—	—	—	—	0	1	1	2	3	3	4	5	5	6	7	7	8	9	9	10
	—	—	—	—	**0**	**0**	**1**	**1**	**2**	**2**	**3**	**3**	**4**	**5**	**5**	**6**	**6**	**7**	**8**	
5	—	—	—	0	1	2	3	4	5	6	7	8	9	10	11	12	13	14	15	16
	—	—	—	—	**0**	**1**	**1**	**2**	**3**	**4**	**5**	**6**	**7**	**7**	**8**	**9**	**10**	**11**	**12**	**13**
6	—	—	—	1	2	3	4	6	7	8	9	11	12	13	15	16	18	19	20	22
	—	—	—	**0**	**1**	**2**	**3**	**4**	**5**	**6**	**7**	**9**	**10**	**11**	**12**	**13**	**15**	**16**	**17**	**18**
7	—	—	0	1	3	4	6	7	9	11	12	14	16	17	19	21	23	24	26	28
	—	—	—	**0**	**1**	**3**	**4**	**6**	**7**	**9**	**10**	**12**	**13**	**15**	**16**	**18**	**19**	**21**	**22**	**24**
8	—	—	0	2	4	6	7	9	11	13	15	17	20	22	24	26	28	30	32	34
	—	—	—	**1**	**2**	**4**	**6**	**7**	**9**	**11**	**13**	**15**	**17**	**18**	**20**	**22**	**24**	**26**	**28**	**30**
9	—	—	1	3	5	7	9	11	14	16	18	21	23	26	28	31	33	36	38	40
	—	—	**0**	**1**	**3**	**5**	**7**	**9**	**11**	**13**	**16**	**18**	**20**	**22**	**24**	**27**	**29**	**31**	**33**	**36**
10	—	—	1	3	6	8	11	13	16	19	22	24	27	30	33	36	38	41	44	47
	—	—	**0**	**2**	**4**	**6**	**9**	**11**	**13**	**16**	**18**	**21**	**24**	**26**	**29**	**31**	**34**	**37**	**39**	**42**
11	—	—	1	4	7	9	12	15	18	22	25	28	31	34	37	41	44	47	50	53
	—	—	**0**	**2**	**5**	**7**	**10**	**13**	**16**	**18**	**21**	**24**	**27**	**30**	**33**	**36**	**39**	**42**	**45**	**48**
12	—	—	2	5	8	11	14	17	21	24	28	31	35	38	42	46	49	53	56	60
	—	—	**1**	**3**	**6**	**9**	**12**	**15**	**18**	**21**	**24**	**27**	**31**	**34**	**37**	**41**	**44**	**47**	**51**	**54**
13	—	0	2	5	9	12	16	20	23	27	31	35	39	43	47	51	55	59	63	67
	—	—	**1**	**3**	**7**	**10**	**13**	**17**	**20**	**24**	**27**	**31**	**34**	**38**	**42**	**45**	**49**	**53**	**56**	**60**
14	—	0	2	6	10	13	17	22	26	30	34	38	43	47	51	56	60	65	69	73
	—	—	**1**	**4**	**7**	**11**	**15**	**18**	**22**	**26**	**30**	**34**	**38**	**42**	**46**	**50**	**54**	**58**	**63**	**67**
15	—	0	3	7	11	15	19	24	28	33	37	42	47	51	56	61	66	70	75	80
	—	—	**2**	**5**	**8**	**12**	**16**	**20**	**24**	**29**	**33**	**37**	**42**	**46**	**51**	**55**	**60**	**64**	**69**	**73**
16	—	0	3	7	12	16	21	26	31	36	41	46	51	56	61	66	71	76	82	87
	—	—	**2**	**5**	**9**	**13**	**18**	**22**	**27**	**31**	**36**	**41**	**45**	**50**	**55**	**60**	**65**	**70**	**74**	**79**
17	—	0	4	8	13	18	23	28	33	38	44	49	55	60	66	71	77	82	88	93
	—	—	**2**	**6**	**10**	**15**	**19**	**24**	**29**	**34**	**39**	**44**	**49**	**54**	**60**	**65**	**70**	**75**	**81**	**86**
18	—	0	4	9	14	19	24	30	36	41	47	53	59	65	70	76	82	88	94	100
	—	—	**2**	**6**	**11**	**16**	**21**	**26**	**31**	**37**	**42**	**47**	**53**	**58**	**64**	**70**	**75**	**81**	**87**	**92**
19	—	1	4	9	15	20	26	32	38	44	50	56	63	69	75	82	88	94	101	107
	—	**0**	**3**	**7**	**12**	**17**	**22**	**28**	**33**	**39**	**45**	**51**	**56**	**63**	**69**	**74**	**81**	**87**	**93**	**99**
20	—	1	5	10	16	22	28	34	40	47	53	60	67	73	80	87	93	100	107	114
	—	**0**	**3**	**8**	**13**	**18**	**24**	**30**	**36**	**42**	**48**	**54**	**60**	**67**	**73**	**79**	**86**	**92**	**99**	**105**

Table D.10 of *Introductory Statistics* by R. E. Kirk. Copyright © 1978 by Wadsworth, Inc. Reprinted by permission of Brooks/Cole, Pacific Grove, California 93950.

TABLE B.10   CRITICAL VALUES OF T FOR THE WILCOXON SIGNED-RANKS TEST*

*To be significant, the obtained T must be *equal to* or *less than* the critical value. Dashes (—) in the columns indicate that no decision is possible for the stated α and n.

	LEVEL OF SIGNIFICANCE FOR ONE-TAILED TEST					LEVEL OF SIGNIFICANCE FOR ONE-TAILED TEST			
	.05	.025	.01	.005		.05	.025	.01	.005
	LEVEL OF SIGNIFICANCE FOR TWO-TAILED TEST					LEVEL OF SIGNIFICANCE FOR TWO-TAILED TEST			
n	.10	.05	.02	.01	n	.10	.05	.02	.01
5	0	—	—	—	28	130	116	101	91
6	2	0	—	—	29	140	126	110	100
7	3	2	0	—	30	151	137	120	109
8	5	3	1	0	31	163	147	130	118
9	8	5	3	1	32	175	159	140	128
10	10	8	5	3	33	187	170	151	138
11	13	10	7	5	34	200	182	162	148
12	17	13	9	7	35	213	195	173	159
13	21	17	12	9	36	227	208	185	171
14	25	21	15	12	37	241	221	198	182
15	30	25	19	15	38	256	235	211	194
16	35	29	23	19	39	271	249	224	207
17	41	34	27	23	40	286	264	238	220
18	47	40	32	27	41	302	279	252	233
19	53	46	37	32	42	319	294	266	247
20	60	52	43	37	43	336	310	281	261
21	67	58	49	42	44	353	327	296	276
22	75	65	55	48	45	371	343	312	291
23	83	73	62	54	46	389	361	328	307
24	91	81	69	61	47	407	378	345	322
25	100	89	76	68	48	426	396	362	339
26	110	98	84	75	49	446	415	379	355
27	119	107	92	83	50	466	434	397	373

Adapted from F. Wilcoxon, S. K. Katti, and R. A. Wilcox, *Critical Values and Probability Levels of the Wilcoxon Rank-Sum Test and the Wilcoxon Signed-Ranks Test*. Wayne, NJ: American Cyanamid Company, 1963. Adapted and reprinted with permission of the American Cyanamid Company.

APPENDIX C Solutions for Odd-Numbered Problems

Note: Many of the problems in the text require several stages of computation. At each stage there is an opportunity for rounding answers. Depending on the exact sequence of operations used to solve a problem, different individuals will round their answers at different times and in different ways. As a result, you may obtain answers that are slightly different from those presented here. As long as these differences are small, they probably can be attributed to rounding error and should not be a matter for concern.

CHAPTER 1 **INTRODUCTION TO STATISTICS**

1. Descriptive statistics are used to simplify and summarize data. Inferential statistics use sample data to make general conclusions about populations.

3. **a.** The population consists of college students.
   b. The sample is the set of 50 students selected for the study.

5. The distinguishing characteristics of the experimental method are *manipulation* and *control*. In the experimental method, the researcher manipulates one variable and observes a second variable. All other variables are controlled to prevent them from affecting the outcome.

7. In an experiment, the researcher manipulates one variable and observes a second variable for changes. The manipulated variable is called the *independent variable* and the observed variable is the *dependent variable*.

9. It is not an experiment because the researcher is not manipulating a variable to create the groups. Instead, the study is comparing preexisting groups (boys versus girls).

11. No. The researcher is simply observing sugar consumption and activity level. Neither variable is manipulated.

13. **a.** An ordinal scale provides information about the direction of difference (greater or less) between two measurements.
   b. An interval scale provides information about the magnitude of the difference between two measurements.
   c. A ratio scale provides information about the ratio of two measurements.

15. **a.** The independent variable is the color of the room, and the dependent variable is the score on the mood questionnaire.
   b. Room color is measured on a nominal scale.

17. **a.** $\Sigma X = 10$
   b. $\Sigma X^2 = 30$
   c. $(\Sigma X)^2 = (10)^2 = 100$
   d. $\Sigma(X - 1) = 6$

19. **a.** $\Sigma X = 10$
   b. $\Sigma X^2 = 42$
   c. $\Sigma(X + 1) = 14$
   d. $\Sigma(X + 1)^2 = 66$

21. **a.** $\Sigma X = 0$
   b. $\Sigma Y = 7$
   c. $\Sigma XY = -1$

23. **a.** Sum the scores, then subtract 4 from the total.
   b. Square each score, then sum the squared values.
   c. Add 4 points to each score, then square the resulting values. Finally, sum the squared values.

25. **a.** First, subtract the constant C from each score. Second, square each of the resulting values. Third, sum the squared values.
   b. $\Sigma(X - 4)^2 = 26$
   c. $\Sigma(X - 4) = 0$

CHAPTER 2   **FREQUENCY DISTRIBUTIONS**

1.

X	f	p	%
10	4	.20	20
9	3	.15	15
8	5	.25	25
7	6	.30	30
6	0	0	0
5	2	.10	10

3. A bar graph leaves a space between adjacent bars and is used with data from nominal or ordinal scales. In a histogram, adjacent bars touch at the real limits. Histograms are used to display data from interval or ratio scales.

5. A grouped frequency distribution is used when the original scale of measurement consists of more scores or categories than can be listed in a regular table. Around 20 or more categories is generally considered too many for a regular table.

7. a.

X	f
6	5
5	6
4	3
3	2
2	1
1	1

b.

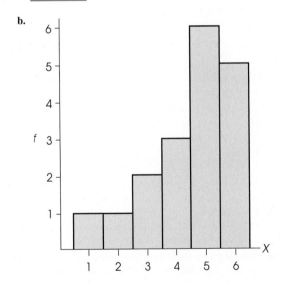

c. The distribution is negatively skewed. If a score of $X = 4$ is considered typical, most of the individuals in this class have reading scores that are higher than those of the typical third-grade student.

9. a. $N = 11$ **b.** $\Sigma X = 30$

11. a. $N = 11$ **b.** $\Sigma X = 32$ **c.** $\Sigma X^2 = 106$

13. a.

X	f
MA	2
ME	1
NH	1
NJ	1
NY	19
OH	2
PA	3
VA	1

(Note: The states are listed in alphabetical order but any order is appropriate.)

b.

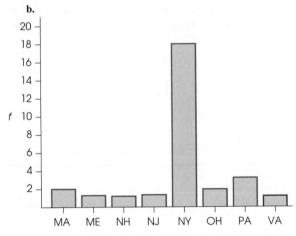

15. a.

X	f
6	4
5	3
4	2
3	2
2	1
1	1

b.

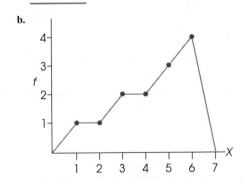

c. (1) The distribution is negatively skewed.
   (2) The scores are centered around $X = 4$ or $X = 5$.
   (3) The scores are piled up around $X = 5$ or $X = 6$, but they are spread across the entire scale.

17. a.

X	f		X	f
85–89	2		80–89	2
80–84	0		70–79	7
75–79	3		60–69	12
70–74	4		50–59	5
65–69	6		40–49	4
60–64	6			
55–59	2			
50–54	3			
45–49	3			
40–44	1			

19.

X	f	cf	c%
5	7	25	100
4	8	18	72
3	5	10	40
2	3	5	20
1	2	2	8

21. a. The 20th percentile is $X = 5.5$
   b. The percentile rank for $X = 11.5$ is 80%.
   c. The 25th percentile is $X = 6.0$.
   d. The percentile rank for $X = 12$ is 83%.

23. a. The 80th percentile is $X = 4.5$
   b. The percentile rank of $X = 2.5$ is 30%.
   c. The 75th percentile is $X = 4.25$.
   d. The percentile rank of $X = 4$ is 70%.
   e. The 25th percentile is $X = 2.25$.

25.

3	599
4	37863790
5	62750
6	61

CHAPTER 3   CENTRAL TENDENCY

1. The purpose of central tendency is to identify the single score that serves as the best representative for an entire distribution, usually a score from the center of the distribution.

3. a. The mean is the *balance point* of a distribution because the sum of the distances above the mean is exactly equal to the sum of the distances below the mean.

   b. The median is the *midpoint* of a distribution because exactly 50% of the scores are greater than the median and 50% are less than the median.

5. The mean, median, and mode are identical for a symmetrical distribution with one mode.

7. With a skewed distribution, the extreme scores in the tail can displace the mean. Specifically, the mean is displaced away from the main pile of scores and moved out toward the tail. The result is that the mean is often not a very representative value.

9. The mean is $50/10 = 5$, the median is 5, and the mode is 5.

11. The mean is $116/20 = 5.8$, the median is 5.5, and the mode is 5.

13. $\Sigma X = 105$

15. The original $\Sigma X = 160$ and the new $\Sigma X = 120$. The new mean is $120/8 = 15$.

17. The original $\Sigma X = 30$ and the new $\Sigma X = 28$. The new mean is $28/4 = 7$.

19. The original $\Sigma X = 55$ and the new $\Sigma X = 72$. The score that was added is $X = 17$.

21. For the combined sample $n = 20$ and $\Sigma X = 450 + 120 = 570$. The mean for the combined sample is $570/20 = 28.5$.

23. a. Mode $= 2$
   b. Median $= 2.5$
   c. You cannot find the total number of absences (ΣX) for this class.

25.

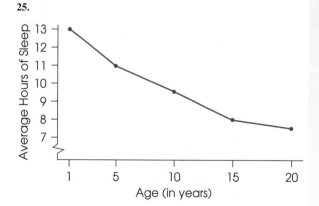

CHAPTER 4 **VARIABILITY**

1. a. *SS* is the sum of squared deviation scores.
 b. Variance is the mean squared deviation.
 c. Standard deviation is the square root of the variance. It provides a measure of the standard distance from the mean.

3. *SS* cannot be less than zero because it is computed by adding squared deviations. Squared deviations are always greater than or equal to zero.

5. An *unbiased* estimate means that on average the sample statistic will provide an accurate representation of the corresponding population parameter. More specifically, if you take all the possible samples and compute the variance for each sample, then the average of all the sample variances will be exactly equal to the population variance.

7. $SS = 20$, $s^2 = 5$, $s = 2.24$

9. a. The definitional formula is easy to use when the mean is a whole number and there are relatively few scores.
 b. The computational formula is preferred when the mean is not a whole number.

11. $SS = 8$, the sample variance is 1, and the standard deviation is 1.

13. $SS = 36$, the population variance is 9, and the standard deviation is 3.

15. a. Your score is 5 points above the mean. If the standard deviation is 2, then your score is an extremely high value (out in the tail of the distribution). However, if the standard deviation is 10, then your score is only slightly above average. Thus, you would prefer $\sigma = 2$.
 b. If you are located 5 points below the mean then the situation is reversed. A standard deviation of 2 gives you an extremely low score, but with a standard deviation of 10 you are only slightly below average. Here, you would prefer $\sigma = 10$.

17. a.

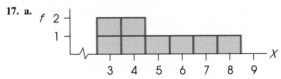

 b. The sample mean is $M = 5$ and the standard deviation appears to be about 2 points.
 c. For this sample, $SS = 24$, the variance is $24/7 = 3.43$ and the standard deviation is 1.85, which is reasonably close to the original estimate of $s = 2$.

19. a.

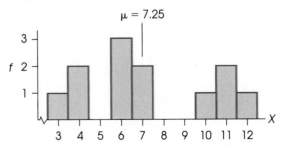

b. The mean is $\mu = 87/12 = 7.25$ and the standard deviation appears to be about 2 or 3.
 c. For this population, $SS = 102.25$, the variance is 8.52, and the standard deviation is 2.92, which is reasonably close to the original estimate of $\sigma = 2$ or 3.

21. a. $SS = 36$, variance = 9, standard deviation = 3
 b. $SS = 36$, variance = 9, standard deviation = 3
 c. $SS = 144$, variance = 36, standard deviation = 6
 d. Adding a constant to each score does not change any of the deviations and does not change the standard deviation.
 e. When each score is multiplied by a constant, the deviations and the standard deviation are also multiplied by the constant.

23. a.

THE NUMBER OF PROBLEMS SOLVED UNDER
DIFFERENT LEVELS OF AROUSAL

AROUSAL LEVEL		
Low	$M = 5.1$	$SD = 1.52$
Moderate	$M = 17.0$	$SD = 2.16$
High	$M = 7.7$	$SD = 1.25$

 b. Performance is best with a moderate level of arousal. If arousal is either too high or too low, performance suffers.

25. a.

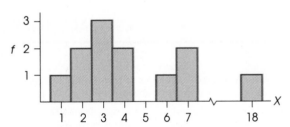

 b. $M = 5$ and $s = 4.53$
 c. Median = 3.5; semi-interquartile range = 2 ($Q1 = 2.5$ and $Q3 = 6.5$)
 d. Most of the scores are clustered around $X = 3$ or 4, so the median and semi-interquartile range provide a better description. The extreme score of $X = 18$ distorts the mean and standard deviation.

CHAPTER 5   *z*-SCORES

1. A z-score describes a precise location within a distribution. The sign of the z-score tells whether the location is above ($+$) or below ($-$) the mean, and the magnitude tells the distance from the mean in terms of the number of standard deviations.

3. **a.** $X = 80$ is in the extreme right-hand tail at $z = 2.00$.
   b. $X = 80$ is slightly above the mean at $z = 0.50$.

5. **a.**

X	z		X	z		X	z
110	0.50		140	2.00		80	-1.00
60	-2.00		105	0.25		130	1.50

b.

X	z		X	z		X	z
150	2.50		90	-0.50		70	-1.50
105	0.25		80	-1.00		115	0.75

7.

X	z		X	z		X	z
66	0.50		48	-1.00		84	2.00
55	-0.42		70	0.83		75	1.25

9. **a.**

X	z		X	z		X	z
75	-0.25		90	0.50		110	1.50
95	0.75		60	-1.00		40	-2.00

b.

X	z		X	z		X	z
130	2.50		70	-0.50		50	-1.50
85	0.25		65	-0.75		100	1.00

11. **a.** $z = 2.00$
    b. $z = -1.00$
    c. $z = 1.50$
    d. $z = -3.00$

13. $\sigma = 6$

15. $\mu = 90$

17. $\mu = 54$ and $\sigma = 8$. The distance between the two scores is 12 points, which is equal to 1.50 standard deviations.

19. **a.** With $\sigma = 2$, the score $X = 86$ corresponds to $z = 3.00$, which is an extreme value.
    b. With $\sigma = 12$, the score $X = 86$ corresponds to $z = 0.50$, which is a central value.

21. $\sigma = 10$

23. Mary's statistics score corresponds to $z = -0.25$ and her biology score corresponds to $z = -2.00$. She should expect the higher grade in statistics.

25.

ORIGINAL SCORE	TRANSFORMED SCORE
$X = 34$	$X = 80$
$X = 36$	$X = 90$
$X = 37$	$X = 95$
$X = 41$	$X = 115$
$X = 46$	$X = 140$

27. **a.** $\mu = 3$ and $\sigma = 2$
    b. & c.

ORIGINAL X	z-SCORE	TRANSFORMED X
0	-1.50	35
4	0.50	55
6	1.50	65
1	-1.00	40
3	0	50
4	0.50	55

CHAPTER 6   **PROBABILITY**

1. **a.** $p = 60/90 = 0.667$
   b. $p = 15/90 = 0.167$
   c. $p = 5/90 = 0.056$

3. The two requirements for a random sample are: (1) each individual has an equal chance of being selected, and (2) if more than one individual is selected, the probabilities must stay constant for all selections.

5. **a.** tail to the right, $p = 0.1587$
   b. tail to the left, $p = 0.0228$
   c. tail to the right, $p = 0.0668$
   d. tail to the left, $p = .2266$

7. **a.** $p(z > 1.50) = 0.0668$
   b. $p(z > -2.00) = 0.9772$
   c. $p(z < 0.50) = 0.6915$
   d. $p(z < -0.75) = 0.2266$

9. **a.** $z = 1.28$ **b.** $z = 0.25$ **c.** -0.84

11. **a.** $p = 0.0440$
    b. $p = 0.1498$
    c. $p = 0.4772$
    d. $p = 0.5319$

13. You cannot use the unit normal table to find proportions or z-scores because the distribution is not normal.

15. a. $z = 2.00, p = .0228$
   b. $z = -0.67, p = 0.2514$
   c. $z = 0.67, p = 0.2514$

17. a. $X = 628$ **b.** $X = 552$ **c.** $X = 525$

19. a. $z = 0.13, p = 0.4483$
   b. $z = -0.87, p = 0.8078$
   c. $z = 0.33, p = 0.6293$
   d. $z = -0.33, p = 0.3707$

21. a. $X = 416$ and $X = 584$ ($z = \pm 0.84$)
   b. $X = 372$ and $X = 628$ ($z = \pm 1.28$)
   c. $X = 304$ and $X = 696$ ($z = \pm 1.96$)

23. a. $z = 1.28, X = 264$
   b. $z = -0.52, X = 174$
   c. $z = 0.67, X = 233.5$

25. a. With four suits possible, the probability of guessing correctly is $p = 1/4$.
   b. With $n = 48$, $p = 1/4$, and $q = 3/4$, it is possible to use the normal approximation to the binomial distribution with $\mu = 12$ and $\sigma = 3$. Using the real limit of 18.5, $p(X > 18.5) = p(z > 2.17) = 0.0150$.
   c. Using the real limit of 17.5, $p(X > 17.5) = p(z > 1.83) = 0.0336$.

27. a. With $n = 50$ and $p = q = 1/2$, you may use the normal approximation with $\mu = 25$ and $\sigma = 3.54$. Using the upper real limit of 30.5, $p(X > 30.5) = p(z > 1.55) = 0.0606$.
   b. The normal approximation has $\mu = 50$ and $\sigma = 5$. Using the upper real limit of 60.5, $p(X > 60.5) = p(z > 2.10) = 0.0179$.
   c. Getting 60% heads with a balanced coin is an unusual event for a large sample. Although you might get 60% heads with a small sample, you should get very close to a 50-50 distribution as the sample gets larger.

CHAPTER 7   THE DISTRIBUTION OF SAMPLE MEANS

1. a. The distribution of sample means consists of the sample means for all the possible random samples of a specific size (n) from a specific population.
   b. The expected value of M is the mean of the distribution of sample means (μ).
   c. The standard error of M is the standard deviation of the distribution of sample means ($\sigma_M = \sigma/\sqrt{n}$).

3. With $n = 36$, the distribution of sample means will be approximately normal even if the population is not. The expected value is $\mu = 65$ and the standard error is $12/6 = 2$.

5. a. Standard error $= 100/\sqrt{4} = 50$
   b. Standard error $= 100/\sqrt{25} = 20$
   c. Standard error $= 100/\sqrt{100} = 10$

7. a. $n \geq 9$
   b. $n \geq 36$
   c. No. When $n = 1$, the standard error is equal to the standard deviation. For any other sample size, the standard error will be smaller than the standard deviation.

9. a. When $n = 1$, the standard deviation, $\sigma = 6$, measures the standard distance between X and μ.
   b. With $n = 9$, $\sigma_M = 2$ points.
   c. With $n = 36$, $\sigma_M = 1$ point.

11. a. Standard error is 4, $z = 0.50$.
   b. Standard error is 2, $z = 1.00$.
   c. Standard error is 1, $z = 2.00$.

13. a. $z = 1.00$. This is a typical sample.
   b. $z = 1.00$. This is a typical sample.
   c. $z = 2.00$. This is an extreme sample.

15. a. $p = 0.9544$
   b. $p = 0.5762$
   c. The middle 95% of all sample means range from 45.13 to 54.87.

17. a. Using $z = \pm 1.65$ for the middle 90%, the sample means should be between 483.5 and 516.5.
   b. The sample means should be between 480.4 and 519.6.
   c. Using $z = \pm 2.58$, the sample means should be between 474.2 and 525.8.

19. a. $\sigma_M = 10$, $z = 0.50$, $p = 0.3085$
   b. $\sigma_M = 5$, $z = 1.00$, $p = 0.1587$
   c. $\sigma_M = 2$, $z = 2.50$, $p = 0.0062$

21. a. The distribution is normal with $\mu = 55$ and $\sigma_M = 2$.
   b. $z = +2.00, p = 0.0228$
   c. $z = +0.50, p = 0.6915$
   d. $p(-1.00 < z < +1.00) = 0.6826$

23. a. The distribution is normal with an expected value (mean) of 80 and a standard error of 4.
   b. $M = 85$ corresponds to $z = 1.25, p = 0.1056$.
   c. The distribution is normal with an expected value (mean) of 80 and a standard error of 2.
   d. $M = 85$ corresponds to $z = 2.50, p = 0.0062$.

25. a. With a standard error of 3.41, this sample mean corresponds to a z-score of $z = 3.14$. A z-score this extreme has a probability of only $p = 0.0008$.
   b. It would be almost impossible to obtain a sample mean this large by random sampling.

27. a.

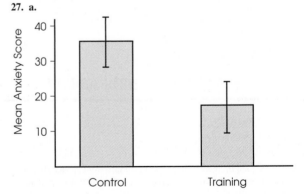

   b. Even considering the standard error for each mean, there is no overlap between the two groups. The relaxation training does seem to have lowered scores, even allowing for the standard error.

CHAPTER 8   **INTRODUCTION TO HYPOTHESIS TESTING**

1. a. $M - \mu$ measures the difference between the sample data and the hypothesized population mean.

 b. A sample mean is not expected to be identical to the population mean. The standard error measures how much difference, on average, is expected between M and μ simply by chance.

3. Lowering the alpha level causes the boundaries of the critical region to move farther out into the tails of the distribution. Lowering α reduces the probability of a Type I error.

5. a. A Type I error is made by rejecting a true H_0. This can occur if you obtain a very unusual sample with scores that are much different from the general population.

 b. A Type II error is made by failing to reject a false H_0. This can happen when the treatment effect is very small. In this case the treated sample is not noticeably different from the original population.

7. The analyses are contradictory. The critical region for the two-tailed tests consists of the extreme 2.5% in each tail of the distribution. The two-tailed conclusion indicates that the data were not in this critical region. However, the one-tailed test indicates that the data were in the extreme 1% of one tail. Data cannot be in the extreme 1% and at the same time fail to be in the extreme 2.5%.

9. a. H_0: $\mu = 40$. The sample mean $M = 42$ corresponds to $z = 1.00$. This is not sufficient to reject the null hypothesis. You cannot conclude that the program has a significant effect.

 b. H_0: $\mu = 40$. The sample mean $M = 44$ corresponds to $z = 2.00$. This is sufficient to reject the null hypothesis and conclude that the program does have a significant effect.

 c. The 2-point difference between the sample and the hypothesis in part (a) was not sufficient to reject H_0; however, the 4-point difference in part (b) is significantly more than chance.

11. a. The dependent variable is SAT score and the independent variable is whether they take the special course or not.

 b. H_0: $\mu = 500$; H_1: $\mu \neq 500$. The critical boundaries are $z = \pm 1.96$. For these data, $z = 2.16$, so our decision is to reject H_0 and conclude that the course did affect SAT scores.

 c. With $\alpha = .01$, the critical region consists of z-score values greater than 2.58 or less than -2.58.

 d. With $\alpha = .01$, the decision is to fail to reject H_0. With $\alpha = .01$, you have less risk of a Type I error. Therefore, the test requires a larger treatment effect in order to reject H_0.

13. a. H_0: $\mu = 100$. The critical boundaries are $z = \pm 1.96$. With $\sigma = 12$, the standard error is $\sigma_M = 2$ and $z = 3.00$. Reject H_0.

 b. With $\sigma = 30$, $\sigma_M = 5$, and $z = 1.20$. Fail to reject H_0.

 c. There is a 6-point difference between the sample mean and the hypothesized population value. In part (a), the standard error is 2 points and the 6-point difference is significantly more than is expected by chance. However, in part (b), the standard error is 5 points and the 6-point difference is not sufficient to reject the null hypothesis.

15. a. H_0: $\mu = 65$. $\sigma_M = 4$ and $z = -1.25$. Fail to reject H_0.

 b. H_0: $\mu = 55$. $\sigma_M = 4$ and $z = 1.25$. Fail to reject H_0.

 c. With a sample mean of $M = 60$, it is reasonable to expect that the population mean has a value near 60. Any value that is reasonably near 60 will be an acceptable hypothesis for μ. The standard error and the alpha level determine the range of acceptable values.

17. a. For this sample, $\sigma_M = 0.32$ and $z = -4.06$. Reject H_0, and conclude there has been a significant change in homework hours.

 b. $d = 0.41$

19. a. H_0: $\mu = 60$. $\sigma_M = 2.5$ and $z = 0.80$. Fail to reject H_0. Cohen's $d = 0.20$.

 b. H_0: $\mu = 60$. $\sigma_M = 1$ and $z = 2.00$. Reject H_0. Cohen's $d = 0.20$.

 c. The hypothesis test is influenced by sample size because the standard error is reduced with a larger sample. However, sample size has no influence on Cohen's d.

21. a. The null hypothesis states that the training program has no effect on the problem-solving scores (H_0: $\mu = 80$). The critical region consists of z-score values beyond ± 1.96. For these data, the standard error is 2.36 and $z = 1.88$. Fail to reject H_0. There is not sufficient evidence to conclude that the training program has any effect.

 b. For a one-tailed test, the null hypothesis would state that training does not improve problem-solving ability; H_0: $\mu \leq 80$. The critical region consists of z-score values greater than $+1.65$. For the one-tailed test, the z-score is in the critical region and the decision is to reject H_0.

23. For the one-tailed test, the null hypothesis states that there is no memory impairment. H_0: $\mu \geq 50$. The critical region consists of z-scores less than -2.33. For these data, $\sigma_M = 1.28$ and $z = -2.34$. Reject H_0, and conclude the alcoholics have significantly lower memory scores.

25. H_0: $\mu \leq 12$ (no increase during hot weather). H_1: $\mu > 12$ (there is an increase). The critical region consists of z-score values greater than $+1.65$. For these data, the standard error is 1.50, and $z = 2.33$, which is in the critical region, so we reject the null hypothesis and conclude that there is a significant increase in the mean number of hit players during hot weather.

27. Power increases.

29. As the size of the treatment effect gets smaller, the treatment distribution moves closer to the null distribution and power decreases. When the difference between the two distributions is extremely small, the size of the critical region (alpha) corresponds almost exactly with the portion of the treatment distribution that is located in the critical region (power). Thus, the smallest possible power is equal to alpha.

CHAPTER 9 INTRODUCTION TO THE t STATISTIC

1. The z-score is used when the population standard deviation (or variance) is known. The t statistic is used when the population variance or standard deviation is unknown. You use the sample data to estimate the variance and the standard error.

3. **a.** As variability increases, t becomes smaller (closer to zero).
   b. As sample size increases, the standard error decreases and the t value increases.
   c. The larger the difference between M and μ, the larger is the t value.

5. As the value of df increases, the t distribution becomes less variable (less spread out) and more like a normal distribution. For $\alpha = .05$, the critical t values move toward ± 1.96 as df increases.

7. **a.** $M = 6$ and $s = 4$ **b.** $s_M = 2$

9. **a.** $M = 5$ and $s = 4$ **b.** $s_M = 2$ points

11. **a.** For sample I, $s^2 = 9$ and $s = 3$. For sample II, $s^2 = 9$ and $s = 3$.
    b. For sample I, and $s_M = 1.50$. For sample II, and $s_M = 1$.
    c. Sample size is one component of the estimated standard error. As sample size increases, the estimated standard error decreases.

13. **a.** With $\alpha = .05$ and $df = 24$, the critical boundaries are $t = \pm 2.064$. For these data, the estimated standard error is 2 and $t = 2.50$. Reject H_0.
    b. With $\alpha = .01$ and $df = 24$, the critical boundaries are $t = \pm 2.797$. For these data, the estimated standard error is 2 and $t = 2.50$. Fail to reject H_0.
    c. The test with $\alpha = .01$ involves less risk of a Type I error but also requires more evidence from the data. In this problem, the data are sufficient to reject H_0 at the .05 level, but are not sufficient to satisfy the criteria for $\alpha = .01$.

15. **a.** With $n = 25$, the standard error is 2 points and the critical boundaries are $t = \pm 2.064$. The 6-point difference corresponds to $t = 3.00$, which is significant.
    b. With $n = 4$, the standard error is 5 points and the critical boundaries are $t = \pm 3.182$. The 6-point difference corresponds to $t = 1.20$, which is not significant.
    c. A larger sample means a smaller standard error and a greater likelihood of rejecting the null hypothesis.

d. Cohen's $d = 6/10 = 0.60$ whether $n = 25$ or $n = 4$. Cohen's d is not influenced by sample size.

17. **a.** For the one-tailed test, the null hypothesis states that the mean will not be greater than $\mu = 2$ ($\mu \leq 2$). The critical value is $t = 1.753$. For these data, the sample variance is 4, the standard error is 0.50, and $t = 4.6$. Reject the null hypothesis, and conclude that the rehearsal training produced a significant increase in memory performance.
    b. With a standard deviation of 2, Cohen's $d = 2.3/2 = 1.15$ and $r^2 = 0.585$.

19. **a.** Because $df = 99$ is not in the table, use $df = 60$ to obtain a critical boundary of -1.671. The sample variance is 4, the estimated standard error is 0.20, and $t = -11.00$. Reject H_0.
    b. Cohen's $d = 2.2/2 = 1.1$ and $r^2 = 0.55$.

21. **a.** With $n = 25$, the critical boundaries are ± 2.064. For these data, the standard error is 2.00 and the t statistic is $t(24) = -2.00$. Fail to reject H_0.
    b. Because $df = 399$ is not in the table, use $df = 120$ to obtain critical boundaries of ± 1.98. With $n = 400$, the standard error is 0.50 and the t statistic is $t(399) = -8.00$. Reject H_0.
    c. The larger the sample, the smaller is the standard error (denominator of the t-statistic). If other factors are held constant, a smaller standard error will produce a larger t statistic, which is more likely to be significant.

23. H_0: $\mu = 21$. Because $df = 99$ is not in the table, use $df = 60$ to obtain critical boundaries of ± 2.00. With a standard error of 0.50, $t(99) = -4.60$. Reject H_0, and conclude that humidity has a significant effect on the rats' eating behavior.

25. H_0: $\mu \leq 20$ (not more than expected by chance). The critical region consists of t values greater than 1.860. For these data, $M = 22.22$, $s^2 = 17.44$, the standard error is 1.39, and $t(8) = 1.60$. Fail to reject H_0. The performance for this sample is not significantly better than expected by chance.

27. H_0: $\mu \geq 15$. The critical region consists of t values less than -1.943. For these data, $M = 11.71$, $s^2 = 12.24$, the standard error is 1.32, and $t(6) = -2.49$. Reject H_0, and conclude that right-hemisphere brain damage does significantly reduce spatial skills.

CHAPTER 10 HYPOTHESIS TESTS WITH TWO INDEPENDENT SAMPLES

1. An independent-measures study requires a separate sample, independently selected, for each of the treatments or populations being compared. An independent-measures t statistic is appropriate when a researcher has two samples and wants to use the sample mean difference to test hypotheses about the population mean difference.

3. The homogeneity of variance assumption specifies that $\sigma_1^2 = \sigma_2^2$ for the two populations from which the samples are obtained. If this assumption is violated, the t statistic can cause misleading conclusions for a hypothesis test.

5. **a.** The first sample has $s^2 = 8$ and the second has $s^2 = 12$.
   b. The pooled variance is $164/16 = 10.25$.

7. **a.** The first sample has a variance of 8, the second sample variance is 12, and the pooled variance is 10 (halfway between).
   b. The first sample has a variance of 4, the second sample variance is 16, and the pooled variance is 7.27 (closer to the variance for the larger sample).

9. **a.** The pooled variance is 120.
   b. The standard error is 4.00.
   c. A mean difference of 8 would produce $t = 8/4 = 2$. With $df = 28$, the critical values are ± 2.048. Fail to reject H_0.
   d. A mean difference of 12 would produce $t = 12/4 = 3$. With $df = 28$, the critical values are ± 2.048. Reject H_0.
   e. With a mean difference of 8 points, $r^2 = 0.125$. With a difference of 12 points, $r^2 = 0.24$.

11. The null hypothesis states that there is no difference in the mother/child relationship between easy-delivery and difficult-delivery daughters, H_0: $\mu_1 - \mu_2 = 0$. With $df = 22$ and $\alpha = .05$, the critical region consists of t values beyond ± 2.074. For these data, the pooled variance is 96, the estimated standard error is 4, and $t(22) = -1.25$. Fail to reject H_0. The data are not sufficient to indicate a significant difference between the two birth conditions.

13. a. For treatment 1 the variance is $420/15 = 28$, and for treatment 2 the variance is $540/15 = 36$. The pooled variance is $960/30 = 32$.

  b. The null hypothesis states that there is no difference between the two treatments, H_0: $\mu_1 - \mu_2 = 0$. With $df = 30$ and $\alpha = .05$, the critical region consists of t values beyond ± 2.042. For these data, the estimated standard error is 2, and $t(30) = 3.00$. Reject H_0. The data are sufficient to conclude that there is a significant difference between the two treatments.

  c. The samples in this problem are much larger than the samples in Problem 12. The larger samples produce a smaller standard error so that the sample mean difference is now significant.

15. a. The null hypothesis states that there is no difference in the performance scores between the students with pets and those without pets, H_0: $\mu_1 - \mu_2 = 0$. With $df = 23$ and $\alpha = .05$, the critical region consists of t values beyond ± 2.069. For these data, the pooled variance is 150, the estimated standard error is 5, and $t(23) = 2.40$. Reject H_0. The data are sufficient to conclude that there is a significant difference between the two groups.

  b. For these data, $d = 0.98$.

17. The null hypothesis states that there is no difference in difficulty between the two problems, H_0: $\mu_1 - \mu_2 = 0$. With $\alpha = .01$, the critical region consists of t values beyond ± 3.355. With the tacks in the box, $M = 119.2$ and $SS = 2746.80$. With the tacks and box separate, $M = 43.2$ and $SS = 1066.80$. The pooled variance is 476.7, the standard error is 13.81, and $t(8) = 5.50$. Reject H_0. The data are sufficient to indicate a significant difference between the two conditions.

19. a. There are a total of 26 subjects in the two samples combined.

  b. With $df = 24$ and $\alpha = .05$, the critical region consists of t values beyond ± 2.064. The t statistic is in the critical region. Reject H_0, and conclude that there is a significant difference.

  c. With $df = 24$ and $\alpha = .01$, the critical region consists of t values beyond ± 2.797. The t statistic is not in the critical region. Fail to reject H_0, and conclude that there is no significant difference.

21. a. For treatment 1, the mean is $M = 4$ and for treatment 2, the mean is $M = 11$. The means are the same as in Problem 20.

  b. For treatment 1, $SS = 96$ and for treatment 2, $SS = 100$. The pooled variance is 32.67 and the estimated standard error is 4.04. The variances and the standard error are much larger than they were in Problem 20.

  c. The null hypothesis states that there is no difference between the two treatments, H_0: $\mu_1 - \mu_2 = 0$. With $df = 6$ and $\alpha = .05$, the critical region consists of t values beyond ± 2.447. For these data, $t(6) = -1.73$. Fail to reject H_0. The data are not sufficient to conclude that there is a significant difference between the two treatments.

  d. The difference between Problem 20 and Problem 21 is the magnitude of the sample variance. In Problem 21, the large variance produces a large standard error, with the result that the difference between sample means is not significant.

23. a. The null hypothesis states that feedback will have no effect on estimation of length, H_0: $\mu_1 - \mu_2 = 0$. With $df = 6$ and $\alpha = .05$, the critical region consists of t values beyond ± 2.447. For the no-feedback group, $M = 6$ and $SS = 6$. For the feedback group, $M = 3$ with $SS = 8$. For these data, the pooled variance is 2.33, the estimated standard error is 1.08, and $t(6) = 2.78$. Reject H_0. The data are sufficient to conclude that there is a significant difference between the two feedback conditions.

  b. For these data, Cohen's $d = 1.96$ (a very large effect) and $r^2 = 0.56$ (56%).

CHAPTER 11   **HYPOTHESIS TESTS WITH TWO RELATED SAMPLES**

1. a. This is an independent-measures experiment with two separate samples.

  b. This is repeated measures. The same sample is measured twice.

3. For a repeated-measures design, the same subjects are used in both treatment conditions. In a matched-subjects design, two different sets of subjects are used. However, in a matched-subjects design, each subject in one condition is matched with respect to a specific variable with a subject in the second condition, so that the two separate samples are equivalent with respect to the matching variable.

5. a. For sample 1, the scores pile up around $M_D = 4$ and the majority of the scores are spread over a range from -6 to 14. A value of zero is located near the center of the distribution. For sample 2, the scores pile up around $M_D = 12$ and the majority of the scores are spread over a range from 2 to 22.

A value of zero is an extreme value, not located near the center of the distribution.

  b. Sample 2 is more likely to reject a hypothesis of $\mu_D = 0$. For this sample there is a clear difference between the center of the sample and the value of zero.

7. a. The estimated standard error is 1.35.

  b. $M_D = 3$ produces $t = 2.22$. The obtained t value is beyond the critical boundaries of ± 2.131. Reject H_0.

9. The null hypothesis says that the magnetic therapy has no effect, H_0: $\mu_D = 0$. With $df = 8$ and $\alpha = .01$, the critical values are $t = \pm 3.355$. For these data, the standard error is 4, and $t(8) = 3.13$. Fail to reject H_0, and conclude that the magnetic therapy has no significant effect on chronic pain.

11. a. The null hypothesis says that the medication has no effect, H_0: $\mu_D = 0$. Because $df = 35$ is not in the table, use $df = 30$ to obtain critical values of $t = \pm 2.042$. For these data, the standard error is 1.33, and $t(35) = 18.05$. Reject H_0, and conclude that the medication produces a significant change in reaction time.

b. For these data, $r^2 = 0.90$ (90%).

13. a. The null hypothesis says that the drug does not increase the pain threshold, $H_0: \mu_D \leq 0$. With $\alpha = .05$, the one-tailed critical region consists of t values greater than 1.860. For these data, $s^2 = 0.16$, the standard error is 0.13, and $t(8) = 12.31$. Reject H_0, and conclude that there has been a significant change.

b. For these data, $r^2 = 0.95$ (95%).

15. a. As an independent-measures design, the pooled variance is 8, the standard error is 2, and $t(6) = 2.00$. The critical values for t are ± 2.447, so we fail to reject the null hypothesis and conclude that there is no significant difference between the treatments.

b. As a repeated-measure design, the variance of the difference scores is 4, the standard error is 1, and $t(3) = 4.00$. The critical values for t are ± 3.182, so we reject the null hypothesis and conclude that there is a significant difference between the treatments.

c. With the repeated-measures design, individual differences are removed, so that the variance and the standard error are smaller.

17. a. The difference scores are 9, 5, 1, and 13. $M_D = 7$.

b. $SS = 80$, sample variance is 26.67, and the estimated standard error is 2.58.

c. With $df = 3$ and $\alpha = .05$, the critical values are $t = \pm 3.182$. For these data, $t = 2.71$. Fail to reject H_0. There is not a significant treatment effect.

19. a. For Problem 17, Cohen's d is 1.36 and for Problem 18, $d = 2.02$. The percentage of variance accounted for is $r^2 = .71$ for Problem 17 and $r^2 = .85$ for Problem 18.

b. In general, the smaller sample variance (Problem 18) produces the larger effect size.

21. The null hypothesis says that the oatmeal has no effect, $H_0: \mu_D = 0$. With $df = 8$ and $\alpha = .05$, the critical values are $t = \pm 2.306$. For these data, $M_D = -7$, $SS = 288$, the standard error is 2, and $t(8) = -3.50$. Reject H_0, and conclude that the oatmeal has a significant effect on cholesterol.

23. The null hypothesis says that the treatment has no effect on agoraphobia, $H_0: \mu_D = 0$. With $df = 6$ and $\alpha = .05$, the critical values are $t = \pm 2.447$. For these data, $M_D = 8$, $SS = 238$, the estimated standard error is 2.38, and $t(6) = 3.36$. Reject H_0, and conclude that the treatment does have a significant effect on agoraphobia.

25. The null hypothesis says that there is no difference between shots fired during versus between heart beats, $H_0: \mu_D = 0$. With $\alpha = .05$, the critical region consists of t values beyond ± 2.571. For these data, $M_D = 2.83$, $SS = 34.83$, $s^2 = 6.97$, the standard error is 1.08, and $t(5) = 2.62$. Reject H_0, and conclude that the timing of the shot has a significant effect on the marksmen's scores.

27. $H_0: \mu_D \geq 0$ (no decrease). For these data, $M_D = -4.20$, $s^2 = 8.70$, the standard error is 1.32, and $t(4) = -3.18$. For a one-tailed test with $\alpha = .05$, the critical boundary is $t = -2.132$. Reject H_0, and conclude that the treatment significantly reduces nightmares.

CHAPTER 12   **ESTIMATION**

1. The general purpose of a hypothesis test is to determine whether or not a treatment effect exists. A hypothesis test always addresses a "yes/no" question. The purpose of estimation is to determine the size of the effect. Estimation addresses a "how much" question.

3. a. The larger the sample, the narrower is the interval.

b. The larger the sample standard deviation (s), the wider is the interval.

c. The higher the percentage of confidence, the wider is the interval.

5. a. Use the sample mean, $M = 7.4$ for the point estimate.

b. Using a single-sample t statistic with $df = 24$, the standard error is 0.60, the t score boundaries for 90% confidence are ± 1.711, and the interval extends from 6.373 to 8.427.

7. a. Using a single-sample t statistic with $df = 3$, the standard error is 3, the t-score boundaries for 90% confidence are ± 2.353, and the interval extends from 26.941 to 41.059.

b. Using a single-sample t statistic with $df = 15$, the standard error is 1.5, the t score boundaries for 90% confidence are ± 1.753, and the interval extends from 31.370 to 36.630.

c. Using a single-sample t statistic with $df = 35$, the standard error is 1, the t score boundaries for 90% confidence are

± 1.697 (using $df = 30$ because 35 is not listed), and the interval extends from 32.303 to 35.697.

d. The larger the sample size, the narrower is the confidence interval.

9. a. Use the sample mean difference, 3.6 seconds, for the point estimate.

b. Using an independent-measures t statistic with $df = 18$, the pooled variance is 5, the standard error is 1, the t score boundaries for 80% confidence are ± 1.330, and the interval extends from 2.27 to 4.93 seconds.

11. a. Use the sample mean difference, 3.5 miles per hour, for the point estimate.

b. Using an independent-measures t statistic with $df = 198$, the pooled variance is 27.5 and the standard error is 0.74. Using $df = 120$ because 198 is not listed, the t score boundaries for 95% confidence are ± 1.980, and the interval extends from 2.035 to 4.965 miles per hour.

13. a. Use the sample mean difference, 5.2, for the point estimate.

b. Using an independent-measures t statistic with $df = 28$, the pooled variance is 30, the estimated standard error is 2.00, the t score boundaries for 95% confidence are ± 2.048, and the interval extends from 1.104 to 9.296.

c. The t score boundaries for 99% confidence are ± 2.763, and the interval extends from 0.326 to 10.726.

d. The change is significant for $\alpha = .05$ because a mean difference of zero is outside the interval and therefore rejected with 95% confidence. However, a mean difference of zero is within the 99% interval and therefore is an acceptable value (fail to reject) with 99% confidence.

15. Use the sample mean difference, 0.72, for the point estimate. Using a repeated-measures t statistic with $df = 24$, the standard error is 0.20, the t score boundaries for 95% confidence are ± 2.064, and the interval extends from 0.3072 to 1.1328.

17. a. $M = 7.36$ and $s^2 = 2.05$

b. Use $M = 7.36$ as the point estimate for μ.

c. With $df = 10$ and an estimated standard error of 0.43, the t score boundaries for 80% confidence are $t = \pm 1.372$, and the interval extends from 6.77 to 7.950.

19. Use the sample mean difference, 5.5, for the point estimate. Using a repeated-measures t statistic with $df = 15$, the standard error is 2, the t score boundaries for 80% confidence are ± 1.341, and the interval extends from 2.818 to 8.182.

21. a. Use the sample mean difference, 24, for the point estimate.

b. Using a repeated-measures t statistic with $df = 35$, the estimated standard error is 1.33, the t score boundaries for 95% confidence are ± 2.042 (using $df = 30$ because 35 is not listed), and the interval extends from 21.284 to 26.716.

23. a. Use the sample mean difference, 2.83, for the point estimate.

b. Using a repeated-measures t statistic with $df = 5$, the estimated standard error is 1.08, the t score boundaries for 95% confidence are ± 2.571, and the interval extends from 0.053 to 5.607.

CHAPTER 13   INTRODUCTION TO ANALYSIS OF VARIANCE

1. When there is no treatment effect, the numerator and the denominator of the F-ratio are both measuring the same sources of variability (individual differences and experimental error). In this case, the F-ratio is balanced and should have a value near 1.00.

3. As the differences between sample means increase, $MS_{between}$ also increases, and the F-ratio increases. Increases in sample variability cause MS_{within} to increase and, thereby, decrease the F-ratio.

5. Posttests are done after an ANOVA where you reject the null hypothesis with 3 or more treatments. Posttests determine which treatments are significantly different.

7. a. Reducing the differences between treatments should reduce the value of $MS_{between}$ and should lower the value of the F-ratio.

b.

SOURCE	SS	df	MS	
Between treatments	32	2	16	$F(2, 9) = 4.00$
Within treatments	36	9	4	
Total	68	11		

With $\alpha = .05$, the critical value is $F = 4.26$. Fail to reject the null hypothesis, and conclude that there are no significant differences among the three treatments. Notice that the F-ratio is substantially smaller than it was in Problem 6.

9. a.

SOURCE	SS	df	MS	
Between treatments	32	2	16	$F(2, 9) = 10.26$
Within treatments	14	9	1.56	
Total	46	11		

With $\alpha = .05$, the critical value is $F = 4.26$. Reject the null hypothesis, and conclude that there are significant differences among the three treatments.

b. For these data, eta squared is $32/46 = 0.70$.

11. a.

SOURCE	SS	df	MS
Between treatments	32	2	16
Within treatments	0	9	0
Total	32	11	

13. The three sample variances are $2/2 = 1$, $8/2 = 4$, and $2/2 = 1$. The average of the three variances is $6/3 = 2$, which is exactly the value of MS_{within} from Problem 12.

15.

SOURCE	SS	df	MS	
Between treatments	24	2	12	$F(2, 33) = 6.00$
Within treatments	66	33	2	
Total	90	35		

17. a.

SOURCE	SS	df	MS	
Between treatments	56	2	28	$F(2, 9) = 7.00$
Within treatments	36	9	4	
Total	92	11		

With $\alpha = .05$, the critical value is $F = 4.26$. Reject the null hypothesis, and conclude that there are significant differences among the three treatments.

b. For these data, eta squared is $56/92 = 0.61$.

19. a.

SOURCE	SS	df	MS	
Between treatments	24	3	8	$F(3, 12) = 6.00$
Within treatments	16	12	1.33	
Total	40	15		

With $\alpha = .05$, the critical value is $F = 3.49$. Reject the null hypothesis, and conclude that there are significant differences among the four presentation rates.

b. Tukey's HSD is 2.42. The presentation rate of $\frac{1}{2}$ second is significantly different from both the 1-second and 3-second rates.

21. a.

SOURCE	SS	df	MS	
Between treatments	70	2	35.00	$F(2, 12) = 26.32$
Within treatments	16	12	1.33	
Total	86	14		

The critical value is $F = 3.88$. Reject H_0. These data do provide evidence of differences among the three therapies.

b. For these data, Tukey's HSD is 1.94. According to this criterion, therapy A is significantly different from both B and C, but therapies B and C are not significantly different from each other.

23. The null hypothesis states that there are no differences among the three treatments, H_0: $\mu_1 = \mu_2 = \mu_3$. The critical value for $\alpha = .05$ is 3.88.

SOURCE	SS	df	MS	
Between treatments	15	2	7.5	$F(2, 12) = 1.50$
Within treatments	60	12	5	
Total	75	14		

Fail to reject H_0. No evidence of any significant differences.

25.

NUMBER OF DAYS TO ADJUST TO JETLAG		
Westbound	$M = 2.50$	$SD = 1.05$
Eastbound	$M = 6.00$	$SD = 1.41$
Same time zone	$M = 0.50$	$SD = 0.55$

The null hypothesis states that the direction of flight has no effect on jeg lag. H_0: $\mu_1 = \mu_2 = \mu_3$. The critical value for $\alpha = .05$ is 3.68.

SOURCE	SS	df	MS	
Between treatments	93	2	46.50	$F(2, 15) = 41.15$
Within treatments	17	15	1.13	
Total	110	17		

With a critical value of $F = 3.68$ you should reject H_0.

CHAPTER 14   REPEATED-MEASURES ANOVA

---

1. A repeated-measures design generally uses fewer subjects and is more likely to detect a treatment effect because it eliminates variability due to individual differences.

3. The second stage of the repeated-measures ANOVA removes the variability due to individual differences from the error term. The individual differences are automatically eliminated from the numerator (between treatments) because the same subjects are used in all treatments. To keep the F-ratio balanced, the individual differences must also be removed from the denominator.

5. The F-ratio has $df = 3, 42$.

7. a. Two treatments
   b. One sample of $n = 39$ participants.

9. a.

SOURCE	SS	df	MS	
Between treatments	32	2	16	$F(2, 6) = 12.03$
Within treatments	14	9		
Between subjects	6	3		
Error	8	6	1.33	
Total	46	11		

With $df = 2, 6$, the critical value is 5.14. Reject the null hypothesis, and conclude that there are significant differences among the three treatments.

b. For these data, eta squared is $32/40 = 0.80$ or 80%.

11. a. The null hypothesis states that there are no differences among the three treatments, H_0: $\mu_1 = \mu_2 = \mu_3$. With $df = 2, 6$, the critical value is 5.14.

SOURCE	SS	df	MS	
Between treatments	8	2	4	$F(2, 6) = 0.27$
Within treatments	94	9		
Between subjects	4	3		
Error	90	6	15	
Total	102	11		

Fail to reject H_0. There are no significant differences among the three treatments.

b. In Problem 9 the individual differences were relatively large and consistent. When the individual differences were subtracted out, the error was greatly reduced.

13. a. Increasing the individual differences should have no effect on the F-ratio for a repeated-measures ANOVA. The variability within treatments will increase, but the variability between subjects will also increase so that the error variance is unchanged.

b.

SOURCE	SS	df	MS	
Between treatments	32	2	16	$F(2, 6) = 8.00$
Within treatments	84	9		
Between subjects	72	3		
Error	12	6	2	
Total	116	11		

The F-ratio is the same as in Problem 12.

15.

SOURCE	SS	df	MS	
Between treatments	24	1	24	$F = 4.00$
Within treatments	120	28		
Between subjects	36	14		
Error	84	14	6	
Total	144	29		

17.

SOURCE	SS	df	MS	
Between treatments	270	3	90	$F = 9$
Within treatments	410	28		
Between subjects	200	7		
Error	210	21	10	
Total	680	31		

19. a. $SS_{within} = 26 + 26 = 52$ **b.** $SS_{between\ subjects} = 52$

21. a. The null hypothesis states that the mean number of errors does not change across sessions, $H_0: \mu_1 = \mu_2 = \mu_3 = \mu_4$. The critical value is 4.76.

SOURCE	SS	df	MS	
Between treatments	18	3	6.00	$F(3, 6) = 8.96$
Within treatments	12	8		
Between subjects	8	2		
Error	4	6	0.67	
Total	30	11		

Reject H_0, and conclude that there is a significant practice effect.
b. For these data, eta squared is $18/22 = 0.82$ (or 82%).

23. a. The means are 6, 4, and 2.
b. The null hypothesis states that there is no change in motivation across the three grade levels, $H_0: \mu_1 = \mu_2 = \mu_3$. The critical value is 4.46.

SOURCE	SS	df	MS	
Between treatments	40.00	2	20	$F(2, 8) = 24.1$
Within treatments	26.00	12		
Between subjects	19.33	4		
Error	6.67	8	0.83	
Total	66.00	14		

Reject H_0. There are significant differences among the three grade levels.

25. a. The means and SS values are

First	Second	Third
$M = 94.6$	$M = 59.0$	$M = 34.8$
$SS = 831.2$	$SS = 1246.0$	$SS = 474.8$

The null hypothesis states that there is no change in looking behavior across the three presentations, $H_0: \mu_1 = \mu_2 = \mu_3$. For $\alpha = .01$, the critical value is 8.65.

SOURCE	SS	df	MS	
Between treatments	9048.4	2	4524.2	$F(2, 8) = 19.39$
Within treatments	2552.0	12		
Between subjects	685.06	4		
Error	1866.94	8	233.37	
Total	11600.4	14		

The data indicate a signficant change in the mean amount of time spent looking as the same stimulus is repeatedly presented.
b. For these data, eta squared is $9048.4/10,915.34 = 0.83$ (or 83%).

CHAPTER 15 **TWO-FACTOR ANALYSIS OF VARIANCE**

1. a. In analysis of variance, an independent variable is called a *factor*.
b. The values of a factor that are used to create the different groups or treatment conditions are called the *levels* of the factor.
c. A research study with two independent variables is called a *two-factor study*.

3. During the second stage of the two-factor ANOVA, the mean differences between treatments are analyzed into differences from each of the two main effects and differences from the interaction.

5. All F-ratios have the same error term, MS_{within}, and therefore have the same df value for the denominator.

7. a. 20   **b.** 0   **c.** 60

9. a.

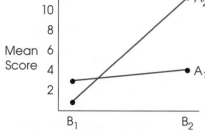

b. Overall the scores in A_1 are slightly lower than the scores in A_2. There may be a small main effect for factor A. There appears to be a large main effect for factor B with scores in B_2 much larger than scores in B_1. The lines are not parallel, so there appears to be an interaction.

c. The null hypotheses state that there is no mean difference between levels of factor A ($H_0: \mu_{A1} = \mu_{A2}$), no difference between levels of factor B ($H_0: \mu_{B1} = \mu_{B2}$), and no interaction. All F-ratios have $df = 1, 16$ and the critical value is $F = 4.49$.

SOURCE	SS	df	MS	
Between treatments	280	3		
A	20	1	20	$F(1, 16) = 1.00$
B	180	1	180	$F(1, 16) = 9.00$
$A \times B$	80	1	80	$F(1, 16) = 4.00$
Within treatments	320	16	20	
Total	600	19		

The A effect is not significant, but factor B does have a significant effect at the .05 level. The interaction does not reach significance at the .05 level.

11.

SOURCE	SS	df	MS	
Between treatments	280	7		
Achievement need	16	1	16	$F(1, 40) = 2.00$
Task difficulty	144	3	48	$F(3, 40) = 6.00$
Interaction	120	3	40	$F(3, 40) = 5.00$
Within treatments	320	40	8	
Total	600	47		

13.

SOURCE	SS	df	MS	
Between treatments	30	3		
Achievement need	12	1	12	$F(1, 40) = 4.00$
Task difficulty	3	1	3	$F(3, 40) = 1.00$
Interaction	15	1	15	$F(3, 40) = 5.00$
Within treatments	132	44	3	
Total	162	47		

15. The null hypotheses state that there is no mean difference between genders ($H_0: \mu_{MALE} = \mu_{FEMALE}$), that the amount of the chemical has no effect ($H_0: \mu_{NONE} = \mu_{SMALL} = \mu_{LARGE}$), and that there is no interaction. For $df = 1, 24$, the critical value is $F = 4.26$. For $df = 2, 24$, the critical value is $F = 3.40$.

SOURCE	SS	df	MS	
Between treatments	70	5		
A	30	1	30	$F(1, 24) = 6$
B	20	2	10	$F(2, 24) = 2$
$A \times B$	20	2	10	$F(2, 24) = 2$
Within treatments	120	24	5	
Total	190	29		

The results indicate that the males are significantly more active than the females. However, the data do not provide sufficient evidence to conclude that the chemical has a significant effect on activity or that the chemical has a different effect on males than on females.

17. a.

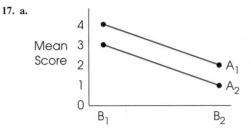

b. The lines are parallel, so there does not appear to be any interaction. There is a small difference between the lines for A_1 and A_2, so there may be a small main effect for factor A. The difference between B_1 and B_2 is larger, so there probably is a main effect for factor B.

c. The null hypotheses state that there is no difference between levels of factor A($H_0: \mu_{A1} = \mu_{A2}$), no difference between levels of factor B, ($H_0: \mu_{B1} - \mu_{B2}$), and no interaction. All F-ratios have $df = 1, 36$ and the critical value is $F = 4.11$.

SOURCE	SS	df	MS	
Between treatments	50	3		
A	10	1	10	$F(1, 36) = 2.00$
B	40	1	40	$F(1, 36) = 8.00$
$A \times B$	0	1	0	$F(1, 36) = 0$
Within treatments	180	36	5	
Total	230	39		

The main effect for factor B is significant, but the A effect and the interaction are not.

19. a. The null hypotheses state that counseling has no effect ($H_0: \mu_{YES} = \mu_{NO}$), the drug has no effect ($H_0: \mu_{YES} = \mu_{NO}$), and there is no interaction. All F-ratios have $df = 1, 36$ and the critical value is $F = 4.11$.

SOURCE	SS	df	MS	
Between treatments	200	3		
A (counseling)	0	1	0	$F(1, 36) = 0$
B (drug)	160	1	160	$F(1, 36) = 32.00$
$A \times B$	40	1	40	$F(1, 36) = 8.00$
Within treatments	180	36	5	
Total	380	39		

b. The data indicate that overall the drug has a significant effect. The main effect for counseling is not significant. However, there is a significant interaction, meaning that the effect of the drug is significantly affected by counseling. Without counseling the drug has a minimal effect, but with counseling the drug has a big effect.

21. a. The cell means are as follows:

	B_1	B_2	B_3
A_1	2.25	4.00	8.00
A_2	7.00	4.00	2.50

b. There appears to be no main effect for factor A (the means are nearly equal) and only a small effect for factor B. There does appear to be an interaction because the means increase across levels of B for A_1, but the means decrease across B for A_2.

c. The null hypotheses state that factor A has no effect ($\mu_{A1} = \mu_{A2}$), factor B has no effect ($\mu_{B1} = \mu_{B2} = \mu_{B3}$), and that there is no interaction. For $df = 1,18$ and $\alpha = .01$, the critical value is $F = 8.28$, and for $df = 2, 18$ and $\alpha = .01$ the critical value is $F = 6.01$.

SOURCE	SS	df	MS	
Between treatments	111.875	5		
A	0.375	1	0.375	$F(1, 18) = 0.09$
B	6.25	2	3.125	$F(2, 18) = 0.78$
$A \times B$	105.25	2	52.625	$F(2, 18) = 13.19$
Within treatments	71.75	18	3.99	
Total	183.625	23		

The A and B main effects are not significant, but there is a significant interaction.

23. The null hypotheses state that gender has no effect ($H_0: \mu_{MALE} = \mu_{FEMALE}$), the drug dosage has no effect ($H_0: \mu_{NONE} = \mu_{SMALL} = \mu_{LARGE}$), and there is no interaction. For $df = 1, 24$, the critical value is $F = 4.26$ and for $df = 2, 24$, the critical value is $F = 3.40$.

SOURCE	SS	df	MS	
Between treatments	130	5		
A	30	1	30	$F(1, 24) = 6$
B	20	2	10	$F(2, 24) = 2$
$A \times B$	80	2	40	$F(2, 24) = 8$
Within treatments	120	24	5	
Total	250	29		

The significant interaction indicates that the hormone affects the eating behavior of males differently than females. For males, small doses appear to increase eating relative to no drug or a large dose. The test of simple main effects indicates that the drug effect is significant for males, $F(2, 24) = 7.00, p < .05$. For females, the drug (in any dose) appears to inhibit eating, but a test of simple mean effects indicates that the effect is not significant, $F(2, 24) = 3.00, p > .05$.

25. a. The null hypotheses state that sleeping has no effect ($H_0: \mu_{ASLEEP} = \mu_{AWAKE}$), the test delay level has no effect ($H_0: \mu_{2\text{-}HOUR} = \mu_{8\text{-}HOUR}$), and there is no interaction. For $\alpha = .05$ and $df = 1, 36$, the critical value is $F = 4.11$.

SOURCE	SS	df	MS	
Between treatments	200.48	3		
A (asleep/awake)	172.23	1	172.23	$F(1, 36) = 73.92$
B (delay time)	7.23	1	7.23	$F(1, 36) = 3.10$
$A \times B$	21.02	1	21.02	$F(1, 36) = 9.02$
Within treatments	83.90	36	2.33	
Total	284.38	39		

The significant interaction indicates that the amount you forget over time (2 hours versus 8 hours) depends on whether you are asleep or awake. For the asleep group, there was no evidence for forgetting over time and the scores were generally high. For the awake group, there was a substantial drop between 2 hours and 8 hours and the scores were generally low.

CHAPTER 16   CORRELATION AND REGRESSION

1. A positive correlation indicates that X and Y change in the same direction: As X increases, Y also increases. A negative correlation indicates that X and Y tend to change in opposite directions: As X increases, Y decreases.

3. The Pearson correlation measures the degree of linear relationship. The Spearman correlation measures the degree to which the relationship is monotonic or one directional independent of form.

5. For set I, $SP = 10$. For set II, $SP = -11$.

7. a.

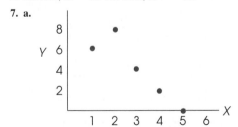

b. Estimate a strong negative correlation, probably $-.8$ or $-.9$.

c. $SS_X = 10$, $SS_Y = 40$, $SP = -18$, $r = -0.90$.

9. a. The correlation between X and Y is $r = +0.60$.
b. The correlation between Y and Z is $r = +0.60$.
c. Based on the answers to part (a) and part (b) you might expect a fairly strong positive correlation between X and Z.
d. The correlation between X and Z is $r = -.20$.

e. Simply because two variables are both related to a third variable does not necessarily imply that they are related to each other.

11. For these data, $SS_{1st} = 64$, $SS_{2nd} = 100$, $SP = 76$, and the Pearson correlation is $r = 0.95$.

13. For these data $SS_{anxiety} = 18$, $SS_{exam} = 72$, $SP = -32$, and the Pearson correlation is $r = -0.889$.

15. With $n = 25$ the correlation must be greater than 0.396 to be significant at the .05 level. Reject H_0, and conclude that these data provide evidence for a nonzero correlation in the population.

17. The ranked values are as follows:

X RANKS	Y RANKS	
2	2	The Spearman correlation is
1	1	$r_s = +0.90$.
3	4	
4	3	
5	5	

19. a. $r_s = +0.907$
b. Yes, there is a strong positive relationship between the grades assigned by the two instructors. The critical value for $\alpha = .05$ is 0.618, so the data indicate a statistically significant Spearman correlation.

21. a.

X	Y
1	9
1	7
1	6
1	10
0	4
0	7
0	3
0	6

  b. $r = 0.688$

23. a. The converted data show 5 individuals with scores of $X = 1$ and $Y = 1$, 3 individuals with $X = 1$ and $Y = 0$, 1 individual with $X = 0$ and $Y = 1$, and 3 individuals with $X = 0$ and $Y = 0$.

  b. $r = 0.354$ (the sign is meaningless)

25. a. $SS_X = 10$, $SS_Y = 108$, $SP = 30$, $r = +0.913$.

  b. The regression equation is $Y = 3X - 3$.

  c. and **d.** For each X, the predicted Y value and the error are

X	Y (PREDICTED)	ERROR	
1	0	2	$SS_{error} = 18$
4	9	−2	
3	6	−1	
2	3	−2	
5	12	2	
3	6	1	

  e. $SS_{error} = (1 - r^2)SS_Y = (0.166)(108) = 17.93$ (within rounding error of 18)

27. a. The standard error of estimate is 4.00.

  b. The standard error of estimate is 0.80.

CHAPTER 17   **CHI-SQUARE TESTS**

1. Nonparametric tests make few if any assumptions about the populations from which the data are obtained. For example, the populations do not need to form normal distributions, nor is it required that different populations in the same study have equal variances (homogeneity-of-variance assumption). Parametric tests require data measured on an interval or ratio scale. For nonparametric tests, any scale of measurement is acceptable.

3. H_0 states that the distribution of automobile accidents is the same as the distribution of registered drivers: 16% under age 20, 28% age 20–29, and 56% age 30 or older. With $df = 2$, the critical value is 5.99. The expected frequencies for these three categories are 48, 84, and 168. Chi-square = 13.76. Reject H_0, and conclude that the distribution of automobile accidents is not identical to the distribution of registered drivers.

5. The null hypothesis states that there are no preferences among the four colors; $p = 1/4$ for each color. With $df = 3$, the critical value is 7.81. The expected frequencies are $f_e = 15$ for all categories, and chi-square = 5.73. Fail to reject H_0, and conclude that there are no significant preferences.

7. The null hypothesis states that there is no preference among the three photographs: $p = 1/3$ for all categories. The expected frequencies are $f_e = 50$ for all categories, and chi-square = 20.28. With $df = 2$, the critical value is 5.99. Reject H_0, and conclude that there are significant preferences.

9. The null hypothesis states that the grade distribution for last semester has the same proportions as it did in 1985. For a sample of $n = 200$, the expected frequencies are 28, 52, 62, 38, and 20 for grades A, B, C, D, and F, respectively. With $df = 4$, the critical value for chi-square is 9.49. For these data, the chi-square statistic is 6.68. Fail to reject H_0, and conclude that there is no evidence that the distribution has changed.

11. a. The null hypothesis states that there is no preference among the three colas; one-third of the population prefers each of the three colas. For a sample of $n = 120$, the expected frequency is 40 for each cola. For $df = 2$ and $\alpha = .05$, the critical value for chi-square is 5.99. For these data, the chi-square statistic is

10.40. Reject H_0, and conclude that there are significant preferences among the colas.

  b. A larger sample should be more representative of the population. If the sample continues to be different from the hypothesis as the sample size increases, eventually the difference will be significant.

13. The null hypothesis states that the distribution of preferences is the same for both groups of students (the two variables are independent). With $df = 2$, the critical value is 5.99. The expected frequencies are:

	BOOK 1	BOOK 2	BOOK 3
Upper half	26.0	20.5	13.5
Lower half	26.0	20.5	13.5

Chi-square = 17.32. Reject H_0.

15. The null hypothesis states that there is no relationship between allergies and childhood pets. With $df = 2$ and $\alpha = .05$, the critical value is 5.99. The expected frequencies are:

	NUMBER OF PETS		
	0	1	2 OR MORE
No allergies	30	45	15
Allergies	10	15	5

The chi-square statistic is 13.16. Reject H_0 with $\alpha = .05$ and $df = 2$.

17. The null hypothesis states that littering is independent of the amount of litter already on the ground. With $df = 2$, the critical value is 5.99. The expected frequencies are:

	AMOUNT OF LITTER		
	NONE	SMALL	LARGE
Litter	31.25	31.25	31.51
Not litter	87.75	87.75	88.49

Chi-square = 22.43. Reject H_0.

19. The null hypothesis states that there is no relationship between group size and helping behavior. With $df = 2$, the critical value is 9.21. The expected frequencies are:

	NUMBER IN GROUP		
	2	3	6
Help	7.75	15.50	7.75
Not help	5.25	10.50	5.25

Chi-square = 7.91. Find to reject H_0, and conclude that there is no relationship between helping and group size.

21. The null hypothesis states that there is no relationship between handedness and eye preference. With $df = 1$ and $\alpha = .01$, the critical value is 6.63. The expected frequencies for left-handed participants are 12 left eye and 18 right eye. For right-handed

participants the expected frequencies are 48 for left eye and 72 for right eye. Chi-square = 11.11. There is a significant relationship between handedness and eye preference.

23. a. The null hypothesis states that the distribution of opinions is the same for those who live in the city and those who live in the suburbs. For $df = 1$ and $\alpha = .05$, the critical value for chi-square is 3.84. The expected frequencies are:

	FAVOR	OPPOSE
City	51.33	48.67
Suburb	102.67	97.33

For these data, chi-square = 16.68. Reject H_0, and conclude that opinions in the city are different from those in the suburbs.
b. Using Equation 17.6, the phi coefficient is 0.236.

CHAPTER 18   THE BINOMIAL TEST

1. H_0: $p = .08$ (still 8% learning disabled). The critical boundaries are $z = \pm1.96$. With $X = 42$, $\mu = 24$, and $\sigma = 4.70$, we obtain $z = 3.83$. Reject H_0, and conclude that there has been a significant change in the proportion of students classified as learning disabled.

3. H_0: $p = .25$ (the general population has the same proportion of belief as the psychotherapists). The critical boundaries are $z = \pm1.96$. With $X = 65$, $\mu = 48$, and $\sigma = 6$, we obtain $z = 2.83$. Reject H_0, and conclude that the proportion of belief is significantly different for the general population and for psychotherapists.

5. H_0: $p = .20$ (correct predictions are just due to chance). The critical boundaries are $z = \pm1.96$. With $X = 15$, $\mu = 10$, and $\sigma = 2.83$, we obtain $z = 1.77$. Fail to reject H_0, and conclude that this level of performance is not significantly different from that expected by chance.

7. H_0: $p = q = 1/2$ (no preference). The critical boundaries are $z = \pm1.96$. With $X = 39$, $\mu = 32$, and $\sigma = 4$, we obtain $z = 1.75$. Fail to reject H_0. There is no evidence of a significant preference for either of the two drinks.

9. a. With $n = 20$ and $p = q = 1/2$, you would need at least $X = 15$ to have a z-score above the critical boundary of $z = 1.96$ (for a two-tailed test).
b. With $n = 40$ and $p = q = 1/2$, you would need at least $X = 27$ to have a z-score above 1.96.
c. With $n = 100$ and $p = q = 1/2$, you would need at least $X = 60$ to have a z-score above 1.96.

11. H_0: $p = .20$ (still 20% after the training). The critical boundaries are $z = \pm1.96$. With $X = 2$, $\mu = 12$, and $\sigma = 3.10$, we obtain $z = -3.23$. Reject H_0, and conclude that there has been a significant change in the proportion of students who fail the exam.

13. H_0: p(female) $= .80$ (still 80% female). The critical boundaries are $z = \pm1.96$. With $X = 43$, $\mu = 60$, and $\sigma = 3.46$, we obtain

$z = -4.91$. Reject H_0, and conclude that there has been a significant change in the proportion.

15. H_0: p(book A) $= .50$ (no preference). The critical boundaries are $z = \pm2.58$. With $X = 19$, $\mu = 12.5$, and $\sigma = 2.50$, we obtain $z = 2.60$. Reject H_0, and conclude that there is a significant preference.

17. H_0: $p = q = 1/2$ (no difference between the two stimulus presentations). The critical boundaries are $z = \pm2.58$. The binomial distribution has $\mu = 15$ and $\sigma = 2.74$. With $X = 22$, we obtain $z = 2.55$. Fail to reject H_0, and conclude that there is no significant difference between the first and second presentations.

19. H_0: $p = q = 1/2$ (any change in grade point average is due to chance). The critical boundaries are $z = \pm2.58$. The binomial distribution has $\mu = 22.5$ and $\sigma = 3.35$. With $X = 31$, we obtain $z = 2.54$. Fail to reject H_0, and conclude that there is no significant change in grade point average after the workshop.

21. H_0: $p = q = 1/2$ (people are just guessing). The critical boundaries are $z = \pm1.96$. (For a one-tailed test, the boundary is $z = -1.65$.) With $X = 27$, $\mu = 22.5$, and $\sigma = 3.35$, we obtain $z = 1.34$. Reject H_0. Performance is not significantly better than expected by chance.

23. H_0: $p = q = 1/2$ (any difference between the two paragraphs is due to chance). The critical boundaries are $z = \pm1.96$. The binomial distribution has $\mu = 18$ and $\sigma = 3$. With $X = 25$, we obtain $z = 2.33$. Reject H_0, and conclude that there is a significant change in performance.

25. H_0: $p = q = 1/2$ (no discrimination). The critical boundaries are $z = \pm1.96$. The binomial distribution has $\mu = 12.5$ and $\sigma = 2.50$. With $X = 15$, we obtain $z = 1.00$. Fail to reject H_0, and conclude that there is no significant evidence of discrimination based on taste.

CHAPTER 19   **TESTS FOR ORDINAL DATA**

1. The ranks for the 10 subjects are as follows:

Subject:	A	B	C	D	E	F	G	H	I	J
Rank:	1	3	3	3	5	6	7	8.5	8.5	10

3. If one or more set of scores is extremely variable, it may violate the homogeneity of variance assumption and can create a very large error term for an F-ratio or t statistic. Also, if the original scores include an undetermined value, it is impossible to compute a mean or variance. In these cases, ranking often can eliminate the problem.

5. The null hypothesis states that there is no systematic difference between the scholarship students and other students. The critical value is 11. For the scholarship group, $\Sigma R = 32.5$ and for the non-scholarship group, $\Sigma R = 103.5$. $U = 11.5$. Fail to reject H_0, and conclude there is not a significant difference between the two groups.

7. The null hypothesis states that there is no systematic difference between the two treatments. The critical value is 5. For treatment 1, $\Sigma R = 51$ and for treatment 2, $\Sigma R = 27$. $U = 6$. Fail to reject H_0, and conclude there is no significant difference between the two treatments.

9. The null hypothesis states that there is no difference between the two species. The critical value is $U = 23$. For species A, $\Sigma R = 73.5$ and for species B, $\Sigma R = 136.5$. $U = 18.5$. Reject H_0, and conclude there is a significant difference in eating behavior.

11. The null hypothesis states that there is no difference between the two sections. The critical value is $U = 36$. For the morning section, $U = 53$ and for the afternoon section, $U = 87$. There is no significant difference between the two sections.

13. The null hypothesis states that there is no difference between the two paints. The critical value is $U = 5$. For the paint company, $\Sigma R = 28$ and for the competitor, $\Sigma R = 50$. Mann-Whitney $U = 7$. Fail to reject H_0. These data do not provide evidence for a significant difference between the two paints.

15. With $n = 12$, the critical value for the Wilcoxon test is $T = 13$. Because only one participant showed a decrease, the sum of the ranks for the decreases cannot be larger than 12 (the single individual must have a rank between 1 and 12). Therefore, the data must produce a Wilcoxon T that is less than 13, and the decision must be to reject the null hypothesis.

17. **a.** The difference scores and ranks are as follows:

SUBJECT	DIFFERENCE	RANK
A	11	8
B	2	2.5
C	18	10
D	7	6
E	−4	4
F	2	2.5
G	14	9
H	9	7
I	5	5
J	−1	1

b. The null hypothesis states that there is no treatment effect. The critical value is $T = 8$. For increases, $\Sigma R = 5$ and for decreases, $\Sigma R = 50$. $T = 5$. Reject H_0.

19. The null hypothesis states that the number of letters has no effect on solving anagrams. The critical value is 0. For the increases, $\Sigma R = 20$ and for the decreases, $\Sigma R = 1$. Wilcoxon $T = 1$. Fail to reject H_0, and conclude that the number of letters does not affect performance.

21. The null hypothesis states that the drug has no effect on maze-learning performance. The critical value is $T = 13$. For the increases, $\Sigma R = 73$ and for the decreases, $\Sigma R = 5$. Wilcoxon $T = 5$. Reject H_0, and conclude that the drug does affect maze-learning performance.

23. The null hypothesis states that there are no systematic differences among the three levels of attractiveness. With $df = 2$, the critical value for chi-square is 5.99. The totals of the ranks for the three levels are $T_1 = 65$, $T_2 = 109$, and $T_3 = 126$. For these data, the Kruskal-Wallis $H = 4.96$. Fail to reject the null hypothesis, and conclude that there are no significant differences among the levels of attractiveness.

25. The null hypothesis states that there are no systematic differences among the three networks. With $df = 2$, the critical value for chi-square is 5.99. The totals of the ranks for the three groups are $T_1 = 48$, $T_2 = 72$, and $T_3 = 90$. For these data, the Kruskal-Wallis $H = 2.19$. Fail to reject the null hypothesis, and conclude that there are no significant differences among the networks.

27. The null hypothesis states that there are no systematic differences among the three personality types. With $df = 2$, the critical value for chi-square is 9.21. The totals of the ranks for the three groups are $T_1 = 116$, $T_2 = 73$, and $T_3 = 136$. For these data, the Kruskal-Wallis $H = 4.58$. Fail to reject the null hypothesis, and conclude that there are no significant differences among the three types.

Statistics Organizer

The following pages present an organized summary of the statistical procedures covered in this book. This organizer is divided into four sections, each of which groups together statistical techniques that serve a common purpose. The four groups are

I. Descriptive Statistics

II. Parametric Tests for Means and Mean Differences

III. Nonparametric Tests

IV. Measures of Relationship Between Two Variables

Each of the four sections begins with a general overview that discusses the purpose for the statistical techniques that follow and points out some common characteristics of the different techniques. Next, there is a decision map that leads you, step by step, through the task of deciding which statistical technique is appropriate for the data you wish to analyze. Finally, there is a brief description of each technique and the necessary formulas.

I DESCRIPTIVE STATISTICS

The purpose of descriptive statistics is to simplify and organize a set of scores. Scores may be organized in a table or graph, or they may be summarized by computing one or two values that describe the entire set. The most commonly used descriptive techniques are as follows:

A. Frequency Distribution Tables and Graphs

A frequency distribution is an organized tabulation of the number of individuals in each category on the scale of measurement. A frequency distribution can be presented as either a table or a graph. The advantage of a frequency distribution is that it presents the entire set of scores rather than condensing the scores into a single descriptive value. The disadvantage of a frequency distribution is that it can be somewhat complex, especially with large sets of data.

B. Measures of Central Tendency

The purpose of measuring central tendency is to identify a single score that represents an entire data set. The goal is to obtain a single value that is the best example of the average, or most typical, score from the entire set.

CHOOSING DESCRIPTIVE STATISTICS: A DECISION MAP

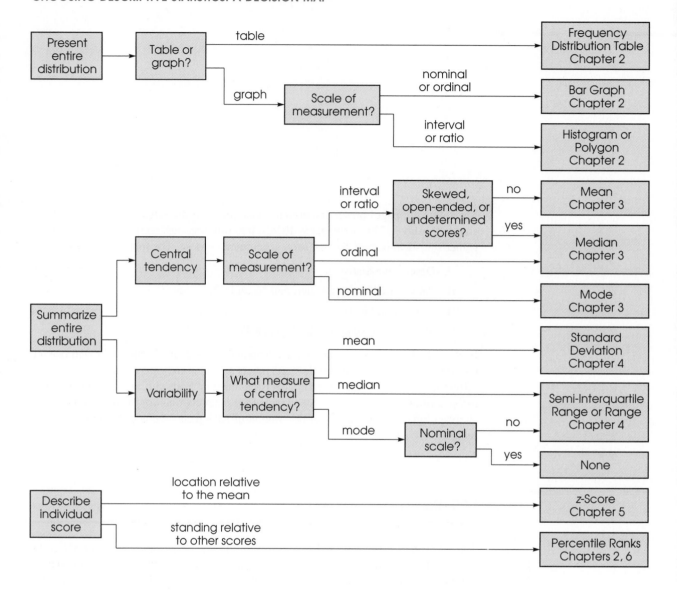

Measures of central tendency are used to describe a single data set, and they are the most commonly used measures for comparing two (or more) different sets of data.

C. Measures of Variability

Variability is used to provide a description of how spread out the scores are in a distribution. It also provides a measure of how accurately a single score selected from a distribution represents the entire set.

D. *z*-Scores

Most descriptive statistics are intended to provide a description of an entire set of scores. However, *z*-scores are used to describe individual scores within a distribution. The purpose of a *z*-score is to identify the precise location of an individual within a distribution by using a single number.

1. The Mean (Chapter 3)

The mean is the most commonly used measure of central tendency. It is computed by finding the total (ΣX) for the set of scores and then dividing the total by the number of individuals. Conceptually, the mean is the amount each individual will receive if the total is divided equally.

$$\text{Population: } \mu = \frac{\Sigma X}{N} \qquad \text{Sample: } M = \frac{\Sigma X}{n}$$

2. The Median (Chapter 3)

Exactly 50% of the scores in a data set have values less than or equal to the median. The median is the 50th percentile. The median usually is computed for data sets when the mean cannot be found (undetermined scores, open-ended distribution) or when the mean does not provide a good, representative value (ordinal scale, skewed distribution).

List the scores in order from smallest to largest.
a. With an odd number of scores, the median is the middle score.
b. With an even number of scores, the median is the average of the middle two scores.
c. With several scores tied at the median, use interpolation to find the 50th percentile.

3. The Mode (Chapter 3)

The mode is the score with the greatest frequency. The mode is used when the scores consist of measurements on a nominal scale.

No calculation. Simply count the frequency of occurrence for each different score.

4. The Range (Chapter 4)

The range is the distance from the lowest to the highest score in a data set. The range is considered to be a relatively crude measure of variability.

Find the upper real limit for the largest score and the lower real limit for the smallest score. The range is the difference between these two real limits.

5. The Semi-Interquartile Range (Chapter 4)

The semi-interquartile range is one-half of the range covered by the middle 50% of the distribution. The semi-interquartile range is often used to measure variability in situations where the median is used to report central tendency.

Find the first quartile (25th percentile) and the third quartile (75th percentile). The semi-interquartile range is one-half of the distance between the two quartiles.

6. Standard Deviation (Chapter 4)

The standard deviation is a measure of the standard distance from the mean. Standard deviation is obtained by first computing SS (the sum of squared deviations) and variance (the mean squared deviation). Standard deviation is the square root of variance.

Sum of Squares

$$\text{Definitional: } SS = \Sigma(X - \mu)^2$$
$$\text{Computational: } SS = \Sigma X^2 - \frac{(\Sigma X)^2}{N}$$

Variance

$$\text{Population: } \sigma^2 = \frac{SS}{N}$$
$$\text{Sample: } s^2 = \frac{SS}{n-1}$$

Standard Deviation

$$\text{Population: } \sigma = \sqrt{\frac{SS}{N}}$$
$$\text{Sample: } s = \sqrt{\frac{SS}{n-1}}$$

7. z-Scores (Chapter 5)

The sign of a z-score indicates whether an individual is above ($+$) or below ($-$) the mean. The numerical value of the z-score indicates how many standard deviations there are between the score and the mean.

$$z = \frac{X - \mu}{\sigma}$$

II PARAMETRIC TESTS: INFERENCES ABOUT POPULATION MEANS OR MEAN DIFFERENCES

All of the hypothesis tests covered in this section use the means obtained from sample data as the basis for testing hypotheses about population means. Although there are a variety of tests used in a variety of research situations, all use the same basic logic, and all of the test statistics have the same basic structure. In each case, the test statistic (z, t, or F) involves computing a ratio with the following structure:

$$\text{test statistic} = \frac{\text{obtained difference between sample means}}{\text{mean difference expected by chance}}$$

The goal of each test is to determine whether the observed sample mean differences are larger than expected by chance. In general terms, a *significant result* means that the results obtained in a research study (sample differences) are more than would be expected by chance. In each case, a large value for the test statistic ratio indicates a significant result; that is, when the actual difference between sample means (numerator) is substantially larger than chance (denominator), you will obtain a large ratio, which indicates that the sample difference is significant.

The actual calculations differ slightly from one test statistic to the next, but all involve the same basic computations.

1. A set of scores (sample) is obtained for each population or each treatment condition.

2. The mean is computed for each set of scores, and some measure of variability (SS, standard deviation, or variance) is obtained for each individual set of scores.

3. The differences between the sample means provide a measure of how much difference exists between treatment conditions. Because these mean differences may be caused by the treatment conditions, they are often called *systematic* or *predicted*. These differences are the numerator of the test statistic.

4. The variability within each set of scores provides a measure of unsystematic or unpredicted differences due to chance. Because the individuals within each treatment condition are treated exactly the same, there is nothing that should cause their scores to be different. Thus, any observed differences (variability) within treatments are assumed to be due to error or chance.

CHOOSING A PARAMETRIC TEST: A DECISION MAP FOR MAKING INFERENCES
ABOUT POPULATION MEANS OR MEAN DIFFERENCES

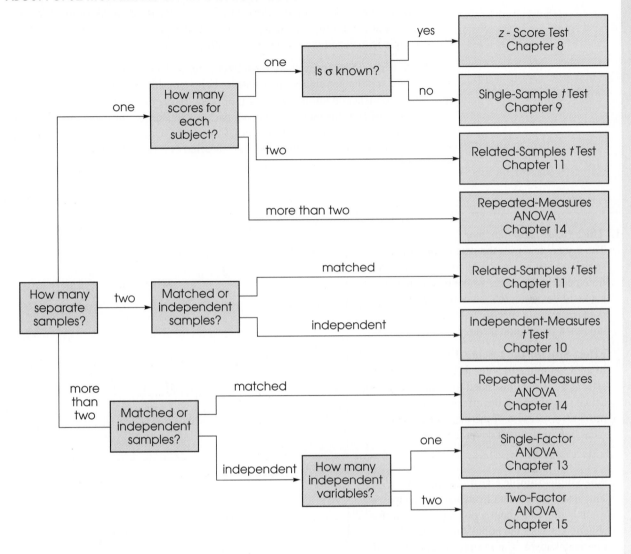

With these considerations in mind, each of the test statistic ratios can be described
as follows:

$$\text{test statistic} = \frac{\text{differences (variability) between treatments}}{\text{differences (variability) within treatments}}$$

The hypothesis tests reviewed in this section apply to three basic research designs:

1. **Single-Sample Designs.** Data from a single sample are used to test a hypothesis
   about a single population.

2. **Independent-Measures Designs.** A separate sample is obtained to represent each
   individual population or treatment condition.

3. Related-Samples Designs. Related-sample designs include repeated-measures and matched-subjects designs. For a repeated-measures design, there is only one sample, with each individual subject being measured in all of the different treatment conditions. In a matched-subjects design, every individual in one sample is matched with a subject in each of the other samples.

Finally, you should be aware that all parametric tests place stringent restrictions on the sample data and the population distributions being considered. First, these tests all require measurements on an interval or a ratio scale (numerical values that allow you to compute means and differences). Second, each test makes assumptions about population distributions and sampling techniques. Consult the appropriate section of this book to verify that the specific assumptions are satisfied before proceeding with any parametric test.

---

1. The z-Score Test (Chapter 8)
The z-score test uses the data from a single sample to test a hypothesis about the population mean in situations where the population standard deviation (σ) is known. The null hypothesis states a specific value for the unknown population mean.

$$z = \frac{M - \mu}{\sigma_M} \qquad \text{where } \sigma_M = \frac{\sigma}{\sqrt{n}}$$

Effect Size for the z-Score Test
In addition to the hypothesis test, a measure of effect size is recommended to determine the actual magnitude of the effect. Cohen's d provides a measure of the mean difference measured in standard deviation units.

$$d = \frac{M - \mu}{\sigma}$$

---

2. The Single-Sample *t* Test (Chapter 9)
This test uses the data from a single sample to test a hypothesis about a population mean in situations where the population standard deviation is unknown. The sample variance is used to estimate the unknown population variance. The null hypothesis states a specific value for the unknown population mean.

$$t = \frac{M - \mu}{s_M} \qquad \text{where } s_M = \sqrt{\frac{s^2}{n}}$$

$$df = n - 1$$

Effect Size for the Single-Sample *t*
The effect size for the single-sample t test can be described by either Cohen's d or by r^2, which measures the percentage of variability that is accounted for by the treatment effect.

$$d = \frac{M - \mu}{s}$$

$$r^2 = \frac{t^2}{t^2 + df}$$

3. The Independent-Measures t Test

(Chapter 10)
The independent-measures t test uses data from two separate samples to test a hypothesis about the difference between two population means. The variability within the two samples is combined to obtain a single (pooled) estimate of population variance. The null hypothesis states that there is no difference between the two population means.

$$t = \frac{(M_1 - M_2) - (\mu_1 - \mu_2)}{s_{(M_1-M_2)}} \quad \text{where } s_{(M_1-M_2)} = \sqrt{\frac{s_p^2}{n_1} + \frac{s_p^2}{n_2}}$$

$$\text{and } s_p^2 = \frac{SS_1 + SS_2}{df_1 + df_2}$$

$$df = df_1 + df_2 = (n_1 - 1) + (n_2 - 1)$$

Effect Size for the Independent-Measures t

The effect size for the independent-measures t test can be described by either Cohen's d or by r^2, which measures the percentage of variability that is accounted for by the treatment effect.

$$d = \frac{M_1 - M_2}{s_p}$$

$$r^2 = \frac{t^2}{t^2 + df}$$

4. The Related-Samples t Test (Chapter 11)

This test evaluates the mean difference between two treatment conditions using the data from a repeated-measures or a matched-subjects experiment. A difference score (D) is obtained for each subject (or each matched pair) by subtracting the score in treatment 1 from the score in treatment 2. The variability of the sample difference scores is used to estimate the population variability. The null hypothesis states that the population mean difference (μ_D) is zero.

$$t = \frac{M_D - \mu_D}{s_{M_D}} \quad \text{where } s_{M_D} = \sqrt{\frac{s^2}{n}}$$

$$df = n - 1$$

Effect Size for the Related-Samples t

The effect size for the repeated-measures (related samples) t test can be described by either Cohen's d or by r^2, which measures the percentage of variability that is accounted for by the treatment effect.

$$d = \frac{M_D}{s}$$

$$r^2 = \frac{t^2}{t^2 + df}$$

5. Single-Factor, Independent-Measures Analysis of Variance (Chapter 13)

This test uses data from two or more separate samples to test for mean differences among two or more populations. The null hypothesis states that there are no differences among the population means. The test statistic is an F-ratio that uses the variability between treatment conditions (sample mean differences) in the numerator and the variability within treatment conditions (error variability) as the denominator. With only two samples, this test is equivalent to the independent-measures t test.

(continues)

Total
$$SS = \Sigma X^2 - \frac{G^2}{N}$$
$$df = N - 1$$

Between Treatments
$$SS = \Sigma \frac{T^2}{n} - \frac{G^2}{N}$$
$$df = k - 1$$
$$MS = \frac{SS}{df}$$

Within Treatments
$$SS = \Sigma SS_{\text{each treatment}}$$
$$df = N - k$$
$$MS = \frac{SS}{df}$$

$$F\text{-ratio} = \frac{MS_{\text{between treatments}}}{MS_{\text{within treatments}}}$$

Effect Size for Analysis of Variance
The effect size for an analysis of variance is described by a measure of the percentage of variability that is accounted for by the treatment effect. In the context of ANOVA, the percentage is called η^2 (eta squared) instead of r^2.

$$\eta^2 = \frac{SS_{\text{between treatments}}}{SS_{\text{total}}}$$

6. Single-Factor, Repeated-Measures Analysis of Variance (Chapter 14)
This test is used to evaluate mean differences among two or more treatment conditions using sample data from a repeated-measures (or matched-subjects) experiment. The null hypothesis states that there are no differences among the population means. The test statistic is an F-ratio using variability between treatment conditions (mean differences) in the numerator exactly like the independent-measures ANOVA. The denominator of the F-ratio (error term) is obtained by measuring variability within treatments and then subtracting out the variability between subjects. The research design and the test statistic remove variability due to individual differences and thereby provide a more sensitive test for treatment differences than is possible with an independent-measures design.

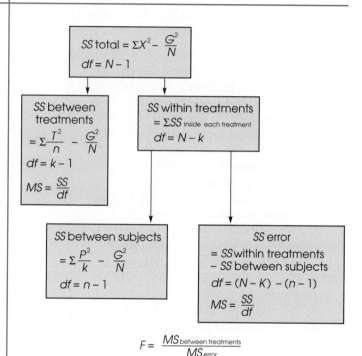

Effect Size for Repeated-Measures Analysis of Variance
The effect size for the repeated-measures ANOVA is described by η^2 (eta squared), which measures the percentage of variability that is accounted for by the treatment effect. The percentage is computed after other explained sources of variability have been removed.

$$\eta^2 = \frac{SS_{\text{between treatments}}}{SS_{\text{total}} - SS_{\text{between subjects}}}$$

7. Two-Factor, Independent-Measures Analysis of Variance (Chapter 15)

This test is used to evaluate mean differences among populations or treatment conditions using sample data from research designs with two independent variables (factors). The two-factor ANOVA tests three separate hypotheses: mean differences among the levels of factor A (main effect for factor A), mean differences among the levels of factor B (main effect for factor B), and mean differences resulting from specific combinations of the two factors (interaction). Each of the three separate null hypotheses states that there are no population mean differences. Each of the three tests uses an F-ratio as the test statistic, with the variability between samples (sample mean differences) in the numerator and the variability within samples (error variability) in the denominator.

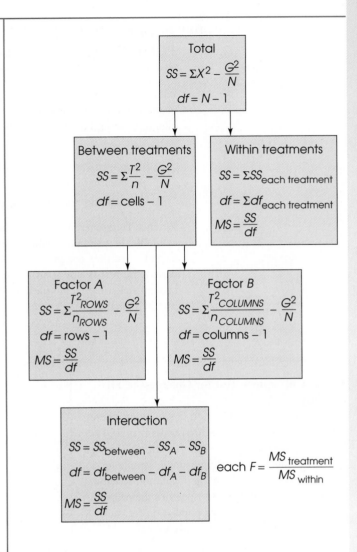

Effect Size for the Two-Factor Analysis of Variance

The effect size for each main effect (A and B) and for the interaction are described by η^2 (eta squared), which measures the percentage of variability that is accounted for by the specific treatment effect. The percentage is computed after other explained sources of variability have been removed.

For example, for factor A:

$$\eta^2 = \frac{SS_A}{SS_{\text{total}} - SS_B - SS_{A \times B}}$$

III   NONPARAMETRIC TESTS: INFERENCES ABOUT POPULATIONS OR POPULATION DIFFERENCES

Although the parametric tests described in the previous section are the most commonly used inferential techniques, there are many research situations in which parametric tests cannot or should not be used. These situations fall into two general categories:

1. The data involve measurements on nominal or ordinal scales. In these situations, you cannot compute the means and variances that are an essential part of parametric tests.

2. The data do not satisfy the assumptions underlying parametric tests.

When a parametric test cannot be used, there is usually a nonparametric alternative available. In general, parametric tests are more powerful than their nonparametric counterparts, and they are preferred over the nonparametric alternatives. However, when a parametric test is not appropriate, the nonparametric tests provide researchers with a backup statistical technique for conducting an analysis and statistical interpretation of research results.

---

1. The Chi-Square Test for Goodness of Fit
(Chapter 17)
This chi-square test is used in situations where the measurement procedure results in classifying individuals into distinct categories. The test uses frequency data from a single sample to test a hypothesis about the population distribution. The null hypothesis specifies the proportion or percentage of the population for each category on the scale of measurement.

$$\chi^2 = \Sigma \frac{(f_o - f_e)^2}{f_e} \qquad \text{where } f_e = pn$$

$$df = C - 1$$

---

2. The Chi-Square Test for Independence
(Chapter 17)
This test serves as an alternative to the independent-measures t test (or ANOVA) in situations where the dependent variable involves classifying individuals into distinct categories. The sample data consist of frequency distributions (proportions across categories) for two or more separate samples. The null hypothesis states that the separate populations all have the same proportions (same shape). That is, the proportions across categories are independent of the different populations.

$$\chi^2 = \Sigma \frac{(f_o - f_e)^2}{f_e}$$

$$\text{where } f_e = \frac{(\text{row total})(\text{column total})}{n}$$

$$df = (R - 1)(C - 1)$$

CHOOSING A NONPARAMETRIC TEST: A DECISION MAP FOR MAKING INFERENCES ABOUT POPULATIONS OR POPULATION DIFFERENCES

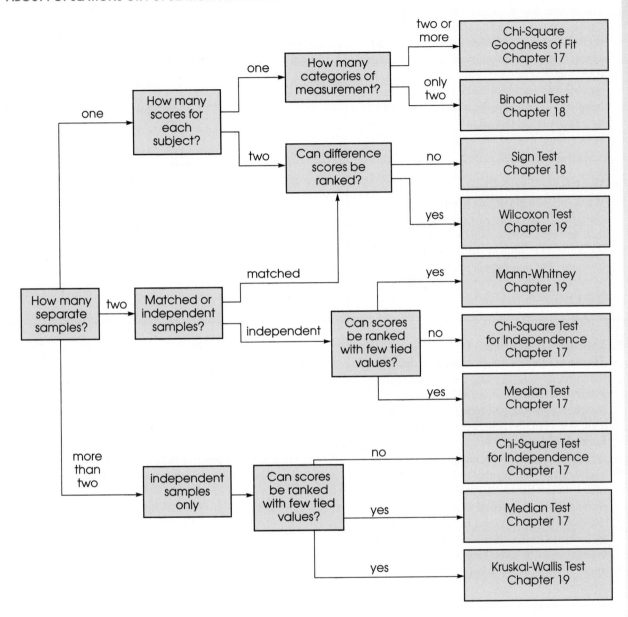

3. **The Median Test** (Chapter 17)

   The median test is a special application of the chi-square test for independence. This test is used to determine whether two or more populations share a common median. The test uses data from two or more separate samples and requires that each individual be classified as being above or below the median. The null hypothesis states that all of the separate populations share a common median.

   Use the chi-square formula and df from the chi-square test for independence.

4. **The Binomial Test** (Chapter 18)

   When the individuals in a population can be classified into exactly two categories, the binomial test uses sample data to test a hypothesis about the proportion of the population in each category. The null hypothesis specifies the proportion of the population in each of the two categories.

   $$z = \frac{X - pn}{\sqrt{npq}} \quad \text{or} \quad z = \frac{X/n - p}{\sqrt{\dfrac{pq}{n}}}$$

5. **The Sign Test** (Chapter 18)

   This test evaluates the difference between two treatment conditions using data from a repeated-measures or matched-subjects research design. The null hypothesis states that there is no difference between the two treatment conditions.

   The sign test is a special application of the binomial test and requires only that the difference between treatment 1 and treatment 2 for each subject be classified as an increase or a decrease. This test is used as an alternative to the related-samples t test or the Wilcoxon test in situations where the data do not satisfy the more stringent requirements of these two more powerful tests.

   Count the number of positive differences (X) in the total sample (n). Then use the binomial test with $p = q = 1/2$.

6. The Mann-Whitney U Test (Chapter 19)
This test uses ordinal data (rank orders) from two separate samples to test a hypothesis about the difference between two populations or two treatment conditions. The Mann-Whitney test is used as an alternative to the independent-measures t test in situations where the data can be rank-ordered but do not satisfy the more stringent requirements of the t test.

Combine the two samples (A and B) and rank-order the complete set of individuals. Then find the sum of the ranks for each sample. Finally, compute the U value for each sample as follows:

$$U_A = n_A n_B + \frac{n_A(n_A + 1)}{2} - \Sigma R_A$$

$$U_B = n_A n_B + \frac{n_B(n_B + 1)}{2} - \Sigma R_B$$

7. The Wilcoxon T Test (Chapter 19)
The Wilcoxon test uses the data from a repeated-measures or matched-samples design to evaluate the difference between two treatment conditions. This test is used as an alternative to the related-samples t test in situations where the sample data (difference scores) can be rank-ordered but do not satisfy the more stringent requirements of the t test.

Rank-order the difference scores, ignoring the signs ($+$ or $-$). Then compute the sum of the ranks for the positive differences and for the negative differences. The Wilcoxon T is the smaller sum.

8. The Kruskal-Wallis Test (Chapter 19)
This test uses ordinal data (ranks) from three or more separate samples to test a hypothesis about the differences between three or more populations (or treatments). The Kruskal-Wallis test is used as an alternative to the independent-measures ANOVA in situations where the data can be ranked-ordered but do not satisfy the more stringent requirements of the analysis of variance.

Combine the separate samples and then rank-order the complete set of individuals. Find the sum of the ranks (T) for each sample. Also, use the number in each sample (n) and the total number in the study (N) to compute a statistic equivalent to chi-square with df equal to the number of samples minus one.

$$H = \frac{12}{N(N + 1)} \Sigma \frac{T^2}{n} - 3(N + 1)$$

IV   MEASURES OF RELATIONSHIP BETWEEN TWO VARIABLES

As we noted in Chapter 1, a major purpose for scientific research is to investigate and establish orderly relationships between variables. The statistical techniques covered in this section all serve the purpose of measuring and describing relationships. The data for these statistics involve two observations for each individual—one observation for each of the two variables being examined. The goal is to determine whether or not a consistent, predictable relationship exists and to describe the nature of the relationship.

CHOOSING A MEASURE OF RELATIONSHIP BETWEEN TWO VARIABLES: A DECISION MAP

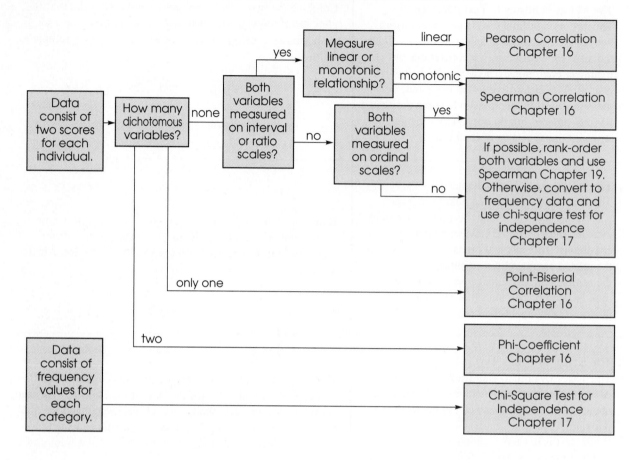

Each of the different statistical methods described in this section is intended to be used with a specific type of data. To determine which method is appropriate, you must first examine your data and identify what type of variable is involved and what scale of measurement was used for recording the observations.

1. **The Pearson Correlation** (Chapter 16)
   The Pearson correlation measures the degree of linear relationship between two variables. The sign ($+$ or $-$) of the correlation indicates the direction of the relationship. The magnitude of the correlation (from 0 to 1) indicates the degree to which the data points fit on a straight line.

$$r = \frac{SP}{\sqrt{SS_X SS_Y}}$$

where $SP = \Sigma(X - M_X)(Y - M_Y) = \Sigma XY - \dfrac{(\Sigma X)(\Sigma Y)}{n}$

2. Linear Regression and Standard Error of Estimate (Chapter 16)

The purpose of linear regression is to find the equation for the best-fitting straight line for predicting Y scores from X scores. The regression process determines the linear equation with the least-squared error between the actual Y values and the predicted Y values on the line. The standard error of estimate provides a measure of the standard distance (or error) between the actual Y values and the predicted Y values.

$$\hat{Y} = bX + a \qquad \text{where } b = \frac{SP}{SS_X} \quad \text{and} \quad a = M_Y - bM_X$$

$$\text{Standard error of estimate} = \sqrt{\frac{\Sigma(Y - \hat{Y})^2}{n - 2}} = \sqrt{\frac{(1 - r^2)SS_Y}{n - 2}}$$

3. The Spearman Correlation (Chapter 16)

The Spearman correlation measures the degree to which the relationship between two variables is one-directional or monotonic. The Spearman correlation is used when both variables, X and Y, are ranks (measured on an ordinal scale).

Use the Pearson formula on the ranked data or the special Spearman formula:

$$r_s = 1 - \frac{6\Sigma D^2}{n(n^2 - 1)}$$

4. The Point-Biserial Correlation (Chapter 16)

The point-biserial correlation is a special application of the Pearson correlation that is used when one variable is dichotomous (only two values) and the second variable is measured on an interval or ratio scale. The value of the correlation measures the strength of the relationship between the two variables. The point-biserial correlation often is used as an alternative or a supplement to the independent-measures t hypothesis test.

Convert the two categories of the dichotomous variable to numerical values 0 and 1; then use the Pearson formula with the converted data.

5. The Phi-Coefficient (Chapter 16)

The phi-coefficient is a special application of the Pearson correlation that is used when both variables, X and Y, are dichotomous (only two values). The value of the correlation measures the strength of the relationship between the two variables. The phi-coefficient is often used as an alternative or a supplement to the chi-square test for independence.

For each variable, convert the two categories to numerical values 0 and 1; then use the Pearson formula with the converted data.

6. **The Chi-Square Test for Independence**
(Chapter 17)
This test uses frequency data to determine whether or not there is a significant relationship between two variables. The null hypothesis states that the two variables are independent. The chi-square test for independence is used when the scale of measurement consists of relatively few categories for both variables and can be used with nominal, ordinal, interval, or ratio scales.

$$\chi^2 = \Sigma \frac{(f_o - f_e)^2}{f_e}$$

where $f_e = \dfrac{\text{(row total)(column total)}}{n}$

$$df = (R - 1)(C - 1)$$

References

American Psychological Association. (2000). *Publication manual of the American Psychological Association* (5th ed.) Washington, DC: Author.

Blest, A. D. (1957). The functions of eyespot patterns in the Lepidoptera. *Behaviour*, 11, 209–255.

Blum, J. (1978). *Pseudoscience and mental ability*. New York: Monthly Review Press.

Bransford, J. D., & Johnson, M. K. (1972). Contextual prerequisites for understanding: Some investigations of comprehension and recall. *Journal of Verbal Learning and Verbal Behavior*, *11*, 717–726.

Byrne, D. (1971). *The attraction paradigm*. New York: Academic Press.

Cerveny, R. S., & Balling, Jr., R. C. (1998). Weekly cycles of air pollutants, precipitation and tropical cyclones in the coastal NW Atlantic region. *Nature*, *394*, 561–563.

Cialdini, R. B., Reno, R. R., & Kallgren, C. A. (1990). A focus theory of normative conduct: Recycling the concept of norms to reduce littering in public places. *Journal of Personality and Social Psychology*, *58*, 1015–1026.

Cohen, J. (1988). *Statistical Power Analysis for the Behavioral Sciences*. Hillsdale, NJ: Lawrence Erlbaum Associates.

Cook, M. (1977). Gaze and mutual gaze in social encounters. *American Scientist*, *65*, 328–333.

Cowles, M., & Davis, C. (1982). On the origins of the .05 level of statistical significance. *American Psychologist*, *37*, 553–558.

Craik, F. I. M., & Lockhart, R. S. (1972). Levels of processing: A framework for memory research. *Journal of Verbal Learning and Verbal Behavior*, *11*, 671–684.

Darley, J. M., & Latane, B. (1968). Bystander intervention in emergencies: Diffusion of responsibility. *Journal of Personality and Social Psychology*, *8*, 377–383.

Davis, E. A. (1937). *The development of linguistic skills in twins, single twins with siblings, and only children from age 5 to 10 years*. Institute of Child Welfare Series, No. 14. Minneapolis: University of Minnesota Press.

Downs, D. S., & Abwender, D. (2002). Neuropsychological impairment in soccer athletes. *Journal of Sports Medicine and Physical Fitness*, *42*, 103–107.

Duncker, K. (1945). On problem-solving. *Psychological Monographs*, *58*, no. 270.

Friedman, M. & Rosenman, R. H. (1974). *Type A behavior and your heart*. New York: Knopf.

Gibson, E. J., & Walk, R. D. (1960). The "visual cliff." *Scientific American*, *202*, 64–71.

Gintzler, A. R. (1980). Endorphin-mediated increases in pain threshold during pregnancy. *Science*, *210*, 193–195.

Hays, W. L. (1981). *Statistics*, 3rd ed. New York: Holt, Rinehart, and Winston.

Hunter, J. E. (1997). Needed: A ban on the significance test. *Psychological Science*, *8*, 3–7.

Katona, G. (1940). *Organizing and memorizing*. New York: Columbia University Press.

Keppel, G. (1973). *Design and analysis: A researcher's handbook*. Englewood Cliffs, NJ: Prentice-Hall.

Keppel, G., & Zedeck, S. (1989). *Data analysis for research designs*. New York: W. H. Freeman.

Knutson, B., Wolkowitz, O. M., Cole S. W., Chan, T., Moore, E. A., Johnson, R. C., Terpstra, J., Turner, R. A., & Reus, V. I. (1998). Selective alteration of personality and social behavior by serotonergic intervention. *American Journal of Psychiatry*, *155*, 373–379.

Levine, S. (1960). Stimulation in infancy. *Scientific American*, *202*, 80–86.

Loftus, G. R. (1996). Psychology will be a much better science when we change the way we analyze data. *Current Directions in Psychological Science*, *5*, 161–171.

Moore-Ede, M. C., Sulzman, F. M., & Fuller, C. A. (1982). *The clocks that time us*. Cambridge: Harvard University Press.

Pelton, T. (1983). The shootists. *Science 83*, *4(4)*, 84–86.

Reifman, A. S., Larrick, R. P., & Fein, S. (1991). Temper and temperature on the diamond: The heat-aggression relationship in major league baseball. *Personality and Social Psychology Bulletin*, *17*, 580–585.

Rogers, T. B., Kuiper, N. A., & Kirker, W. S. (1977). Self-reference and the encoding of personal information. *Journal of Personality and Social Psychology*, *35*, 677–688.

Sachs, J. (1967). Recognition memory for syntactic and semantic aspects of a connected discourse. *Perception and Psychophysics, 2*, 437–442.

Scaife, M. (1976). The response to eye-like shapes by birds. I. The effect of context: A predator and a strange bird. *Animal Behaviour, 24*, 195–199.

Schachter, S. (1968). Obesity and eating. *Science, 161*, 751–756.

Schleifer, S. J., Keller, S. E., Camerino, M., Thorton, J. C., & Stein, M. (1983). Suppression of lymphocyte stimulation following bereavement. *Journal of the American Medical Association, 250*, 374–377.

Schmidt, S. R. (1994). Effects of humor on sentence memory. *Journal of Experimental Psychology: Learning, Memory, & Cognition, 20*, 953–967.

Segal, S. J., & Fusella, V. (1970). Influence of imaged pictures and sounds on detection of visual and auditory signals. *Journal of Experimental Psychology, 83*, 458–464.

Shrauger, J. S. (1972). Self-esteem and reactions to being observed by others. *Journal of Personality and Social Psychology, 23*, 192–200.

Siegel, J. M. (1990). Stress life events and use of physician services among the elderly: The moderating role of pet ownership. *Journal of Personality and Social Psychology, 58*, 1081–1086.

Siegel, S. (1956). *Nonparametric statistics for the behavioral sciences.* New York: McGraw-Hill.

Snyder, S. H. (1977). Opiate receptors and internal opiates. *Scientific American, 236*, 44–56.

Tryon, R. C. (1940). Genetic differences in maze-learning ability in rats. *Yearbook of the National Society for the Study of Education, 39*, 111–119.

Tukey, J. W. (1977). *Exploratory data analysis.* Reading, MA: Addison-Wesley.

Tulving, E., & Osler, S. (1968). Effectiveness of retrieval cues in memory for words. *Journal of Experimental Psychology, 77*, 593–601.

Tversky, A. & Kahneman, D. (1973). Availability: A heuristic for judging frequency and probability. *Cognitive Psychology, 5*, 207–232.

Tversky, A., & Kahneman, D. (1974). Judgment under uncertainty: Heuristics and biases. *Science, 185*, 1124–1131.

Walcott C. (1972). Bird navigation. *Natural History, 81*, 32–43.

Weinberg, G., Schumaker, J., & Oltman, D. (1981). *Statistics: An Intuitive Approach,* 4th ed. Belmont, CA: Wadsworth.

Wilkinson, L., and the Task Force on Statistical Inference. (1999). Statistical methods in psychology journals. *American Psychologist, 54*, 594–604.

Winget, C., & Kramer, M. (1979). *Dimensions of dreams.* Gainesville: University Press of Florida.

Yapko, M. D. (1994). Suggestibility and repressed memories of abuse: A survey of psychotherapists' beliefs. *American Journal of Clinical Hypnosis, 36*, 163–179.

Zigler, M. J. (1932). Pressure adaptation time. A function of intensity and extensity. *American Journal of Psychology, 44*, 709–720.

Index

Two-factor, independent-measures ANOVA,
  (*cont.*)
    extending to higher-order factorial
      designs, 505
    graphs, 486–487
    hypothesis tests, 495–499
    interactions, 482–486, 500–502
    main effects, 480–482, 500–502
    matrix, 479
    notation and formulas, 488–495
    overview of, 478–479
    reporting results of, 500
    stage 1 of analysis, 491–492
    stage 2 of analysis, 492–493
Two independent samples. *see* Independent-
  measures *t* test
Two-tailed hypothesis tests. *see* Hypothesis
  tests
Type I errors, 242–246, 401, 426
Type II errors, 243–244

U

Unbiased statistics, 121–122
Undetermined values, 91
Unit normal table, 170–171, 194–195
Unknown population, 232–233, 287
Unplanned comparisons, 426–427
Upper real limit, 19

V

Validity tests, 526
Variability, 104–135
  applications of, 105–106
  biased and unbiased statistics, 121–122
  computing, 113–115, 132–133
  defined, 104
  degrees of freedom and, 119–121
  equations, 113
  factors affecting measurements of,
    128–129
  graphs of, 115–116
  inferential statistics and, 126–128
  interquartile range, 107–109
  range, 106–107
  sample, 116–119
  semi-interquartile range, 107–109
  standard deviation and, 109–111,
    122–123, 124–126

statistical measures and, 129
sum of squared deviations, 111–113
transformations of scale, 124
between treatments. *see* Between-
  treatments variance
within treatments variance. *see* Within-
  treatments variability
Variables
  confounding, 14–15
  continuous, 18–19
  correlational method and. *see* Correlation
  defined, 10
  dependent, 14
  dichotomous, 549
  discrete, 18–19
  experimental method and, 12–13
  independent, 13, 397
  measuring. *see* Measurement, of variables
  quasi-independent, 17, 398
Variance. *see also* Pooled variance
  computing, 113–114
  defined, 110
  effect size and, 292
  factors affecting measurements of,
    128–129
  formulas for, 113
  independent-measures *t* test and, 330
  as mean square, 411
  population variance, 118–119
  relationship to other statistical measures,
    129
  sample size and, 315
  scores, 314
  t tests and, 292–296
Visual cliff, 617

W

Walk, R. D., 617
Weighted mean, 76–77
Wilcoxon signed-ranks test, 645–650
  calculating, 646
  hypothesis tests, 647
  normal approximation (binomial),
    648–650
  repeated measures studies, 645
  reporting results of, 648
  tied scores and zero scores, 646–647
  when to use, 636
Within-subjects design. *see* Repeated-
  measures *t* test

Within-treatments variability
  degrees of freedom (*df*), 410
  repeated-measures ANOVA, 449–456
  single-factor, independent-measures
    ANOVA, 403
  sum of squares (*SS*), 408
  two-factor ANOVA, 491–492

X

X-axis (abscissa), 43
X values, 36, 156–157
X variables, 523

Y

Y-axis (ordinate), 43
Y variables, 523

Z

z-score transformations, 144–148
z-scores
  binomial, 620
  comparisons with, 148, 238–239
  converting to/from *X* values, 156–157
  decision making and, 239–240
  definition, 140
  distribution, 283–284
  formulas, 142–143
  hypothesis tests and, 281
  introduction to, 138–140
  limitations of, 271, 281
  locations within a distribution, 140–144
  probability and, 171–173
  as ratio, 241
  raw score from, 143–144
  as recipe, 240–241
  research studies and, 152–154
  sample means, 210–211
  standardizing distribution with, 144–152
  t test as alternative to, 281–283
  as test statistic, 240, 248
Zedeck, S., 462
Zero differences, 626, 646–647
Zero, interval scale and, 21–22
Zero probability, 164